ISBN 978-0-266-90156-3
PIBN 10906972

FB &c Ltd, Dalton House, 60 Windsor Avenue, London, SW19 2RR.
Company number 08720141. Registered in England and Wales.

For support please visit www.forgottenbooks.com

DEPARTMENT OF THE INTERIOR.

REPORT

OF THE

UNITED STATES GEOLOGICAL SURVEY

OF

THE TERRITORIES.

F. V. HAYDEN,
UNITED STATES GEOLOGIST-IN-CHARGE

VOLUME III.

WASHINGTON:
GOVERNMENT PRINTING OFFICE.
1884.

NOTE.

DEPARTMENT OF THE INTERIOR,
UNITED STATES GEOLOGICAL SURVEY.

On the 27th of September, 1882, at the request of Dr. F. V. Hayden, the completion of the publications of the United States Geological and Geographical Survey of the Territories, formerly under his charge, was committed to the charge of the Director of the Geological Survey by the following order from the honorable the Secretary of the Interior:

DEPARTMENT OF THE INTERIOR,
Washington, September 27, 1882.

Maj. J. W. POWELL,
Director U. S. Geological Survey, City:

SIR: The letter of Prof. F. V. Hayden, dated June 27th, bearing your indorsement of July 20th, relating to the unpublished reports of the survey formerly under his charge, is herewith returned.

You will please take charge of the publications referred to in the same, in accordance with the suggestions made by Professor Hayden.

It is the desire of this office that these volumes shall be completed and published as early as practicable.

Very respectfully,

H. M. TELLER,
Secretary.

Of the publications thus placed in charge of the Director of the Survey, the accompanying volume is the first to be issued. It is understood that its preparation was begun in 1879, by the transmission of a part of the manuscript to the Public Printer.

At the time when the work was turned over to the Director of the Geological Survey an important portion of the manuscript was yet unprepared; but, through the energy of Professor Cope, the volume has been rapidly brought to completion. The work constitutes a valuable contribution to paleontology, and is a monument to the labor and genius of the author and to the administrative talent of Dr Hayden.

The yet unpublished volumes will be pushed to completion at an early day.

J. W. Powell

Director U. S. Geological Survey.

WASHINGTON, *September* 11, 1883.

LETTER OF TRANSMITTAL.

WASHINGTON, *January* 1, 1883.

SIR: I have the honor to transmit for your approval the third volume of the series of final reports of the United States Geological Survey of the Territories, which during its existence was under my charge.

The present volume, which has been prepared by the eminent paleontologist, Prof. E. D. Cope, of Philadelphia, represents the labor of several years, both in the field and in the study, and may be regarded as one of the most important contributions to the rich field of vertebrate paleontology of the western Territories ever made in this country.

It was the original purpose to include all the material in the author's possession from the Cenozoic and Mesozoic formations in the third and fourth volumes of the series, but they accumulated to such an extent that it became necessary to limit them to the Cenozoic alone. Therefore, the two volumes are essentially one in subject matter.

This volume consists of 1002 pages of text, illustrated with more than one hundred plates, and the fourth volume, which is to follow, may be regarded as a continuation of the present one, both comprising the material in the author's possession from the Cenozoic formations of the West.

The two volumes are divided into four parts, viz:

Part I, Puerco, Wasatch, and Bridger Faunæ (Eocene);

Part II, White River and John Day Faunæ (Lower and Middle Miocene);

Part III, Ticholeptus and Loup Fork Faunæ (Upper Miocene); and

Part IV, Pliocene.

The present volume includes Part I, and the first portion of Part II as far as the Ungulates; including, therefore, the Marsupials, Bats, Insectivores, Rodents, and Carnivora of the Miocene.

Part I includes the following most important contributions to paleontology and evolution:

1. The discovery of the fauna of the Puerco Group, of thirty genera and sixty-three species. This includes many important details, such as the discovery and definition of three new families, with many species of a new order, the *Taxeopoda*, as the *Periptychidæ*, *Meniscotheriidæ*, and a new suborder, the *Taligrada*, represented by the genus *Pantolambda;* also the discovery of the Plagiaulax type (of the Jurassic) and other Marsupials, and the Laramie Saurian genus Champsosaurus in the Puerco Group.

2. The new classification of the Ungulata rendered possible by the discovery of the complete remains of the Wasatch types of *Phenacodus* and *Coryphodon*, especially the former, from Wyoming Territory. The light thrown on the phylogeny of the Ungulata by this discovery exceeds that derived from all other sources together.

3. The new classification of the lower clawed mammals, based on the analyses of fifteen new genera and forty-seven new species of flesh-eaters, and six new genera and sixteen new species of allied forms, all discovered since the publication of the author's volume in connection with the Wheeler Survey.

4. The restoration of Hyracotherium, the four-toed horse of the Wasatch Group.

5. The restoration of the genera Triplopus and Hyrachyus of the Bridger Fauna.

6. The determination of the systematic relation of the Dinocerata as seen in the genera Loxolophodon and Bathyopsis.

The whole number of genera described in this volume is 125, and of species 349, of which 317 species were determined by Professor Cope.

The explorations that furnished the materials for these volumes began in 1872, and are still being continued. It will therefore be readily seen that the amount of new matter towards the origin and history of the Mammalian group brought together by the author in these two volumes is most extraordinary, and will probably never be surpassed.

The plates which illustrate this volume were engraved by Thomas Sinclair & Son, of Philadelphia, and the figures were drawn on stone from the specimens themselves, under the immediate supervision of Professor Cope.

At the close of the fiscal year ending June 30, 1882, at my request, the Secretary of the Interior placed the printing of the uncompleted volumes of the quarto series in the care of Maj. J. W. Powell, the Director of the United States Geological Survey, and I desire him to accept my cordial thanks for his very kind attention and for many personal courtesies.

Very respectfully, your obedient servant,

F. V. HAYDEN,

United States Geological Survey of the Territories.

The Hon. the SECRETARY OF THE INTERIOR,

Washington, D. C.

UNITED STATES GEOLOGICAL SURVEY OF THE TERRITORIES.

THE VERTEBRATA

OF THE

TERTIARY FORMATIONS

OF

THE WEST.

BOOK I.

By EDWARD D. COPE,
MEMBER OF THE NATIONAL ACADEMY OF SCIENCES.

WASHINGTON:
GOVERNMENT PRINTING OFFICE.
1883.

CONTENTS.

THE PUERCO, WASATCH, AND BRIDGER FAUNÆ—Continued.

THE PUERCO, WASATCH, AND BRIDGER FAUNÆ—Continued.

THE PUERCO, WASATCH, AND BRIDGER FAUNÆ—Continued.

THE PUERCO, WASATCH, AND BRIDGER FAUNÆ—Continued.

THE PUERCO, WASATCH, AND BRIDGER FAUNÆ—Continued.

PART SECOND.

THE WHITE RIVER AND JOHN DAY FAUNÆ—Continued.

THE WHITE RIVER AND JOHN DAY FAUNÆ—Continued.

~~Reptilia:~~

LIST OF WOOD-CUTS.

LETTER OF TRANSMISSION.

JANUARY 1, 1879.

SIR: I send herewith a report on the Tertiary Faunæ of the United States as represented by collections made in various Territories and States west of the Mississippi River, embraced within the boundaries of your survey. The explorations from which the collections have been derived cover portions of the States and Territories included between British America on the north, the western boundaries of Minnesota and Missouri on the east; the northern borders of the Indian Territory and Arizona, and the middle of New Mexico on the south; and the Sierra Nevada on the west. The amount of material which I have procured through these explorations is large, and is but partially represented in the following pages. I trust that you will find the results a useful contribution to the records of your Geological Survey and to the science to which you have devoted your life; and that you may find in this report some compensation for the arduous official duties which have recently withdrawn you to some degree from your chosen field of research.

The preface gives an account of the methods pursued in conducting the investigation; while the introduction embraces a general view of the stratigraphy of the Tertiary formations of the West. The system adopted is that proposed by yourself and Mr. King, with a few additions; while several correlations with the horizons of the Old World are based on my own paleontological studies. The order of succession of faunæ is observed in the following sections of the work; that is, part first, the Puerco, Wasatch, and Bridger formations; part second, the White River and the John Day beds; and part third, the Loup fork and Equus beds. The second half of the second part, the third part, and faunal lists, will constitute the succeeding volume, No. IV, of your series.

I desire to express here the obligations under which I have been placed through the important aid and hospitality rendered me by the following gentlemen:

In 1872, at Fort Bridger, Wyo., I was assisted by Capt. E. O. Clift, in command of the post, and by Lieutenant Rogers, quartermaster, and Dr. Joseph Corson, surgeon; also by Judge W. Carter and Dr. J. V. Carter. In Montana, in 1876, I received important aid from General E. O. C. Ord, commanding the Department of the Missouri, and Major Ilges, in command at Fort Benton; and in my explorations in Washington Territory, in 1879, I was under obligations to Dr. George H. Sternberg, U. S. A.* In 1880–'81 the military authorities at Fort Washakie, Wyo., rendered me much assistance, particularly Col. J. W. Mason, commanding, and Dr. W. H. Corbusier, post surgeon.*

I have received important aid from Professor Baird, of the Smithsonian Institution, and from Professor Condon, of Eugene City, Oreg. I wish here to place on record the names of my assistants, who have contributed greatly to the success of my expeditions, viz: William G. Shedd, Charles H. Sternberg, Jacob Boll, J. C. Isaac, Russell S. Hill, Frank Hazard, Jacob L. Wortman, and D. Baldwin.

I have been also favored by special rates by the general officers of the Union and Central Pacific, and Pennsylvania Railroad companies. I wish here to express my thanks to Messrs. Kimball and Stebbins of the Union, and Governor Stanford of the Central Pacifics, and to Presidents Scott and Roberts of the Pennsylvania Railroad.

The lithographic work of Messrs. T. Sinclair & Son maintains the well-known reputation of their house, and will prove satisfactory to students generally.

I am, with respect,

E. D. COPE,

Paleontologist, United States Geological Survey of the Territories.

Dr. F. V. HAYDEN,

Director of the United States Geological
and Geographical Survey of the Territories.

---

*In my explorations of formations other than those treated of in this volume, I have been assisted by other gentlemen, generally officers of the Army, to whom I will refer in the appropriate place.

PREFACE.

1. Sources of Collections.—The localities which yielded the fossils described in the following pages are the following:

In 1872 I conducted an exploring party in Southwestern Wyoming. I left Fort Bridger July 19, and followed the road to Cottonwood Creek, southeast eighteen miles, whence we made our first excursions into the bad lands. After this our route laid along Cottonwood Creek to Smith's Fork of Green River, thence along Black's Fork, and thence to Green River City. We then followed Bitter Creek to Black Buttes, and, leaving the line of the Union Pacific Railroad, traveled south toward the headwaters of the Vermillion. Before reaching this point we explored the Mammoth Buttes, which form the water-shed between South Bitter Creek and Vermillion, and examined the bad lands of the Washakie Basin carefully. In reaching this point we crossed a portion of the Cretaceous formation, and I took especial pains to determine the relations of the strata at these points.

We returned from this region and struck Green River seventeen miles above Green River City. We proceeded northward to the mouth of Labarge Creek, and, returning a short distance, ascended Fontanelle Creek to near its source in the outlying ranges of the Ham's Fork Mountains. The relation between the lake-deposits and the older strata here claimed special attention. We then descended Ham's Fork to the Union Pacific Railroad and returned to Fort Bridger.

Special expeditions were made to the region round Evanston, and to Elko, Nev., with gratifying success.

We obtained, in round numbers, one hundred species of vertebrated animals of the Eocene period, of which about sixty were new to science. We obtained material for the addition of two orders of mammals to those previously represented in this fauna in the United States, viz, the *Mesodonta* and *Amblypoda*, the latter in several types of remarkable interest.

In 1873 I fitted out an expedition at Greeley, Colo., and traversed the Plains eastward toward Julesburg as far as the eastern branches of the Cedar or Horse Tail Creek. Our route was parallel to the line of the so-called Chalk Bluffs, which extend from west to east, forming a break in the southern slope of the surface of the country from the dividing of the waters of the North and South Platte. It consists of the Loup Fork sandstones resting on a basis of the upper beds of the White River formation. The country between the foot of the bluffs and the South Platte River is composed in its northern part of the White River formation, which presents exposures at various points, and nearer the river consists of the Laramie formation. On this part of the expedition I obtained seventy-five species from the White River beds, and twenty-one from the Loup Fork.

We then turned to the southwest, crossing the South Platte, and moved up the valley of Bijou Creek towards the highlands of Colorado east of the mountains, known as the Colorado divide. On this part of the expedition, which was in charge of William G. Shedd, a number of interesting reptiles of the Laramie period were discovered. The party then entered the South Park and obtained a fine collection of the fishes of the Florissant shales During this time I had made an excursion to Fort Bridger, Wyo., and had supplemented the collections of the previous year.

In 1877 I sent my assistant, J. C. Isaac, to Montana for the purpose of examining the valley of Deep River for the fossiliferous deposits previously reported to exist there by Captain Ludlow, United States Engineers, and examined by Messrs. Dana and Grinnell of his party. The results were satisfactory, a considerable number of fine specimens having been secured. Mr. Isaac then passed southeastward into Wyoming, and explored the White River beds of the southern parts of that Territory and the adjacent borders of Wyoming.

The same year I employed Charles H. Sternberg to conduct an exploration of the Cretaceous and Tertiary formations of Kansas. After a successful search I sent Mr. Sternberg to Oregon, and Russell S. Hill took charge of the expedition. Under his management an excellent collection of the

Mammalia and Reptilia of the Loup Fork formation of Northern Kansas was made, Mr. Hill discovering several new species of *Mastodon*, rhinoceroses, tortoises, &c.

The Tertiary formations explored in 1878 were the John Day, Loup Fork, and Equus beds, of Oregon. These were examined by Mr. C. H. Sternberg, who received important aid from his brother, Dr. George M. Sternberg, U. S. A. The John Day formation was chiefly examined on the John Day's River, and the Loup Fork beds at various points in the same region. These yielded about fifty species, many of them represented by specimens in an admirable state of preservation. The Equus beds were examined both in Washington and Oregon; in the former near to Fort Walla Walla, and in the latter in the desert east of the Sierra Nevada. The basin of an ancient lake, originally discovered by Governor Whitaker, of Oregon, was found to be strewn with the bones of llamas, horses, elephants, sloths, and smaller mammals, with birds; and all were collected by Mr. Sternberg and safely forwarded to Philadelphia. I examined this locality myself in 1879, and obtained further remains of extinct and recent species of mammalia found mingled with numerous worked flints.

In 1879 Mr. J. L. Wortman took charge of my party exploring in Oregon, and made extensive and valuable collections of the fossils of the John Day and Loup Fork beds of the eastern part of that State. In 1880 Mr. Wortman explored the deposits of the Idaho Pliocene lake of the Snake River Valley, and made a valuable collection.* The same year he examined the Eocene beds of the Wind River Basin previously discovered by Dr. Hayden, and sent east forty-five species of vertebrata, of which twenty-four were new to science. In the following year Mr. Wortman pushed his explorations northwards, and discovered that the basin through which the lower part of the Big Horn River flows is filled with deposits of Wasatch Eocene age. These he examined for vertebrate remains, and succeeded in obtaining sixty-five species, of which twenty-seven were previously unknown. Most important additions to our knowledge of the structure of various types were made, owing to the remarkably perfect condition of some of the specimens.

*Proceedings Academy, Philadelphia, 1883, p. 153.

In 1881 I employed Mr. D. Baldwin to collect fossils in the Puerco formation of New Mexico, which I discovered in 1874. Mr. Baldwin's success has had a very important bearing on the science of paleontology. He has obtained more than sixty species from that formation, nearly all of which were new to science.

The expeditions have not been conducted without risks. My exploration in Western Kansas was made during a state of hostility of the Cheyenne Indians, and in a region where they were constantly committing murders and depredations. During my expedition of 1872 I was abandoned by some of my party, who robbed me of mules and provisions, and placed me in some bodily peril. My expedition of 1873 was in the Cheyenne country, and constant vigilance was necessary. The year following my visit the whites were driven from the region, or murdered, by the Indians of that tribe. In 1876 I entered the Sioux country with my party on the Upper Missouri while the Indians were engaged with General Custer on the Little Big Horn and the Rosebud. My guide and camp tender abandoned me, and before leaving the country we passed a point a day's ride from Sitting Bull's camp on the Dry Fork of the Missouri. Mr. Sternberg's expedition of 1877 was interrupted by the Bannock war, and both himself and Mr. Wortman were compelled to leave their camp and outfit in the field and fly to a place of safety on their horses. In attempting to cross the Wind River in 1880 Mr. Wortman's horses and wagon were carried away by the current and the greater part of his baggage and provisions lost. His exploration of 1881 was conducted under circumstances of much risk from the absence of water. All the water necessary to the existence of his animals and men had to be carried a distance of twenty miles on the backs of mules

It is evident that an enthusiastic devotion to science has actuated these explorers of our western wilderness, financial considerations having been but a secondary inducement. And I wish to remark that the courage and regardlessness of physical comfort displayed by the gentlemen above referred to in the pursuit of the idea of progress, are qualities of which their country may be proud, and are worthy of the highest commendation and of imitation in every field.

I have also received miscellaneous collections from G. W. Marnock, of Texas, from the late Tertiary formation of the southern part of that State, and from various persons in Nebraska, Dakota, &c. A few small collections received through the office of the United States Geological Survey of the Territories are mentioned in the proper places.

2. Mode of Preservation of Collections.—Sinee the value of determinations in vertebrate paleontology depends greatly on the condition of the collections, I give here some explanation of the methods I have employed in this direction.

Prior to the publication of the descriptions of *Elasmosaurus* and various species of *Pythonomorpha*, from Kansas, in 1869–'70, complete skeletons from the western deposits were unknown in eastern collections, or, if existing the fragments of different animals were so commingled as to be unavailable for purposes of determination. As it is self-evident that the science can make little progress without the discovery of entire skeletons, I have made every effort to secure them, commencing with my exploration of the Cretaceous beds of Kansas in 1871.

In the field entire skeletons are not rare, as animals have often been entombed in soft deposits more or less uninjured. To obtain them in an entire condition, however, requires an unusual conjunction of circumstances. The skeleton must be visible, but not so far exposed to the weather as to have suffered injury from frost and rain, and it must not penetrate a hard matrix so deeply as to be inaccessible. As is the matrix, so is usually the fossil; friable fossils belong to a soft rock, and hard ones to a hard rock. The exceptions to this rule are fossils found in dry sand, which are hard.

In collecting, the first precaution to be observed, is to trace weathered fragments to their proper source in the adjacent deposit. This will of course be done, if at all, by following up the line of descent, either of escarpment or of water wash. If the remainder of the skeleton be found in place, the true correlation of the fragments will soon be discovered. The difficulty of extricating bones from the matrix depends on the hardness or softness of the latter. The most favorable condition is an intermediate one, neither hard nor soft. The chalk of the Niobrara Cretaceous presents the most favorable conditions; next in order the matrix of the Bridger and

John Day formations preserves the bones best for extrication. The White River formation of the Plains is only inferior in being a little softer, while the material of the Laramie formation varies between too great hardness or too great softness. The same difficulty, though in a less degree, is met with in the Loup Fork beds, softness predominating, while the least favorable of all for the preservation of fossils are the Puerco and Wasatch formations, where concretionary hardness prevails.

In all of my expeditions great care has been exercised in preserving the relations in which the fossils have been found by placing marks on the same and by preserving notes and drawings made on the ground. These precautions are of course absolutely necessary to secure accuracy in the reference of the various fragments into which a skeleton is often broken. On the arrival of the collections in Philadelphia the labels on the packages insure their correct classification, and the work of reuniting the broken pieces commences. In many cases crania, bones, and skeletons having been taken out in a more or less entire condition, inclosed in rock masses, much time is consumed in dressing them out with mallet and chisel. The amount of labor required for the preparation of the material of the present report alone, is easily seen to have been very great. I here refer to the skill of my assistant, Mr. Jacob Geisman, to whom the excellent character of this work is largely due.

3. Publication of Results.—The media of publication of the results of the investigations embraced in the present volume have been the following:

1. Bulletin of the United States Geological and Geographical Survey of the Territories, F. V. Hayden in charge, Washington.

2. Annual Reports of the United States Geological and Geographical Survey of the Territories, Washington.

3. Proceedings of the American Philosophical Society, Philadelphia.

4. Proceedings of the Academy of Natural Sciences, Philadelphia.

5. Paleontological Bulletins. By E. D. Cope, Philadelphia.

The last-named series consists in large part of reprints of papers which have appeared in the serials, Nos. 3 and 4; principally in No. 3. These reprints have averaged 200 copies each, but have sometimes amounted to

300 copies; in a few cases but 100 copies were issued. They have mostly appeared in advance of the number of the serial which contains them, owing to the long intervals at which the latter were or are issued. Thus the Proceedings of the American Philosophical Society were, up to a recent date, published but once in six months, and those of the Academy of Natural Sciences three times in the year. In some instances the Paleontological Bulletins have not appeared in any serial. In the earlier part of my investigations the reading of the proofs of these and other papers was sometimes intrusted to other persons, owing to my absence from Philadelphia while conducting explorations. These persons at times allowed important typographical errors to escape them, and in a few instances introduced alterations of the text, for which I wish to disclaim responsibility. This experience led me to avoid such confidence thereafter, so far as practicable.

The literature of the Paleontology is given under the head of the separate divisions of the subject in which it appropriately falls.

4. Rules of Nomenclature.—I have adhered to the law of priority, as generally understood, in the use of names both in the biological and stratigraphical aspects of the subject. I take this opportunity of noting what appears to have been at times forgotten by a few students of vertebrate paleontology—although fully recognized by biologists generally—that a name, unaccompanied by a definition or a precise reference to an existing definition, has no status in scientific nomenclature. A word so introduced is meaningless, and cannot be used, because that which it represents is unknown. Thus, names of classes and orders which refer only to popular definitions, such as "flesh-eaters," "insect-eaters," "whales," "worms," &c., have no scientific existence. These divisions of recent animals having been, however, by this time, well established by true analysis, the names proposed for extinct groups which are now being discovered claim our attention.* The progress of paleontology has been retarded by the publication of numerous names, supposed to refer to family and generic divisions, which are not accompanied by descriptions or by any statement of the reasons why their author has created them. Characters of the species described

* See Proceedings American Philos. Society, 1873, p. 73. Report U. S. Geol. Survey Terrs., 4to, II, p. 113. Report of Lieut. G. M. Wheeler, U. S. Geogr. Surv. W. of 100 Mer., IV, Pt. II, p. 148.

under the proposed generic name are usually given, and in some instances characters which really belong to the definition of the genus to which it belongs may be found mingled with them. In these cases it is left for the reader to discover these characters. Should he do so, he becomes the real discoverer of the genus, and as such is entitled to name it. The publication of names in the manner objected to is, from every point of view, pernicious, and is very properly forbidden by well-known rules. It matters not if it be ascertained at a subsequent date, and by some circumstantial evidence, what the author of such names referred to as to species and specimens. Such information cannot habilitate a *nomen nudum;* nor is such circumstantial evidence accessible to students generally, especially to those who live at some distance from the locality whence it may be obtained.

I now append the most important rules of nomenclature, as adopted by a majority—in most instances, a very large majority—of forty-five of the leading biologists of North America. They are included in the report of a Committee of the American Association for the Advancement of Science, appointed in 1876, of which Capt. W. S. Dall, U. S. N., was chairman.*

1. The reading of a paper before a scientific body does not constitute a publication of the descriptions or names of animals or plants contained therein.

2. A name applied to a group of species without a specification of any character possessed by them in common (that is, without any so-called generic diagnosis or description), is not entitled to recognition as an established generic name by subsequent authors.

3. A generic name applied to a single (then or previously) described species without a generic diagnosis or description of any kind, is not entitled to recognition as above, by subsequent authors.

4. A subsequent author shall not be permitted in revising a composite genus (of which no type was specified when it was described) to name as its type a species not included by the original author of the genus in that latter author's list of species given when the genus was originally described.

5. When an old genus without a specified type has been subdivided by a subsequent author, and one of the old species is retained and specified

*See American Naturalist, August, 1878.

to be the type of the restricted genus bearing the old name, it is not competent for a third author to discard this and select another of the original species as a type, when by so doing changes are necessitated in nomenclature.

6. When a generic name has lapsed from sufficient cause into synonymy, it need not be thenceforth entirely rejected from nomenclature, and may still be applicable to any new and valid genus.

The earlier pages of this volume were printed between two and three years prior to the greater part of it, hence some of the earlier statements will be found to be modified in the more detailed discussions which follow. One such point is the distinction which should be maintained between the John Day and White River epochs; another point is the great distinction which should be recognized to exist between the Puerco and later Eocene periods. The faunæ of the Puerco and Wasatch epochs are as diverse from each other as are those of the Bridger and White River.

Some inequalities in the text, and the intercalation of numerous plates which carry letters attached to their numbers must be explained. These peculiarities are due to the fact that the discovery of the Puerco fauna was made after the first pages of the volume had been struck off, and the greater number of the plates had been numbered and printed.

The present volume includes the vertebrata of the Eocene and of the Lower Miocene, less the Ungulata. There are described three hundred and forty-nine species, of which I have been the discoverer of all except thirty-two. They are referred to one hundred and twenty-five genera.

The most important results which have accrued to paleontology through the researches here set forth, are the following:

(1.) The discovery of the Laramie genus *Champsosaurus* in Tertiary beds.

(2.) The discovery of *Plagiaulacidæ* in Tertiary beds.

(3.) The discovery of the characters of five families and many genera and species of the Creodonta.

(4.) The discovery of the characters of the *Periptychidæ* and its included genera; and

(5.) Of the *Meniscotheriidæ;* and

(6.) Of the *Phenacodontidæ* and its genera.

(7.) The discovery of the characters of the suborder of Condylarthra and of the phylogenetic results of the same.

(8.) The discovery of the characters of the *Pantolambdidæ;* and

(9.) Of the suborder *Taligrada* and its implications in phylogeny.

(10.) The discovery of the *Anaptomorphidæ* of the Prosimiæ.

(11.) The reconstruction of *Hyracotherium;* and

(12.) Of *Hyrachyus.*

(13.) The discovery of numerous *Marsupialia* in the Lower Miocene.

(14.) The discovery of the phylogenetic series of the Canidæ; and

(15.) The same of the ancestors of the Felidæ.

ERRATA.

Page 167. For *Diplæthra* read *Diplarthra.*

Page 168. In table of genera, correct definition of genus Polymastodon to read: Fourth inferior premolar conical, simple; inferior molars with two, superior molars with three, rows of tubercles. Strike out genus Catopsalis and its definition, as they refer to the inferior dentition of Polymastodon.

Page 169. In phylogenetic table of Plagiaulacidæ erase Catopsalis, and place Polymastodon on a side branch out of the line between Thylacoleo and Ctenacodon; one which is derived from such mesozoic forms as Tritylodon and Stereognathus of Owen.

Page 240. The genus Necrolemur is erroneously included in the Anaptomorphidæ and should be placed in the family Mixodectidæ. The arrangement of the genera of these families is then as follows (see American Naturalist, 1884, p. 60):

I. Canine teeth large and lateral; well separated.

First superior premolar without internal lobe; superior true molars tritubercular with cingula.. *Tricentes.*

II. Canine teeth median in position or much reduced in size.

α. Last inferior premolar without internal tubercle.

nferior premolars all one-rooted. Canine and incisor small .... *Necrolemur.*

First premolar only one-rooted; canine small; incisor very large .... *Mixodectes.*

αα. Last inferior premolar with internal tubercle.

A very large ? canine; first premolar only one-rooted .... *Microsyops.*

A very large ? canine; first and second premolars both one-rooted .... *Cynodontomys.*

The new genus Tricentes includes three species, and perhaps four. It will be described in the last part of Volume IV of this series with the genus Indrodon of the next family.

The genera of Anaptomorphidæ differ as follows:

α. Incisors three.

First superior premolar without inner lobe; posterior inner tubercle present on first and second superior true molars .... *Indrodon.*

αα. Incisors two.

First superior premolar with inner lobe; no posterior inner tubercle on superior molars.. *Anaptomorphus.*

Page 260. Line 7, the genus Diacodon should probably be removed to the section of the family where the fourth premolar is different in character from the first true molar.

Page 391. In synonymy of PERIPTYCHUS RHABDODON, for *Catathtœrhus* read *Catathlœus.*

Page 739. Omit from the definition of the Prosimiæ, in the table at the bottom of the page, the words "superior true molars quadritubercular."

Page 920. Eighth line from bottom, for "fishes" read fisher.

Explanation of Plate XXIX d (p. 42), at bottom, correct fig. 2 by stating that the anterior two teeth figured belong really to the lower jaw, which is very robust, and that the second upper premolar has an internal lobe. Add that the *Periptychus ditrigonus* does not belong to Conoryctes, but to a genus of Periptychidæ, which differs from all those known by the presence of a conic cusp external to the usual external tubercles of the superior molars. It may be called Ectoconus m. To an allied genus must be referred the second specimen figured and described as *Conoryctes comma* m.

INTRODUCTION.

SECTION I.

THE TERTIARY FORMATIONS OF THE CENTRAL REGION OF THE UNITED STATES.

The principal Tertiary formations of the region between the Mississippi River and the Sierra Nevada are the following, as mainly determined by Dr. Hayden: The Puerco, the Wasatch, the Bridger, the Uinta, the White River, the Loup Fork, and the Equus beds. Several of these are again distinctly subdivided, and in a few instances such divisions have been regarded by authors as of equal importance with those above mentioned; as, for instance, the Green River portion of the Wasatch. But the evidence of vertebrate paleontology is not as yet clearly favorable to further primary subdivision than is indicated by the above names. I will briefly describe the character and distribution of these formations before entering on the description of the fossils which they contain.

The general history of the succession of the Tertiary Lakes of the interior of the North American continent and their deposits has been developed by the labors of various geologists, prominent among whom must be mentioned Hayden, Newberry, and King. It may be synoptically stated as follows:

The Laramie-Cretaceous period witnessed a great difference in the topography of the opposite sides of the Rocky Mountain range. To the east were extensive bodies of brackish and nearly fresh water, with limited ocean communication, studded with islands and bordered by forests. On the west side of the range was a broad continent, composed of mostly marine Mesozoic rocks, whose boundaries are not yet well ascertained. Towards the close of the Laramie the bed of the great eastern sea began to emerge from the waters, and the continent of the western side of the great range descended. The relations of the two regions were changed; the east became the continent and the west became the sea. The latter, receiving the drainage of the surrounding lands, was a body of fresh water, whose connection with the ocean permitted the entrance of a few marine fishes only. This was the great Wasatch Lake, whose deposits extend from the headwaters of the Yellowstone far south into New Mexico and Arizona, between the Rocky Mountains on the east and the Wasatch range on the west. Its absence from the east side of the former range indicates the continental condition of that area at the time. The only locality where the Wasatch deposits are extensively deposited on the Laramie, is in the region intermediate between the two districts in Wyoming and New Mexico. Here the sediments of the former are seen to have succeeded those of the latter, and to have been coincident with an entire cessation of brackish conditions. Elevations of the continent northward and southward contracted the area of the great Wasatch sea, and perhaps deepened it, for at this time were deposited the fine limestones and silico-calcareous shales of the Green River epoch. There is no evidence that these beds had a greater eastern extension than that of the parent Wasatch Lake. King has given distinct names to these ancient lakes. I think it better to pursue the usual course of using for them the names already given to their deposits, as involving less strain on the memory; the more as the number of these lakes will be probably enlarged by future discoveries. The only known region covered by this lake west of the Wasatch range, is represented to-day by the calcareous strata in Central Utah, which I have called the Manti beds. The exact equivalency of these is, however, not quite certain. Further contraction reduced this area to perhaps two lake basins, whose deposits now form two isolated tracts in Southern Wyoming, and are known as the Bridger formation. Continued elevation and drainage caused the desiccation of these basins also, leaving only, so far as present knowledge extends, a body of water on the south of the Uinta Mountains, in Northeastern

Utah. The sediments of this lake form the Uinta formation, which is the latest member of the series now found in the region lying between the Rocky and Wasatch Mountains.

About the time that the elevation of the present drainage basin of the Colorado River was completed, a general subsidence of level of the great region east of the Rocky Mountains commenced. Extensive lakes were formed in the depressions of the Laramie and older beds which formed the surface, which were probably connected over a tract extending from near the Missouri River to Eastern Wyoming and Colorado. At the same time a similar body of fresh water occupied a large part of what is now Central Oregon and certain areas in Northwestern Nevada, according to King. The sediments now deposited constitute the White River formation, and the faunal distinctions which I have discovered to characterize the eastern and western basins have led me to employ for them the subdivisional names of White River beds for the former and Truckee (King) for the latter. It may have been during the early part of this period, or during the Uinta, that there existed two contemporary bodies of water, separated by a wide interval of territory. One of these extended over a considerable tract in Northern Nevada, and deposited a coal bed near Osino. A formation probably the same has been found by Professor Condon in Central Oregon underlying the Truckee Miocene beds. The other lake left its sediments near Florissant, in the South Park of Colorado. This formation I have named the Amyzon beds,* from a characteristic genus of fishes which is found in it. It has been referred to the Green River formation by King, but without the necessary paleontological evidence, as it appears to me.

The oscillations of the surface which brought the White River period to a close are not well understood. Suffice it to say here, that after an interval of time another series of lakes was formed, which have left their deposits at intervals over a wider extent of the continent than have those of any other epoch. These constitute the beds of the Loup Fork period, which are found at many points between the Sierra Nevada and the Rocky Mountains from Oregon to New Mexico, and over parts of the Great Plains

* American Naturalist, May, 1879.

of Colorado, Kansas, and northward, and in the valleys of the Rocky Mountains. A probably continuous succession of lakes has existed from this period to the present time in ever-diminishing numbers. The most important of these were in the Great Basin in Oregon, in Washington, and in Nebraska, and their deposits enclose the remains of a fauna entirely distinct from that of the Loup Fork period and of more modern character. They are known as the Equus beds. This fauna was probably contemporaneous with that which roamed through the forests of the eastern portion of the continent, whose remains are enclosed in the deposits of the caves excavated from the ancient limestones.

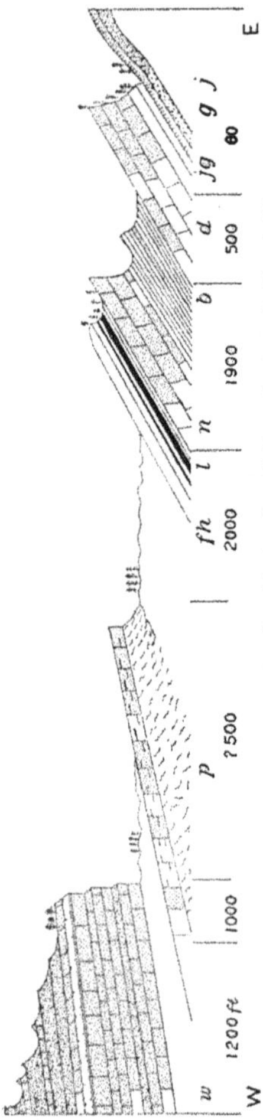

FIG. 1.—Section west of the Gallinas Mountains, New Mexico, from Gallinas Creek to the Eocene Plateau. (Cope.) Letter *j*, Jurassic; *jg*, Jurassic gypsum; *g*, Gallinas Creek; *d*, Dakota; *b*, Benton; *n*, Niobrara; *l*, lignite; *fh*, Fox Hills; *p*, Puerco; *w*, Wasatch. (From the Report of Capt. G. M. Wheeler, vol. iv, pt. ii, p. 1.)

A more detailed account of the formations is now given. Faunal lists are reserved for the close of the volume.

THE PUERCO.

This formation, having furnished numerous Mammalian fossils, is known to belong to the Tertiary rather than the Post-Cretaceous series. It is regarded by Dr. Endlich as a subdivision of the Wasatch, but the characteristics of its fauna are so marked as to constitute it a distinct horizon.

The most southern locality at which it has been observed, the one from which I named it, and where its characters are distinctly displayed, is west of the Jemez and Nacimiento Mountains, in New Mexico, at the sources of the Puerco River. At this place its outcrop is about 500 feet in thickness, and has an extent of several miles on both sides of the river. From this point the strike is northwards, keeping at the distance of a few miles to the eastward of an escarpment of Wasatch formation. It contracts in depth to the northward, and beyond the Gallinas Mountains I have not observed it.

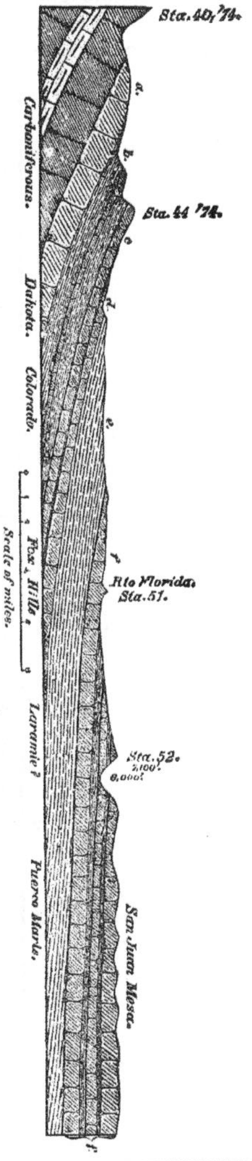

It is well developed in Southern Colorado, where Dr. F. M. Endlich* and William H. Holmes,† of Dr. Hayden's Survey, detected it in 1876. Its mineral character is there similar to that seen in New Mexico, and its thickness is much greater. On the Animas River it is 1,000 to 1,200 feet; on the San Juan River, near the Great Hog Back, 700 feet. The general characters of the formation are expressed in the following description, extracted from my report to Lieut. G. M. Wheeler.‡

South of the boundary of the Wasatch, the varied green and gray marls formed the material of the country, forming bad-land tracts of considerable extent and utter barrenness. They formed conical hills and flat meadows, intersected by deep arroyos, whose perpendicular walls constituted a great impediment to our progress. During the days of my examination of the region heavy showers of rain fell, filling the arroyos with rushing torrents, and displaying a peculiar character of this marl when wet. It became slippery, resembling soap in consistence, so that the hills were climbed with difficulty, and on the levels the horses' feet sank at every step. The material is so easily transported that the drainage channels are cut to a great depth, and the Puerco River becomes the receptacle of great quantities of slimy looking mud. Its unctuous appearance resembles strongly soft soap, hence the name *Puerco*, greasy These soft marls cover a belt of some miles in width, and continue at the foot of another line of sandstone bluffs, which bound the immediate valley of the Puerco to a point eighteen miles below Nacimiento.

Fig 2. Section along the east side of the Animas in Colorado; Endlich in Hayden's Annual Report, 1875, p. 18.

*Annual Report U. S. Geol. Surv. Terrs., 1875, p. 189. †Loc. cit., 247.

‡Annual Report of Chief of Engineers, 1875, p. 89. Appendix L L.

The Puerco marls have their principal development at this locality. I examined them throughout the forty miles of outcrop which I observed for fossil remains, but succeeded in finding nothing but fossil wood. This is abundant in the region of the Gallinas, and includes silicified fragments of dicotyledonous and palm trees. On the Puerco, portions of trunks and limbs are strewn on the hills and ravines; in some localities the mass of fragments indicating the place where some large tree had broken up. At one point east of the river I found the stump of a dicotyledonous tree which measured 5 feet in diameter.

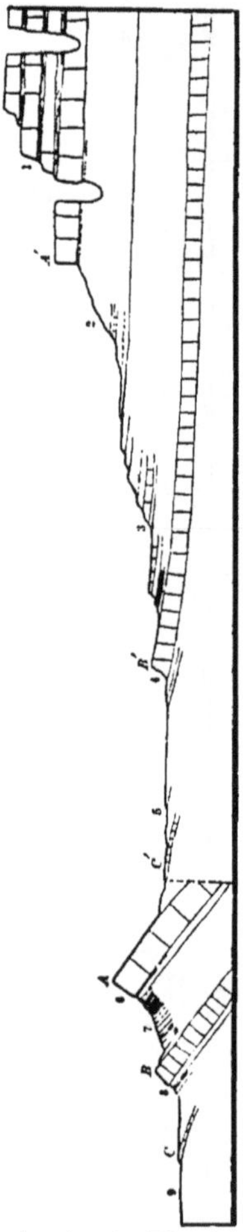

The fauna of this formation is different from that of the other Eocenes in the presence of a saurian, *Champsosaurus*, which is characteristic of the Laramie Cretaceous, and of marsupial Mammals (*Ptilodus* and *Catopsalis*) which are remnants of a type known otherwise from the Jurassic. Characteristic genera are *Catathlæus*, a many-toed omnivore, *Psittacotherium*, a gnawing Tillodont, and various flesh-eaters with primitive teeth. *Coryphodon* is, so far, unknown.

FIG. 3.—Section on the San Juan River, Colorado. (From Dr. Hayden's Annual Report, 1875, p. 252.) 1, Wasatch; 2, Puerco; 3?, Laramie (with lignite vein); 5–8, Nos. 4 and 5, Cretaceous; 9, Benton Cretaceous. The deep notches in the Wasatch bed represent the cañons of (left) the La Plata and (right) the Animas Rivers.

THE WASATCH.

The Wasatch Group is the lowest of a series of these fresh-water Tertiary groups, all of which are intimately connected, not only by an evident continuity of sedimentation throughout, but also by the passage of a portion of the molluscan species from one group up into the next above. Not only were those three groups, aggregating more than a mile in thickness, evidently produced by uninterrupted sedimentation, but it seems equally evident that it was likewise uninterrupted between the Laramie and Wasatch epochs, although there was then a change from brackish to fresh waters, and a consequent change of all the species of invertebrates then inhabiting those waters.

In his annual report for 1870, Dr. Hayden proposed the name "Wasatch Group" for a series of strata that are extensively developed in Southern Wyoming and adjacent parts of Utah and Colorado. I regard the series of strata to which Mr. King has given the name "Vermilion Creek Group," and Professor Powell that of "Bitter Creek Group," as geologically equivalent with the Wasatch Group of Dr. Hayden, and I therefore use that name in this report in accordance with the recognized rule in such cases.

The preceding remarks I have quoted from the report of Dr. C. A. White to Dr. Hayden,* as expressive of the position of this important formation. In lithological character, the Wasatch consists of a mixed arenaceo-calcareous marl, alternating with beds of white or rusty sandstone. The more massive beds of sandstone are in New Mexico, Colorado, and Wyoming, at the base of the formation. The marls readily weather into the fantastic forms and canyon labyrinths of bad-land scenery. The marls often contain concretionary masses of a highly silicious limestone, which cover the banks and slopes of the bluffs with thousands of angular fragments. It is characteristic of this formation that the marls contain brightly colored, usually red, strata; and in many localities the colors are various, giving the escarpments a brilliantly banded appearance.

Petrographically, this formation has two divisions, the Wasatch proper and the Green River beds; the latter name having sometimes been given to the entire formation as well as the former. Dr. White thus describes it:†

> Resting immediately and conformably upon the Wasatch are the strata of the Green River Group. Although intimately connected with the former by continuous sedimentation and specific identity of molluscan species, they differ considerably from those of that group in general aspect, and in composition also. The group is, lithologically, at least, separable into two divisions, but they are not regarded as severally of co-ordinate value with the other recognized Tertiary groups. The lower division consists mainly of silicious and sandy shales and laminated and thin-bedded sandstones, with, in some places, especially in the western part of this district, frequent layers of hard, dark-colored carbonaceous shales. In some places the strata are also quite calcareous, occasional layers being nearly pure, compact, finely-laminated limestone. Others of the calcareous layers are sometimes oölitic in texture. The general aspect of the strata, as seen exposed at a distance, is light gray.
>
> The upper division consists mainly of sandstones that are coarser, as well as less thinly and distinctly bedded, than those of the lower division. In some parts it is shaly and in others carbonaceous. Much of its sandstone is ferruginous in aspect, instead of having the gray tint that the lower division has. Sometimes certain beds of its sandstones are earthy and easily disintegrated, often leaving, weathered out of the mass, spherical concretions of hard sandstone that vary in size from a fraction of an inch to two or three feet in diameter. Other beds sometimes present buttress-like masses in the brow of bluffs, which form conspicuous and somewhat remarkable features in the landscape. Such features are very characteristic of this division in the bluffs of Green River, in the vicinity of Green River City, Wyo., and, to a less extent, they also appear in the bluffs which border the cañon and valley of White River, in the southwest portion of this district.

*Annual Report, 1876 (1878), p. 35.

†Annual Report U. S. Geol. Surv. Terrs., 1876 (1878), p. 35.

The invertebrate fossils which this group affords are similar to those that are found in the fresh-water portion of the Wasatch Group, some of the species being

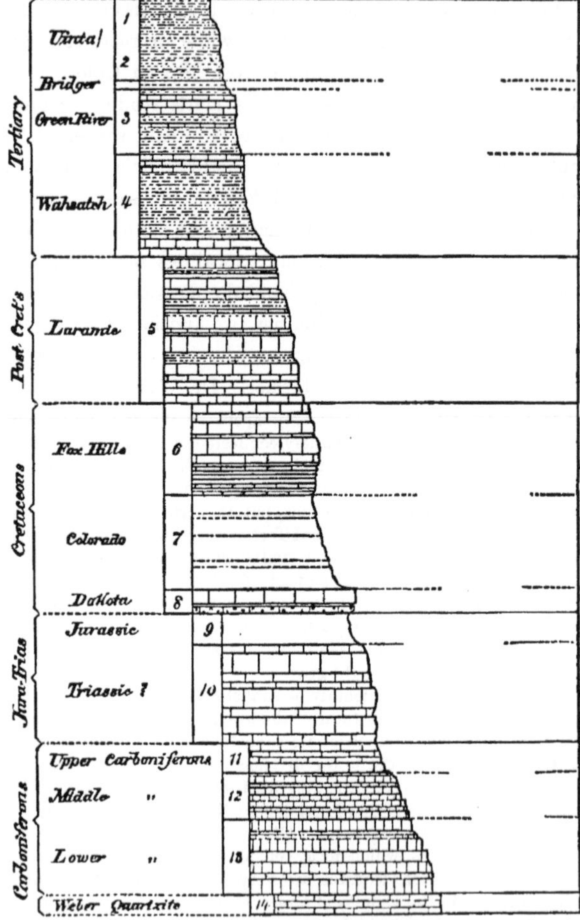

FIG. 4.—General section in the Yampa district. White in Annual Report United States Geological Survey Territories, 1876.

identical, and indicate a purely fresh-water condition throughout. They are almost wholly molluscan, and belong to the branchiferous genera *Unio*, *Viviparus*, and *Gonio-*

basis, besides several genera of pulmonate gastropods, including both the limnophile and geophile divisions. The Green River Group has become somewhat noted for the fossil fishes that have been discovered in its strata in Wyoming, and, like the Wasatch Group, it has at various localities also furnished considerable collections of fossil vertebrates and plants.

Of the few vertebrate fossils known from the Green River division, some are identical with those of the Wasatch, while at least one genus of fishes is common to the Bridger.

The Wasatch beds proper are much more widely distributed than those of the Green River. They appear first in the south in Northwestern New Mexico, and extend thence into the adjacent parts of Colorado. They are exposed over extensive areas of Colorado west of the Rocky Mountains, and reappear in Southwestern Wyoming. They extend along the western portion of the Green River Valley, whose northern portion they entirely occupy. On the eastern side of the Wind River Mountains it has, according to Hayden, an exposure of from one to five miles in width for a distance of one hundred miles, from the source of the Wind River to the Sweet Water River. North of this point it fills the extensive basin of the Big Horn River to the borders of Montana. It does not occur east of the Rocky Mountain range. The thicknesses given by geologists are the following:

Northwestern New Mexico (Cope).

	Feet.
Red-striped marls	1,500
Reddish-brown sandstone	1,000
	2,500

Rio San Juan, Colorado (Holmes).

Coarse yellowish sandstones, alternating with variegated marls	1,200

White and Yampa Reservations (Endlich and White).

Chiefly yellow and reddish sandstones, alternating with shales	1,500

Bear River, Wyoming (Hayden).

Red-banded marls	700
Sandstones and shales	800
	1,500

Wind River Valley (Hayden).

Variegated marls and sandstones	5,000

The Green River division of the Wasatch is much less extensively distributed than the Wasatch proper. Its exposures are confined to the valley of Green River, particularly the regions between its affluents both north and south of the Uinta Mountains. In the Bridger Basin it forms a wide rim around the Bridger formation, and is especially developed on Fontanelle Creek, and on Bitter Creek, and the region to the south of it. I here found its thickness to be 1,200 feet.* Farther south, in Western Colorado, near the Yampa River, Dr. White gives its depth at 1,400 feet.† Farther south, in Western Colorado, Dr. A. C. Peale‡ gives the united thickness of this formation and the Wasatch at 7,670 feet; but how much of this is to be referred to the Green River proper we are not informed. It does not appear to exist on the San Juan, according to Endlich and Holmes, and I did not find it in New Mexico.

According to King, the deposits of the Green River formation rest unconformably on those of the Wasatch.§ He also believes that it has a considerable extent west of the Wasatch Mountains, over parts of Utah and Nevada. Under the head of the Bridger formation I show that the paleontological evidence is opposed to the identification of these "Amyzon" beds with the Green River, and that they are probably of later origin. There is, however, a series of calcareous and silico-calcareous beds in Central Utah, in Sevier and San Pete Counties, which contain the remains of different species of vertebrates from those which have been derived from either the Green River or Amyzon beds. These are *Crocodilus* sp., *Clastes* sp., and a fish provisionally referred to *Priscacara*, under the name of *P. testudinaria*. There is nothing to determine to which of the Eocenes this formation should be referred, but it is tolerably certain that it is to be distinguished from the Amyzon beds. In its petrographic characters it is most like the Green River.‖

The writer first referred the Wasatch to the Eocene division of the Tertiary, it having been previously regarded as Miocene. (Proceedings American Philosophical Society, February, 1872.)

The vertebrate fauna of the Wasatch is rich, and presents many pecu-

* Annual Report U. S. Geol. Surv., 1873, pp. 436, 437.
† Annual Report U. S. Geol. Surv., 1876, p. 36.
‡ Annual Report 1874, p. 156.
§ U. S. Survey of the Fortieth Parallel, i, p. 377.
‖ See American Naturalist, April, 1880.

liarities. It is nearly identical with that of the Suessonian of Western Europe, which is at the base of the Eocene series. The fullest account of it is that which I have given in the Report of Captain Wheeler of Explorations and Surveys West of the 100th Meridian, vol. IV.

THE BRIDGER.

This is one of the more important of the groups among those that, in Western North America, are referred to the Tertiary period, especially as regards the vertebrate remains that have been obtained from its strata. It is most fully and characteristically developed in the region known as the Green River basin, north of the Uinta Mountains, only the southeastern portion of the formation, so far as it is now known, extending into Northwestern Colorado. In its typical localities it is found resting conformably upon the Green River Group, into which it passes without a distinct plane of demarkation among the strata.

Its molluscan fossil remains correspond closely with those of the Green River Group, some of the species being common to both, all indicating a purely fresh condition of the waters in which the strata of both groups were deposited. At the typical localities the group is composed in great part of soft, variegated, bad-land sandstones, a peculiar greenish color often predominating over the others, which are reddish, purple, bluish, and gray. Limestone strata, marly and clayey beds, and cherty layers are not uncommon, and grits and gravelly layers sometimes occur.

To the above general remarks of Dr. C. A. White I add, that the material of this formation consists of indurated clays more or less arenaceous, which display various degrees of hardness. The harder beds are, however, thin, and the intervening strata yield readily to meteoric influences. They are frequently quite arenaceous, and rather thin beds of conglomerate are not uncommon. The colors that predominate are greenish-gray and brownish-green, with frequent ash-colored beds. The peculiar condition of hardness of most of the strata render it one of the formations which most generally present the bad-land scenery; it permits the erosive action of the elements without general breaking down, great numbers of fragments of the strata remaining in spaces between the lines of erosive action. The result is the extraordinary scenery of Black's Fork, Church Buttes, and Mammoth Buttes, of which mention will be made in the section of this volume especially treating of the Bridger formation.

The distribution of the Bridger formation is limited, and is, so far as I am aware, restricted to three areas, whose mutual connection is as yet uncertain. Its principal mass is in the Bridger basin, which extends from the northern base of the Uinta Mountains to the latitude of the mouth of the Big Sandy River northward. In this area it reaches a depth, according to King,* of 2,000 or 2,500 feet. A second district is also in Wyoming, and lies east of Green River between Bitter Creek and the northern boundary of Colorado, in what is called by King the Washakie basin. The depth of the formation there reaches 1,200 feet.† The third region is in Western Colorado, where it loses much of its importance. Dr. C. A. White found it only 100 feet in thickness near the White River.‡ Dr. Peale found it near the Gunnison River, as he discovered vertebræ of *Pappichthys*, a genus which belongs to this horizon only; but he did not distinguish it from the underlying formations, so that I do not know its thickness at that point. South of this locality it is unknown.

As already pointed out, this period is especially characterized by a peculiar and rich vertebrate fauna.

THE UINTA.

Resting directly, but by unconformity of sequence, upon all the Tertiary and Cretaceous groups in the region surrounding the eastern end of the Uinta Mountain range is another Tertiary group, that has received the name of "Uinta Group" from Mr. King, and "Brown's Park Group" from Major Powell.§ It is possible that this group was deposited continuously, at least in part, with the Bridger Group, but at the places where the junction between the two groups has been seen in this region there is an evident unconformity, both by displacement and erosion.

The group consists of fine and coarse sandstones, with frequent layers of gravel, and occasionally both cherty and calcareous layers occur. The sandstones are sometimes firm and regularly bedded, and sometimes soft and partaking of the character of bad-land material. The color varies from gray to dull reddish-brown, the former prevailing north of the Uinta Mountains and the latter south of them.

The only invertebrate fossils that are known to have been discovered in the strata of this group are some specimens of a *Physa*, very like a recent species. Therefore, invertebrate paleontology has furnished no evidence of its assumed Tertiary age and lacustrine conditions of its deposition. Its fresh-water origin, however, seems unquestionable, because of its intra-continental position, its limited extent, and the fact that none but fresh-water deposits are known in this part of the continent that are of later date than the close of the Laramie period.

To these remarks of Dr. White I add, that several species of *vertebrata* have been obtained from this formation by Professor Marsh, who has deter-

* Geol. Explor. Fortieth Parallel, ii, p. 245.

† Annual Report U. S. Geol. Surv. Terrs., 1873 (1874), pp. 436–437.

‡ Annual Report, 1876, p. 36.

§ Annual Report U. S. Geol. Surv. Terrs., 1874, pp. 157, 158.

mined from it a few genera of Tertiary and Upper Eocene character. Such are of *Mesodonta*, the genus *Hyopsodus*, and of *Ungulata*, the Perissodactyle form *Amynodon*.

THE WHITE RIVER.

The materials of which the beds of this formation are composed in their eastern division, are calcareous clays and marls, alternating with a few unimportant strata of light-colored sandstone. They are white and gray, with occasionally a pink and red, and sometimes greenish tinges. The beds of the western deposit in Oregon consist of a more or less indurated mud, which is, according to King, of trachytic origin, which is rarely hard, and frequently rather soft. Its predominating color is light green, but is frequently olive and light brown. The depth of the formation on the

Fig. 5.—Scene in the Bad Lands of the White River formation in Nebraska. From Dr. Hayden.

White River of Nebraska is, according to Hayden,* about 150 feet; and on Crow Creek, Colorado, according to King,† 300 feet. Sixty miles east of Crow Creek I estimate its thickness as somewhat greater. The Truckee beds of Oregon have, according to Marsh, a depth of from 3,000 to 4,000 feet, and King estimates the deposit exposed in the Hawsoh Mountains, Nevada, at 2,300 feet.‡ An extensive deposit exposed in the region of the Cajon Pass, Southern California, is suspected by King to belong to the same horizon.

The following section by Dr. Hayden exhibits the strata of both this formation and the overlying Loup Fork epoch, as displayed on the White River, Nebraska. The bed marked E, and those above it, belong to the last-named formation.

		Subdivisions.	Localities.	Estimated thickness.
Miocene Tertiary.	Bed H.	Gray and greenish-gray sandstone, varying from a very fine compact structure to a conglomerate.	Bijou Hills, Medicine Hills, Eagle Nest Hills.	20 feet.
	Bed G.	Yellowish-gray grit, passing down into a yellow and light-yellow argillo-calcareous marl, with numerous calcareous concretions and much crystalline material, like sulphate of baryta. Fossils: Hippotherium, Protohippus, Steneofiber, &c.	Bijou Hills, Medicine Hills, Eagle Nest Hills, and numerous localities on south side of White River, also at the head of Teton River.	50 feet.
	Bed F.	Grayish and light-gray rather coarse-grained sandstone, with much sulphate of alumina? disseminated through it.	Along White River Valley, on the south side.	20 feet.
	Bed E.	Yellowish and flesh-colored indurated argillo-calcareous bed, with tough argillo-calcareous concretions, containing "Testudo, Hippotherium, Steneofiber, Oreodon, Rhinoceros," &c.	Seen along the White River Valley, on the south side.	30 feet.
	Bed D.	Yellow and light-yellow calcareous marl, with argillo-calcareous concretions and slabs of silicious limestone, containing well-preserved fresh-water shells.	On the south side of White River. Seen in its greatest thickness at Pinao's Spring.	40 feet.

* Proceedings Academy Philada., 1857, p. 153.
† Report of Geol. Survey of 40th Parallel, i, 410.
‡ L. c., p. 423; l. c., p. 415.

		Subdivisions.	Localities.	Estimated thickness.
Miocene Tertiary.	Bed C.	Light-gray silicious grit, sometimes forming a compact fine-grained sandstone.	Seen on both sides of White River. Also at Ash Grove Spring.	20 feet.
	Bed B.	A reddish flesh-colored argillo-calcareous indurated material, passing down into a gray clay, containing concretionary sandstone, sometimes an aggregate of angular grains of quartz, underlaid by a flesh-colored argillo-calcareous indurated stratum, containing a profusion of mammalian and chelonian remains. Turtle and Oreodon bed.	Revealed on both sides of White River and throughout the main body of the Bad Lands.	80 feet.
	Bed A.	Light-gray calcareous grit, passing down into a stratum composed of an aggregate of rather coarse granular quartz, underlaid by an ash-colored argillaceous indurated bed with a greenish tinge. Menodus bed.	Best developed at the entrance of the basin from Bear Creek. Seen also in the channel of White River.	50 feet.

The following diagram represents without much detail the section in Eastern Colorado, along the Horse Tail Creek, from the Chalk Bluffs southward. (See fig. 6, p. 16.)

At both localities the lower beds carry the bones of the gigantic *Chalicotheriidæ*, *Menodus* in Nebraska, and *Symborodon* with *Menodus* in Colorado. But few other types occur in this bed in Colorado, the great number of genera and species being found in bed B, in which I did not discover any fragments of *Chalicotheriidæ* among a large quantity of remains of *Ungulata*, *Carnivora*, *Rodentia*, etc. The lithology is as follows: Bed A is a white calcareous soft clay rock, breaking into angular fragments. Bed B has a similar mineral character, with frequently a red color of different obscure shades. Bed C is a sandstone of varying persistence. Bed D is a white argillaceous rock like that of bed A. Fossils are less numerous than in bed B, and included no *Symborodons* nor other *Chalicotheriidæ*.

The eastern area of this formation is the true White River epoch of Hayden; the western deposits form the Truckee epoch of King. I named this formation the Oregon, but Mr. King's name is the older, and must be retained.*

According to Professor Condon, the Truckee formation of Oregon, on the John Day River, rests unconformably on the laminated beds containing *Toxodium* and fish remains, which, as I have suggested on a previous page, may be an extension of the Amyzon shales. These in turn rest on a formation of hard laminated beds, which contain an abundance of *Calamites*, which doubtless belong to the Triassic or Jurassic period. The Truckee beds are, like the true White River, overlaid by the Loup Fork, and this in turn by heavy beds of basalt.

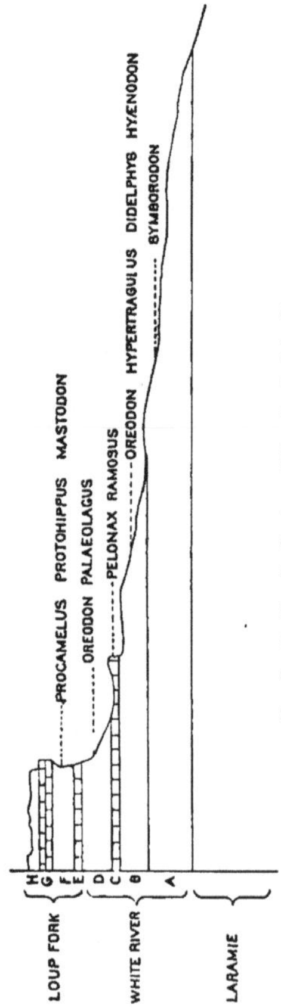

Fig. 6.—Section in Eastern Colorado (original).

The fauna of the Truckee presents some characters which distinguish it from that of the White River. These are, the absence of *Hyænodon* and *Ischyromys* and most of the *Menodontidæ*, and the presence of several genera of *Canidæ*, *Felidæ*, and *Rodentia*. Many genera, and apparently several species, were common to the two epochs.

THE LOUP FORK.

This formation has now been studied in many widely-separated localities in the region west of the Mississippi River. It was discovered by Dr. Hayden, whose collections furnished the basis of Dr. Leidy's determination in 1858.† It was next observed by myself in Colorado in 1873,‡ and twenty-one species were determined; and, in the following year, I identified the Santa Fé marls of New Mexico, already observed by Dr. Hayden, with the

* Bulletin U. S. Geol. Surv. Terrs., v, p. 52.

† See Proc. Acad. Nat. Sci. Phila., 1858, p. 20, and Extinct Mammalia of Dakota and Nebraska.

‡ Bulletin of the U. S. Geol. Surv. Terrs., No. 1, Jan., 1874.

same horizon.* Messrs Hayden and King have discovered it west of the Wasatch Range in Utah and Nevada, and Marsh has observed it in Oregon. Messrs. Dana and Grinnell found it occupying the valley of Deep River in Montana, and Professor Mudge and myself have seen it in Northern and Western Kansas. There is a near lithological resemblance between the strata at these localities, and the fauna presents a common character as distinguished from those which preceded and followed it; but sufficient care has not always been exercised to distinguish its upper members from the *Equus* beds above them. The latter contain a distinct fauna.†

According to King, about 1,500 feet of beds are included in this formation. Hayden found 300 to 400 feet on the Loup Fork and Niobrara Rivers. The following is his section, beginning at the top:

It consists of, "1st. Drab-gray or brown sand, loose, incoherent, with remains of mastodon, elephant, etc. 2d. Sand and gravel, incoherent. 3d. Yellowish-white grit, with many calcareous arenaceous concretions. 4th Gray sand with a greenish tinge; contains the greater part of the organic remains. 5th. Deep yellowish-red arenaceous marl. 6th. Yellowish-gray grit, sometimes quite calcareous, with numerous layers of concretionary limestone from 2 to 6 inches in thickness, containing fresh-water and land shells, *Succinea*, *Limnæa*, *Paludina*, *Helix*, etc., closely allied and perhaps identical with living species; also much wood of a coniferous character." The White River section appears at the upper part of the table, on page 14.

The water-shed between the South Platte River and the Lodge Pole Creek is composed superficially of formations of the Loup Fork epoch, as defined by Hayden. On its southern side is an abrupt descent in the level of the country, which generally presents the character of a line of bluffs varying from 200 to 900 feet in height. This line bends to the eastward, and extends in a nearly east and west direction for at least sixty miles.

The upper portion of this line of bluffs and buttes is composed of the Loup Fork sandstone in alternating strata of harder and softer consistency. It is usually of medium hardness, and such beds, where exposed, on both

* Ann. Rep. Chief of Engineers, 1874, ii, p. 603.

† See Bulletin U. S. Geol. Surv. Terrs, iv, p. 389, and v, p. 47.

the Lodge Pole and South Platte slopes of the water-shed, appear to be penetrated by numerous tortuous friable silicious rods and stem-like bodies They resemble the roots of the vegetation of a swamp, and such they may have been, as the stratum is frequently filled with remains of animals which have been buried while it was in a soft state. No better-preserved remains of plants were seen. The depth of the entire formation is not more than 75 feet, of which the softer beds are the lower, and vary in depth from one foot to twenty. The superior strata are either sandstone, conglomerate, or a coarse sand of varying thickness and alternating relations; the conglomerate contains white pebbles and rolled Loup Fork mammalian remains.

This formation rests on a stratum of white friable argillaceous rock of the White River epoch, as represented in Fig. 6.

The lithological characters above described are precisely those presented by the same formation in New Mexico.*

Mr. King employs the name Niobrara for this formation, but Dr. Hayden's name † was introduced many years previously. The new name has also the disadvantage of being already in use for a horizon of the Cretaceous, which is well distinguished paleontologically.

I have divided the Loup River formation into two divisions, on paleontological grounds,‡ under the names of the *Ticholeptus* bed and the *Procamelus* bed. The former occur in the valley of Deep River, Montana, on the White River in Northern Nebraska, and in Western Nebraska, where it has been found by Mr. Hill. Its fauna presents in Montana a mixture of fossils of the *Procamelus* horizon; while in Nebraska, according to Hayden, its typical genera are accompanied by White River mammalia. In the former region, *Hippotherium*, *Protohippus*, and *Blastomeryx* are mingled with genera allied to *Leptauchenia* and with *Merycochœrus*. In Nebraska, *Leptauchenia* is said to be accompanied by *Ischyromys*, *Palæolagus*, *Hyracodon*, and even *Oreodon*, genera which do not extend to the *Procamelus* bed. There is, however, a question in my mind whether this collocation is entirely correct. It is bed D of Hayden's section in Leidy's Extinct Fauna, Dakota and Nebraska, p. 20.

---

* See Report Lieut. G. M. Wheeler's Explorations West of 100th Meridian, vol. iv, p. 283.
† See Dana's Manual of Geology, edit. 1864, p. 511.
‡ Bull. U. S. Geol. Surv. Terrs., v, pp. 50–52.

The material of the *Ticholeptus* horizon is a more or less friable argillaceous sand; not so coarse and gritty as the *Procamelus* bed, nor so calcareo-argillaceous as the White River.

The *Procamelus* bed is extensively distributed. It is found in Kansas, Nebraska, Colorado, New Mexico, Utah, Nevada, and Oregon.

THE EQUUS BEDS.

I can give little information respecting the depth and stratigraphy of the beds of this period as they occur on the plains west of the Mississippi River, for although sections of them as they occur in Nebraska and elsewhere have doubtless been published by authors, their paleontological status has not been determined for the localities described. My own knowledge of the deposits is based on localities in California and Oregon. In Nebraska they have probably been confounded with the Loup Fork, beds. They represent the latest of all the Tertiary lakes, and include a fauna which consists of a mixture of extinct and living species, with a few extinct genera.

I have received fossils of this age from Idaho, Washington, Oregon, and California. The most important locality in Central Oregon is from thirty to forty miles east of Silver Lake.* The depth of the formation is unknown, but it is probably not great. It consists, first, of loose sand above, which is moved and piled into dunes by the wind; second, of a soft clay bed a few inches in thickness; third, by a bed of sand of 1 or 2 feet in depth; then a bed of clay mixed with sand of unknown depth. The middle bed of sand is fossiliferous. In Northern and Middle California the formation is chiefly gravel, and reaches a depth in some localities of several hundred feet. Here, as has been proven by Whitney, it contains human remains, associated with *Mastodon*, *Equus*, *Auchenia*, etc. I have obtained *Mylodon* from the same gravel.

Traces of this fauna are found over the eastern United States, and occur in deposits in the caverns excavated in the Lower Silurian and Carboniferous limestones, wherever the conditions are suitable. This deposit is a red or orange calcareous mud, varied with strata of stalagmite and

* See American Naturalist, 1878, p. 125.

gypsum. Remains of the fauna are found in clay deposits along several of the Atlantic Rivers, as the Delaware and Potomac.

It is probable that the formation in the western localities mentioned is mostly sand. Near Carson City, Nevada, it consists of a light-buff friable calcareous sandstone.

This is the Upper Pliocene of King, and the Post-Pliocene of various writers.

FIG. 6a.—Sand Hills Northwestern Nebraska From Hayden.

SECTION II.

THE HORIZONTAL RELATIONS OF THE NORTH AMERICAN TERTIARIES WITH THOSE OF EUROPE.

Uniformity of system requires that an identical scale of stratigraphy be employed by all geologists. In accomplishing this object, the students of distinct regions necessarily rely on paleontology as the guide in making determinations of the relations of strata, since determination by observation of continuity is impossible.

The progress of the vertebrate paleontology in North America of late years has been such as to permit of comparisons with the extinct faunæ of Europe and other continents, which give definiteness to our knowledge of the relations between their geologic periods. Comparisons made by Morton and Leidy had nearly determined the position of some of the eastern Cretaceous strata, and those of Leidy had approximately fixed those of the White River beds. Lyell and Conrad early determined the positions of the Eastern Tertiaries. My own views as to the European equivalency of our Keuper* and Laramie† were first expressed, and I later established the ages of the Wasatch,‡ Bridger,|| Loup River,§ and Permian¶ formations in America. A more detailed comparison being very necessary, I visited Europe in 1878 for the purpose of examining the rich collections of vertebrate fossils, and read a general synopsis of results before the Congress of Geologists of Paris of that year. The present section embraces a summary of that paper, with some additional matter.**

The history of the succession of life upon any one portion of the earth's surface is replete with matter for speculation. It shows us a series of faunæ succeeding each other, each of which, in many instances, commences without previous announcement in the forms of older periods, and

* Proceedings Academy Phila., 1866, p. 249.

† Report on U. S. Geol. Surv. Fortieth Parallel, iii, p—. Transac. American Philosophical Soc., xiv, 1869, pp. 40, 98, 243.

‡ Proceedings Amer. Philos. Society, February, 1872.

|| Loc. cit., August, 1872.

§ Annual Report U. S. Geol. Surv. Terrs., 1873, p. 402; Proceed. Academy Phila., 1875, p. 257.

¶ American Naturalist, 1878, p. 327; Proceedings Amer. Philos. Society, 1878, p. 530.

** This synopsis was first published in the Bulletin U. S. Geol. Surv. Terrs., v, p. 33. See Comptes Rendus Sténographiques, Congrès Internationale de Géologie 1880, p. 144.

disappears without leaving representatives in later ones. With this basis of fact, which naturally enough has been furnished by the longest explored and best-known portion of the earth, Europe, we turn to other lands, with the hope of obtaining further light upon a subject so full of mystery. These types of life, did they originate in a single centre, from which they disseminated themselves; and, if so, did each form originate in a region of its own, or not? Or, did the same types of generic structure appear at different points on the earth's surface independently; and, if so, whether cotemporaneously or at different times?

For a solution of these and similar questions, we naturally look to a comparison of the facts first established, with those obtained more recently by exploration in other regions. In this quest, no portion of the earth offers greater promise of results than America. As the second great continent, separated from the other by the greatest possible water surface, we anticipate the widest diversity in the character of its life-history. If the types of life have originated independently, we will find evidence of it by studying American paleontology; if their origin has been through gradual modification, America should furnish us with many intermediate faunæ.

Let us first consider the nature of the evidence on which we should rely in classifying faunæ and the deposits which contain them. We are accustomed, at present, to rely for our definitions upon all the faunal peculiarities upon which we can seize: the period of appearance of certain types; the duration of certain types; and the disappearance of certain types, depending on orders, families, and genera for the major divisions, and species at a given locality for the lesser. It is, of course, evident that either of the above-mentioned three criteria are variable quantities, since discovery is constantly extending our knowledge of the distribution of types. Hence the definitions are empirical and temporary. We must, then, if we desire a stable system, examine the principles involved, and endeavor to discover definitions which stand on stronger foundations than those which we now possess.

As a matter of fact, the old definitions of epochs and periods are continually invalidated by new discoveries. As a matter of theory, this should be the case.

To the believers in the doctrine of derivation, the obliteration of faunal distinctions is not a cause of surprise. Such await with confidence the day when complete phylogenies will be possible, and at present regard the interruptions in the succession of life as local only. Will the result then, be, that paleontology will cease to be available in the definition of ages and of deposits? I answer no, on various grounds. Interruptions in the succession of life in any given locality, due to various causes, have doubtless often occurred, and have left traces in the crust of the earth which are ineffaceable by discovery. But apart from this, one fact in this history is patent both to the friends and to the opponents of the doctrine of derivation: It is known that the world has witnessed, at every stage of its history, the extinction of some important type of life. Familiar examples are the *Placodermi* of Paleozoic time, various reptilian groups of Mesozoic time, and the *Amblypoda* of the Tertiary. Each minor subdivision of time offers its own record of persistences and extinctions of particular families and genera.

Now, all departments of biology compel us to recognize the law of classification, that the order of forms is from the less to the more generalized, from the simple to the more complex, and *vice versa*, whether the lines of succession be those of descent or of creative order; and this law is true in time as well as in classification. It follows from this, that all types of life are, at the time of their appearance, less distinct and more general in their characters than they are later in their history.

It also follows, as a consequence of the principle of descent, which states that the types of one age have taken their origin from generalized types of preceding ages, that there is no descent from the most specialized types; which is to say, conversely, that the genera, families, and orders whose extinction has been a marked feature of every geologic age have been the specialized types of those ages.

We now have a clue to a basis of a definition for faunæ, and hence for epochs, which discovery can safely build upon. The successive increments of structure by which an important modification of animal type is introduced, preclude the possibility of exact determination of the time at which such type may be said to have *appeared*. Even where such a point may be arbitrarily fixed, the type must then be less characteristically represented than it is at the other limit of its existence, viz, the period of its disappearance.

For these reasons I must regard the latter criterion as the true one in the discrimination of the subdivisions of geologic time, while the point of the appearance of types must be looked upon as of provisional use only, and this quite independently of the changes which discovery will from time to time compel us to make in our knowledge of the distribution of life in time and space. It must, however, be borne in mind that disappearance may be due to two causes: first, to extinction; and secondly, to modification; a distinction which is entirely essential. The case of disappearance by modification is identical with that of appearance by modification, and cannot be used otherwise in classification. It is, then, the period of extinction of types to which I have reference.

With these principles in view, we attempt the comparison of the extinct faunæ of Europe and North America, employing principally the nomenclature of D'Orbigny for the former, and Hayden for the latter, in the Mesozoic and Tertiary series.

It is well known that no remains of *Vertebrata* have yet been discovered in North America in strata of Silurian age, while several species are known from the Upper Silurian of Europe. The latter are *Placodermi* and sharks, and are not very numerous in species. They have been derived from England, Germany, and Russia. In America, the first fishes appear in the Corniferous limestone at the base of the Devonian. Professor Newberry, who has devoted much attention to this department, points out important distinctions as existing between the Devonian fish faunæ of North America and Europe, and also to important coincidences. The first of these is the occurrence of the genus *Macropetalichthys* in both continents; in Germany in the Eifel limestone, and in America in the Corniferous limestone of Ohio. The other examples are furnished by the Catskill beds of New York and Pennsylvania, which contain a part of the fauna of the old Red Sandstone of Scotland, including the genera *Holoptychius* and *Bothriolepis*.*

The structure of the *Batrachia* of the Coal-Measures is not yet sufficiently well known to enable the most exact comparisons to be made, but

* Geological Survey of Ohio, i, pp. 264, 271.

close parallels, if not identities, of genera exist. Such are the *Oëstocephalus* and *Ceraterpeton* of Ohio as compared with the *Urocordylus* and *Ceraterpeton* of Great Britain.

The Permian vertebrate fauna which I discovered in Illinois and Texas exhibits close parallels, but not yet generic identity, in the two continents. Thus, the American *Clepsydrops* and *Dimetrodon* are near to the *Deuterosaurus* of Perm, in Russia, and the *Lycosaurus* of the mountains of South Africa. The Texan genus *Pariotichus* may, with further information, prove to be identical with *Procolophon* Ow. from the Tafelberg. Humeri of the type discovered by Kutorga in Russia, and by Owen in South Africa, are found in North America, and the same remarkable type has been recently discovered by Gaudry in France. The peculiar type of *Ganocephalous* vertebræ described by me under the genus *Rhachitomus* from Texas, has been discovered by Gaudry in France. The even more remarkable *Cricotus* (Cope), type of the *Embolomera*, is paralleled by the *Diplovertebron* (Fritsch) of Bohemia. *Ectosteorhachis* represents in Texas the genus *Megalichthys*. The present indications are that close similarity between the faunæ of this period in Europe and America will be discovered. Nevertheless, up to the present time no representatives of the striking American forms *Bolosaurus*, *Diadectes*, and *Empedocles* have yet been found in any other continent.

As regards the Triassic fauna, it differs from that of the Permian in being better known in Europe than America. As marine Trias is little developed in North America, so the vertebrate fauna of the Muschelkalk has not been discovered in the latter country. It is otherwise with the Keuper. The characteristic genus of that epoch, *Belodon*, existed in America, and parallels, if not identity, are seen in the genera *Thecodontosaurus* and *Palæosaurus*. These are known in America from teeth only. The reptiles are accompanied in North America, as in Europe, by Stegocephalous *Batrachia*, mostly Labyrinthodonts, but their generic affinities are yet unknown.

The great Jurassic faunæ are as yet but sparsely represented in North American paleontology. The marine *Vertebrata* of the Lias are either unknown or are represented by a few provisional identifications of unsatisfactory fragments. We do not yet know any deposits in North America which contain the typical reptilian genera *Plesiosaurus*, *Ichthyosaurus*, *Plio-*

saurus and *Dimorphodon*, or the fishes of the *Dapediidæ*. This formation, so important in Europe, is almost omitted from the North American series. Several characteristic fossils of the Rocky Mountain region represent the Oölite, particularly the Upper Oölite. Such is a genus not yet distinguishable from *Megalosaurus*. This genus has not been identified beyond doubt from above the Oölite in England. *Teleosaurus* and *Steneosaurus*, and their allies, are not yet known from North American beds. From the same beds in the Rocky Mountain region come genera which nearly resemble the one from the English Oölite (Forest Marble) called by Phillips *Cetiosaurus*, and the genus from the Oxfordian of Honfleur, called by von Meyer *Streptospondylus*. Beyond this no comparisons can be made, and we therefore pass to the rich fauna of the Kimmeridge. North America cannot show such records of this epoch as have been found in Europe. There are no *Archæopteryx*, *Rhamphorhynchus*, nor *Pterodactylus;* no *Leptolepis*, *Thrissops*, nor other of the numerous fishes of Solenhofen. The *Omosaurus* has, however, some very close relatives in the *Camarasaurus* beds of the Rocky Mountains. Remains of the primitive Marsupial fauna which occurs in the Purbeck have been recently detected in the Western Continent. A partial representation of the Wealden fauna of Europe is found in the beds of the Rocky Mountains mingled with the types of the Oölite and Kimmeridge already mentioned. The important genus *Camarasaurus* represents the *Ornithopsis* of Europe, and with *Amphicœlias* included the most gigantic of land animals. The relationships of this fauna to those of the European Jurassic series may be thus exhibited:

American.	**European.**
CAMARASAURUS BEDS.	WEALDEN.
?	*Iguanodon.*
? *Hypsilophodon.*	*Hypsilophodon.*
	Hylæosaurus.
? *Cetiosaurus.*	*Cetiosaurus.*
Camarasaurus.	*Ornithopsis.**
Amphicœlias.	
? *Goniopholis.*	*Goniopholis.*

* *Chondrosteosaurus* Owen.

American.	**European.**
	KIMMERIDGE.
Hypsirhophus.	*Omosaurus.*
Caulodon.	? *Caulodon.**
	OXFORD.
Epanterias.	*Streptospondylus.*
	OÖLITE.
	"*Cetiosaurus.*"
? *Megalosaurus.*	*Megalosaurus.*

From the above table it will be seen how difficult it is to parallelize the related beds of the Jurassic periods of the two continents at the present time. All that can be said is, that many types resembling† nearly those of different horizons of the European Jurassic are found to have lived together or near together in the Rocky Mountain region of North America.

That the Cretaceous fauna of North America was the richest in the cold-blooded *Vertebrata* is indicated by the present state of discovery. The ocean of the interior of the continent deepened from the beginning of the period until the epoch of the Niobrara, and then gradually shallowed until the elevations of the bottom began to divide the waters. The closing scenes of this great period were enacted amid a labyrinth of lagoons and lakes of brackish and fresh water, whose deposits form the beds of the Laramie epoch.

The fauna of the deep-sea epoch, the Niobrara, is the best known. Here the remains of *Pythonomorpha* constitute its prevailing characteristic, while *Elasmosaurus* and *Polycotylus*, with but few species, represent the numerous *Sauropterygia* of Europe. Crocodiles were apparently wanting, while turtles and a peculiar group of *Pterosauria* were only moderately abundant. The fish fauna was very rich and varied. Here the *Saurodontidæ*, like the molluscous family of the *Rudistes*, appeared and as soon dis-

* *Iguanodon præcursor* Sauv.

† A near affinity has been shown by Professor Owen to exist between *Eucamerotus* and *Camarasaurus*. Professor Owen believes these genera to be identical; but the neural spines of the anterior dorsal vertebra are very different, being single in the former and double in the latter.

appeared, accompanied by the peculiar form *Erisichthe*, and the family of *Stratodontidæ*. The genera of Mount Lebanon, *Leptotrachelus* and *Spaniodon*, occur in this bed in Dakota; but the closest parallelism is exhibited with the Lower Chalk or Turonian of Western Europe. The general *facies* of the reptilian fauna is that of the Lower Chalk, and there is little doubt that several genera are identical in the two continents, *e. g. Elasmosaurus*. The apparent peculiarity of the Chalk in America is the abundance of forms (four genera) of *Pythonomorpha*, with numerous species, while but two genera have yet been found in Europe, and the presence of birds with *biconcave* vertebræ and teeth. This interesting type, which was first discovered by Seeley in the genus named by him *Enaliornis*, and afterwards found by Marsh to possess teeth, has been found at a lower horizon in England, the Upper Greensand. But in England, France, and Westphalia occur the genera of fishes above mentioned, as *Portheus, Ichthyodectes, Saurodon, Saurocephalus, Erisichthe, Empo, Pachyrhizodus, Enchodus, Leptotrachelus*, etc. This close relationship of the horizons permits an identification, and it is the first instance which appears to me to be susceptible of satisfactory demonstration.

The next horizon of the Cretaceous which has yielded many vertebrate remains in North America is the Fox Hills formation (including the Fort Pierre bed). Here the genus *Mosasaurus* appears in America, and is accompanied by the earliest crocodiles with procœlous vertebræ, and by numerous marine turtles which partake of the characters of both *Chelydridæ* and *Cheloniidæ*, which I have called the *Propleuridæ*. *Beryx* appears first here in America. The predominant genus of fishes is *Enchodus*, and the principal *Dinosauria* are *Lælaps* and *Hadrosaurus*. This horizon has been parallelized with the Maestricht of Europe, and several genera are common to the two beds; such are *Mosasaurus* and *Enchodus*. The genus *Hadrosaurus*, and the family of turtles I have called the *Adocidæ*, remain undiscovered in Europe; hence the identity of faunæ cannot be established.

The lacustrine beds, or summit of the American Cretaceous series, the Laramie of Hayden, present the remains of a populous fauna and a rich flora. The students of the palæobotany have declared this flora to be of Eocene, and the later portions of Miocene, character, while the lacustrine constitution of the strata has influenced the stratigraphic geologists to concur in the view that the formation should be arranged with the Tertiary epochs. That the fauna was of a mixed character is the result of a study of its vertebrate fossils. The predominant type in North America was the *Dinosauria*, which were abundant in species and individuals, and this fact alone will suffice most paleontologists as a reason for referring the epoch to the Cretaceous series. The genera of *Dinosauria* (*Palæoscincus*, *Cionodon*, *Diclonius*, *Monoclonius*, *Dysganus*, etc) have not yet been found in any other part of the world. Mingled with them were species of crocodiles and turtles of indifferent character, while a number of other forms existed which had a limited range in time, and hence are important indicators of stratigraphic position. Such are the genera of fishes, *Myledaphus* Cope and *Clastes* Cope, which have been found also near Reims, France, by Dr. Lemoine, in the Sables de Bracheux, which are regarded as the lowest Tertiary. Such is the curious Saurian type *Champsosaurus* (Cope) (*Simædosaurus* Gerv.), and the turtle genus *Compsemys* Leidy, which Lemoine finds a little higher up in the series, in the conglomerate of Cerny, which is in the lower part of the Suessonian. In France, a genus of the Laramie, *Polythorax*, extends into the Lignite or Upper *Coryphodon* bed of the Suessonian Thus the Laramie is intercalated by its characters between the Cretaceous period on the one hand and the Tertiary on the other, and its fauna includes genera and orders of both great series. These relations may be exhibited in tabular form as follows. I here include the faunæ of the Sables de Bracheux and of the conglomerate of Cerny as one, since both possess the types of the Laramie, while the horizon of the Lignite of Meudon, or the Suessonian, does not.

Sables de Bracheux and Conglomerat de Cerny.	Laramie.
α. Tertiary.	
Lophiochœrus.	
Plesiadapis.	
Pleuraspidotherium.	
Arctocyon.	
Clastes.	*Clastes.*

β. PECULIAR.

Champsosaurus.	*Champsosaurus.*
Compsemys.	*Compsemys.*
Myledaphus.	*Myledaphus.*
	Scapherpeton.
SABLES DE BRACHEUX AND CONGLOMERAT DE CERNY.	LARAMIE.

γ. CRETACEOUS.

	Palæoscincus.
	Dysganus.
	Monoclonius.
	Diclonius.
	Cionodon.
	Lælaps.
	Aublysodon.

If the Conglomerate of Cerny be the same horizon as the Conglomerate of Meudon, we must add *Coryphodon* to the upper left-hand column, and probably *Gastornis* also. The result is clear that the French and American formations together bridge most completely the interval between the Cretaceous and Tertiary series, as has been anticipated by Hayden, in America, on geological grounds. It is also evident that another formation must be added to the series already recognized in France, viz, the Laramie or Post-Cretaceous. This will be defined as the beds of the genera *Champsosaurus* and *Myledaphus*. In France, the presence of mammalia will characterize the formation as a subdivision, for which it is probable that the name Thanetian must be retained; while to the American division, which is characterized by the presence of *Dinosauria*, the name of Laramie beds has been given.

In arranging the Laramie Group, its necessary position is between Tertiary and Cretaceous, but on the Cretaceous side of the boundary, if we retain those grand divisions, which it appears to me to be desirable to do. The reasons for retaining it in the Cretaceous are two, viz: (1) because

Dinosauria are a Mesozic type, not known elsewhere from the Tertiary; (2) because *Mammalia* (should they be found in the future in the Fort Union) are not equal as evidence of Tertiary age, since they have been also found in Jurassic and Triassic beds.

The Eocene fauna is so varied, especially in Europe, that it is necessary to compare the divisions separately, as in the case of the Cretaceous. Thus, the fauna of the Suessonian is quite as distinct from that of the Calcaire Grossier and Gypse (Parisian and Tongrian) in France as are those of the Wasatch and Bridger epochs in North America.

I have already identified the Wasatch with the Suessonian or Orthrocene, on account of the community of the following genera in the two continents: *Coryphodon*, *Hyracotherium*, *Amblyctonus*, *Clastes*, and a form of birds close to *Gastornis*. I can now add *Phenacodus*, *Orotherium* (Cope), and very probably *Hyopsodus*, *Adapis*, *Opisthotomus*, and *Prototomus*. But, as above mentioned, in the lower beds of the Suessonian in France occur genera which are, so far as yet known, wanting in the Wasatch of America, but present in the beds of the Laramie.

The parallelism of the American Wasatch with the Upper Suessonian of France is the second identification which may be regarded as provisionally established. The only important discordant element at present known is the *Tæniodonta* of the Wasatch, which have not so far been found in Europe.

Above the Suessonian, a divergence in the characters of the European and North American faunæ commences, and continues to be marked throughout the remainder of Tertiary time. So far as the *Mammalia* are concerned, the diversity between the continents was greater during the periods of the Upper Eocene and Miocene than at the present era. During these periods, a limited number of genera, common to the two continents, was associated with numerous genera in the one which did not exist in the other. As a consequence, our paleontological means of identification of the horizons are limited to a restricted list, and the task of applying a uniform nomenclature is, under the circumstances, difficult. Another difficulty in the way of determining the place of the American beds in the European scale consists in the fact that the physical history of the two continents during the Terrary period appears to have been different. In America, the changes of evel appear to have been more uniform in character over large areas. Each deposit has a wider geographical extent, and the fauna presents less rregular variation. In Europe we have a great number of comparatively estricted deposits, each of which differs from the others in possessing more or less peculiarity of fauna. After a study of these faunæ, their natural arrangement in Europe into three series—Eocene, Miocene, and Pliocene—does not appear to rest on any solid basis. This is especially true of the distinction between the first two; and authors are at variance as to the point of demarkation between the last two. Thus, the Tongrian is the summit of the Eocene according to Renevier, while Gaudry, with Filhol and others, places it at the base of the Miocene. One opinion is as well supported by facts, as now interpreted, as the other.

As an essential aid in the estimation of the limits of the formations, I appeal to the criterion adopted at the opening of this chapter, viz, the period of extinction of animal groups.

If we take a general view of the Tertiary faunæ, we find that the following well-marked types representing families and higher groups have become extinct, and have left no living descendants or successors: Among *Bunotheria*, the American groups *Tæniodonta* and *Tillodonta;* also the *Mesodonta* of both continents; of *Edentata*, *Macrotherium*, and *Ancylotherium* in Europe, and the *Megatheriidæ* in North America; among the *Carnivora*, the *Hyænodons* and *Proviverræ*, with the *Drepanodons;* of *Ungulata*, the entire order of *Amblypoda*, which, however, doubtless disappeared in some of its members by modification; but its only known suborders, the *Pantodonta* and the *Dinocerata*, become absolutely extinct. Among *Perissodactyla*, both continents lost by extinction the *Chalicotheriidæ*, which terminated in a great development in North America; and the *Rhinoceridæ*. Of *Artiodactyla*, two great divisions, representative of each other in the two continents, totally disappeared, viz, the *Oreodontidæ* and the *Anoplotheriidæ;* to which must be added the *Hyopotamidæ*. Of true ruminants, the most important type which has disappeared from both continents is that of the *Camelidæ*. Of Suilline genera, *Anthracotherium* and *Elotherium* may be looked upon as having left no persistent successors. Last of all, the *Proboscidea* retreated to the continents of the south.

In view of the complexity of the European record, I first present the relations of the above-mentioned phenomena as displayed in the simpler American system. As the present essay commences with the earliest periods, I exhibit the succession in descending order on the page. The horizons of the Tertiary which present distinct terrestrial faunæ in North America have been named the Wasatch, the Bridger, the Uinta, the White River, the Loup Fork, the *Equus* beds, and the Champlain. The types which became extinct* with the close of each of these epochs are the following:

WASATCH.
- *Gastornithidæ.*
- *Pantodonta.*

BRIDGER.
- *Baënidæ.*
- *Tillodonta.*
- *Stypolophus.*
- *Dinocerata.*

UINTA.
- *?Mesodonta.*
- *Amynodon.*

WHITE RIVER.
- *Leptictidæ.*
- *Hyænodon.*
- *Chalicotheriidæ.*
- *Hyopotamidæ.*

LOUP RIVER.
- *Rhinoceridæ.*
- *Hippotherium.*
- *Oreodontidæ.*

EQUUS BEDS.
- *Megatheriidæ.*
- *Drepanadon.*
- *Tapiridæ.*
- *Elephas.*
- *Camelidæ.*

The above table exhibits the present state of our knowledge; it will doubtless be much extended by future discovery, but not otherwise modified.

The numerous able writers on European vertebrate palæontology have more frequently recorded the appearance of types in defining their faunal

* This means, as already mentioned, the forms which left no direct successors in the Nearctic and and Palæarctic faunæ.

divisions than their disappearance. The following table is compiled from the writings of Gervais, Gaudry, Pomel, Filhol, Renevier, and others, but is not as complete as I would desire.

SUESSONIAN.
Pantodonta.
PARISIAN (Bruxellian, Bartonian, and Sestian).
Palæophis (Bruxellian).
Proviverra.
Pterodon.
Mesodonta.
Lophiodon (Bruxellian).
TONGRIAN.
Palæotheridæ.
Chalicotherinm.
Anoplotheridæ.
Elotherium.
AQUITANIAN.
Hyænodon.
Hyopotamus.

FALUNIAN.
Anchitherium.
Anthracotherium.
Palæochœrus.
Cænotherium.
OENINGIAN.
Ancylotherium.
Dinotherium.
Hippotherium.
Aceratherium.
SUBAPENNINE.
Mastodon.
Tapiridæ.
DILUVIAL.
Hyæna.
Drepanodon.
Elephas.
Rhinocerus.
Hippopotamus.

The above tables show that the history of mammalian life in the two continents presents many points of resemblance; but that there is a great difficulty in correlating the epochs represented by the known faunæ. As regards the two primary divisions, Eocene and Miocene, they have no special *raison d'être*, as such faunæ as the Tongrian and Oeningian are absolutely transitional in their character. More detailed comparisons of the European and American faunæ bring out many relationships not displayed by the above tables, and which I will now briefly consider.

In the American Bridger, various genera of *Mesodonta* represent the few *Adapidæ* of the Parisian, the genus *Adapis** Cuv. being probably common to the two continents. A near ally of the American *Anaptomorphus*, a

* *Notharctus* is undistinguishable from *Adapis* in inferior dental characters.

true Lemur, has been found by M. Filhol in the Phosphorites, and named *Necrolemur*. The characters of the numerous *Carnivora* of the Bridger are as yet unknown. The *Stypolophus* of the Bridger is perhaps the *Prototomus* of the Wasatch, and this again has been discovered by M. Filhol* in France; while a very similar genus has been discovered in the Swiss Siderolitic, and named *Proviverra*. *Hyænodontidæ* probably occur in the Bridger. Nowhere in Europe do we find the *Dinocerata* and *Tillodonta* of the Bridger. *Palæosyops* is also unknown in Europe, but it plays the part in America of the *Palæotherium*, from which it does not greatly differ in structure. The latter genus is most largely developed in the Parisian, but is also characteristic of the Tongrian. *Hyrachyus* is the American *Lophiodon*, the difference between them being but slight; both are found in France; the former in the Lower Parisian, the latter in the Phosphorites. *Tapirulus*† Gerv. is a genus common to the Bridger and to more than one horizon of the Parisian. The squirrel-like rodents of the Bridger are like those of the Parisian, but they are not confined to either epoch. The character which distinguishes the Parisian most widely from the Bridger, besides the absence of the *Dinocerata* and *Tillodonta*, is the presence of numerous Selenodont *Artiodactyla*, as *Xiphodon*, *Cænotherium*, *Amphimeryx*, *Anoplotherium*, etc. These are of primitive type, it is true; the *Anoplotheriidæ* especially having probably four toes in the very short manus (*Eurytherium*), including the pollex, and three behind. They also display the character of a fifth crescent of the superior molars, which is wanting in the higher Selenodont types. But even these genera are absent from the Bridger. The *ensemble* is then that the latter displays relationships backwards, or to the Suessonian, while the Parisian has a later *facies*, constituting an approach to the Tongrian and White River. ‡

The following table presents the relations of the Bridger fauna succinctly, but it is much less complete than we hope to make it when its numerous species are fully described. The Parisian is here regarded as including the divisions Bruxellian, Bartonian, and Sestian (Gypse).

---

* It is described as *Cynohyænodon* with two species.

† Gervais, 1850; *Helaletes* Marsh, 1872, vide Scott, Spier, and Osborne.

‡ See Ann. Rept. U. S. Geol. Surv. Terrs. 1873, pp. 461–462, where this view is proposed.

Parisian.	Bridger.
Didelphys.	? *Didelphys.*
Vespertilionidæ.	*Vespertilionidæ.*
Plesiarctomys.	*Plesiarctomys.*
	Tillodonta.
Hyænodontidæ.	*Hyænodontidæ.*
Adapis.	*Adapis.*
Necrolemur (Phosph).	*Anaptomorphus.*
	Dinocerata.
	Palæosyops.
Palæotherium.	
Lophiodon.	
Hyrachyus (Phosph.).	*Hyrachyus.*
Tapirulus.	*Tapirulus.*
Anthracotherium.	*Achænodon.*
Chæropotamus.	
Dichobune.	
Anoplotherium.	
Xiphodon.	
Amphimeryx.	

The rich Tongrian (Stampian) fauna is, according to authors, represented in the Sables de Fontainebleau, Puy en Velay, Ronzon, Hempstead, and Cadibona in Italy. We find here *Didelphys* in abundance, *Hyænodon*, *Amphicyon*, *Canis*, *Palæotherium*, *Paloplotherium*, *Chalicotherium*, and *Aceratherium*. Of *Artiodactyla*, the Suillines are *Anthracotherium* and *Elotherium;* the Selenodonts, *Hyopotamus* and *Gelocus*. This list is the nearest known counterpart of that of the fauna of the White River epoch of North America. To reproduce the latter we must omit from the above catalogue the genera of *Palæotheriidæ*, and replace them by the allied Chalicotheroid *Menodus* and *Symborodon*, subtract *Anthracotherium*, and add the great body of the *Oreodontidæ*. Then there are included in the White River fauna the higher Selenodont Artiodactyles of the *Poëbrotheriidæ* and *Hypertragulidæ*, the corresponding types of which belong to the fauna of St. Gerand le Puy in France, or the Aquitanian epoch, which directly succeeded the Stampian. In Europe, we have here *Dremotherium*, *Amphitragulus*, *Lophiomeryx*,

Dorcatherium; in America, *Leptomeryx*, *Hypertragulus*, *Hypisodus*, and *Poëbrotherium*. It is curious that while *Leptomeryx* is also European,* it has not yet been found above the Phosphorites. Among Suillines, the *Palæochærus*† of the Oregon White River beds has also not been found below the Aquitanian in Europe. But the American *Didelphys*‡ *Hyænodon*, *Amphicyon*, *Elotherium*, and *Hyopotamus*, with the numerous Chalicotheroid species, show clearly that the White River fauna may be looked upon as a mixture of those of the Stampian and Aquitanian, the former of which is sometimes referred with reason to the Upper Eocene, while the latter is always left in the lowest Miocene. And the solution of this question of position as regards the White River beds appears to me to be at present by no means easy.§ According to the system of Naumann, it should be called Oligocene.

Although Artiodactyles with Selenodont molars are far more abundant in both continents during this period than the last, a remarkable difference is to be observed between them. Those of Europe still largely consist of the types with five crescents, as represented by the numerous *Hyopotami* and *Cænotheria*, while in America the modern four-crescent-bearing molar characterizes almost the entire suborder, the only exception being two species of *Hyopotamus*.

The following table will represent the relations of the White River fauna:

STAMPIAN AND AQUITANIAN.	WHITE RIVER.
Didelphys.	*Didelphys.*
	Leptictidæ.
Protomyidæ	*Protomyidæ.*‖
	Saccomyidæ.¶
STAMPIAN AND AQUITANIAN.	WHITE RIVER.
Steneofiber.	*Steneofiber.*
Leporidæ.	*Leporidæ.*

* I think M. Filhol's *Prodremotherium* is identical with *Leptomeryx*.

† *Thinohyus* Marsh appears to be the same.

‡ *Herpetotherium* Cope; *Peratherium* Aym.

§ See Ann. Report U. S. Geol. Surv. Terrs. 1878, p. 462, where the White River beds are determined as Lower Miocene.

‖ *Ischyromys* Leidy.

¶ *Entoptychus* and *Pleurolicus* Cope.

STAMPIAN AND AQUITANIAN.	WHITE RIVER.
Hyænodon.	*Hyænodon.*
Amphicyon.	*Amphicyon.*
Canis.	*Canis.*
	Temnocyon.
	Enhydrocyon.
Gulo. *	*Gulo.* *
Proælurus.	
Aelurogale (Phosp.).	*Archælurus.*
	Nimravus.
	Dinictis.
Palæotheriidæ.	*Hoplophoneus.*
	Chalicotheriidæ.
	Hyracodon.
Aceratherium.	*Aceratherium.*
	Anchitherium.
Elotherium.	*Elotherium.*
Palæochœrus.	*Palæochœrus.*
Anthracotherium.	
	Oreodontidæ.
Anoplotheriidæ.	
Hyopotamidæ.	*Hyopotamidæ.*
	Poëbrotherium.
Lophiomeryx.	*Hypertragulus.*
Amphitragulus.	
Leptomeryx (Phosph.).	*Leptomeryx.*
Dremotherium.	
	Hypisodus.

The Falunian epoch includes in the large sense the Langhian, Helvetian, and Tortonian divisions, embracing the rich deposits of the Orléanais, of Simorre, and of Sansan. We have here the true Miocene fauna, of which the following genera are characteristic: Edentata, *Macrotherium;* Proboscidea, *Dinotherium*, *Mastodon;* Perissodactyla, *Anchitherium*, *Listriodon;*

---

* *Amphictis* Pom.

† Cope 1874; *Procervulus.* Gaudry, 1878; *Dicrocerus* Cope, 1874 (not Lartet); *Merycodus* et *Cosoryx* Leidy, nomina nuda.

Artiodactyla, *Palæomeryx*, *Dicrocerus*, *Cosoryx;*† Carnivora, *Amphicyon*, *Hyænarctos*, *Drepanodon;* Quadrumana, *Pliopithecus*. The ancient genera *Anthracotherium* and *Cœnotherium* continue throughout, and the existing genera *Arvicola*, *Lutra*, and *Sus* appear. The succeeding epoch, the Oeningian, including with it the horizons of Epplesheim and Pikermi, presents the additional genera *Dorcatherium*, *Helladotherium*, several genera allied to *Antilope*, with *Hippotherium*, the huge edentate *Ancylotherium*, and the monkey *Mesopithecus*.

It is from these materials that we must determine by comparison the American Loup Fork epoch, whose deposits are widely spread, and whose fauna is of well-marked character. Although called by my predecessors Pliocene in age, I have insisted that it should be referred to the Miocene series, and I think that the evidence to that effect which I have produced will be found conclusive. Nevertheless here, as in other American Tertiary horizons, the element of geographical peculiarity enters, and diminishes the number of identical types.

FALUNIAN.	LOUP FORK.
Steneofiber.	*Steneofiber.*
	Mylagaulus.
Macrotherium. } *Ancylotherium.* }	*Morotherium.*
Amphicyon.	*Amphicyon.**
Dinotherium.	*Cœnobasileus.*
Tetralophodon.	*Tetralophodon.*
Aceratherium.	*Aphelops.*
Anchitherium.	
Listriodon.	
Hippotherium (Oeningian).	*Hippotherium.*
	Protohippus.
	Hippidium.†
Cœnotherium.	*Oreodontidæ.*
Anthracotherium.	
Palæomeryx.	*Blastomeryx.*
Dicrocerus.	

* *Canis ursinus* Cope. † *Pliohippus* Marsh.

FALUNIAN.	LOUP FORK.
Cosoryx.	*Cosoryx.*
	Protolabis.
	Procamelus.

The existing genera mentioned as found in the Falunian fauna are paralleled by the *Dicotyles*, *Hystrix*, and *Mustela* of the Loup Fork beds. It is evident that this latter horizon retains in its *Oreodontidæ* the same traces of antiquity that the Falunian does in its *Cænotherium*, but shows a more modern aspect in the omission of *Anchitherium* and its replacement by *Hippotherium* and *Protohippus*, and in the still more modern type *Hippidium*. Although but six genera of the two continents are determined as identical in the above table, yet others, which are facing on the same line, are very nearly allied. Other differences are geographical. The *facies* of the Loup Fork horizon is then a compound of that of the Falunian and Oeningian, or Middle and Upper Miocene.

In commenting on the above-described fauna in 1874,* I remarked that "the proper discrimination of the American Pliocene remains to be accomplished." It was not long after that date that material for making the identification of this horizon on this continent first came into my hands. This was derived from the superior Tertiary of Oregon, and includes a considerable number of species of fishes, birds, and *Mammalia*. I published a list of some of the species in 1878.† The character of the fauna from that region coincides with that which has from time to time been unearthed in the caves and other Eastern deposits, to such an extent, as to lead us to suspect that the differences between them are geographical only. In Europe the Pliocene, or Subapennine, includes, according to D'Orbigny (1855) and Gaudry (1878), the Plaisancian and Astian, which are represented at the following localities:

SUBAPENNINE.

Plaisancian.—Montpellier; Casino (Tuscany).

Astian.—Perrier, near Issoir, Coupet, Vialette (Upper Loire), Chagny; English Crag; part of deposits of the Val d'Arno.

The characteristic of this fauna is the fact that the species belong

* Report Lieut. G. M. Wheeler, iv, Paleontology of New Mexico, 1874, p. 364.

† Bull. Hayden's U. S. Geol. Surv. Terrs. iv, 1878, p. 389.

mostly to existing genera, the chief exception being *Hippotherium.* The horses are chiefly represented by *Equus.* Common genera are *Arctomys, Lepus, Elephas, Mastodon, Tapirus, Sus, Cervus, Antilope, Bos, Canis, Drepanodon, Felis, Ursus.* In the *Equus* beds of Oregon a few extinct genera in like manner share the field with various recent ones, while not a few of the bones are not distinguishable from those of recent species. I give the following list, the extinct species being in italics:

Mylodon sodalis.	Canis latrans.	*Auchenia magna.*
Thomomys (nr.) clusius.	*Elephas primigenius.*	*Auchenia vitakeriana.*
Thomomys talpoides.	*Equus occidentalis.*	*Cervus fortis.*
Castor fiber.	*Equus major.*	
Lutra near *piscinaria.*	*Auchenia hesterna.*	

The species derived from the cave formations of the Eastern States are more numerous, and differ from the Oregon fauna in many respects; yet the parallelism is close in the genera with the *Equus* beds on the one hand and the Pliocene of Europe and South America on the other. The differences distinguishing it from the *Equus* beds of Oregon are, however, such as compel me to regard it as a distinct division of the Pliocene, under the name of the *Megalonyx* beds.*

Megatherium (p).	*Castoroides.*	*Drepanodon Smilodon* (vel).
Mylodon (p).	Lagomys (s).	*Mastodon* (sp).
Megalonyx (p).	Lepus (s).	Equus (sp).
Sciurus (s).	*Anomodon.*	? *Hippotherium* (s).
Arctomys (s).	Scalops.	Tapirus (s).
Jaculus (s).	Arctotherium (p).	Dicotyles (p),
Arvicola (s).	Procyon.	Cariacus (p).
Erethizon.	Canis (sp).	Bos (s).
Hydrochœrus (p).	Mustela (sp).	

In the above list the extinct genera are marked in italics. There exists, as a marked feature of the North American Pliocene, to which I called attention several years ago,† a considerable representation of the fauna of

* Bull. U. S. Geol. Surv. Terrs. v, p. 52, 1879.

† Proc. Acad. Phila. 1857, 156; Proc. Am. Philos. Soc. 1869, 178.

the Pampean formation of South America; such are twelve genera, of which six are extinct genera, and four are peculiar to that formation and fauna. The genera found in the Pampean are marked (*p*), and those of the Subapennine (*s*). In the list from the Oregon localities, *Mylodon* and *Auchenia* were observed to be the only distinctively Pampean genera. As a conclusion of the comparison of the American *Equus* beds in general with those of Europe, it may be stated that the number of identical genera is so large, that we may not hesitate to parallelize them as stratigraphically the same. On the other hand, the agreement with the South American Pampean formation is so marked in some respects as to induce us to believe that the distinction is geographic rather than stratigraphic. Believing that the Pampean formation contains too large a percentage of extinct genera to be properly regarded, as it has been, as Post-Pliocene or Quaternary, its characters, both essentially and as a result of the comparison which I have been able to make, refer it properly to the Pliocene. It appears, then, that the term Pliocene or Subapennine is applicable to the horizon of this fauna in Europe and North and South America.

RÉSUMÉ OF COMPARISONS.

The conclusions to be derived from the facts enumerated in the preceding pages are as follows:

I. Portions of all the faunæ of all the primary divisions of geologic time have been recognized on both the European and North American continents.

II. Parallels requiring general identification of principal divisions of these faunæ may be detected. These are: the Coal Measures; the Permian; the Laramie; the Maestrichtian; the Eocene; the Miocene.

III. Exact identifications of restricted divisions may be made in a few instances only; such are the Turonian and the Niobrara; the Suessonian and the Wasatch; the *Equus beds* and the Pliocene.

It is not impossible that some of the relations mentioned in II will be by the accession of further information, referrible to the list of exact comparisons in III. In all cases of identification it will be necessary to employ the name first proposed with definition for the horizon, other names taking places as synonymes. But in the majority of strata it will be necessary to preserve the special names: thus those of Bear River, Bridger, White River, and Loup Fork, applicable to beds having no exact equivalents in Europe, cannot be set aside for older ones, but must themselves be applied to corresponding faunal horizons elsewhere, should any such be found in future. And it will rarely happen that the minor subdivisions of such faunæ will be found to have an extent sufficient to warrant their having other than special names.

In the accompanying diagram the series of strata of Europe and North America, as determined by their paleontology, are placed side by side for the purpose of comparison. Complete parallelism can only be predicated of divisions of the first order, separated by horizontal lines. Such relation is indicated by exact opposition of the areas representing the epochs in question. In giving the minor divisions of the European epochs I have generally restricted myself to those of the epochs which have American equivalents. Where there is no equivalent on one side or the other, the vacancy is represented by a diagonal line. In employing names for epochs and their divisions, I have adhered to the law of priority as far as my knowledge of the literature allows.* I have given a few names to American formations, but only in instances where such had not been previously given. In such cases I have preferred employing the name of some characteristic genus of fossils, rather than one of local origin.

COMPARISON WITH THE SCALE DERIVED FROM PALEOBOTANY.

I now consider another kind of relation presented by the American and European horizons. I allude to the floræ, for my knowledge of which I am necessarily dependent on the labors of others. I first exhibit the determinations of the ages of the American formations, already discussed, made by Mr. Lesquereux on the basis of the vegetable remains which they contain. I place by the side of these my own determinations of the ages of the same beds, as already related. The former are derived from the full memoir of Mr. Lesquereux in the Annual Report of the United States Geological Sur-

*In the European system I have been much aided by the writings of Woodward, Gervais, Hébert, Pomel, Gaudry, etc., and by the atlas of Professor Renevier, of Lausanne.

vey of the Territories for 1872, pp. 410–417. It will be observed that there is a constant discrepancy between the two tables.

Lesquereux.	Formation.	Cope.
....	Loup Fork	Miocene.
....	White River	Oligocene.
Upper Miocene..................	Bridger	Middle Eocene.
Miocene	{ Wasatch .... { Green River....................	Lower Eocene.
Lower Eocene...................	Laramie	Upper Cretaceous.

If the determinations of Mr. Lesquereux be correct,* it is evident from the above that the vegetable life of North America reached its present condition one epoch or period earlier than the higher *Vertebrata*, and that the nomenclature is thus thrown back by so much. It would appear that the recent flora of North America is a period older than the fauna, *i. e.*, has persisted longer than the latter by a certain length of geologic time. Applying the same reasoning to the past, I embodied the idea in reference to the Laramie period ("Fort Union") in the statement that "a Cretaceous fauna was then contemporary with a Tertiary flora";† and, later, that "an Eocene fauna was contemporary with a Miocene flora." It may have to be added that a Miocene fauna was contemporary with a Pliocene flora. Since Mr. Lesquereux has the support of the best paleobotonists of Europe in his conclusions, it is useless to take the ground assumed by a few of my colleagues, that the former gentleman has simply erred in his determinations. He gives us grounds for believing that he has not done so, by giving us the European standard by which his identifications are governed.‡ It is as follows:

Pliocene ...	Lower limits not positively fixed; largely developed in Italy. †(Subapennine, E.D.C.)
Miocene ...	Oeningian; Mayencian; Aquitanian.
Oligocene..	Tongrian.
Eocene	Gypse of Aix; Alum Bay; Mt. Bolca; London Clay; Sheppey; Grès of the Sarthe.
Paleocene..	Upper Landenian; Sezanne (=Paniselian). Suessonian (Lignitic Soissonais; Sables de Bracheux); Lower Landenian. Hersian; Gelinden. Limestone of Mons, overlying unconformably the Maestrichtian.

* The above parallels are well presented by Dr. Peale in his report to Dr. Hayden, Ann. Rept. U. S. Geol. Surv. Terrs. 1874, p. 141 *et seq.* † Bulletin U. S. Geol. Surv. Terrs. I, art. 2. p. 16, April, 1874 ‡ Ann. Report U. S. Geol. Surv. Terrs. 1874, p. 285.

This system, it will be observed, is almost exactly identical with that employed in the preceding pages as the standard of comparison for the *Vertebrata*. Yet it has resulted, from a most careful comparison of both faunæ and floræ of America with this standard scale, that two distinct paleontological series have to be adopted, the one for the vertebrate life and the other for the plants, of the Western Continent. If this result be accurate, and there appears to be no avoiding it, an explanation must be sought. There are only two possible ones; either the animal life of North America has lagged behind that of Europe by one period during past geologic time; or, secondly, the vegetable life of America has been equally in advance of that of Europe during the same period. In other words, if the plant-life of the continents was contemporaneous, ancient types of animals remained a period longer in North America than in Erope. If animal life was contemporaneous, plant-life had advanced by one period in Europe beyond that which it had attained in North America. In any case, either the faunal or the floral standard of estimation of geologic age of strata for North America is a false one, since there can be but one standard of comparison for anything. But this great fact being understood, the evidence of each of the great departments of life possesses its own intrinsic value.

In conclusion, it may be observed that the lacunæ in the series as presented by one continent, render us dependent on the other for the evidence necessary for the complete elucidation of the laws of the creation of animal life. Phylogenies can be thus constructed which would otherwise be impossible, and the results of researches into the earliest types of *Vertebrata* become intelligible. Thus I have been able to prove, in support of a thesis published in 1874, that the earliest Ungulate *Mammalia* were pentadactyle and plantigrade. I have also shown that the ankle-joint had not, in the primitive *Mammalia*, the hinge-like character that it has in the later ones, but that it is without the interlocking superior articulation. The small size of the brain of early *Mammalia*, already pointed out by Lartet, has received extensive confirmation by the researches of Marsh, who has also shown the progressive increase in size of the whole body in various mammalian lines. To these results I have added another, which is derived from the study of numerous Permian *Vertebrata*, viz, that the earliest land vertebrates had a persistent chorda dorsalis.

PART FIRST.

THE PUERCO, WASATCH, AND BRIDGER FAUNÆ.

PART FIRST.

THE WASATCH AND BRIDGER FAUNÆ.

PISCES.

The remains of fishes are abundant in the lacustrine Eocene formations of the United States, and several principal groups are represented. These pertain to the *Elasmobranchi* and the *Hyopomata; Dipnoi* and *Holocephali* are unknown. Of Hyopomatous fishes, indications of the *Crossopterygia* and the *Chondrostei* have not yet been found, but of the third group, the *Actinopteri*, we have several distinct orders, commencing with the more generalized *Ginglymodi* and ending with the specialized *Percomorphi.* The facies of the Eocene fish fauna is that of the existing fresh waters of the United States, exclusive of the great order of the *Plectospondyli* (unless the Amyzon beds are Eocene), and with the addition of two families, *Osteoglossidæ* and *Chromididæ* (aff.), at present confined to the southern hemisphere.

ELASMOBRANCHI.

XIPHOTRYGON Cope.

American Naturalist, 1879, p. 333.

Family *Trygonidæ;* that is, the tail furnished with a serrate spine and the pectoral fins united in front of the rostral cartilage. Teeth closely placed in a few rows, the crowns developed into a triangular cusp, which is directed backwards, as in *Raja.* Pelvic arch without anteriorly directed inferior processes. No superficial ossification of the rostral cartilage. No caudal fins observed.

This genus is *Trygon,* with the teeth of *Raja.* It further differs strikingly from the typical *Trygones* in the form of the caudal spines. These are trigonal in section, and bear a ventral keel, and a serrate edge on each side. The extinct species of the Monte Bolca Eocene, *Trygon muricata*, has, according to Gazzola, the spine of the true *Trygones*.

XIPHOTRYGON ACUTIDENS Cope.

American Naturalist, May, 1879 (April).

Plate I, figs. 1-5.

This species is of graceful proportions, having no great transverse expansion, and possessing a long and slender tail. The size is inferior to that of a fully-grown skate, but much exceeds that of the *Cyclobatis oligodactylus* of the Lebanon.

The ossification of the superficial part of the cranial cartilage is wide, extending to the branchial fissures below. It terminates anteriorly, in a slightly concave truncation, a short distance in front of the orbital openings. Two convex lobes immediately behind the mouth, divided by a median fissure, resemble the labial flaps. They are marked by rather larger hexagons than the other surface. The least hexagons form a longitudinal oval patch on the middle line behind these flaps, which corresponds in position to the superior fontanelle. Posterior to the scapular arch the ossification forms a band on each side of the vertebral column, and, gradually narrowing, disappears near the origin of the caudal spines.

The propterygia extend well forward, giving outline to an acute snout. They are segmented to the extremity. The outline of the fin expands gradually from this apex. The metapterygial border is very stout, and is not so long as the propterygial. The posterior border of the pectoral fin does not extend quite so far posteriorly as the posterior border of the ventral fin. The latter, in turn, extends for about three-fourths the length of the claspers from the base of the fin. Pectoral rays; metapterygial, 31; mesopterygial, 10; propterygial, 41.

The vertebræ are fully ossified; the caudal series becomes very slender distally, and measures nearly twice as long as from the pelvic arch to the anterior border of the cephalic ossification. In the specimen described there are three caudal spines situated near together, whose origins are a little posterior to the middle of the length of the caudal series of vertebræ. They are all depressed at the base and triangular in section beyond, and have an acute ventral edge. The lateral edges are finely and rather remotely serrate, the serration being obsolete on the smallest or anterior spine. In all, the infero-lateral faces of the spine form a shallow groove, like that of a bayonet. These spines are very different from those of the Trygons of the American and European seas, where they are depressed, oval in section, and have the teeth on each side much more closely placed.

The teeth of this species are small. Viewed from below, those of the upper jaw form a very few series of triangles, with their bases approximated and their acute apices directed backwards. Their bases are of different form, and are expanded and probably bifurcate, as several sections or anterior views of teeth are preserved, which exhibit two divergent roots and a flat summit. The functional surfaces of the triangular crowns are flat.

The greater number of the segments of the fin-rays are shown by the sections to have been hollow cylinders, with a fibrous axis

Measurements.

	M.
Total length (entire)	.515
Total width at middle of abdomen	.230
Length of head (without muzzle) to scapular arch above	.100
Length of abdomen to pelvic arch	.064
Length of tail	.351
Length to origin of spine	.160
Length of spine	.040
Depth of spine	.003
Width of base of muzzle	.023
Width between propterygia (greatest)	.060
Width between metapterygia (greatest)	.065
Width of pelvic arch in front	.043

This species is so far known to me from a single specimen. This was obtained from Twin Creek, in the Bear River region of Southwestern Wyoming, by Leslie A. Lee, of Bowdoin College, Maine, who very liberally placed it at my disposal for study and description.

Its presence in the Green River shales, adds to the evidence offered by other anadromous types of fishes, in favor of the view that the Green River Lake had communication with the ocean.

The first information as to the existence of rays in the Green River formation was furnished by Professor Marsh, who obtained a specimen from

Twin Creek. He gave a meager description of the species, but quite neglected to describe the genus, on which account its affinities remain unknown. The presence of several tubercles mentioned by Marsh indicates that it is a different species at least from the *Xiphotrygon acutidens.*

GINGLYMODI.

Cope, Proceedings American Association for the Advancement of Science, 1871, p. 330.

Physostomous *Actinopteri*, with a præcoracoid arch and a coronoid bone of the mandible. Vertebral centra opisthocœlous. Parietal bones in contact; pterotic and opisthotic bones absent; pectoral fin with mesopterygium and five other basal elements.

This order contains as yet but one family, the *Lepidosteidæ*, which embraces three genera, the recent *Litholepis* and *Lepidosteus*, and the extinct *Clastes.* The existing species, as is well known, are confined to the rivers and lakes of North America, while the extinct forms occur in both Europe and North America. The earliest appearance of this type in geological history yet known was in the Laramie or Upper Cretaceous epoch in North America. Individuals of one species, *Clastes occidentalis* Leidy, were very numerous at that period. During the Wasatch or Suessonian in North America, they were equally abundant, and I have described two species from this horizon in New Mexico. Species and individuals are plentiful in the Bridger beds, as indicated in the following pages, but in the various tracts of the White River epoch they are absolutely wanting.* They do not occur in the Loup River deposits east of the Rocky Mountains, and only reappear in the interior of the Continent in the present period. On the other hand, the family is represented in the marine Miocene beds of the Atlantic seaboard by the genus *Pneumatosteus* Cope, of which a single species† has been found in North Carolina. In Europe the *Lepidosteidæ* make their appearance at nearly the same horizon as in America, Lemoine having obtained *Clastes* from the lowest Suessonian (Chalons-Sur-Vesle), and Gervais having determined it from the Upper Suessonian.

* Proceedings American Philosophical Society, 1877, p. 9.

† Proceedings American Philosophical Society, 1869, p. 242.

CLASTES Cope.

Annual Report U. S. Geol. Surv. Terrs., 1872 (1873), p. 633. American Naturalist, 1878, p. 761.

Mandibular ramus, without or with reduced fissure of the dental foramen, and without the groove continuous with it found in *Lepidosteus*. One series of large teeth, with small ones exterior to them in the dentary bone, the inner superior aspect of that bone without prominent dentiferous or rugose rib.

An inspection of French specimens, probably belonging to this genus, has shown that the maxillary bone is much less segmented than in *Lepidosteus*, if it be divided at all. The characters of *Clastes* were originally derived from the under jaw, and I have observed them in two species, one which I suppose to be the *Lepidosteus glaber* Marsh, and the other *L. atrox* Leidy.

The species of this genus resemble in many ways the *Lepidostei* of the present day. Their scales are rhombic and pierced by a duct on the lateral line. The cranial bones are ornamented by tubercles of ganoine, distributed variously, according to the species. Some of these fishes reached a large size, exceeding any now living, others resemble the true *Lepidostei* in this respect.

The first indication of the occurrence of gars in our Western Tertiaries was furnished by Professor Marsh, who announced his discovery of them before the Academy of Philadelphia (Proceedings 1871, p. 105). He named two species, but did not give any descriptions, excepting so far as the statement that one of them has "unusually short vertebræ," and that the other has them "proportionally longer," may be regarded as such. Under these circumstances I have been unable to identify the species referred to, and think that the names proposed for them by Marsh cannot be used.

CLASTES ANAX Cope.

Annual Report U. S. Geol. Surv. Terrs., 1872 (1873), p. 633.

Plate II, figs. 50–52.

Represented by some cranial bones and especially by a post-temporal, which indicate a very large species of gar, two or three times as large as the alligator gar of the Mississippi, (*Litholepis ferox*). The bone has a free, ovate posterior outline, and its superior surface is covered with a thick layer of dense bone, which has not the brilliant surface of ganoine. This substance is thrown into elevated, corrugated ridges, which are generally transverse to the long axis of the bone, and inosculate and are interrupted frequently. The spaces between are as wide as the bases of the ridges.

Measurements.

	M.
Width of bone	.042
Thickness of bone	.012

Found in the Bridger bad lands of Ham's Fork, Wyoming.

CLASTES ATROX Leidy.

Cope, Annual Report U. S. Geol. Surv. Terrs., 634, 1872 (1873).

Lepidosteus atrox Leidy, Proc. Acad. Nat. Sci. Phila., 1873, 97.—Report U. S. Geol. Surv. Terrs., i, 1873, p. 189, Pl. xxxii, figs. 14–15.

Abundant, and represented by both rough and smooth scales, the former from the anterior part of the body. As this species has been already described by Leidy, I only refer to my Plate II, figs. 1–24.

CLASTES CYCLIFERUS Cope.

Annual Report U. S. Geol. Surv. Terrs., 1872 (1873), p. 634.

Plate II, figs. 25–45.

Established on numerous remains of a small species, in which the scales are rather wide, and generally with obtuse extremital angles, and frequently in certain regions of the body entirely rounded at the posterior border. Fragments of the cranial bones are ornamented with scattered tubercles of ganoine of rounded form, and not distributed in lines as in some species. In a fragment from the posterior part of the mandible, there is a single row of large teeth, with a series of minute ones between them on the outer edge of the bone. The external face presents a smooth, superior surface, and a rugose inferior portion which is marked by irregular lines of points of ganoine.

Measurements.

	M.
Depth of dentary bone	.0070
Width of above	.0055
Length of a scale (exposed face)	.0060
Width of a scale (exposed face)	.0060

From the Mammoth Buttes, Washakie Basin.

Clastes cuneatus Cope.

Proceedings American Philosophical Society, 1877, p. 9; (name only).

Plate I, fig. 6.

This gar is represented by numerous specimens from the Manti Shales of Central Utah, some of which are preserved almost entire. None of them exceed a foot in length. I describe the best specimen accessible to me, a small one, kindly lent me through Dr. Hayden by Mr. Barfoot, director of the museum of Salt Lake City.

The proportions are rather stout, and the base of the ventral fin is a little nearer the base of the tail than the end of the snout. The head is not perfectly preserved, but its outline, as clearly defined on the slab of limestone, is wedge-shaped, not longer than in *Lepidosteus platystomus*, but narrower. This view is, however, partly profile. The posterior and inferior borders of the operculum form a continuous segment of a circle, and the depth of the suboperculum is .66 the horizontal width of the operculum. The preoperculum is superficially divided by transverse grooves into four scuta, of which the superior is the largest; they are ornamented with small tubercles of ganoine. All the other bones, including the frontals, present radiating lines of the tubercles which are capped with ganoine, excepting on the operculum and suboperculum, where they form scarcely interrupted ridges. The scales are smooth, even near the scapular arch, and are arranged in eighteen or nineteen longitudinal series, as seen in an oblique row directed obliquely upward from the ventral fin. Fulcra of the ventral fin rather long and slender. The region of the dorsal fin is somewhat disturbed; the fin is in any case situated very far posteriorly; anal and extremity of caudal fins wanting.

Measurements.

	M.
Axial length from end of muzzle to base of caudal fin	.215
Axial length from end of muzzle to base of dorsal fin	.182
Axial length from end of muzzle to base of ventral fin	.125
Axial length from end of muzzle to edge of operculum	.063
Length of skull to preopercular border	.049
Depth of skull at preopercular border	.036
Depth of body at ventral fin	.050

This species differs from some of those of the Bridger formation in the smoothness of the scales on the anterior part of the body. It is a smaller species than most of those of both that formation and the Wasatch. The characteristic vertebræ are exhibited by various specimens.

HALECOMORPHI.

PAPPICHTHYS Cope.

Annual Report U. S. Geol. Surv. Terrs., 1872 (1873), p. 634.

Family *Amiidæ*. Vertebræ short, the abdominal with prominent diapophyses, and with each neurapophysis articulating with two centra; sides of the centrum not pitted. Maxillary bone supporting a single series of teeth and with a supplementary bone on its distal upper border. Dentary bone deeply grooved on the inner side and with but one series of teeth. Surface of cranial bones sculptured.

This genus differs from the existing *Amia* in the presence of only one series of teeth instead of several, on the bones about the mouth. The vertebral centra possess a smaller anteroposterior diameter and relatively greater transverse diameter in the anterior part of the column; but the value of these characters is not yet certainly understood.

The maxillary bone overlaps the premaxillary extensively by its proximal extremity, and presents no condylar facets (*P. plicatus*). The symphysis of the dentaries is not sutural. The condyle of the inferior quadrate is rough (*P. lævis*, *P. plicatus*). Its posterior grooves show the position of a symplectic; while the inferior anterior portion shows a coarse sutural serrate junction with the ectopterygoid (in the above species). The centra of the vertebræ are most transverse anteriorly; in the posterior abdominal region they become subround; in the anterior caudal region, higher than wide; and in the greater part of the notocaudal region are subround They all have a minute notochordal performation. The neurapophysial facets of the anterior and posterior positions are distinct in the anterior abdominal vertebræ, and confluent on the caudals of all the species; the point at which they become confluent is different in the different species. On a few anterior abdominal centra the inferior surface is entire, or displays a slight depression; soon two parallel fissures, one on each side of the median line, appear, which become oblong fossæ. These continue until they become narrowed again, anterior to the caudal series. In the latter they are as well developed as the neurapophysial pits and are much like them.

Several species of this genus have been found in the Bridger formation in Wyoming and Colorado, but it does not occur in the Wasatch. While some of them were first reported by Marsh, who referred them to *Amia*, they were first described by Leidy (Report U. S. Geol. Surv. 4to vol, I., p. 184, Plate xxxviii). One generic and one subgeneric names were used by Dr. Leidy in this connection, but without diagnoses. Among the specimens at my disposal I have found but one genus, to which I gave the name now used, with a characteristic diagnosis

This genus is the earliest known representative of the order *Halecomorphi*, which consists at present of but two genera; the present one and *Amia*. The latter first appears in America in the ? Eocene Amyzon shales of Florissant in the South Park of Colorado, where two species have been found; *A. scutata* Cope, and *A. reticulata* Cope.* *Pappichthys* first occurs in the Bridger formation, and constitutes one of the faunal distinctions between that epoch and the Wasatch, which immediately preceded it. However, it occurs in the Wind River beds, mixed with Wasatch Mammals. In Europe this genus is found at lower horizons than in America, having been discovered by Dr. Lemoine† near Reims, in the Suessonian conglomerate, which answers partly to the lowest Wasatch and partly to the Upper Laramie epochs.

PAPPICHTHYS SCLEROPS Cope.

Annual Report U. S. Geol. Surv. Terrs., 1872 (1873), p. 635.

Plate III, fig. 1.

Established on a ramus of the mandible of one, and other similar specimens of other individuals. These indicate a large fish, equal in size to the alligator gar of the Mississippi. The dentary bone is more compressed and deeper than in *P. plicatus*. The longitudinal groove runs above the middle line, and the portion of the bone below it thins to an edge. The upper portion is thickened, and the alveolar border is wide and bounded by an angle

* See Bulletin of the U. S. Geol. Surv. Terrs., 1875, p. 1.
† Recherches s. l. Oiseaux Foss. Tert. Infer. des Env. Reims, 1878, p. 65.

on the inner side. The alveoli are large and shallow; in .025 m. scarcely three find place. Near the symphysis is a smaller one which is separated by a considerable diastema from the succeeding one (perhaps abnormally). The external face of the bone is rough and somewhat tubercular.

Measurements.

	M.
Depth of dentary at symphysis	.023
Depth of dentary at middle	.038
Depth of dentary at eleventh tooth	.048
Thickness of dentary at eighth tooth	.018

Pappichthys lævis Cope.

Annual Report U. S Geol. Surv. Terrs., 1872 (1873), p. 636.

Plate III, figs. 2–11.

Represented by various fragments, including dentary and vertebral bones. The former differs from that of the species just described in the smaller size of its teeth, there being six in a space occupied by but four in it, at a point where the dentaries have equal depth. In other words, there are four in .0250 m. The alveolar face is also much more oblique, being in fact continuous with the inner face of the bone. The external face of the dentary is smooth and thus different from that of *P. sclerops.* The premaxillary bones of opposite sides are not coössified with each other, and they are narrowed fore and aft at their anterior suture. The posterior side of the distal half of the bone is beveled for the anterior process of the maxillary. Its alveolar face is marked for the bases of six teeth. The posterior face of the inferior quadrate shows the symplectic to have been a large bone, and to have descended nearly to the condyle of the former. Some fragments of the top of the skull show that it was roughened with low, obtuse ribs and lines.

A dorsal vertebra is but little wider than deep and is truncate below, presenting a prominent infero-lateral angle.

Measurements.

	M.
Depth of dentary near middle	.037
Thickness of dentary near middle	.012
Depth of centrum of vertebra	.029
Width of centrum of vertebra	.038
Length of centrum of vertebra	.009

From the bluffs of Cottonwood Creek.

Pappichthys plicatus Cope.

Annual Report U. S. Geol. Surv. Terrs., 1872 (1873), p. 635.

Plate III, figs. 12–19, and Plate IV, figs. 1–5.

Established on a series of bones of the skull and vertebræ. Some of the cranial bones are deeply grooved and with parallel ridges between. The outer face of the dentary is roughly grooved on the inferior half of its posterior two-thirds. The inner face is marked by a strong groove near its middle to the symphysis, above which it is very convex; below it extends to a thin edge. The dental alveoli are shallow and in close contact; there are six in .025 m. at its middle, where it is also .019 deep. The teeth become smaller at the symphysis. The maxillary bone is rod-like proximally, but flattens out much distally, and is there slightly rugose in parallel lines on the outer face. The teeth are smaller than the mandibulars, there being at the middle fourteen in .025 m. The alveoli are larger proximally. The depth of the bone at the beginning of the suture for the supplementary maxillary is .020 m. The superior extremity of the hyomandibular is broad and flat. The inferior quadrate is thickened behind, and has a broadly oval condyle. The coarsely serrate suture of the pterygoid adjoins it closely. Number cranial ridges in .010 m., ten. The vertebræ preserved are from an anterior position and are quite short and have sessile diapophyses; they are broader than deep; width, m., .026; depth, .019; thickness, .005. The articular surfaces for the neural arches are confluent, so as to have a subquadrate outline.

Another specimen is represented by numerous fragments, one of which is the right maxillary bone. The proximal extremity of this element rises beak-like upward, and shows on its inferior surface the face for contact with the premaxillary. A portion of this bone passed behind the tooth-bearing part of the maxillary for a short distance, and in front of the proximal end of the beak. The external face of the maxillary is nearly flat, and is delicately grooved distally only; the inner face is strongly convex. The teeth are close together, and gradually increase in size, and become more cylindric at the base to the anterior extremity. The crowns are lost.

A fragment of supposed palatine bone exhibits a series of large marginal teeth, and numerous smaller ones within them, as in *Amia calva.* Its superior face exhibits a deep, longitudinal groove, which opens out posteriorly. The proötic bone is a half disk, thickened on the straight edge and with concave sides with a round, flat tuberosity on one of them. On some of the cranial bones the ridges are interrupted.

Six adjacent abdominal vertebræ of this individual are less transverse than the corresponding ones of the *P. medius* (Leidy) and the anterior ones found with the specimen of *P. plicatus* above described. The diapophyses occupy nearly the middle of the sides of the centrum, and are rather elongate. The outline of the articular faces above the level of these processes is a regular wide arch; below them it consists of three sides, two long laterals which are nearly straight, and a short median inferior one, which is slightly concave. The apices of the articular cavities are above the middle. The inferior surface is marked by two parallel angles, each of which bears a narrow longitudinal undivided fossa, which are separated by a concavity. The neurapophysial facets are nearly divided by a median constriction, and are of unequal size.

PAPPICHTHYS CORSONI Cope.

Annual Report U. S. Geol. Surv. Terrs., 1872, p. 636. *P. symphysis* Cope, l. c., 636.

Plate IV, figs. 21-36.

Established on a number of vertebræ of an individual which was of much smaller zize than any of the preceding, and about equal to the largest growth of *A. calva.* The form of the dorsal centra is a little wider than deep; the caudal deeper than wide. What distinguishes these from the vertebræ of the species above described is the lack of distinction between the articular facets of the adjacent neurapophyses. These are almost confluent instead of nearly or quite isolated, as in the *P. lævis* and *P. plicatus.*

Measurements.

	M.
Length of centrum dorsal	.006
Depth of centrum dorsal	.014
Width of centrum dorsal	.018
Depth of centrum caudal	.0115
Width of centrum caudal	.0105
Length of centrum caudal	.0040

The dorsals of the above specimen have short diapophyses, and might be regarded as posterior, and the anterior might be anticipated to present a different type of articulation with the neurapophyses, as in *P. plicatus*. But a vertebra of the same size and form but with long diapophyses, from another locality (Upper Green River), presents the same subquadrate articular faces, slightly constricted in the middle. Hence I suspect this character to be characteristic of the species.

Another specimen is rather smaller than the last. A dorsal vertebra, with inferior diapophyses, is but little wider than deep. The articular surfaces for the neurapophyses are 8-shaped, the areas confluent. A marked peculiarity is seen in the dentary bone. It is much curved in the vertical plane as well as in the horizontal, and must have inclosed a wide mouth. The groove is median, and the inferior and superior surfaces reach it by a nearly equal slope. The former leaves the alveoli without horizontal border, though the latter themselves open on a horizontal plane. There are four and a fraction in .010 m.

Measurements.

	M.
Depth of ramus at middle	.013
Thickness of ramus at middle	.006
Length of posterior dorsal vertebra	.006
Depth of posterior dorsal vertebra	.012
Width of posterior dorsal vertebra	.013

From Upper Green River, dedicated to Dr. Joseph Corson, formerly at Fort Bridger, Wyoming, whose researches among the vertebrata of the Bridger basin were attended by rich results.

NEMATOGNATHI.

Gill; Cope, Proceedings American Association for the Advancement of Science, 1871, p. 330.

Parietal and supraoccipital bones confluent. Four anterior vertebræ coössified, and with ossicula auditus. No mesopterygium. Basis cranii and pterotic simple; no coronoid bone. Third superior pharyngeal bone wanting, or small and resting on the fourth; second directed backwards. One or two pairs of basal branchihyals; two pairs of branchihyals. Suboperculum wanting; premaxillary forming mouth border above. Interclavicles present.

This order, so extensively developed in recent times, first appears in geological history in a single genus in the Bridger Eocene. It has not yet been found at a lower horizon than this. These earliest forms do not differ widely from recent ones, so far as appears.

RHINEASTES Cope.

Proceed. Amer. Philos. Soc., 1872, p. 486 (published August 20, 1872). Annual Report U. S. Geol. Surv. Terrs., 1872 (1873), p. 638.

This genus differs from those at present inhabiting North America in the presence of teeth on the vomer (*R. calvus*). The teeth are everywhere coarsely villiform. The occipital bone exhibits a pit on the middle line below, and a surface for attachment for the inferior branch of the post-temporal on each side (*R. calvus*, *R. smithii*). The modified anterior vertebræ mass is deeply grooved below (*R. smithii*). The cranium is covered with a rugose exostosis (*R. peltatus*, *R. calvus*, *R. smithii*), and has a strong closed groove in the position of the usual fronto-parietal fontanelle. The vertebræ (*R. smithii*) are short, and the sides of the centra only striate with the circumference. There are no lateral pits, but a pair above and a pair below, with a coössified apophysis at the base of one of them.

The spines preserved belong chiefly to the pectoral fin. They are strongly striate and weakly dentate, and have the usual hinge with superior recurved flange above, and two embracing processes below at the base. The dorsal spine is weaker in *R calvus*, but strong in *R. peltatus*. In the former species it stands on the transversely expanded summit of the intert neural bone, which presents a median process upwards and an articular face on each side upwards and backwards. The median process is divided from above, and the excavation receives a subglobular enlargement of the middle of the base of the spine. This, with the two lateral facets of the basal expansions of the spine, constitute the hinge on which the latter moves.

This genus differs from those at present inhabiting the fresh waters of North America, not only in the presence of vomerine teeth, but also in the exostosis of the superior and lateral surfaces of the skull. The anterior part of the cranium being absent from my specimens, I am not able to determine whether *Rhineastes* should be referred to the *Pimelodina* or the

Ariina of Günther's system. In the former case the genus resembles the *Phractocephalus* or the *Piramutana*. In the latter case it will fall into the immediate neighborhood of *Arius*. These three genera are at present existing in South America, so that it appears that the *Nematognathi* of the Eocene of the Rocky Mountains present the same neotropical resemblances to be traced in the *Dapedoglossus* and *Priscacara*.

My expedition obtained remains of four or five species of this genus from the Bridger beds, and one from the Amyzon beds of Colorado; but none have as yet been discovered in the shales of the Green River formation.

The species are distinguished as follows:

I. *Rhineastes;* a large, massive nuchal shield.
Cephalic ossification pappilliform .... *R. peltatus.*

II. *Astephus;* nuchal shield narrow and short.
Cephalic ossification in smooth lines; one basioccipital pit; pectoral spines serrate on both edges .... *R. smithii.*
Three basioccipital pits; pectoral spines serrate on both edges...... *R. calvus.*
Pectoral spines serrate behind only; curved...... .... *R. arcuatus.*

The cephalic bones of the *R. arcuatus* are unknown.

Rhineastes peltatus Cope.

Proceedings American Philosophical Society, 1872, 486.

Plate V, figs. 1-2.

Established on cranial and other bones with spines of a siluriform fish of the size of, perhaps, the *Amiurus lophius*. The form and the excessive rugosity of the external bony surfaces, reminds one of some of the Brazilian *Dorades*. The frontal fontanelle is closed, though very distinctly marked by a deep groove, with its fundus smooth. The rugosity consists of innumerable well distinguished osseous papillæ. The cranial ossification is continued posteriorly as a shield, which is strongly convex from side to side. The spine is symmetrical and probably dorsal. It is compressed and curved antero-posteriorly, and is deeply grooved behind Laterally it is closely striate-grooved; the anterior face is narrowed, obtuse, and minutely serrate, with cross ridges; each side of it is rugose, with several irregular series of pronounced tubercles arranged transversely.

Measurements.

	M.
Width frontal bone near front of fontanelle	0.012
Thickness at same point	.004
Thickness of casque	.004
Width spine	.005
Depth spine	.009

The single individual of this species whose remains are preserved shows that it was the most robust, though not the largest of the genus. I found it on South Bitter Creek in the Washakie basin of the Bridger formation.

RHINEASTES SMITHII Cope.

Proceed. Amer. Philos. Soc., 1872, p. 486 (August 20). Annual Report U. S. Geol. Surv. Terrs., 1872, p. 639 (1873).

Plate V, figs. 5–11.

Represented by remains of several individuals, including one with vertebræ, basioccipital, opercular and other cranial bones, with spines. They indicate a fish of the size of the large catfishes of the Ohio River. The pectoral spines are quite compressed and distinctly striate-grooved on the sides. The posterior groove is occupied by short, spaced, recurved teeth; the anterior by an acute edge, bounded by a groove on each side, which has a fine, close serration. The surface of the modified vertebral mass is striate ridged; that of the basioccipital still more strongly ridged. In the latter there is a median pit behind, and the points of attachment of the inferior limb of the post-temporal is in front of it, smooth, and without reverted edges. The operculum has a large compressed, sessile cup, and its external surface is strongly ridged and grooved, radiating from above in front.

Measurements

	M.
Diameter of a vertebra	.021
Length of centrum	.009
Diameter of modified vertebra	.013
Diameter of groove of vertebra	.005
Diameter of occipital articulation	.015
Length of cup of operculum	.013
Diameter of spine at base	.008
Diameter of spine at .004 from base	.0037

Another pectoral spine is larger; diameter at base .010.

From the Mammoth Buttes and Laclede, on South Bitter Creek.*

* Named for my respected friend, Daniel B. Smith, of Germantown, Philadelphia, many years principal of Haverford College, and a student and lover of natural sciences.

Rhineastes calvus Cope.

Annual Report U. S. Geol. Surv. Terrs., 1872 (1873), p. 640.

Plate V, figs. 3-4.

Represented by several specimens, including most parts of the cranium, spines, etc.

One of these shows the supraoccipital production to have the form of an equilateral triangle, with a sinus of the posterior border on each side of it, which advances in front of the epiotic bone below. Shortly in front of this point the deep groove representing the fontanelle commences. The cranial rugæ are lines parallel to the fontanelle, which diverge to the margins of the occipital prolongation, and are frequently connected by cross-ridges. The frontal portion of the skull is much expanded laterally, and the part beneath inclosed by the prefrontals, particularly wide. The fontanelle in this region does not appear to have been entirely closed. The surface is here also strongly rugose. The vomer has a T-shaped anterior extremity, which is immediately followed by two transverse parallelogrammic patches of premaxillary brush-teeth in several rows. They are about twice as long as wide, and in contact medially. The anterior margin of the premaxilla projects their length beyond them, and is perfectly smooth, and has a smooth, rounded border. The basioccipital has a subcordate cotylus. In front of the median inferior pit are three groove-pits; the articular face for the post-temporal is opposite the former and is rugose and has strongly reverted edges.

Measurements.

	M.
Diameter occipital articulation	.0082
Diameter base supraoccipital shield	.0130
Width front above orbits	.0043
Length from vomer to premaxillary border	.0110
Length of both tooth patches	.0120
Diameter pectoral spine at base	.0031

The pectoral spine is serrate on both edges. The base of the dorsal is symmetrical and articulates with its interneural bone by two lateral flat, and one convex median anterior surfaces, whose surfaces are curiously rugose. The interneural has a rugose median superior keel, which terminates in a point which is received into a pit of the base of the spine; there is a similar production on the posterior side for a similar purpose. The basis of the spine proper is smaller than that of the pectoral, and is about as wide as deep.

In a number of fragments of another individual found together, the basioccipital has the characters already described. The dentary bone is curved inward, and is acute below, widening regularly to the alveolar border. There is no groove on the inner face, while the outer is striate-grooved and has a series of pits along its lower middle.

Measurements.

	M.
Diameter occipital articulation	.009
Width alveolar face	.004
Depth ramus at middle	.008

A part of the operculum of a third individual (with similar spines) displays great rugosity and elevated radiating ridges; length of articular cup ^{m}.0065.

The specimens are chiefly from the bad lands of the Upper Green River.

RHINEASTES ARCUATUS Cope.

Plate V, fig. 12.

Annual Report U. S. Geol. Surv. Terrs., 1872 (1873), p. 641. *Pimelodus antiquus* Leidy, Final Report U. S. Geol. Surv. Terrs., i, 1873, p. 193, Pl. xvii, figs. 9, 10. Proceedings Academy Phila., 1873, p. 99, name only.

There are numerous spines about the size of those of the last species, which differ in the want of the fine serrated anterior edge. I select one as the type, which belonged to the pectoral fin of the right side. It is unbroken, and is curved from base to apex. The latter is acute by an oblique posterior truncation. The surface is strongly striate and the teeth of the posterior edge are closely set; the proximal point distally, the distal proximally. In this specimen there is a trace of anterior serration; in many specimens none whatever. The external surfaces of the epiclavicular and coracoid bones are strongly rugose-striate, as is the case in all the species of this genus, and the most characteristic fragment is that portion of the scapular arch at the base of the pectoral spine.

Measurements.

	M.
Length of spine on curve	.052
Diameter at base; long	.006
Diameter at base; short	.004

The recurved plate of the base is rugose as in other catfishes. The spines themselves are less compressed than in *R. calvus.*

It is probable that the name applied by Dr. Leidy to this species was published a short time before my own, but as it was not accompanied by a description it cannot be used.

From the Bridger beds of the Upper Green River.

? Rhineastes radulus Cope.

Annual Report U. S. Geol. Surv. Terrs., 1872 (1873), p. 639.

Plate V, figs. 14-17.

This species rests on a number of broken cranial bones. I referred it formerly to this genus, but now regard the reference as purely provisional. It is likely that it does not belong to *Rhineastes*, but what its proper generic position is, I am not at present able to determine.

The cranial bones present a pattern of exostosis quite distinct from that observed in the known species of *Rhineastes.* This consists of closely placed crenate ridges, which radiate from various points, and are sometimes broken up, but always rough or serrate on the edges. The bones are not so thick as in the *R. peltatus; i. e.*, .0025 m.

From bad lands of Cottonwood Creek, Wyoming.

ISOSPONDYLI.

Cope, Proceedings American Association for the Advancement of Science, 1871, p. 33.

Actinopterous fishes with physostomous characters, having the scapular arch suspended to the cranium; a præcoracoid arch, and a symplectic bone, but no coronoid bone, and with the anterior vertebræ unmodified and without ossicula auditus.

Two families of this order are represented in the Green River and Bridger beds by numerous individuals. These are the *Osteoglossidæ* and the *Clupeidæ*, which are distinguished by the following characters of the skeleton:

Tail, homocercal; pterotic bone, normal; basis cranii, double; superior pharyngeals four, distinct, third largest and directed forwards; basal branchihyals three; parietals separated by supraoccipital; one vertebra included in the caudal fin. (Psuedobranchiæ and pyloric appendages). *Clupeidæ.* Tail, homocercal; pterotic, normal; basis cranii, simple; basal branchihyals and superior pharyngeals, each three; (no pseudobranchiæ) *Osteoglossidæ.*

To the first named family belongs the genus *Diplomystus;* to the last named, *Dapedoglossus.*

DAPEDOGLOSSUS Cope.

Bulletin U. S. Geol. Surv. Terrs., 1877, p. 807 (August 15). *Phareodus* (*nomen nudum*) Leidy. Proceed. Acad. Nat. Sci. Phila., 1873, p. 99. Final Report U. S. Geol. Surv. Terrs., i, p. 193.

A single row of elongate acute teeth on the premaxillary, maxillary, and dentary bones, vomer, tongue, and (?) basihyal bones closely studded with short conic grinding teeth. Mouth rather short. Pectoral fin with the anterior ray elongated; dorsal fin not elongate, with the anal well separated from the caudal. No beards. Two vertebræ included within the caudal fin.

This interesting genus presents the characters of the family to which I refer it, in its segmented scales, posterior dorsal fin, etc., and does not differ widely in essentials from *Osteoglossum.* The principal differences between the two genera are, the small mouth in *Dapedoglossus,* the absence of barbels, and the generally abreviated form. From *Arapæma,* it differs in proportions, and in the abundance of teeth on the bones of the roof and floor of the mouth.

The peculiar structure of the scales characteristic of this family is well displayed in this genus. The whole of the scale is composed, between the inferior and superior surface layers, of subhexagonal or diamond-shaped cells, which are arranged in spirals trending to the center. Their contents are more thoroughly calcified on the exposed than in the concealed portion of the scale. No radial grooves. Tube of the lateral line issuing by a round pore.

The discovery of this genus, in the Green River shales, is one of the most interesting in the history of this department of paleontology which has been made.

Osteoglossum is known only in a recent state, and with a range of distribution quite unparalleled among Teleostean fishes. Thus one species—*O. bicirrhosum,* Vand., occurs in Brazil; *O. formosum,* Schl. Müll., in Borneo, etc., and *O. leichardtii* Gthr., in New Zealand, all in the southern hemisphere, or near the equator. Two other genera, *Vastres* and *Heterotis*, have been associated with it, and these belong to the same hemisphere, or to those faunæ which characterize it, in their extensions north of the equator. It is, therefore, interesting to note that the first representative of the type found in any of the northern faunal regions, belongs to an age evidently Eocene.

Our first knowledge of this family and genus as it occurs in North America, was based on a fragment of shale from the railroad cut at Green River, Wyoming, which bears part of the scales of one side of the body of the fish. The specimen is without any portion of the head or fins. In consideration of the structure of the scales I was induced to refer the species to the genus *Osteoglossum*. The next addition to our knowledge of this form was furnished by Dr. Leidy, who gave a brief description of portions of the jaws supporting teeth, of a species found in the Bridger formation. This fish he termed *Phareodus acutus*. No diagnosis of the genus was given, nor were any grounds for creating it assigned. I have not, therefore, been able to use the name. Subsequently I attempted to define the genus thus named by Dr. Leidy,* inadvertently writing it *Phareodon*. The specimens of jaws then in my possession were found by myself, mingled with the bones of *Rhineastes*, and it was from the latter that my diagnosis was drawn up. *Phareodon* Cope must then be regarded as a synonym of *Rhineastes*, and as it had its origin in error, naturalists will probably agree with me in sinking it altogether.

Four species of this genus are enumerated in the following pages, but I am not able to give the distinguishing characters of three of them. It is probable that the *D. acutus* Leidy is distinct from the other two, which have been found in the Green River shales, but as only its jaws are known, these offer but few characters. At present the type specimen of *D. encaustus* remains considerably larger than any of those of the *D. testis*, which have been found, and comes from a locality distant from where the latter is found.

* Annual Report U. S.Geol. Surv. Terrs.,1872 (1873), p. 637.

DAPEDOGLOSSUS ENCAUSTUS Cope.

Bulletin U. S. Geol. Surv. Terrs., 1877, p. 808. *Osteoglossum encaustum* Cope, Annual Report U. S. Geol. Surv. Terrs., 1870, p. 430.

Plate VI, fig. 1.

Represented by a portion of the side of a large individual, including the series of scales bearing the lateral lines, and three series above and three below it, more or less perfectly preserved. The longitudinal extent of the fragment includes seventeen transverse series. These scales are of large size, the included portions are smooth to the naked eye, but rugose under the microscope, and with but few and faint traces of concentric lines. Exposed portion with entire margin, bearing a large lenticular rugose surface. This rugosity consists of elevated portions of an enamel-like material, between small pits and grooves. The septa between the cells are distinctly visible on the smooth part of the scale; on the rugose surface they are represented by grooves. The cells are in curved series, which extend to the center of growth, growing smaller as they converge. The rugose part of the exposed surface diminishes in relative extent towards the anterior part of the body. The tubes of the lateral line are in this species concealed beneath the external layer of the scale. The opening is nearer the margin than the center of the scale, is round, and is frequently accompanied by a smaller one above and in front of it.

Measurements.

	M.
Length of fifteen consecutive scales	.23
Depth of six longitudinal series scales	.127
Vertical diameter of a scale	.035
Transverse diameter of a scale	.025
Diameter of a submarginal scale cell	.003
Width of rugose area of scale	.011

As compared with the species of *Osteoglossum*, whose scales have been figured, the present offers clear distinction. In *O. bicirrhosum*, figured in Agassiz and Spix Brazilian Fishes, Tab. XXV, the scales have distinct concentric grooves, and the rugosity consists of a few points or projections. In *O. formosum*, figured in Solomon Mueller's travels in Borneo, &c., the rugosity is uniform on the exposed surface, and very minute, and there are no concentric grooves; the cells are smaller. In *Vastres* the exposed surfaces are still more rugose; in large examples quite honeycombed.

The specimens represent an individual of 3 or 4 feet in length. Discovered at the fish slate cut on the Green River, on the line of the Union Pacific Railroad by Lucius E. Ricksecker, C. E.

DAPEDOGLOSSUS TESTIS Cope.

Bulletin U. S. Geol. Surv. Terrs., 1877, p. 807 (August 15).

Plate VII, fig. 1; Plate VIII, figs. 1-2.

Form oval, contracting subequally to the muzzle and caudal peduncle. The front is gently convex and the mouth is terminal. The depth is little less than half the length minus the caudal fin, and the length of the head enters the same 3.4 times. The dorsal fin is shorter than the anal, and its first ray stands over the sixth of the latter. The ventrals are small, and extend about one-half the distance from their base to the first anal ray, which equals the distance to the base of the pectoral. The latter is elongate, especially the first ray, which, although jointed, as in *Osteoglossum bicirrhosum*, reaches nearly to the end of the ventral. Radii: D. 22–23; A. 27–30. The caudal fin is slightly concave. Scales five or six series above the vertebral column and seven below it. Their exposed surface is rather wide, and is minutely granulated and without grooves. The cells are invisible except when this surface is removed, and they are rather large. Vertebræ, 18 dorsal, 24–25 caudal.

The orbit is rather large, and is reached by the end of the maxillary bone. The suborbital bones are not much enlarged, as is the case in the recent genera. Preoperculum entire; suboperculum very narrow. Branchiostegals slender, rather numerous; coracoid wide, forming a vertical keel, which is not produced. Length of the longest specimen, $0^{m}.230$; of the shortest, $0^{m}.165$.

The numerous specimens of this which I have seen differ in size, and are three-fifths and less, of the dimensions of the *D. encaustus*.

As compared with the *D. acutus*, I notice that the dentary bone does not support quite so many long teeth. I count twenty-three in the former and seventeen or eighteen in the *D. testis*.

From the Green River shales at Twin Creek, Wyoming.

Dapedoglossus acutus Leidy.

Phareodus acutus Leidy, Proceed. Acad., Phila., 1873, p. 99; Final Report U. S. Geol. Surv. Terrs., i, p. 193, Plate 32, figs. 47–51.
Phareodon acutus Leidy, Cope, part. Annual Report U. S. Geol. Surv. Terrs., 1872 (1873), p. 637.
Phareodon sericeus Cope, l. c., p. 638.

Plate V, figs. 18–20.

Represented by numerous remains. The teeth, as preserved, are black with white translucent slightly incurved apices. The dentary bones are deep, incurved, and with an erect elevated point at the symphysis; their outer surface is rugose with deep longitudinal grooves and pits of irregular sizes. They are narrow transversely, and support a single series of twenty two or three closely placed slender teeth, which together form a comb. The bases of these teeth are rugose-striate, and the apices abruptly acuminate.

Measurements.

	M.
Depth of dentary at symphysis	.009
Depth of dentary at fourteenth tooth	.015
Length of eighth tooth	.0056
Diameter of eighth tooth at base	.0015
Six teeth in	.0100

The palatine bones support a mass of teeth, there being one external series of large ones rather abruptly pointed, and several series of small ones of little elevation, whose size diminishes inwards. On two teeth of the external series of a large individual, I proposed the species *Phareodon sericeus*, with the following description: "They differ from those of the *P. acutus* in their large size and stout conic form; also in having the basal striation finer, parallel, and extending over half the length of the crown. The basal portions as preserved are black, the apex white, and with a slightly abrupt contraction."

Measurements.

	No. 1.	No. 2.	No. 3.
Length of crown	.0050	.0090	.0070
Dameter of crown at base	.0035	.0040	.0040

The specimens of *Dapedoglossus acutus* which I have seen, I obtained from the upper valley of Green River, from the marls of the Bridger epoch.

DAPEDOGLOSSUS ÆQUIPINNIS Cope.

Bulletin U. S. Geol. Surv. Terrs., 1878, p. 77 (February 5).

Plate VII, fig. 2.

Two specimens present the principal character of this species, viz, the equality in number of rays in the dorsal and anal fins and the near equality in their size. The radii are in one, D. 23; A. 22: in the other, D. 22; A. 22. In *D. testis*, the formula is D. II—18; A. II—26. The vertebræ in one of the specimens of *D. æquipinnis* number D. 19; C. 27: while in *D. testis* there are, D. 18; C. 24–25. (The number, 21 dorsal, originally given, must be corrected, as based on an imperfect specimen.) In *D. æquipinnis*, the first pectoral ray is not so largely developed as in *D. testis*, not being of unusual size The hyoid apparatus and vomer are closely studded with teeth, as required by the generic character.

Measurements.

	M.
Length of No. 1	.051
Axial length of head of No. 1	.014
Axial length to line of anal fin	.030
Axial length to line of dorsal	.028
Axial length to origin of caudal	.040
Depth of head	.012
Depth at first dorsal ray	.008
Depth of caudal peduncle	.004
Length of No. 2	.092
Depth at middle of dorsal line	.032
Depth at base of dorsal fin	.024
Depth of caudal peduncle	.008

The specimens described are much smaller than those of the *D. testis* yet known, and No. 1 is probably young. This fact will not account for the peculiarity of the radial formula, etc.

Green River shales at Twin Creek, Wyoming.

DIPLOMYSTUS Cope.

Bulletin U. S. Geol. Surv. Terrs., 1877, p. 808 (August 15).

Family *Clupeidæ*, and nearly related to the genus *Clupea*. It differs from *Clupea* in the presence of a series of dorsal scuta, which extend from the supraoccipital region to the base of the dorsal fin, corresponding in position with those of the ventral surface. Unlike these, they have no lateral processes. The dorsal fin originates in front of the anal. In the typical forms, teeth are well developed in a single series on the dentary, premaxillary, and maxillary bones; but, in the small forms, they are invisible. Mouth moderate.

There are two sections of this genus, the species of which differ in the form of the dorsal scuta. In section I, these shields are transverse, and their posterior borders are pectinate, a median tooth being especially prominent. In section II, the scuta are not wider than long, and have but one, a median tooth, which is the extremity of a median longitudinal carina. The species of section I are *D. dentatus*, *D. analis*, *D. theta*, and *D. pectorosus*; those of section II are *D. humilis* and *D. altus*. The species of section I display a longer anal fin than those of section II.

The species of this genus were more numerous in individuals than all others combined, during the period of the Green River Lake.

Diplomystus dentatus Cope.

Bulletin U. S. Geol. Surv. Terrs., 1877, p. 809.

Plate X, fig. 1.

Fin-radii: D. I—13; A. I. 35. Vertebræ: dorsal, 18; caudal, 21. The greatest depth enters the length without the caudal fin two and a half times, and the head enters the same nearly three and one-third times. The eye is large, its horizontal diameter a little exceeding the length from its border to the inferior edge of the premaxillary bone, and is a little greater than one-fourth the length of the head. The premaxillary and dentary bones are short and deep, the latter with a deep notch on the anterior border; both are directed upwards. The maxillary bone is long and narrow, and curved backwards at its lower end, which reaches a point below the anterior border of the orbit. The profile behind the premaxillary bone is nearly horizontal; above the posterior part of the orbit, it rises, and a compressed supraoccipital crest carries it to the gently convex dorsal line. The abdomen is convex, and is about as long as the caudal region. The last dorsal ray rises above a point anterior to the first anal ray. The caudal is deeply forked. The ventrals originate at a point barely in advance of a vertical line from the first dorsal ray. The pectoral fins are short. The scuta of the inferior median line are large and acute. The scales are rather small, and are delicately grooved; twenty rows may be counted between the vertebral column and the dorsal fin.

Measurements.

	M.
Total length	.365
Length of head	.083
Length (axial) to below first dorsal ray	.145
Length to above first anal ray	.185
Length to base of external caudal rays	.285
Depth at orbit	.055
Depth at occiput	.093
Depth at first dorsal ray	.118
Depth at middle anal ray	.050
Depth at base of caudal fin	.030

Not rare at Twin Creek; specimens occur of fifty centimeters length, or about the size of the shad.

Diplomystus analis Cope.

Bulletin U. S. Geol. Surv. Terrs., 1877, p. 809.

Plate VII, fig. 4; Plate VIII, fig. 3; Plate X, fig. 2.

Radial formula: D. I. 11; A. I. 40. Vertebræ: dorsal, 17–18; caudal, 23–24. This species is more elongate in proportion to its depth than either of the other species, the length being three times the greatest depth. The anal portion of the body is considerably longer than the abdomen, and the anal fin is long and with short rays. The ventral fin commences well in front of the dorsal, whose last ray is considerably in advance of the first anal ray. The pectoral fin reaches the ventral, and contains thirteen rays. The greatest depth is at the pectoral region, the outlines contracting to the base of the anal fin. The dorsal outline is convex. The profile descends gently. The muzzle is half as long as the diameter of the orbit, which enters the length of the head three times. The latter enters the length without the caudal fin three and three-fourths times There is a row of short, conical teeth along the middle line of the mouth, which is not on the vomer, but is on the parasphenoid or axial hyal bones. Similar teeth exist in the mouth of *D. dentatus*. The jaws may be furnished with minute teeth, or they may be wanting. The scales are thin and difficult to count; there are fifteen rows between the vertebral column and the anterior anal rays.

Measurements.

	M.
Total length	.195
Length of head	.040
Axial length to below first dorsal ray	.073
Axial length to above first anal ray	.092
Axial length to base of external caudal rays	.149
Depth at orbit	.030
Depth at occiput	.044
Depth at first dorsal ray	.047
Depth at middle anal ray	.027
Depth at base of caudal fin	.016

This Herring is represented by a great number of well-preserved specimens, and was, next to the *D. humilis*, the most abundant fish of the waters of the ancient Green River Lake basin. It is distinguished from the *D. dentatus* by the large number of anal and smaller number of dorsal radii, and by the shorter head and relatively more slender body. The specimen measured represents the average size; the largest obtained is half as large again, and much smaller than the type of *D. dentatus*.

From Twin Creek, Wyoming.

DIPLOMYSTUS PECTOROSUS Cope.

Bulletin U. S. Geol. Surv. Terrs., 1877, p. 810.

Plate X, fig. 3.

This Clupeoid is represented by small specimens of a deeper form than that seen in the two preceding species. It is also characterized by a smaller number of dorsal radii than either of them. Formula: D. I. 8–9; A. I. 40–44. Vertebræ: dorsal, 16–17; caudal, 22. The greatest depth is in the pectoral region, and enters the length minus the caudal fin a little less than three times. The outlines contract from the ventral fins, and the anal region is longer than the abdominal. The eye is a little more than one-fourth the length of the head, and the latter enters the total minus the caudal fin three and a half times. The ventral fins are small, and commence well in advance of the line of the dorsal. The last dorsal ray is nearly above the first anal; the caudal is deeply forked. As in the two preceding species, the neural spines in front of the interneurals present a laminar antero-posterior expansion. The dorsal scuta are furnished in the *D. pectorosus* with an especially prominent median keel.

Measurements.

	M.
Total length	.090
Length (axial) to below D. I	.038
Length (axial) to above A. I	.043
Length (axial) to base of caudal fin	.070
Length of head	.022
Depth at orbit	.017
Depth at pectoral fin	.026
Depth at dorsal fin	.024
Depth at caudal peduncle	.008

This species is represented by several specimens, from Twin Creek, Wyoming.

DIPLOMYSTUS THETA Cope.

Bulletin U. S. Geol. Surv. Terrs., 1877, p. 811. *Clupea theta* Cope, Annual Report U. S. Geol. Surv. Terrs., 1873, p. 461.

Represented by a specimen from the Green River shales near the mouth of Labarge Creek, in the upper valley of Green River. It is a larger species than the *C. humilis* Leidy, which is also found at the same locality, and it has much longer anal fin. Its radii number 26, possibly a few more, as the end appears to have been injured. The dorsal fin is short; the last ray in advance of the line of the first of the anal. The body is deep. Number of vertebræ from the first interneural spine to the last interhæmal, 29. Depth at first dorsal ray, .0485; depth at last anal ray, .0170; length of 29 vertebræ, .0780.

The posterior part of the body having been lost, the number of anal rays is unknown. It is quite possible that further investigation may show that the *D. analis* is identical with this species.

DIPLOMYSTUS HUMILIS Leidy.

Proceed. Acad. Phila., 1856, p. 256. Final Report U. S. Geol. Surv. Terrs. i., p. 195, Plate xvii, fig. 1. *Clupea pusilla* Cope, Proceed. Amer. Philos. Soc., 1870, p. 382. Annual Report U. S. Geol. Surv. Terrs., 1870, p. 429.

Plate IX, fig. 8; Plate X, fig. 4.

This and the following species, already referred to a distinct section of the genus *Diplomystus*, differ from those above described in several points. They have a much shorter anal fin, and the caudal part of the vertebral column is thus shorter. The anterior neural spines do not present the antero-posterior laminar expansion. The ventral fin commences a little behind the origin of the dorsal. The formulæ for the *D. humilis* are as follows:—Radii: D. I—11; A. I—14. Vertebræ: D. 21; C. 13. Depth to length as 3 : 8. 5.

Specimens of this fish are equally abundant at the Green River and Twin Creek localities. A rather small specimen from the former place was described by me as the type of another species, but I think it represents merely a young individual. When I described it I was under the impression that the *D. altus* was the true *D. humilis* of Leidy. This view was justified by Dr. Leidy's description of the *D. humilis*, where the measurements given are those of the *D. altus*. The figures given (3½ inches long by 16 lines deep) are, however, partly erroneous, as they do not agree with those subsequently given, nor with the plate above cited. The description of a Green River specimen is as follows:

Greatest depth contained four times in the total length or 3.5 times to basis of caudal fin. Length of head 3.2 to basis caudal. This measurement may require revision, as the end of the muzzle is slightly injured. Orbit large, contained twice in length of head behind it. Middle of dorsal fin near the middle of the length, and about over the origins of the ventrals. D. II, 11, v. 7. Pectoral extending half way to ventrals. Vertebræ, 29–30; dorsals, 16–20. Ventral keeled ribs, 18. Anal fin lost. Caudal peduncle slender caudal fin deeply furcate. Length, ${}^{m}.044$; greatest depth, ${}^{m}.011$.

A second specimen exhibits the characters of the species more distinctly in some respects. There are 30 vertebræ, of which 13–14 are caudal. The general shape is regularly fusiform, and the head rather acuminate.

Measurements.

	M.
Total length	.054
Length to preopercular edge	.01
Length to opercular edge	.013
Length to posterior margin dorsal	.0255
Length to anterior margin anal	.034
Length to base caudal	.044
Depth at occiput	.011
Depth at middle of dorsal	.0115
Depth at caudal peduncle	.0046

The largest specimen of this species which I possess is ${}^{m}.150$ in length.

As at the Green River locality, so at Twin Creek, this Herring is the most abundant species One-third the entire number of specimens are referable to it.

DIPLOMYSTUS ALTUS Leidy.

Cope, Bulletin U. S. Geol. Surv. Terrs., 1877, p. 811. *Clupea alta* Leidy. Final Report U. S. Geol. Surv. Terrs., i, p. 196

Plate XVII, fig. 2.

This small Herring is abundant in the Green River shales, both at Green River and at Twin Creek. It is distinguished from the *D. humilis* by the greater relative depth of the body, resembling in this respect the *D. pectorosus.* The difference which it presents in this respect is rather too great to permit its union with *D. humilis.* Nevertheless intermediate specimens occur, but their characters are sometimes found to be due to distortion.

Formulæ:—Radii: D. I. 11; A. I. 13–15. Vertebræ: D. 22; C. 12. Depth to length (without caudal fin) as 4 to 8. Size that of the *D. humilis.*

PERCOMORPHI.

Cope, Proceedings American Association for the Advancement of Science, 1871, p. 341.

Physoclystous Actinopteri, with the shoulder-girdle attached to the skull, and thoracic or jugular ventral fins. Maxillary and dentary bones distinct; cranium symmetrical; epiotics normal; no interclavicles; post-temporal not coössified with cranium. Basal pectoral radii not enlarged; femora suspended (generally) from the scapular arch. Basibranchials three; superior pharyngeals with the third usually the largest; sub and interoperculum present, plate-like.

Three families are represented in the Eocene Tertiary beds, two of which certainly belong to this order, and the third very doubtfully. The former are the *Percidæ* and ?*Pomacentridæ*, representing the suborders *Distegi* and *Pharyngognathi*, respectively. The third group, represented by the genera *Amphiplaga*, *Trichophanes*, and *Erismatopterus*, is related to the *Aphododiridæ;** and as I know of no characters as yet by which to distinguish it, shall for the present consider it under that head. This family lies on the extreme verge of the order towards the *Haplomi*, to which the genus *Erismatopterus* almost affords a transition.

* This name is variously spelled, and I am not yet sure as to the orthography to be adopted.

ERISMATOPTERUS Cope.

Annual Report U. S. Geol. Surv. Terrs., 1870, p. 427.

Dorsal and anal fins short, with two or three strong appressed supporting spines in front; no other interhæmal spines than those supporting those of the anal fin. Dorsal fin above the anterior median or posterior abdominal region. Ventrals originating in front of or opposite to the origin of the dorsal. Pubes sending a limb upwards, which is in contact with the inferior post-clavicle. Teeth minute or (?) wanting. Caudal fin bifurcate.

I originally referred a species of this genus to the *Cyprinodontidæ*, and many of the characters are similar to those of that family. The arc of the mouth is formed by the premaxillary bone, and the ventral fins have a rather anterior position, which is neither pectoral nor ventral, and the caudal is furcate; the scales are cycloid. The strength of the spinous fin radii and supporting interhæmal spines attracted my attention, and on careful examination I observe other approximations to the type of *Asineops* and the *Aphredodiridæ*. The inferior post-clavicle is very long and styliform, as in the latter genus, and the pubic bones are slender and directed upwards, so as to rest on the post-clavicles. In one specimen there appears to be an anteriorly directed pubic limb, but this does not exist in other specimens. The pubes do not reach the clavicles, as in true *Physoclysti*. Vertebræ hour-glass shaped. Ventral radii seven, in the species *E. rickseckeri* and *E. endlichi.*

ERISMATOPTERUS LEVATUS Cope.

Annual Report U. S. Geol. Surv. Terrs., 1870, p. 428. *Cyprinodon levatus* Cope. Proceed. Amer. Philos. Soc., 1870, p. 382.

Plate IX, fig. 6–7.

Anterior margin of anal fin commencing a little behind, opposite the posterior margin of the dorsal. Vertebræ: 11–14–5, seven between the interneural and interhæmal bones of the dorsal and anal fins. Radii: D. 8, A. II. 8, v. 8. Caudal fin deeply furcate; first anal ray strong. General form elongate, the greatest depth contained three times in the length between the scapular arch and the basis of the caudal fin. Scales preserved, small; seven longitudinal series above and seven below the vertebral column, probably two rows concealed by it. The caudal peduncle is but little contracted. Length from scapular arch to extremity of caudal, m.0335; depth at origin dorsal fin, m.008.

Measurements.

	M.
Total length No. 2	.055
Length of cranium	.013
Length to basis D. I	.0232
Length to basis A. I	.033
Length to basis V. I	.0205
Length to basis caudal	.0466
Depth at D. I	.01
Depth of caudal peduncle	.0058

There are many individuals on the slabs of Green River slate, some of them perfectly preserved. Many of these slabs represent that portion of the stratum which is highly carbonaceous, portions of it thrown into the fire burning freely. Dr. Hayden, who has brought numerous specimens from this locality, informs me that the laminæ exhibit greater numbers of these little fishes. No doubt the carbonaceous character of the shales is due to the decomposition of their bodies. The nature of the deposit, and mode of preservation, remind one strongly of the *Cyprinodon meyeri* of Agassiz, from the neighborhood of Frankfort-on-the-Main. That species differs specifically in presenting 18 anal radii.

Some of the specimens above described were obtained from the Green River Cut, and preserved for scientific study, by L. E. Ricksecker.

Erismatopterus rickseckeri Cope.

Annual Report U. S. Geol. Surv. Terrs., 1870, p. 427.

Plate VI, fig. 2.

Length, three to four inches; head large. Vertebræ: D. 13; C. 16; = 29, ten between the interneural bone supporting the first dorsal ray, and the first interhæmal supporting the first anal ray. There are only seven in this position in *E. levatus*. Anterior dorsal ray anterior to the point half way between end of muzzle and end of vertebral column. Branchiostegal radii fin distinguishable. Head stout, mouth terminal, orbit equal length of muzzle; maxilliary bone reaching line of middle of orbit. Scales small, with numerous concentric and no radiating grooves. Fin radii: D. II. —8 (last split); C. 8—19—8; A. II—9. V. 7, p. 15.

Measurements.

	M.
Total length No. 1	.0743
Cranium to supraclavicle	.018
Length to base D. I	.029
Length to end vertebral column	.06
Length of A. II	.003
Length of cranium No. 2	.0175
Length to preoperculum	.012
Length to D. I	.0275
Length to A. I	.043

Five more or less complete specimens of this fish were obtained by Lucius E. Ricksecker from the Green River shales, and I dedicate it to him in recognition of his interesting discoveries in this department.

Its difference from *E. levatus* is seen in the more anterior position of the dorsal fin, more numerous vertebræ, etc.

ERISMATOPTERUS ENDLICHI Cope.

Bulletin U. S. Geol. Surv. Terrs., 1877, p. 812.

Plate XII, fig. 5.

The radial formula in this fish is: D. III—11; C. 6–19–6; A. III—7. V. 7. The vertebræ are: D. 13; C. 17; Centra between the lines of the first interneural and first interhæmal spines, 10. Ten rows of small scales visible above the vertebral column.

The general form of the fish is stout, and the caudal peduncle is deep. The top of the head is convex, and the eye large. The front descends abruptly to the rather projecting muzzle in the specimen, but whether this is a distortion or not is uncertain. The coracoid is wide and well produced backward, while the clavicle is, as usual, directed forward. The femur is slender, and connected with its fellow by a posterior transverse bar. The greatest depth is a little less than one-fourth the length without the caudal fin. The diameter of the eye is one-fourth the length of the head. The origin of the ventral fin is in advance of the first dorsal ray; the origin of the anal is below the penultimate dorsal ray. The caudal fin is openly forked.

Measurements.

	M.
Total length	.061
Length of head	.016
Length to line of ventral fin	.020
Length to line of dorsal fin	.022
Length to line of anal fin	.031
Length to base of caudal fin	.048
Depth at caudal peduncle	.008
Depth at dorsal spine	.011

The more numerous rays of the dorsal fin, and more numerous scales are among the characters which distinguish this species from the two above described. It is dedicated to Dr. Frederick M. Endlich, geologist in charge of one of the parties of the United States Geological Survey of the Territories under Dr. F. V. Hayden.

AMPHIPLAGA Cope.

Bulletin U. S. Geol. Surv. of the Terrs., 1877, p. 812 (August 15).

Generally as in *Erismatopterus*, but with strongly ctenoid scales. The dorsal fin is over the abdomen, and is supported by a few strong, adherent spines in front, which rest on stout interneurals; the soft rays have no interneurals, either in this fin or the anal. They are present in *Erismotopterus.* The ventrals originate a little in advance of the line of the dorsal, and the caudal fin is deeply forked. This genus approximates *Aphrodedirus.*

The scales in this genus are thin and like those of *Trichophanes.* In other respects *Amphiplaga* resembles that type, and I have only distinguished it on account of the absence of interneural bones below the soft dorsal radii. It has occurred to me that this may be abnormal or due to accident, but the bases of the dorsal radii, as well as the anterior interneural bones, are perfectly preserved, so that the accidental removal of the posterior interneurals seems improbable. It must also be remembered that the interhæmal bones are absent from the soft anal rays in both this genus and in *Erismatopterus.*

But one species of this genus is yet known.

AMPHIPLAGA BRACHYPTERA Cope.

Bulletin U. S. Geol. Surv. Terrs., 1877, p. 812.

Radii: D. II—8; A. III—4. Vertebræ of the caudal series 15. Scales: transverse row, 22; longitudinal row behind first interneural bone, 40. The only specimen I possess lacks the head, so that various characters cannot be ascertained. The depth of the body at the first dorsal spine enters the length from that point to the base of the caudal fin two and a half times, giving a general form of medium proportions. Caudal peduncle stout. The vertebræ are contracted medially, and not shortened; they have two or three longitudinal keels, which are somewhat irregular in their connections. This species is larger than any of the *Erismatopteri* yet known.

Measurements.

	M.
Length from first dorsal spine	.073
Length from first anal spine	.051
Length of anal fin	.023
Length of second dorsal spine	.015
Length of third anal spine	.013
Depth at first anal spine	.018

From the Green River shales at Twin Creek, Wyoming.

ASINEOPS Cope.

Proceed. Amer. Philos. Soc., 1870, p. 380. Annual Report U. S. Geol. Surv. Terrs., 1870, p. 425.

Branchiostegal radii, seven; ventral radii I. 6–7. Opercular and other cranial bones unarmed; scales cycloid. Spinous and cartilaginous dorsal fins continuous; caudal rounded; anal with two spines. Lateral line distinct. Operculum with regularly convex posterior border. Teeth coarsely villiform, without canines. Both spinous and soft portions of dorsal and anal fins moderately scaly.

This well marked genus is established on the remains of numerous individuals, in various states of preservation, so that the characters undistinguishable in one can be discovered in another. Thus the lateral line is preserved in one only, and the teeth in another. In none can I be entirely sure that I see the vomer.

The scales are preserved in many specimens, and I cannot find a ctenoid margin in any, nor any radiating sculpture, but delicate concentric ridges continued round the central point proximally, distally forming parabolic curves, the less median not completed, but interrupted by the margin of the scale. Near the margin all the ridges become gently zigzagged.

There is no depression between the two portions of the dorsal fin, though the cartilaginous portion is the more elevated. Laid backwards, the latter is in line with the extremity of the anal, and both extend beyond the basis of the caudal.

The affinities of this genus are rather obscure, but are in some degree to that aberrant family of Physoclysti, the *Aphredodiridæ*. This is indicated by the increased number of ventral radii, the slender separated pubes, and the reduced number of interneural spines. The *Aphredodiridæ* betray Physostomous tendency in the same characters, with still greater reduction of the spinous dorsal and anal fins, though its ctenoid scales and spinous orbital and preopercular bones are of Physoclyst significance. In *Asineops* the scales are cycloid, and the cranial bones unarmed. The ventral fins occupy nearly the same position as in the extinct genus *Erismatopterus* Cope, which accompanies it. There is at least in these genera another illustration of the approximation of forms now very distinct, in past periods. The pubes are, however, supported by the clavicles in *Asineops*, and by the post-clavicles in *Erismatopterus*, though the latter bones are very long in *Asineops* also. *Asineops* will thus constitute a family *Asineopidæ* differing from the *Aphredodiridæ* in the simple pubes. I suspect that the genus *Pygæus* of Agassiz will be found also to belong to it, though the increased number of ventral radii is not assigned to it in the Poissons Fossiles. Some of its species may even be found to belong to *Asineops*. Nine species are described by Professor Agassiz, all from Monte Bolca, in Italy, from an Eocene stratum.

ASINEOPS SQUAMIFRONS Cope.

Proceed. Amer. Philos. Soc., 1870, p. 381. Annual Report U. S. Geol. Surv. Terrs.,, 1870, p. 426. *Asineops viridensis* Cope. Annual Report 1870, p. 426.

Plate IX, fig. 5; Pl. x.

General form suboblong, the greatest depth just behind the head, and contained two and a half times in the length exclusive of caudal fin. Radii D. VIII—IX, 14; A. II, 11–12; C. 14; V. I, 7; P. ?11 ?13. Scales 5—

?30—10, vertical line counted a little behind the ventral fins. The line of the extremities of the second dorsal and anal fins, marks the basal third of the caudal fin. The dorsal spines are subcylindric, slightly curved, and of nearly equal length; the length equals the depth of the body at the middle of the second dorsal fin.

The external series of villiform teeth are stout of their kind, conic, and a little incurved. I cannot see the pharyngeal bones or teeth.

The number of vertebræ which extends between the caudal fin and the superior margin of the operculum, where one or more are concealed, is twenty-five, of which fifteen are of the caudal portion (in two I can only count fourteen).

The mouth is directed obliquely upwards and is rather large; the mandible, when closed, does not project beyond the premaxillary border. The maxillary, where preserved, is narrow distally, and does not project beyond the posterior line of the orbit. The latter is rather small, and though not well defined in any specimen, is not more than one-eighth the length of the head, and 1.5 to 1.75 times inside of muzzle. The margins of all the opercular bones are entire and smooth. The interoperculum is narrow, and lies obliquely upwards, narrowing the operculum. The greatest width of the latter is more than two-thirds its depth. The pelvic supports of the ventral fins are simple, slender and in contact anteriorly, their length about half that of the fin. The pectorals are not elongate.

The scales extend over the top of the head, to or beyond the orbits. They also extend along the ramus of the under jaw. Those of the fins are quite small, they extend to a considerable distance on the unpaired and on the caudal fins.

Measurements.

	M.
Total length of the largest specimen	.19
Do. No. 2, smaller example (with caudal)	.12
Length of head of do	.044
Depth of do. posteriorly about	.036
Length of base of spinous dorsal	.0265
Length of posterior spinous ray	.017
Length of operculum	.0125
Length of maxillary bone about	.0145
Depth No. 3, at base 1st dorsal	.045
Depth No. 3, at base anal, 1st ray	.0325
Length of basis anal=basis caudal	.0162
Length of caudal fin	.034

Tertiary strata of Green River, Wyoming; Dr. F. V. Hayden, Coll. Mus. Smithsonian.

In the original specimen of this species, but nine soft anal rays were preserved; in a more perfect one, subsequently obtained, I find eleven or twelve. I characterized a supposed species under the name of *A. viridensis* on small specimens, one of which was stated to possess fourteen soft anal rays. A re-examination of this fish leads me to believe that the separate rods which represent this number are parts of but twelve rays.

I have not seen the *Asineops squamifrons* from the Twin Creek locality, where another species takes its place.

ASINEOPS PAUCIRADIATUS Cope.

Bulletin U. S. Geol. Surv. Terrs., 1877, p. 813.

Pl. XIV, fig. 1.

This Perch is represented by a single specimen, which is larger than any of those of the *A. squamifrons*, which have yet been found, and which is of more robust proportions. It differs materially in the radial and vertebral formulæ, and in the greater relative shortness of the dorsal spines. I observe at the base of these, a series of short subhorizontal basilar interneural bones.

Formulæ:—Radii: D. IX—12; A. II. 7. Vertebræ: D. 9; C. 13. One or two vertebræ may be concealed behind the epiclavicle, but these, as in the description of *A. squamifrons*, are uncounted. The depth enters the length 2.25 times, the caudal fin being omitted. The length of the head is little less than the depth. The dorsal spines are not very robust, and are (excepting the first) of subequal length. The longest equals only half the depth of the body at the middle of the second dorsal fin. The caudal is rounded, and the ventrals are below the pectorals. The origin of the latter is a little in advance of that of the first dorsal spine. Its base is attached to four short basilar bones, of which the inferior two are stout in proportions There are about ten rows of cycloid scales below the vertebral column. Scales extend on the top of the head as far as the orbits. The mouth is terminal. The total length of the type-specimen is $^{m}.243$, of which the head constitutes $^{m}.075$. The longest (ninth) dorsal spine measures $^{m}.027$, and the second anal spine $^{m}.024$.

Twin Creek, Wyoming.

MIOPLOSUS Cope.

Bulletin U. S. Geol. Surv. Terrs., 1877, p. 813.

Allied to *Labrax* and *Perca*. Branchiostegal rays, 7 or 8; ventral rays, I. 5; scales ctenoid. Two dorsal fins slightly connected at base; only two anal spines. Operculum rounded, without spines or emargination. Preoperculum without spine, and smooth on the posterior border; inferior border with teeth. Premaxillary and dentary with small uniform teeth in a narrow series. Clavicle unarmed. Vertebræ with two lateral fossæ. Caudal fin emarginate.

The discovery of this genus in the Green River shales is of no small importance to fossil ichthyology, proving the existence, at that early period, of the type which is one of the highest among the true fishes. It probably belongs to the *Percidæ*, although I have not ascertained the presence of teeth on the vomer, and there may be eight branchiostegal rays. As compared with the genera, recent and extinct, which are allied to *Perca*, it differs in the unarmed operculum, and the preoperculum with teeth only on the lower limb, and in the presence of but two anal spines. It is therefore a weaker form than they, and though of a higher type, less strongly protected by spines than the cotemporary *Asineops*. *Mioplosus* embraces the largest Physoclystous fishes yet known from this formation, and specimens are not rare at the locality from which they have been procured. They are often in a state of excellent preservation. The type of the genus is the *M. labracoides*.

Mioplosus abbreviatus Cope.

Bulletin U. S. Geol. Surv. Terrs., 1877, p. 814.

The *M. abbreviatus* is represented by but one specimen, from which the muzzle has been broken away. It is the stout species of the genus, and the others succeed it in this enumeration in the order of their greater elongation of form. The depth at the first dorsal fin enters the total length (including caudal fin) three and a half times; and the depth at the first anal ray enters the length of the vertebral column two and eight-tenths times. Vertebræ visible behind clavicle: D. 9; C. 14. Radii: D. IX—I.

11; A. II—11; P. 14. Ventral with a very weak spine. The last dorsal spines, as in all the other species, are very short, the anterior ones slender and moderately long; in this species they are curved. The anal spines are short and slender, the first a rudiment. There are six rows of scales above and six below the vertebral column on the caudal peduncle.

Measurements.

	M.
Length of vertebral column	.125
Length of third dorsal spine	.025
Length of ninth dorsal spine	.007
Depth at middle of first dorsal fin	.060
Depth of caudal peduncle	.025

Twin Creek, Wyoming.

Mioplosus labracoides Cope.

Bulletin U. S. Geol. Surv. Terrs., 1877, p. 814.

Plate XII, fig. 1.

This Perch is represented by five specimens, mostly in good preservation. They have much the proportions of the Rock-fish. The origins of the pectoral and ventral are in nearly the same vertical line, and that of the first dorsal is not far behind them. That of the first ray of the anal is below the second or third ray of the second dorsal. The rays of none of the fins are prolonged; the dorsal spines are slender and nearly straight; the longest (third), when depressed, reaches but four-tenths the distance to the first ray of the second dorsal. The last dorsal spine is very short. The soft dorsal rays are rather longer than the spinous. Formulæ:—Rays: D. IX—I. 12; C. 8—17—8; A. II—14; V. I. 5. Vertebræ: D. 10; C. 15.

The depth at the first dorsal fin enters the total four times; the depth at the first anal ray enters the length of the vertebral column three times. The length of the head enters the total four times, and that of the orbit enters the head 4.66 times, and into the length of the muzzle one and one-third times. The profile of the top of the head is slightly convex, and the dorsal line is also slightly convex. The mouth opens somewhat obliquely upward. The end of the maxillary bone reaches a point below the middle of the orbit. The teeth of the inferior border of the preoperculum are strong, and are directed forward; they number five. The angle of the lower jaw is not produced, but the inferior edge of the ramus is laminar and acute; the symphysis is shortly truncate. The superior edge of the maxillary bears a supernumerary bone at its distal portion. There are six branchiostegal rays preserved, with impressions of two others: the anterior three are slender; the others wide, as in allied genera. There is a low supraoccipital crest. The abdomen bears fourteen rows of scales below the vertebral column, and six rows may be counted above it; on the caudal peduncle I count 5—5.

Measurements.

	M.
Total length	.280
Axial length of head	.070
Axial length to line of first dorsal spine	.085
Axial length to line of first ray of second dorsal	.143
Axial length to line of first anal spine	.152
Axial length to base of caudal	.232
Depth at orbit	.051
Depth at first anal ray	.055
Depth of caudal peduncle	.030
Length of third dorsal spine	.030
Length of second anal spine	.016

Abundant at Twin Creek, Wyoming. A single specimen from Green River, Wyoming, from L. E. Ricksecker.

MIOPLOSUS LONGUS Cope.

Bulletin U. S. Geol. Surv. Terrs., 1877, p. 815.

Plate XII, fig. 3.

I have questioned the right of the form to which the above name is given to be maintained as a species distinct from the *M. labracoides.* It is represented by two individuals of much smaller size than those of the latter, and which are of a more elongate form. They have also two anal radii fewer.

The formulæ are: D. IX—12; A. II. 12. Vertebræ: D. 10; C. 15. The depth at the first dorsal fin enters the total length five times, and the depth at the first anal ray three and one-half to three and eight-tenths times. The dorsal spines are straight and slender, the posterior ones very short. The caudal is forked. The teeth of the inferior border of the preoperculum are strong and acute; there are three large and two small ones.

Measurements.

	M.
Total length	.175
Length of head	.042
Length to line of first dorsal	.054
Length to line of second dorsal	.085
Length to line of anal	.091
Length to caudal	.140
Depth at orbit	.025
Depth at first dorsal	.037
Depth at second dorsal	.034
Depth of caudal peduncle	.020

The scales are similar to those of the *M. labracoides*.

Twin Creek, Wyoming.

Mioplosus beani Cope.

Bulletin U. S. Geol. Surv. Terrs., 1877, p. 816.

Plate XII, fig. 2.

This, the most slender species of the genus, is represented by one specimen, which is the smallest obtained which is referable to this genus. The depth enters the total length six times, and the depth at the first anal spine enters the length of the vertebral column a little more than four times. Radii: D. IX—I. 13; A. II—12; P. 13. Vertebræ: D. 10; C. 15. The general characters are as in *M. labracoides*, but the scales are not preserved The form of the head is that of a younger fish, but its proportions as compared with the body are not those of immaturity. The length enters the total 4.2 times, and the orbit enters it 4.5 times. The profile of the front is descending. The teeth of the inferior limb of the preopercle are obtuse and not well defined. There are impressions of seven branchiostegals preserved.

This Perch is named in honor of my friend Dr. T. H. Bean, of the United States Fish Commission.

Measurements.

	M.
Total length	.131
Length of head	.031
Length to line of first dorsal	.040
Length to line of second dorsal	.064
Length to line of anal fin	.070
Length to line of caudal fin	.109
Depth at orbit	.020
Depth at first dorsal ray	.023
Depth at first anal ray	.019
Depth of caudal peduncle	.011

From Twin Creek, Wyoming.

MIOPLOSUS SAUVAGEANUS Cope.

This fish is known to me by a single specimen of the average size of the *M. labracoides*. It is distinguished from that species by its more slender proportions, and especially by the elongation of its muzzle, which projects a little beyond the lower jaw. In *M. labracoides* it is the inferior jaw which projects. *M. sauvageanus* also differs from all the known species of the genus in the small number of its soft anal rays, and larger number of its vertebræ.

Radial formula; D. VI +, 14; A. I, 9. Vertebræ, D. XI, C. XVI. Depth at middle of first dorsal fin (the greatest) enters the length to the extremity of the vertebral column, 3 5-6 times, that is, nearly five times in the total length. In *M. labracoides* this dimension enters the same length three times or a little over. The dentary bone is quite shallow, and the chin and inferior crests are less prominent than in *M. labracoides*. The inferior preopercular teeth are distinct.

From the Green River shales of Twin Creek, Wyoming, from Mr. Leroy, of the Central Pacific Railroad.

Dedicated to my friend Dr. H. E. Sauvage, of Paris, author of various important works on ancient and modern fishes.

PRISCACARA Cope.

Bulletin U. S. Geol. Surv. Terrs., 1877, p. 816.

This type might be included in the *Pomacentridæ*, but it differs from the genera now known in the possession of vomerine teeth, and apparently in having eight branchiostegal rays.

In general, *Priscacara* may be characterized as *Pharyngognathi*, with ctenoid scales and well-developed spinous rays. The preoperculum is, in the typical species, sharply serrate on both free borders. There are three anal spines, and the lateral line is well developed, not extending near the dorsal line. The caudal fin is rounded. The jaws are toothless. The pharyngeal bones, both superior and inferior, are closely studded with short, sessile, conical, teeth; a row of small ones stands on the external border of the inferior pharyngeal. One dorsal fin.

The species of *Priscacara* are referable to two sections. In the first, the ventral spine is very strong, and there are but ten or eleven soft dorsal radii: here belong *P. serrata*, *P. cypha*, *P. oxyprion*, and *P. testudinaria*. In the second, the first ventral spine is weak and slender, and there are thirteen or fourteen radii of the second dorsal fin; in this division belong *P. liops*, *P. pealei*, and *P. clivosa*.

A pair of superior pharyngeal bones from the Washakie basin of the Bridger formation strongly resemble those of this species.

PRISCACARA SERRATA Cope.

Bulletin U. S. Geol. Surv. Terrs., 1877, p. 816.

Plate XIII, fig. 1.

Form a regular wide oval, with a subequal contraction at both extremities. The spinous dorsal rays become longer than the soft ones, but the posterior spines are shorter than the anterior soft rays, so as to produce a wide emargination in the superior outline. The spines are very robust, especially those of the pectoral and anal fins. The first anal spine is near two-thirds the length of the second. The pectoral fin does not extend to the anal, and the soft parts of the anal and dorsal, which are equal, do not overlap the base of the caudal. Radii: D. X—11; A. III—10; C.?—17—?. Vertebræ: D. 9; C. 14. The centra have a strong median lateral ridge, which separates two fossæ.

The greatest depth is at the base of the ventral fins, or the third dorsal spine; it enters the total length (with caudal fin) two and four-tenths times. The length of the head enters the same three and four-tenths times. The orbit is large, its diameter exceeding the muzzle and entering the length of the head a little over four times. The mouth is terminal, and the premaxillary extends obliquely downward and backward; the maxillary reaches the line of the anterior border of the orbit.

The scales are longer than deep, and the rough surface has but a small extent, and is finely granulated. The remainder of the scale is marked with strong concentric grooves. Those on the gular region are small. On the belly, there are from twenty to twenty-five rows (about) below the vertebral column. A row of scales extends along the postero-inferior edge of the operculum. This part is well preserved in only one of my three specimens which represent the species.

Measurements.

	M.
Total length	.217
Length of head	.064
Length to line of first spine of first dorsal	.070
Length to line of first spine of second dorsal	.121
Length to line of anal	.122
Length to base of caudal	.173
Depth at first dorsal spine	.093
Depth at first dorsal soft ray	.070
Depth of caudal peduncle	.027
Length of fourth dorsal spine	.030
Length of second anal spine	.027

This species is about the size of the Crappie, *Pomoxys annularis*. Not rare at Twin Creek, Wyoming.

Priscacara cypha Cope.

Bulletin U. S. Geol. Surv. Terrs., 1877, p. 817.

Plate XIII, fig. 2.

This species is nearly related to the last, but presents a number of differences which require its separate consideration. These are: (1) The more arched or convex dorsal outline; (2) The relatively longer head; (3) The presence of an additional dorsal spine; (4) The entire covering of the operculum with scales. There is also probably a smaller number of dorsal vertebræ, but this is not certain, as that region has been somewhat disturbed. Formulæ:—Rays: D. XI—10—11; A. III—9; P. 15. Vertebræ, 6–14.

The greatest depth enters the total length 2.6 times; the length of the head enters the same 3.3 times. The spines are more robust, and the serrature of the preopercle more produced in the individual now described than in any of those of the *P. serrata* in my possession. The size is about the same as that of the latter species.

Twin Creek, Wyoming.

Priscacara oxyprion Cope.

Bulletin U. S. Geol. Surv. Terrs., 1878, p. 74 (February 5).

Plate XIV, fig. 5.

Five specimens in nearly complete preservation represent this species in our collections. It is more nearly allied to the *P. serrata* than to the other species, as the spine of the ventral fin is large and robust. It differs from that and from all the other known species of the genus in the small number of the radii of its anal fin. It agrees with *P. serrata* in the small number of the rays of the second dorsal. It is a smaller species than the *P. serrata*, being intermediate in size between it and the *P. pealei.* It is especially marked by the long, acute serræ of the entire posterior and inferior margins of the preoperculum. The operculum, suboperculum, and cheek are scaled; the preoperculum is naked.

Formula: Br. VIII; D. X—11; V. I—5; A. III—8; Vert. D. 10; Caud. 14. The form is an elongate oval, rather more elongate than any other species of the genus. The mouth is terminal and the front gently convex and descending. The length of the head enters the total, less the caudal fin, two and a half times, and the greatest depth is half of the same. The dorsal spines are long and strong, the longest equaling the soft rays in length. The anal spines are very robust, the second or longest not equaling the longest soft rays of the same fin. The origin of the first spine is below the first ray of the soft dorsal. There are three long and one short interneural bones in front of the dorsal fin. The origin of the ventral is below the third (or fourth) dorsal spine. The vertebræ have two fossæ on each side, separated by a ridge. The jaws are edentulous. The scales are small and the specimens very well preserved.

In the largest specimen, I count, in a vertical line drawn from the first dorsal soft ray to the middle of the abdominal line, fifteen longitudinal rows of scales above, and twenty-five below, the vertebral column. On the opercular flap of a smaller, the typical specimen, I count nine vertical and fourteen transverse rows of scales.

Measurements.

	M.
Length of type-specimen	.137
Length to base of caudal fin	.109
Length to apex of first interhæmal	.067
Length of head	.040
Length of third dorsal spine	.024
Length of second anal spine	.018
Length of pectoral spine	.019
Depth at first dorsal spine	.050
Depth at first anal spine	.041
Depth of caudal peduncle	.019

The lateral line is visible in the largest specimen. It extends parallel to the dorsal border, marking at its greatest convexity less than one-third the distance from the vertebral column to the dorsal outline. It disappears behind the vertebral column below the seventh soft dorsal ray, and does not reappear.

This fish came from a deposit of the Green River Shales on Twin Creek, Wyoming.

PRISCACARA CLIVOSA Cope.

Bulletin U. S. Geol. Surv. Terrs., 1878, p. 76.

Plate XIII, fig. 3.

In the last-named fish, there are eight dorsal and fourteen caudal vertebræ. Radii: D. X—13; A. III—11. The ventral fin appressed, nearly reaches the base of the anal, a point in which it differs materially from the two allied species. Another characteristic is the form of the profile, which resembles that of some of the species of *Geophagus*. This descends steeply from a point just anterior to the base of the dorsal fin, giving an obliquity to that part of the outline, and an inferior position to the mouth. The vertebral column is more arched anteriorly, appropriately to the prominence of the anterior dorsal region. The depth at the base of the first dorsal fin enters the total length (with caudal fin) 2.6 times, and the length of the head 3.6 times in the same.

Measurements.

	M.
Total length	.115
Axial length of head	.032
Axial length to line of first dorsal	.032
Axial length to origin of ventral fin	.041
Axial length to origin of anal fin	.057
Axial length to origin of second dorsal fin	.056
Axial length to origin of caudal fin	.082
Depth of caudal peduncle	.016

The preopercular border is not visible in the only specimen of this species known to me. The operculum is scaly. There are 11–13 rows of scales on a line from the vertebral column to the abdominal border.

Twin Creek, Wyoming.

PRISCACARA PEALEI Cope.

Bulletin U S. Geol. Surv. Terrs., 1878, p. 75.

Plate VIII, fig. 4; XIV, fig. 4.

Outline elliptic, with the extremities contracting equally or symmetrically to the head and tail. Depth at ventral fins entering length (with caudal fin) 2.60 times. Mouth rather small; length of head entering total length 3.8 times. Short conic teeth *en brosse.* Preorbital and preopercular bones finely serrated on their free margins. Vertebræ: D. 7; C. 14. Radii: D. X—14; A. III—11; V. I. 5 or 6. The dorsal spines are rather slender; the anal spines are stouter, but shorter; the ventral spine is weak and slender. The ventral fin when appressed against the belly fails to reach the anal fin by a space a little greater than the length of the ventral spine; its origin is beneath the third dorsal spine. The scales are difficult to observe on the specimens, but there are not less than fifteen to seventeen longitudinal rows along the abdomen in front of the anal fin.

Measurements.

	M.
Total length	.130
Axial length of head	.035
Axial length to first dorsal spine	.038
Axial length to first dorsal soft ray	.062
Axial length to first anal spine	.070
Axial length to base of caudal fin	.103
Depth at orbit	.025
Depth at first anal spine	.041
Depth of caudal peduncle	.016
Length of fifth dorsal spine	.018

This species is similar in size and proportions to the *Priscacara liops*, but differs in having constantly but seven dorsal or abdominal vertebræ, while that species presents nine. I have not observed any serratures on the preoperculum of the *P. liops*, but the typical specimens are imperfect in that region, although good impressions of it remain on the matrix.

Two complete specimens present all the characters of this species, while in two others all the more important ones can be seen. Two additional specimens may be referred to it with the greatest probability, and I have found it abundant in various collections. Some were obtained by Dr. A. C. Peale, in charge of one of the parties under Dr. F. V. Hayden, from the shales of the Green River formation of Wyoming. The species is dedicated to Dr. Peale, in recognition of his services to geological science.

PRISCACARA LIOPS Cope.

Bulletin U. S. Geol. Surv. Terrs., 1877, p. 818.

Plate XIV, figs. 2–3.

A smaller fish than either of the preceding is referred to this genus, although it differs in one feature, regarded as important among the *Poma-*

centridæ, i. e., the preopercular border is entire. It conforms closely to the *P. serrata* in other respects, as the form of the dorsal fin, three anal spines, form of caudal fin, character of scales and lateral line, edentulous jaws, and, indeed, in form to such an extent as to lead me to suspect that in this genus, as in *Lepomis*, etc., the serration of the preopercle is not of much systematic value. One character by which the *P. liops* may be distinguished from *P. serrata*, in addition to the smooth preoperculum and small size, is the constantly larger number of rays in the second dorsal fin.

Formulæ:—Rays: D. X—13–14; C. 5—19—6; A. III. 10–11. Vertebræ: D. 9; C. 13. The pectorals originate below the first dorsal spine, and the ventrals a little behind it. The spines are moderately stout, and the emargination of the dorsal fin is not deep. There are twenty-five rows of ctenoid scales traversed by a vertical line from the middle of the spinous dorsal, and smaller scales cover the operculum and more or less of the preoperculum.

Measurements.

	M.
Total length	.113
Length of head	.032
Length to first dorsal spine	.034
Length to first dorsal soft ray	.057
Length to first anal spine	.057
Length to base of caudal	.086
Depth at orbit	.030
Depth at first dorsal spine	.043
Depth at first dorsal soft ray	.035
Depth of caudal peduncle	.014

Two specimens of this fish have been received from Twin Creek. They are somewhat injured, and it is possible that better specimens will show minute serrations of the preopercle.

PRISCACARA TESTUDINARIA Cope.

Plate I, fig. 7.

My best specimen of this fish is without the greater part of the skull; otherwise it is nearly complete. Under the circumstances it is difficult to make a final generic reference, but as the parts preserved are identical with those included in the definition of the genus *Priscacara*, I refer it here for the present.

The specimen is larger than those which I have seen of the other species of the genus, agreeing in this respect with another which I have seen in the museum at Salt Lake City. It is also more elongate in its form, and is further characterized by its large scales. The more general characters are, the undivided dorsal fin; the wide interneural spines, and the well developed basilar interneurals. Also the hæmal spines of the caudal fin retain the division into three or four parts. The vertebræ have a flat median, lateral rib, bounded by a fossa above and below. The lower border of the inferior fossa on the dorsal vertebræ is deflected. The anterior base of the neural spine is excavated on the second, third, and fourth vertebræ, behind the scapular arch. The lateral rib of the second is oblique vertical. On the fifth and sixth it has an excavated, down-looking superior border. The scales, which are beautifully preserved, though much disarranged, are about as deep as long, with the anterior border subtruncate. The borders, excepting the posterior, are marked with four lines of growth, and from five to ten grooves radiate from the center, across the posterior border lines. There are numerous minute elevated points on the center of the scale, and the same reappearing on the posterior border, give the ctenoid character.

Radii: D. XI—12; A. + 21 +. The number of the rays of the inferior fins cannot be ascertained, but there may be counted between the first and last, the impression of eleven interhæmal spines. The dorsal spines are very stout, and the first is very short. The ventral spine is not very long, but is very robust. The interhæmals that support the anal spines are extraordinarily robust. The outline of the body is an elongate oval, the depth at the ventral fin entering the length without the caudal fin (or head) two and two-fifths times. Vertebræ: D. from scapular arch, 8; C. 16.

Measurements.

	M.
Length without head and caudal fin	.255
Depth at posterior base and fin	.044
Length of series of caudal vertebræ	.140
Depth from vertebra to ninth dorsal spine	.026
Vertical diameter last dorsal vertebra	.011
Length of fifth dorsal spine	.052

From the calcareous shales of the Lower Eocene, near Manti, Utah.

A small specimen, apparently of the young of this species, was sent me by Dr. A. S. Packard. It is of a more elongate form than any of the other species assigned to this genus, the depth at the first dorsal (the length of the head) entering the total four times. The muzzle is short and the eye large, apparently in consequence of the immaturity of the fish. Radii: Br. 5; D. X—8; A III—12. Vertebræ from edge of operculum, D. IX, C. XIV, or from two to four more than in any other *Priscacara*. The three spinous anals are robust as in other species of the genus, while the first ventral spine is long and strong and deeply ground on the inner side. The numbers of soft rays above given is subject to revision owing to the condition of the specimen. Total length, M, .058.

From the shales of City Creek Cañon, near Salt Lake City, Utah. This specimen indicates a great northern extension of the Manti shales.

BATRACHIA.

Remains of *Batrachia* are rare in North American formations later than the Permian. There are two or three species of *Stegocephali* known from the Trias, above which formation that order is not known to extend in any country. No Batrachians have been obtained from the Jurassic or Cretaceous systems excepting from the top of the latter, in the Laramie. Here occur the genera *Scapherpeton* and *Hemitrypus* Cope. A single specimen from the Eocene is mentioned below, and then we miss them until the Loup Fork or Upper Miocene, where *Anura* and salamanders have been found.

The vertebral column and part of the cranium of a probably incompletely developed tailless Batrachian, were procured by Dr. F. V. Hayden, from the fish shales of the Green River epoch, from near Green River City, Wyoming. They are not sufficiently characteristic to enable me to determine the relation of the species to know forms. It is the oldest of the order *Anura* yet discovered, the fossil remains of the known extinct species having been derived from the Miocene and later formations of Europe.

REPTILIA.

The Eocene period, was, of the divisions of the Tertiary, the most prolific of reptilian life. It is true that the orders of reptiles which characterized the Mesozoic periods no longer existed. The *Dinosauria* had perished from the land; the *Ichthyopterygia*, *Sauropterygia* and *Pythonomorpha* no longer inhabited the sea, and the *Pterosauria* had disappeared from the air. The Eocene reptiles were not a new creation, nor a new evolution, but a remnant of the types that had coexisted with those monarchs of life during previous ages. We must except from this statement the serpents, which first appear at this time.* The crocodiles, tortoises, and lacertilians represent orders already abundant in the Mesozoic faunæ. Their decadence in central North America did not commence until the Miocene period, when the crocodiles and nearly all the tortoises disappeared. From the Loup Fork or Upper Miocene, only a few traces of lizards have been obtained, and snakes were apparently not very numerous. On the eastern coast regions crocodiles existed, and tortoises were more numerous during the Miocene period; but here also they were less abundant and varied than during the Eocene.

The *Crocodilia* did not differ in important respects from those now existing. I have distinguished five species from the Wasatch beds, and six different ones from the Bridger.

The *Testudinata* include a great variety of forms. I have seen sixteen species from the Wasatch formation, and thirty-two from the Bridger and Washakie. Of these, six are common to the two formations, as indicated by imperfect material, leaving a total of forty-two. Three genera, *Emys*, *Trionyx*, and *? Plastomenus* hold over from the Cretaceous period, while six appear for the first time. Of these, five gènera are not known to continue later than the Eocene period.

Of lizards I have obtained the remains of a half dozen of species, but none of them in a complete state of preservation. Professor Marsh has been more fortunate, as he has described from his material from the Bridger beds, twenty-one species.† He arranges these under five generic heads, as fol-

* Since the above was written, it is reported that Dr. Sauvage, of Paris, has discovered the remains of a serpent in the cretaceous formation.

† American Journal of Science and Arts, 1871, June, and October, 1872.

lows: *Thinosaurus* Marsh, five species; *Glyptosaurus* Marsh, eight species; *Xestops* Cope (1873, *Oreosaurus* Marsh, not Peters), five species; *Tinosaurus* Marsh, two species; and *Iguanavus* Marsh, one species. As Profesor Marsh does not give us any clue to the affinities of these forms, they cannot be further considered here. In Lieutenant Wheeler's Survey Report* I have pointed out that the dermal scuta and a few other fragments which I obtained in the Wasatch beds of New Mexico, were probably referable to the *Placosauridæ*, a family created by Gervais to receive certain *Lacertilia* of the Eocene of France. To this family no doubt some of the species described by Marsh from the Bridger horizon are to be referred.

The snakes of the Eocene are not very numerous as to species. The first known American species (*Palæophis littoralis* and *P. halidanus*) were determined by myself from New Jersey specimens. None have been procured from beds lower than the Bridger, and in that formation I found a single form. Professor Marsh has described five species.*

The whole number of species of reptiles thus far discovered in the Eocene of the central region of North America is as follows:

Crocodilia	12
Testudinata	42
Lacertilia	22
Ophidia	6
	82

OPHIDIA.

PROTAGRAS Cobe.

Paleontological Bulletin No. 3, p. 3, August 7, 1872. Annual Report U. S. Geol. Surv. Terrs., 1872 (1873) p. 632. Proceed. Amer. Philos. Soc., 1872, p. 471.

Transverse processes large, the extremity entirely occupied by the costal articular surface. This consists of a superior and an inferior convex portions, which are separated by a constriction, which is most profound on the posterior border. Zygosphene wider than articular cups, and giving rise to a low ridge which extends along the side of the neurapophysis. Articular ball and cup wider than deep, the former looking very obliqely upwards, its surface extending to the bases of the neurapophyses. A prominent ridge connects the pre- and postzygapophyses. A strong hypapophysial keel, and a latero-inferior ridge extending posteriorly from the base of the transverse process.

* Vol. iv, pt. ii, p. 42, pl. xxxii, figs. 26–36.

The only extinct genus with which it is necessary to compare the present one, is the *Boavus** of Marsh, which was described more than a year previously. There are various points in which Professor Marsh's full description corresponds with my specimens, but I observe two important differences: One is, that in *Boavus* the diapophyses are said to be "convex throughout," while here they present a median constriction, giving a figure eight outline. The other is, that the cup and ball are "more nearly vertical" than in *Boa;* the ball is very oblique in *Protagras.*

The modern affinities of *Protagras* will be fully considered in connection with the *Ophidia* of the Miocene period in a later portion of the present work.

But one species is known as yet. It was found in the Bridger beds of Wyoming.

PROTAGRAS LACUSTRIS Cope.

Paleontological Bulletin No. 3, 1872, p. 3. Proceed. Amer. Philos. Soc. loc. sup. cit. Annual Report U. S. Geol. Surv. Terrs., 1872 (1873), p. 632.

Plate XXIII, figs. 17–18.

A serpent of about the size of the existing pine snake (*Pityophis melanoleucus*).

A vertebra before me has the longitudinal hypopophysial keel horizontal, and terminating in a very obtuse point. The ball looks extensively upwards. The upper articular extremity of the diapophysis is short and obtuse, and the inferior equally so, and directed shortly downwards, their articular surface being continuous with each other. It sends an obtuse latero-inferior keel backward, which terminates distinctly in front of the ball. Its inferior angle stands below the inferior margin of the articular cup. The angle connecting the diapophysis and zygapophysis is strong, while the former is narrow.

Measurements.

	M.
Length of centrum with ball, below	.0090
Elevation behind, total	.0135
Elevation before, total	.0119
Width between diapophyses, below	.0055
Width of articular cup	.0054
Depth of articular cup	.0043
Depth of inferior keel	.0010

Found by myself in the Bridger bad lands of Cottonwood Creek, Wyoming.

* American Journal Science and Arts, 1871, May.

? LACERTILIA.

CHORISTODERA.

Cope, Proceedings Academy Phila. 1876, p. 350.

Vertebral centra amphiplatyan. Processus dentatus free from axis. Neural arches separate from centrum during maturity.

CHAMPSOSAURUS Cope.

Loc. cit. p. 348, published Jan. 30, 1877. Paleontological Bulletin No. 23, p. 9, Jan. 10, 1877. *Simædosaurus* Gervais, Journal de Zoölogie, 1877, No. 1, p. 76, ? February.

This genus was established on species found in the Laramie Cretaceous formation. It has been found to be abundant in the Puerco of the Tertiary series, and is hence introduced here.

The characters presented by the vertebral column are the following: The ribs have a single head, which articulates with a prominent tuberculum, excepting those of the cervical vertebræ. On these there is a small capitular tubercle below the diapophysis. It commences very small, and inferior in position, being removed, in fact, but a short distance from the inferior middle line in the first vertebra in which it appears. It rises rapidly in the succeeding centra until it is merged in the tuberculum of the diapophysis. The latter projects from the neural arch, which is free from the centrum, but in none does the base of the diapophysis rise from a point above the floor of the neural canal. On the dorsals it is vertically compressed. One of the anterior cervicals, probably the axis, is obliquely truncated below its anterior articulate face, for a free hypopophysis or *os odontoideum*. This vetebra has no parapophysis, and the articular faces for the neurapophysis are superior. The few vertebræ in each of several series, probably from the sacral region, are more depressed than the others, and the facets for the diapophyses present a greater antero-posterior extent, but none are coössified. The caudal vertebræ are distally quite compressed. In all, except the anterior ones, the neural arch is coössified with the centrum, and in such there are no diapophyses. In those with free neural arch, the facets for the neuropophyses turn down on the sides of the centrum.

The articular extremities of the centra are plane, those of the caudal series slightly concave. There are no hypapophyses behind the axis, excepting a longitudinal carina, which ceases to exist on the dorsal vertebræ. The zygapophyses are simple. The chevron bones are free.

The relations of the atlas and axis, though not fully elucidated by my specimens, are peculiar. The former has separate neurapophyses, which have nearly the shape of those of the Streptostylicate *Reptilia*, resembling much those of the *Pythonomorpha*. Although I procured numerous cervical vertebræ, there are but few which exhibit the antero-inferior facet for supposed hypapophysis, already described. The position of this vertebra was in front of the first cervical which displays a parapophysis, and is, on this account, likely to be the axis or the third cervical vertebra. It is the more probably the axis, as there is no other among the large number of vertebræ in my collection which can be referred to that position. Its anterior articular face is smooth and like the posterior, showing that the odontoid bone was not coössified with *it*. Now in the *Crocodilia* the odontoid bone is united with the anterior extremity of the axis by suture, which may become coössified with age, while the free hypapophysis is wanting. In the streptostylicate orders the hypapophysis is present, and the odontoid is above it, but united to the axis by suture. On the other hand, in the *Rhynchocephalia*, the axis is coössified with both odontoid and hypapophysis, and a few succeeding vertebræ possess free hypapophyses.

A few entire ribs and the heads of many others have been obtained. The cervical ribs are long, and the dorsals are relatively stout and short. The head of an anterior dorsal is truncate and compressed, its articular face contracted, forming a narrow figure eight. The shaft is obliquely flattened. The extremities are separated from the lateral surfaces by a narrow angle, as though capped with cartilage in life, as in the *Pythonomorpha*.

Bones of the extremities are very rare. One fragment resembles the proximal end of a crocodilian tibia, and another is like the distal half or more of the tibia of the same type.

The above characters were derived from the Laramie species, and those of the Puerco agree with them exactly in those respects. The latter enable me to add, that the jaws are slender, and that the splenial bone of the mandible is well produced anteriorly. The teeth are set in shallow alveoli, and are replaced from the inner side as in *Lacertilia* and *Pythonomorpha*.

Dr. Lemoine has found this genus in the Suessonian formation near Reims, France, and his material has enabled him to furnish some characters in addition to the above mentioned. He states* that the quadrate bone is "non soudé," and that the limbs resemble both those of Crocodilia and Lacertilia, and are apparently adapted to aquatic habits.

Ignorance of the structure of the skull has prevented a definite conclusion as to the true position of this genus and its allies. Dr. Lemoine's observation makes it appear that they belong to the Streptostylicate division, and that they form an aberrant division of the *Lacertilia* or *Pythonomorpha*. For the present I refer them to the former, but they will constitute a distinct sub-order with the definition given on a preceding page. Besides *Champsosaurus*, the *Champsosauridæ* include the genus *Ischyrosaurus* Cope, which differs from it in the heavy subfusiform ribs, and the flat articulation between the centrum and neural arch of the vertebræ.

This genus was named by Professor Gervais at nearly the same time with myself. His publication was made in the February no. of the Journal de Zoölogie for 1877, and mine in the Proceedings of the Philadelphia Academy for December, 1876. My description did not appear until January 10, 1877, and although I do not know of the precise date of the publication of the Journal de Zoölogie, it was presumably not until some weeks later.

There have been four species of this genus described from the American Laramie formation, viz: the *C. profundus*, *C. annectens*, *C. brevicollis*, and *C. vaccinsulensis*. The species from Reims is called *C. lemoinei* Gerv. I distinguish three species from the Puerco bed of New Mexico, which differ from the Laramie species in obvious ways. Vertebræ of a species from the Laramie were figured without name by Leidy in the Transactions of the American Philosoph. Society, 1860.

The Puerco species differ as follows:

Small; dorsal centra with semicircular faces, much wider than deep; anterior dorsal keeled below .... *C. australis.*

Large; dorsals with cordate faces a little wider than deep; none known to be keeled below.. *C. puercensis.*

Medium; length, width, and depth of dorsal centra equal; faces subround; not keeled below; axis not keeled below... .... *C. saponensis.*

*Communication sur les Ossements fossiles des Terr. Tertiaires Infs. des. Env. de Reims; Assoc. Franç. pour l'Avanc't des Sciences, 1880, p. 15.

CHAMPSOSAURUS AUSTRALIS Cope.

American Naturalist, 1881, p. 670 (July).
Plate XXIII b; figs. 1–4.

Eleven vertebræ, probably of one individual, were found by Mr. Baldwin, mingled with jaw fragments, with teeth of the Eocene Mammal *Catathlœus rhabdodon*. All the pieces are enclosed in the black ferruginous matrix, in which the mammalia of the Puerco epoch are found embedded.

The vertebræ are of about the same size and form, and all belong to the dorsal series. They are characterized by their large width as compared with their depth, differing in this respect from all of the known species. The centra are regularly rounded below, and the borders are scarcely at all flared. One of the dorsals, probably an anterior one, has a prominent angular keel in the middle line below. The outline of the articular faces for the neural arches is pyriform, the wide portion concave, with its external edge decurved, and on the anterior half of the side of the centrum. The decurvature is sometimes sufficient to resemble part of a rib-facet. Articular faces of centra nearly plane. Sides of centra very little concave, a shallow fossa below the base of each diapophysis. Non-articular surfaces of centrum marked with a delicate thread-like sculpture, and there is no coarse sculpture near the edges of the articular surfaces.

Diameters of keeled dorsal centrum: anteroposterior M. .012; vertical .014; transverse .017. Diameters of a rounded dorsal: anteroposterior .013; vertical .012; transverse .015. The dorsal vertebræ are wider and more transverse than in either of the four known American species. They are longer than those of the *C. vaccinsulensis*, and lack the marginal wrinkles of the *C. saponensis*. From near Canyon Largo, San Juan River, New Mexico.

CHAMPSOSAURUS PUERCENSIS Cope.

Proceeds. American Philosoph. Society, 1881, Dec. p. 195 (1882). Paleontological Bulletin No. 34, 1882, p. 195, Feb. 20.
Plate XXIII b; figs. 5–10.

This species is represented by a number of fragments, which include three dorsal and four caudal vertebræ of apparently one individual. They represent an animal of larger size than any of those heretofore referred to *Champsosaurus*, excepting the *C. vaccinsulensis*. In all of the vertebræ the neural arch is more or less coössified with the centrum, and the animal had probably reached its full size.

One of the dorsal centra is split vertically and longitudinally, and shows the structure already figured by Leidy in the *Ischyrosaurus antiquus** Leidy. The surface exposed displays two diagonal lines of fissure crossing each other at right angles. They indicate clearly the mode of origin of this amphiplatyan type of centrum. The centrum is first deeply amphicoelous, as in the Theromorphous reptiles of the Permian. The conical cavities are filled by the ossification of the remaining portion of the notochord, forming a conical body which always remains distinct from the remainder of the centrum.

The articular faces of the dorsal centra are a little wider than deep, and the depth about equals the length of the body. They are not nearly so depressed as those of *C. australis*, and their outline is different. This is wider above and narrows below; in both *C. australis* and *C. saponensis* the inferior outline is part of a circle. None of the dorsals preserved are keeled below. There is a fossa below the diapophysis which has a subvertical posterior boundary. The general surface (somewhat worn) does not display wrinkles near the articular faces. An anterior dorsal has a short compressed diapophysis with a narrow figure 8 articular surface, and its superior border in line with the roof of the neural canal. The anterior caudals have subround articular faces; the posterior are more oval, and the bodies compressed. With greater compression the length increases.

Measurements.

		M.
Diameters of an anterior dorsal	anteroposterior	.025
	vertical	.025
	transverse	.030
Height of costal facet of do		.021
Diameters neural canal do.	vertical	.007
	transverse	.009
Diameters anterior caudal	anteroposterior	.024
	vertical	.021
	transverse	.021
Diameters posterior caudal	anteroposterior	.025
	vertical	.018
	transverse	.018

The typical specimen was found by Mr. Baldwin near the Puerco River west of the Nacimiento Mountain, New Mexico, in the typical locality of the Puerco formation.

*Trans. Amer. Philos. Soc. 1860.

Champsosaurus saponensis Cope.

Paleontological Bulletin No. 34, p. 196; Feb. 20, 1882. Proceedings Amer. Philos. Society, Dec. 1881, p. 196, 1882.

Plate XXIII b; figs. 11-22.

Represented in my collection by six cervical and several dorsal vertebræ; one only of the latter with well preserved centrum, parts of ribs, and various other bones, whose reference is not yet certain.

The cervical vertebræ include the os dentatum or centrum of the atlas. This shows its streptotylicate character in its distinctness from both the centrum and the free hypapophysis of the axis. Nevertheless, it is more Crocodilian than Lacertilian in form. Its anterior face is transverse, with a little lip carrying forwards the floor of the neural canal, below which the face is bevelled posteriorly. The inferior surface is narrow and transverse, as though adapted for the anterior part of the hypapophysis of the axis. At each side it terminates in a prominent tuberosity, as though for the attachment of a cervical rib as in the Crocodilia. The anterior face is bounded posteriorly by a transverse groove which terminates in a down-looking fossa on each side. The posterior articular face of the os dentatum is wider than deep. The lateral angles of the superior face are rounded, and its median portion is concave.

The axis displays a large facet for the hypapophysis. Behind it the inferior middle line is not keeled, but is coarsely wrinkled longitudinally. The posterior edge of the hypapophysial facet is the most prominent part of the inferior surface. The posterior articular face is deeper than wide. This is true of the faces of all the cervical vertebræ. The latter gradually increase in size posteriorly, and the dorsals become larger. The articular faces of all the centra are regularly rounded and not contracted below. The five cervicals are strongly keeled below, the keel of the third centrum being split up anteriorly into narrow ridges On the sixth the keel is more prominent and acute. The dorsal is not keeled. A trace of the parapophysis appears low down on the fourth cervical; it rises and becomes prominent as a rounded tuberosity on the fifth and sixth. It appears on the superior edge of the centrum of the dorsal vertebra, where it is connected with the diapophysis. It is near the middle of the length of the centrum, and not near the anterior border, as in *C. australis*.

The surfaces of the vertebræ are very smooth, excepting where thrown into coarse wrinkles near the border of the articular faces and near the hypapophysis. The edges of the articular faces are somewhat revolute on the sides in the cervicals, but not on the dorsal. They are impressed in the centre to a point, most strongly so as we pass forward in the series. There is a fossa below the space anterior to the parapophysis of the dorsal vertebra, which is abruptly bounded below by a horizontal angle. A separate neural spine, perhaps of a cervical vertebra, has the following form. It is stout and contracts rather abruptly at the apex from behind forwards. The section is broadly lenticular, angulate in front and truncate behind. The posterior face has several longitudinal wrinkles, including a median raised line, and there are some more irregular wrinkles on the sides.

Measurements of vertebræ.

			M.
Anterior face of os dentatum	width		.025
	depth (oblique)		.012
Posterior face of os dentatum	width		.020
	depth		.018
Length os dentatum above			.014
Diameters axis	posterior face	depth	.022
		width	.020
	length		.0185
Hypopophysial facet os dentatum	depth		.008
	width		.014
Diameters fourth cervical	length		.022
	anterior	depth	.0225
		width	.022
Diameters sixth cervical	length		.0215
	anterior	depth	.0245
		width	.0235
Space between parapophysis and diapophysis of do			.0040
Diameters of dorsal	length		.0265
	anterior	depth	.0260
		width	.0265
Height of neural spine of ? from posthypapophysis			.0210
Anteroposterior width of do. at base			.0100

The portions of ribs are separated heads and shafts. The former are double and therefore cervical, and are quite large. If the shafts belong to them, the neck of this species must have been wide. The shafts are slender and of dense bone. The section is oval at the middle, but towards the distal extremity becomes flattened and grooved, and delicately lined on one side. The extremities of the long bones are without condyles, but have concave surfaces like those of the ribs. The bodies are robust and angular. They may be abdominal ribs of unusual stoutness.

TESTUDINATA.

As the Eocene forms of this order are of unusual interest, I give an analysis of the extinct genera of the Cryptodire division, which have been found in North America and Europe up to the present time.

In the check-list of the North American *Batrachia* and *Reptilia*,* I enumerated nine families of this division of the *Testudinata*, three of which are extinct. Subsequently another extinct family, the Baënidæ, was added. I now define all of these families.

I. Plastron not articulated to the carapace, but presenting to it more or less open digitations. *Dactylosterna.*

Phalanges of anterior limb without condyles, and covered by a common integument; eight pairs of costal bones .... *Cheloniidæ.*

Phalanges of anterior limb without condyles; nine or more costal bones. *Propleuridæ.*

Phalanges of anterior limb with condyles; digits inclosed in distinct integuments; eight costal bones; sternal elements united by digitations and inclosing fontanelles; caudal vertebræ procoelous. *Trionychidæ.*

Phalanges of anterior limbs with condyles; digits distinct; eight costal bones; sternal elements united by suture and inclosing no fontanelles; caudal vertebræ opisthocoelous .... *Chelydridæ.*

II. Plastron uniting with the costal bones of the carapace by suture, with ascending axillary and inguinal buttresses. (Feet, ambulatory.) *Clidosterna.*

A Intersternal bones present.

No intergular scuta .... *Pleurosternidæ.*†

Intergular scuta; caudal vertebræ optisthocoelous .... *Baënidæ.*

A A No intersternal bones.

a Intergular scuta.

A mesosternal bone .... *Adocidæ.*

a a No intergular scuta.

A mesosternal bone; three series of phalanges .... *Emydidæ.*

No mesosternal bone; three series of phalanges .... *Cinosternidæ.*

A mesosternal bone; two series of phalanges .... *Testudinidæ.*

III. Plastron uniting with the marginal bones of the carapace by straight suture only. (Feet, ambulatory.) *Lysosterna.*

No intersternal bone nor intergular scutum; a mesosternal bone and three series of phalanges .... *Cistudinidæ.*

The extinct species of the *Cryptodira* of this continent belong to eight of the above families. I give diagnoses of the genera to which they are referred. Names of existing genera are in Roman type.

---

* Bulletin U. S. National Museum, No. 1, 1875. p. 16.

† There are two genera of this family, neither of them yet found in America; *Pleurosternum* Ow. with smooth shell, and *Helochelys* Meyer, with sculptured shell.

CHELONIIDÆ.

Postabdominal bones distinct from each other .... Chelonia Brong.
Postabdominal bones united with each other by suture .... *Puppigerus* Cope.

PROPLEURIDÆ Cope.*

Transactions of American Philosophical Society, xiv, 1870, p. 235.

Ten costal bones; first two marginals united with carapace by suture; shell smooth, flattened .... *Osteopygis* Cope.
Nine costal bones; first two marginals united to carapace by suture; shell sculptured; a high dorsal keel .... *Peritresius* Cope.
Nine costal bones; one marginal united with carapace by suture; second by costal gomphosis; shell not keeled nor sculptured. *Propleura* Cope.
? Nine costal bones; first united with carapace by suture; second without costal gomphosis; shell not sculptured .... *Catapleura* Cope.
? Nine costal bones; marginals all free; shell not sculptured ...... *Lytoloma* Cope.

TRIONYCHIDÆ.

a Surface of bones smooth.
Postabdominal suture digitate .... *Axestus* Cope.
a a Surface of bones sculptured.
β Sutures of plastron digitate.
A dermal flap protecting posterior legs below; marginal bones .... Emyda Gray.
A dermal flap; no marginal bones .... Cyclanosteus Peters.
No dermal flap nor marginal bones; muzzle much abbreviated .... Chitra Gray.
No dermal flap nor marginal bones; muzzle elongate .... Trionyx Geoffr.
β β Suture for post-abdominal coarsely serrate.
Postabdominal recurved in front .... *Plastomenus* Cope.

CHELYDRIDÆ.

a Bridges of plastron wide; ? caudal vertebræ.
One row of marginal scuta; six pairs of scuta of the plastron ..... *Idiochelys* Myr.
One row of marginal scuta; scuta of plastron ? not distinct .... *Hydropelta*† Myr.
a a Bridges of plastron very narrow.
β Carapace smooth, not sculptured.
Two rows of marginal scuta; five pairs of scuta of the plastron .. Macrochelys Gray.
One row of marginals; five pairs on plastron .... Chelydra Schw.
One row of marginals; four pairs of scuta on plastron .... Claudius Cope.
β β Carapace sculptured.
One row of marginal bones .... *Anostira* Leidy.

BAËNIDÆ.

Cope, Annual Report U. S. Geol. Surv. Terrs., 1872 (1873) p. 621.

Supramarginal scuta (Rütimeyer) and intermarginal scuta; no interhumerals .... *Platychelys* Myr.
No interhumeral scuta; no supramarginals .... *Baëna* Leidy.
Interhumeral scuta; no supramarginals .... *Polythorax*‡ Cope.

---

* *Palæochelys novemcostatus* Geoffr. belongs to this family, but not *Palæochelys* Myr.
† *Eurysternum* Wagn. (*Palæomedusa* et *Acichelys* Myr. fide Rütimeyer) is nearly allied to *Hydropelta*.
‡ Possibly one of the *Adocidæ*; see Proceed. Acad. Phila., Oct., 1876.

ADOCIDÆ.

Cope, Proceedings American Philosophical Society, 1870, p. 559.

a Vertebral bones and scuta normal.
One intergular scutum entirely separating the gulars............. *Adocus* Cope.
Either two intergulars, or the gulars meeting behind intergular.... *Amphiemys* Cope.
a a Vertebral bones wedge-shaped, widening upwards; vertebral scuta not wider than the bones.
Elements of carapace early coössified .... *Homorhophus* Cope.

EMYDIDÆ.

a No scutal sutures.
Surface sculptured .... *Apholidemys* Pom.
a a Scuta including intermarginals and two anals.
Lobes of sternum narrow......... .... Dermatemys Gray.
Lobes of sternum wide .... *Agomphus* Cope.
a a a Scuta; two anals, no intermarginals.
Surfaces of carapace sculptured; plastron fixed................. *Compsemys* Leidy.
Surfaces of carapace smooth; plastron fixed; recent *Emydidæ** and the genus .... Emys Brong.
Posterior lobe of plastron moveable; surfaces smooth............. *Ptychogaster* Pom.
a a a a Scuta; one anal, no intermarginals.
Carapace smooth; pectoral plates entire........................ *Stylemys* Leidy.
Carapace smooth; pectoral plates small, widely separated from each other .... Manuria Gray.

TESTUDINIDÆ.

a Two anal scuta.
Ten abdominal scuta . ... *Hadrianus* Cope.
a a One anal scutum.
β Claws 4–4.
Jaws with aveolar grooves.................................. Testudinella Gray.
Lower jaw smooth; upper with a ridge......................Homopus Dum. Bibr.
β β Claws 5–4.
γ Hinder part of carapace mobile.
Gulars distinct.. Kinixys Wagl.
γ γ Hinder part of carapace fixed.
δ Plastron moveable in front.
Gulars distinct.. Pyxis Bell.
δ δ Plastron fixed.
Gular plates united .... Chersina Gray.
Gular plates distinct.
Lower jaw with two cutting edges.............................. Xerobates Agass.
Lower jaw with one cutting edge. .... Testudo Linn.

The extinct tortoises of the Cretaceous and Eocene throw considerable light on the probable origin of various existing genera,† and while much

* Gray has distinguished several genera among existing species, on cranial characters.

† See on the Extinct Tortoises of the Cretaceous of New Jersey. Proceed. Amer. Ass. Adv. Sci., 1871, p. 344.

remains obscure, the following observations may be derived from the study of the forms in question.

The order makes its appearance in the Triassic period, for I am assured by Dr. F. Endlich, of Reading, Pa., that the species obtained by Professor Quenstedt, in Würtemburg, belong undoubtedly to the *Testudinata*. With their special structure we are not yet fully acquainted. A number of genera appear in the Jurassic, and there is a successive increase in the number of species in the Cretaceous and Tertiary formations. *Sphargis*, which is without carapace, and has a greatly reduced plastron, may be regarded as nearest the primitive types of the order. One species, *S. pseudostracion*, has been found in the European Jurassic, and the genus exists in recent seas. *Protostega* of the Kansas Cretaceous is its nearest extinct ally known. *Protostega* is superior in the well-developed marginal bones, and prepares the way for consideration of the genera with incomplete shields, of the present period (*Cheloniidæ*), which also possesses the natatory extremities. Two structural features of importance mark many of the Jurassic forms. First, the incomplete union and ossification of the elements of the plastron and carapace; second, the reduction in size of the lobes of the plastron. Genera retaining some or all of these peculiarities persist to the present day, but the ossified types with wide plastron are far more abundant, and are comparatively rare in the period of the Jura.

Trionyx appears to represent another point of departure. Its plastron presents a grade of development near to that of *Propleura*, and its eight costal bones ally it to other types. In its half-ossified carapace wanting the marginals, it is inferior to both. It leads us at once to the existing *Chelydra*, the closing of the sternal fontanelles being accompanied by a contraction of its extent in respect to the bridges and lobes. In *Propleura* of the Cretaceous we have a state of things intermediate between some of the Jurassic *Chelydridæ*, as *Idiochelys*, and *Chelone*. These genera had a common origin near the Jurassic predecessor of *Protostega*. The peculiar sculpture of *Trionyx* is seen in the Eocene *Anostira* (which is much like *Chelydra* in form), and in an obsolete condition, in *Adocus* of the Cretaceous, which adds Chelydrine and Pleurodire characters in a remarkable manner. It is closely joined by *Plastomenus*, which, in turn, presents points of resemblance to *Emydidæ*.

The Jurassic genus *Aplax* Myr., is nearly as deficient in ossification of carapace and plastron as *Protostega*, and is allied to the Chelydrid series, which existed cotemporaneously and during the Cretaceous. *Idiochelys* represents a rather more advanced form, with distinct marginal bones, and with affinities to *Chelydra* of a decided character. It was probably its ancestor. Allied to it we have such forms as *Adocus* and *Baëna*, which, while furnished with fully ossified shell, still approach *Chelydra* in the contracted form of plastron, and have several points of affinity to the Pleurodire series. From some common ancestor of these, sprang also the true Pleuroderes of the Cretaceous, as *Taphrosphys*, while, by the omission of most of the tendencies towards that series, we have *Dermatemys*, the genus of *Emydidæ* nearest to *Adocus*. From this point we pass to true *Emydidæ*, and thence, by the loss of a series of phalanges, to *Testudo*. From *Taphrosphys* we pursue the Pleurodire series to the similarly modified type *Pelomedusa*.

The accompanying table expresses the relations indicated, supposed to be genetic, and in accordance with the theory of evolution. (Types beginning in the Jurassic in *italics;* Cretaceous to recent, SMALL CAPITALS); Eocene to recent, s p a c e d ; recent only, Roman. The apparent reversal of the order of time displayed by the Jurassic and Cretaceous families is an indication of our ignorance of the Jurassic *Testudinata*.

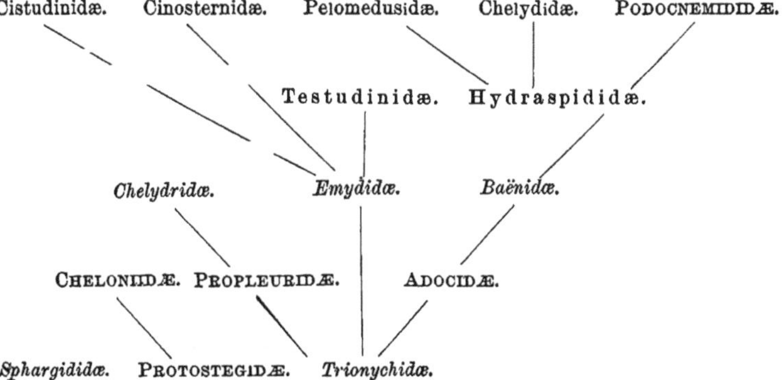

AXESTUS Cope.

Proceed. Amer. Philos. Soc., 1872 p. 462, (published July 29). Annual Report U. S. Geol. Surv. Terrs., 1872 (1873), p. 615.

This is a genus of Trionychidæ which is represented by a species not fully known. The type specimen is represented by bones of the limbs and various vertebræ, with the post-abdominal bone of the left side.

The general characters are those of *Trionyx*. The scapula is elongate, the procoracoid long and narrow, and the coracoid of medium width. The humerus is sigmoid, with widely separated tuberosities, and flattened extremity, with marginal groove. The femur is also curved, but less strongly than the humerus, and has a median anterior low angular ridge. The claws are large, some curved, and some entirely straight. The cervical vertebræ are relatively large and elongate. The two sacrals are free from the carapace above, have broad articular surfaces for diapophyses, and flattened centra. The caudals are procœlian, and have short diapophyses. The post-abdominal bone has the form seen in existing *Trionyx*. It presents two dentate processes forwards for the hyposternal, and two inwards to its mate in front. It is prolonged backwards and inwards into a flat process. It is especially distinguished by its tenuity, and the entire absence of the superficial sculpture of *Trionyx*. The usual dense layer is present, but is quite thin, and exhibits the peculiar decussating pattern of lines of deposition characteristic of the same layer of the dermal scuta of *Crocodilia*. No portions were obtained which can with certainty be referred to the carapace. The ilium is short, stout, and recurved, and the pubis is largely expanded.

AXESTUS BYSSINUS Cope.

Loc. cit.

Plate XV, figs. 1-12.

The procoracoid and scapula are of equal lengths, and the coracoid is much dilated distally.

The portions of plastron preserved are thin for the size of the animal, and all the bones are especially dense and smooth. The post-abdominal has the free margins acute and serrulate. There is an external, gently convex edge, with a long process extending backwards; and one long narrow one inwards. The dense layer is marked with decussating lines of osseous deposits, as in woven linen The cervical vertebra is without spine; it is not depressed in the middle, and is without any pneumatic foramen.

	M.
Length of cervical vertebra	.068
Diameter at middle	.020
Diameter at end	.035
Diameter of caudal vertebra at ball	.010
Length of caudal vertebra	.023
Length of an ungual phalange	.043
Proximal depth of an ungual phalange	.013
Length of post-abdominal, broken	.180
Width of post-abdominal	.120

A hyposternal bone of a large Trionychoid turtle displays the characters of this genus in its absence of superficial sculpture, and in the decussating bone structure of its borders and processes. It belongs to a species of the size of the *A. byssinus*, but is so much more massive than the post-abdominal bone of that species that I suspect that it belongs to a distinct one. Its median surface presents a few faint traces of tubercular roughening.

Occasionally the superficial layer of the bones of the plastron of species of *Trionyx* of this formation, are found nearly smooth, but they do not display the decussating bone structure, nor the thin edges of the species of *Axestus* It has often occurred to me that these peculiarities may be the result of erosion, and that the animals possessing them should be referred to the genus *Trionyx*. This may be the case, but there are two objections to such a view. First, the middle, or more prominent parts of the bone, where the attrition must be greatest, displays the characters least; second, if we imagine that movements of the limbs have caused the attrition, we are met by the fact that the decussating structures appear on the inner borders of the bones where the limbs do not reach.

The typical specimen was found on Black's Fork of Green River; the second specimen on Upper Green River.

TRIONYX Geoffr.

Turtles of this genus were very abundant during the Eocene period in North America. They disappeared from the interior basin with the close of this period and did not reappear; but they continued on the Atlantic slope, and are to-day abundant in the tributaries of the Mississippi and in the streams that flow into the Gulf of Mexico.

There were evidently several species of *Trionyx* during both the Wasatch and Bridger epochs; but the specimens found are generally so fragmentary, that it is difficult to ascertain what characters can be relied on to distinguish them. In the Paleontology of New Mexico (Wheeler's Report) I enumerated five as occurring in the Wasatch formation of that region, but I am not sure that more than three of them will ultimately be found to deserve that distinction. In my collection from the Bridger beds I have three manifestly distinct species, and I am compelled to admit two others. Of the five, I regard two as identical with New Mexican Wasatch species, and one as common to the Bridger and Wasatch beds of Wyoming. The proper definition of the doubtful species must be left to future discovery of better material.

The Eocene Trionyches may be distinguished in tabular form as follows:

I. Sculpture of the extremities of the costal bones thrown into ridges.	
a Superficial layer of costal bones overhanging rib extremities.	
Ridges close together	*T. leptomitus.*
Ridges widely separated	*T. cariosus.*
a a Superficial layer of costal bones, sloping into free rib ends.	
Ridges widely separated	*T. radulus.*
Ridges close together, interrupted and vermiform	*T. ventricosus.*
II. Sulpture of the extremities of the costal bones honeycombed or punctate.	
Carapace with sculpture all honey-combed; six vertebral bones ...	*T. uintaënsis.*
Carapace covered with pits which are little wider than their interspaces; seven vertebral bones	*T guttatus.*
Carapace with longitudinal ribs crossing the ends of the costals; pits small	*T. concentricus.*
Carapace with longitudinal ribs along its middle; seven vertebral bones; pits not large	*T. heteroglyptus.*
III. Sculpture of extremities of costal bones, consisting of small tubercles formed of the broken ridges.	
Carapace honeycombed with large fossæ; no ribs; seven vertebral bones	*T. scutumantiquum.*

The Wasatch species are: *T. leptomitus*, *T. cariosus*, *T. radulus*, *T. ventricosus*, *T. guttatus*, and *T. scutumantiquum.* The Bridger species are: *T. radulus*, *T. uintaënsis*, *T. guttatus*, *T. concentricus*, *T. heteroglyptus*, and *T. scutumantiquum.*

TRIONYX RADULUS Cope.

Systematic Catalogue of the Vertebrata of the Eocene of New Mexico; U. S. Geo. Geol. Expl. W. of 100th Mer., G. M. Wheeler, 1875, p. 35. Report Paleontology, do. iv. ii, p. 45.

Plate xxvi, figs. 11–16.

This turtle is nearly allied to the *T. cariosus.* As in it, the proximal portions of the costal bones and the vertebral bones are honeycombed, while the distal parts of the former are parallel-ribbed. Five to nine of these ribs can be counted from the free end. They are not closely placed, being narrower than their intervals. The size of the species is the same as that of *T. cariosus*, but the costal bones are more uniformly thinner.

A specimen from the Bridger beds shows a rather wide, smooth band, along the front of the carapace.

Measurements.

	M.
Thickness at front of carapace at middle	.010
Thickness of a costal bone at middle of border	.006
Three ridges at end of costal in	.010

I have but one specimen of this species, in many fragments, from the Bridger formation.

TRIONYX GUTTATUS Leidy.

Trionyx guttatus Leidy, Report U. S. Geol. Surv. Terrs. (4to), p. 176, pl. ix, fig. 1. Cope, Report Expl. Surv. W. of 100th Mer., G. M. Wheeler, iv. pt. ii, p. 46. *T. uintaënsis* Leidy, Cope, Syst. Catal. Vert. Eocene New Mexico, p. 35 (not of Leidy).

This is the most abundant species of the Bridger formation, and I obtained parts of many individuals. Its characters are expressed in the definition already given. To it must be added that the distal ends of the costal bones are beveled regularly to the free rib-extremity.

Almost the entire carapace of one of the individuals of this species was obtained by myself in New Mexico. The pitting is uniform and without interruption, extending even to the sutural edges of costal bones. It is strong on the vertebral bones, but, near the distal ends of the costals, becomes obscure; the border itself being smooth. The ribs separating the pits are coarse, but not so wide as the pits.

The posterior part of the carapace of a Bridger specimen has the last two pairs of costals in contact, showing that there were only six vertebral bones. The pits are rather small, having, in some places, interspaces as wide as themselves; this is, however, not generally the case. The dividing ridges are always wider and more obtuse than in *T. scutumantiquum* and *T. uintaënsis*.

I did not obtain the last-named species in Wyoming, so far as I know, and I refer to Dr. Leidy's work for a full description of it and of the *T. guttatus*.

TRIONYX HETEROGLYPTUS Cope.

Annual Report U. S. Geol. Surv. Terrs., 1872 (1873), p. 616.

Plate XVI, fig. 2.

Carapace broad, flat, concavely truncate behind. Free portion of costal bones short. The last pair of costal bones are in contact by a common suture by about two-thirds their width, the anterior portion being separated by the last vertebral bone. There is a great difference between the sculpture of the middle of the carapace and its lateral portions. The former region is coarsely ribbed longitudinally, the intervening grooves being mostly uninterrupted. On the middle portions of the costals the ridges are more or less broken up, and distally they are very delicate, forming an inosculating pattern, inclosing small pits. On the last costal they retain their ridge-like character. The posterior vertebrals are marked by a single groove down their middle.

Measurements.

	M.
Width of carapace at antepenultimate costal bone	.235
Length from front of carapace at antepenultimate costal bone backwards	.095
Width of carapace at antepenultimate costal distally	.048
Length of last two vertebrals	.037

This is a handsome species, and appears to be rare, as I have but two specimens that I can definitely refer to it. It is, however, difficult to distinguish separated ends of costal bones from those of *T. guttatus*. I dug one of the specimens from the summit of the Church Butte, Wyoming.

TRIONYX CONCENTRICUS Cope.

Proceed. Amer. Philos. Soc., 1872, p. 469 (published July 29).

Plate XVI, figs. 3-6.

This species reposes on various fragments, in one case representing numerous portions of a carapace. The sculpture is intermediate between those of *T. heteroglyptus* and *T. guttatus*. The costals have subequal and subround pits throughout the entire length of the bones, but their interspaces are raised into longitudinal ribs at intervals of from one to three rows of pits. These ribs are equally developed at both ends of the costals.

Measurements.

	M.
Width of a costal bone near middle	.020
Thickness of costal bone near middle	.003

The type specimen is smaller than that of the last. From Cottonwood Creek, Wyoming.

TRIONYX SCUTUMANTIQUUM Cope.

Annual Report U. S. Geo. Geol. Surv. Terrs. 1872 (1873), p. 617.
Plate XVI, fig. 1.

Established on a nearly perfect carapace and part of the plastron from the bad lands of Cottonwood Creek. These indicate the largest species of the genus yet found in North America.

The carapace is a longitudinal oval, broadly rounded in front. The median line forms a marked depression, and the costal bones rise and descend again, forming an arch on each side. The free portion of the ribs is not very long The sculpture consists of numerous honeycomb-like pits separated by rather narrow ridges. On the middle parts of the carapace these are subequal, but on the middle of the length of the costals all the ridges run together longitudinally, and on their distal parts, these are broken up so as to produce innumerable irregular tubercles and pits. The lines of the intercostal sutures are smooth. Eight costal bones, and an anterior marginal coössified with the first costal by its entire width, and sending out a broad costal extremity, which curves backwards, its anterior margin smooth. Eight vertebrals, the last separating the anterior portions of the last costals.

Measurements.

	M.
Length of carapace	.425
Greatest width of carapace, axial	.410
Thickness of fifth costal	.0075
Thickness of fourth vertebra	.034
Thickness of centrum of vertebra	.010

Several fragmentary individuals from the Wasatch beds, near Black Butte, Wyoming, as also one from the corresponding formation on Bear

River, resemble this species more nearly than any other. They display similarity in the fineness and acuteness of the ridges between the fossæ, and their disposition to break into small tubercles on the distal parts of the costal bones. More perfect specimens will be neccessary to decide whether this species is common to the two horizons or not.

PLASTOMENUS Cope.

U. S. Geol. Surv. Terrs. 1872, p. 617. Report Expl. Surv. W. 100th Mer., G. M. Wheeler, iv, pt. ii, p. 47, 1877.

The structure of the skeleton in this genus remains incompletely known in spite of the abundance of specimens which I have procured in the Eocene beds of the West. As already stated, it is allied to the genus *Trionyx*, but differs in some important points in the bones of the plastron. The hyosternal bones which I have seen in *P. multifoveatus* are generally like those of *Trionyx*, while the hyposternals, if I have correctly identified them, differ materially. These elements are preserved in the species named, and in *P. corrugatus*, and here they display a transverse width behind the inguinal region more like an Emydoid than a Trionychoid genus. The inguinal border is thickened, and at the bridge somewhat recurved. The inguinal buttress is in all three of the species more robust and more vertically directed than in *Trionyx*. The post-abdominal suture is closer and less digitate in the *P. trionychoides*. In *P. corrugatus* there is a fontanelle at the supposed post-abdominai suture, as in *Trionyx*, while there is no indication of one in the *P. trionychoides*. The hyposternals also display a more completed ossification than in *Trionyx*, in the fullness of the borders between the internal and external digitations. Thus, in *P. multifoveatus*, the internal border is regularly convex, and the processes for the episternal bone scarcely project beyond it. The external digital process projects more extensively, while the free ends of the ribs extend little or not at all beyond the border of the carapace. Among the various remains from Wyoming and New Mexico, no marginal bones have been found, nor dermal scutal sutures.

Portions of the skeletons of the species of this genus are very abundant in the Eocene of New Mexico and Wyoming. Though one seldom obtains an entire carapace or plastron, the form, size, and sculpture indicate that the remains belong to several species. The figures, composed of ridges, pits, etc., variously distributed, are often quite elegant. The species do not attain the average size of the Trionyches of the same era; but the *P. communis*, *P. lachrymalis*, and *P. multifoveatus* exceed in dimensions the living species of that genus of North American waters.

The three species above named, in which the sternal characters are evident, are the only ones which can certainly be referred to the genus; but several others from the Eocene beds can with much probability be referred here also, the whole number being eight. Four species from the Fort Union Cretaceous beds have been referred to *Plastomenus*, but, as already remarked, as a provisional arrangement until their structure is better known. The *P. thomasii* is also of uncertain reference to this genus.

I. Surface without welts, or with the sculpture thrown into ridges:

a. No ridge-lines:

Surface with sharp, fine wrinkles .... *P. corrugatus.*
Surface with more remote wrinkles, little inosculating .... *P. trionychoides.*
Surface honeycombed with thick, inosculating ridges .... *P. multifoveatus.*

a a. Sculpture thrown into ridge-lines:

Surface coarsely honeyeombed with fine ridges .... *P. fractus.*

II. Sculpture interrupted with solid welts; pits small or reduced to punctæ:

Surface with transverse ribs separated by one or two rows of pits ...... *P. serialis.*
Welts on all the *thin* costals, and separated by numerous pits .... *P. molopinus.*
Welts only on the posterior costals, which are all thick; numerous punctæ between them .... *P. communis.*
Welts broken up into short ridges behind; intervening surface punctate. *P. lachrymalis.*
Welts represented posteriorly by tubercles separated by smooth surface, anteriorly unbroken; the surface punctate .... *P. œdemius.*

Of these species, *P. corrugatus*, *P. multifoveatus*, *P. fractus*, *P. serialis*, *P. communis*, and *P. lachrymalis* have been found in the Wasatch beds of New Mexico; and the *P. trionychoides*, *P. multifoveatus*, *P. molopinus*, and *P. œdemius* in the Bridger beds of Wyoming.

Plastomenus trionychoides Cope.

Annual Report U. S. Geol. Surv. Terrs., 1872 (1873), p. 619. *Anostira trionychoides* Cope, Proceed. Amer. Philos. Soc., 1872, p. 461 (published July 29).

Plate XVI, fig. 1.

The original specimen of this species was found mingled with one of *Anóstira ornata*, and being of about the same size, the two were supposed to pertain to a single species. I subsequently distinguished the fragments clearly, and find portions of another individual from another locality to pertain to the same.

The sculpture of the costal bones consists of reticulated ridges which inclose coarser pits than in the next species, and show no tendency to run into ribs extending obliquely acrossthe bones. The first costal exhibits a greatly beveled suture for the nuchal, and its alar portion behind its costal rib is twice as wide as the latter. The last costal differs from that of *P. thomasii* in being angulate instead of truncate at the rib-extremity, and the latter projects strongly beyond the angle.

The sculpture of the costal bones is somewhat like that of *Trionyx scutumantiquum*. It can hardly be regarded as the young of that species, for, although of small size, the complete ossification of the costal bones indicates that the specimens are of mature age.

Bad lands of Cottonwood Creek, near Fort Bridger, Wyoming.

PLASTOMENUS MULTIFOVEATUS Cope.

Annual Report U. S. Geol. Surv. Terrs., 1872 (1873), p. 619. Report U. S. Geo. Geol. Surv. W. of 100th Mer., iv, pt. ii, p. 49. *P. thomasi* pars. Annual Report U. S. Geol. Surv. Terrs., 1872 (1873), p. 618.

Plate XVII, figs. 2–8.

Represented by various parts of four individuals, a sufficient number of identical pieces being present in all to insure their specific unity. The bones of both carapace and plastron have a honeycomb pattern of reticulation, with shallow pits, which on weathering become punctæ. The intervening ribs tend to connect into ridges running diagonally across the costal bones. The pits tend to form linear series parallel to the borders, on some of the bones of the plastron. The latter are gently convex at the transverse suture. The last costal is very wide, and is in contact with its fellow on the median line as in other species of the genus, except a sutural margination behind, apparently for a pygal bone. The outer border is straight, truncating the last rib-extremity.

Measurements.

	M.
Thickness of a costal	.004
Width last costal, distally	.048
Width of hyosternal	.018
Thickness of hyosternal	.005
Length of a vertebral	.018
Width of a vertebral	.014

Both the hyosternal and hyposternal bones are more convex than in any of the *Trionyches* of the Eocene period. The inguinal edge is thinned, and is very gently concave. The hyosternal is much thicker on the outer part of its posterior suture than at the internal part Of course the same is true of the hyposternal. The latter is characterized by the steepness of the ascent of its external buttress, which is also situated nearer to the hyosternal suture than in *Trionyches* generally. The external sculptured layer rises on its base and forms a narrow rim below the inguinal edge of the posterior part of the hyposternal for a short distance only. The pits of the inferior surface of the hyosternal are more or less parallel with the borders of that bone, while those of the hyposternal are irregular.

In a costal bone of a large specimen, the sculpture is a shallow, but sharply impressed honeycomb pitting, smaller than in the preceding species Thus there are seventeen or eighteen pits across the middle, to seven or eight in *P. trionychoides.* No ribs whatever

Measurements.

	M.
Width of costal at middle	.240
Width of costal at end	.350
Thickness at middle	.0035

I included the specimens of this species in my descriptions of *P. thomasi* in the Annual Report of the United States Geological Survey for 1872. When I stated there that *P. thomasi* is the type of the genus *Plastomenus*, I referred to these specimens; it is therefore to be observed that the type of this genus is really the *P. multifoveatus*. The true *P. thomasi* was founded on sternal bone perhaps of a small species of *Trionyx.*

Some of the specimens of this species were found on Cottonwood Creek, Wyoming, and others on the Upper Green River.

PLASTOMENUS MOLOPINUS Cope.

Annual Report U. S. Ceol. Surv. Terrs., Hayden, 1872, p. 602. *Anostira molopina* Cope. Proceed. Amer. Philos. Soc., 1872, p. 461. *Plastomenus communis* Cope, var. ii, Cope. Report Expl. Surv. W. of 100th Mer., Lieut. G. M. Wheeler, iv, pt., ii, p. 50.

Plate XVII, figs. 9–14.

This tortoise was common in the Bridger epoch in Wyoming, and also in the Wasatch in New Mexico. My collection from the beds of the former includes eight individuals in a fragmentary condition. I have already mentioned those from the latter horizon as a thin variety of *P. communis*, and figured some costal bones on Plate xxv, figs 5–6, of the Report of Lieutenant Wheeler, above quoted. It is probable that Dr. Leidy has figured part of a costal bone on Plate xvi, fig. 12, of the final Report of Dr. Hayden, vol. i.

The costal bones of this species are generally ribbed towards the distal ends; some of them at the proximal also. The ribs are not close together as in *P. serialis*, and their directions are somewhat irregularly transverse to the length of the costal bone. The sculpture of the surface between them is punctate rather than reticulate, since the impressed fossæ are not wider than the intervals between them. The difference between this species and the *P. communis* is found in the much thinner bones of the carapace.

A portion of the last costal bone of one of the specimens is without impressed punctæ. Its ribs are thickened, and run parallel to the median line. Were they broken up into tubercles I would refer the specimen to *P. œdemius*. I do not possess the corresponding part in any other specimen.

The size of the *P. molopinus* is about that of the *P. œdemius*, and is less than that of my examples of *P. multifoveatus*.

Measurements.

	M.
Width of a costal bone, proximally	.018
Thickness of a costal bone, proximally	.004
Width of another costal bone, distally	.020
Width of the same costal bone, distally	.003

The specimens are from various portions of the Bridger basin.

PLASTOMENUS ŒDEMIUS Cope.

Annual Report U. S. Geol. Surv. Terrs., 1872 (1873), p. 619. *Anostira œdemia* Cope. Proceed. Amer. Philos. Soc., 1872, p. 461 (July 29).

Plate XVII, figs. 15–17.

Represented by parts of three specimens. These all display the last and middle costals, and two of them the second costals. Sternal bones are wanting (except, perhaps, in one).

From these it appears that the anterior costals have a distantly punctate sculpture, with rib-like swellings running diagonally across them. On the middle costals the punctæ disappear and the ribs grow thicker; on the last costals the ribs are broken into a number of smooth tubercular swellings whose axes are nearly at right angles to that of the carapace. The second costal has its posterior alar portion twice as wide as the rib portion; its suture with the first costal is very oblique, and is bounded behind by a rabbet-edge. The last costals are peculiar in their union throughout their entire length without emargination for pygal, and in the gently convex posterior outline (with projecting rib end), differing in these respects markedly from *P. multifoveatus* and *P. trionychoides*.

Measurements.

		M.
No. 1.	Length of last costal common suture	.045
	Length of last costal anterior suture	.063
	Length of last costal exterior border	.052
	Width of middle costal	.021
	Thickness of middle costal	.004
No. 2.	Width of first costal, proximally	.026
	Width of first costal behind rib, distally	.014
No. 3.	Width of middle costal	.021

This handsomely marked turtle is quite peculiar in its sculpture, which departs more from ordinary patterns than any of those referred to the present genus.

Two specimens from Cottonwood Creek, Wyoming.

ANOSTIRA Leidy.

Proceedings of Academy of Natural Sciences, 1871, p. 102.

In this genus the epidermis was thin and adherent to the bones, and not divided into scuta. The carapace is composed, as in *Emydidæ*, of costal vertebral, and marginal bones, the last united to the first by suture and gomphosis. The series of vertebrals does not continue to the caudal except by the intervention of a pygal. The sternum is cruciform, with narrow longitudinal prolongations or lobes, and narrow bridges. It appears not to have possessed any fontanelles, but the presence of mesosternum is not yet fully made out. The cranium and limbs are unknown.

This genus must be regarded as an interesting intermediate type connecting *Plastomenus* and *Chelydra* or *Dermatemys*. In skin and sculpture it is identical with the first; in carapace and plastron most like *Chelydra*.

Two species, a large and a small, are known.

ANOSTIRA RADULINA Cope.

Proceedings American Philosophical Society, 1872, p. 555 (published October 12).

Plate XVII, figs. 18–19.

Based on two marginal bones, one from the front the other from the rear of the carapace of an animal of twice the bulk of the largest *Anostire* previously found. Apart from size, the sculpture is peculiar. It consists in the anterior marginal costal bone, of closely packed vermicular ridges which run out flat on the posterior and upper edge. In the posterior, it consists of only closely placed minute tubercles over the whole surface, which are more or less confluent on the proximal part of the bone.

Measurements.

	M.
Length of front marginal on free edge	.025
Width of front marginal on free edge	.028
Length of posterior marginal on free edge	.025
Width of posterior marginal on free edge	.025

The specimen on which this species reposes cannot well be regarded as an overgrown *A. ornata*, since the sculpture of the bones is not enlarged in proportion to the size of the elements of the skeleton. The tubercles and ridges are not larger than those on a small *A. ornata.*

One specimen from the Bridger bad lands of Hams Fork, Wyoming.

ANOSTIRA ORNATA Leidy.

Proceed. Acad. Phila., 1871, p. 102. Report U. S. Geol. Surv. Terrs., 1872 (1873), p. 174. Plate xvi, figs. 1–6.

This species has been so fully described by Leidy, that I only give a brief synopsis of its characters.

The outline is a broad oval with an open emargination in front. The median dorsal line is keeled posteriorly, as far as the posterior border. The posterior marginal bones are thickened so as to have in part a triangular section. The margin is acute and not or but little recurved. The sculpture of the costal bones is in obsolete ridges running parallel with the middle line, and close together. That of the marginals is in small tubercles which run together at the proximal part of the bone above, and generally on the inferior surface.

The branches of the plastron are all narrow, and the transverse ones quite long. The external borders of the latter are coarsely dentate, but not digitate. The sculpture of the plastron is obsolete.

The size of the *Chelopus guttatus* or *Chysemys pictus* of our streams.

Specimens were found by myself and party on Cottonwood Creek, and on the Upper Green River, Wyoming.

EMYS Brong.

The species of the Eocene formation which have been referred to this genus are evidently members of the family *Emydidæ;* but owing to the absence of descriptions and specimens of the crania, it is not certain to which genus of the family they should be referred. As in similar cases in paleontology, they are retained in the genus *Emys* until reasons for distinguishing them shall be discovered.

As already remarked by Leidy, the species so referred, have left more numerous remains in the Bridger beds than those of any other genus. The same is true of the Green River and Wasatch formations, the genus *Trionyx* only having left more abundant traces in the latter. During my explorations in Wyoming, in 1872, I detected three species in the Wasatch and Green River beds, one in the Washakie, and four in the Bridger formation; in 1874 I obtained two species in the Wasatch strata.

These species may be briefly distinguished in tabular form, as follows:

I. The bridge sutures on not, or moderately, elevated axillary and inguinal buttresses.	
a Dorsal line with projecting keel.	
Vertebral bones smaller, thicker	*E. polycypha.*
Vertebral bones larger, thinner; gular plates not reaching mesosternal bone; grooves moderate	*E. terrestris.*
Vertebral bones larger, thinner; grooves very deep and wide	*E. megaulax.*
a a No dorsal keel.	
β Gular scuta not extending on mesosternum.	
Bones massive, with lines of growth on some; costals swollen at proximal end of costal scuta	*E. testudinea.*
Bones thinner; no lines of growth; costal bones flat	*E. euthneta.*
β β Gular scuta extending on mesosternal bone.	
γ Vertebral bones wide.	
Shell thin; lip of plastron not very wide	*E. lativertebralis.*
γ γ Vertebral bones elongate.	
Shell thin, little arched; vertebral scuta wide; sutures of plastron straight; its lip narrower	*E. vyomingensis*

Shell very thick, little arched; vertebral scuta wide; sutures of plastron straight; its lip narrow .... *E. shaughnessiana.*
Shell thin, little arched; vertebral scuta wide; lip of plastron very wide .... *E. latilabiata.*
Shell thin coössified, much arched above; sutures of plastron irregular; vertebral scuta longer than in other species............ *E. haydeni.*
II. Axillary and inguinal buttresses very prominent.
Shell thin, carapace convex, not keeled; scutal sutures not deep.... *E. septaria.*

The distribution of these species is as follows:

Wasatch, Wyoming—*E. testudinea*, *E. euthneta*, *E. megaulax*; New Mexico—*E. lativertebralis.*

Bridger, Wyoming—*E. polycypha*, *E. terrestris*, *E. wyomingensis*, *E. shaughnessiana*, *E. latilabiata*, *E. haydeni.*

Washakie—*E. septaria.*

It is true that in many *Emydidæ* the young stages are characterized by a dorsal carina and greater width of the dermal scutal grooves. Dr. Leidy has suggested* that the immature stages of *Emys wyomingensis* are represented by certain keeled specimens in his possession; and also states that the mesosternal bone is more elongate in such specimens than in the larger ones. I have suspected that the forms I have named, *Emys polycypha* and *E. terrestris*, might be the young of *E. shaughnessiana* and *E. wyomingensis* respectively, but I have not identified them on account of the lack of specimens displaying intermediate characters, and also because of the shorter gular scuta of *E. terrestris.* If the mesosternal bone is longer in the young than in the adult *E. wyomingensis*, it should bear more rather than less of the gular scuta. The *Emys megaulax* of the Wasatch beds presents the characters of immaturity in the low median keel and the deep and wide sutural grooves. It is much larger than either of the two species just named, and its bones are stout. It cannot be the young of its cotemporary *E. euthneta*, for that does not exceed it in size. I have parts of several individuals of both for comparison. It is true that in all three of the species presenting these characters of immaturity, the shells are, so far as known, without fontanelles, and that in the smallest, *E. polycypha*, the vertebral bones are relatively the thickest.

* Report U. S. Geol. Surv., i, p. 148.

I have not included the *E. cibollensis* (Cope Report Expl. Surv. W. of 100th Mer., Wheeler, iv, pt., ii, p. 57), as it may have been founded on a larger *E. euthneta*.

EMYS POLYCYPHA Cope.

Annual Report U. S. Geol. Surv. Terrs., Hayden, 1872, pp. 625, 630. *Palæotheca polycypha* Cope, Proceed. Amer. Philos. Soc., 1872, p. 463.

Plate XVII, figs. 20–22.

This species of tortoise is indicated by vertebral costal and marginal bones of very small individuals. These bones are, however, not only thoroughly ossified, but are very stout, indicating the adult age of the animal. The deeply impressed scutal sutures, and heavy proportions, as well as the elevated carina of the carapace, indicate affinity with *Cistudo* or, perhaps *Testudo*. As a generic character, it may be noted that the vertebral bones are subquadrate, and support the neural canal without intervening lamina.

The carina of the carapace is abruptly interrupted at intervals; sometimes with, sometimes without, a pair of pits, one on each side. The marginal bones are well secured and the scutal sutures are deeply impressed on them.

Measurements.

	M.
Length of vertebral bone	.009
Width of vertebral bone	.0085
Length of marginal bone	.01

This is the last of the tortoises of the Bridger formation.

EMYS TERRESTRIS Cope.

Annual Report U. S. Geol. Surv. Terrs., Hayden, 1872, p. 629. *Palæotheca terrestris*, Cope, Proc. Amer. Philos. Soc., 1872, p. 464.

Plate XVII, figs. 23–25.

In this species and the following, the lip only is inclosed by the gular scuta, which only reach the apex of the mesosternal. In neither are the articulations of the bridge with the costals known. Represented by three individuals, one of which may be regarded as the type. They are all thinner than the *P. polycypha*, and larger, being about equal to the *Aromochelys odoratus* of our ponds.

In the type specimen the carina of the vertebral bones is interrupted by a deep sutural groove, which is less pit-like than in *P. polycypha*. The bone itself is broader than long, being perhaps, from the hinder part of the carapace. The clavicular (episternal) bone is preserved. It is characterized by the considerable and abrupt projection of that part inclosed by the gular scutum, which resembles what is sometimes seen in *Testudo*. The edge of this part is entire and acute. The posterior part of the projection forms a step-like prominence behind, on the superior or inner face. The bone is almost as wide as long, and the mesosternal causes a very slight median truncation, but overlaps much on the inner side. The gular dermal suture does not reach it.

Measurements.

	M.
Length of vertebral bone	.009
Width of vertebral bone	.018
Length of episternal	.02
Width of episternal (transverse to axis of body)	.017
Width of a costal	.011
Thickness proximally	.003

In the second specimen, a strong groove is seen to bound the lip of the front lobe of the plastron. In it the marginal is seen to be stout, a little recurved, and sharp-edged. A vertebral differs from those of other species in being longer than wide. In a third individual the gular lip is not so prominent as in the type, and the mesosternal bone truncates the clavicular extensively, giving it thus a more elongate form. The gular scuta expand to its front margin. The marginal bone is stout and sharp-edged, and is not so deeply impressed by the dermal suture as in *P. polycypha*.

EMYS MEGAULAX Cope.

Annual Report U. S. Geol. Surv. Terrs., 1872, p. 628. *E. pachylomus* Cope, *loc. cit.*, 629. Plate XVII, figs. 26–33.

Represented by remains of five or six specimens. They pertained to a species of about the size of the salt-marsh terrapin *Malacoclemmys palustris*. The marked peculiarity consists in the broad and abruptly sunken sutures which separate the dermal scuta of the carapace. This is visible on vertebral, costal, and marginal bones, where the areæ between the sutures are abruptly separated. The sutures partially interrupt the dorsal carina. This is wide and low. The surface is otherwise smooth. The scutal sutures are not so much impressed on the plastron, and those of the gular scutes extend on the mesosternal bone.

Measurements.

	M.
Length of a marginal	.016
Width of a marginal	.023
Width of a vertebral	.018
Length of a vertebral	.017

The vertebrals are subquadrate in form. Neither carapace or plastron are thick. The mesosternal is transversely diamond-shaped, and angular in front.

Measurements of mesosternal.

	M.
Length	.023
Width	.034

Other fragmentary specimens similar in size to the last, have the scutal sutures strongly marked, but not so widely and deeply impressed. Though they are finer, they interrupt the dorsal carina, which swells up from it, and they divide the flat proximal portion from the much swollen marginal part of the marginal bones. The mesosternal bone is similar in form to that of the last specimens; the only specimen is obtusely rounded in front, and bears part of the gular scuta.

From the Wasatch or Green River beds at Black Buttes. A third but uncharacteristic series of fragments from the first Eocene lignite bed above the Cretaceous, probably belong to this species.

Emys euthneta Cope.

Annual Report U. S. Geol. Surv. Terrs., 1872, p. 628.
Plate XVII, figs. 34–42.

Represented by numerous portions of several specimens. These pertained to a species of about the size of the salt-water terrapin, *Malacoclemmys palustris.* There is no dorsal keel, and the scutal sutures, though distinct, are not very much impressed nor the interspaces swollen. The lip of the plastron is narrow, thick, and not notched; the sutures of the gular scales do not extend on to the mesosternum. The margins of the lobes of the plastron are a little thickened and the sutures of the bones coarse and at the hypoxiphisternal junction, etc., with gomphosis. (It is fine and close at this point in *E. testudinea.*)

The costal sutures for the bridge are projecting and curved, in one position; in the other, straighter, and very near the margin of the costal bone. Surfaces smooth.

Abundant in the red beds which lie between those identified as belonging to the Green River and Bridger epochs at Black Buttes, Wyoming.

EMYS TESTUDINEA Cope.

Annual Report U. S. Geol. Surv. Terrs., Hayden, 1872, p. 627, *Notomorpha testudinea* Cope. Proceed. Amer. Philos. Soc., 1872, p. 475.

Plate XXIII, figs. 12–13.

Represented by portions of four or more individuals. In one of these the anterior lobe of the plastron is in part preserved. The mesosternum is a transverse oval, the posterior margin regularly convex, the anterior with three equal borders. The median of these is concave. The sutures are radiating, and the groove separating the humeral scuta, appears to traverse the entire length of the bone. The outer surface is gently convex. The free margin of the episternal and hyposternal bones is acute, and with an internal thickening, as in *Cistudo*, *Testudo*, *&c.*, forming a ridge with abrupt inner face. This face extends backwards as a groove, to the axillary process of the hyosternal, forming a characteristic mark. Although the extremity of the episternal bone is lost, and the mesosternal exhibits no trace of the intergular scute, the outer sutures of the gular scuta are so far posterior as to render it highly probable that the intergular plate existed. At the point where this suture reaches the margin, the latter is openly emarginate. The posterior suture of the humeral suture crosses the margin half way between the axilla and the episternal suture, and is not marked by a notch. The last-named suture is transverse. On the xiphisternal bones the groove of the anterior suture of the anals is plainly visible. It is regularly convex forwards, and in one specimen is double.

In a second specimen of about the same size, parts of two costal bones are preserved They are thick, and display the usual costal and vertebral scute-sutures, the latter in a groove; for the middle of the vertebrals is elevated, and the costals project shoulder-like just outside the groove.

In a third specimen a little larger, xiphisternals with several marginals are preserved. A free posterior marginal is regularly recurved, and the scute-sutures are deeply impressed. The marginal scuta have evidently been marked with concentric grooves within their margins. The first marginal bone of the bridge has a very obtuse edge.

In none of the specimens are the surfaces sculptured.

Measurements.

No. 1.	M.
Width of plastron at axilla	.086
Length of plastron from axilla (approximate)	.05
Thickness of hyosternal at mesosternal	.099
Thickness of hyosternal at hyposternal	.0005
Width of mesosternal	.037
Length of mesosternal	.026
Thickness of a vertebral	.006
Thickness of xiphisternal (normal)	.004
Thickness of xiphisternal at pubis	.007
No. 2.	
Thickness of costal at hump	.0075
Width of costal	.0175
No. 3.	
Width of posterior marginal	.027
Length of posterior marginal	.019

The mesosternal, though found with No. 1, does not fit it exactly and does not belong to it.

From Green River formation near Evanston, Wyoming.

EMYS VYOMINGENSIS Leidy.

Annual Report U. S. Geol. Surv. Terrs., Hayden, 1871, p. 367, Montana. Proc. Acad. Phila., 1869, p. 66. *E. stevensonianus* Leidy, *loc. cit.*, 1870, p. 5, fide Leidy. *E. jeanesi* Leidy, *loc. cit.*, 1870, p. 123, fide Leidy.

Plate XXIII, figs. 9–11.

Of this, the most abundant tortoise of the Bridger Eocene, I obtained numerous specimens on my expedition of 1872. I refer especially to three as most characteristic, one a chelonite entire, but with plastron crushed in; a second broken up, but including portions of most of the shell; and, thirdly, a nearly perfect plastron. These all show that the species had no intergular scute, as finally decided by Leidy. They also show that the notch on each side of the lip of the plastron is not uniformly present. As Dr. Leidy has given a very full account of this species, with good figures,* I do not redescribe it here, but refer to it in the descriptions of the allied species.

EMYS SHAUGHNESSIANA sp. nov.

Plate XXIII, figs. 3–6.

I describe under this name a species, which is represented by the greater part of one individual in a dislocated condition. The separated elements are in excellent preservation, so that the characters can be readily ascertained.

---

*Report U. S. Geol. Surv. Terrs., i, 1873, p. 140.

The species has the linear dimensions of the *E. vyomingensis*, and differs primarily in the much greater thickness of the bones of the carapace, especially of the vertebral bones. The plastron is thick, but does not so much exceed the corresponding parts of the thicker examples of *E. vyomingensis*. Another character is the great thickness of the costal parts of the axillary and inguinal buttresses. In an antero-posterior diameter they are twice as large as those of *E. vyomingensis*, but not more prominent transversely. The vertebral dermal scuta are a little longer than wide, and the lateral sutures are strongly bracket-shaped, while the anterior and posterior one present an angle forwards on the median line, thus differing from the sutures in *E. haydeni*. The vertebral bones are longer than wide anterior to the sixth; the latter, with the seventh, are wider. The costal capitula are quite weak. The posterior marginal bones are not recurved, and the anal is not notched, and its border is a little convex. The lateral and anterior marginals are not grooved nor recurved at the margin. The dermal grooves of the carapace are generally strong.

The plastron is of elongate proportions as compared with the ordinary *E. vyomingensis*. The lip is moderately wide, and has a notch on each side, as in the species just named. The mesosternum is wider than long, and is marked anteriorly by the gular scuta. The humero-pectoral groove reaches but does not cross it posteriorly, in which it differs from that of most of the Bridger species. The smooth border of the anterior lobe of the plastron is very wide anteriorly; that of the posterior lobe is even wider at the position of the ranges of movement of the thigh, where it is bounded within by a sharp groove. The posterior notch is well marked but open. The dermal sutures are straight and not sinuous, as in *E. haydeni*.

Measurements.

	M.
Length of carapace	.380
Width of carapace at second costal bone	.145
Length of third vertebral bone	.040
Width of third vertebral bone	.025
Thickness of third vertebral bone	.016
Length of seventh vertebral bone	.023
Width of seventh vertebral bone	.033
Thickness of seventh vertebral bone	.011
Width of axillary buttress	.018

		M.
Anterior lobe of plastron	length	.106
	width at axilla	.144
Length of bridge		.144
Posterior lobe of plastron	length	.105
	inguinal width	.148
Width of anterior lip		.068

There is a peculiarity in the form of the lip of the plastron of the specimen on which this species rests, which may be a specific character. Instead of having an abrupt lateral prominence and truncate or concave anterior border, its outline is regularly convex, only interrupted by the notch, which is half way between the median and lateral gular sutures.

I dedicate this species to my friend Arthur O'Shaughnessy, of the British Museum, who has published a number of important papers on herpetological subjects.

The specimen above described was found by myself on Cottonwood Creek, near Fort Bridger, Wyoming.

EMYS HAYDENI Leidy.

Proceedings of the Philadelphia Academy, 1870, p. 123. Annual Report U. S. Geol. Surv. Terrs., 1871, p. 366. *E. wyomingensis* Leidy, part Report U. S. Geol. Surv. Terrs., Hayden, i, p. 14, Pl. ix, fig. 6.

The central parts of the carapace and plastron of an *Emys* from Cottonwood Creek, Wyoming, belong to a species distinct from the *E. wyomingensis*, as it appears to me, and agrees very nearly with the figure and description of the specimen on which Dr. Leidy established his *E. haydeni.* In fact, the general appearance of the specimen is that of a Baëna, a resemblance produced by the density of the tissue and general coössification of the parts as well as the fineness of the sutures where apparent.

The species differs from the *E. wyomingensis* in the marked and regular convexity of the carapace, both longitudinally and transversely, resembling no little a portion of the shell of an egg. The dermal sutures are straighter and less undulating on the carapace. The vertebral scuta are relatively longer, and their borders are not bracket-shaped laterally, and are very little or not angulate before and behind. On the other hand the grooves of the plastron are irregular, crossing and recrossing the median line at various points. The humero-pectoral suture crosses the mesosternum well in advance of its posterior border.

The carapace is of moderate thickness, and the capitula are robust, much more so than in the *E. shaughnessiana*. The anterior marginals are robust; a posterior is much more thickened inferiorly, and is consequently more recurved than in *E. shaughnessiana*, and displays narrower marginal scuta.

Measurements.

	M.
Length of third vertebral bone	.045
Width of third vertebral bone	.030
Length of third vertebral scute	.080
Width of third vertebral scute	.056
Thickness of third marginal bone on suture for second	.021
Thickness of plastron at middle of hyposternal	.017
Width of anterior lobe at axilla	.140

This specimen resembles the one observed by Dr. Leidy, in having the fourth vertebral bone octagonal, a character I have not met with in any other species.

EMYS LATILABIATUS Cope.

Proceedings of American Philosophical Society, 1872, p. 471.

Represented by a perfect specimen of a tortoise of a broadly oval form, and somewhat terrestrial habit. Its prominent characters are to be seen in the plastron, of which the posterior lobe is deeply bifurcate. The anterior lobe is peculiar in the unusual width of the lip-like projection of the clavicular ("episternal") bone, which is twice as wide as in *E. wyomingensis*, and not prominent. Bones all smooth; margins of lobes of plastron thickened.

There are three scars, perhaps, of muscular insertions near the posterior margin of the plastron, one oval one opposite to each lobe, and one round one opposite to the notch.

As compared with *E. septarius* this species has no such septa nor sculpture; the emargination of the plastron is more open, and the lip much shorter and wider.

Measurements.

	M.
Length of carapace	.255
Width of carapace	.250
Width of lip of plastron	.06
Depth of posterior notch	.02

The temporary misplacement of the typical specimen of this species prevents my giving other than my original description.

From near Black's Fork of Green River.

Emys septaria Cope.

Annual Report U. S. Geol. Surv. Terrs., 1872 (1873), p. 625.

Plate XVIII, figs. 9-13.

Established on a nearly complete specimen of the size of *Ptychemys rugosa*. The carapace is rather thin and the sutures not obliterated. The vertebræ are sessile on the vertebral bones. The form is quite convex. The plastron is flat and rather stout. The mesosternum is rhombic, the longer angle anterior on the outer side, but posterior on the inner side. Its anterior angle is embraced by the gular scuta. The anterior lobe of the plastron is contracted near the axillæ, and flared with a thin edge in front of it, then contracted to the rather narrow lip of the middle front. The posterior lobe is somewhat flared and has a wide beveled margin, and is deeply notched behind, the notch being close, and the lobes projecting.

The surface is delicately sculptured with obsolete ridged lines across the axis of the costal bones. The vertebral region is somewhat swollen between the cross-sutures, which present an obtuse angle in the same direction, both before and behind. The scuta are longer than wide, and have bracket-shaped outlines. The surface has the obsolete ridges, which diverge in every direction from the inlooking angle of one end, but are mostly longitudinal.

In old specimens this delicate sculpture might become obsolete.

Measurements.

	M.
Length of plastron	.325
Width of plastron at groin	.150
Width of lip	.054
Length of lip	.020
Width of clavicular bone behind	.041
Width of mesosternal, externally	.058
Length of mesosternal, externally	.045
Thickness of hyosternal behind	.015
Length of vertebral scutum	.072
Width of vertebral scutum	.068
Width of a costal bone	.029
Thickness of a costal bone	.006

Found in the bad lands of the Washakie basin, on South Bitter Creek, by the writer.

HADRIANUS Cope.

Proceedings of the American Philosophical Society, 1872, p. 468.

This genus resembles *Testudo* in form, but has two anal scuta, as in most *Emydidæ*. The claws are short and stout; one ungual phalange is a long oval viewed from above, and is oval in section, with obtuse edges. The articular surface is subinferior. A cervical vertebra is of moderate length and has a very prominent anterior zygapophysis. The centrum presents two distinct convex articular surfaces anteriorly, and one transverse one behind. A sacral is free from the carapace above; it presents two surbround articular cups posteriorly and outwardly; the anterior are broken off. These characters are observed in a large specimen of *H. corsoni.*

HADRIANUS ALLABIATUS Cope.

Proceedings of the American Philosphical Society, 1872, p. 471.

Plate XIII, figs. 13–15.

This large land tortoise differs from both the *H. quadratus* and the *H. octonarius* in the absence of the projecting lip of the anterior lobe of the plastron, which is thus simply truncate. The mesosternum is not cordate, but has much the shape of that of *H. quadratus;* that is, rhombic. The scutal sutures are deeply impressed. The plastron is strongly concave. Carapace without irregularities of the surface. Length eighteen inches.

From the bad lands of Cottonwood Creek, Wyoming.

HADRIANUS OCTONARIUS Cope.

Proceedings of the American Philosophical Society, 1872, p. 468.

Plate XX.

The *H. octonarius* is distinguished from its congeners as follows. It is of elongate form, strongly contracted at the bridges, but expanded and arched above the limbs. The carapace is very convex. The plastron has the posterior lobe emarginate rather than bifurcate, as seen in *H. corsoni.* Each projection represents a right-angled triangle rather than a wedge. The anterior lobe presents an elongate lip, which is expanded and slightly emarginate, at the end. The mesosternal bone is heart-shaped, the posterior emargination being wide and not very deep.

The anterior margin of the carapace is somewhat flared above the limbs. The nuchal scutum is very narrow transversely, but elongate. The carapace descends steeply and is incurved in the middle of the posterior margin. The superior portions of the anal scuta cover an ovate projection of the surface.

Measurements.

	M.
Length (below)	.730
Width at middle	.437
Width at hind limbs	.525

This species is perhaps the largest of our extinct land tortoises, and is founded on a beautifully perfect male specimen from the bluffs of Cottonwood Creek.

HADRIANUS CORSONI Leidy.

Geological Survey Montana, 1871, p. 365. *Testudo hadrianus* Cope, Proceed. Amer. Philos. Soc., 1872, p. 463. *Hadrianus quadratus*, *loc. cit.*, p. 468.

Indicated by several individuals, one nearly perfect, the others represented by all parts of the skeleton. This proves the existence of a very massive species of the terrestrial genus *Hadrianus*. The plastron presents a short, wide lip in front, which is turned outwards, forming a strong angle with the plane of the upturned front of the lobe. This lobe is bordered by a thickening of the upper surface, which cuts off the basin from the lip, as a high ridge. The posterior lobe is deeply bifurcate, each post-abdomnal projecting as a triangle. There is a notch at the outer angle of the femoral scute. The hyposternal bone is generally thickened within the margin above, and an elevated ridge bounds the basin of the plastron behind, as anteriorly. The middle of the plastron is thin. The carapace is without marked keel or serrations. It is remarkable for its expanded and truncate anterior outline, which is nearly straight between two lateral obtuse angles, thus giving a quadrate outline when viewed from above or below. Length of carapace, $^{m}.750 = 29$ inches, width $^{m}.630$. The marginal scuta are narrow, and there is a large nuchal plate.

Abundant in the Bridger beds.

DERMATEMYS Gray.

Annals Mag. Nat. His., 1847, p. 60. Catal. Shield Reptiles British Museum, 1855, p. 49. *Baptemys* Leidy, Proceed. Acad. Phila., 1870, p. 4. Report U. S. Geol. Surv. Terrs., i, 173, p. 157.

This genus is similar to Emys in the structure of the carapace and plastron, except that the lobes of the latter are narrowed and shortened. The scuta are similar, excepting that there is a series of intermarginals on the bridge on each side. There are thirteen marginals on each side, those of the last pair in contact throughout. In a specimen of the extinct species of the Bridger, I find a trace of an intergular scute, as is sometimes seen in *D. berardi* now living in Mexico.

DERMATEMYS VYOMINGENSIS Leidy.

Baptemys wyomingensis, Report of the United States Geological Survey of Territories, i, 1873, p. 157. Plates xii, and xv fig. 6.

This tortoise is not uncommon in the Bridger beds, but generally in a dislocated or fragmentary condition. It is readily distinguished as a species by the elevated keel of the posterior vertebral and pygal bones. The mesosternal is large and emydiform, and is not marked by the humero-pectoral dermal suture. It is extensively occupied by the large gular scuta, which exceed in size those of any other species of the formation. The size rather exceeds that of the average *Emys vyomingensis*. The Wasatch species *D. costilatus** Cope differs in having an obtuse keel on the costal bones, parallel with the median line of the carapace, as in *Staurotypus triporcatus.*

NOTOMORPHA Cope.

Proceedings of the American Philosophical Society, 1872, p. 474.

This genus reposes on a clavicular or episternal bone, which gives characters not seen in any other genus known to me. The sutures are distinct and fine, and the form of the mesosternum is emydoid. The gular scuta are small, and occupy an angular space between the large intergular and humeral, which are extensively in contact. It is uncertain whether there are two or only one intergular. The general characters of the other

*Report Expl. and Surv. W. 100th Mer., G. M. Wheeler, iv, pt. ii, p. 62.

bones are those of the less aquatic types of the Emydes. It is likely, as already remarked, that this genus belongs to the *Adocidæ*.

The following were the characters with which I commenced the original description of the genus: "The elements of the carapace and plastron are massive, and the former was well arched; both exhibit well-defined grooves for the sutures of the dermal scuta. The mesosternum is broad ovate, and the bones of the plastron are united by immovable sutures. The elevated lateral processes of the hyo and hyposternal bones are not broad and unite by suture with the lower plate of the first and last bridge-marginal bones. They are thus recurved in both cases, but none of the ribs indicate any sutural union as is seen in various genera. The costal bones unite with the marginals by serrate suture. In one species a large intergular scutum has left its impression, the gulars being lateral and rather small. The anterior lobe of the plastron is emarginate."

I then added that the pubis was united by suture with the post-abdominal (xiphisternal) bone, and inferred that the genus should, therefore, be referred to the *Pleurodira*. I subsequently became convinced that the bones showing this sutural union are really costals, bearing sutures for the buttresses, and that there is no evidence to show that the sutural union of the pubis and ischium which characterizes the *Pleurodira* exists in this genus. At the same time, having doubts as to the homologies of the dermal scuta observed, I referred the species which displays the supposed intergular bone to *Emys*. While I believe this course to be the proper one in the case of one of the species (*Emys testudinea*) referred to *Notomorpha*, I now believe that the characters displayed by the other species (*N. gravis*) justify the retention of the genus *Notomorpha*.

The only species known to me was obtained from the Wasatch formation of Wyoming.

Notomorpha gravis Cope.

Proceed. Amer. Philos. Soc., 1872, p. 476. *N. garmani* loc. cit., p. 476. *Emys gravis* Annual Report U. S. Geol. Surv. Terrs., 1872 (1873), p. 626.

Plate XXXIII, figs. 14–16.

This species is known from a number of separated bones which were found together. It is probable that the pieces of carapace and plastron belong to the same individual. The dermal sutural grooves are well marked. There is a large intergular scutum, which evidently encroached considerably on the mesosternal (a piece not preserved), and was probably subtriangular in shape. The gulars are reduced to triangular areas on the outer anterior angles, the suture with the humeral being in front of the middle point between the angle and the hyosternal suture. The margin is less distinctly emarginate at this suture than in *Emys testudinea.*

The marginal bones belong to both bridge and free edge. They are all much thickened medially, but with thin proximal sutural margins. The free ones are well recurved, and with regular rather thickened margins. The bridge marginals have very obtuse margins. Their general massiveness is in contrast to the thinness of the costals, of which there are numerous fragments. Portions of vertebral bones are intermediate in thickness.

There is no thickening or ridge on each side of the vertebral scuta. The scutal grooves are everywhere well marked. The surface of the marginals and episternal is obsoletely rugose, somewhat as in some species of Taphrosphys from the Cretaceous.

The posterior marginal bones are stout and more thickened inferiorly than those of the *E. vyomingensis* and *E. shaughnessiana*, and are more recurved.

Measurements.

	M.
Length of episternal (approximate)	.04
Length from posterior suture (approximate) to gular scute	.02
Thickness of episternal, behind	.011
Length of a marginal bone	.042
Width of same marginal bone	.045
Thickness of same marginal bone	.015
Width of a bridge-marginal	.04
Thickness of a verbetral	.007

From the Wasatch beds of Wyoming, six miles north of Evanston, near Bear River.

BAËNA Leidy.

Geological Survey Wyoming, 1870, p. 367. Survey Montana, 1871, p. 368. Report U. S. Geol. Surv. Terrs., Hayden, i, p. 160. Cope, Annual Report U. S. Geol. Surv. Terrs., 1872 (1873), p. 621. *Chisternum* Leidy, Proceed. Acad. Phila., 1872, p. 162.

This genus agree with the *Adocidæ** in the presence of intergular scuta and the absence of coössification of the ischium and pubis with the plastron,

* Proceedings of the American Philosophical Society, 1870, p. 547.

but differs in the presence of an intersternal bone on each side, as in the *Pleurodira*. As generic characters it possesses two marginal intergular plates, which resemble the gulars of *Emydidæ*. It has a series of intermarginal scuta. The free lobes of the sternum are narrowed and shortened and the bridge is very wide. The dermal scuta are everywhere distinct. The mesosternal bone is in form between T-shaped and sagittate. The last pair of marginals, instead of being in contact, are separated by a wide emargination.

The preceding characters were first noticed by Leidy. Another one appears in my specimens of *B. arenosa*, *B. undata*, and *B. hebraica*, which Dr. Leidy does not mention, viz, the presence of five costal scuta instead of four. The accessory one is anterior, and is taken from the usual first costal and first vertebral, both of which are contracted in consequence. Leidy's specimens are damaged in the region in question, and do not display anything. The character is unique in the order *Testudinata*, unless it be found in the *Platychelys* of the European Jurassic, which is one of the *Baënidæ*.

The affinities of this genus are complex and interesting. It would be a pleurodire, but for the fact that the pelvis is not coössified with the plastron; nevertheless there are rudiments of this union in the form of a shallow pit on each side. The posterior or ischiadic is near the posterior end, and on the lateral margin of the post-abdominal bone; it is of a narrow, oval form. The anterior is shallow and sublaterally impressed into the side of the upright septum which supports the carapace. Whether it received the pubis or not is uncertain.

The double intergular scute is not found in any existing genus of *Pleurodira*, but exists in *Tropidemys* Rütim. of the Jurassic.

The posterior margin of the carapace is excavated as in *Chelydra*, but the margin is more arched in this position. This form in *Baëna* suggests the presence of a large tail, and the serrate margin of the carapace posteriorly reminds one again of *Chelydra*. There are in *B. arenosa* fourteen marginal scuta without the nuchal; in *Chelydra serpentina*, as in *Emydidæ*, but thirteen.

There are prominent axillary and inguinal septa, as in some *Emydidæ*.

They are composed of the produced edges of two coössified costal bones, united with the ascending buttresses of the plastron.

The affinities appear to be to *Adocus* on the one side and *Hydraspididæ* on the other, perhaps as descendant of the former and ancestor of the latter. It also possesses traces of the other relationships of *Adocus*, *i. e.*, to *Dermatemys*, and more remotely to *Chelydra*.

Baëna hebraïca Cope.

Baëna hebraica, Proceed. Amer. Philos. Soc., 1872, p. 463 (published July 29). Annual Report U. S. Geol. Surv. Terrs., 1872, p. 621.

Plate XIX, figs. 1–2.

General form depressed and discoid, as wide as long. Bridge wider than long, but the length equal to the width of the bases of the sternal lobes. Anterior lobe longer than wide at the base, and narrowed at the extremity. The inguinal and axillary septa are very prominent. The edge of the carapace from the front to the inguinal region, is without emargination. All the osseous elements are coössified.

The scuta are well distinguished. The nuchal is very small and wider than long; the first marginal is shorter but more prominent. The second and third are larger but narrow; the fourth and fifth are wider, but the sixth widens by an inward projection of its border so as to meet the intercostal suture between the second and third costal scuta. From this one to the ninth (as far as preserved) the inner margins are produced so far as to make the scuta nearly twice as wide as long when viewed from above. The first costal is small; its posterior border is curved. The first vertebral is pyriform, truncate in front. It is (perhaps abnormally) divided by a transverse suture into a quadrate anterior and cordate transverse posterior portion. The other vertebrals are somewhat longer than broad, and are separated by sutures convex anteriorly.

The intermarginal scuta are all wider than long; their number is normally four, but a narrow one is intercalated behind the inguinal on one side. The longitudinal suture of the scutes of the plastron is exceedingly tortuous, winding between points more than an inch apart. The gulars and intergulars are transverse and bounded by transverse sutures. They cross the median suture (which is straight on the anterior lobe) some distance apart. The humerals are long, and the humero-pectoral scutal suture is convex backwards, its extremities reaching the margin in front of the axillæ. The anterior extremity of the anterior sternal lobe has a quadrilobate outline.

The surface is smooth except along the lines of intercostal sutures, where short grooves parallel to the general axis alternate with protuberances having the same direction, the whole having somewhat the appearance of sculptured characters.

Measurements.

	M.
Length of carapace (axial) .... (19 inches)..	.500
Width of carapace (axial) .... (19 inches)..	.500
Length of plastron from groin	.295
Width of base of anterior lobe	.155
Width of extremity of anterior lobe (at gulars)	.080
Length of anterior lobe (at gulars)	.123
Width of nuchal scute	.011
Length of nuchal scute	.024
Length of third marginal	.038
Width of third marginal	.015
Width of eighth marginal	.090
Length of eighth marginal	.063

This species, when compared with its nearest ally, *B. undata*, differs in the greatly wider marginal scuta; in the latter the corresponding ones (6–7–8–9) are much longer than wide, as in most other tortoises. The intermarginal scuta are of more elongate form, and the normal number is five in *B. undata* instead of four. The sculpture in the longer known species is entirely distinct, consisting of pits and tubercles scattered generally over the surface; while the peculiar sculpture of the suture lines is wanting. *B. hebraïca* is relatively wider.

Bad lands of Cottonwood Creek, S. W. Wyoming.

BAËNA UNDATA Leidy.

Annual Report U. S. Geol. Surv. Montana, 1871, p. 369. Cope, Annual Report U. S. Geol. Surv. Terrs., 1872, p. 622. *Chisternum undatum* Leidy, Report U. S. Geol. Surv. i, p. 169, pl. xiv.

Plate XIX, figs. 3–5.

A specimen of this species presents the following characters: The anterior lobe of the plastron is as wide as that of *B. hebraica*, but little more than half as long. The posterior lobe is truncate at the extremity. The nuchal scute projects beyond the first marginal; the reverse is the case in the type of *B. hebraica*. The posterior sutures of the intergular and gular scuta have a common center, and that of the gular has a rectangular curvature, the nearly transverse middle portion slightly convex forwards. The suture separating the femoral and anal scuta is similar, but reversed in direction, presenting two obtuse right angles, two portions being transverse and one longitudinal on each side.

The *Baëna undata* is quite similar to the *B. arenosa* in most respects. As in Leidy's specimens, the sutures of the plastron in the *B. arenosa* are obliterated in my single specimen, while in several of the *B. undata* they leave distinct traces even when coössified. As the latter are of larger size than the former, the difference in this respect cannot be due to age. Besides, the plastron is smoother and presents no median carina. It is more roughened posteriorly with small irregular tubercular ridges. Traces of the grooves seen in *B. arenosa* are found on the anterior median region. The peculiar fifth or anterior costal scute is similar to that of *B. arenosa*, as are the gular and intergular scutes. In fact, the resemblances between this species and the *B. arenosa* are so close that I suspect that when we come to know the younger stages of the latter we will find that the intersternal bones are present, as in the *B. undata*. On this ground I have not adapted the genus *Chisternum* proposed by Dr. Leidy for the latter on account of the presence of the intersternals.

Baëna undata is more abundant in the Bridger beds than any other species of the genus. I found it on Black's Fork and elsewhere.

BAËNA ARENOSA Leidy.

Proceed. Acad. Phila., 1870, p. 123. Report U. S. Geol. Surv. Terrs., i, p. 161, pl. xvi and xx, figs. 1–5 and xiii. *B. affinis* Leidy, Annual Report U. S. Geol. Surv. Terrs., 1871, p. 367.

Plate XVIII, figs. 1–2.

A perfect specimen of this species is of smaller size than those of the preceding, and about equal to the *Pseudemys rugosa*, and not dissimilar in form.

The carapace is strongly convex, and all its component parts, as well as those of the plastron, are coössified. The sutures of the intersternal bones are visible. The posterior end of the carapace is arched upwards and smoothly excavated; the postero-lateral borders are thin, and deeply notched at the ends of the scutal sutures. Similar but shallower emarginations mark the borders of the marginal scuta. The anterior margin is slightly concave. The lobes of the plastron are narrow, the posterior wider and slightly emarginate. The bridge is wide, and not more than half as long as the width of the base of the posterior lobe.

The general surface is minutely rugose or shagreened, on the plastron strongly so, and without other sculpture. The carapace is marked by strong grooves disposed in a regular manner. A double groove extends along the median line of the second, third, and fourth vertebral scuta. Other grooves are nearly parallel to this one, whose extremities diverge to the angles of the vertebral scuta. At the anterior angles of the costal scuta oblique grooves converge towards the vertebrals, and are continued backwards as parallel to the median line. They are separated by parallel tuberosities. On the first and last vertebral scuta there are transverse grooves next the adjacent vertebrals, and longitudinal ones towards the margins of the carapace.

The scuta are well marked. The marginals are all longer than wide except the four preceding the last, which are all wider than long. The last is suboval, and is very small, while the anal is altogether wanting. The nuchal is divided (it is single in *B. hebraica*), the first marginal is very small and projecting, the third is longer, while the fourth, fifth, and sixth are rather short. The vertebral scuta are all longer than wide, and the fourth is deeply emarginate to receive the last scute. The first is a broad triangle with anterior angle truncate, and the two basal ones cut off to a less degree.

The scutal sutures of the plastron are but little sinuous. The intergulars have precisely the form of gulars of *Emydes*. The posterior gular suture crosses the median line a short distance posterior to those of the intergulars, and each half consists of an obtuse **V** directed backwards. The posterior humeral suture originates in front of the axilla. There are four intermarginal scuta on the one side and three on the other, the additional one being a small one behind the left axillary. The femoro-anal suture is nearly straight.

Measurements.

	M.
Length of carapace (axial)	.450
Width of carapace (axial)	.240
Length of plastron	.290
Length of anterior lobe	.082
Length of posterior	.085
Width of extremity of anterior lobe	.038
Width of extremity of posterior lobe	.057
Length of nuchal scuta	.030
Length of third marginal	.023
Width of third marginal	.020
Width of fourth marginal	.024
Length of fourth marginal	.028
Length of eighth marginal	.050
Width of eighth marginal	.035

This species differs in many details from the preceding species, notably in the form of the marginals. The anterior are wider than in either species, while the median are narrow as in *B. undata*. The sculpture is very distinct from that of either.

From the bad lands of Ham's Fork, Wyoming.

BAËNA PONDEROSA Cope.

Annual Report U. S. Geological Survey of the Territories, 1872 (1873), p. 624.

Plate XVIII, figs. 3–8.

Established on numerous fragments of a specimen of a species which I cannot refer to this genus with certainty, but which agrees with the species already known in some particulars of structure. Thus, the last marginal plates were separated by an excavation of the posterior border; at least this is the only position to which I can refer a portion of the margin of the carapace where the marginal scutes suddenly cease. The lateral ribs of the bridge are received into a deep pit between two costals.

The marginal and other bones are very massive, much more so than in any other known water tortoise of this formation; the margins of the former are thickened, especially at the last marginal scute, which is on a massive protuberance. The sutures are entirely regular. The lateral marginal scuta are about as long as broad. The surface of the shell is marked with irregular impressions, which are sometimes like rain-drop pits. A posterior vertebral bone possesses a median rib similar to that in *Dermatemys vyomingensis*.

Measurements.

	M.
Length of an anterior marginal scute	.045
Width of an anterior marginal scute	.039
Thickness of bone at anterior marginal scute	.023
Length of a free marginal bone	.050
Width of a free marginal bone	.057
Length of first marginal of bridge	.060
Thickness at simple end	.023

From the bad lands of Ham's Fork, Wyoming.

CROCODILIA.

The fauna of the Eocene periods of the United States included a number of species of *Crocodilia*, some of which were represented by great numbers of individuals. They were equally numerous in the Wasatch and Bridger epochs, but none have been found in the Green River formation proper. They are moderately abundant in the Wind River beds, and a species is known from the Manti beds of Utah. None are known from the Miocene formations west of the Rocky Mountains, and but one species from that formation to the east of them; but they are not rare in the marine Miocene of the Atlantic coast. All the species belong to two genera, *Diplocynodus* Pomel, and *Crocodilus* Linn. One species of the former is found in the Wasatch beds, with three or four species of *Crocodilus*. In the Bridger beds I know of six species of the latter genus.

CROCODILUS Linn.

The Eocene species of true crocodiles differ much in size and characters, ranging from the *C heterodon*, which is not larger than an *Iguana*, to the *C. clavis*, which rivals the existing species of the East Indies.

The species are divided into two sections, which are distinguished by the form of the frontal bone. In the one it is thin, and has low lateral olfactory crests. Such species are as yet only known from the Wasatch formation. They are the *C. grypus* Cope and *C. wheeleri* Cope. The species of the second section have massive frontal bones with strong lateral olfactory crests. The *C. chamensis* and *C. heterodon* of the Wasatch belong here; also the *C. elliottii* of the Bridger and the *C. clavis* of the Washakie formation. The frontal bones of *C. subulatus* and *C. polyodon* of the Bridger are unknown. I distinguish the species described in the succeeding pages by the following characters, among others:

Teeth with coarse sulci at the base only; crowns long .... *C. subulatus.*
Teeth with finer deep sulci extending towards the apex of the robust crown .... *C. sulciferus.*
Teeth acute, compressed; crowns with numerous shallow sulci; muzzle slender; no distinct ledge or smooth space on frontal bone......... *C. acer.*
Teeth without sulci, crowns robust; frontal bone with a transverse ledge, with a smooth space in front of it; premaxillaries conical .... *C. affinis.*
Teeth without sulci; frontal without ledge, honeycombed; premaxillary teeth conical... ..... *C. clavis.*
Teeth without sulci; front plane, honeycombed; premaxillary teeth compressed, trenchant; size small.................................. *C. heterodon.*
Crowns of teeth unknown; sizes very irregular, numerous small teeth with a few large ones interspersed .... *C. polyodon.*

The *C. affinis* and *C. elliottii* belong to the section or genus *Thecachampsa* Cope, which is characterized by the concentric structure of the teeth. They are composed of layers, which form a cone-in-cone structure, each cone being distinct from the one which it incloses. I do not know the structure in the other species above named.

I have formerly referred two of the species, the *C. subulatus* and *C. polyodon,* to *Diplocynodus* Pom, a genus characterized by the presence of two similar and adjacent canine teeth in the ramus of the mandible. *C. subulatus* has two such canines in the upper jaw, but there is no evidence that there are such present in the lower, as the mandible is broken off anterior to the canine which is present.

The typical specimen of *C. polyodon* is broken in the same manner, so that I leave it also provisionally in *Crocodilus.*

Crocodilus subulatus Cope.

Crocodilus (*Ichthyosuchus*) *subulatus* Cope, Proceed. Amer. Philos. Soc., 1872, p. 554 (October 12). *Diplocynodus subulatus* Cope, Annual Report U. S. Geol. Surv. Terrs., F. V. Hayden in charge, 1872 (1873), p. 613.

Plate XXIV, figs. 5–19.

Some of the cervical vertebræ without hypapophyses. Their cups round, with smooth bordering surface of the sides of the centrum. The jaws only are preserved from the cranium; the premaxillary is strongly pitted, but the dentary has remote shallow pits on the outer face, and shallow grooves below. Dentition very characteristic. There are two very long canine-like teeth in the premaxillary bone near its posterior margin, directed somewhat backwards; these are preceded, after a space, by a medium-sized tooth, which, after a similar space, is preceded by another long tooth. Anterior to this the alveoli are lost. Two very long, smooth, compressed straight teeth in the front of the ramus mandibuli. These are followed abruptly by a distantly set series of subequal teeth of not one-fourth the size, varying little to the back of the jaw. All the long teeth have subcompressed crowns with opposed cutting edges, and are smooth except at their bases. These are sulcate with wide grooves, the separating ridges being acute. The smaller teeth are cones with cutting edges. There are fourteen alveoli and one pit in the dentary bone from the posterior end to the beginning of the short symphysis.

Measurements.

	M.
Length of alveolar series to beginning of symphysis	.130
Diameter of alveolus of seventh tooth	.008
Elevation of eighth tooth	.017
Diameter of eighth tooth at base	.0065
Depth of dentary at base	.025
Elevation of first lower canine	.018
Length of crown of second upper canine	.017
Diameter of crown at base	.007
Length of third cervical (with ball)	.037
Diameter of cup, vertical	.016
Diameter of cup, transverse	.018
Length of a posterior dorsal	.041
Diameter of cup, transverse	.026
Diameter of cup, vertical	.022

Found on the bluffs of the Upper Green River, of the Bridger epoch.

This species agrees in some respects with the very brief description given by Marsh for his *Crocodilus liodon.* He does not mention the fluting of the base of the crown so remarkable in this species; and states the vertebræ to be "strongly rugose" near the extremity, a character not seen in the present animal.

The *Crocodilus subulatus* was about as large as the Mississippi alligator

CROCODILUS POLYODON Cope.

Diplocynodus polyodon Cope, Annual Report U. S. Geol. Surv. Terrs., F. V. Hayden in charge, 1872 (1873), p. 614.

Plate XXIV, figs. 1–2.

Represented by portions of cranium and teeth, with probably some vertebræ found close to them. This crocodile is similar in size to the *C. subulatus* or our existing alligator. It differs much from the last in the arrangement of the teeth. There is one prominently large canine opposite the symphysis (in *C. subulatus* this tooth is opposite the posterior end of the same), which is followed by nine very small teeth, whose round alveoli are only separated by very thin walls. Following the last of these immediately, is another very large tooth, with nearly round alveolus, which is closely succeeded by other smaller teeth of larger size than those in front of it, and not differing in this respect among themselves. The crowns of the teeth are cylindric at the base, and have a double ridge on the anterior outer aspect. The enamel is obsoletely rugose striate at the base. The external surface of the dentary bone is deeply and coarsely pitted; at its anterior part the pits are close, deep, and small; on the inferior face they are deep short grooves. There is a series of close small foramina along the inner side of the alveoli.

Measurements.

	M.
Depth of symphysis	.014
Diameter of "anterior canine tooth"	.008
Distance of same from median "canine"	.030
Depth of dentary bone at latter	.027
Width of ramus at anterior canine	.025

This species differs in many respects from the one last described. The teeth anteriorly are much more closely placed, and the anterior and middle canines are less separated, and more numerous small teeth occupy the interval. The splenial bone has a larger share in the symphysis, and the sculpture is much more profound. The teeth are not fluted.

The type specimen was found on the bluffs of Upper Green River by the writer.

CROCODILUS ACER Cope.

Plate XXIII, figs. 1–2.

This species is represented by a perfect skull which lacks the lower jaw. In its general form this skull resembles the existing *Crocodilus acutus*, and is narrower than the *C. elliottii* and *C. affinis*. It belongs with the latter in the group with robust frontal bone with strong lateral ridges.

The top of the muzzle is absolutely flat, transversely and longitudinally. In this respect it differs from the *C. americanus*, which is characterized by the presence of a strong convexity of the posterior part of the nasal bones, and the parts adjoining. The table of the skull is wider than long; the orbits are convex inwards but not regularly, so that the outline of the inter-orbital part of the frontal bone contracts forwards. Anteriorly the orbits are angulate by the union of two oblique borders, the malar and prefrontal. The angle which is in the lachrymal bone, is continued as a shallow gutter for a short distance forwards. There are no crests on the head. The anterior extremities of the nasal bones are prolonged a short distance into the external nares. The postero-external angle of the squamosal bone is compressed. The undulation of the superior alveolar line is moderate, The external edge of the pterygoid bone is thickened and truncate.

A considerable triangular area of the supraoccipital bone appears on the superior face of the skull. The premaxillary bone measured to its posterior apex, enters the length to the extremity of the quadrate bone, three and two-thirds times, or a little more than three times, to the posterior border of the cranial table. The palatine bones extend very little beyond the anterior border of the inferior orbital openings, a character in which the *C. acer* resembles the *C. affinis*, and differs from the *C. americanus*.*

The pitting of the surface of the skull is strongly marked everywhere, except on a very small space at the junction of the frontal and nasal bones. Five pits may be counted across the middle of the interorbital front of the frontal bone. On this bone they are subround and not deep nor confluent, but are separated by ridges narrower than themselves.

There are five premaxillary and thirteen maxillary teeth on each side. They present characters which readily distinguish them from those of any other species known to me. Their sizes are graduated, and the larger ones do not present an abrupt contrast of size, as in *C. polyodon*. Their crowns are all more or less compressed, and have distinct acute cutting edges. The compression is most marked on the last six of the maxillary

* For fine specimens of this species I am indebted to the Smithsonian Institution at Washington.

series, and the last four are short lancet-shaped. The crowns anteriorly are long and acute, but they begin to shorten with the fifth maxillary and diminish regularly posteriorly. The basal portion of a fully protruded crown is smooth; the greater portion is, however, longitudinally grooved. There are eight ridges on the narrower and ten on the larger teeth. The grooves are not so deep, nor the ridges as acute as in *C. sulciferus* Cope, of the Bridger beds, and the crowns are less robust and not so incurved. The teeth of the present species have more acute edges. On these grounds I have been obliged to regard the *C. acer* as distinct from the *C. sulciferus*.

Measurements.

	M.
Length of skull to line of extremities of quadrates	.390
Length of skull to posterior border of cranial table	.345
Length of skull to line of anterior border of orbit	.238
Width of premaxillary bones	.060
Width at premaxillary notch	.046
Width at fifth maxillary tooth	.072
Width at anterior angle of orbits	.058
Width at posterior border of quadrates	.192
Width of interorbital space	.021
Width of posterior table at { temporal fossæ	.090
Width of posterior table at { squamosal angles	.126
Vertical diameter of skull { above occipital condyle	.032
Vertical diameter of skull { below occipital condyle	.034
Width of extremity of os quadratum	.033
Width of posterior nares	.014
Width of inferior orbit	.031
Length of inferior orbit	.095

This species is smaller than the *C. affinis*, to which it is generally allied. It has a slight trace of the ledge between the anterior borders of the orbits seen in several of the Bridger species. The forms and sculpture of the teeth are entirely different from those of the *C. elliottii* and allies. In life the species had about the size and form of head of the narrow-nosed caiman now living in South America, *Jacare punctulata* Natt.

My only specimen of the *Crocodilus acer* was obtained by Charles H. Sternberg, from the white limestone near Manti, in central Utah. Other specimens have been subsequently found. This formation belongs to the Eocene period, but its exact relation to those on the east side of the Wasatch Mountains is yet uncertain. I have called it the Manti formation.

Crocodilus sulciferus Cope.

Proceed. Amer. Philos. Soc., 1872, p. 555 (October 12). Annual Report, *loc. cit.*, 612.

Plate XXIV, fig. 23.

The specimen on which this species was established included various portions of the skull and skeleton, which were of about the size of corresponding parts of the *C. elliottii*. The characteristic teeth were removed for purposes of description and illustration, and it has unfortunately become impossible to identify the specimens with which they were originally associated. I can therefore at present only reproduce the description originally published.

A medium-sized species with cranium deeply and roughly pitted. The chief character is at present visible in the teeth. The larger of these are of subcylindric and short conic crown, which is superficially grooved from basis to near apex; sulci coarse, open, but close together, and separated by strong narrow ridges.

From the Bridger bluffs of Upper Green River.

Crocodilus clavis Cope.

Proceed. Amer. Philos. Soc., 1872, p. 485. Paleontological Bulletin No. 6, p. 3 (August 20).

Plate XXI, figs. 4–9; Plate XXII.

This large species is represented primarily by a portion of a cranium with vertebræ, of a specimen which I found in the Mammoth Buttes of the Washakie basin, near South Bitter Creek, Wyoming. A portion of this specimen has been, through the vicissitudes of a moving, mislaid; but enough remains to furnish determinative characters. Some time subsequently I obtained from the same region a nearly complete skull, accompanied by various parts of the skeleton, which I refer to the same species. I now give the original description of the type The muzzle of this species is of narrowed proportions, and sufficient depth to give it a broad, oval section. The nasal bones appear to have reached the nareal orifice. The anterior superior teeth are very large, especially the canine. The inferior tooth corresponding is large, and occupies an emargination which approaches near to the nasal suture. The pitting of the muzzle is fine, and the swollen interspaces much wider. The teeth have stout conic crowns, with well developed cutting edges and coarse striate sculpture. The mandible is acuminate to the narrow extremity, and has a long symphysis which extends to opposite the third tooth behind the notch. The cervical vertebræ preserved have round cups; they have a simple elongate hypapophysis, with a pit behind it; shoulder very prominent.

Measurements.

	M.
Length of ramus with teeth	
Length of symphysis	.135
Width of ramus at end of symphysis	.083
Width of ramus at end of mandible	.020
Width of maxillary at third tooth above	.020
Width of maxillary at notch above	.020

Portions of a few vertebræ are preserved. An anterior cervical has a massive hypapophysis which connects the two parapophyses. Its quadrate base is thickened posteriorly; as the apex is broken off it is impossible to ascertain its length. Immediately behind it on the middle line is a deep fossa like a foramen. On the middle of the posterior shoulder below, is a low, acute tuberosity; the shoulder is very prominent, and its sides are slightly rugose. The cup is subround. A posterior dorsal has a capitular costal surface extending vertically. The cup is a little wider than deep, and the base indicates a strong hypapophysis. The face of the posterior shoulder of another dorsal is much roughened by closely placed ridges and small tubercles.

Measurements of vertebræ.

		M.
Length of cervical (28 lines)		.061
Diameters of cup of cervical	vertical	.032
	transverse	.032
Diameters of cup of dorsal	vertical	.035
	transverse	.040
Width of femoral condyles		.064

The skull of the second specimen has a resemblance in general to that of the *Crocodilus affinis*, but differs materially. The three most prominent points of distinction are the following: There is no distinct transverse ledge of the frontal bone between the orbits, and the space between the anterior parts of the latter is honey-combed like the posterior frontal region, and not smooth. Secondly, the posterior part of the squamosal bone, where it rests on the quadrate, is broadly truncate, instead of acuminate, and its postero-interior surface is subhorizontal, instead of vertical. Thirdly, the basal part of the angular process of the mandible is expanded inwards into a shelf with convex border. One or the other, particularly the first and second, of these characters have been verified on several individuals of the *C. affinis*.

The general form of the cranium is much like that of the wider forms of the *Crocodilus americanus.* The front and top of muzzle are flat, and there are no crests or ridges on either. The maxillary border is strongly convex to the position of the posterior canine tooth, and is deeply notched to accommodate the inferior canine. The muzzle is shortened in front of the nares, since the premaxillary border descends steeply from their anterior margin. In consequence of the mature age of the individual, the sutures of the skull are obliterated. The united nasal bones project into the nareal opening for about one-third the long diameter. The orbits are somewhat narrowed by the convexity of the internal border of the malar bones. The interorbital space is plane, of course, excepting the sculpture, but there is a slight tendency to a transverse ridge about opposite the middle of the orbits. The superior border of the quadrate condyle is deeply notched near the middle, to receive a corresponding angle of the mandible. The projecting angular process is very wide at the base, its superior surface having two concavities, of which the inner is nearly twice as wide as the external. The inner convex border contracts rapidly distally, leaving the obtuse free end a little wider than long, and directed inwards.

The sculpture is roughly honeycombed on the superior surfaces, especially on the squamosal, post-frontal, frontal, and top of muzzle. On the middle line, posterior to the middle of the latter, the sculpture is reduced to a few longitudinal grooves, closely placed. The pits are much smaller on the borders of the maxillary and front of the maxillary. The rami of the mandible have longitudinal grooves on the external sides, and the anterior part of the chin has small pits.

The symphysis of the mandible is like that of *C. affinis*, somewhat elongate for this genus, and produced and rather narrowed to the apex. The rami separate opposite the second tooth behind the notch of the upper jaw. The lateral posterior foramen terminates anteriorly opposite the posterior border of the orbit. The anterior border of the palatine orbital foramen marks the posterior third of the distance between the orbit and external nares, on the top of the skull. The posterior border marks the posterior edge of the orbit. The pterygoid bones are produced and present backwards an acute external angle. The posterior nares are well behind the posterior border of the orbital foramina. The palatine bones are narrowed posteriorly between the orbito-palatine foramina, being at the middle of the latter, a little narrower than either foramen.

The posterior maxillary and mandibular teeth are concealed in the specimen, owing to the closure of the jaws. There are four teeth in the maxillary between the canine and the notch for the lower canine. In front of the latter, in the maxillary bone, there are five teeth, of which the third and fourth, counting from the front, are largest. The teeth are all separated by short interspaces, and are graduated in sizes, the large teeth not being abruptly larger, as in *C. polyodon.* The crowns are robust, and with round section at the base. They have a low cutting edge in front and rear. The enamel is roughened by numerous short crowded filiform ridges, as in many other crocodiles, which are worn off in old crowns.

Measurements of skull.

	M.
Total length with angles of mandible	.700
Length on superior surface; parietal bone estimated	.530
Width between external angles of quadrates	.300
Width at orbits	.230
Width at superior canines	.168
Width at superior notch	.095
Width at middle of nares	.130
Width between orbits	.050
Width of nares	.047
Length from end of muzzle to orbit	.360
Length from end of muzzle to line of canine	.180
Length from end of muzzle to line of notch	.100
Diameter of canine tooth at base	.020
Length of symphysis mandibuli	.115

The only difference to be noted between the fragments of the upper jaw of the type specimen and the corresponding parts of this one, is to be seen in the premaxillary teeth. In the former they are larger and are not separated by as distinct interspaces; the third and fourth appear to have been in contact. While the form of the symphysis of the mandible is the same in both, the ramus of the type is stouter.

The vertebræ are mostly injured. A cervical has a simple anterior hypapophysis with a concavity on each side of its base, and an obtuse keel on the middle line behind it. The ball is nearly round, and is bounded by a strong shoulder. The external slope of this shoulder is marked by a few ridges, and by considerable rugosity at the base of the neural arch. In a dorsal, the ball is a little deeper than wide, and the middle line behind the hypapophysis is a keel. The ball of a lumbar is wider than deep, and the external border surface of its shoulder, as well as that of the cup, are rough with short ridges. The first caudal has a robust diapophysis, and a fossa on the median line below. The chevron facets are large.

Measurements of vertebræ.

		M.
Length of a cervical		.160
Diameters of base of ball	vertical	.035
	transverse	.033
Length of a lumbar		.061
Diameters of a lumbar	vertical	.032
	transverse	.041
Length of first caudal		.060
Diameters of ball	vertical	.030
	transverse	.028

The right half of the pelvis is preserved, wanting the distal extremities of the pubis and ischium. It has the typical crocodilian character of the perforation of the acetabulum open anteriorly by the failure of contact between the pubis and ilium. The ilium is much like that of the Mississippi alligator, so much so as to render description superfluous.

Measurements of pelvis.

Length of ilium	.160
Depth of ilium	.092
Width of contact with ischium	.035
Width of ischium at base	.080
Width of ischium at middle	.033
Width of pubis at base	.040
Width of pubis at middle	.019

Two dermal bones preserved are probably from the lateral dorsal region, although this is not certain. One of them is large, the other small; both are oval in shape, the larger rather narrowly so, and neither have a distinct keel. Both have a slight median elevation in the short diameter. The fossæ are rather far apart; edges smooth. Length of the larger $^{m}.085$: width .045.

There are traces of the sutures of the neural arches of the lumbar vertebræ, showing that the individual was adult, but not aged, at the time of death

The only species with which the present one can be confused, is the *C. aptus* Leidy, which was founded on a cervical vertebra from South Bitter Creek, Wyoming. In that locality the beds of the Wasatch and Green River formation occur, and probably the Bridger; those of the Washakie group are not many, perhaps fifteen miles distant. This vertebra belongs, according to Leidy, to the cervical series of an adult animal, and measures only 16 lines long. A vertebra of the *C. clavis*, which must correspond in position very nearly with the one described by Leidy, measures 27 lines in length, and is therefore between half as long again and twice as long. This indicates an animal of so much greater size as to render their specific identity highly improbable. A crocodile occurs in the Washakie beds with the *C. clavis*, of which I possess a fragmentary skull. It is of a size appropriate to the vertebra typical of *C. aptus*.

Crocodilus affinis Marsh.

American Journal of Science and Arts, 1871, June.
Plate XXI, figs. 1–3.

This is the most abundant species of the beds of the Bridger basin. I took a nearly complete cranium with some vertebræ from a bad land bluff on Smith's Fork of Green River; and my friend George Wilson, of Cheyenne, Wyoming, presented me with a considerable part of the skeletons of two individuals, including two nearly complete skulls from the Church Buttes. Fragments of others were found by various members of my party on Black's and Ham's Forks of Green River.

I have pointed out the characters which distinguish this species from the *C. clavis*. Under the description of *C. elliotii*, Dr. Leidy *loc. cit.*, has given a pretty full description of another near ally, so far as his material permitted.

One readily observes that the frontal and parietal regions of the skulls of this species are less rugose than those of the *C. clavis*, especially in the plane in front of the interorbital ledge. It is there absolutely smooth. Posterior to the bridge, the fossæ are frequently no wider than their interspaces, which is not the case in *C. clavis*. The middle line of the posterior half of the muzzle is nearly smooth. The sculpture of the malar bones is very strong, and that of the superior middle of the maxillaries nearly as much so; that of the lower jaw is distinct.

The form of the skull is wedge-shaped, and it is flat above, without keels or crests. The "interorbital ledge" is an abrupt change of level, with an outline concave forwards. It is somewhat like the corresponding locality in the existing jacares of South America. The extremity of the premaxillary drops off abruptly from the nares. The outline of the upper jaw is sinuous; the orbits are vertical in direction. The posterior part of the squamosal bone is narrow, with nearly vertical interno-posterior side, and acuminate extremity. The posterior edge of the quadrate condyle is emarginate. The angular process is rather narrow, and is obtusely rounded at the extremity. The lateral mandibular foramen extends as far forwards as the line of the middle of the orbit. The symphysis of the lower jaw is of medium length, reaching the seventh tooth from the front. The chin is wedge-shaped.

The teeth of the upper jaw are: premaxillary, 5; maxillary, 4, the canine, and 11. The anterior teeth are elongate conic, with somewhat compressed crowns, and weak fore-and-aft cutting-edges; the posterior ones have very short crowns. The enamel is finely and roughly striate. In the mandible the first tooth is larger than the two succeeding; the fourth is the very large canine; those following the canine have about half its diameter. In this part of the dentition the *C. affinis* is like the *C. clavis*.

The sutures are well preserved in one of the crania. The posterior part of the parietal is nearly as wide as that of the frontal. The anterior part of the latter is much produced between the prefrontals. The nasals extend backwards behind the apices of the prefrontals and lachrymals, and are continued forwards as an acute process into the external nostrils to the third of their long diameter, as in *C. clavis*.

Measurements of skull.

	M.
Length to angles of mandible	.500
Length to posterior border of parietals	.370
Width between external angles of quadrates	.218
Width at orbits	.180
Width of superior canines	.120
Width of superior notch	.060
Width of middle of nares	.085
Width between orbits	.033
Width of nares	.030
Length from end of muzzle to orbit	.250
Length from end of muzzle to canine	.120
Length from end of muzzle to notch	.069
Diameter of canine tooth at base	.015
Length of symphysis mandibuli	.080

The vertebræ preserved are five lumbars; four with centra nearly complete. They have nearly round cups, and the shoulder at the base of the ball is not so prominent as in the lumbars of *C. clavis* described. The edge of this and of the cup, is marked with distinct short longitudinal ridges. What characterizes these vertebræ as different from the lumbars of *C. clavis* is the presence of a wide open groove of the inferior median line of the centrum. The sides bounding these grooves are regularly rounded and not angulated. This fact, with the absence of chevron facets, satisfies me that these vertebræ are not caudals, which are always grooved below.

Measurements of a lumbar vertebra.

		M.
Length of centrum, including ball		.041
Diameters of cup	vertical	.022
	transverse	.022
Elevation with neural spine		.063
Expanse of prezygapophyses		.049

In some of these lumbars the neurapophysial suture is obliterated, indicating the maturity of the individual.

Professor Marsh distinguishes his *C. affinis* from the *C. elliotti* of Leidy by the shorter premaxillary bones and a few other characters. I find my crania to agree nearly with the former in the characters in question.

CROCODILUS HETERODON Cope.

Systematic Catalogue Vertebrata, Eocene of New Mexico; U. S. G. G. Survey W. 100 Mer., by G. M. Wheeler, 1875, p. 34. *Alligator heterodon* Cope, Proceed. Amer. Philos. Soc., 1872, p. 544. Annual Report U. S. Geol. Surv. Terrs., F. V. Hayden, 1872 (1873), p. 614.

Plate XXIV, figs. 11-18.

The anterior and posterior teeth of this species differ exceedingly in shape; the former are flattened, sharp-edged, and slightly incurved; the edges not serrate. Those of the premaxillary bone are subequal in size, while one behind the middle of the maxillary is larger than the rest. The posterior teeth have short, very obtuse crowns with elliptic fore-and-aft outline. They resemble some forms seen in Pycnodont fishes, and are closely striate to a line on the apex. The upper surface of the cranium is pitted, the frontal and parietal bones with large, deep, and closely-placed concavities. The former is perfectly plane and the latter is wide. The squamosal arch is also wide, and the crotaphite foramina are large and open.

The dermal scuta are very large for the size of the animal, and were not united by suture. They are keelless and deeply pitted, with smooth margins.

The vertebral centra found with other specimens are round. The coössified neural arches indicate the adult age of the animal.

Measurements.

	M.
Height of crown of premaxillary tooth	.004
Width of crown of premaxillary tooth at base	.0035
Long diameter of crown of a maxillary	.005
Short diameter of crown of a maxillary	.0035
Width of parietal	.009
Width of frontal, posterior	.020
Width of frontal, interior orbital	.010
Width of malar below the eye	.008

The variation in the form of the teeth is a slight exaggeration of that seen in the dentition of various species of crocodilians.

The axial portion of the basioccipital bone is a transverse vertical plate with vertical carina on the distal half. The frontal bone exhibits no ledge or crests, and the crotaphite foramina are open. The quadratojugal arch is stout. The dermal scuta are not united, and with the cranium, are deeply pitted. They are very abundant in some of the beds of the Green River epoch. Some of them exhibit a faint trace of keel. Vertebræ associated with them have subround articular extremities.

This is the smallest North American species, and is as small as any member of the genus that is known. It did not probably exceed three feet in length. I only found it in the beds of the Wasatch or perhaps Green River epoch, south of Black Butte, Wyoming. A species of similar proportions left its remains in the Bridger beds, judging from vertebræ which I found on Black's Fork of Green River.

A somewhat similar small species is found in the Wasatch beds of New Mexico, the *C. chamensis* Cope. In that species the dermal scuta are articulated together by suture.

MAMMALIA.

The lacustrine Eocene strata have been found in all parts of the world, where existing, to contain remains of an abundant mammalian life. The character of this mammalian fauna has been found to be particularly interesting, and for the following reasons.

Much light is thrown on the history of the *Mammalia* by the researches into the structure of those of the Eocene formation, and I deem it demonstrated to a certainty that the case with the mammals of this formation is the same as with the reptiles of the Permian, *i. e*, that the family types are all more generalized, and the orders not nearly so widely distinguished as in later periods of the world's history.

The recent orders of fishes were in existence in the Cretaceous period, and probably earlier. Their period of evolution was in the Devonian and the Carboniferous periods. The existing orders of reptiles were all established prior to the Eocene; the period of evolution was the three Mesozoic ages, but especially the Permian. The orders of birds were inchoate in the Cretaceous, but when they were fully differentiated is unknown. The existing orders of *Mammalia* were already established in the Miocene period; during the Eocene they were in process of differentiation, and were less, or scarcely at all distinctly defined.*

The characters of the Placental Mammalian orders which existed during Eocene time are as follows:

I. Ungual phalanges claws (unguiculate).

α Cerebral hemispheres small; cerebellum and olfactory lobes large and uncovered.

β Teeth sheathed in enamel.

Glenoid cavity longitudinal; mandibular condyle round; anterior limbs ambulatory . . . . *Rodentia.*

*See Annual Report U. S. Geol. Surv. Terrs. 1872, p. 645.

Glenoid cavity and mandibular condyle transverse; anterior limbs constructed for flight.. *Chiroptera.*
Glenoid cavity and mandibular condyle transverse; anterior limbs ambulatory .... *Bunotheria.*

II. Ungual phalanges hoofs (ungulata).

a Os magnum supporting the lunar and not articulating with the scaphoid.
The astragalus articulating with the navicular only, and the cuboid with the calcaneum only .... *Taxeopoda.*
(The astragalus articulating with the navicular only; cuboid articulating with distal faces of calcaneum and navicular. *Proboscidia.**)
The astragalus articulating with the cuboid and navicular.. *Amblypoda.*

aa Os magnum supporting the scaphoid, and more or less of the lunar.
Astragalus articulating with both cuboid and navicular..... *Diplaethra.*

MARSUPIALIA.

Although many of the Mammalia of the Lower Eocene formation resemble the *Marsupialia*, few of them present characters which are unquestionably those of that order. They appear in many instances to possess characters of the Insectivorous and Carnivorous orders as well, so that it has been thought best to refer them to a single order in combination with the *Insectivora*, the *Bunotheria.* A few species, however, present the marsupial facies so decidedly, as to leave no alternative but to refer them to that order, until further evidence shall confirm or set aside such a conclusion.

The two genera now to be treated of are not very nearly related to any existing form of Marsupials. The nearest ally of one of them at least is characteristic of the Jurassic age, and has been referred by Professor Marsh to a distinct order under the name of the *Allotheria.* As Professor Marsh does not offer any characters by which this group can be distinguished as an order from either the *Marsupialia* or the *Bunotheria*, I have not been able to adopt it. As Falconer has suggested, their nearest ally is perhaps *Hypsiprymnus* among the existing Marsupials, and *Thylacoleo* has perhaps an equal affinity. As the only part of the structure of these genera which is well known is the dentition, I define them as follows. The family of the *Plagiaulacidæ* differs from that of the *Macropodidæ* in the possession of but two inferior true molars. Most of the genera have the fourth pre-

* Not known as Eocene.

molar trenchant, and generally those anterior to it also, while in the *Macropidæ* the trenchant premolar, if present, is the third. The genera differ as follows:

a Several large cutting premolars.

Premolars four, sides not ridged .... *Ctenacodon.*

Premolars typically three, with oblique lateral ridges .... *Plagiaulax.*

a a One large cutting premolar.

β Inferior molars large, with several tubercles.

Large premolar without posterior cusp; edge directed upwards; sides ridged .... *Ptilodus.*

a a a Fourth premolar rudimental or wanting.

Large premolar with posterior cusp; edge directed forwards; molars with two rows of tubercles .... *Catopsalis.*

Fourth premolar? wanting; molars with three rows of tubercles ....*Polymastodon.*

β β Inferior molars small, with few lobes; the last rudimental.

Large premolar without posterior cusp; edge directed upwards; sides not ridged .... *Thylacoleo.*

Of the above genera, *Plagiaulax* is represented by two species in the English Jurassic; *Ctenacodon* by two species in the North American Jurassic; *Ptilodus* by two species, from the Lower Eocene, one from France and one from North America; *Thylacoleo* by one species from the Pliocene of Australia; *Catopsalis* by two species from the Lower Eocene of North America; and *Polymastodon* by one species from the Lower Eocene of North America.

The phylogeny of these forms in connection with that of the kangaroos may be expressed as follows: It is evident that such forms as *Thylacoleo*, *Ptilodus*, and *Catopsalis* are more specialized than *Plagiaulax* and *Ctenacodon*, inasmuch as the number of teeth is reduced, and the cutting function of the premolars is concentrated in a single large tooth, or is obsolete. This is quite the same kind of specialization as that which has taken place in the history of the descent of the Carnivora. *Ctenacodon*, as having the largest number of premolars, which have the least amount of sculpture, is the least specialized of all the genera. *Thylacoleo*, with the rudimental character of the true molar teeth, is the most specialized, as it is the latest in time. The *Macropodidæ* retain the full series of true molar teeth of the primitive Mammalia, and present only a cutting third premolar in the lower jaw, the fourth resembling the true molars. Thus the cutting tooth of *Thylacoleo* is not the homologue of the cutting tooth of *Hypsiprymnus* as supposed by Professor Flower;* since the latter corresponds with the cutting tooth of *Ptilodus*, which is the fourth premolar of *Plagiaulax.* We must therefore regard *Hypsiprymnus* as the descendant of a type from which the *Plagiaulacidæ* were also derived, in which some of the premolars, as far as the third only, were trenchant, and in which the fourth premolar possessed the tubercular character of the three true molars. Such a type would belong to Jurassic and perhaps even to Triassic times, and might well have continued to the Eocene. I call it provisionally by the name *Tritomodon.* The lines of descent will appear as follows:

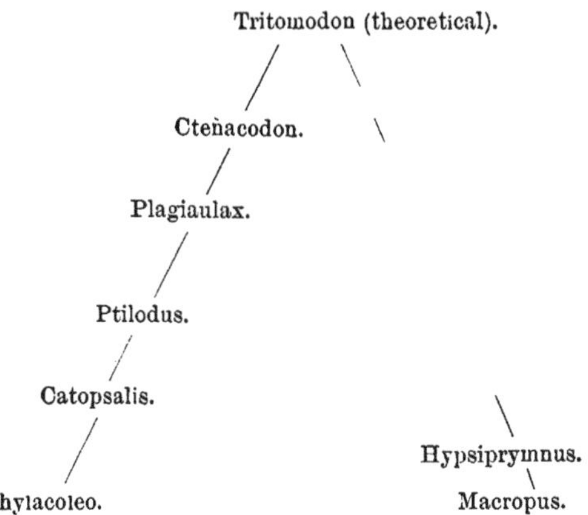

The discussion between Professor Owen on the one hand, and Messrs. Falconer, Krefft, and Flower on the other, as to the nature of the food of *Thylacoleo*, is known to paleontologists. From the form of the teeth alone Professor Owen inferred the carnivorous nature of the food of this genus, while his opponents inferred an herbivorous diet from the resemblance between the dentition and that of the herbivorous *Hypsiprymnus.* As the result of the discussion affects in some degree the genera *Catopsalis* and *Ptilodus*, I recall it here. The comparison of *Thylacoleo* with *Hypsiprymnus* is weakened by two considerations: first, the fact that the cutting tooth of the former is not homologous with the cutting tooth of the latter; and second, that the grinding series of the former is rudimental, and in the latter it is complete. It evidently does not follow that because *Hypsiprymnus* is herbivorous, *Thylacoleo* is so also. Professor Flower refers to the absence of molars in *Thylacoleo* as slightly complicating the problem, and concludes that the food of that animal may have been fruit or juicy roots, or even meat. It is difficult to imagine what kind of vegetable food could have been appropriated by such a dentition as that of *Ptilodus* and *Thylacoleo.* The sharp thin serrate or smooth edges are adapted for making cuts, and for dividing food into pieces. That these pieces were probably swallowed whole, is indicated by the small size and weak structure of the molar teeth, which are not adapted for crushing or grinding. It is not necessary to suppose that the dentition was used on the same kind of food in the large and the small species. In *Ptilodus mediævus* the diet may have consisted of small eggs which were picked up by the incisors and cut by the fourth premolar. In *Thylacoleo* it might have been larger eggs, as those of crocodiles, or carrion, or even the weaker living animals. The objection to the supposition that the food consisted of vegetables, is found in the necessity of swallowing the pieces without mastication. In case it could have been of a vegetable character, the peculiar teeth would cut off pieces of fruits and other soft parts as suggested by Professor Flower, but that these genera could have been herbivorous in the manner of the existing *Macropodidæ*, with their full series of molars in both jaws, is clearly inadmissible.

CATOPSALIS Cope.

American Naturalist, May, 1882 (April 24), page 416.

This genus is known from a part of a mandibular ramus with a few other bones associated. The jaw is broken off in front of the fourth premolar, and the fracture displays the shaft of a large incisor tooth. It is impossible to state how many premolars there are. The fourth is of large size, and is exceedingly compressed. The alveolar border descends abruptly from its posterior root, having the outline of the diastema of the jaw in various rodents, where, however, it is edentulous. The result of this form is, that the crown is presented forwards in an acute edge. The inferior two-thirds of this edge is broken off, so that it is not possible to state whether it is grooved or serrate. The superior part is neither, and rises into a cusp posteriorly. The two molar teeth are very peculiar, and the first is much larger than the second. The arrangement of the cusps is alternating on opposite sides of a median groove. The grooves are deep, and resemble the impression of a simply pinnate leaf with alternating leaflets.

The coronoid process rises opposite the second molar. The inferior face of the posterior part of the ramus is flat, owing to both internal and external inflections. Both are well marked, the latter bounding the masseteric fossa, which is open in front, and without foramen. The internal inflection bounds a deep fossa, like that seen in *Hypsiprymnus* and *Macropus* to terminate in the dental foramen.

The only species of this genus known to me is from the Puerco horizon.

CATOPSALIS FOLIATUS Cope.

American Naturalist, 1882, p. 416, April 24.
Plate XXXIII c; fig. 2.

The mandibular ramus which represents this animal is robust and deep. The alveolar line rises from behind forwards, as in *Elephantidæ* and various rodents, and then suddenly descends. The inner side of the ramus is concave, while the external side, anterior to the masseteric fossa, is convex. The incisive alveolus is thus thrown inside the line of the molars in front. There is a large fossa exposed by weathering below and behind the last molar, which is identical with that seen in *Hypsiprymnus* and *Macropus*, and indicates a large dental foramen.

Below the middle of the fourth premolar tooth the incisor tooth is quite large, suggesting whether it had not a persistent growth, as in the Rodentia. The posterior cusp of the fourth premolar is triangular in profile, the anterior edge descending steeply. It is uncertain whether the edge of the crown rises again, forming another lobe. The apex of the cusp is conic. The first true molar is of large size and remarkable form. The crown viewed from above is a long oval. It has a deep median longitudinal groove, which sends out branch grooves alternately and at right angles to the edge. The spaces between the grooves form block-shaped tubercles, four on the inner and five on the outer sides, whose transverse diameter generally exceeds their anteroposterior. The median groove is open at its anterior extremity.

The posterior is closed by an elevated convex margin. The apices of the lobes are obtuse where not distinctly worn. The last (second) true molar is much shorter and a little wider than the first, and has the same character of surface. There are two large tubercles on the inner side and four small ones on the external side. The posterior end of the crown is narrower than the anterior. The anterior base of the coronoid process is opposite the posterior extremity of the first true molar tooth. The jaw, with its dentition, in its present condition, has a curious resemblance to that of a tubercular-toothed *Mastodon*, with the order of size of the molars reversed.

Measurements.

		M.
Length of base of true molars		.0166
Length of base of fourth premolar		.0108
Vertical diameter of root of incisor		.0070
Diameters M. i	anteroposterior	.0107
	transverse	.0050
Diameters M. ii	anteroposterior	.0060
	transverse	.0060
Depth of ramus at front of P-m. iv		.0120
Depth of ramus at front of M. i		.0190
Depth of ramus at posterior edge of M. ii		.0150
Width of ramus below P-m. iv		.0070
Width of inferior face of ramus below M. ii		.0110

Found by my assistant, D. Baldwin, in the Puerco Formation of Northwestern New Mexico.

PTILODUS Cope.

American Naturalist, 1881, November, p. 921 (October 28).

Dental formula of inferior series; I. 1; C. 0; Pm. 2; M. 2. Incisor occupying a deep alveolus and probably growing from a persistent pulp. It forms an arc of a circle, with an anterior enamel band. First (third) premolar rudimental. Second (fourth) premolar disproportionately large, with compressed crown with a convex cutting edge, and lateral ridges directed upwards and posteriorly. Molars small; the first longer than wide, its crown divided by a deep longitudinal median groove into two lateral ridges, which are divided into lobes by transverse fissures. The masseteric fossa is well marked. The internal pterygoid fossa is very deep and terminates in the foramen dentale, and is bounded below by a horizontal inflection of the inferior border of the mandibular ramus.

The characters of the fossæ of the mandibular ramus are those of the marsupial order, and much like those of the family of the kangaroos. The absence of connection between the masseteric and pterygoid fossæ at the dental foramen distinguishes *Ptilodus* from that family. The differences in the dentition have been already discussed.

The announcement of the discovery of this genus in the Eocene formation was a circumstance of much interest, and it has shown how persistent the type of the *Plagiaulacidæ* has been. It is true that no representative of the *Plagiaulacidæ* has yet been obtained from the beds of the Cretaceous period, which represent the long interval of time which elapsed between the Jurassic and the Eocene. We will not, however, on this account permit the supposition that they did not exist at that time.

The existence of the *Plagiaulacidæ* in the Eocene period was first ascertained by Dr. Victor Lemoine, of Reims, France, and was announced to the Geological Society of France at its meeting of January 10, 1881, and published in its Bulletin for 1881, p. 168, for May. This announcement had escaped my observation when six months later (October) I published the account of its discovery in New Mexico. I, at that time, gave the genus the name of *Ptilodus*, and I am not aware that Dr. Lemoine has yet printed any name, either generic or specific, for the form discovered by him.

PTILODUS MEDIÆVUS Cope.

American Naturalist, 1881, p. 921, November; (published October 28).
Plate XXIII c; fig. 1

This species is represented by two mandibular rami of probably the same animal, one of which lacks the part anterior to the premolars, and the other the part posterior to the premolars; by a single fourth premolar tooth of a second individual; and by all the inferior molars of a fourth.

The ramus is short, and is deep posteriorly; anteriorly its depth is reduced by the concavity of the surface at the diastema, as in most rodents. The inferior border is rather thick anterior to its posterior expansion. The symphysis is not coössified. The incisor tooth is quite slender, and its section is a vertical oval, a little flattened on the inner side. The enamel band covers less than two-fifths of the face of the tooth, beginning with the inner edge of the anterior face and extending externally. Its surface is entirely smooth. The diastema is moderately long. The first premolar is a very small tooth and has a single root. The crown is conic, and is pressed against a narrow truncate face of the base of the anterior edge of the fourth (second) premolar.

The vertical section of the fourth premolar is wedge-shaped, and the profile of the crown is regularly and strongly convex. The anterior root has a greater anteroposterior diameter than the posterior root, and the enamel extends further downwards on it than on the posterior root, on the external side; on the internal side this expansion is not so marked. The enamel of the sides of the crown is thrown into parallel ridges, which are gently curved, and which extend upwards and posteriorly to the edge, except at the posterior border, where they fall a little short of the edge. When they reach the edge the latter is angulate, forming a serrate outline. There are twelve ridges from front to rear, and the anterior ones are closer together than the posterior, the widths of the interspaces increasing in regular progression. The first true molar is very small, absolutely and relatively, its length being one-third of that of the two premolars together. The outline of the crown is elongate oval longitudinally placed. The median groove is wide, and is open at both extremities. The lateral lobes have continuous longitudinal acute edges, and number four on each side. The posterior lobe is twice as long as any of the others on both sides, and its edge is weakly notched. No cingula. The alveolus of the anterior root of the second molar indicates a rather wider tooth than the first molar.

The anterior border of the masseteric fossa is a strong ridge, and passes the alveolar border at the posterior edge of the first true molar. The fundus of the internal pterygoid fossa is deeper or more external than the line of the molars, falling below the external edge of the anterior masseteric ridge.

Measurements.

		M.
Length of ramus to last true molar inclusive		.0205
Diameters M. i	anteroposterior	.0040
	transverse	.0020
Diameters P-m. iv	anteroposterior	.0085
	transverse at base	.0030
	vertical at middle	.0045

	M.
Length of diastema	.0060
Diameters incisor { anteroposterior	.0026
Diameters incisor { transverse	.0020
Depth of ramus at diastema	.0070
Depth of ramus at middle of P-m. iv	.0090

The length of the skull of this animal was about equal to that of the Norway rat *Mus decumanus*, but the large proportions of the fourth premolar indicate that the cranium was much deeper than in that animal, and had probably the form of that of the *Thylacoleo carnifex*.*

RODENTIA.

Remains of species of this order are not abundant in the beds of the Wasatch epoch, and are rather common in those of the Wind River and Bridger. They are not very various as to type, and the greater number are apparently allied to the squirrels.

PLESIARCTOMYS Bravard.

Plesiarctomys Bravard, Ossemens fossiles de Desbruge, 1850, p. 5.—Gervais, Paléontologie française explic., tab. 36, p. 4.
Paramys Leidy, Report U. S. Geol. Surv., 4to, i, 1873, p. 109; Proc. Phila. Acad., 1870 (name only).
Pseudotomus Cope, Paleontological Bulletin No. 2, p. 2, August 3, 1872, nomen nudum; Annual Report U. S. Geol. Surv. Terrs., 1872 (1873), p. 610 (defined).

The inferior molars by which this genus has been generally known resemble much those of existing *Sciuridæ*, but there are cranial characters which distinguish it from the existing forms of that family.

The crowns of the inferior molars support four rather small and strictly marginal tubercles, which inclose a median valley. The anterior inner tubercle is more elevated than the others, and the posterior two tubercles are connected by a low ridge on the posterior border, which may be more or less tubercular on the last molar. In some of the species, the marginal tubercles are merely elevations of the margin, while, in others, the adjacent tubercles of a pair approximate, so as to form a pair of interrupted cross crests.

There are five superior molars, of which the anterior is of small size. They resemble those of *Sciurus*, but the transverse crests are obsolete or

*A restoration of the skull of this animal is given by Flower, Quarterly Journ. Geolog. Society, vol. xxiv, 1868.

wanting. The positions corresponding to their external extremities are marked by more or less distinct cusps. There is a single internal tubercle of the crown. In the third and fourth molars of *P. delicatissimus* I observe rudiments of a second internal tubercle.

The incisor teeth are compressed, with narrow anterior face. The enamel is not grooved, and is little or not at all inflected on the inner side of the shaft, while it is extensively so on the external face. There is a large, round *foramen infraorbitale exterius*, like that of *Ischyromys* and *Fiber*, and entirely unlike that of *Gymnoptychus* and *Sciurus*, conforming in this respect to the forms of the extinct group of the *Protomyidæ* of Pomel.

The cranium of the specimen originally described by me as *Pseudotomus hians*, exhibits the following characters: The superciliary margin is short, and without post-frontal process. The temporal fossæ are large, and contract the brain case behind the orbits to a striking degree. Their anterior margins rise from the post-frontal angles and converge backwards, meeting in a sagittal ridge opposite the anterior part of the squamosal bone. The parietal bones increase rapidly in width to the squamosal, which are largely inferior at their zygomatic portion. They do not extend very far on the superior aspect of the skull, nor backwards beyond the auditory meatus. The occipital region is concave, and surrounded by a prominent crest.

The *foramen infraorbilate exterius* has an inferior position, being a little above the alveolar border. There is a prominent tuberosity on the under side of the basal front of the malar bone, just exterior to the position of the second molar of *Arctomys;* its inferior face is truncate.

The pterygoid laminæ are prolonged, inclosing a trough. Their sphenoid alæ descend steeply from their posterior base, and have an external ridge, which marks out a pterygoid fossa. The otic bulla is not large. Paroccipital process distinct. The *foramen ovale* is large, and is divided by a thin bridge of bone. The two external foramina resulting are also the alisphenoids. There are no additional foramina in this region. The space for the otic bulla is moderately large; the basicranial axis is grooved at the junction of the basioccipital and sphenoid bones. The zygomatic arch is deep and thin. The glenoid cavity is wide but longitudinal.

The cast of the brain indicates smooth oval hemispheres, which leave the cerebellum and olfactory lobes entirely exposed. The latter are ovoid and expanded laterally.

The coranoid process of the mandible is large and high, as in *Arctomys*. The condyle is small and compressed. The angle is produced, so that the posterior border of the ramus is concave.

Associated with the skull of a *P. delicatissimus* are various parts of the skeleton. A lumbar vertebra is not elongate; its anterior articular face is slightly convex and the posterior plane. The outline of the former is a little more than half a circle. The diapophyses are large, and are opposite the floor of the neural canal. The prezygapophyses are subvertical, and the superior exterior edges are developed into well marked metapophyses. The middle line below has a low narrow keel, which separates two large nutritious foramina.

The proximal part of the scapula is preserved. It resembles in general that of a squirrel, having a recurved coracoid process, and a well developed acromion. The latter is quite flat, and is continuous with a horizontal expansion of the spine. The humerus has a subround head, and the tuberosities are little prominent, and enclose but a shallow bicipital groove. The deltoid crest is prominent, and extends to the middle of the shaft. Its inferior portion is more prominent and compressed than the superior portion. The teres facet is well marked. The distal end is transversely extended by the large size of the internal epicondyle. The condyles are simple hour-glass-shaped, and without crest. The epitrochlear foramen is distinct. The ulna and radius are long and slender, and their carpal articular surfaces are of subequal size. The head of the radius is nearly round. Two metacarpals are preserved, and they are rather short; there is an inferior trochlear keel of the distal articular extremity.

The pelvis is much like that of a squirrel. The ilium is not much expanded towards the crest, but cannot be called prismatic. Its external rib is near the anterior border, and the posterior edge is thin, and bounds a concavity of the external face. The anterior inferior spine is prominent. The ischium is two-thirds the length of the ilium, and is moderately expanded distally in a vertical plane. Its spine is distinct Fragments of femora, associated with the other specimens, have the characters of those of the *Sciuridæ*. The great trochanter is about as high as the head, and bounds a large fossa. The little trochanter is well developed. The rotular groove is moderately elongate, with nearly equal bounding keels. The condyles are subequal and present posteriorly. The distal extremity of the tibia supports an internal malleolus which is flat on its inner side, and is without distal facets. The external trochlear groove of the peronealy is larger than the internal. The posterior border is produced downwards in a subangular process as in other *Sciuridæ*, which is as long as the malleolus, and is openly grooved to carry the tendon of the *flexor longus pollicis* muscle. It is separated by a deep notch from the internal malleolus, through which passed the tendons of the *flexor longus digitorum*, and the *tibialis posticus* muscles.

The *astragalus* also resembles that of a squirrel. The head is directed inwards from the anteroposterior axis of the troachlea, and has a depressed and convex distal extremity. The trochlea is wide, and the groove is well marked. Its external and internal faces are vertical. The calcaneum is large, and the free portion is compressed. The sustentaculum is small.

The species from which most of the characters of the genus as above stated have been derived are the *P. delicatior* and *P. delicatissimus*. They display the following general points. The anterior limbs are relatively longer than in recent species of squirrels. The head of the radius is rounder, indicating an unusual power of rotation of the anterior limb. The pelvis is larger, being as long as the skull, and it is probable that the posterior limb is larger. These points indicate approximation to the *Mesodonta*.

No characters have yet been offered by which to distinguish the American species as representing a genus distinct from the *Plesiarctomys gervaisii* of the French Eocene. Bravard briefly distinguishes the genus as distinct from *Arctomys* in the greater thickness of the angles of the molars, which thus become tubercles. Only the mandible and mandibular teeth of the *P. gervaisii* are known. It has been found in the Upper Eocene, near Perreal, Apt, France.

I have seen five species of this genus, of which one, *P. hians* belongs to the Bridger beds; one, *P. leptodus*, to the Washakie; one, *P. buccatus*, to the Wasatch and Wind River, and two, *P. delicatior* and *P. delicatissimus*, to all the Eocenes except the Washakie.

PLESIARCTOMYS BUCCATUS Cope.

Report U. S. Geol. & Geog. Expl. Surv. W. 100th Meridian, G. M. Wheeler, iv, pt. ii, p. 171; pl. xliv, fig. 8.

Plate xxiv *a*, fig. 14.

Jaws of four specimens which agree in proportions with those of this species found in New Mexico, were obtained by Mr. Wortman in the bad lands of the Big Horn basin, Wyoming. The species was established on a maxillary bone bearing teeth, while the present specimens are all mandibles. I do not detect any difference between these and the lower jaws of *P. delicatissimus*, excepting the inferior size of the former, from which, however, the first and last molars are wanting. I give the following measurements:

Measurements.

	M.
Length of inferior molar series	.012
Length of base of last molar	.0038
Diameter of inferior incisor { antero-posterior	.0033
Diameter of inferior incisor { transverse	.0008
Depth of ramus at second molar	.0088

PLESIARCTOMYS DELICATISSIMUS Leidy.

Cope, Report U. S. Geog. & Geol. Expl. Surv. W. of 100th Mer., G. M. Wheeler, iv, pt. ii, p. 172, pl. xliv, figs. 9–12. *Paramys delicatissimus* Leidy, Proceed. Acad. Phila. 1871, 231; Report U. S. Geol. Surv. Terrs. ii, p. 111, pl. vi, figs. 28–9.

Plate xxiv *a*, figs. 1–10.

Portions of the mandibles of four specimens of this rodent were brought by Mr. Wortman from the Wind River Basin; also, the skull of another individual, and the greater part of the skeleton with the skull of a sixth individual.

The specimen last mentioned furnishes the following characters. The skull has much the form of the large arboreal squirrels of the present day. The muzzle is of moderate length, and the zygomata are not very widely expanded. The skull is contracted just behind the eyes, for the orbits are not defined posteriorly. Above the eyes the superciliary border is angular, but not prominent, and each one is continued as a delicate anterior temporal ridge. The ridges converge backwards and unite into a low sagittal crest opposite the anterior angle of the squamosal bone. The anterior superior angle of the malar bone forms the inferior half of the anterior border of the orbit, but its inferior part is supported by a peduncle of the maxillary bone. Its posterior angle extends as far posteriorly as the external border of the glenoid cavity. The occiput is vertical, and is wider than deep. It is convex in the position of the vermis, and concave on each side above the occipital condyle. The condyles are rather small, and are widely separated, bounding a very large foramen magnum. Inferiorly the basioccipital is flat in front of the condyles. Between the otic bullæ it has a median keel, with a concavity on each side, bounded externally by a prominent descending border in contact with the bulla.

The massateric fossa of the mandible reaches the line of the posterior border of the penultimate molar tooth. The articular service of the condyle is nearly round, and is strongly convex, and projects as far backwards as a vertical line drawn from the angle. The coronoid process is much higher, and has a wide summit with a convex border.

Of the superior molars, the apex of the minute first is bifid, one cusp being the larger; the second is a little narrower than the third, and the fifth is a little wider antero-posteriorly. The enamel is smooth on all the teeth in both jaws, including the incisors, and there are no cingula.

The extremities of the coracoid and acromion are about in one horizontal line. The former is abruptly turned backwards and truncate on the external face.

The ulna and radius appear to have lost their epiphyses; allowing for this, they are a little longer than the humerus. The shaft of the ulna is rather compressed, while that of the radius in round in section. The distal expansion of each is about equal. The trochanteric fossa of the femur is very deep, and the little trochanter is very large. The third trochanter is represented by a low ridge. The distal extremity of the tibia has nearly equal anteroposterior and transverse diameters, and the shaft at the distal two-fifths the length is nearly round.

Measurements of No. 6.

	M.
Length of skull to lateral border of external nares to paroccipital process	.0700
Length from lateral edge of nares to orbit	.0220
Length from anterior border of orbit to posterior edge of zygomatic fossa	.0300
Length from posterior border of superior incisor to first molar	.0170

	M.
Length of superior molar series	.0145
Length of molars II and III together	0065
Length of molars IV and V together	.0070
Transverse diameter of M. II	.0042
Transverse diameter of M. V	.0035
Width at eyes	.0165
Width behind eyes	.0100
Length of mandibular ramus from incisor to condyle	.0400
Length of diastema	.0060
Length of molar series	.0135
Depth of ramus at second molar	.0100
Depth of ramus at coronoid process	.0310
Depth of ramus at condyle	.0235
Length of humerus (Nos. 5 or 6)	.0600
Proximal diameter of humerus { anteroposterior of head	.008
Proximal diameter of humerus { transverse, with tuberosities	.012
Distal, with humerus	.016
Width of distal condyles of humerus	.011
Diameter of head of radius { vertical	.006
Diameter of head of radius { transverse	.007
Depth of radius (?minus epiphysis)	.048
Depth of ulna at coronoid	.006
Anteroposterior diameter of head of femur	.009
Transverse diameter at little trochanter	.0135
Transverse diameter below little trochanter	.009
Transverse diameter at distal end	.016
Vertical diameter at distal end (greatest)	.015
Diameter shaft of tibia at distal two-fifths of length	.006
Diameter at distal end of tibia { anteroposterior	.010
Diameter at distal end of tibia { transverse	.010
Diameter of trochlea of astragalus { anteroposterior	.008
Diameter of trochlea of astragalus { transverse	.007
Length of head of astragalus	.005
Angle of axis of head with trochlea	9°
Diameters of navicular face { vertical	.0045
Diameters of navicular face { transverse	.0066
Length of a metatarsal	.029
Length of ilium to acetabulum	.0785
Widths of ilium { at crest	.0150
Widths of ilium { at acetabulum	.0105
Vertical diameter of acetabulum	.0105
Length of ischium	.0128
Widths of ischium { at acetabulum	.0075
Widths of ischium { at distal border	.0150

From the above measurements the following comparisons with the *Sciurus niger* may be made: The pelvis is longer as compared with the bones of the fore leg. The humerus is longer as compared with the length of the ulna and radius. The species exceeds the *S. niger* in size, one-fourth linear.

PLESIARCTOMYS DELICATIOR Leidy.

Cope, Report U. S. Geog. & Geol. Expl. Surv. W. of 100th Mer. G. M. Wheeler, iv, pt. ii, p. 172. *Paramys delicatior* Leidy, Proceed. Acad. Phila. 1871, p. 231. Report U. S. Geol. Surv. Terrs. 1873, i, p. 110, pl. vi, figs. 26–7, pl. xxvii, figs. 16–18.

Plate xxiv *a*, figs. 11–13.

This squirrel is represented by the jaws of at least four individuals in Mr. Wortman's Wind River collections. Two of these are further represented by other portions of the skeleton, one of them including a maxillary bone with four molars in place.

The bones of the skeleton coincide with the teeth in their superior dimensions as compared with those of the *P. delicatissimus*. The pieces of the last-named individual which are preserved are portions of the femur and tibia, with astrogali and calcanea; also, portions of ulna and metatarsals and ribs. The only difference of form I observe between these and corresponding parts of *P. delicatissimus* is in the distal extremity of the tibia; This has a relatively greater transverse diameter as compared with the anteroposterior than in the species just named. The calcaneum of the right side is nearly perfect. The free portion is strongly compressed; the anterior portion appears depressed on account of the extent of the sustentaculum on the inner side, and the well-developed corresponding process on the external side opposite to it. The cuboid facet is, however, as deep as wide, and truncates the calcaneum transversely. The external astragaline facet is very convex, and presents inwards and a little forwards.

The superior molars accompanying this specimen are of the size and proportions of the inferior molars, but they belong to the Mesodont, *Pelycodus frugivorus*, described later in the present work, where the distinction between the molar teeth of this genus and those of *Pelycodus* are pointed out.

Measurements.

		M.
Length of inferior molar series		.016
Length of last inferior molar		.005
Length of diastema		.0085
Depth of ramus at penultimate molar		.0115
Diameter of shaft of tibia two-fifths distance from extremity		.0073
Diameters of extremity of tibia	anteroposterior	.0095
	transverse	.0110
Diameters of trochlea of astragalus	anteroposterior	.0090
	transverse	.0080

	M.
Length of head of astragalus	.0055
Diameters of navicular face of astragalus { vertical	.0042
Diameters of navicular face of astragalus { transverse	.0080
Depth of heel of calcaneum at base	.0080
Depth of cuboid facet of calcaneum at base	.0057
Width at susterotaculum	.0132

PLESIARCTOMYS LEPTODUS Cope.

Paramys leptodus Cope, Paleontological Bulletin, No. 12, p. 3, March 8, 1873.

Plate xxiv, fig. 1.

Established on a right mandibular ramus with all the teeth preserved. It indicates an animal of about the size of the *P. delicatus* Leidy, but with smaller incisors, which have little more than half the diameter of the same tooth in those species. The molars have two anterior separate, and three posterior contiguous cones, the median smallest, the anterior and posterior of both sides separated by a deep excavation. The anterior tooth is peculiar in its greater compression. The posterior tubercles are not separated, and the anterior inner is situate behind the outer, and is connected with the posterior inner by a concave ridge.

Measurements.

	M.
Length molar series	.0221
Length M. IV	.0060
Width M. IV	.0055
Length M. I	.0060
Width M. I	.0048
Diameter lower incisor, transverse	.0024
Diameter lower incisor, anteroposterior	.0038

From the South Bitter Creek, Wyoming, from the Washakie Basin.

PLESIARCTOMYS HIANS Cope.

Pseudotomus hians Cope, Paleontological Bulletin No. 2, p. 2, 1872. Annual Report U. S. Geol. Surv. Terrs., 1872, p. 611.

Plate xxiv, figs. 3–5.

This species was established on a nearly complete cranium, from which some parts of the walls are broken, and only the anterior portion of the mandible remains. It belonged to an old individual which had lost its molar teeth, and whose inferior incisors are very much worn.

The cranium is of depressed form, and with considerably expanded zygomata. The muzzle is broad and but little elevated, so that the nasal meatus is between the alveolæ of the superior incisors. The enamel of the superior incisors is not grooved, but has a delicate striate sculpture. The inferior incisor does not project as far as the alveolar border of the jaw; its surface worn by the upper incisor is horizontal and anterior. The inferior diastema is a thin edge, and the ramus is deep there. The temporal surface of the parietal bones is rugose. The cranium is depressed, and has a trace of sagittal crest. The anterior margin of the temporal fossa is marked by a curved angle on each side of the frontal bone. The supra-orbital arch is very short. The mandibular incisors are narrowly separated by a narrow prolongation of the symphysis. The exposure of the tooth is lateral, its direction nearly anterior. It projects anteriorly very little beyond the symphysis, and has a horizontal triturating surface below the level of the latter. There are alveolæ for but three molar teeth, each with three roots. The teeth themselves are not contained in them, but were apparently lost before the cranium was entombed in the Eocene mud. The position of the first molar is occupied by spongy bone in both maxillaries, and appears as though such teeth might have existed earlier in life, and been shed.

Measurements.

	M.
Length of cranium (3.75 in.)	.095
Width of cranium (without zygomas)	.040
Width of cranium (with zygomas)	.072
Width of occiput	.032
Width of occiput near end of nasals	.027
Width of upper cutting tooth	.007
Depth of upper cutting tooth	.0085
Length exposed part lower tooth	.009
Width exposed part lower tooth	.006

The species differs from the *P. delicatus* Leidy, in its superior size, being, in fact, the largest species of the genus.

The typical specimen was found by myself on Cottonwood Creek, Wyoming, in the Bridger beds.

BUNOTHERIA.

This division embraces the unguiculate *Mammalia* of low cerebral development, which have the transverse articulation of the lower jaw, and ambulatory limbs. It embraces a series of types which present a great range of variation in dental characters, but which at the same time pass into each other by sensible gradations. It is possible that some of the types which I have referred here may turn out to be *Marsupialia*, but the number of such cases is probably small. The following is the definition of this order:

Cerebral hemispheres small, leaving the olfactory lobes and cerebelum exposed; the surface smooth, or nearly so. Limbs ambulatory, armed with a greater or less number of compressed ungues. Articulation of the mandible transverse. Molar teeth of the superior series (and usually ot the lower) tubercular, and without continuous crests. Incisor teeth present in the premaxillary bone. Teeth invested with enamel. Feet with five digits (with a few exceptions). Usually a third trochanter of the femur.

I once applied to this order the name *Insectivora*, so as to avoid the creation of a new one, but I subsequently concluded to adopt the latter course as the preferable one. The name *Insectivora* has acquired currency as applied to the well-known modern group of that name, and its application to types of such apparent diversity as those now associated under a single head is not a convenience. I therefore proposed the name *Bunotheria* for the order, and included under it the suborders *Creodonta*, *Mesodonta*, *Insectivora*, *Tillodonta*, and *Taeniodonta*.* I suspect that the *Prosimiæ* must also be included in this order. The suborders are characterized as follows:

Superior incisors normal, not growing from persistent pulps; canines much enlarged; premolars compressed; molars more or less sectorial; astragalus generally not grooved above, articulating with cuboid and navicular; scaphoid and lunar bones (?always) distinct .... *Creodonta.*

Incisors not growing from persistent pulps; molars tubercular, never sectorial; third trochanter elevated; astragalus not grooved above .... *Mesodonta.*

Incisors enlarged, simple, not growing from persistent pulps, canines reduced; astragalus concave above .... *Insectivora.*†

---

* Report of Lieut. G. M. Wheeler of the Expl. Surv. W. 100th Mer., iv, 1877, p. 72.

† The typical *Insectivora*, Linn., Bonap., Gill.

Incisors much enlarged, growing from persistent pulps, and faced with enamel in front only; therefore scalpriform .... *Tillodonta.*
Incisors much enlarged growing from persistent pulps, the superior with enamel in anterior and posterior bands, and hence truncate .... *Tæniodonta.*

The order of *Bunotheria* with these subdivisions, is not more heterogeneous than that of the Marsupialia, and presents a great similarity in its component parts. Thus the *Creodonta* resemble *Sarcophaga*, the *Insectivora* the *Entomophaga*, and the *Tillodonta* the *Rhizophaga*. *Phascolomys*, the type of the last suborder presents several points of resemblance to the *Tillodonta.*

The affinities of the groups here combined under one ordinal caption are very divergent. The order is generalized, and, as such, does not present the peculiar features of the *Chiroptera*, *Rodentia*, and *Edentata*, but is so far negative in its character as to preclude more than subordinate subdivision. While the existing division *Insectivora* maintains the typical characters, the *Dermoptera*, also existing, are doubtless relics of the group from which the *Chiroptera* derive their ancestry. The *Tillodonta* exhibit some kind of affinity to the *Rodents*, while the *Tæniodonta* present us with a point of connection with the *Edentata.* The discovery of this fact was particularly welcome, as we had not previously had any hint of the relations between that anomalous order and the remainder of the *Mammalia.* So far the relationships indicated are to smooth-brained (lissencephalous) orders only. The connections with the *Gyrencephala* (or *Educabilia*) are quite as close; namely, as already pointed out, through *Mesodonta* to the *Prosimiæ* and the *Quadrumana*, and through the *Creodonta* to the *Carnivora.* Standing in this structural relation to different existing types, and in an antecedent relation as to time, it is easy to look on the *Bunotheria* as ancestral to some of them. In the first place, the *Insectivora* represent them in the existing fauna. The *Creodonta* are probably the ancestors of the *Carnivora*, and the *Mesodonta* of the *Prosimiæ.* This ancestry is rendered almost certain by the discovery, by Drs. A. Milne Edwards and Grandidier, of the affinity existing between the *Prosimiæ* and the *Carnivora.*

Before the discovery of the species and genera which form the subjects of this report, I wrote as follows: "I trust that I have made it sufficiently obvious that the primitive genera of this division of Mammals must have been *Bunodonts* with pentadactyle plantigrade feet. We may anticipate the discovery of such a genus, and believe that it will not be widely removed from the Eocene *Hyopsodus*, or perhaps *Achænodon*. But it will be more than this: it cannot be far removed from the primitive *Carnivora* and the primitive *Quadrumana*. The *Carnivora* are all modified *Bunodonts*, and the lower forms (Ursus, Procyon, etc.) are pentadactyle and plantigrade; as to the *Quadrumana*, man himself is a pentadactyle plantigrade *Bunodont*.*

Such a hypothetical type might be expressed by the name *Bunotheriidæ*, with the expectation that it will present subordinate variations in premolar, canine, and incisor teeth. The premolars might be expected to differ in the degree of development of the internal lobes, the canine in its proportions, and the incisors in their number."

The history of discovery of the *Eocene* forms of this order is briefly told. Professors Leidy, Marsh, and myself had described *Creodonta* as *Carnivora*, until I pointed out, in some remarks before the Philadelphia Academy (published December 22, 1875) their true relations. The first species of *Tillodonta* was described by Leidy from an inferior molar from New Jersey, in 1868. Dr. Leidy next described the dentition of the mandible of the same genus from Wyoming. Subsequently, Marsh described the superior molars of an allied genus, from Wyoming. In 1874, the writer described the dentition of the *Tæniodonta* from specimens collected by Lieutenant Wheeler's Survey in New Mexico. In March, 1875, Marsh proposed the *Tillodonta* as an order of Mammals, giving its dental characters, and stating the brain was small. In December, 1875, in his remarks on *Creodonta*, I referred this group to the *Insectivora* as a suborder. In March, 1876, Professor Marsh gave a full description of the cranial characters of the genus *Tillotherium*, describing the characters of the brain from a cast of the cranial cavity. In the same month of 1876 the writer characterized the suborder *Tæniodonta*, referring to it the genera *Ectoganus* and *Calamodon*.

TÆNIODONTA.

The *Tæniodonta* agree with the *Tillodonta* in the possession of a pair of inferior incisors of rodent character, but it adds several remarkable pecu-

*On the Homologies and Origin of the Types of Molar Teeth of Mammalia Educabilia, Journal Academy Philadelphia, 1874, p. 20.

liarities. Chief among these is the character of the inferior canines. In the *Tillodonta* they are either wanting, as in *Erinaceus*, according to the Cuvierian diagnosis, or they are insignificant. In *Calamodon* they are of large size, and though not as long-rooted as the second incisors, grow from persistent pulps. They have two enamel faces, the anterior and posterior, the former like the corresponding face of the rodent incisors. The function of the adult crown is that of a grinding tooth. This character distinguishes *Calamodon* as a form as different from *Tillotherium* as the latter is from *Esthonyx*. There are, however, other characters. The external incisors, wanting in *Tillotherium*, are here largely developed, and though not growing from persistent pulps, have but one, an external band-like enamel face. Their function is also that of grinders The fact that the rodent teeth in the lower jaw are the second incisors, renders it probable that those of the *Tillodonta* hold the same position in the jaw. This is to be anticipated from the arrangement in *Esthonyx*, where the second inferior incisors are much larger than the first and third. The superior dentition of the *Tæniodonta* is unknown.

Two families represented this suborder in the Eocene period of North America. The first or *Ectoganidæ*, with two species, possess molar teeth with several roots. In the *Calamodontidæ*, with five species, each lower molar has a simple conic fang. The great reduction in the extent of the enamel investment is an interesting approximation to the *Edentata*, where this substance is altogether wanting. The reduction is greatest on the adjacent sides of the molars; it has a little greater extent on the inner side, while it extends as a band on the exterior side, so that in worn teeth this surface alone remains. In addition, there are a heavy cementum investiture and undivided roots in the genus *Calamodon*, features essentially characteristic of the *Edentata*.

Thus we have in the *Tæniodonta* the first hint as to the relations of the *Edentata* in early Tertiary time.

CALAMODON Cope.

Report Vert. Foss. New Mexico, U. S. Geog. Survs. W. of 100th M., 1874, p. 5; *Id.*, Ann. Report U. S. Geol. Survs. W. of 100th M., 1874, p. 117; System. Cat. Vert. Eocene New Mexico, U. S. Geog. Survs. W. of 100th M., 1875, p. 24; Report U. S. Geog. Survs. W. of 100th M., vol. iv, pt. ii, 1877, p. 162.

Formula of inferior dentition, I. 3; C. 1, M. 5; without distinction into premolars and molars. First incisors small, with conic roots; second incisors rodent-like, very large; third incisors large, with conic roots, faced with enamel in front only. Canines placed obliquely to the long axis of the ramus, their long horizontal axis extending inwards and backwards. Enamel confined to their narrow anterior and posterior faces. Molars subcylindric and with conic roots, one or more of them within the base of the coronoid process. Form of the jaws deep and robust.

A characteristic feature of the dentition in this genus is the thick coating of cementum, which invests those portions of the molars and inferior canines which are not protected by enamel. In these teeth it is thicker than the enamel, and forms thickened raised borders surrounding the latter, producing a characteristic appearance not known in the other genera. It is not observable in the large inferior incisors.

A part of the skeleton of one of the species is preserved. It shows that the humerus was robust, and was pierced distally by a large arterial foramen The condyles are not very convex, nor the internal epicondyle, so prominent as in some of the *Creodonta.* The head of the radius is flat and incapable of rotation, and the shaft is rather slender, while the ulna is deep and thin. An ungual phalange is stout and compressed, and but little curved, and without the basal sheath seen in some *Carnivora* and *Edentata.*

The exact homologies of the seven mandibular teeth are obscure, and it is uncertain to how many the expression molar should apply, since the wear of all in my specimens is nearly equal.

The symphysis is solid and long; it projects wedge-like between the large incisors, whose anterior borders are closely approximated. There is a large mental foramen.

Two species, the *C. arcamænus* and *C. simplex,* were found by myself in New Mexico. A third one, discovered by Mr. J. L. Wortman in the Wind River beds, has been described by myself under the name of *C. cylindrifer;* it is not sufficiently well preserved to settle finally the question of its generic position.

Calamodon simplex Cope.

Report Vert. Foss. New Mexico, U. S. Geog. Surv. W. of 100th Meridian, Capt. G. M. Wheeler, 1874, p. 5. Report U. S. Geog. Survs. W. of 100th Meridian, Capt. G. M. Wheeler, 1874, iv, ii, p. 166. Pl. xlii, figs. 6–8; xliii; xliv, figs. 2–5. Pal. Bulletin No. 34, p. 147, 1882.

Plate XXIV c, Fig. 1.

The nearly complete mandible of this species was obtained by J. L.

Wortman in the Wasatch Bed-lands of the Big Horn Basin of Northern Wyoming. It has furnished the information necessary to complete our knowledge of the inferior dentition of the genus. It indicates an animal of the size of a tapir, and with a very peculiar physiognomy. This is due to the shortness and depth of face, as indicated by the lower jaws.

If the anterior border of the coronoid be held vertically, the masticating surface of the molars is horizontal. The inferior border of the ramus descends steeply to below the second molar, and then rises to the level of the grinding surfaces in a curve which is the arc of a circle. This curve follows the external border of the second incisor, which terminates at the fundus of its alveolus, below the third molar. The ascending inferior border is gently concave below the coronoid process, and passes into the widely convex angle, whose middle is about in the line of the alveolar edge of the jaw. The condyle is pretty well elevated. Its articular face descends gently inwards, and its convexity looks upwards and a little backwards. The basis of the coronoid is extended anteroposteriorly. The distance from its anterior edge to the posterior border of the ramus is equal to the length from the former point to the anterior edge of the third incisor. The anterior edge is thickened inferiorly and extends downwards to below the alveolar border, and externally to it, and graduates into the surrounding surface. The summit of the coronoid is obtuse. Its posterior border descends to near the condyle, and its superior half is bevelled from an angle of its inner face. The alveolar ridge extends within the base of the coronoid to a point below its apex, and then sinks. Below it the inner face of the ramus is concave continuously with the internal pterygoid fossa. The edge of the angular curve is bevelled on the external side. The angle separating the external face from the bevel is continuous with the posterior margin below the condyle, while the bevelled portion projects beyond it in an angle, half way between the condyle and the inferior border. The latter is not in the least incurved.

The crowns of the first incisors in their present condition display no enamel, and the grinding surface of each is a rather wide oval with the long axis anteroposterior. The second incisor has the usual shoulder or ledge behind the scalpriform portion, as though it could be used for grinding as well as for cutting purposes. From this ledge the anteroposterior diameter is equal throughout. The exposed dentinal portion is more concave on one side than the other, though both are slightly concave. The enamel is strongly convex all round, excepting a shallow longitudinal groove on the more concave side. Its surface is marked with obsolete parallel ridges. The third incisor is a large prismatic tooth of triangular section. The inner face, which is applied to the second incisor, is flat, and is the widest. The anterior is convex, and is the next in width. The posterior external is a little narrower, and is flat. Its enamel has some obsolete longitudinal ridges, and several transverse undulations or obsolete ribs. The external face of the canine is narrower than that of the third incisor, and is wider than the posterior face. Anteriorly there is a faint longitudinal groove and several transverse undulations. The posterior face has transverse wrinkles.

The grinding surfaces of the molars are subquadrate in outline, with rounded angles. The posterior three are rather narrower posteriorly than anteriorly. They are of about equal size, the first only being a little smaller than the others. The roots are conic in this adult specimen. The enamel extends well down on the external faces, although concealed in the alveolus. This looks as though the tooth enjoyed a considerable period of growth before the dentigerous pulp disappeared and the apex of the root closed. This is probably true in a more marked degree of the *C. cylindrifer*. In wearing, the anterior external border of the crown is a little more elevated than the remainder.

Measurements.

			M.
Length of ramus from condyle to I. ii inclusive			.206
Length of ramus from condyle to front of base of coronoid process			.100
Length of dental line			.130
Length of molar series			.080
Diameters of M. ii	anteroposterior		.0145
	transverse		.0145
Diameters of canine	longer		.021
	shorter		.011
Diameters of I. iii	anteroposterior		.021
	transverse		.010
Diameters external face I. iii	transverse		.017
	vertical from alveolus		.020
Diameters I. ii (extremity)	anteroposterior		.030
	transverse	anteriorly	.016
		posteriorly	.010
Depth of ramus at condyle			.083
Depth of ramus at apex of coronoid			.115

A young specimen apparently of this species is represented by the crowns of four molar teeth and by a canine. All are unworn and display the character of the unmodified grinding surfaces. The external and internal enamelled faces of the canine approach each other at the apex, as in *Ectoganus*, and are separated by a well-marked notch. The crowns of the molars are a good deal like those of *Psittacotherium*. They support two transverse crests which are separated by a deep open valley. The posterior crest is the shorter, and in most of the molars is divided into three tubercles, of which the median is the smallest. The anterior crest is divided by a deep fissure into two tubercles, each of which has a subacute transverse edge. Enamel smooth. From the Big Horn. J. L. Wortman.

CALAMODON CYLINDRIFER Cope.

Bulletin U. S. Geol. Surv. Terrs. F. V. Hayden, 1881, p. 184.
Plate XXIV a; figs. 15–16.

The only individual of this species discovered by Mr. Wortman is represented by fragments of the jaws, with several teeth, both loose and imbedded in matrix. The former show that the molars have but one root. The latter include the large rodent-like incisors in a fragmentary condition, and a nearly complete tooth intermediate in character between the flat banded teeth and the molar teeth of the known species of *Calamodon*. It may occupy an intermediate position in the jaw, but I do not know of any appropriate place for it in the mandible of *Calamodon arcamœnus*. I think there is little doubt the individual belongs to a species with narrower teeth than any of those of the two species already named.

The characteristic tooth in question is nearly cylindric, and the part preserved is quite long and slender. Its grinding surface is worn concavely, as in the flat teeth of the known species of *Calamodon*. The enamel is in two bands, one wider than the other, and each of equal width throughout. The space of cementum separating them on one side is nearly twice as wide as that on the other. The cementum layer is not so thick as in the species of the genus hitherto described. The shaft of the tooth is slightly curved, and the wider band of cementum is on the inner side of the curve.

Measurements.

		M.
Width of enamel of large incisor		.018
Length of shaft of cylindric tooth		.041
Diameters of grinding surface of cylindric tooth	anteroposterior	.011
	transverse	.010

TÆNIOLABIS Cope.

American Naturalist, July (June), 1882.

Established on a tooth whose position is on the arc of the alveolar line which connects the molar and middle incisor regions. It is probably either the third incisor of the superior or inferior series, or canine of the inferior series. In either case it differs from the corresponding tooth of any of the known genera of *Tillodonta* or *Tæniodonta*. The long diameter of the root being placed anteroposteriorly, that of the crown makes with it an angle of 30°.

Section of crown oval; the grinding surface scalpriform in the manner of a rodent incisor; but bevelled on one side of the long diameter instead of on the end, as in that order. Enamel consisting of a wide band on the external side of the tooth, which embraces more of the circumference near the apex than elsewhere. Apex grooved behind.

If this be an inferior canine tooth it differs from that of the *Tillodonta* in its large size and incisor-like form. It most resembles the external or third inferior incisor of *Calamodon*. From this it differs in the scalpriform wear, and the oval instead of triangular section, and in the absence of cementum layer.

TÆNIOLABIS SCALPER Cope.

American Naturalist, 1882, July (June).

Plate XXIII d; fig. 7.

The enamel band does not cover the entire width of the external face, but leaves a part of the dentinal surface anterior and posterior to it, except at the apex. At the latter point there are seven coarse shallow grooves of the enamel surface; and the posterior of these split up below, and become narrower, while the anterior run out at the more oblique anterior edge of the enamel band. The posterior apical groove has a flat bottom. At the front of the apex the enamel is involute to the inner side for a short distance. The inner face of the tooth displays five facet-like bands of the dentinal surface which soon disappear inferiorly.

Measurements.

		M
Length of tooth (root restored)		.058
Length of enamel band		.031
Width of enamel band at middle		.0095
Diameters of middle of tooth	anteroposterior	.0130
	transverse	.009
Long diameter of apex of tooth		.008

This tooth indicates a new and interesting type, perhaps, of *Calamodontidæ,* and one of which more information will be awaited with interest. Judging from the size of the tooth its possessor was as large as a sheep. From the Puerco Eocene of New Mexico; from D. Baldwin.

TILLODONTA.

There are three allied groups represented by the genera *Esthonyx, Tillotherium,* and *Calamodon* of the American Eocenes, which are equally unlike each other. *Esthonyx,* as I long since showed, is related to the existing *Erinaceus;* very nearly, indeed, if the dentition alone be considered. Its anterior incisor teeth are usually developed and have, as in *Erinaceus,* long roots. One pair at least in the lower jaw has enamel on the external face only, and enjoys a considerable period of growth. The genus *Tillotherium* is (fide Marsh) quite near to *Esthonyx.* Its molars and premolars are identical in character with those of that genus, the only important difference being found in the incisors. Here, one pair above and one pair below are faced with enamel in front only, and grow from persistent pulps as in the *Rodentia.* This character has been included by Marsh in those he ascribes to his "order" of *Tillodontia,* but as he includes *Esthonyx* in that order,* which does not possess the character, it is not very clear on what the supposed order reposes. The rodent character of the incisors is the only one I know of which distinguishes *Tillotherium* from the *Insectivora.* I have on this account retained the *Tillodonta* as a suborder, and referred *Esthonyx* to the *Insectivora.*

There are three genera of this suborder: *Psittacotherium* Cope, *Anchippodus* Leidy, and *Tillotherium* Marsh. The last one I only know from the

*Report of U. S. Geol. Surv. 40th parallel, by Clarence King, vol. i, page 377.

descriptions and figures of its describer. There are three species ascribed to *Tillotherium*, while the other genera have two each. The genera are defined as follows:

First inferior incisors present; six inferior molars with cross-crests.. *Psittacotherium.*
First inferior incisors present; ?six inferior molars; true molars with V's .... *Anchippodus.*
First inferior incisors wanting; seven inferior molars; true molars with V's .... *Tillotherium.*

The first-named genus approaches most nearly to *Calamodon* in the structure of the crowns of the molar teeth, and in the shortness of its dental series. It is only known from the Puerco formation. The two remaining genera have been found in the beds of the Bridger epoch. *Tillotherium* is nearest to *Esthonyx*, while *Anchippodus* is between the former and *Psittacotherium.*

PSITTACOTHERIUM Cope.

American Naturalist, 1882, Feb., p. 156 (Jan. 25).

This genus differs widely from the two genera hitherto known, *Anchippodus* and *Tillotherium.* Owing to the absence of the superior dental series it is not possible to be sure which tooth is the canine. The inferior dental formula may be therefore written: I. 2; C. 1; Pm. 3; M. 3; or I. 3; C. 0; Pm. 3; M. 3; or I. 3; C. 1; Pm. 2; M. 3. The first and second incisors are large and rodent-like, growing from persistent pulps; the second are the larger. The third, or canines, are small and probably not gliriform. There is no diastema. The first premolar (or canine) has a compressed crown with two cusps placed transverely to the jaw axis, and has a complete enamel sheath, and probably two roots. The succeeding tooth is also transverse, and is two-rooted, judging from the alveolus. The first and second true molars are rooted, and the crown consists of two transverse separated crests, each partially divided into two tubercles. On wearing, the grinding surface of each assumes the form of a letter B with the convexities anterior The last inferior molar is injured. The rami are short, and the symphysis deep and recurved.

Two species are known, a larger and a smaller. In both the incisor teeth are very powerful, and the symphysis of the lower jaw is very deep, for their accommodation. They may have been adapted for breaking nuts and seeds, as well as cutting roots.

PSITTACOTHERIUM MULTIFRAGUM Cope.

American Naturalist, 1882, p. 157 (Feb.).
Plate XXIV c, fig. 2.

This animal is represented by an almost entire mandible, which indicates an animal of about the size of the capybara (*Hydrochoerus capybara*). The specimen has been subjected to pressure which has pressed the symphysis backwards, and given it an angle with the ramus rather steeper than the normal one.

The base of the coronoid process is opposite the junction of the second and third true molars. The ramus is deep and moderately stout. The enamel of the first incisor does not extend below the alveolar border, at the internal and external faces, and does not reach it at the sides. It has a few wrinkles on the anterior face. The anterior enamel face of the second incisor is thrown into shallow longitudinal grooves with more or less numerous irregularities, from the low dividing ridges. There is a deeper groove on each side of the tooth, and there are about a dozen ridges between these on the anterior face. Both cusps of the first premolar are conic, and the external is the larger. The second true molar is a little smaller than the first. The enamel of the premolars and molars is smooth, and there are no cingula.

Probable length of dental series .0750; diameters of I. 1: anteroposterior .0120, transverse .0066; diameters I. 2: anteroposterior .0160, transverse .0115; diameters Pm. 1.: anteroposterior .0072, transverse .0130; diameters of M. 11: anteroposterior .0090, transverse .0090. Length of true molars .0038; depth of ramus at M. 11. .0360.

The short deep jaws of this animal must have given it a very peculiar appearance, not unlike that of a parrot in outline.

PSITTACOTHERIUM ASPASIÆ Cope.

Paleontological Bulletin No. 34, p. 192, Feb. 20, 1882; Proceed. Amer. Philos. Soc. 1881, p. 192.
Plate XXIV c, figs. 3-4.

Represented by two mandibular rami of two individuals, one adult, the other nearly so, but with the last inferior molar not fully protruded. The latter specimen must be used for description, as it presents two molar teeth, while the other specimen has lost them.

The most obvious difference from the *P. multifragum* is its inferior size, which can be readily perceived from the measurements given. The posterior crest of the molars appears to have less transverse extent than in the larger species. This crest in the last inferior molar has a curved crenate edge, with a small conic tubercle at its external extremity. The anterior crest consists of two conic tubercles, whose apices converge, but whose bases are closely appressed, and only distinguished by a superficial fissure. The valley between the crests is uninterrupted. The preceding molar is larger, and its posterior crest is like that of the lost molar. The apex of the anterior crest is broken off.

The ramus deepens rapidly forwards, and contains the enormous alveolus for the incisors. The coronoid process leaves the alveolar border at the line separating the last two molars, or, in the smaller specimen, a little anterior to this point, and is quite prominent. The masseteric fossa is well marked, but shallows gradually anteriorly and inferiorly.

Measurements.

No. 1.

	M.
Depth of ramus at penultimate molar	.027
Width of last molar anteriorly	.008
Length of crown of do	.009

No. 2.

Depth of ramus of penultimate molar	.029
Depth of ramus at P-m. ii	.043
Length of five consecutive alveoli	.047

From the Puerco bed of N. W. New Mexico. Discovered by my assistant, Mr. D. Baldwin.

INSECTIVORA.

To this suborder I refer the genus *Esthonyx*, on account of the near resemblance of such parts of the dentition as are known, to some of the genera now existing. It is not unlikely that other genera of the Eocene which have been referred to the *Insectivora* belong here. *Esthonyx* exhibits an approximation to the *Tillodonta* in the restriction of the enamel-layers of the incisors of one of the jaws to the anterior face only. The inferior molars have much the constitution of those of *Anchippodus*, and in their details resemble also those of *Erinaceus*. As compared with the *Creodonta*, there is a near resemblance between these teeth and the tubercular molars of *Didymictis*, and through them to the tubercular sectorials of the *Oxyænidæ*, with which they agree in essential composition.

On the other hand, resemblances between the dentition of *Esthonyx*, and the supposed Lemurine genus *Pelycodus* are not wanting, and the rodent-like anterior teeth of the Lemuroid *Chiromys* suggest still further affinities between the Eocene members of that group and the *Tillodonta*.

There are two genera of this group known to me from the American Eocenes, which differ as follows:

Anterior lobe of fourth inferior premolar a cone; two inferior incisors..... *Conoryctes*.
Anterior lobe of fourth inferior premolar with a concave edge; three inferior incisors...... *Esthonyx*.

As the scaphoid and the lunar bones are separate in *Esthonyx*, that genus cannot be placed in the *Erinaceidæ*.

CONORYCTES Cope.

Paleontological Bulletin No. 33, p. 486, Sept. 30, 1881; Proceedings Amer. Philos. Soc. 1881, p. 486.

Dental formula, I. $\frac{?}{2}$; C. $\frac{1}{1}$; P-m. $\frac{?3}{4}$; M. $\frac{3}{3}$. Incisors small; canines large, with conic crowns. First two superior and inferior premolars with conic crowns; first inferior one-rooted, the second two-rooted, the fourth with an anterior conic cusp and a posterior grinding heel. Superior true molars transverse, with two external cusps. The inferior consist of two lobes, of subcylindric section, separated by deep vertical grooves. Enamel developed on internal and external faces of crowns

CONORYCTES COMMA Cope.

Loc. sup. cit.

Plate XXIII d, figs. 1-5; XXV c, figs. 3-4.

Founded on a mandibular ramus which lacks the last molar, and has the crowns of the others worn. The external faces of the molars are much more exposed than the internal, and are somewhat contracted upwards. In the unworn crown there is a distinct anterior inner cusp, which is soon confounded on attrition. The heel of the last premolar has a crescentic section, the internal horn the narrower. The anterior lobe is a robust cone. The base of the second and third premolar is oblique to the axis of the ramus outwards and forwards. First premolar small, filling the short space between the second and the canine No cingula; enamel obscurely plicate, ramus robust. Length of inferior molars minus the last .0465; length of base of first true molar, .010; width of do. .009; elevation of crown do. .0055; length of base of fourth premolar .011; width of do. .008; elevation of crown of do. .0065. Anteroposterior diameter of base of crown of canine .010. Depth of ramus at first true molar .023; width of do. at do. .013. This genus differs from *Esthonyx* in the form of the fourth premolar. In the latter the anterior lobe is compressed and trenchant. The species is larger than any of that genus, and nearly equal to the *Ectoganus gliriformis*.

Since the preceding was written, I have received from the same region a much more complete specimen. It includes the greater part of the dentition of both jaws, with mandibles, parts of cranium, limbs, etc.

The mandible shows that the first true molar is the largest tooth, and that the crowns diminish in size in both directions. The third premolar has a nearly regular conic crown, with an oblique anteroposterior diameter a little the longer. Each of the preceding premolars has a single root. The inferior canines are very robust, and their crowns are strongly recurved. The external incisor is of good proportions, while the interior incisors are only half as large. The ramus is robust, and the symphysis is coössified. The inferior border of the ramus is slightly convex, and then rises, to the angle, commencing below the last molar. The angle is rounded, and is quite prominent, the posterior border being strongly incurved below the condyle. The latter is situated in the plane of the grinding surfaces of the molars. Its articular surface presents equally superiorly and posteriorly. The edge of the angle is incurved, and rises as a support to the internal extremity of the condyle. The coronoid process rises immediately in advance of the condyle, and its base has a wide anteroposterior extent; apex lost. The masseteric fossa is deeply impressed, but has no distinct inferior boundary. The anterior border is prominent, forming a flattened front of the process, which is oblique, projecting forwards as well as outwards. The ridge extends to below the anterior lobe of the last molar. On the inner side of the angular region there are four ridges for the insertion of the internal pterygoid muscle, etc. The two inferior are sublongitudinal and subparallel.

The crown of the superior third premolar is a slightly trihedral cone, with a low cingulum on the inner side. The crowns of the last four molars are transverse, narrowing to the internal rounded border. The fourth premolar has two external conic tubercles, of which the posterior is the smaller, and no cingula. The true molars have a low, interrupted, external cingulum, and two external cusps. Of these the anterior is large, and behind it is a small cingular cusp. The surfaces of the crowns are too much worn to display the detailed structure. As in the mandibular series, the first true molar is the largest tooth, from which the proportions diminish in both directions. The crown of the third molar is not larger than that of the third premolar. The molar bone is robust at the base, but becomes rapidly attenuated and compressed. It has no postorbital angle.

Measurements of jaws.

		M.
Length of last five superior molars		.040
Length of true molars		.024
Diameters crown P–m. iii	anteroposterior	.006
	transverse	.007
Diameters crown P–m. iv	anteroposterior	.007
	transverse	.011
Diameters crown M. i	anteroposterior	.008
	transverse	.011
Diameters crown M. iii	anteroposterior	.006
	transverse	.008
Length of mandible from incisive border to edge of angle		.122
Length from incisive border to anterior basis of coronoid process		.074
Length of base of coronoid process		.033
Length of symphysis (oblique)		.043
Depth at P–m. iii		.023
Depth at M. iii		.023
Depth at condyle		.037
Width between inferior canines at base		.007
Diameter of base of inferior canine (long)		.010
Length of molar series from canine		.058
Length of premolar series		.029
Diameters crown P–m. iii	anteroposterior (oblique)	.0075
	transverse (oblique)	.006
Diameters crown P–m. iv	anteroposterior	.010
	transverse	.007
Diameters M. i	anteroposterior	.011
	transverse	.0085

The internal roots of the superior molars are produced inwards, giving the crowns a strong internal support and extension at the base. The enamel is best developed on the internal side, while on the inferior molars it extends further on the external side.

A portion of the frontal bone preserved shows that the interorbital region is flat, and that there is a strong sagittal crest. The chambers for the large olfactory lobes fall below the anterior part of the sagittal crest. The glenoid cavity of the squamosal bone is wide anteroposteriorly, and has no anterior crest. The postglenoid foramen is rather large. The anteorbital foramen issues above the posterior part of the third superior premolar.

The proximal end of the humerus displays a well-developed, rather flat, greater tuberosity (with a distinct teres facet), which is continued as a prominent bicipital crest to the lower part of the shaft. The distal extremity exhibits an epitrochlear foramen. A part of the shaft of the ulna is very robust; both sides are grooved. The shaft of the tibia is quite slender, with the long diameter anteroposterior, the posterior edge acute on the inferior third. The edge turns outwards, forming the posterior edge of the wide fibular face. Internal malleolus prominent distally, most so posteriorly. The astragalar face is oblique and nearly flat, or slightly concave, having even less excavation than that of most creodonta.

Measurements of limbs.

		M.
Least diameters of humerus	anteroposterior	.009
	transverse	.014
Depth shaft of ulna at middle		.012
Diameters shaft tibia	anteroposterior	.009
	transverse	.006
Diameters extremity tibia	anteroposterior	.011
	transverse	.015

This species was robust in its characters, and evidently lived on hard food; its strong and worn canines show that they had more than the usual use. In its dentition it stands nearer the creodonta than does any other member of the group. It was probably a burrower.

CONORYCTES CRASSICUSPIS Cope.

Plate XXIII d; fig. 6.

The posterior part of a mandibular ramus supporting the last two molar teeth indicates a second and larger species of the genus The ramus is one-half deeper than that of the *C. comma*, and the second true molar is much larger than in that species. The last true molar is much smaller than the penultimate, and consists of three anterior cusps and a longer heel. The former are obtuse, the external the larger, the internal equal, the anterior on the inner edge of the crown. The heel sustains a low conic tubercle. Measurements will be given in the explanation of the figure above cited.

From the Puerco beds of northwestern New Mexico.

ESTHONYX Cope.

Report Vert. Foss. New Mexico, U. S. Geog. Survs. W. of 100th Meridian, 1874, p. 6; *Id.*, Ann. Report U. S. Geog. Survs. W. of 100th Meridian, 1874, p. 118; System. Cat. Vert. Eocene, New Mexico, U. S. Geog. Survs. W. of 100th Meridian, 1875, p. 23; Report U. S. Geog. Survs. W. of 100th Meridian, vol. iv, pt. ii, 1877, p. 153.

The dental formula is, I. $\frac{2}{3}$; C. $\frac{1}{1}$; P-m. $\frac{?\,3}{3}$; M. $\frac{3}{3}$.

Incisors of two forms, the inferior subgliriform, but not growing from persistent pulps; the enamel covering a long and narrow external vertical face, and terminating above the alveolus, thus distinguishing crown and root. The superior incisors with the apex incased in enamel, which extends much farther on the outer than the inner side; the crown compressed, not wider than the root. The first superior incisor is large, and the crown is somewhat spoon-shaped. The second incisor is as robust as the first, but the crown is shorter. The second premolar has one external and one internal lobe, in the third (fourth) premolar these lobes are much enlarged, and the tooth is transverse. The true molars have two external cusps, which are flattened, close together, and well within the margin of the base of the crown. There is one internal lobe at the junction of two ridges, which inclose a triangular area with the connected bases of the external two cusps; and a strong posterior ledge, as in the opossums. Of the inferior incisors, the median is large and half gliriform, while the first and third are small. The inferior, like the superior canines, are large. The first and second (third) premolars have no internal lobes, but the second (third) has a heel. The fourth is more or less like the first true molar.

The inferior molars support two V's, with rounded apices directed outward, the posterior soon wearing into a triangle lower than the anterior. The anterior is elevated and transverse, only distinguished from a triangle by a notch on the inner side. Last lower molar with this anterior transverse triangle, a diagonal ridge, and a heel with raised border. The fourth premolar has a V-shaped crest on its anterior half, the angle being an elevated apex of the external face, the limbs descending inward.

This genus differs from *Anchippodus* and *Ectoganus* in the far less gliriform character of the incisor teeth, although the composition of the molar teeth exhibits a true resemblance to that seen in those genera. The incisor is annectant to the form usual in Mammals, betraying the rodent character in the absence of enamel from the posterior face, and the oblique bevel posteriorly from the apex to the shank. The ? canine or superior incisor (second form) is elongate, and without distinction between crown and root, but is straight, and not gliriform. A resemblance to the superior incisor of *Ectoganus* can be observed in the deep emargination of the enamel to near the apex on the inner side, and the convexity of the opposite side.

A strong resemblance can be discovered between some characters of this genus and *Tomitherium*, which is described under the *Mesodonta*. The composition of the inferior molars in the latter is essentially the same in the two genera, but the anterior cusps and yokes are relatively less developed in *Tomitherium*. An obvious resemblance is seen in the last premolar, which is somewhat sectorial in the form of its anterior half in both genera. There is no enlarged external incisor in *Tomitherium*, but either arrangement is consistent with mesodont affinities, and even incisors of rodent-like character, in view of the structure of *Chiromys*. Its resemblances to *Erinaceus* are, however, so many that I leave it in the neighborhood of that genus. The premolars of the superior series are nearly alike in the two genera, and so are the inferior molars, excepting the last, which is much smaller in *Erinaceus*. The superior molars of the hedge-hogs only differ in the development of the posterior cingulum into a posterior internal cusp, which is connected with the posterior external cusp by a transverse ridge.

It is probable that this genus represents the group ancestral to the existing *Erinaceidæ*.

The following points may be derived from parts of a skeleton of *E. burmeisteri*. The processus dentatus of the axis is short and obtuse, and has an oval section. The dorsal vertebræ are much smaller than the cervicals, and the tubercular articulation is on a long diapophysis. The caudals are very large. The scapula has a prominent coracoid hook. The manus has five digits. The scaphoid is larger than the cuneiform, and is distinct from the lunar. The trapezium is large and distinct; unciform large. The phalanges are like those of *Creodonta*. The ilium has a narrow plate with strong external ridge, which makes the section of its peduncle an equilateral triangle. There is a large anterior inferior spine.

The distinctness of the scaphoid and lunar bones and the five digits of the manus show that this genus cannot be referred to the family of the *Erinaceidæ*, but to belong rather near to *Solenodon*, and perhaps within the boundaries of the *Centetidæ*.

The specimens show that my original determinations of the incisors based on loose teeth were correct. They also show that this genus is not far removed from the *Creodonta*.

There are several species of the genus, which I define as follows:

I. Fourth inferior premolar like first true molar.

Larger; third superior premolar larger; fourth premolar with the external cusp bilobate .... *E. acutidens.*

Medium; third superior premolar smaller; fourth premolar with external cusp simple; superior incisors wide; large inferior narrower .... *E. burmeisteri.*

Medium; superior incisors narrow; large inferior wider .... *E. bisulcatus.*

II. Fourth inferior premolar with anterior V open and cutting.

Smallest; incisors unknown .... *E. acer.*

A species of the size of *E. acer* has been named *E. spatularius*, but I cannot place it in the above key, as the premolar and incisor teeth are unknown. The section II approximates nearer the genus *Conoryctes* than section I.

ESTHONYX BURMEISTERI Cope.

Report Vertebrate Foss., New Mexico, 1874, p. 7: Report U. S. G. G. Surv. W. of 100th Mer., G. M. Wheeler, iv, ii, p. 156, pl. xi, fig. 26.

Plate XXIV*b*, figs. 1–10.

A fractured cranium exhibits the entire dentition of this species, and gives the characters satisfactorily. The anterior premolars are wanting.

The third premolar consists of an external strongly compressed lobe and an internal keel or wide cingulum. The posterior third of the external lobe is reflected outwards, inclosing a groove with the anterior part. The latter rises into an apex, which is separated by an open notch from the posterior portion. The latter has a superior convex edge, and both lobes together present a convex face inwards. The internal lobe is low and sends a ledge continuous with its apex along the posterior base of the crown. Its prominence inwards gives the base of the crown a triangular outline.

The fourth premolar is generally similar to the true molars in its transversely elongate form, and the production outwards of the anterior external base of the crown as well as the posterior external angle. It differs from them in that the principal external cusp is single and not double, and in the absence of a well-defined posterior internal ledge. On this account the internal angle is more nearly median than in the true molars. There are rudimental anterior and posterior cingula on the inner part of the crown. The principal cusp is compressed and acute, and stands well within the ear-like external expansions of the base of the crown. In the true molars there is a rudimental anterior cingulum. The principal cusps are compressed and well separated, though connected at their bases; those of the first not more so than those of the second, contrary to what is seen in *E. acutidens.* The posterior external ear-like lobe is a little longer than the anterior in the first molar; a little smaller in the second, and much smaller in the third. Enamel smooth.

The inferior molars do not differ much from the corresponding teeth of *E. bisulcatus.* The anterior limb of the anterior V is not so elevated as the posterior limb, but more so than the posterior V. On the last molar the intermediate lobes are well developed, but the external is much the larger. The anterior V of the fourth premolar has not as great transverse extent as the anterior V of the true molars, but it is well developed, and not a cutting lobe, as in *E. acer.* The third premolar has an anterior cutting lobe and a posterior keel with a cutting edge continuous with that of the principal lobe. The anterior edge of the latter is convex and rises to the acute apex. The second (first) premolar is two-rooted; crown lost on both rami. The inferior canine has a robust base, and issues close to the first premolar. Crown broken.

off. The third incisor is a small tooth with small truncate crown, and enamel on the external face only. The second incisor is compressed, and has an elongate crown with the anterior face convex in both directions. The posterior edge is beveled like that of a rodent, but has a thin investiture of enamel, which is worn away by use on a low median ridge. The sides of the crown are covered with enamel. The external side is regularly convex; the inner one is flattened. The first incisor is much smaller than the second, and the section of the crown is sub round, and narrowed to an angle posteriorly. It is as long as the second incisor, and therefore much longer than the third.

The symphysis is coössified and is nearly horizontal in its direction, its plane making an acute angle with the alveolar line. As a consequence the incisor teeth are directed forwards at the base, curving slightly upwards at their summits.

Measurements.

		M.
Length of posterior five molars		.047
Length of true molars		.026
Diameters p.-m. iii	anteroposterior	.008
	transverse	.008
Diameters p. m. iv	anteroposterior	.008
	transverse	.010
Diameters m.-ii	anteroposterior	.009
	transverse	.013
Length of inferior dental series (axial)		.074
Length of inferior true molars		.028
Length of inferior premolars		.021
Diameter of base of canine		.007
Length exposed anterior face m. ii		.0115
Width posterior face I. ii		.0035
Width posterior face I. iii		.0020
Diameters p.-m. iii	anteroposterior	.0070
	transverse	.0040
Diameters m. ii	anteroposterior	.008
	transverse	.0065
Diameters m. iii	anteroposterior	.0110
	transverse	.0060
Diameters crown superior I. i	long (oblique)	.0080
	short	.0055
Length of alveolus of I. ii		.0090
Depth of mandibular ramus at m. iii		.024
Depth of mandibular ramus at p. m. ii		.014
Length of alveolar border of premaxillary bone		.020

This species is well distinguished from the *E. acutidens* in the characters of the posterior premolar teeth of the superior series. In the latter, the external lobe of the third superior premolar is much longer, while the internal lobe is smaller than in the *E. burmeisteri*. The external lobe of the fourth has two lobes instead of one. I mention here that the loose tooth of the *E. acutidens*, which I figure Plate XXIV*a*, figs. 17–17*a*, as a first superior premolar, may not really be such, but may be a second superior incisor.

From *E. bisulcatus*, which I discovered in the Wasatch region of New Mexico, this species is separated by the much longer superior incisors and smaller inferior second incisors.

Part of the skeleton of a second specimen includes the superior premaxillary bone with the second incisor of the right side; parts of the mandible with three incisors and the posterior four molars; vertebræ from various parts of the column; parts of the scapula humerus, manus, pelvis, ribs, and fibula. The generic characters observable in this specimen have been already recited.

The atlas is characteristic. The transverse process is on the posterior edge of the vertebra, and its inferior edge is narrow. It is perforated from before upwards and backwards by the vertebral canal. The lateral canal pierces the neural arch at the middle of its external border. The anterior border of the arch is notched medially. No neural spine. The axis is quite robust and the neural arch is large. The neural spine is elevated and has not an elongate base. The paradiapophysis is small. The inferior median line is keeled. The processus dentatus is constricted at the base above by a groove; its superior face is quaquaversally convex. The posterior articular face is wider than deep, and is oblique. A more posterior cervical centrum has oblique transverse articular faces about as wide as the body is long. The inferior face has a median keel with a concavity on each side. The centra of two dorsals, an anterior and a posterior, are depressed. The inferior median line is a flat band, which widens posteriorly. The two venous foramina of the neural floor are large. The caudal vertebra associated with the other specimens has no neural arch. It is very large, more than twice as long as the longest dorsal and one half longer than the axis with odontoid process. It has a ridge on each side, which terminates in a short process posteriorly, and a median ridge below, which terminates behind in a double tuberosity. If this vertebra belongs to this skeleton, the animal had an unusually long tail.

Measurements of vertebræ.

			M.
Diameters of atlas	vertical		.018
	anteroposterior at side of arch		.009
	transverse without diapophysis		.021
Diameters of axis	longitudinal,	with odontoid	.0205
		without odontoid	.016
	vertical to base neural spine		.016
Diameters posterior face centrum axis.	transverse		.010
	vertical		.007
Diameters of anterior dorsal centrum.	anteroposterior		.009
	vertical		.0055
	transverse		.009
Diameters posterior dorsal centrum.	anteroposterior		.0095
	vertical		.006
	transverse		.0085
Diameters caudal centrum.	anteroposterior		.0285
	posteriorly,	vertical	.008
		transverse	.009

The glenoid cavity of the scapula is an oval, narrower next the coracoid, and not produced at that extremity into a tuberosity. In the manus the cubito-carpal articulating surface of the carpus is not very convex in any direction. The proximal surfaces of the scaphoid and lunar bones are concave. The two proximal surfaces of the cuneiform are about equal; that for the pisiform the longer by a little. The scaphoid's greatest diameter is transverse. The transverse is the shortest of the lunar. The sizes of the anterior faces of the carpals of the second row are, in diminishing order, the unciform, the trapezium, the trapezoides, and least the magnum. The trapezium is wider than long, and extends distadt over the proximal extremity of the second metacarpal. The proximal extremity of the second metacarpal is concave; that of the fourth is convex. The fifth has a proximal external tuberosity. Its shaft is proximately as stout as those of the others, which are of subequal proportions, but the second and fifth contract more rapidly.

Measurements of fore limb.

		M.
Width of extremity of scapula with coracoid		.019
Diameters glenoid cavity	anteroposterior	.013
	transverse	.009
Anteroposterior diameter humerus below head		.013
Width of carpus		.021
Length of carpus alt trapezoides		.008
Diameters scaphoid	anteroposterior	.0055
	vertical in front	.0045
	transverse	.0085

		M.
Diameters lunar	anteroposterior	.007
	vertical in front	.006
	transverse	.005
Diameters cuneiform	anteroposterior	.006
	vertical in front	.004
	transverse	.008
Proximal widths of metacarpals	second	.005
	fourth	.005
	fifth	.0055

The ilium is slightly concave longitudinally on the external face. The inner border is abruptly contracted near the origin of the os pubis. The anterior inferior spine is prominent and extended with the length of the ilium; it is four times as long as wide. The groove of the acetabulum is within the anterior margin of the ischium. The posterior edge of the pelvis opposite to the acetabulum is not thinned nor expanded. A bone which may be the radius, or possibly the fibula, has marked peculiarities. The distal portion has for a considerable distance on the inner side a median keel with a gutter on each side. The keel expands into the triangular articular facet. The rest of the shaft has an oval section. The external face of the head has two flat tuberosities, of which the external is much the larger, and has a concave posterior border. This bone is apparently too stout for the fibula and too slender for the radius. As one extremity is lost I cannot be positive as to its position. Its articular face fits the scaphoid and lunar bones. The inner face of the radius in *Erinaceus* is excavated, but has no keel.

Measurements.

	M.
Width of neck of ilium	.011
Width of acetabulum	.015
Width of neck of ischium	.015
Width of (?) distal facet of (?) radius	.011
Diameter of shaft of (?) radius near middle	.006

The teeth are like those of the specimen first described. The second superior incisor has three sides of the crown; a large extero-anterior convex, a wider interior, and narrower posterior, plane. The true molars measure .026 on their bases, and the ramus is .023 deep at the anterior part of the last molar.

Several fragments of jaws were found by Mr. Wortman, besides the type specimen described, which are probably referable to it. All are from the Big Horn bad-lands of Northern Wyoming. No specimens of the *E. acutidens* have been identified from that region, but several present the measurements of the *E. spatularius*.

ESTHONYX ACUTIDENS Cope.

Bulletin U. S. Geol. Surv. Terrs. F. V. Hayden, 1881, p. 185.

Plate xxiv *a*, figs. 17–21.

The largest species of the genus, and represented by two individuals. The first of these includes the last molars of both series and an anterior true molar; the second includes most of the dentition of one maxillary bone, the last two true molars being probably the only teeth missing. Four of the molars of this specimen are in place, and three are loose. Under the circumstances, I estimate six molars, of which the fourth premolar is like the first true molar, and the third premolar has its internal lobe very much reduced. The two preceding premolars have one root, and short, compressed, and acute crowns. The second is abruptly very much smaller than the third, and is close to it; the first is close to the second, and is a little larger. The canine is larger still, and is somewhat compressed. Externally viewed, it looks like the canine of a carnivorous mammal; but viewed from within, it displays marked peculiarities. It has here a median rib, separated from the fore and aft edges of the crown by a groove on each side. This ridge is without enamel, and the edges are produced and very sharp. The enamel of the external face extends twice as far towards the base as on the interior side. The enamel of this tooth, with that of the premolars, is wrinkled; that of the molars is smoother. The two external cusps of the last premolar are closer together than those of the true molars. The posterior part of the external basal cingulum of the second and third true molars is more prominent than the anterior part.

The details of the inferior teeth preserved do not differ much from those of the *E. bisulcatus*, excepting that the heel of the last true molar is much more produced.

The *E. acutidens* is considerably larger than either of the species of the genus heretofore described.

Measurements.

No. 1.		M.
Diameters of last inferior molar	vertical	.0065
	antero-posterior	.0130
	transverse	.0064
Diameters of a true molar	antero-posterior	.0095
	transverse	.0074
Diameters of last superior molar	antero-posterior	.0097
	transverse	.0130

No. 2.

	M.
Length of five superior molars preserved	.0410
Length of premolar series	.0325
Length of bases of Pm. I and II	.0125
Diameters of Pm. III { antero-posterior	.0097
Diameters of Pm. III { transverse	.0098
Diameters of first true molar { antero-posterior	.0086
Diameters of first true molar { transverse	.0133
Antero-posterior width of base of crown of canine	.0080
Transverse width of base of crown of canine	.0050

Wind River Basin, Wyoming. J. L. Wortman.

ESTHONYX SPATULARIUS Cope.

American Naturalist, 1880 (Nov. 25), p. 908.

Plate xxiv *a*, figs. 22–25.

Represented by five molar and premolar and two incisor or canine teeth, apparently belonging to one individual. These are about the size of those of *E. bisulcatus*, but present several differences of detail. Thus the basin of the heel of the last inferior molar is not obliquely cut off by a crest which extends forwards from the heel, but is surrounded by an elevated border, which rises into a cusp on the external side. The incisor-canine teeth are more robust than those of *E. bisulcatus*, one of them especially having a spoon-shaped crown, with the concave side divided by a longitudinal rib, on which the enamel is very thin. The enamel descends much further down on the external than the internal side of these teeth. The rodent-like tooth does not accompany the specimen. Length of base of last inferior molar, .009; width anteriorly, .005; length of crown of canine incisor No. 1, .009; width at base, .005; length of crown of second canine incisor at base, .012; width, .006.

Basin of the Big Horn River, Wyoming. J. L. Wortman.

MESODONTA.

Since 1872 the Eocene formations of the Rocky Mountains have been known to contain the remains of numerous species of Mammals which possess greater or less proportions of characteristics of the order *Quadrumana*. Some of these were referred by their first describers to the *Insectivora*, and others to the *Ungulata*. In October, 1872, the writer described a genus,

Anaptomorphus, represented by a jaw found in the Bridger beds of Wyoming, in whose dentition Quadrumanous characteristics are so marked as to have induced me to compare it with such typical forms as *Simia*. The characters of the mandibular dentition then recorded are those of the true monkeys, but the permanent separation of the mandibular rami distinguishes the genus from these and from the marmosets, constituting a resemblance to the lemurs. The dental formula is I. 2; C. 1; P. M. 2; M. 3; the crowns of the premolars with a single, undivided, compressed tubercle. In the following year I published (May 6, 1873) a second paper, in which the characters of *Anaptomorphus* and of the earlier described *Tomitherium* (Cope) were more fully elaborated. In this essay I referred* the latter genus also to the *Quadrumana*, but as expressing a type even more aberrant than the Lemurs, and therefore well separated from the true monkeys. I cited, as reasons for this reference, the flat ilium, the long femur, the round head of the radius, the form of the distal end of the radius, with the coössified symphysis and four transverse incisors of the lower jaw. I pointed out that the forms of the molars are similar to those of the *Quadrumana*, and to animals of some other orders as well, while the number of molars is greater than in the Lemurs, or any other known group of the order. The formula of the mandible is I. 2; C. 1; P. M. 4; M. 3. I also pointed out the resemblance between this genus and *Hyopsodus*, which was then estimated as Ungulate, but which has since been stated to be Lemurine. Finally, I added to this series, in the same year,† the genus described by Leidy as *Notharctus*, and a fourth species, which belongs to the genus *Pantolestes* Cope.

In the Actes of the Linnæan Society of Bordeaux for 1873,‡ M. Delfortrie published a description of the cranium of a Mammal which he named *Palæolemur betillei*, which he referred to the Lemurs, pointing out certain differences. He gave a number of characters which he deemed sufficient reasons for such a course. Chief among these are the completed orbits, directed partially forwards, which are associated with elongate nasal bones, large petrous bone, and acutely tubercular molars. M. Delfortrie also points

*On the Primitive Types of the Mammalia Educabilia.

†Annual Report U. S. Geol. Surv. Terrs., 1872 (pub. 1873), p. 549.

‡Ann. Sc. Géol., t. iv, No. iv, p. 18, pl. vii, viii, 1874, and Journal de Zoologie, iv, p. 464.

out that the dentition differs from that of the known *Lemuridæ* in the more numerous premolars, giving the following formula: I. ?2; C. 1; P. M. 4; M. 3; or the same as that of *Tomitherium*.

At the time of the discovery of *Anaptomorphus*, Prof. O. C. Marsh expressed the opinion* that some of the forms noticed by him in the Bridger formation of Wyoming are allied to the Lemurs. He, however, did not state the characters which led him to entertain this opinion, nor did he give such descriptions as would enable the anatomist to judge of its correctness. Up to the present date no more complete account of these animals has appeared.

The history of discovery of the European forms of this group is similar to that of our own, in respect to the difficulty at first experienced by paleontologists in referring them to their proper systematic position. The investigations conducted by Cuvier during the early part of this century into the extinct *Vertebrata* of the Eocene of the neighborhood of Paris revealed, among other types, the genus *Adapis* (Cuv.). This he referred to the Ungulates and to the neighborhood of *Anoplotherium*. Laurillard and Blainville believed that its affinities were to the *Insectivora*. The above-mentioned discovery by M. Delfortrie, of Bordeaux, of the greater part of the cranium, at Bebuer (Department of Lot), of his *Palæolemur betillei*,† led him to announce that Lemurs inhabited France during early Tertiary times. This was in confirmation of the opinion of M. Rütimeyer, who had already described a *Coenopithecus lemuroides* from the Eocene of Switzerland. But MM. Gaudry and Gervais, on further investigation, came to the conclusion that the *Palæolemur* is the *Adapis* of Cuvier, and that *Aphelotherium* Gerv., and *Cœnopithecus*, are also synonyms of it. And they are disposed to accede to the conclusion of Delfortrie as to its affinities. Subsequently M. Filhol established for this genus, and a new one which he called *Necrolemur*, a family, the *Pachylemuridæ*, adding a new species, *Adapis magnus*. In this paper he recognizes the characters pointed out by previous authors as allying this family to the *Lemuridæ*, as well as the higher dental formula which

*Amer. Jour. Sci. and Arts, Oct. 8, 1872.

†Actes de la Société Linnéenne de Bordeaux, xxix, 1873. The separate copies of this paper are dated May 25, 1873, while a supplement attached to the last page is dated September 4, 1873.

distinguishes it, and adds some important characters which are strongly marked in the genus *Adapis*. He finds the cranium to be strongly contracted just behind the orbits and at the pterygoid plates, in a manner unknown to existing *Lemuridæ*.

Subsequent to the above dates the number of known species of these puzzling Eocene *Mammalia* has been increasing, and the Wheeler expedition of 1874 added a number of genera and species to those previously known. An illustrated account of these was published in the final report of that organization, Vol. IV.

I have seen no reason to modify the view originally expressed as to the Quadrumanous affinities of *Anaptomorphus;* on the contrary they have been placed beyond doubt by the discovery of the entire cranium by M. Filhol, of his genus *Necrolemur* above mentioned, which is very similar to *Anaptomorphus*, in the parts of both which are known, the mandible and its dentition. Additional light was thrown on the structure of *Tomitherium* by my researches in New Mexico, conducted in 1874. The fragments of the skeletons of two species of a closely allied genus, *Pelycodus*, were found, which include numerous bones of the tarsus, and these are identical with corresponding parts in the *Creodonta* and different from those of the *Lemuridæ*. The astragalus extends anterior to the shortened calcaneum, and the navicular is short and the cuboid not elongate. The superior aspect of the astragalus presents two oblique surfaces, one for the internal malleolus, the other for the transverse facet of the tibia. The portions of femur, including the third trochanter, the proximal part of the ulna, and the distal portion of the humerus, are all closely similar to those of the *Creodonta*. The type specimen of *Tomitherium* includes some parts of the skeleton not present in the New Mexican species. Thus the ilium of *T. rostratum*, while furnished with the prominent anterior inferior spine of the *Creodonta*, is flattened toward the crest, and is not angulate on the external face. The femur is furnished with a very elevated third trochanter, which is opposite to the little trochanter, as in *Chiromys* and *Talpa*, and not low down, as in *Creodonta*. The head of the radius is rounder than in *Creodonta*. The skeleton of *Tomitherium*, in fact, bears a strong resemblance to that of *Chiromys*, leaving the skull out of view.

The skeleton of the New Mexican form includes an entocuneiform, like that of *Stypolophus hians*, which indicates a non-opposable hallux.

It is apparent that the supposed Lemurine *Mammalia* of the type of *Tomitherium*, which have the formula of the molar teeth 4–3, cannot be separated by ordinal distinction from the *Creodonta*. They differ from them, it is true, in their wholly tubercular molar teeth, but in this relate to them as the Bears and *Procyonidæ* do to other *Carnivora*. I have, therefore, proposed* to constitute these a distinct group or suborder, intermediate in position between the *Creodonta* and the *Prosimiæ*, under the name of the *Mesodonta*. I cannot now find characters by which to distinguish this division from the *Insectivora* as an order.

In my report to Dr. Hayden on the paleontology of the Bridger Eocene of Wyoming,† I included six species, viz, *Tomitherium rostratum*, *Pantolestes longicaudus*, *Sarcolemur furcatus*, *S. pygmæus*, *Hyopsodus vicarius*, and *H. paulus* Leidy, which belong to this suborder. As many species of *Mesodonta* referred to various orders are described by Dr. Leidy in his quarto report in the same series. In my report to Lieutenant Wheeler on the Vertebrata of the Eocene of New Mexico obtained by the expedition of 1874, eleven species are included, none of which had been certainly obtained from the Bridger beds.

A synopsis of the genera is given below, in which the characters are derived from the dentition of the lower jaw, the part usually preserved. While considerable variety is to be observed in the structure of the teeth, they furnish also close approximations, so that their discrimination requires careful scrutiny.

The genus *Anaptomorphus* Cope, although included in the synopsis, probably belongs to the *Prosimiæ*.

I. Last true molar with cusps in opposing pairs.
  A. Anterior inner cusps, two or a double one, on some of the molars.
    * Three premolars.
      "Last premolar without inner tubercle" (Leidy), like the other premolars .... *Omomys*.
      Last premolar with inner tubercle like the molars .... *Microsyops*.
    * * Four premolars; last molar heeled.
      Last premolar without inner tubercle; premolars two-rooted.. *Pantolestes*.

* Proceed. Acad. Phila., 1876, p. 88. Report U. S. Geol. Surv. West of 100th Mer., iv, p. 80, 1877.
† Annual Report U. S. Geol. Surv. Terrs. for 1872, 1873, pp. 546, 607.

Last premolar with inner cusp; all the molars with basin-shaped heel behind; second premolar one-rooted .... *Tomitherium.*
Like *Tomitherium,* but second premolar two-rooted .... *Pelycodus.*
Last premolars with inner cusp; all the molars with elevated cusps behind; second premolar two-rooted .... *Sarcolemur.*

AA. The anterior inner cusp undivided on all the molars.
Last molar with heel; last premolar with inner cusp; four premolars .... *Hyopsodus.*
Last molar without heel.
Premolars four, last large; sectorial, without inner cusp...... *Apheliscus.*
Three or two premolars; last without inner cusp.......... *Anaptomorphus.*

II. Last molar with a longitudinal series of alternating cusps, including a heel.
Tubercles of molars 1–2, alternating .... *Adapis.*
Tubercles of molars 1–2, opposite .... *Opisthotomus.*

MICROSYOPS Leidy.

Proceed. Acad. Phila. 1872, p. 20 (characterized). Report U. S. Geol. Surv. Terrs. i, p. 82, 1873. Cope, Report Expl. and Surv. W. of 100th Mer., G. M. Wheeler, iv, pt. ii, p. 134, 1877.

This genus is easily distinguished from its allies by the absence of the first inferior premolar, and probably of the superior first premolar also. The canine tooth of the lower jaw of the *M. scottianus* and *M. gracilis* is very large, so that I suspect that when the dentition is fully known there will be found to be a deficiency in the number of the inferior incisors. The dentition is otherwise as in *Sarcolemur, i. e.,* the last premolar with internal cusps, and two anterior internal cusps of the true molars.

The three species are distinguished as follows:

I. Heel of last inferior molar very short.
Length of true molar series m.0077............................... *M. spierianus.*
Length of true molar series m.0172 .... *M. gracilis.*

II. Heel of last inferior molar long.
Length of true molar series m.0135........................... *M. scottianus.*

MICROSYOPS SPIERIANUS Cope.

American Naturalist, 1879, p. 908.

Plate xxv *a*, fig. 8.

Established on a portion of a mandibular ramus which contains the three true molars in perfect preservation. As the number of premolar teeth is unknown, its reference to this genus is provisional only. The last true molar has the form of that of the *M. gracilis.* It is distinguished by its small size, since it is considerably less than the *Hyopsodus vicarius* (*H.?*

minusculus), and by the equality in size of the molars. The heel of the third molar is very small, and the two cones of the inner side of the crowns of all the molars are acute. The external crescents are very well defined, the anterior sending a horn round the anterior extremity of the crown. The posterior is connected with the corresponding internal tubercle by a median conic posterior tubercle. Length of true molar series, .008; length of second molar, .0026; width of second molar, .0022; length of last true molar, .0025; width of last true molar, .0016; depth of ramus at second molar, .0043. Dedicated to my friend Mr. Francis Spier, of Princeton, N. J., who, in connection with Messrs. Scott and Osborne, has made important additions to our knowledge of the Eocene *Vertebrata.*

Valley of the Big Horn River, Wyoming. J. L. Wortman.

MICROSYOPS ELEGANS Marsh.

Limnotherium elegans Marsh, Amer. Journ. Sci. and Arts, 1871, ii, p. 43, *fide* Leidy.
Microsyops gracilis Leidy, Proceed. Acad. Phila. 1872, p. 20.

This, the largest species of the genus, left abundant remains in the beds of the Wind River, Wyoming. The specimens in my possession agree closely with the descriptions and figures given by Leidy. The cusps of the molars are well developed, and the angles bounding the heels are distinct. The fifth lobe or heel of the last true molar is quite small, and is on the inner side of the median line.

Measurements.

		M.
Length of five last molars on bases		.0206
Length of last two premolars on bases		.0070
Diameter of base of fourth premolar	anteroposterior	.0030
	transverse	.0030
Diameters of second true molar	anteroposterior	.0040
	transverse	.0035
Diameters of last true molar	anteroposterior	.0050
	transverse	.0035
Depth of ramus at third premolar		.0090
Depth of ramus at third true molar		.0100

MICROSYOPS SCOTTIANUS Cope.

Bulletin U. S. Geological Survey Terrs., 1881, p. 188.

Plate xxiv *a*, fig. 26.

A nearly entire left mandibular ramus is all that I have seen of this species. The crowns of the fourth and sixth molars furnish the only dental characters available, but the number and forms of the bases of the others are readily ascertainable.

The ramus of the jaw is more slender than in *M. gracilis*, and the last true molar has quite a different form. Instead of being shorter than in allied species, this tooth is rather longer, evidently in consequence of a well-developed heel. The fourth premolar has a strong inner tubercle, and no anterior cusp or cingulum. Its heel has an elevated posterior border, inclosing a fossa with the principal cusps. No external or internal cingula. Third premolar with two roots. Alveolus of the second, large and apparently simple; it is filled with matrix. Canine large, directed forwards, and occupying all the space between a short diastema and the symphysis. The latter extends posteriorly to below the anterior part of the third premolar. The ramus is compressed and maintains an equal depth to the end of the molar series. Its inferior border descends below the coronoid process, and is not incurved, but the external face is convex. The anterior masseteric ridge is well marked, descending to below the middle of the ramus. Masseteric fossa flat. Mental foramen below the third premolar.

Measurements.

	M.
Length of the fragment of ramus	.0435
Length of dental series without incisors	.0280
Diameters of canine { antero-posterior	.0040
Diameters of canine { transverse	.0025
Length of premolar series	.0100
Length of fourth premolar	.0040
Width of fourth premolar behind	.0027
Length of true molar series	.0136
Length of last true molar	.0052
Width of last true molar anteriorly	.0030
Depth of ramus at third premolar	.0090
Depth of ramus at last molar	.0090

This species is dedicated to my friend Prof. William B. Scott, of the College of New Jersey.

TOMITHERIUM Cope.

Third Account of New Vertebrata from the Bridger Eocene of Wyoming, p. 2, August 7, 1872; Proceedings American Philosophical Society for 1872 (published January, 1872); On the Primitive Types of Mammalia Educabilia, 1873, May 6, p. 2; Annual Report U. S. Geol. Surv. Terrs., 1872, p. 456 (1873).

Dental formula of the inferior series: I. 2; C. 1; Pm. 4; M. 3. The last molar has an expanded heel. The third premolar consists of a cone with posterior heel; the fourth premolar exhibits, besides its principal cone, an interior lateral one, and a large heel. The true molars support two anterior tubercles, of which the inner is represented by two distinct cusps in one or more of them, and the external is crescentoid in section. The posterior part of the crown is wide and concave, and is bordered at its posterior angles by an obsolete tubercle on the inner and an elevated angle on the outer side. In the *T. rostratum*, the type of the genus, the middle incisors have transverse cutting-edges.

This genus is allied to *Adapis* Cuvier, of the French Eocene, but differs in the possession of but two incisors on each side; in *Adapis*, there are three, according to Filhol. From that genus and *Opisthotomus*, it differs also in the structure of the last inferior molar, as exhibited in the analytical table.

An account of the osteology of this genus, so far as indicated by my material, was given in the papers above referred to. It was shown that the hind limbs, especially the femur, are quite elongate, more so than the fore limbs, and that the proportions of both fore and hind limbs are slender. The head of the radius is subround, and its distal extremity a subequilateral triangle. The humerus is distally expanded, with large inner and outer epicondyles and an arterial foramen. The tuberosities of its head are small. The ilium is rather narrow and flat, except at the acetabulum, where it supports a large anterior inferior spine.

The first impression derived from the appearance of the lower jaw and dentition, and from the humerus, is that of an ally of the coati (*Nasua*). The humerus, indeed, is almost a fac-simile of that of *Nasua*, the only difference being a slight outward direction of the axis of the head. The same bone resembles also that of many marsupials, but the flat ilium, elevated position of dental foramen, and absence of much inflection of the angle of the lower jaw, etc., render affinity with that group highly improbable. The length of the femur indicates that the knee was entirely free from the body as in the Quadrumana, constituting a marked distinction from anything known in the *Carnivora*, including *Nasua*. The round head of the radius indicates a complete power of supination of the fore foot, and is different in form from that of *Carnivora*, including *Nasua;* and, finally, the distal end of the radius is still more different from that of *Nasua*, and resembles closely that of *Semnopithecus*.

We have, then, an animal with a long thigh free from the body, a manus capable of complete pronation and supination, and details of lower jaw and teeth quite similar to that of the lower monkeys. The form of the humerus and its relative length to the femur are quite as marked as in some of the lemurs. The most marked difference is seen in the increased number of teeth; but in this point it relates itself to the other *Quadrumana*, as the most ancient types of *Carnivora* and *Ungulates* do to the more modern; *e. g.*, *Hyænodon* to the former and *Palæosyops* to the latter.

Through the great kindness of Dr. Filhol I came into possession of a mandibular ramus of the *Adapis parisiensis* Cuv., with nearly entire dentition. The specimen was derived from the Phosphorite beds of central France, whose remarkable fauna has been so fully elucidated through the labors of M. Filhol. I cannot distinguish the characters presented by this jaw from those of the genus *Notharctus* of Leidy, although the species on which the latter was based is distinct from *A. parisiensis*. Until some distinctive character is discovered, I use the oldest name, *Adapis*, for the genus.

In the American Journal of Science and Arts for July, 1871, Prof. O. C. Marsh gave a generic name, *Limnotherium*, which was accompanied by a description in which the characters of species and genus were not distinguished, nor were the grounds of separation from other genera previously described, set forth. For these reasons I have been unable to identify the genus, or use the proposed name. Some years later Professor Marsh stated that the genus he thus referred to, is the one I have called *Tomitherium.** As the name proposed by Marsh was not accompanied by a distinct and separate diagnosis I cannot adopt it, although in this instance its author includes in his description a greater number of generic characters than customary with him. Some of these characters are not applicable to *Tomitherium*, if the language of the description is to be literally understood. Thus, the external tubercles of the inferior molars are called *cones*, a term only applicable to the inner tubercles in the present genus. Nothing is said also of the third inner tubercle found in one or more of the molars, a character I think to be generic in this group. Professor Marsh also ascribes large canine teeth to the typical species he describes. They are

*American Journal of Science and Arts, 1875, p. 239.

quite small in the type of *Tomitherium*, but the character may be one of specific value only.

The additional species referred to this genus in my report to Lieut. G. M. Wheeler on the vertebrate fossils of New Mexico belong to the allied form *Pelycodus*.

TOMITHERIUM ROSTRATUM Cope.

Paleontological Bulletin No. 3, p. 2, Aug. 7, 1872. Proceed. Amer. Philos. Soc., 1872, p. 470. Annual Report U. S. Geol. Surv. Terrs., 1872 (1873), p. 548.

Plate xxv, figs. 1-9.

The portions of the skeleton of the type species preserved are: the entire dentition of the lower jaw, minus the crowns of the outer incisor, canine, and first premolar, the left ramus nearly complete, the extreme angle being wanting; the right humerus complete, with right ulna and radius, the latter lacking the distal extremity; a large part of the left ilium; the right femur nearly entire; part of the left humerus, metatarsals, etc.

This species was about the size of the *Cebus capucinus*. The first and second premolar have but one root, the base of the second being about the size of the base of the canine. The latter are cylindric at base. The incisors form a parabolic outline, and have entire edges, the middle pair transverse ones. Enamel generally smooth, premolars somewhat striate; an indistinct inner cingulum.

The mandibular rami are quite stout, but not very deep; the symphyseal portion long and oblique, and the coronoid and condylar portions elevated, with axis at right angles to that of the horizontal portion. The condyle is well elevated, and the coronoid process small; the dental foramen is half way between the margins of the ascending ramus, and opposite the bases of the crowns of the molars. The inferior margin of the jaw shows no tendency to inflection at a point immediately below this foramen, where it is broken off. The mental foramen is divided, the exits being at points opposite the fissures between the premolars 1–2 and 2–3.

The humerus has a round head directed backwards and a little outwards. The tuberosities are rather small, of about equal size, and obtuse; they inclose a short bicipital groove. The bicipital crests are very largely developed, and extend to the middle of the shaft, inclosing an open groove between them. The external is narrow and most elevated, the internal more obtuse and directed inwards. The shaft is thus subtriangular in section. The distal extremity is nearly at right angles to the axis of the proximal, and is much expanded transversely. A large part of this expansion is caused by the truncate internal tuberosity, and by the less prominent external one. The latter is continued in a thin ala, which only sinks into the shaft at its middle. The condyles are small, the external the most prominent. There is a shallow olecranar fossa, and no coronoid, and hence no supercondylar foramen. There is a large arterial foramen above the internal tuberosity.

The *ulna* is compressed, and contracts rapidly towards the distal extremity. The olecranon is broad and obtuse and the humeral cotylus oblique to the long axis. The coronoid process is low. The shaft is curved from right to left (inwards), perhaps by distortion. The *radius* has a discoidal head with central depression, and it was evidently capable of complete rotation. It exhibits a tuberosity and slight flexure below the head. The distal extremity has a horizontal triangular section with the apex internal and truncate; the shaft near it is quite flat. The carpal articulation is a simple, shallow concavity.

The left *ilium* is obspatulate and flat, widest at the convex crest, and slightly concave on the outer side. It is rather thin, and the impression for the sacral diapophyses is elongate. The inferior border thickens gradually to the acetabulum; the superior is excised so as to form an open concavity.

The right *femur* is remarkable for its length. Its shaft is flattened from before backwards, and without flexure. The great trochanter is large, and embraces a deep in-looking fossa. There is a flat tuberosity (third trochanter) looking outwards just below, and the little trochanter is a little below opposite to the latter. The condyles are subsimilar in size, the trochlear surface wide, but not flat, and the inner border thickened and considerably elevated. The femur is 1.75 times as long as the humerus; it was scarcely longer, though a small piece is wanting from the shaft of our specimen.

Measurements.

	M.
Length of entire dental series (straight)	.044
Length of symphysis mandibuli	.020
Depth of ramus at second molar	.010

	M.
Length of crown of second molar	.006
Width of crown of second molar	.0045
Width between two second molars	.014
Width between canines	.005
Width of ascending ramus above dental foramen	.016
Length of humerus	.083
Diameter of head	.013
Diameter of shaft at middle	.0085
Diameter of distal end, transverse	.023
Diameter of distal end, antero-posterior	.0078
Depth of olecranon	.009
Depth of ulna at coronoid	.010
Diameter extremity of radius, proximally	.009
Diameter extremity of radius, distally	.010
Length of ilium from acetabulum	.042
Width near crest	.017
Length of femur preserved	.137
Width just below neck	.017
Width at middle	.011
Width at extremity	.019
Width of trochlea	.009
Longest chord of condyles and trochlea	.019

The following points may be gained by comparison with the skeleton of *Lemur collaris* (catalogue Verreaux). There is considerable resemblance in the details of structure of the molars from the third to the sixth, inclusive. Of course the anterior teeth differ widely in the two, and the last true molar of the *Lemur* has no heel. The principal difference in the humeri is seen in the superior size of the epicondyles of the *T. rostratum*, and the rather more robust character of the shaft. The proximal half of the ulna is deeper, and the olecranon is not so wide in *T. rostratum*. The proximal part of the radius is very similar in the two species, but the distal extremity is in the *T. rostratum* less transversely extended, and thicker anteroposteriorly. There is also much similarity in the ilia. The crest is more extensive in *T. rostratum*, and the inferior border is thinner at its proximal part. Towards the acetabulum the increase in width of this border is similar, and the anterior inferior spine is as prominent. The resemblance between the femora amounts to identity of character; that of the *T. rostratum* is more robust

The remains of this species were found together by the writer in the Bridger beds in an isolated spot on Black's Fork, Wyoming.

Professor Marsh states (*loc. cit.*) that this species is the one he named *Limnotherium affine* in a paper in the American Journal of Science and Arts, the advance copies of which bear date August 7, 1872. This is also the day of publication of the paper in which the name *Tomitherium rostratum* was proposed. Professor Marsh's description is extremely brief, consisting of five lines and six measurements. No fuller description from Professor Marsh's pen has appeared since. An elaborate description of my own specimens appeared May 6, 1873, in a paper entitled "On the Primitive Types of the Orders of the Mammalia Educabilia," which was included in the Proceedings of the American Philosophical Society for that year. My original description was fuller than that of Professor Marsh, consisting of seventeen lines and seven measurements.

PELYCODUS Cope.

Systematic Catalogue of the Vertebrata of the Eocene of New Mexico, Explorations and Surveys west of 100th Meridian, G. M. Wheeler, 1875, p. 13.

Having received considerable accessions to my material, representing species which I have referred to this genus, I am of the opinion that the latter is distinct from *Tomitherium*, with which I formerly united it. The character on which I originally proposed it, the two-rooted second premolar, is constantly present in several species. I may add that the third trochanter of the femur in *Tomitherium* has the elevated position seen in *Lemuridæ*, while in *Pelycodus* it has the position on the middle of the shaft, as in the *Creodonta*.

The superior molars of *Pelycodus tutus* have the following characters: First, one-rooted; second, two-rooted; third, three-rooted. Crown of fourth transverse, with one external cusp, with base antero-posteriorly extended, and one internal cusp. The true molars possess two exterior cusps which are flat on the external face. The first and second have two internal cusps, and the third has but one. Of the inferior molars, the last has a well-developed heel. The anterior inner cusps of one or more of the molars are double. The fourth premolar has an internal tubercle and a heel. The third has no internal tubercle. The heels of the true molars are not bounded by elevated cusps behind. I have pointed out in my report to Captain Wheeler that the tibio-tarsal articulation in this genus is, like that of most of the *Creodonta*, without trochlea.

Comparison of the superior molars of four species which I have referred to *Pelycodus*, with two of *Hyopsodus*, reveal the following characteristic differences. In both species of *Hyopsodus* there are two distinct internal tubercles, and there is no distinct V extending from the intermediate tubercles. That is, the posterior internal tubercle is not connected by a ridge with the anterior inner tubercle. There is no internal cingulum. In *Pelycodus tutus* and *P. frugivorus*, and presumably in *P. jarrovii*, there are two internal tubercles, of which the posterior is distinct from the internal cingulum, and the anterior inner tubercle is the apex of a V, which includes the intermediate tubercles. In *P. angulatus* and *P. pelvidens*, the apex of the V is the only distinct internal tubercle, the anterior. The posterior is a part of a cingulum which extends round the inner base of the crown. In these two species there is no median external tubercle, while in the two *Pelycodi* first mentioned the external cingulum sends up such a lobe between the external cusps. The species of *Pelycodus* may be distinguished as follows:

a. Posterior internal tubercle of superior molars distinct from the posterior cingulum.

Length of true inferior molars on base, $^{m}.019$ .... *P. jarrovii.*
Length of true inferior molars on base, $^{m}.017$ .... *P. tutus.*
Length of true inferior molars on base, $^{m}.015$ .... *P. frugivorus.*

aa. Posterior internal tubercle of superior molars, small, and a process of the posterior cingulum.

Length of true inferior molars on base, $^{m}.024$ .... *P. pelvidens.*
Length of true inferior molars on base, $^{m}.012$ .... *P. angulatus.*

Remains of species of this genus are very common in the Wind River and Big Horn bad lands. They were originally found in the Wasatch beds of New Mexico, and have not yet been announced from the Bridger formation.

PELYCODUS PELVIDENS Cope.

Lipodectes pelvidens Cope, American Naturalist, 1881, p. 1019, Nov. 29.
Plate XXIV d; fig. 3.

The largest species of the genus, represented by a single right mandibular ramus which supports the posterior four molars. The species is readily distinguished from the other members of the genus by the great projection of the heel of the last molar and its constricted form. The fourth premolar is also larger than in the other species, and has a large anterior basal tubercle, which is much better developed than in the other species.

In the true molars the three anterior tubercles are well distinguished, and the anterior, though the smallest, is well on the inner side. The posterior inner cusp is a little larger than the anterior outer. In the first and second true molars the inner margin of the heel is elevated, inclosing a basin-like fossa, and rises into a flat cusp posteriorly. There is a small median posterior marginal tubercle, which runs into a posterior cingulum, and is wanting from the *Deltatherium fundaminis.* The tubercular has the three anterior cusps distinct as in *Didymictis* sp., while the heel is longer than in the known species of that genus. Its external border rises into a prominent cusp with triangular base. The fourth premolar has a small heel on the inner posterior side, and an acute anterior basal cusp. The principal cusp is robust and the basal portion is widely grooved posteriorly (apex lost). True molars with an external cingulum. Enamel obsoletely wrinkled. Length of true molar series, .024; of fourth premolar, .0075; length of last molar, .008; width of heel of second true molar, .005; length of crown of do., .007.

A second specimen of this species includes a mandibular ramus which supported the last five molars, and a maxillary which supported the last four molars, both evidently parts of the same animal. The third inferior premolar, which is wanting in the type specimen, is present here. It consists of an elevated acute simple cusp, which has median anterior and posterior ridges, and a low internal ridge, which separates a lateral plane from a postero-internal plane. There is a rudimental anterior basal lobe and a short heel with transverse posterior edge. No lateral cingula.

A third specimen consists of parts of the maxillary bones of one individual, which support the last four superior molars in better preservation than those of the specimen described above, where two of the four are broken. The two individuals clearly represent one species. The fourth premolar is transverse, and consists of a principal large external cusp and a smaller but well developed acute internal cusp. The external cusp is flush with external base of the crown, and is flanked by a small basal lobe both anteriorly and posteriorly. These lobes are connected with each other by a weak external cingulum, and with the internal cusp by an anterior and a posterior cingula of greater strength. No intermediate tubercles. The true molars may be readily distinguished from those of most of the *Creodonta*, *Deltatherium*, for instance, by the presence of intermediate tubercles. The external cusps are low and well separated, and are not so far within the base as in *Deltatherium* and *Didelphys*, though they are bounded externally by a low cingulum. The latter terminates in two angles, one anterior and one posterior, which send inwards cingula, which meet in a prominent angle interior to the middle of the crown. These cingula support also the intermediate tubercles. Another cingulum arises below the intermediate tubercles, and passes round the inner base of the crown, where it has a truncate outline, owing to the development of tubercular angles anteriorly and posteriorly. The differences between the true molars are as follows: The first is smaller than the second and larger than the third. The first and second have a strong tubercle at the posterior inner angle of the cingulum; in the third the cingulum only is present. The anterior inner angle of the cingulum is much more pronounced in the second than the first molar, and in the third it is not angulate, the inner extremity of the tooth being narrowed oval in outline. In this tooth the posterior external cingular angle is wanting, and the posterior external cusp reduced in size.

Measurements.

		M.
Length of bases of superior true molars		.0185
Diameters P-m. iv	anteroposterior	.0060
	transverse	.0070
Diameters M. ii	anteroposterior	.0070
	transverse	.0085
Diameters M. iii	anteroposterior	.0050
	transverse, at middle	.0062

This species was found by my assistant, Mr. D. Baldwin, in the Lower Eocene, probably Puerco beds of Northwest New Mexico.

PELYCODUS JARROVII Cope.

Systematic Catal. Vert. Eocene, New Mexico. U. S. Expl. Surv. W. of 100th Mer., 1875, p. 13. *Prototomus jarrovii* Cope, Ann. Rep. U. S. Geog. & Geol. Expl. Surv. W. 100th Mer., in report of Chief of Engineers, 1874, p. 126. *Tomitherium jarrovii*, Rep. Expl. Survey W. of 100th Mer., G. M. Wheeler, iv, pt. ii, p. 137, 1877.

Part of a mandibular ramus supporting the last two true molars from the Wind River beds has the dimensions of this species, and is probably to be referred to it.

PELYCODUS TUTUS Cope.

Bulletin U. S. Geol. Surv. Terrs. 1881, p. 187. *Tomitherium tutum* Cope, Report Expl. Surv. W. of 100th Mer., under Capt. G. M. Wheeler, iv, pt. ii, p. 141.

Plate XXV*a*, figs. 1–3.

Jaws of six individuals of this species from the Wind River beds are contained in my collection. Two of these include both rami of the mandible, and two others a greater or less part of the superior dentition.

In one of the latter most of the right premaxillary and left maxillary bones are preserved. The former is light, and supports only two teeth, and although the apex is broken away I do not believe in the existence of a third incisor. My belief is partly based on the wide spaces which separate the posterior incisor from both the anterior incisor and from the maxillary suture. The crown of this tooth, the only one preserved, is small, and is directed obliquely forwards; its posterior face is concave, and the cutting edge is thin. The fourth superior premolar and all the true molars have anterior and posterior basal cingula, which on the last two true molars extend round the inner base of the crown. It extends round the external base of the four molars mentioned. The exterior border of the last molar is rounded, the posterior cusp having consequently a more interior position. Between each pair of external lobes a small tubercle rises from the cingulum, as in most ungulates with external crescents. The anterior internal tubercle is connected with the anterior basal cingulum by an oblique ridge. The posterior inner tubercle is independent, and between it and the posterior external cusp is a low, angular tubercle. Like the posterior inner tubercle, this one is wanting from the fourth premolar. A loose canine from the same specimen has a sharp edge on the posterior border of the crown, which is terminated at the base by a small, acute tubercle. A longitudinal concavity bounds one side of the edge. The infraorbital foramen opens above the posterior root of the third premolar.

On the inferior molars of the same specimen, the second cusp of the anterior inner pair is rudimental on the last tooth, and distinct on the penultimate. The heel of the last molar is large, and its internal border is crenate, and the posterior border is notched. The three other lobes are robust. The lateral angles of the heel of the penultimate molar are prominent. No cingula on inner bases, traces only externally; enamel smooth.

The last two premolars are preserved in the rami of another specimen. Both are rather robust and have wide heels; that of the third very short. The latter has no anterior basal lobe; in the fourth it is a rudiment on the anterior cutting-edge. The latter has an internal tubercle; the third has none.

Measurements.

No. 1.

		M.
Length of superior molar series		.0280
Length of superior premolars		.0150
Length from second (first) incisor to maxillo-premaxillary suture		.0060
Diameters of fourth premolar	anteroposterior	.0048
	transverse	.0070
Diameters of second true molar	anteroposterior	.0060
	transverse	.0080
Diameters of last true molar	anteroposterior	.0050
	transverse	.0065

No. 2.

Length of bases of Pm. III, IV		.0095
Diameters of Pm. IV	anteroposterior	.0050
	transverse	.0040
	vertical	.0038

From collections made by Jacob L. Wortman.

PELYCODUS FRUGIVORUS Cope.

Systematic Catal. Vert. Eocene New Mexico, U. S. Geog. & Geol. Surv. W. of 100th Mer., 1875, p. 14. *Tomitherium frugivorum* Cope, Report U. S. Geog. & Geol. Expl. Surv. W. of 100th Mer., Capt. G. M. Wheeler, iv, pt. ii, p. 144. *Pelycodus nunienum* Cope, Bulletin U. S. Geol. Surv. Terrs. 1881, p. 187.

Plate xxv *a*, figs. 4–5.

This species was not rare during the Wind River epoch of the Eocene in Wyoming. Fragmentary jaws of six individuals of this species were found by Mr. Wortman.

The best preserved ramus supports all the teeth posterior to and including the third premolar. The last-mentioned tooth has an elevated acute crown, without any anterior basal tubercle, and a very short posterior heel. The fourth premolar is very stout; its cusps are not much elevated, and the heel is short. The anterior basal tubercle is quite small. All of the true molars have a second cusp in front of the anterior tubercle, but it is quite small, excepting on the first, where it is more distinct. The external crescents of all the molars are well defined, but the posterior does not inclose the crown behind with an extension of its horn. The last molar is a little longer than the others, and its posterior border is produced into two cusps. A simple raised border is found here in the typical specimen of *P. frugivorus*.

Measurements.

		M.
Length of molar series from third premolar, inclusive		.0228
Length of true molars		.0150
Diameters of first true molar	anteroposterior	.0050
	transverse	.0038
Diameters of last true molar	anteroposterior	.0065
	transverse	.0040
Depth of ramus at Pm. III		.0095
Depth of ramus at last true molar		.0095

The character of the last molar above mentioned distinguishes the four specimens of this species where that tooth is preserved, from the type of *P. frugivorus*. I originally looked on these as representing a distinct species on this account, and called it *P. nunienum*, but further investigation will be necessary in order to ascertain whether this course was justifiable or not.

A series of superior molar teeth accompanying the lower jaws and bones of *Plesiarctomys delicatior*, so strongly resembles those of *Pelycodus tutus* that I am induced to refer them to the *P. frugivorus*, with which they agree in size. They differ from those of *P. delicatissimus* in possessing an external basal cingulum and only one principal external cusp of the fourth premolar. The posterior molar is also abbreviated posteriorly, as in *P. tutus*. As compared with that species, the intermediate tubercles are less distinct, though present, and the inner basal cingula are weaker. The following measurements show the smaller dimensions:

Measurements.

		M.
Length of last four superior molars		.0160
Diameters fourth premolar	anteroposterior	.0036
	transverse	.0055
Diamteters second true molar	anteroposterior	.0050
	transverse	.0070
Diameters third true molar	auteroposterior	.0050
	transverse	.0058

The infraorbital canal is contracted and long, and issues above the third premolar. This proves the fragment not to be rodent.

This species is abundant in the Big Horn bad lands. Mr. Wortman obtained there two entire mandibles and seven separate rami which agree in all respects with the typical specimens. He obtained three mandibular rami in which the molar teeth measure .016 in length; that is, intermediate between the *P. frugivorus* and the *P. tutus*. Of the latter species Mr. Wortman discovered four mandibular rami. One ramus shows a length of .018 for the true molars, which are therefore intermediate in size between the *P. tutus* and *P. jarrovii*. It will be necessary to study other parts of the skeleton in order to ascertain the status of these individuals.

PELYCODUS ANGULATUS Cope.

Systematic Catal. Ent. Vert. Eocene, New Mexico. U. S. Geog. Surv. W. of 100th Mer., 1875, p. 14. Report U. S. Geog. Surv. W. 100th Mer. iv, pt. iii, p. 144, pl. xxxix, fig. 15.

Plate XXIV d; fig. 4.

The *P. angulatus*, originally known from New Mexico, is represented in the Big Horn collection by four mandibular rami, and a portion of a maxillary bone with teeth.

The inferior molars present much the appearance of those of the larger species. The anterior inner cusp is well developed, and nearly on the inner border of the crown, though not so large as the other anterior cusps. The posterior basin-like heel is wide, and is bounded on the external side by an angular crest. A cusp rises from the posterior inner angle, which is separated from the anterior cusp by a deep notch. There is a small posterior marginal tubercle at its external base. The heel of the fourth premolar is wide, and has a low cusp at each posterior angle. There are but two anterior cusps, the external of which is the larger, and is situated anteriorly to the well developed inner one. The last inferior molar is lost from this specimen. The mandibular ramus is compressed and rather deep, and becomes deeper anteriorly. The symphysis begins below the fourth premolar and was not coössified, although the animal is clearly adult.

Measurements of mandible.

		M.
Length of posterior four molars		.0160
Length of true molars		.0120
Length of base of last molar		.0046
Diameters M. ii	anteroposterior	.0040
	transverse	.0030
Diameters Pm. iv	anteroposterior	.0036
	transverse	.0028
Depth ramus at Pm. iv		.0078
Depth ramus at M. iii		.0086

Associated with some of the mandibular fragments, but not known to belong with any in the same skeleton, is the fragment of upper jaw above mentioned. It supports the first and second true molars, in perfect preservation. They differ from those of *Hyopsodus vicarius* (Plate —, fig. —), which are of about the same size, in their rather more triangular form. This is due to the reduction of the posterior inner tubercle to a mere ledge supporting a small cusp. The anterior cusp is prominent, and is connected by ridges with two small angular tubercles on the anterior and posterior margins of the crown. The external lobes stand on the external border of the base, are distinct, acute, angular, and with lenticular section. Their external base is bounded by a cingulum which rises into a low cusp opposite the interspace between them. Slight anterior and posterior cingula.

Measurements.

		M.
Length of base of M. i and M. iii		.0070
Diameters M. ii	anteroposterior	.0040
	transverse	.0045
	vertical	.0020

SARCOLEMUR Cope.

Proceed. Acad. Nat. Sci. Phila., 1875, p. 256.

Molars 4–3, the last with heel; crowns of true molars of four opposite or slightly alternating tubercles, the external pair slightly crescentic in section; anterior inner tubercle bifid. The premolars are compressed, the last acute and with an acute inner tubercle. This form differs from its nearest ally, *Pelycodus*, in the development of acute cusps on the heels of the true molar teeth. The same character distinguishes it from the other genera here enumerated, excepting *Hyopsodus*. But this genus has the anterior inner tubercles simple. The type is *S. pygmæus* Cope (*Antiacodon furcatus* Cope, Annual Report U. S. Geol. Surv. Terrs., 1872, p. 608); another species is the *S. mentalis* Cope (Systematic Catal. Vert. New Mexico, 1875, p. 17). *S. crassus, loc. cit.*, belongs to some other genus, Mesodont or Creodont.

SARCOLEMUR PYGMÆUS Cope.

Proceed. Acad. Phila., 1875, p. 256; *Lophiotherium pygmæum* Cope, Proceed. Amer. Philos. Soc., 1872, extras July 29; *Hyopsodus pygmæus* Cope, *loc. cit.*, p. 461; *Antiacodon furcatus* Cope, On some Eocene Mammals, etc., p. 1, March 8, 1873. Annual Report U. S. Geol. Surv. Terrs., 1872, p. 608, 1873; *Sarcolemur furcatus* Cope, Proc. Phila. Acad., 1875, p. 256.

Plate XXIV, figs. 18–19.

Represented by a portion of the right mandibular ramus, with the penultimate and antepenultimate molars in perfect preservation. These teeth present four cusps, of which the outer are crescentoid in section, the inner conic. They are all elevated, and the inner anterior is in both teeth compressed and bifid. It receives an oblique ridge from the outer posterior crescent, which also sends a ridge to the posterior inner. Enamel smooth.

Measurements.

	M.
Length of penultimate molar	.0045
Width of penultimate molar behind	.0040
Depth of ramus at posterior margin of penultimate molar	.0070

The typical specimen of the *S. pygmæus* is a part of the right ramus mandibuli, with the three molars and last premolar in perfect preservation. The crowns of the molars are composed of two external chevron-shaped tubercles, the apices rising as acute cusps, and two internal cones, the interior of which is flattened and strongly bifid, both points being more elevated than any of the others. The cusps are nearly opposite to each other, and behind the interval between the two posterior rises another, not so elevated as the others, except on the posterior molar. Here it is elevated, and nearly equidistant from the two in front of it: The enamel is smooth, and there is no cingulum on either side. The premolar consists of a principal sectorial cusp, and has a smaller but stout acute anterior cusp, with a small rudiment of another behind; a stout cusp rises from the inner posterior margin of the principal one, giving it a subbifid appearance.

Measurements.

	M.
Length of four molars	.0195
Length of three true molars	.0149
Length of last true molar	.0055
Length of first true molar	.0043
Width of first true molar front	.0025
Width of first true molar posteriorly	.0031
Depth of ramus at front of M. 3	.0075
Depth of ramus at front of Pm. last	.0055

This species differs from the following in the presence of the posterior tubercles on the M. 2–3, and the absence of external cingulum. The sizes are not very differnt.

From the bluffs of the Upper Green River.

This Mammal is about equal in size to a weasel.

From Cotonwood, Wyoming, near Fort Bridger.

HYOPSODUS Leidy.

Proceed. Acad. Nat. Sci. Phila., 1871; Geol. Surv. Montana, 1871, p. 362.

The dentition of the mandible is in this genus characterized by the simplicity of the anterior inner tubercle, and the development of cusps at the angles of the heels of the molars. The fourth premolar has the usual internal cusp at the side of the external. The dentition of the maxillary bone is characterized by the presence of the full number of premolars, of which the last two at least have a well developed simple internal lobe. The internal lobe of the true molars is double, and there are two small intermediate tubercles. The external cusps are two, distinct and compressed, and stand on the external base of the crown. The third true molar does not differ materially from the second.

Though the species of this genus are not numerous, individuals of two of them are exceedingly common in the Eocene beds of North America.

The species of this genus known to me by their mandibles are five, and these differ chiefly as follows:

a. Posterior inner cusps of inferior molars, elevated.
Length of true molars M. .0175; last molar elongate .... *H. powellianus.*
Length of true molars M. .0165; last molar elongate .... *H. lemoinianus.*
Length of true molars M. .0140; last molar longer than second .... *H. paulus.*
Size as last; last molar shorter than second .... *H. miticulus.*
Length of true molars M. .0115; last molar elongate .... *H. vicarius.*
aa. Heel of true molars i and ii, basin-shaped, without posterior inner cusp.
Length of inferior true molars M. .010; last molar as long as penultimate.. *H. acolytus.*

Hyopsodus powellianus Cope.

Plate XXIIId, figs. 3–4.

The largest species of the genus is represented by more or less imperfect mandibular rami of eleven individuals, none of which unfortunately support premolar teeth. The characters of the true molars are those of the other species of the genus.

The cusps alternate with each other, and are quite acute. The sections of the inner cusps are nearly round, while those of the externals are crescentic. This is due to the fact that they send out descending ridges to the inner side, one anteriorly, the other posteriorly. There is a small median posterior tubercle, rather better marked than in *H. paulus.* A low ledge connects the anterior cusps anteriorly, but there are no other cingula. The last inferior molar is narrowed and produced posteriorly, and the edge of the heel is elevated. Enamel entirely smooth.

The anterior border of the masseteric fossa is distinct to about the middle of the depth of the ramus, where it disappears.

Measurements.

		M.
Length of premolars		.0175
Diameters of M. ii	anteroposterior	.0055
	transverse	.0042
Diameters of M. iii	anteroposterior	.0062
	transverse	.0045
Depth of ramus at M. ii		.0115

I formerly referred the specimens of this species to the *Phenacodus zuniensis*, which is found in the Puerco horizon, but from which it is clearly distinct. It is dedicated to Major J. W. Powell, director of the United States Geological Survey.

From the Wasatch epoch of the Big Horn River, Wyoming, J. L. Wortman.

HYOPSODUS LEMOINIANUS Cope.

Paleontological Bulletin, No. 34, p. 148, 1882; Proceed. Amer. Philos. Soc., 1881, p. 148 (1882).

Plate XXIVe, figs. 8-9.

This Mesodont is distinguished from the known species of the genus by its superior size, and the fully developed heel of the inferior third molar. The anterior inner cusps of the inferior molars are simple, though robust, and the same teeth have a weak external and no internal cingulum. The cusps are elevated, and are not strictly opposite, the external one being a little in advance of the corresponding internal one. The posterior external cusp is connected by a low ridge with the two internal cusps, respectively. On the first and second true molars there is a well-marked posterior median cusp. The fourth premolar is a robust tooth, with a short, wide heel, and a mere rudiment of an anterior basal tubercle. The heel has a principal submedian keel and small marginal cusp. There is another and rudimental lobe on the posterior border. The third premolar has neither internal nor anterior tubercle. Its heel is short and wide, and has a low angular median marginal lobe. Enamel smooth.

Measurements.

No. 1.	M.
Length of third and fourth premolars	.0088
Diameters fourth premolar { anteroposterior	.0045
Diameters fourth premolar { transverse	.0030
Diameters second true molar { anteroposterior	.0050
Diameters second true molar { transverse	.0040
Depth ramus at Pm. iv	.0080
No. 2.	**M.**
Length of M. iii	.0056
Depth ramus at M. iii	.0090

Mr. Wortman found in the bad lands of the Big Horn, Northern Wyoming, nine more or less fragmentary mandibles of this species. It is dedicated to Dr. Victor Lemoine, of Reims, France, well known for his many important discoveries in the lower Eocene formation, and his investigations in various departments of zoölogy.

Hyopsodus paulus Leidy.

Proc. Acad. Phila., 1870, p. 110; 1872, 20. Report U. S. Geol. Surv. Terrs., I., p. 75.

Plate V, figs. 1–9 and 18–22.

This species is abundant in the Bridger, Washakie, and Wind River Basins, and I have thirty-eight more or less broken mandibular rami from the Big Horn. With all these jaws there is not a single skull or skeleton. I am therefore unable to add anything to Dr. Leidy's description.

Hyopsodus vicarius Cope.

Microsyops vicarius Cope. On some Eocene Mammals obtained by Hayden's Geol. Surv. of 1872, p. 1, March 8, 1873. Annual Report U. S. Geol. Surv. Terrs., 1872 (1873), p. 609. *H. minusculus* Leidy. Report U. S. Geol. Surv. Terrs., I, 1873.

Plate XXIV, figs. 20–21; XXVa, fig. 7.

Founded on portions of the mandibular rami of two individuals from the bad lands of Cottonwood Creek, Wyoming. These represent an animal considerably smaller than the *Hyopsodus paulus*. The most complete specimen was obtained from the Wind River region of Wyoming. It includes the dentition of both jaws, excepting the superior canine.

The three superior incisors are of equal size and have angular spatulate crowns in contact with each other. The external cusps of the premolars are simple. All the superior molars have a weak anterior basal cingulum. The last true molar is but little smaller than the others, and only differs in form in the rather more oblique external border. The inferior incisors are closely packed together. The root of the canine is not larger than that of the I. iii or Pm. i. The premolars follow the canine and each other without diastemata. The crown of the third premolar is rather obtuse, and has no anterior nor internal tubercle, but has a rather wide heel. The anterior edge of the fourth premolar is curved round abruptly so as to embrace the base of the inner cusp, and does not support an anterior cusp. Heel wide, with a submedian and a lateral keel. The cusps of the true molars are alternate, and the internal intermediate of the last molar is near the fifth lobe or heel of the same.

The symphysis of the mandible is not coössified, in contrast to the condition in *Tomitherium.* There are three mental foramina, one below the second and one below the anterior part of the fourth premolars.

Measurements.

		M.
Length of superior dental series		.0262
Length of superior incisors		.0060
Length of superior true molars		.0100
Diameters M. ii	anteroposterior	.0036
	transverse	.0050
Length of inferior incisors, oblique on bases		.0030
Length of inferior true molars		.0110
Diameter of root of inferior canine		.0020
Diameters Pm. iv	anteroposterior	.0027
	transverse	.0024
Diameters M. ii	anteroposterior	.0039
	transverse	.0030
Diameters M. iii	anteroposterior	.0040
	transverse	.0030

Eleven mandibular rami were procured from the Big Horn Basin, and a smaller number from the Wind River. Among the former, a few specimens are intermediate between this species and the last in dimensions, the inferior true molars measuring M. .0120 and .0125 in length.

Hyopsodus acolytus Cope.

Proceed. Amer. Philos. Soc., 1882, p. 462. Paleontological Bulletin No. 35 (Nov. 11, 1882).

Plate XXIIId, figs. 5–6.

This, the least species of the genus, is also the oldest, being derived from the Puerco horizon. Parts of the two individuals furnish the characters of the inferior and superior true molars, and the fourth superior premolars. The species differ from those hitherto described in other characters than the minute size. One of these is the absence of posterior interior cusp, the heels of the first and second true inferior molars being bounded by a ridge only at this point, as in most of the species of *Pelycodus.* The last inferior molar is not smaller than the second, nor longer. The anterior cusps of all the molars are robust, so that on the first and second true molars they are separated by a shallow notch only. There is a rudiment of the anterior inner cusp on the first true molar, but more on the second and third. The posterior external is obtuse and has a triangular section on all the molars. A crest is continued from the heel of the third molar on the inner side of the crown half way to the anterior inner cusp.

Measurements.

		M.
Length of the inferior true molars		.0100
Diameters M. ii	anteroposterior	.0038
	transverse	.0034
Diameters M. iii	anteroposterior	.0038
	transverse	.0028
Depth of ramus at M. ii		

The *Microsyops spierianus* differs from this species in its smaller size (true molars .008) and in the presence of posterior internal cusps of the true molars.

The *Hyopsodus acolytus* was found by Mr. D. Baldwin, in Northwestern New Mexico.

PROSIMIÆ.

The suborder may be differentiated from the Mesodonta by the possession of an opposable hallux of the posterior foot. This character is, however, not yet demonstrated in the genera of the American Eocene, which I provisionally give to it, nor is the absence of the character known to belong to any of the genera of Mesodonta excepting *Pelycodus.* It is, however, very probable that the other genera referred to the Mesodonta agree with *Pelycodus.* It is also possible that some of the genera here referred to the Prosimiæ agree with *Pelycodus.*

In the uncertainty which exists as to the reference of the genus *Cynodontomys* and its immediate allies, I compare the genera of the Eocene lemuroids as follows. I premise by observing that the genus *Chiromys* clearly represents a primary division of the Bunotheria, which occupies a position between the Prosimiæ and the Tillodonta. The rodent-like incisors with permanent pulps are those of the Tillodonta, but the opposable hallux of the posterior foot is not found as yet in that suborder. The suborder has been named by Gill* the superfamily Daubentonioidea.

*Arrangement of the families of Mammals, 1872, p. 54.

We can distinguish three families among our Eocene forms of lemuroids, in the dental characters, as follows:

Inferior premolars, four .... *Adapidæ*.
Inferior premolars, three .... *Mixodectidæ*.
Premolars, two, with internal lobes in the upper jaw .... *Anaptomorphidæ*.

The genera of the *Adapidæ* have already been considered (p. 215).* I therefore compare the genera of the two remaining families.

MIXODECTIDÆ.

a. Last premolar without inner tubercle.
A very large incisor; canine smaller; first premolar, only one-rooted ...... *Mixodectes*.
aa. Last premolar with internal tubercle.
A very large ? canine; first premolar only one-rooted .... *Microsyops*.
A very large ? canine; first and second premolars each one-rooted .... *Cyodontomys*.

ANAPTOMORPHIDÆ.

Inferior incisors two; canine small; premolars two-rooted .... *Anaptomorphus*.
Inferior premolars one-rooted .... *Necrolemur*.

The correct nomenclature of the large tooth in the front of the mandible of the genera of *Mixodectidæ* is not yet ascertained. It may be a canine or an incisor. I will also remark that in the genus *Necrolemur* of Filhol the characters of the superior premolars are not yet completely known.

MIXODECTES Cope.

Proceed. Amer. Philos. Soc., 1883, p. 447.

Char. Gen.—The position of this genus is uncertain, but may be near to *Cynodontomys* Cope, which I have provisionally placed among the Prosimiæ † It is known from mandibles, which have presumably the following dental formula: I. 0; C. 1; Pm. 4; M. 3. An uncertainty exists as to the proper names of the anterior teeth, which cannot be decided until the dis-

* From that synopsis must be omitted the genera now placed in other families, *Microsyops* and *Anaptomorphus*, and *Pantolestes* (Artiodactyla). *Omomys* must be placed in the section with four premolars, in the place of *Pantolestes*.

† Paleontological Bulletin, No. 34, p. 151.

covery of the superior series. For instance. the formula may be: I. 1; C. .1; 'Pm. 3.

The supposed incisor is a large tooth, issuing from the ramus at the symphysis like a rodent incisor, and has an oval section, with long diameter parallel to the symphysis. The crown is lost from all the specimens. The second tooth is similar in form to the first, but is much smaller. It is situated posterior and external to the first. The next tooth is still smaller, and is one-rooted. The third and fourth premolars have simple conic crowns, and more or less developed heels without cusps. The true molars are in general like those of *Pelycodus; i. e.*, with an anterior smaller, and a posterior, triangle or V. The supplementary anterior inner cusp is quite small, while the principal anterior inner is elevated. The posterior inner is much more elevated than in the species of *Pelycodus*. Last inferior molar with a fifth lobe.

This genus cannot be referred to its place without additional material, but the parts discovered indicate it to be between *Pelycodus* and *Cynodontomys*, either in the *Mesodonta* or the *Prosimiæ*. I may here remark that in defining the latter genus I was in doubt as to the number of the inferior premolars. The discovery of the present genus renders it probable that it has three such teeth, and that the anterior two are each one-rooted.

Mixodectes pungens Cope.

Proceed. Amer. Philos. Soc., 1883, p. 447.
Plate XXIVf; fig. 1.

The mandible of the *Mixodectes pungens* is about the size of that of the mink. Its inferior outline is straight to below the second premolar, whence it rises upwards and forwards like that of a rodent. The anterior masseteric ridge is very prominent, but terminates below the middle of the ramus. Inferior masseteric ridge much less pronounced. The inferior part of the ramus is robust below the base of the coronoid process, but there is no indication of recurvature of the edge. Mental foramina two; one below the front of the first true molar, and one below the second premolar.

The oval base of the canine is not flattened on either side; that of the second tooth is flattened on the inner side. There is a great difference between the sizes of the last three premolars. The fourth is twice as large as the third, and the second, judging from the space and the size of its alveolus, was much smaller than the third, and the crown was probably a simple acute cone. The crown of the third is of that form, with the addition of a short heel. The long axis of the base of the crown is diagonal to that of the jaw. The fourth premolar has a relatively larger heel than the third, but it is shorter than the diameter of the base of the cusp. Its posterior edge is elevated. The cusps of the anterior pair of the true molars are elevated, but the interior is the most so. The supplementary one is not exactly in the line of the interior border of the crown. Each of the inner cusps are connected with the base of the external by a ridge, which together form a V. The posterior base is nearly surrounded by a raised edge, which rises into cusps at the posterior lateral angles. Of these the internal is the more prominent. The edge connecting these cusps is slightly convex backwards, and evidently bears a part in mastication. The lateral borders of the last molar are somewhat expanded, and the fifth lobe is very short. No cingula on any of the teeth.

Measurements.

		M.
Length of dental series from "canine" exclusive		.0265
Length of true molar series		.0140
Diameters "canine"	longitudinal	.0040
	transverse	.0030
Long diameter of base of "Pm. i"		.0028
Long diameter of base of Pm. ii		.0017
Diameters Pm. iv	vertical	.0055
	anteroposterior	.0050
Diameters M. ii	transverse	.0038
	anteroposterior	.0050
Length of crown of M. iii		.0060
Depth of ramus at Pm. iii		.0090
Depth of ramus at M. iii		.0100

From the Puerco epoch of New Mexico; D. Baldwin.

MIXODECTES CRASSIUSCULUS Cope.

Proceed. Amer. Philos. Soc., 1883, p. 417.

Plate XXIVf; fig. 2.

This mammal is represented by fragments of three mandibles from different individuals, one less and the other more worn by mastication. The species differs from the last in its greater size and in the relatively greater length of the last inferior molar. The length of the posterior four molars of the *M. pungens* equals that of the three true molars of the *M. crassiusculus;* and the last true molar of the latter is half as long again as the penultimate, while in *M. pungens* it exceeds it but little.

The best preserved true molar is the second. Its most elevated cusps are the anterior and posterior inner, of which the anterior is subconic and more elevated. The anterior external cusp is crescentic in section, and sends crests to the supplementary, anterior inner, and the posterior anterior inner, both of which descend inwards. The posterior crest reaches the posterior base of the anterior inner cusp.

The posterior external cusp is an elevated angle, sending crests forwards and backwards. The former reaches the base of the anterior external cusp (not reaching the inner), while the latter passes round the posterior edge of the crown. As in *M. pungens*, it is convex posteriorly, and rises to the posterior internal cusp. In both species its appearance indicates that it performs an important masticatory function in connection with the superior molar. No cingula.

Measurements.

		M.
Length of bases of M. ii and iii (No. 2)		.0125
Length of base of M. iii (No. 2)		.0070
Diameters crown M. ii (No. 1)	anteroposterior	.0056
	transverse	.0050
Depth of ramus at M. ii (No. 1)		.0100

From the Puerco Eocene of New Mexico; D. Baldwin discoverer.

CYNODONTOMYS Cope.

Paleontological Bulletin, No. 34, p. 151, Feb., 1882. Proceed. Amer. Philos. Soc., 1881, p. 151 (1882).

The characters of this genus are derived from mandibular rami. Dental formula I. ? 0; C. 1; Pm. 3; M. 3. The premolars are counted as three, on the supposition that the anterior two are one-rooted; should it prove to be single and two-rooted, then the number will be two. The canines (or incisors ?) are very large and close to the symphysis, so that there do not appear to have been any other incisors. The true molars have the frequently occurring three tubercles in front and a heel behind; but the arrangement is peculiar in that the three tubercles are but little more elevated than the borders of the heel, and occupy a small part of the crown. The last molar is lost from both jaws, but the space for it is about as large as that occupied by the penultimate. The fourth premolar has but two anterior cusps, and these are more elevated than those of the true molars, and the heel is narrower. The mandibular rami are not coössified.

The dental characters of this genus resemble considerably those of *Anaptomorphus* and *Necrolemur*, but the large size of the inferior canine or incisor tooth distinguishes it from both. The double anterior cusps of the fourth premolar equally distinguish it from them.

CYNODONTOMYS LATIDENS Cope.

Proceed. Amer. Philos. Soc., 1881, p. 151 (1882).

Plate XXIVe, fig. 2.

This species is known from a pair of mandibular rami which are bound together by matrix.

The inferior true molars are subquadrate in horizontal outline, somewhat narrowed anteriorly. The concave heel is the larger part of the crown; it is only elevated into a low cusp at the posterior external angle. The anterior cusps are conic, and are in contact at the base. The external and posterior internal are of about the same size; the anterior inner is smaller and does not project so far inwards as the posterior. The fourth premolar has the posterior border of its heel serrate. The anterior cusps are elevated and moderately acute; the internal is a little less elevated than the external, and is separated from it by a deep notch. The alveoli for the anterior premolar are not so close together as to render it probable that they belong to but one tooth. They are placed somewhat obliquely to the long axis of the jaw. There is no diastema. The section of the base of the crown of the canine is a regular oval, the long diameter coinciding with the vertical diameter of the ramus

The ramus is rather slender, but is shortened anteriorly. The boundaries of the masseteric fossa are well marked, the anterior ridge descending to below the middle line of the ramus. The mental foramen is large, and is situated below the contact of the two premolars. The inferior edge of the ramus is rather thick.

Measurements.

	M.
Length of dental series, including canine	.0240
Length of premolars	.0062
Length of molars	.0114
Long diameter base canine	.0036
Diameters Pm. iv { anteroposterior	.0035
Diameters Pm. iv { transverse	.0026
Diameters Pm. M. ii { anteroposterior	.0042
Diameters Pm. M. ii { transverse	.0038
Depth of ramus at Pm. i	.0060
Depth of ramus at M. iii	.0068

Big Horn bad lands, Northern Wyoming; J. L. Wortman.

ANAPTOMORPHUS Cope.

Proceed. Amer. Philos. Soc., 1872, p. 554. Paleontological Bulletin, No. 8, p. 1. Oct. 12, 1872. ?*Washakius* Leidy. Report of the U. S. Geol. Surv. Terrs. i, p. 123, 1873.

The genus *Anaptomorphus* was characterized by me in 1872, from a mandibular ramus which exhibited the alveoli of all the teeth, three of them occupied by the teeth, viz, the Pm. iv, and the M. i and M. ii. From the specimen the inferior dental formula was ascertained to be I. 2, C. 1, Pm. 2, M. 3. The Big Horn collection contains a nearly entire cranium of what is probably a species of the same genus. From it the superior dentition, exclusive of the incisors, is determined to be C. 1, Pm. 2, M. 3. The premaxillary bones are mostly broken off, but a part of the alveolus of the external incisor of one side remains.

The indications are that the external incisor was a small tooth, not exceeding the canine in size, and it was situated close to the latter. The canine is also small, and its simple crown is not more prominent than those of the premolars. The latter are separated from it by a very short diastema. The long diameter of their crowns is transverse to the long axis of the jaw; and each one consists of a larger external and smaller internal cusp. The true molars are also wider than long, and support two external and only one internal cusps.

This genus was founded on the left ramus mandibuli of a single species. The posterior portion is broken away, and the teeth remaining perfect are the Pm. 4 and M. 1 and 2. The ramus, though small, is stout, and deeper at the symphysis than at the last molar. What appears to be the dental foramen is nearly opposite the bases of the crowns of the molars.

Dentition of the ramus mandibuli, In. 2, C. 1, Pm. 2, M. 3; total 16. There is no interruption in the series near the canine, and the symphysis, though massive, is not coössified. The third (first) premolar is two-rooted. Further details are, the last molar is three-lobed and elongated behind. The composition of the crowns of the preceding molars consists of four opposed lobes, which are very stout, and connected transversely by a thin ridge behind, and are in close contact in front. The premolar tooth, which is best preserved, is a perfect second, which, while having two roots, possesses a crown which stands almost entirely on the anterior, presenting a curved sectorial crest forwards and upwards.

The orbits are large and are entirely inclosed behind. The frontal bone does not send inwards to the alisphenoid a lamina to separate the orbit from the temporal fossa, as is seen in *Tarsius*. There is no sagittal crest, but the temporal ridges are distinct. The occipital region protrudes beyond the foramen magnum, or at least beyond the paroccipital process, which is preserved, the condyles being lost. The otic bulla is large, extending anteriorly to the glenoid cavity. The pterygoid fossa is large, the external pterygoid ala being well developed, and extending well upon the extero-anterior side of the bulla, as in *Tarsius*. As in that genus, the foramen ovale is situated on the external side of the bulla, just above the base of the external pterygoid ala. The carotid foramen, as I suppose it to be, is situated at the apex of the bulla. The lachrymal foramen is situated anterior to and outside of the orbit, as in *Lemuridæ* generally.

The cast of the anterior part of the left cerebral hemisphere is exposed. This projects as far anteriorly as the middle of the orbits, leaving but little room for the olfactory lobes. The relations of the latter, as well as of other parts of the brain, will be examined at a future time. The part exposed does not display fissures, and gentle undulations represent convolutions.

The characters of this genus now known warrant us in thinking it one of the most interesting of Eocene Mammalia. Two special characters confirm the reference to the *Lemuridæ* which its physiognomy suggests. These are the external position of the lachrymal foramen and the unossified symphysis mandibuli. Among *Lemuridæ* its dental formula agrees only with the *Indrisinæ*, which have, like *Anaptomorphus*, two premolars in each jaw. But no known *Lemuridæ* possess interior lobes and cusps of all the premolars, so that in this respect, as in the number of its teeth, this genus resembles the higher monkeys, the *Simiidæ* and *Hominidæ*,* more than any existing number of the family. Of these two groups the resemblance is to the *Hominidæ* in the small size of the canine teeth. It has, however, a number of resemblances to *Tarsius*, which is perhaps its nearest ally among the lemurs, although that genus has three premolars. One of these points is the anterior extension of the otic bullæ, which is extensively overrun by the external pterygoid ala. A consequence of this arrangement is the external position of the foramen ovale, just as is seen in *Tarsius*. Another point is the probably inferior position of the foramen ovale. Though this part is broken away in the cranium of *Anaptomorphus homunculus*, the paroccipital process is preserved, and has the position seen in *Tarsius*, as distinguished from the *Indrisinæ*, *Lemuridæ*, *Galaginæ*, etc. In this it also resembles the true *Quadrumana*.

When we remember that the lower *Quadrumana*, the *Hapalidæ*, and the *Cebidæ* have three premolar teeth, the resemblance to the higher members of that order is more evident. The brain and its hemispheres are not at all smaller than those of the *Tarsius*, or of the typical lemurs of the present period. This is important in view of the very small brains of the flesh-eating and angulate Mammalia of the Eocene period so far as yet known. In conclusion, there is no doubt but that the genus *Anaptomorphus* is the most simian lemur yet discovered, and probably represents the family from which the true monkeys and men were derived. Its discovery is an important addition to our knowledge of the phylogeny of man.

I find on examination of the specimen on which Dr. Leidy based his *Washakius insignis*, which he kindly permitted me to make, that the corresponding parts preserved, the last two inferior molars, do not differ from

---

* In an early description of *Anaptomorphus*, Proc. Amer. Philos. Soc., 1873, the types make me say "this genus * * * might be referred decidedly to the *Lemuridæ*, were it not for the unossified symphysis." It is scarcely necessary to state that *Simiidæ* should be read in place of *Lemuridæ*.

those of the present genus. Dr. Filhol's beautiful collection, made in the Phosphorite beds of Southeastern France, was, through his liberality, thrown open to me, and I used the opportunity to study the extinct lemurs. There is much resemblance between the inferior jaw, with its dentition, of the genus *Necrolemur*,* and that of *Anaptomorphus*, but the two inferior premolars of his *Necrolemur antiquus* have but one root. This character constitutes a basis for its generic separation from the *A. aemulus.*

ANAPTOMORPHUS AEMULUS Cope.

Paleontological Bulletin, No. 8, p. 1. Oct. 12, 1872. Proceed. Amer. Philos. Soc., 1872, p. 554. Annual Report U. S. Geol. Surv. Terrs., 1872 (1873), p. 549.

Plate XXV, fig. 10.

This species was about the size of a marmoset, or of a red-squirrel. The teeth are all closely packed. The alveolus supposed to be that of the canine, is round and a little larger than that of the external incisor. The anterior premolar has two roots; the anterior one much smaller than the posterior. The second root of the last true molar is inserted in the lower part of the ascending ramus, so that the tooth was obliquely placed. The anterior cusps of the true molars preserved are a little higher than the posterior. There are no cingula, and the enamel is entirely smooth. There is a small mental foramen, which is below the anterior alveolus of the anterior premolar.

Measurements.

	M.
Length of dental line	.0140
Length of last molar	.0030
Length of antepenult	.0030
Width of antepenult	.0020
Length of three molars preserved	.0070

From the Bridger beds of the upper valley of Green River.

The mandibular ramus of the *Necrolemur antiquus* is longer and relatively less robust than that of the American animal, according to Filhol.

*Recherches s. les Phosphorites du Quercy, 1877, p. 275, fig. 216. Annales des Sciences Naturelles, 1874.

Anaptomorphus homunculus Cope.

Paleontological Bulletin, No. 34, p. 152, 1882. Proceed. Amer. Philos. Soc., 1881, p. 154 (1882).

Plate XXIVe, fig. 1.

This species was founded on a cranium without lower jaw. It is distorted by pressure, but its form is normally nearly round, when viewed from above or below. The extremity of the muzzle is broken away, but the alveolus of the external incisor indicates that it is short and not prolonged as in *Tarsius spectrum*. The mandibular ramus, already described, proves the same thing. The orbits are large, but not so much so as in *Tarsius spectrum;* their long diameter equals the width of the jaws at the last superior molar teeth inclusive. The supra-orbital borders project a little above the level of the frontal bone, which is concave between their median and anterior parts. The cranium is wide at the postorbital region, in great contrast to its form in the *Adapidæ*, resembling the *Necrolemur antiquus* Filh. in this respect. The postfrontal processes are wide at the basal portion, and flat. From their posterior border the temporal ridges take their origin. These converge posteriorly and probably unite near the lambdoidal suture, but this part of the skull is injured. The anterior lobes of the cerebral hemispheres are indicated externally by a low boss on each frontal bone.

The paroccipital process is short and wide at the base, and it is directed downwards and forwards. The alisphenoid descends so as to form a strong wall on the anterior external side of the otic bulla. This is also the case in *Tarsius spectrum*, but in the extinct species the descending ala is more robust and has a thickened margin. On the latter the external pterygoid ala rests by smooth contact of its thickened superior edge. The ala is twice as prominent as the internal pterygoid ala. The posterior nareal opening is not wide, and its anterior border is parallel with the posterior border of the last superior molar teeth. The palate is wide, and its dental borders form a regular arcade as in man, being quite different from the form usual in monkeys and lemurs, including *Tarsius*. Perhaps the form is most like that of *Microrhynchus laniger*. The proximal parts of the malar bone are prominent, and overhang the maxillary border, as in *Tarsius*.

The *foramina ovale* and *lachrymale* are rather large. There are two infraorbital canals, lying beside each other, and issuing by two foramina externa. The external appearance justified this conclusion, but the fact was demonstrated when I accidently broke away the anterior border of one of the orbits. This displayed the two canals filled with matrix their entire length. The anterior foramen externum is anterior to and above the posterior, and both are above the first (third) premolar tooth. The lachrymal foramen is above the space between that tooth and the canine.

The crown of the canine tooth is a cone with a very oblique base, and a convex anterior face. The base rises behind, and the posterior face has on the median line a low angular edge. The internal cone of the third (first) premolar is not so prominent as that of the second, though large. The external cusps of both premolars rise directly from the external base. They are flattened cones, with anterior and posterior cutting edges. The crowns are a little contracted at the middle, so as to be narrower than the inner lobe of the tooth, which is narrower than the external portion. Both premolars have delicate anterior, posterior, and external cingula. The external cusps of the true molars rise directly from the external base, and, like those of the premolars, have a regularly lenticular section. At the internal base of each one is a small intermediate tubercle, which is connected by an angular ridge with the single internal cusps. There are delicate anterior, posterior, and external cingula, but no internal. The posterior cingulum shows a trace of enlargement at its inner part, which is well marked on the second molar, but it is not as prominent as in many Creodont genera. The posterior external cusp of the last true molar is reduced in size. Taking the molars together, the first true molar is the largest, and they diminish in size both anteriorly and posteriorly. The third true molar is a little smaller than the first (third) premolar. Enamel smooth.

Measurements.

	M.
Length of cranium to occipital prominence above paroccipital process, and minus premaxillary bone	.0280
Total width at posterior border of orbit, below	.0240
Length of palate from front of canine tooth	.0116
Width of palate and penultimate molars	.0125
Length of superior molar series	.0095
Length of superior true molars	.0060

Diameters of crown of canine	anteroposterior	.0018
	vertical	.0018
Diameters crown of Pm. iii	anteroposterior	.0020
	transverse	.0026
Diameters crown of Pm. iv	anteroposterior	.0020
	transverse	.0035
Diameters M. ii	anteroposterior	.0032
	transverse	.0040
Diameters M. iii	anteroposterior	.0016
	transverse	.0028
Diameters of orbit	anteroposterior	.0110
	vertical (? depressed)	.0078
Interorbital width (least)		.0050

The *Anaptomorphus homunculus* was nocturnal in its habits, and its food was like that of the smaller lemurs of Madagascar and the Malaysian islands. Its size is a little less than that of the *Tarsius spectrum.* The typical specimen was found by Mr. J. L. Wortman in a calcareous nodule in the Wasatch formation of the Big Horn Basin, Wyoming Territory.

As compared with the *A. æmulus* it is smaller in the dimensions of its teeth.

CREODONTA.

Cope, Proceed. Acad. Phila., Dec. 1875. Report Capt. Wheeler Expl. Surv. W. of 100th Mer., iv, pt. ii, pp. 72 and 87.

Unguiculate (?) placental *Mammalia*, with separate scaphoid and lunar bones; narrow cerebral hemispheres, and very large and exposed olfactory lobes; and the ankle-joint generally not trochlear.

The above definition was derived from the flesh-eaters of the Suessonian formation of New Mexico and France. The characters of the brain have been demonstrated in three genera: in *Arctocyon* by Gervais; *Oxyæna* by Cope, and *Stypolophus* by Filhol. The peculiar ankle-joint was shown to be present in four genera, *Amblyctonus*, *Oxyæna*, *Stypolophus*, *Didymictis*, and *Heteroborus*. The uniform absence of the characteristic carpal bone of the *Carnivora*, the scapho-lunar, gave ground for inferring its division into its primitive pieces in the same genera; and the view was supported by the existence of this division in the Bridger genus *Mesonyx*.

How far these characters are common to the flesh-eaters of the Bridger formation is yet uncertain. The genus *Mesonyx* differs widely in its ankle-joint, but its dentition is so near that of *Amblyctonus* that they probably belong to the same greater division. In the same way *Miacis* must probably follow *Didymictis*. For the present, then, I refer the genera of the Bridger formation known to me, to the *Creodonta*.

The affinities of the *Creodonta* may be estimated as follows:

The glenoid cavity of the squamosal bone is transverse, and well defined anteriorly and posteriorly, as in the *Carnivora*. In all the genera of the Suessonian or Wasatch, the ilium has a well-marked external anterior ridge, which continues from the acetabulum to the crest, distinct from the internal anterior ridge. The ilium has therefore an angulate or convex external face, as in *Insectivora* and *Marsupialia*, and does not display the usual expansion in a single plane of most of the placentals. In all the genera there is a strong tuberosity in the position of the anterior inferior spine, which is wanting in the *Mammalia*, excepting certain *Insectivora* and *Prosimiæ*, although it marks the position of the origin of the rectus femoris muscle in all types.

In *Amblyctonus*, *Didymictis*, *Protopsalis*, and three undetermined forms, the femur supports a third trochanter.

In some species, where the cuboid bones are preserved, it is evident that the distal end of the astragalus articulated with this as well as with the navicular bone, although the facet of the astragalus is single and continuous. As the extensive transverse distal astragalar face is characteristic of all the species where it is preserved, the contact of the cuboid and astragalus is probably common to all of this division. There is no elongation of the navicular; it is, on the contrary, very short, since the astragalus projects beyond the calcaneum (in the genera where they have been observed). The cuboid is, on this account, rather elongate, but not remarkably so. There were five toes in the hind feet of some of the species. The ungues in some of the genera are compressed and acute. In the genus *Mesonyx*, from the Bridger, I found one of the claws to be broad and flat, so as to be subungulate. I found an ungual phalange in New Mexico, probably belonging to a species of this group, which presented a similar, though less expanded, form. I have every reason for believing that there were five toes on the hind foot of *Stypolophus hians* and a second species.

The characters now adduced lead to the following conclusions as to the systematic position of these animals.

The small size of the cerebral hemispheres and the rare occurrence of convolutions, refer this group to the Lissencephalous or Lyencephalous *Mammalia*. The characters presented by our crania are borne out by those exhibited by the *Arctocyon primævus*, De Blainv., from the Lower Eocene or Suessonian beds of France. Professor Gervais* has discovered that the olfactory lobes are large, and project far beyond the hemispheres, while not only the cerebellum but also probably the corpora quadrigemina were exposed behind. We are therefore restricted, early in the inquiry, to comparisons with a few orders. These are the *Insectivora*, *Marsupialia*, and some of the *Prosimiæ*, which have small brains. Other characters, however, exist, which add to the reasons for separating them from the *Carnivora*.

There is nothing in the dentition inconsistent with the orders *Carnivora*, *Insectivora*, and *Marsupialia*. It resembles that of some *Viverridæ* of the first, *Mythomys* of the second, and the *Sarcophaga* of the third. Nevertheless, in the often limited number of incisor teeth, it approaches most nearly to the *Insectivora*.

The transverse glenoid cavity is that of the three orders named, and distinguishes the group from the *Rodentia*.

So far as known, the coössification of the scaphoid and lunar bones, the distinguishing character of the *Carnivora*, is wanting. The angulate shape of the ilium is that of *Insectivora* and *Marsupialia*. It is less apparent in *Chiromys*, and is not characteristic of the higher *Mammalia*. The large anterior inferior tuberosity is especially a character of the Lemurs, other than *Nycticebinæ* (Mivart),† the *Chiromys*, and of certain *Insectivora*, especially *Solenodon*. It is figured by Mivart in *Indris* and *Loris*, by Owen in *Chiromys*,‡ and by Peters in *Solenodon*.§ It is absent in *Carnivora*, the true *Quadrumana*, *Marsupialia*, and many *Insectivora*. Allman ‖ does not represent it in *Mythomys*. The third trochanter of the femur is wanting in

* Nouvelles Archives du Muséum, 1870, p. 150.

† In a memoir in the Philosophical Transactions, vi, p. 421.

‡ Proceedings of the Zoölogical Society of London, v, pl. xxi.

§ Abhandlungen der königlichen Academie der Wissenschaften, 1863, pl. 3, Ueber *Solenodon cubanus*.

‖ Transactions of the Zoölogical Society of London, vi, pl. 2, On (*Potamogale*) *Mythomys velox*.

the Gyrencephalous orders generally, characterizing only the *Perissodactyla.* Among Lissencephalous orders, it is very common in the *Edentata,* and still more usual in the *Insectivora* It does not occur among Marsupials. But in the *Prosimiæ,* there is often a third trochanter (Mivart, *loc. cit.; e. g., Lemur, Galago*). In *Talpa* and some other *Insectivora,* and also in *Chiromys,* it is situated high up, nearly opposite the little trochanter.

The peculiar character of the ankle-joint already mentioned is not certainly characteristic of all the genera of this division. In *Mesonyx* it is as perfect a tongue and groove as in the most specialized of existing *Carnivora.* In the *Creodonta* of the Wasatch it is on the contrary flat, and resembles more than any other existing type the ankle-joint of the otter. It probably indicates aquatic habits of the species possessing it.

The comparison of this group brings out principally affinities to the *Insectivora* and *Prosimiæ* Besides the differences from the *Marsupialia,* already pointed out, in the genera *Oxyæna* and *Didymictis,* the posterior part of the inferior border of the mandibular ramus is not inflected, as in *Marsupialia;* in *Stypolophus* (*viverrinus*) the lachrymal canal is within the orbit, and not exterior to it. The frequently reduced number of incisors in the lower jaw and the normal number above, are a further ground of distinction from the Carnivorous *Marsupialia.*

Comparison with the *Prosimiæ* shows that the differences consist in the sectorial character of some molar teeth, and large development of the canines, in the Eocene forms; in the short tarsal bones, and peculiar tibio-tarsal articulation; with convex external face of the ilium. This *ensemble* of characters can hardly be regarded as ordinal; and there only remains, to give character to such a distinction, the difference in the size and form of the cerebral hemispheres. This character, in some of the smaller living *Lemuridæ,* is not strongly marked, and in them the approximation to the Lissencephalous Mammals is at its closest.

The differences from the *Insectivora* are less numerous. The only trenchant distinctive character upon which I can seize, in comparison with *Mythomys* and *Solenodon,* is the peculiar tibio-tarsal articulation. On this account, and because of the rather more marked carnassial characters of the molar teeth, I have proposed to place the genera *Amblyctonus, Oxyæna,*

Stypolophus, and *Didymictis* in a suborder of *Insectivora*, under the name of *Creodonta*.* They stand also in relationship to the Lemurs, and more remotely to the *Carnivora*.

History. MM. Laurillard, Pomel, and others have referred the European *Creodonta* to the *Marsupialia*, on account of the great similarity of the dentition. MM. De Blainville and Gervais have, on the other hand, regarded them as placental, a view which I have assigned reasons† for believing to be the correct one. M. Filhol has recently shown that the replacement of the dentition in *Hyænodon*, which has some affinities with the *Creodonta*, is quite as in true placental *Carnivora*. Professor Gaudry has expressed the opinion that the *Creodonta* are the descendants of the *Marsupialia*.‡ I have proposed another view.§

If we suppose that the *Creodonta* are the descendants of the *Marsupialia*, we must suppose that the *Insectivora*, to which they are related, are also the descendants of the *Marsupialia*, and this is on various grounds not very probable. The lower forms of unguiculate Mammalia with small cerebral hemispheres are very much alike in important characters, and to these I have given the name of *Bunotheria*. I suspect that this group is as old as the *Marsupialia*, and may even have given origin to it. That it developed contemporaneously with it in various parts of the world, is evident.

Restoration. The Wasatch beds of New Mexico have yielded remains of more than a dozen species, which ranged from the size of a weasel to that of a jaguar. The Bridger beds of Wyoming probably contain as many species, which range from small size to the dimensions of a bear.

In general appearance the *Creodonta* differed from the *Carnivora*, in many of the species at least, in the small relative size of the limbs as compared with that of the head, and in some instances as compared with the size of the hind feet. The feet are probably plantigrade, and the posterior ones capable of some degree of horizontal rotation. The probable large size of the rectus femoris muscle indicates unusual power of extension of the hind limb. This may indicate natatory habits, a supposition further

* On the Supposed Carnivora of the Eocene of the Rocky Mountains, by E. D. Cope. 8vo. Philadelphia, Dec. 22, 1875.

† Proceedings Academy Phila., 1875. Paleontological Bulletin, No. 20, Dec., 1875.

‡ Enchainements du Monde Animal, 1878, p. 24.

§ Proceedings American Philos. Soc. 1880, p. 76.

justified by the flat tibio-tarsal articulation. They were furnished with a long and large tail. Probably some of the species resembled in proportions the *Mystomys* and *Solenodon*, now existing in Africa and the West Indies, but they mostly attained a much larger size. The habits of many of them were probably aquatic.

Classification. To the *Creodonta* I have referred,* on the information which we possess, the genus *Arctocyon* of Blainville. Professor Gervais has discovered that it possessed the very small cerebral hemispheres characteristic of the *Creodonta* The olfactory lobes are large, and project far beyond the hemispheres, while not only the cerebellum, but probably the corpora quadrigemina, were exposed behind. The tarsal articulation and the posterior part of the mandibular bones are unknown; hence this reference is not certain. Professor Gervais† regards it, after Laurillard,‡ as a marsupial, and establishes an especial family of the order for its reception. It is, however, more probable that its affinities are with the contemporary genera of flesh-eaters, *Palæonyctis* Blv., and *Pterodon* Blv., genera which have near allies among the American forms. *Palæonyctis* was the contemporary of the Coryphodons in the Suessonian period of Western Europe, and presents a strong resemblance to *Amblyctonus* in its mandible, the only part of the skeleton known. The posterior part of the ramus is not inflected according to Gervais, and he therefore does not refer it to the *Marsupialia.*§ The nearest European representative of *Oxyæna* is *Pterodon*, in which the form of the mandible also forbids a reference to the *Marsupialia*, as Gervais has remarked. Both genera are doubtless members of the suborder of *Creodonta.* The genus *Hyænodon*, on the other hand, is not referable to the same group, for I find in a specimen of the *H. requieni* from Desbruges, preserved in the Museum of the Jardin des Plantes, that the scaphoid and lunar bones are coössified. Moreover the figure given by Professor Gervais,‖ representing the brain of the originally-described type, *H. leptorhynchus* of the Miocene period, displays characters of the true *Carnivora.* The anterior part of the cranial cavity of the specimen molded, is broken away.

---

* Report Capt. G. M. Wheeler's Expl. Surv. W. 100 Mer., 1877, iv, pl. ii, p. 38.
† Nouv. archives du museum, 1870, p. 150.
‡ Dict. univ. d'hist. naturelle, ix, p. 400.
§ Nouv. archives du museum, 1870, 151.
‖ Loc. cit., pl. vi, fig. 5.

It is possible that the genus *Diacodon* Cope belongs here also; its species resemble some *Marsupialia* in the inferior dentition, and are of small size.

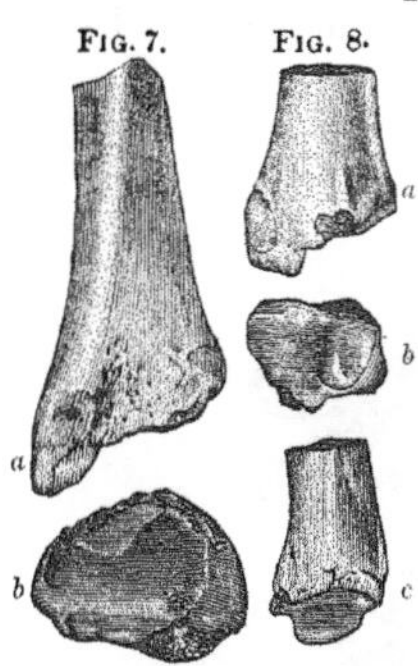

FIG. 7. Distal extremity of tibia of *Amblyctonus sinosus* Cope. FIG. 8. Distal extremity of tibia of *Oxyæna morsitans* Cope. Both two-thirds natural size. From Report Expl. and Surv. W. of 100th Mer., G. M. Wheeler, iv, pt. ii.

The genus *Mesonyx*,* which I discovered in the Bridger beds of Wyoming, has the trochlear face of its astragalus completely grooved above as in the true *Carnivora*, and its distal end presents two distinct facets, one for the cuboid and the other for the navicular bones. It represents on this account a peculiar family, the *Mesonychidæ*.

There are various degrees of development of the sectorial structure of the molars in this suborder. In some of them, as *Didymictis*, only one of the inferior molars presents this structure; in others two, and in others three. In one type, the last superior molar is longitudinal; in others, it is transverse. In *Arctocyon* the superior true molars are tubercular.

The glenoid cavity of the squamosal bone presents differences in the various genera of this suborder. In *Arctocyonidæ* (fide De Blainville), *Oxyænidæ*, and *Mesonychidæ* it is bounded by a transverse crest anteriorly, as well as by the postglenoid posteriorly, while in the *Leptictidæ* it is plane and open anteriorly. In *Amblyctonidæ* its condition is unknown. In existing *Carnivora* this character is not very constant as a family definition; it is best marked in the *Felidæ* and least marked in the *Canidæ*. Nevertheless there is a group of genera allied to the *Oxyænidæ*, which are very marsupial in character, which have been called the *Leptic-*

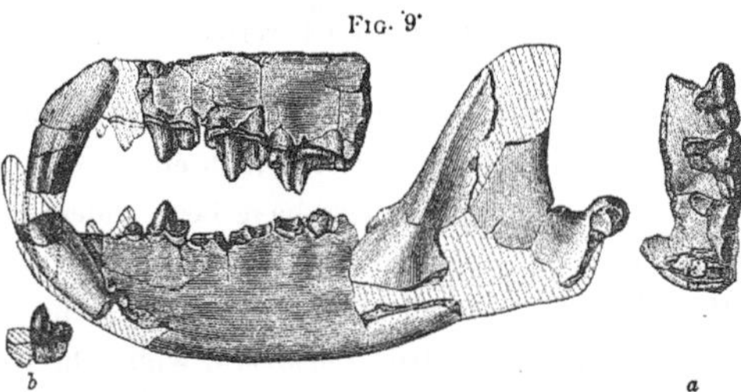

FIG. 9. Portions of maxillary and mandibular bones of *Oxyæna lupina* Cope, one-half natural size; *a*, maxillary bone from below; *b*, last superior molar. From Report Expl. Surv. W. 100th Mer., G. M. Wheeler, vol. iv, pt. ii.

* Annual Report U. S. Geol. Surv. Terrs., 1872, p. 550.

tidæ, and which differ, so far as known, from *Oxyæna* in the absence of the preglenoid crest. I suspect that these forms constitute a family by themselves, and for the present, until our knowledge of them is fuller, I define it by this character.

FIG. 10.

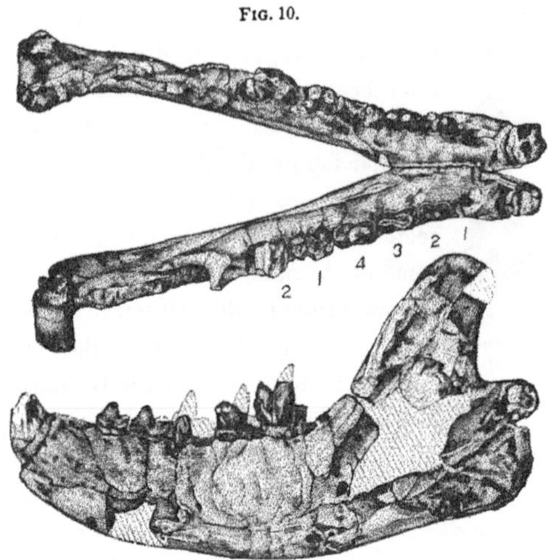

FIG. 10.—Mandible of *Oxyæna forcipata* Cope, one-half natural size; *a* from above, *b* from left side. From Report Expl. Surv. W. 100th Mer., G. M. Wheeler, vol. iv, pt. ii.

Shortly after the publication of my arrangement of the Credonta in 1880,* I obtained a good deal of additional material, which enabled me to improve it in several respects. A number of genera have been added, and the characters which distinguish the *Miacidæ* and *Oxyænidæ* have been more fully brought out. The *Miacidæ* differ from all other families in having the fourth superior premolar sectorial as in the true *Carnivora*, while the true molars are tubercular. In *Oxyæna*, the fourth superior premolar displays no indication of sectorial structure, the first true molar assuming that character. In *Stypolophus* and allies, the second superior true molar is more or less sectorial, and the first true molar and even the fourth premolar in some of the genera, develop something of the same character. But there is every

* Proceedings Amer. Philos. Society, p. 76.

gradation between the triangular *Didelphys*-like and the sub-sectorial *Pterodon*-like forms of the superior molars in this group of genera.

The definitions of the families will then be as follows:

I. Ankle-joint plane transversely, or nearly so.

True molars above and below, tubercular; last superior not transverse.. *Arctocyonidæ.*

Superior true molars, tubercular; last superior premolar sectorial; first inferior "tubercular sectorial" .... *Miacidæ.*

Superior molars triangular, last one transverse; inferior molars tubercular-sectorial, or with reduced anterior cusp; no preglenoid crest .... *Leptictidæ.*

Last superior molar trenchant, transverse; first superior true molar sectorial; inferior true molars tubercular-sectorial; a preglenoid crest .... *Oxyænidæ.*

Last superior molar longitudinal; inferior true molars without developed sectorial blade .... *Amblyctonidæ.*

II. Ankle-joint tongued and grooved, or trochlear.

Molar teeth in both jaws consisting of conic tubercles and heels; none sectorial; a preglenoid crest .... *Mesonychidæ.*

I now give the characters of the genera. All these are derived from examination of typical specimens. The opportunity of doing this I owe to the kindness of Messrs. Leidy, Gervais, Gaudry, Filhol, and Lemoine.

ARCTOCYONIDÆ.

a. Superior molars with two internal tubercles.

Premolars, $\frac{4}{4}$; the first inferior one-rooted; the last inferior well developed. *Arctocyon* Blv.

Premolars $\frac{3}{4}$; first two below one-rooted; superior molars with two internal cusps. *Achænodon* Cope.

aa. Superior molars unknown.

Premolars below, 4; the first two-rooted, the last true molar much reduced (fide Lemoine) .... *Hyodectes** Cope.

Premolars below, 3; first two-rooted; true molars normal .... *Heteroborus*† Cope.

aaa. Superior molars with one internal tubercle.

Premolars below, 4, without internal tubercles; first one-rooted; superior molars with an internal V, and intermediate tubercles .... *Mioclænus* Cope.

LEPTICTIDÆ.‡

I. Fourth inferior true molar like the true molars, or with three anterior cusps.

β. Third superior premolar with internal cusp; anterior cusp of inferior molars small, median.

Third premolar with one external and one internal cusps .... *Mesodectes* Cope.

Third premolar with two external and one internal cusps...... ....*Ictops* Leidy.

---

* Type *Arctocyon gervaisii* Lemoine, Oss. Foss. des Envir. de Reims, 1878, p. 8.

† Type *Arctocyon duelii* Lemoine, loc. cit., p. 9.

‡ The genera referred here resemble considerably the family *Didelphidæ*. The species of *Leptictis* and *Mesodectes* have thus far only been found in the White River beds; see Part Second of this volume.

ββ. Third superior premolar without internal cusps; anterior cusps of inferior molars present.

Cusps of superior molars marginal; two superior incisors........ ...*Leptictis* Leidy.

Cusps of superior molars median in position; anterior cusps of inferior molars well developed; angle of mandible not inflected .... *Peratherium* Aym.

βββ. Anterior cusps of inferior molars wanting.

Fourth inferior premolar like true molars ....*Diacodon* Cope.

II. Fourth inferior premolar different from true molars in a simpler constitution.

α. Inferior true molars not (or not all) tubercular sectorial.

Last inferior molar tubercular; cusps of other true molars well developed; three premolars above and below ....*Deltatherium* Cope.

Inferior true molars alike, with anterior inner cusps little developed; three premolars (?).. .... *Triisodon* Cope.

αα. Inferior true molars all tubercular sectorial.

Inferior true molars alike, with cusps well developed; four compressed premolars below, three above.... ... .*Didelphodus* Cope.

Premolars four below, robust, conic ....*Quercitherium* Filh.

Premolars four above and below, compressed; the fourth superior with a conic cusp and heel externally ....*Styolophus* Cope.

Premolars four below, compressed; fourth superior with a simple blade externally. *Proviverra* Rütim.

MIACIDÆ.

Inferior tubercular molars two, premolars four..........................*Miacis* Cope.

Inferior tubercular molars one, premolars four.... ....*Didymictis* Cope.

OXYÆNIDÆ.

I. Inferior molars without internal tubercles.

Molars, $\frac{4}{3}$ $\frac{3}{3}$; three sectorials in the lower jaw........................*Pterodon* Blv.

II. Inferior molars with internal cusps.

a. Posterior heel of one or more superior molars elongate and trenchant.

Last inferior molar truly sectorial, without internal tubercle; second tubercular-sectorial.................................. ....*Protopsalis* Cope.

Molars, $\frac{4}{4}$ $\frac{2}{2}$; two last inferior molars tubercular-sectorial ... . *Oxyæna* Cope.

AMBLYCTONIDÆ.

Fourth inferior premolar with a broad heel supporting tubercles; an anterior and no internal tubercles...... ....*Amblyctonus* Cope.

Dental formula below, 3, 1, 3, 3. Fourth inferior premolar with a cutting edge on the heel; both internal and anterior tubercles.........................*Palæonyctis* Blv.

MESONYCHIDÆ.

a. Inferior molars seven.

Cones of inferior and superior molars simple..........................*Mesonyx* Cope.

Cones of last two inferior molars with lateral cusps ....*Dissacus* Cope.

aa. Inferior molars ?six.

Internal lobes of penultimate superior molar v-shaped.... ....*Sarcothraustes* Cope.

aaa. Inferior molars five.

Inferior molars with strong anterior lobe........................?*Patriofelis* Leidy.

Of the preceding genera it may be remarked that the structure of the feet of *Pterodon* being unknown, it may be found hereafter to be necessary to remove it from the *Oxyænidæ*, although I do not anticipate that such a course will be necessary. *Palæonyctis* is only known by the mandibular dentition, which is very near to that of *Amblyctonus*. So also it is not certain, but only possible, that *Patriofelis* belongs to the *Mesonychidæ* of the same horizon and locality. The horizontal and geographical distribution of the species of these twenty-seven genera is as follows:

	Lower Eocene.		Middle Eocene.	
	N. A.	Eur.	N. A.	Eur.
Arctocyon primævus Blv		*		
Hyodectes gervaisi Lem		*		
Heteroborus duelii Lem		*		
Mioclænus turgidus Cope	*			
minimus Cope	*			
subtrigonus Cope	*			
ferox Cope	*			
baldwini Cope	*			
protogonioiides Cope	*			
bucculentus Cope	*			
mandibularis Cope	*			
Achænodon insolens Cope			*	
Quercitherium tenebrosum Filh				*
Diacodon celatus Cope	*			
alticuspis Cope	*			
Ictops bicuspis Cope	*			
didelphoides Cope	*			
Peratherium comstockii Cope	*			
Deltatherium fundaminis Cope	*			
interruptus Cope	*			
baldwini Cope	*			
Didelphodus absarokæ Cope	*			
Stypolophus viverrinus Cope	*			
secundarius Cope	*			
multicuspis Cope	*			
strenuus Cope	*			
minor Filh				*
caylusi Filh				*

	Lower Eocene.		Middle Eocene.	
	N. A.	Eur.	N. A.	Eur.
Stypolophus pungens Cope			•	
brevicalcaratus Cope			•	
whitiæ Cope	•		•	
aculeatus Cope			•	
hians Cope	•			
Proviverra typica Rütim		•		
Triïsodon quivirensis Cope	•			
heilprinianus Cope	•			
levisianus Cope	•			
conidens Cope	•			
Miacis parvivorus Cope			•	
edax Leidy			•	
vorax Leidy			•	
canavus Cope	•			
brevirostris Cope	•			
Didymictis altidens Cope	•			
protenus Cope	•			
leptomylus Cope	•			
dawkinsianus Cope	•			
haydenianus Cope	•			
massetericus Cope	•			
curtidens Cope	•			
Pterodon dasyuroides Blv		•		
biincisivus Filh				•
Protopsalis tigrinus Cope			•	
Oxyæna morsitans Cope	•			
lupina Cope	•			
forcipata Cope	•			
Amblyctonus sinosus Cope	•			
sp. no. 2*		•		
Palæonyctis gigantea Blv		•		
Sarcothraustes antiquus Cope	•			
Dissacus navajovius Cope	•			
carnifex Cope	•			
Mesonyx obtusidens Cope			•	
lanius Cope			•	
ossifragus Cope	•			
Patriofelis ulta Leidy			•	

*Represented by a mandible with teeth, from Meudon, associated with the specimens of *Palæonyctis* in the Mus. Jardin des Plantes.

Phylogeny. It is among the genera above enumerated that we are to look for the ancestors of the existing *Carnivora*, excepting, perhaps, the seals, and even these were probably contemporaries. The genera with developed inner cusps and tubercles of the molars, are probably modifications of the *Leptictidæ*, which are also nearest to the *Marsupialia*. In those genera without developed internal tubercles of the molars, we may look for the ancestors of the *Hyænodontidæ*, a family which early attained specialization at the expense of strength of structure, and did not survive the Lower Miocene period. Such genera may be found in the *Mesonychidæ* as the later, and the *Amblyctonidæ* as the earlier types.

In distinguishing between the ancestors of the *Felidæ* and *Canidæ*, we naturally seek to recognize in each an anticipation of the leading characters in the dentition which distinguish those families to-day. This consists, in the *Felidæ*, in the successive abbreviation of the true molar series from behind, so that ultimately two molars are lost, and the remaining or anterior one becomes transverse; also in the development of a preglenoid cross-ridge which embraces the mandibular condyle in front. On the other hand in the *Canidæ*, firstly, the full number of true molars is retained in some genera, as *Amphicyon*, and only one is lost in *Canis*. Secondly, the tubercular character of the posterior molars in both jaws in the *Canidæ* is distinguished from their sectorial character in *Felidæ*. Estimated by these tests the *Miacidæ* are clearly the forerunners of the *Canidæ*, and the *Oxyænidæ*, of the *Felidæ*. In *Miacis* we have in fact a near approach to the dentition of *Canis*, in the lower jaw; while in the same part of *Didymictis*, posterior abbreviation has commenced, reminding one of *Viverra*. In the *Oxyænidæ*, one degree of posterior abbreviation is seen in *Stypolophus*, where the last superior molar is narrowed and turned at right angles to the others. In *Oxyæna*, the process had advanced a step, for there are but two superior true molars, and the last of these is driven in, transversely. The first true molar is functionally sectorial in this genus, while the last premolar is the true sectorial of the superior series in existing *Carnivora*. In the inferior series there are only two true molars in *Oxyæna*, both primitive, or "tubercular-sectorial" in character. In existing *Felidæ* the second is lost, while the first undergoes great changes in becoming a specialized sectorial. The forms of the *Felidæ*, which are nearest, are the *Cryptoprocta*, and the *Proælurus* of Filhol, but they only follow after a wide interval. I have elsewhere discussed the successive steps in the evolution of the sectorial itself.* I have also pointed out† the successive shortening of the anterior part of the dental series in the *Felidæ* and other groups of existing *Carnivora*, which came later in time.

The following table will give an idea of these affinities, and the phylogeny to be derived from them:

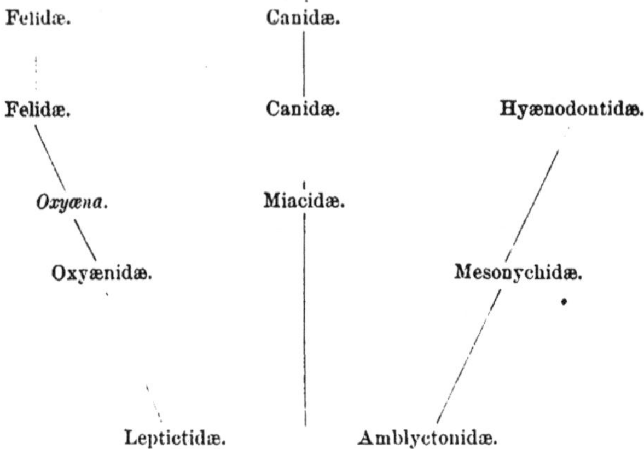

Synonymy. Professor Gaudry has united *Stypolophus* (*Cynohyænodon* Filhol) with *Proviverra.* After an examination of casts of Rütimeyer's types preserved in the Museum of the Jardin des Plantes, I retain them as distinct for the reasons given above. Mr. Bose, in an interesting paper on this subject, published in the London Geological Magazine for May and June, 1880, unites *Didymictis* with *Palæonyctis.* Having examined the types of both genera, my conclusion, as expressed in the preceding pages, is a different one. On the other hand, I have good reason for believing the species to which the name *Synoplotherium* was given, *S. lanius*, is really a second species of *Mesonyx*, of larger size than the *M. obtusidens*, and otherwise

* Proceedings Acad. Phila., 1875, p. 21. † Felidæ and Canidæ, loc. cit., 1879, p. 169-170. On the Extinct Cats of America, American Naturalist, December, 1880.

different. It is likely that some of the species of the Bridger formation, to which Marsh has applied generic names, belong to the *Creodonta*, and may belong to some of the genera described by myself. The fact that no generic definitions accompanied the publication of those names, renders their use impracticable.

ICTOPS Leidy.

Extinct Mammalia of Dakota and Nebraska, 1869, p. 352. Proceedings Philadelphia Academy, 1868, p. 316.

Dental formula (derived from *I. bicuspis* where unknown in *I. dakotensis*) I. $\frac{3}{2}$; C. $\frac{1}{1}$; Pm. $\frac{4}{4}$; M. $\frac{3}{3}$. Third superior premolar tooth with two external and an internal cusp; fourth premolar like the true molar, with two external tubercles, an internal tubercle, and a posterior cingulum. Fourth inferior premolar with an internal and a well developed anterior tubercle; the anterior tubercle of the true molars median in position, and much smaller than the internal tubercle. Heels of molars with elevated cusps. Orbit not closed posteriorly. Cronoid process of the mandible well developed. Inferior margin of mandible not inflected.

The genus *Ictops* was determined by Leidy from a species, the *I. dakotensis* Leidy, from the White River formation. The animals now mentioned are identical with it in generic characters, so far as they are ascertained. The *I. dakotensis* is established on a specimen which does not contain all the teeth, but the parts preserved indicate that those which are wanting are like the corresponding parts of *Leptictis* Leidy and *Mesodectes* Cope, with which the present species also agree. It is unexpected to identify a genus found in the White River horizon with one from the Wasatch *Ictops* agrees very closely with *Didelphys*. The fourth superior premolar has an internal cusp, which is wanting in *Didelphys*, and the inferior border of the mandible is not inflected. There are also but three superior incisors on each side. Under these circumstances I prefer to refer this genus to the *Bunotheria* rather than to the *Marsupialia*, but whether its proper place is in the Creodont or Insectivorous subdivisions I cannot yet determine.

Ictops bicuspis Cope.

Bulletin U. S. Geol. Surv. Terrs., vi, 1881, p. 192. *Stypolophus bicuspis* Cope, American Naturalist, 1880, p. 746, October.

Plate xxix *a*, figs. 2-3.

This species is represented by a skull with mandible, from which the occiput is broken away. From it I have developed the entire dentition of one side, and nearly all of that of the other. The skull is about the size of that of a mink, and has the form of that of a civet.

The premaxillary bone is remarkably extended anteroposteriorly, opposite the nares. The muzzle is moderately elongate, and is contracted laterally. The orbits are not defined posteriorly. The anterior temporal ridges are very obscure, and early unite into a low sagittal crest. The zygoma is proportionately very slender. The glenoid cavity is quite wide, and the postglenoid process is well extended transversely. The infraorbital foramen issues above the anterior border of the first true molar. There are short diastemata behind the posterior incisor, and behind the first and second premolars. The first premolar is situated close to the canine.

The crowns of the canines are compressed and rather short, with the anterior edge subvertical, and the posterior oblique. The latter is also acute. Crowns of second and third superior premolars compressed, with a prominent cusp behind the principal one. First and second true molars with two distinct external cusps and a strong external basal cingulum. The posterior basal cingulum is strong, almost developing a posterior internal cusp in the first and second. There are no external cingula on the third and fourth premolars, excepting a short one opposite the posterior cusp of the fourth. Inferior first premolar one-rooted, third with a posterior heel, and fourth with strong anterior and especially posterior heels. Heels of true molars well developed, cuspidate, the external lobe having a crescentic section, and separated from the internal by a small median tubercle. The latter is strongest on the last molar, giving the heel greater length. The anterior internal cusp of the crown is much larger on the fourth inferior premolar than on the true molars, where it is absolutely median in position.

The mandibular ramus is slender, and the inferior border is gently convex, descending again to the angle. The condyle is not very wide, presents upwards, and is on a line one-third way above the molars in the line to the summit of the coronoid process. The coronoid process is both high and wide at the base, there being no emargination between it and the condyle. The masseteric fossa has a distinct border anteriorly, which rises nearly to the summit of the coronoid process, and turns posteriorly to its posterior margin. Below, it fades out to the general surface. The inferior border at the angle is not inflected. An inferior pterygoid fossa of the inner surface has a marked superior boundary extending horizontally a little below the tooth line. A lower ridge extends downwards and backwards parallel with the inferiorly decurved border of the jaw.

Measurements.

		M.
Length of skull to posttympanic region		.054
Length of skull to orbit		.031
Length of skull to end of molar series		.032
Length from I. 1 to canine		.0065
Length from canine to end of molars		.0220
Length of premolar series		.0148
Elevation of crown of canine		.0040
Length of base of Pm. iii		.0045
Diameters of penultimate molar	anteroposterior	.0040
	transverse	.0045
Length of mandibular ramus from Pm. 1 to condyle		.0425
Elevation of coronoid process from inferior border		.0200
Width of coronoid process at condyle		.0110
Length of inferior molars		.0235
Length of inferior true molars		.0095
Depth of ramus at second true molar		.0070
Depth of ramus at canine		.0035

This species is smaller than any of the known *Stypolophi*, and is about equal to the *Ictops dakotensis*. Besides the characters already mentioned, the two-lobed external wall of the superior fourth premolar will readily distinguish it from any of the species of the former genus.

ICTOPS DIDELPHOIDES Cope.

Bulletin U. S. Geol. Surv. Terrs., vi, p. 192, February 26, 1881.
Plate xxv *a*, fig. 9.

Established on a left mandibular ramus, which supports the last three molars. This demonstrates the former existence of a species of larger size than any of the *Leptictidæ* hitherto known. The general form of the inferior true molars is a good deal like that of *Stypolophus*, but they may be distinguished by three characters in which they at the same time agree with the *Ictops bicuspis:* First, the elevated border of the heel, with a strong external cusp and weaker posterior and internal elevations; second, the small development of the anterior cusp; third, the posterior production of the heel of the third true molar, giving an indication of the fifth lobe. The external anterior cusp of the third molar is elevated; on the first molar it is less so, and the anterior cusp is small. The enamel is smooth, and there are no internal nor external cingula. The mandibular ramus is compressed and deep.

Measurements.

		M.
Length of bases of three true molars		.0165
Diameters of first true molar	anteroposterior	.0055
	transverse behind	.0038
Diameters of last true molar	anteroposterior	.0060
	transverse	.0050
Depth of ramus at anterior root of last true molar		.0095

The jaw fragment described indicates a skull about the size of that of the common opossum.

PERATHERIUM Aymard.

Herpetotherium Cope, Paleontological Bulletin No. 16, p. 1, August, 1873. Annual Report U. S. Geol. Surv. Terrs., F. V. Hayden, 1873 (1874), p. 465.

Dental formula: I. $\frac{?}{4}$; C. $\frac{1}{1}$; Pm. $\frac{3}{3}$; M. $\frac{4}{4}$. Fourth superior premolar like the true molars, with two external cusps and an internal one; third premolar compressed, with one lobe. Inferior molars with heel supporting cusps at the angles, and the three cusps anteriorly, of which the exterior is most elevated and the two inferior subequal. Fourth premolar with the anterior internal cusp well developed; the other inferior premolars simple.

The dental characters of the species included in this genus, so far as they are known, are identical with those of *Didelphys*, and authors have generally regarded the name as a synonym of the latter. Species are numerous in the Upper Eocene of France and in the White River beds of the United States. I retain Aymard's name provisionally, until the number of superior incisor teeth of the species concerned is known. The *Leptictidæ* that are known, do not have so many of these teeth as does *Didelphys*, there being only two in *Leptictis* (fide Leidy) and three in *Ictops* (*bicuspis*) on each side. The genus *Leptictis* is quite near this one, as is also *Ictops*. The reduction of the anterior inferior cusp of the inferior molars, already seen in the latter, is carried nearly to extinction in *Diacodon*.

Only one, a small species of this genus, has yet been found in the Eocene beds of North America.

PERATHERIUM COMSTOCKI Cope.

Plate xxv *a*, fig. 15.

Portions of the mandibles of two individuals of this species were found by Mr. Wortman in the bad lands of the Wind River, Wyoming. They indicate animals a little larger than the *P. fugax* Cope of the White River beds of Colorado, or about equal to the cave-rat (*Neotoma floridana*).

The generic characters include most of those displayed by the dentition. The heels of the molars support an acute tubercle anterior to the posterior border, and their external angles are elevated and have a crescentic section. The anterior internal cusp is not quite so elevated as the median internal cusp, but both are in the same longitudinal plane of the jaw. The posterior internal tubercle is as high as the anterior. The enamel is smooth, and there are no cingula.

Measurements.

			M.
Length of two inferior molars			.0063
Diameters of first true molars	anteroposterior	total	.0037
		of heel	.0019
	transverse at heel		.0017
Depth of ramus at first true molar			.0048

Dedicated to Prof. Theodore D. Comstock, of Cornell University, New York, who explored the Wind River region as geologist of the expedition under Captain Jones, United States Engineers.

TRIÏSODON Cope.

American Naturalist, 1881, p. 667, August (July 27). Proceed. Amer. Philos. Soc., 1881, p. 485.

The type of this genus is only known from portions of lower jaws; some of these include the entire dentition, with unimportant omissions.

True molars alike, consisting of three anterior cusps and a heel. The cusps are relatively small and the heel large. Of the former the internal is much smaller than the external, and the anterior is rudimental, being merely a projection of the cingulum. The cutting edges of the large external cusp are obtuse. The heel is basin-shaped, and its posterior border is divided into tubercles, of which the external is a large cusp. The fourth premolar has no anterior or posterior inner tubercles, so that the anterior part of the crown consists of a compressed cutting cusp. The heel has two well-developed posterior cusps. The third premolar has a similar principal trenchant cusp, but a smaller heel. Canines large.

This genus differs from *Herpetotherium* and *Ictops* in the simplicity of its fourth inferior premolar, and from *Stypolophus* and *Deltatherium* in the rudimental character of the accessory anterior cusps of the true molars, as well as in the three premolars. The rudimental anterior cusp of the true molars, with the three similar true molars, separates it from *Palæonyctis*, and the presence of a conic inner cusp of the same indicates it as different from *Amblyctonus* and *Periptychus*. It is not possible to state whether *Triïsodon* must be placed in the *Amblyctonidæ* or not, on account of the absence of the superior molar teeth of the typical species *T. quivirensis*.

This specimen of the type species of this genus is instructive as showing the succession of premolar teeth. Both the third and fourth premolars have temporary predecessors. The predecessor of the fourth premolar differs much from it in form, and is essentially identical in all respects with the true permanent molars. The crown of the predecessor of the third premolar is wanting, the roots only remaining in the jaw.

The permanent third premolar was protruded before the permanent fourth. Which temporary tooth of *Triïsodon* is homologous with the single one of the *Marsupialia* pointed out by Professor Flower?[1] As the addi-

[1] Transactions of the Royal Society, 1867, p. 631.

tional permanent teeth of the placental *Mammalia* must have appeared later in time than the one already found in the implacentals, they must be those later protruded; hence the fourth tooth in the jaw of *Triïsodon* must be regarded as homologous with the fourth premolar of a placental, which is the last of that series to appear. If this be true, the tooth which follows the shed tooth of the marsupials is not the fourth premolar, as supposed by Professor Flower, but the third premolar. This view is confirmed by the fact that the milk-tooth displaced by the fourth tooth in *Triïsodon* resembles in all respects the true molars, just as the permanent tooth occupying the same position does in *Didelphys* and some extinct Eocene genera. This goes to show that this tooth, permanent in marsupials, is temporary in placentals, and that, in spite of its form in the former group, it is the fourth premolar and not the first true molar, as supposed by Professor Flower. Thus the posterior milk-molar of diphyodonts is a permanent tooth in the *Marsupialia*.

This observation confirms my conclusion that the *Creodonta* form a group intermediate between the *Marsupialia* and *Carnivora*.[1] I may add that in *Triïsodon* the inferior border of the lower jaw is not inflected posteriorly.

Four species of this genus are known, which differ in some points of dental structure as well as in size. Their characters are as follows:

I. Internal anterior cusp of inferior true molars very small and well separated from external anterior cusp.

Length of inferior true molars .044 .... *T. quivirensis.*

II. Internal anterior cusp of inferior true molars larger and nearly connected with the external.

Heel of inferior true molars simple; length of true molars .023; smaller .. *T. levisanus.*

Length of inferior second true molar .011; heel with several lobes; larger. *T. heilprinianus.*

Length of inferior true molars .052; heel with two tubercles; largest..... *T. conidens.*

The superior molar teeth show a resemblance to those of *Mesonyx* and also to those of *Deltatherium*. Among the *Mesonychidæ*, *Triïsodon* approaches *Sarcothraustes* in the form of the inferior molars, in the expanded heel. On the other hand, the appearance of the anterior cusp of the inferior molars approaches what is seen in *Amblyctonus*. The small transverse posterior superior molar of *Triïsodon* further distinguishes it from *Amblyctonus*. A

[1] Proceedings Academy Philadelphia, 1875 (November 30).

series of modifications of the dental characters, proceeding from the simple to the more complex, may be constructed as follows: 1. *Mesonyx;* 2. *Dissacus;* 3. *Sarcothraustes;* 4. *Triïsodon;* 5. *Amblyctonus;* 6. *Deltatherium.* The first three belong to the *Mesonychidæ*, as distinguished by the form of the tarsal articulations. Whether *Triïsodon* must be arranged with *Amblyctonus* or not, cannot be ascertained until the foot structure is known.

As the number of the inferior premolars in three of the species now referred to this genus is unknown, it is possible that some of them may be hereafter referred to *Ictops*.

Triïsodon quivirensis Cope.

Loci supra citati.

Plate XXV. c; fig. 2.

Represented by both rami of the mandible, which exhibit alveoli or crowns of all the teeth excepting the incisors. Size about that of the wolf. Inferior canine directed upwards, its section nearly elliptic; a faint posterior, no anterior cutting edge. Fourth premolar rather large, with an anterior basal cingulum which is angulate upwards, and is not continued on the inner side of the crown. Cusps of the heel each sending a ridge forwards, the internal lower, obtuse, and descending to base of inner side of large cusp; the external larger, with an acute anterior cutting edge continuous with the cutting edge of the large cusp. True molars with an external, but no internal basal cingulum. Border of heel with one large and three smaller tubercles, the former with, the latter without, anterior cutting edge. Enamel of all the teeth nearly smooth. All the cusps are rather obtuse.

Measurements.

	M.
Length of inferior molar series	.080
Long diameter of base of canine	.013
Length of true molar series	.044
Lenth of base of Pm. iv	.016
Elevation of crown of do	.014
Length of base of M. ii	.016
Width of do. in front	.011
Elevation of do	.014

The measurements of the jaw are not given, as the animal is not adult, the last molar not being yet fully protruded.

From the lower Puerco Eocene beds of New Mexico, near the Canyon Largo, a branch of the great canyon of the San Juan River.

TRIÏSODON HEILPRINIANUS Cope.

Proceedings American Philosophical Society, 1881, p. 193. Paleontological Bulletin, No. 34, p. 193, Feb. 20th, 1882.

Plate XXVIII a; fig. 2.

This species may be readily recognized as smaller than the *T. quivirensis*, and as having the anterior inner cusp of the inferior true molar of larger proportions than in the corresponding teeth of the latter species. It is only represented in my collection by a portion of a lower jaw, which supports only one well-preserved molar. As the fourth premolar is not present, it is not positively ascertained that the species does not belong to *Ictops*.

The anterior cusp is very low, and is nearer the inside than the middle of the anterior border. The principal anterior cusps are opposite, and the external is a little the larger. The heel is larger than the basis of the anterior cusps, and has convex borders. Its internal border supports three tubercles, and the external border rises into a cutting lobe with lenticular section. Enamel smooth. No cingula, but the external base is injured.

Measurements.

			M.
Diameters of inferior molar;	vertical	of cusps	.0070
		of heel	.0052
	anteroposterior		.0110
	transverse		.0065

Puerco beds of New Mexico.

Dedicated to my friend, Professor Angelo Heilprin, of Philadelphia, who has contributed to our knowledge of the fauna and geology of the Eocene period.

TRIÏSODON LEVISANUS Cope.

Proceedings American Philosophical Society, 1883, p. 446.

Plate XXIV f; fig. 3.

This creodont is represented by a part of a right mandibular ramus, which contains the fourth premolar minus its principal cusp, and the first and second true molars, with the alveoli of the third. The ramus is deep, and probably belonged to an animal of about the size of the red fox. The molars have the structure most like that of the *T. heilprinianus*, especially anteriorly. The principal anterior cusps are united together for most of their elevation, while the anterior inner is much smaller and lower, and is situated between the middle and inner side of the anterior cusp. The heel is rather wide, and has a raised border. The external part of it is angular, and is somewhat within the vertical line of the base of the crown. The fourth premolar differs from that of the type of the genus, *T. quivirensis*, in having two acute longitudinal tubercles situated close together on the heel.

The anterior masseteric ridge is very prominent The masseteric fossa is strongly concave, but shallows gradually inferiorly. Its inferior border presents a low thickened ridge, which is recurved in front. This may be an individual character only. The inferior outline of the ramus is generally convex, and does not rise much below the masseteric fossa.

Measurements.

		M.
Length of last four inferior molars		.0315
" " true molars		.0230
Diameters of M. i.	anteroposterior	.0085
	tranverse	.0055
Length of Pm. iv. on base		.0090
Depth of ramus at M. i		.0200
Thickness " "		.0085

This *Triïsodon* is not only materially smaller than the *T. heilprinianus*, but differs in the characters of the heel of the inferior molars. In that species the internal border is tubercular; in this one it is entire. The *T. conidens* and *T. quivirensis* differ in the arrangement of the anterior cusps.

Dedicated to my friend, Henry Carvill Lewis, professor of mineralogy and geology in the Academy of Natural Sciences, Philadelphia.

From the Puerco epoch of New Mexico, discovered by D. Baldwin.

TRIÏSODON CONIDENS Cope.

Proceedings Academy Natural Sciences, Philadelphia, 1882, p. 297.

Plate XXIII d; figs. 9-10.

A right maxillary bone and corresponding mandibular ramus represent this species in my collection. The former sustains the last five molars, and the latter the last three, with alveoli of the others and of the canine tooth. The pieces indicate a skull of the size of that of the wolf, and a good deal more robust in its vertical measurements.

The third superior premolar has a base of triangular outline, the external side longer than either of the internal, which are connected by a broadly rounded angle. The external cusp is of lenticular section at the base, and circular section near the apex. An internal cusp is represented by a strong cingulum, as in *Periptychus*, which connects with the posterior base of the external cusp. The crown of the fourth superior premolar has a triangular base of which the anterior side is shorter than either of the other two, which are subequal. The external cusp is large, simple, and subconic. The internal is distinct but smaller, and is continued posteriorly as a cingulum to the posterior base of the external cusp. No internal cingulum. The crown of the first true molar is worn to the roots. The second true molar is the longest of the series. Its base is a triangle, placed transversely to the axis of the jaw, of which the external side is the shortest, the anterior the next longer, and the posterior the longest. The apex or internal extremity of the crown is obtusely rounded. There are two subequal external cusps, which are injured in the specimen. The internal cusp is the apex of a V whose limbs form the anterior and posterior edges of the grinding face of the crown, extending outwards to near the bases of the external cusps. Posterior to the posterior one is a strong basal cingulum. No internal, and a faint anterior cingulum. There is probably an external cingulum, but it is broken away. The last molar is of an oval outline placed transversely to the cranial axis, both the external and internal extremities contracted, the latter a little more so. There is a large anterior external conical cusp. The posterior external is small, and is situated at the posterior third of the posterior border of the crown. The internal cusp is well developed, and has a subcircular section. There are strong external and posterior cingula, and a weak anterior one, but no internal cingulum. The posterior extremity of the maxillary bone within the zygoma, is immediately above the posterior border of the last superior molar.

Measurements of superior molars.

		M.
Length of bases of posterior five		.069
Diameters base, Pm. iii	anteroposterior	.013
	transverse	.009
Diameters base, Pm. iv	anteroposterior	.0145
	transverse	.014

		M.
Length base of true molars		.039
Diameters base of M. ii	anteroposterior	.0175
	transverse	.021
Diameters base, M. iii	anteroposterior	.010
	transverse	.0175
Elevation of base of zygoma above base of M. iii		.018

The ramus of the lower jaw is, as usually with the Creodonta, deeper and less robust than that of Carnivora of corresponding size. It is also more compressed than that of the *Triïsodon quivirensis.* It retains its depth to below the canine teeth, and does not shallow below the middle of the coronoid process, where also there is no tendency to inflection. The anterior masseteric ridge is not very prominent, and the masseteric fossa is not defined below, nor is the inferior edge of the ramus prominent or ridged at that point.

The premolar teeth are lost, but they occupied but a short space, and were probably only three in number. The first and second true molars are subequal, while the third is a little smaller than either. Each consists of an anterior higher and a posterior lower portion, the lower region being at the junction of the two. The anterior part has a nearly circular section, and contracts towards the apex. The latter is divided into three cusps, a larger external and two lesser internal. The external and posterior internal soon fuse on wearing, and their combined section is a crescent. The anterior inner is small, and stands near the inner edge of the crown, and not at the middle as in *T. quivirensis*, and is circular in section. The heel of the tooth rises to its posterior border, which is divided into two cusps. Each of these sends a ridge forwards towards the base of the anterior cone of the tooth. The external is the larger, and reaches that base. The internal is smaller, and falls short of it. The posterior inferior molar differs from the others in form as well as in size. There is no posterior inner anterior cusp, the large external cusp being supplemented by a small anterior internal only, which sends a little ridge downwards and posteriorly. The heel is narrowed, and supports the two cusps on its posterior border in contact, and not separate as on the other teeth. The external is the larger, and extends forwards to the base of the anterior cone near its middle. Some remnants of hard matrix leave it uncertain whether there is a small median posterior marginal tubercle on the first and second molars or not.

The first inferior true molar has a strong external cingulum; the second has none; the third has one, which is most evident between the cusps, is weaker at the base of the posterior lobe, and faint at the anterior lobe. No internal cingula.

Measurements.

		M.
Length of true molar series		.052
Length from M. iii to anterior masseteric ridge		.013
Diameters of M. i	anteroposterior	.017
	transverse	.0115
Diameters of M. ii	anteroposterior	.018
	transverse	.011
Diameters of M. iii	anteroposterior	.016
	transverse	.0105
Depth of ramus at M. iii		.047
Width of ramus at M. iii inferiorly		.013

The molar teeth of this species are more like those of the *T. heilprinianus* than those of the *T. quivirensis.* This is seen in the more conic character of the anterior lobe of the tooth, and the better development of the anterior inner cusp. The species is a good deal larger than the *T. quivirensis.*

From the Puerco beds of N. W. New Mexico, D. Baldwin.

DELTATHERIUM Cope.

American Naturalist, 1881, March 25, p. 337, *Lipodectes* Cope, l. c., 1881, p. 101-9.

Dental formula: I. $\frac{3}{?}$; C. $\frac{1}{1}$; Pm. $\frac{3}{3}$; M. $\frac{3}{3}$.

Superior premolars, the first and second with simple crowns, the third with one large external cusp and an internal small one. The fourth premolar with a large, simple external cusp and a prominent internal one. The first and second true molars with triangular bases, supporting two external compressed conic cusps and a subtriangular internal one. Last molar similar in its internal portions, the external part narrow. A wide diastema in the lower jaw. Inferior premolars simple, two rooted. True molars with anterior inner cusp well developed, forming with the anterior external a sectional edge, as in *Stypolophus.* Heels well developed, much produced, and supporting a special tubercle in the last molar.

The superior molars of this genus may be distinguished from those of *Pelycodus* by the absence of the intermediate tubercle and of the posterior internal tubercle. They differ from those of *Esthonyx* in the absence or weakness of the posterior inner tubercle, and in the absence of the ear-like expansion of the external angles.

The number of possible combinations of tubercular and tubercular-sectorial molar teeth is considerable, and many of them are represented in the genera of the *Creodonta*. *Deltatherium* is a genus which has, in the lower jaw, two tubercular sectorials, and a third with a long heel posterior to them. The genus thus stands between *Stypolophus* and *Didymictis*, but is nearer the former than the latter, since it has three true molars. It differs further from both in having but three premolars and a wide diastema. The canine is well developed. Although there is a tubercular tooth, the cutting apparatus is well developed, and indicates more than usually rapacious habits. There is as yet but one species known. It is allied to *Leptictis*, and agrees with *Ictops* and *Mesodectes* in possessing an internal tubercle of the third superior premolar, but differs from both in having but one external cusp of the fourth superior premolar, resembling in this respect the more typical *Oxyænidæ*.

With this genus we enter a series of forms in which the dentition is more decidedly opossum-like than in those previously considered. Besides *Deltatherium* there are species of the Eocene which I have referred to the Miocene genera *Ictops* and *Peratherium*, and there are the as yet purely Miocene genera *Leptictis* and *Mesodectes*. The Eocene *Diacodon* must be referred to the same category. *Leptictis* differs from *Didelphys* in having but two superior incisors on each side instead of four; *Ictops bicuspis* has three on each side. *Diacodon* differs in not having the anterior inner cusp of *Didelphys*, and in *Ictops* it is present, but small. This character will serve to distinguish these genera empirically from *Stypolophus*, as will also the development of cusps on the heels of the molars.

DELTATHERIUM FUNDAMINIS Cope.

American Naturalist, 1880, March, p. 338. *Lipodectes penetrans* Cope l. c. 1881, p. 1019.
Plates XXIII c, figs. 8–11; XXV a, fig. 10; XXV d, fig. 3.

Represented by specimens which display the dentition of both maxillary bones minus the canines, and by several mandibles. The most instructive specimen includes the cranium anterior to the sagittal crest with dentition, with several fragments of the posterior part of the skull, with parts of both mandibular rami supporting several teeth, with parts of humerus and ulna. These specimens show that the *Deltatherium fundaminis* and *Lipodectes penetrans* are one and the same species. Besides this, there are separate mandibles of two other individuals, making five in all.

In the first-named specimen the second premolar is convex on the inner face; the base of the third is a nearly equilateral triangle. The bases of the true molars are triangles, with bases external. The internal angle supports an acute cusp, and has a posterior basal cingulum, which is very strong in the last three molars. The two external cusps of the first and second molars are situated well within the base, which is folded into a strong cingulum. This cingulum develops strong anterior and posterior angles. This is the largest species of the family yet discovered.

Measurements.

		M.
Extent of series of last six molars		.045
Extent of series of true molars		.026
Diameters of fourth premolar	anteroposterior	.0074
	transverse	.0076
Diameters of second true molar	anteroposterior	.0087
	transverse	.0100

The type specimen of the *Lipodectes penetrans* is a left mandibular ramus, with three of the molars preserved The last has a long heel; the first and second true molars are alike, and resemble those of *Triïsodon*, but the appendicular cusps are better developed. The anterior inner cusp is, however, smaller than the others and is nearly median in position. The heel is elevated on its external border into a strong triangular cusp. The posterior border rises into an acute cusp, which is internal to the middle line. The internal border of the heel is not elevated, and the surface is the oblique inner face of the external cusp. The anterior cusps are only moderately elevated and the cusps are acute. The enamel is smooth, and there is a low cingulum on the external base. The first (second) premolar is two-rooted, and has a large base. The second (third) consists principally of an elevated cusp with a subtriangular section. The heel is very small and acute, and there is no anterior basal tubercle. The internal face is strongly grooved in front. Canines directed upwards, with robust base. Symphysis short. Length of molar series, .043; of premolars, .019; of diastema, .012; length of base of last molar, .010; do. of canine, .007; depth of ramus at last molar, .018; of diastema, 015. As large as, but more robust than, the red fox.

The third specimen above mentioned is somewhat injured by pressure, but exhibits the following characters The sagittal crest rises from an ascending frontal region, so that the profile is concave. The muzzle is short, and is contracted behind the alveolæ of the canine teeth. The latter are prominent, and the premaxillary region is short and rather wide. The superciliary ridges are rather prominent and terminate in postorbital angles, which are rather prominent. This is due to the abruptness of the convergence of the anterior temporal borders, which are angles, and not ridges. The anterior, and probably the posterior, part of the brain cavity is very narrow. The postglenoid process is prominent, and there is no trace of preglenoid ridge. There is a large postglenoid foramen, and the infraorbital foramen issues above the middle of the third premolar. The lachrymal foramen is small, and is entirely within the preorbital border. The posterior nareal opening is small, not exceeding in width the space which separates it on each side from the internal border of the last molar.

Of the three incisors the external is the least, and it is separated by a wide diastema from the canine. The latter is large and the crown is directed vertically downwards. The crowns are mostly broken off, but enough remains to show that the posterior edge is acute, and is bounded within by a wide, shallow groove, and by a less marked groove externally. There is a distinct but short diastema behind the canine. The first (second) premolar is a flattened acute cone, with an acute posterior edge. The base of the third premolar is triangular in section, but the internal projection does not support a cusp. The anterior and posterior basal cusps are rudimental. The fourth premolar has an internal cusp which sends a ridge downwards and outwards on the anterior side of the crown. There is an anterior but no posterior basal lobe, which does not rise into a cusp. Posterior or basal cingulum weak. In the true molars the posterior external cusp is connected with the corresponding external angle of the crown by a ridge, while the anterior cusp is not connected with the anterior angle. The external cusps are set in further on the second than on the first true molar. The last true molar is cut off obliquely on the external side and posteriorly, so that the posterior external cusp stands on the posterior external angle of the crown. The posterior cingulum is strong on all the true molars, and extends round the inner base of the crown. It is strongest on the first and second, but does not rise into a cusp on either, as it does in *Pelycodus* and other forms.

The probable inferior canine has a characteristic form. The crown is not elongate, and the section of its base is a half circle. Above the base the inner face becomes concave, a broad median ridge dividing the concavity into two grooves, which are less marked near the apex. There is a shallow groove external to the posterior edge, which is thus acute; apex, obtuse. The inferior molars, except the first and last true molars, have been described above The first true molar does not differ from the second. The anterior part of the third is just like that of the second, but the heel is much longer. I cannot give the details of its form, as the surface is injured.

Measurements.

		M.
Length of superior dental series from canine		.043
Length of diastema		.005
Length of premolar series		.017
Diameters of canine	anteroposterior	.009
	transverse	.006
Length of precanine diastema		.007
Diameters, Pm. iv	transverse	.008
	anteroposterior	.007
Diameters, M. ii	transverse	.009
	anterposterior	.008
Diameters, M. iii	transverse at middle	.009
	anterposterior	.006
Length from incisive border to postorbital angle		.056
Width between superciliary edges		.030
Length of true inferior molars		.023
Length of last inferior molar		.0098
Depth of ramus at M. ii		.0155

This species was a half larger than the common opossum, and was much more robust. The typical specimen was found by Mr. D. Baldwin at the mouth of Canyon Largo, on the San Juan River, Northwestern New Mexico, in the Lower Eocene formation. Mr. Baldwin informs me that it came from below all the Wasatch Sandstones, which would place its horizon in the Puerco formation. The other specimens came from about the same position.

DELTATHERIUM BALDWINI Cope.

Proceedings American Philosophical Society, 1882, p. 463 (Nov.)
Plate XXIII d; fig. 12.

This Creodont is known only from a portion of a right mandibular ramus, which supports the two last premolars and the first true molar, with part of the second. It differs from the *D. fundaminis* in its materially smaller size, and in the forms of the teeth. The first true molar is a more robust tooth, and the basis of the posterior or heel crest is more rounded and less angulate. The anterior inner cusp projects less anteriorly. The fourth premolar has a distinct anterior basal lobe, which is wanting in the *D. fundaminis.* Its heel is short and wide, and the posterior face of the principal cusp is flat, and there is a rudiment of an internal tubercle on its side. The second premolar is elevated and acute, has no anterior basal lobe, and has a very short, wide heel. Enamel slightly roughened. The animal was rather aged.

Measurements.

		M.
Length of Pm. ii and iii and M. i		.0160
Diameters M. i	anteroposterior	.0058
	transverse	.0040
Elevation of crown of Pm. iii		.0052
Depth of mandible at M. i		.0180

From the Puerco beds of Northwestern New Mexico. Dedicated to Mr. D. Baldwin, the discoverer of the mammals and fauna of the Puerco beds, which is one of the most important in the history of American paleontology.

DELTATHERIUM INTERRUPTUM Cope.

Proceed. Amer. Philos. Soc., 1882, p. 463.
Plate XXIII d; fig. 13.

The smallest species of *Deltatherium* is, like the *D. baldwini*, only represented by the anterior part of a right mandibular ramus, which supports the last premolar and the first true molar, with the bases of the other premolars and part of the canine. The canine is small, and the first premolar, in accordance with the generic character, is wanting. The second premolar is two-rooted. The fourth has an elevated principal cusp and a narrow heel on the inner side of the posterior base; anterior base injured. The first true molar has very little sectorial character, and resembles the corresponding tooth of a *Pelycodus*. It differs entirely from that of the *D. fundaminis* in the possession of a well-marked posterior internal cusp, which is connected by a ridge with the large internal lateral cusp of the heel. The anterior cusps of opposite sides subequal. A weak external basal cingulum on the anterior half of the crown; no internal cingulum. Enamel of the teeth wrinkled.

Measurements.

	M.
Length of premolar series	.0140
Elevation of Pm. iv	.0040
Diameters of M. i { anteroposterior	.0055
Diameters of M. i { transverse	.0042
Depth of ramus at Pm. i	.0090
Depth of ramus at M. i	.0113

On comparison with the *D. fundaminis*, the first molar tooth has the same dimensions, but the premolars are considerably smaller. The ramus is also shallower. Found by Mr. Baldwin in the Puerco beds of Northwestern New Mexico.

DIDELPHODUS Cope.

American Naturalist, June, 1882, p. 522.

Dental formula: I. $\frac{?}{3}$; C. $\frac{1}{1}$; Pm. $\frac{?3}{4}$; M. $\frac{2}{3}$. Second superior molar a simple cusp; third with an internal cusp; fourth with a simple external cusp without heel. True molars with two external cusps, set well in from the external border. Last superior molar narrowed and transverse. Inferior dentition like that of *Stypolophus*, except that the fourth premolar has a small internal tubercle. Canine large. Symphysis not coössified.

This genus differs from *Deltatherium* in the presence of an additional premolar tooth in the lower jaw. It is uncertain whether there is a first superior premolar, but I cannot find an alveolus for it. From *Stypolophus* it differs in the triangular character of the superior molars and the simple trenchant form of the fourth superior premolar. In the latter respect it is identical with *Proviverra* of Rütimeyer. I should refer the American species to this genus but for the fact that it appears to have but three superior premolars, while *Proviverra* has four.

But one species is known.

Didelphodus absarokæ Cope.

American Naturalist, 1882, June; *Deltatherium absarokæ* Cope, l. c. 1881, p. 669.
Plate XXIVe; fig. 13.

This animal repeats very closely the characters of the *D. fundaminis*, but is much smaller in all its proportions. Both branches of the lower jaw accompany the anterior part of the skull, so that the dentition is well displayed. There are three inferior tubercular-sectorial molars, as in *Stypolophus*, but the fourth premolar has an internal tubercle, which is not found in that genus. The same tooth has a rudimental heel, which sends an angle up on the inner side of the crown, which is then deflected and terminates below in the rudimental anterior basal tubercle. The apical angle is little produced, and constitutes the internal cusp above mentioned. The third inferior premolar is large, has a rudimental heel, and no inner lobe; the first premolar is two-rooted. There are only three inferior incisors. The inferior canines have an open groove on the inner side and a rib on the external side. The latter is bounded anteriorly and posteriorly by a shallow open groove. The symphysis is long and oblique; it extends posteriorly to below the front of the third premolar. The posterior mental foramen is below the posterior side of the same tooth, and the anterior one is below the first premolar tooth.

The superior molars are triangular, and the external posterior angle is not produced. The external cusp of the fourth premolar is compressed and simple, as in *Proviverra;* in *Stypolophus* (*Prototomus*) *viverrinus* and *S. multicuspis*, that tooth has a conic cusp and large posterior heel. The crowns of the second and third premolars are quite elongate and acute, and the third has a small posterior basal lobe. On the external base of the fourth are very small anterior and posterior lobes. The principal cusps of all the true molars are double, and stand on the middle of the transverse diameter of the crown. This leaves an extensive obliquely sloping external face, which terminates externally by a narrow cingulum. The two cusps are well separated from each other on all the true molars. The internal cusp is the apex of a V, each limb of which is continued into a cingulum along the anterior and posterior base of the crown. The external base of the first true molar is subequally bilobate, on the second the anterior lobe is produced, and on the third the posterior lobe is only represented by a right angle between the external and posterior borders of the crown, which are of equal length. The second principal cusp is also reduced. Canines well developed. Enamel smooth.

Measurements.

		M.
Length of superior molar series		.0216
Length of superior true molars		.0107
Diameters of second true molars	anteroposterior	.0033
	transverse	.0055
Width of jaws at same tooth		.022
Width between bases of canines		.008
Depth of ramus mandibuli at Pm. i		.005
Depth of ramus mandibuli at M. iii		.009

From the Wasatch Eocene of the Big-Horn River; J. L. Wortman.

This species was an opossum-like animal of the size of the American weasel, *Mustela americana*. Its delicately acute teeth indicate a diet of insects, which no doubt abounded during the Wasatch epoch.

STYPOLOPHUS Cope.

Stypolophus Cope, Second Account New Vertebrata Bridger Eocene (Paleontological Bulletin, No. 2), p. 1, Aug. 3, 1872, Proceed. Amer. Philos. Soc., 1872, p. 446; An. Rep. U. S. Geol. Surv. Terrs., 1872, p. 559. *Prototomus* Cope, Rep. Vert. Foss. New Mexico, 1874, p. 13; System. Cat. Vert. Eocene New Mexico, U. S. Geol. Surv. W. of 100th Mer., 1875, p. 9. *Cynohyænodon* Filhol, Comptes Rendus de la Société Philomathique, Paris, 1873. Recherches sur l. Phosphorites du Quercy, 1877, 227.

Molars seven below, *i. e.*, four premolars and three true molars, and probably the same number above. Inferior true molars, consisting of three elevated cusps in front and a low horizontally-expanded heel behind; the external cusp largest, the internal smallest, and the anterior intermediate, forming with the external a short sectorial blade. The inferior premolars two-rooted (the first only seen in *S. caylusi*, *S. whitiæ*, and *S. hians*); the crown consisting of a compressed cusp and short trenchant heel. Of the superior molars the last is narrow, transverse, and with a blade-like crown. The two preceding have crowns forming right-angled triangles in horizontal section, the right angle being the antero-external. The antero-posterior cutting-edge consists of two cusps in the middle and a short blade at the posterior angle of the crown. The internal angle supports a cusp. The last premolar has a trilobate section at the base, and supports a median subconic cusp, a short posterior blade, and an internal tubercle. The second premolar is compressed without internal heel, and with a rudimental posterior one. (The first premolar is two-rooted in *S. caylusi* and *S. hians.*)

The species of *Stypolophus* of which I have obtained the best preserved remains is the *S. viverrinus*, an animal about the size of a domestic cat, from the Wasatch beds of New Mexico. Its mandibular bones and teeth are unknown, but I have derived from it the characters of the dentition of the maxillary bone, as above stated. The maxillary teeth of the *S. multicuspis* and *S. aculeatus* are similar in generic characters, and of these species I know almost the entire dentition of the lower jaw. The posterior part of the cranium of *S. viverrinus* displays a low sagittal crest. The supraoccipital bone has a moderate extent on the upper surface of the cranium, supporting part of the sagittal crest, as well as the prominent oblique ones of the inion. The front is rather wide, and the nasal bones are flat, and but little narrowed posteriorly. The lachrymal foramen is large, and entirely within the prominent anterior margin of the orbit; it is of a vertically oval form. A suture extends from it postero-externally to the rim of the orbit, and then returns forward and upward on the facial surface, inclosing what I suppose to be the lachrymal bone. On cleaning the surface, I cannot trace any lachrymal bone posterior to the foramen, as is usual in *Carnivora* (*Canis*, *Felis*), and must therefore suppose that this genus presents an external and anterior position of the lachrymal bone, as in *Ungulates*. The evidence of this arrangement is seen on both sides of the head. The foramen infraorbitate is large, and issues above the third premolar.

The characters presented by the vertebræ are those of the *Creodonta* in general, with the following modifications: A cervical is of medium length, possesses a hypapophysial heel, which is produced downwards behind, and has but little trace of a neural spine. The neural arch is wide and flat above, and it is pierced on each side by a foramen not far from the lateral border. Two anterior lumbars from just behind the flying ribs, have no diapophyses unless a small, narrow, broken area indicates the base of a very rudimental one This is at the anterior end of a strong longitudinal ridge, which marks the inferior part of the side of the centrum The anapophyses* are strong, inclosing the anterior zygapophyses of the succeeding vertebra on the lower side. In *Ursus arctos*, *Canis familiaris*, and *Felis catus* there is no vertebra intervening between the last bearing a rib and the first bearing a diapophysis. In *Ursus arctos*, the centra are short, and the diapophysis occupies an elevated position. In *Stypolophus viverrinus*, the centrum is moderately elongate, and the ridge representing the diapophysis has an inferior position, resembling rather *Canis* and *Felis* in these particulars. A portion of the sacrum preserved shows it to have been of robust proportions. Besides the superior intervertebral foramina, there is a small one each side of the neural arch in front of the posterior zygapophysial ridge. A caudal vertebra is relatively large in all its dimensions. A fragment of the femur shows that both the great and little trochanters are well developed, the former inclosing the usual fossa. The distal halves of both tibiæ are preserved, one of them adhering to a mass of the vertebræ. The shaft below the middle is subcylindric, while the distal end presents the peculiarity common to all the flesh-eaters of the Wasatch *Eocene* epoch. The astragalar surface is without groove, and is oblique, both transversely and longitudinally. The inner extremity of the bone is produced downwards, fitting the inner oblique face of the astragalus, as well as the concavity of the side of the neck by its end. There are no strong ligamentous grooves. The bones of the feet are unknown. A comparison of such portions of the limb bones as I have observed (those of *S. viverrinus*) with those of *Felis catus* (*domesticus*), *Canis familiaris*, and *Ursus arctos*, has the following result: In the humerus the tuberosities are not so pronounced; especially is the great tuberosity more produced upward and outward in the recent genera, whence the bicipital groove is deeper. In *Ursus arctos* the greater tuberosity is also produced more posteriorly, and in all of the species named, its posterior bounding ridge is more pronounced on the shaft than in *P. viverrinus*. The great trochanter of the femur has the elevated position of that of *Felis* and *Canis* rather than the depressed form of that

* In my description of this genus in the Wheeler Reports, these processes were called metapophyses.

of *Ursus*, and the compressed and moderately elevated distal end is that of the former two, rather than like the same region in the latter genus. The distal end of the tibia is unlike that of either of the three genera named, but resembles most that of *Ursus*. The entirely distinct character of the astragalar articular extremity has been already described. The anterior end of the shaft is convex in *S. viverrinus*, flat in *Felis* and *Canis;* flat behind in the former, convex in the latter. The external end of the shaft is transverse in *S. viverrinus*, oblique in *Canis* and *Felis;* especially so in the former, being more or less parallel with the inner astragalar groove, while in *S. viverrinus* it diverges from the angle which represents the groove. The tendinous groove is wider and better defined than in *C. familiaris*, more resembling that in *Felis*. The inner malleolus is more anterior in position than in the two genera named, and bears a distal articular facet, which is wanting in *Felis* and *Canis*. As compared with *Ursus arctos*, the inner malleolus is more produced, and the outer distal border quite different, the truncate outline of *Stypolophus* being represented by a tuberosity. The anterior face of the shaft is convex in *Stypolophus*, concave in *Ursus arctos;* the posterior flat in the former, convex in the latter. The entire distal end of the tibia is more transversely expanded in *Ursus*.

This genus, as now defined, is identical with that called by me in previous papers on the paleontology of New Mexico, *Prototomus*. It may be found to be proper to use this name, but for the present I use an older one, which I proposed for similar *Carnivores* of the Bridger Eocene of Wyoming. Unfortunately, I am not able to state the number of the tubercular-sectorial molars of the *S. pungens* Cope, the type of the genus, as my specimens have only the last two in place. The structure of the separate molar teeth of both jaws is identical in the species from the two regions, and the generic characters of the dentition, so far as known, are the same in the best preserved species, *S. multicuspis* and *S. viverrinus* of the Wasatch, and *S. aculeatus* of the Bridger epochs. The three tubercular-sectorials in the lower jaw, and the two bicuspid molars in the upper, distinguish this genus from the allied *Oxyæna*.

M. Filhol has described very fully beautiful specimens of species of *Stypolophus* from the *Phosphorites* of southern central France. He names them *Cynohyænodon caylusi* and *C. minor.** Through the courtesy of M. Filhol I am in possession of specimens of the former, and I have examined the types of his descriptions. There can be no doubt of their pertinence to *Stypolophus.*† *S. caylusi* differs from *S. multicuspis* and *S. aculeatus* in the reduced elevation of the cusps of the first inferior true molar, and the greater obliquity of the superior molars.

Professor Gaudry‡ has identified this genus with the *Proviverra* of Rütimeyer, which was proposed to receive a species from the Swiss Eocene. I have examined casts of Rütimeyer's type, which includes the dentition of both jaws, and which are preserved in the laboratory in charge of Professor Gervais, in the Jardin des Plantes, to whom my acknowledgments are due. I find enough difference to induce me to believe that *Proviverra* and *Stypolophus* cannot be united, excepting by the discovery of species which shall show transitional features in the characters. The difference is, that while in *Stypolophus* the fourth superior premolar has an internal cusp and an external conical cusp flanked anteriorly and posteriorly by a basal heel, in *Proviverra* the external part of the crown is a single triangular trenchant cusp, and the internal heel is low, forming with the rest of the base of the crown, a right-angled triangle. This difference is significant, when we recollect that this tooth is the homologue of the sectorial in true *Carnivora.*

Dr. Leidy has applied the name *Sinopa* to some flesh-eaters of the Bridger epoch without distinctive generic description. An examination of the typical specimen of the *S. vorax*, which Dr. Leidy kindly permitted me, shows that it differs from *Stypolophus* in the rudimental character of the heel of the last molar, if the specimen is not deceptive. It is otherwise identical in the last four inferior molars.

Species. I have referred five species of the Wasatch formation to this genus and a fourth provisionally (*S. hians*). Three species of the Bridger epoch probably belong to it, which, with the two French species, make a total of ten. They may be distinguished as follows, with the imperfect material at my disposal. The molars measured are those of the inferior series; in the case of the *S. viverrinus* their length is estimated from those

* Recherches sur les Phosphorites du Quercy, 1877, p. 227.
† Cope, Bull. U. S. Geol. Surv. Terrs., 1879, p. 43.
‡ Enchainements du Monde Animal, 1878, p. 20.

of the superior series. The depth of the mandibular ramus is taken at the penultimate true molar.

I. Length of true molars less than .0140.

Length of true molars, .0135; premolars less extended............. *S. viverrinus.*
Length of true molars, .0112; premolars much larger.............. *S. minor.*

II. Length of true molars from .0180 to .0220.

True molars, .0183; depth of ramus, .0080; first true molar small... *S. caylusi.*
True molars, .0200; depth of ramus, .0100; heels of molars large, wide.. *S. pungens.*
True molars, .0220; depth of ramus, .0125.......................... *S. secundarius.*
True molars, .0210; depth of ramus, .0150; heels large; first molar not reduced.. *S. multicuspis.*
True molars, .0220; depth of ramus, .0150; heels very narrow and short... *S. brevicalcaratus.*
True molars, .0220; depth of ramus, .0180; heels elongate, basin-shaped.. *S. whitiæ.*

III. Length of true molars exceeding .0230.

Length of true molars, .0235; depth of ramus, .0210, robust........ *S. strenuus.*
Length of true molars, .0250; premolars large; ramus slender; depth, .0140.. *S. aculeatus.*

There is some diversity in the form of the astragalus in the species above named. There may be two generic forms included here, the one having the astragalus without trochlear groove, represented by *S viverrinus*, the other with the tibial articular face of that bone slightly concave, represented by the *S. whitiæ.* As the latter structure is the later in its appearance in time, it is presumably that of *Stypolophus pungens*, the type of the genus which is from the Bridger formation. For the older form, the name *Prototomus* (Cope, 1874) must be used. *P. viverrinus* and *P. hians* are certainly referable to it.

STYPOLOPHUS INSECTIVORUS Cope.

Proceed. Amer. Philos. Soc., 1872, p. 469; Paleontological Bulletin No. 3, p. 1, published August 7, 1872.

Plate xxiv, figs. 10–11.

Represented by a posterior molar and a premolar of the right side of an animal less than half the size of the *S. pungens* Cope. The molar presents three anterior trihedral acute tubercles, of which the exterior is more elevated than the others. Its posterior plane forms one transverse face with that of the inner posterior. The posterior tubercular heel is low, and supports an oblique ridge which bounds a deep groove behind the outer cusp, no doubt to receive the molar of the upper jaw. This arrangement is not seen in *S. pungens*. The premolar is a flat cone with faint traces of a tubercle behind, and cingulum on inner side.

	M.
Length of crown molar	.0050
Height of inner cusp	.0040
Length of heel	.0025
Width of crown	.0030
Height of crown premolar	.0040
Length of crown premolar	.0040

Found in the Eocene bad lands of Black's Fork by the writer.

STYPOLOPHUS PUNGENS Cope.

Loc. cit., 1872, p. 466; Paleontological Bulletin No. 2, p. 1, August 3, 1872.

Plate xxiv, fig. 8.

This is the type of the genus, and is represented by the posterior part of a mandibular ramus which supports the last two molars.

The species is of medium size in the genus, and has a rather shallow mandibular ramus. The heels of the molars are well developed and have a raised border. The inner cusps are not much elevated, the external one much exceeding them. The masseteric fossa is well defined anteriorly, and a well-marked angle bounds it below. There are no cingula on the molars, and the enamel is smooth.

The measurements are:

	M.
Depth of ramus at last molar	.011
Length of last molar	.0072
Width of last molar, posteriorly	.0040
Height of inner tubercle	.0062
Height of external tubercle, anterior	.0040

This species was about the size of the gray fox.

From the bluffs of Cottonwood Creek, Wyoming.

STYPOLOPHUS BREVICALCARATUS Cope.

Loc. cit., p. 459; Paleontological Bulletin No. 3, p. 1, August 7, 1872.

Plate xxiv, fig. 9.

Established on a portion of the left mandibular ramus, containing the penultimate and antepenultimate molars, of an animal of larger size than the type of the genus, *S. pungens.* The molars have the general characters of the corresponding ones of that species, but differ in their greater elevation in comparison with their length, and the greater convexity of the outer side. The shortness is occasioned by the abbreviation of the heel, which, in the last molar present, is very small and flat, and without keel or tubercle on its surface. That of the molar preceding it is larger, and presents in its elevated outer margin a trace of the keel seen in the smallest species, but it is also of reduced size. Enamel smooth.

	M.
Length of two molars	.016
Length of penultimate crown	.008
Width of penultimate crown	.0047
Length of penultimate heel	.002

STYPOLOPHUS WHITIÆ Cope.

Stypolophus strenuus Cope, Bull. U. S. Geol. Surv. Terrs., 1881, vi, p. 192, not of 1875.

Plate xxv *b*, figs. 8–14.

This species is represented by a right mandibular ramus which supports all the molar teeth, and displays the alveolus of the canine, and lacks all posterior to the coronoid process. Also by a portion of the frontal bone, two vertebræ, fragments of scapula, humerus, ulna, radius, ilium, and tibia, and the greater part of both tarsi They represent a species larger than the Virginian opossum, and intermediate between the *S. brevicalcaratus* and *S. strenuus* in proportions. It has not the rudimental heels of the molars of the former species, nor the robustness of the latter.

The inferior outline of the mandible is gently curved from the canine to below the last molar. The anterior border of the masseteric fossa is well marked, but not the inferior border. The ramus is compressed and deep. The canines have stout roots and narrow curved crowns. The first premolar is separated by a short space from the canine, and by a longer from the second premolar. It has either a single compressed root or two roots confluent within the alveolus. The crown is truncated obliquely behind. The second premolar is two-rooted and the crown is elevated anteriorly and depressed posteriorly. The third premolar is more symmetrical, but the heel is produced. It is narrow and keeled medially. The fourth premolar is abruptly larger than the third. Its crown is simple, except a low tubercle at the anterior base and a short trenchant heel at the posterior base. Of the three tubercular-sectorials the first is the smaller. The heels of all three are rather narrowed and elongate. The margin is raised all round, inclosing a basin; a notch in the external margin cuts its anterior part into a tubercle. The two internal tubercles are rather obtuse, and are considerably shorter than the external cusp.

Measurements.

	M.
Length from canine to end of last molar	.060
Length from canine to first true molar	.037
Length from canine to second premolar	.015
Length of base of fourth premolar	.009
Elevation of fourth premolar	.007
Length of base of second true molar	.007
Length of heel of second true molar	.0026
Elevation of second true molar	.009
Depth of ramus at third premolar	.015
Length of superior canine	.028
Length of crown of superior canine with enamel	.012

A portion of the frontal bone shows weak anterior temporal ridges uniting early into a sagittal crest, which is low as far as preserved. The parietal bones overlap the frontal as far forwards as the temporal ridges. Anterior to the latter the front is concave in transverse section. Viewed from below, the spaces for the olfactory lobes are large and entirely anterior to those which received the anterior lobes of the hemispheres; each one is about as wide as long. In the small part of the cerebral chamber wall left there is no indication of convolutions, which would be visible in a gyrencephalous brain; two air-chambers in front of each olfactory lobe.

The base of the transverse process of the atlas is perforated from behind to the middle of its inferior side: from the latter opening a foramen penetrates directly into the neural canal. A posterior dorsal vertebra has the centrum longer than wide and much depressed. Its inferior face is regularly convex in section. The proximal end of the scapula shows that its posterior border is much thickened, and that the spine arises abruptly and near to the glenoid cavity. There appears to have been scarcely any coracoid; the surface adjoining it is, however, injured. The humerus lacks the proximal portion, and the inner half of the condyles with the epicondyles. The deltoid crest is not very prominent, so that the shaft is rather slender.

The external distal marginal crest is thin, and is continued well up on the shaft. The external part of the condyle displays no intertrochlear ridge. Olecranar and coronoid fossæ well marked. The olecranon is robust and deep, and is truncate posteriorly and below. The head of the radius is a regular transverse stout oval.

A fragment of the ilium from near the acetabulum displays a prominent "anterior inferior spine." The best preserved tarsus includes calcaneum, astragalus, cuboid, and navicular bones. The tibial face of the astragalus is strongly convex anteroposteriorly and slightly concave transversely. The head is prolonged some distance beyond the distal extremity of the calcaneum, and presents a convex internal border and a concave external one. Its long axis is parallel to that of the tibial portion, but is not in the same axis, owing to its lateral position. The external face of the trochlear portion is vertical, and is interrupted by a deep fossa behind. The internal face is very oblique, and passes into the superior face of the head. The posterior face of the trochlea is grooved with a wide and shallow groove, which just reaches the superior face, terminating on the external side. The superior face is not grooved, but is shallowly concave in transverse section. The head is a transverse oval, and is convex; it has a small facet for the cuboid on the outer side.

The heel of the calcaneum is large and expands distally, so as to be as wide as deep. The convex astragalar facet is very oblique to the long axis of the calcaneum; the sustentaculum is rather small. Below the latter is a narrow tuberosity looking downwards and forwards. On the external side, close to the cuboid facet, is a depressed crest. The cuboid facet is as deep as wide. The cuboid bone is a little longer than wide proximally, and narrows distally. It has a narrow astragaline facet and a deep fossa below proximally. The hook inclosing the groove for the tendon of the flexor muscle is prominent. The navicular is rather small, and has three inferior facets, which diminish in size outwards. It has a strong posterior knob-like process, with a narrow neck.

When the tarsal bones are in position, and the tibia stands vertically on the astragalus, the cuboid bone is turned inferiorly. This indicates that this species walked on the outer edge of the hinder foot.

Broken metapodial bones are slender and straight. The proximal end of a metacarpal does not display the interlocking lateral articulation. Two phalanges are depressed in form.

Measurements.

			M.
Diameters of a dorsal centrum	anteroposterior		.0145
	vertical		.0075
	transverse		.0115
Diameters of glenoid cavity scapula	anteroposterior		.0145
	transverse		.0090
Depth of olecranon			.0110
Width of head of radius			.0110
Width of neck of ilium anteroposteriorly			.0120
Diameter of shaft of tibia at middle			.0085
Diameters of astragalus	anteroposterior		.0180
	greatest transverse	of trochlea	.0140
		of head	.0100
Length of head			.0070
Length of calcaneum			.0300
Width of calcaneum at sustentaculum			.0140
Width of cuboid facet			.0066
Length of cuboid			.0120
Diameters of cuboid	anteroposterior	distal	.0070
		proximal	.0075
	transverse proximal		.0098
Diameters of navicular	vertical		.0050
	transverse		.0100
	anteroposterior	with tuberosity	.0100
		without tuberosity	.0070

As already remarked, it is possible that the semi-grooved trochlea of the astragalus of this species is an indication that the genus *Prototomus* must be retained as distinct from *Stypolophus*, to which the present species probably truly belongs.

The specimen described, together with the mandibular ramus of another, supporting the last two molar teeth, were found in the bad lands of Wind River, Wyoming, by Mr. J. L. Wortman. A third specimen was found by Mr. Wortman in the true Wasatch bed of the Big Horn River region. This includes a large part of a skull, with one mandibular ramus almost perfect, with an incomplete ulna and fibula. With the aid of the lower jaw I have restored the skull (see Plate XXV d; fig. 1), the occipital region only being inferential. This I have modeled after that of *Stypolophus caylusi* Filh., of which I possess a cast given me by the kindness of Dr. Filhol.

The skull is larger than that of the red fox, and resembles in general proportions that of the opossum. The brain-case is narrow, and the sagittal crest is elevated. The muzzle is contracted but not short, and the palate is wide posteriorly. One character of the *Creodonta*, in which they resemble the opossums, is seen in the relatively small size of the molar teeth. There is no trace of preglenoid crest. The premaxillary bone is narrow, and its superior process does not reach near to the frontal. Its inferior lateral aspect is excavated for the apex of the inferior canine tooth. The nasal bones have a short free extremity, together forming an angulate semicircle, and do not extend beyond the vertical line of the anterior border of the canine teeth. The front is very wide, much exceeding the proportions in *S. caylusi*, and equaling one-half the length from the premaxillary border to the union of the temporal ridges. The latter are low and rather transverse. There are no postorbital processes, and the angle is very obtuse. The sagittal crest grows higher posteriorly. The posterior border of the palate is transverse, and a little concave, and is thickened. Between the processus triangularis and the alveolus of the M. iii there is first a notch and then a short process. The maxillary bone is excavated between the true molar teeth. The malar bones are thin and shallow. They have no postorbital angle. The glenoid cavity extends on the zygomatic process of the squamosal, and terminates in a rounded border.

The incisor foramina are short and rather wide. The infraorbital foramen is large, vertical, and above the posterior root of the fourth premolar. There is no postzygomatic foramen.

There are three closely placed superior incisors on each side, of which the external is separated from the canine by an interspace equal to the widths of two of them. The canine is large, and has a robust root; the crown is lost. A short diastema separates it from the first premolar, which has two well-developed roots. A very short space separates these from the anterior root of the second premolar. The roots of the remaining molars are adjacent. The crowns of all the true molars, and that of the fourth premolar, are preserved; the others are lost. The fourth premolar has the character of other species of the genus. Externally there is a median cone, a posterior distinct heel, and a low anterior basal lobe. The internal lobe is prominent inwards, but is not elevated. A weak external cingulum. In the first and second true molars the external cusp is double, and stands considerably within the external border, less so on the first than in the second tooth. In the first the anterior border is transverse, and the posterior border considerably longer and oblique. The internal angle is prominent, but without cusp; the anterior external basal lobe is very small, while the posterior is elongate, and has an acute edge, which forms with the posterior side of the posterior cone a sectorial blade. The latter feature is seen in a more striking degree in the second true molar; the transverse diameter is greater, so that the posterior border is less oblique. The anterior border is transverse, and the external border is openly emarginate, which is not the case with the first true molar. The last true molar is entirely transverse and narrow. It has one median cusp, which is connected by a cutting anterior edge with the internal tubercle and the external border.

The mandibular ramus is compressed, and the horizontal portion is rather deep, most so below the last molars. The inferior border is gently convex downwards, and then rises below the coronoid process. It is slightly decurved again below the condyle, and is then recurved, terminating in the apex of the hook-like angle. This hook is larger than in most *Creodonta* and *Carnivora*, projecting a short distance beyond the condyle. Between them the posterior border is deeply excavated. The condyle has its superior border nearly straight. Its posterior articular face extends to the inferior side at the internal extremity, and is cut off obliquely from the middle below to the external superior extremity above. The coronoid process is wide anteroposteriorly, and has a regularly convex superior border. This terminates in an angle looking downwards and backwards above the condyle. The masseteric fossa is well marked, but is not defined below. There are two mental foramina, one below the third, the other below the anterior part of the premolar.

The inferior canine is a robust tooth, with rather short crown. Both internal and external faces display a median longitudinal angle. There are short spaces before and behind the first premolar tooth. The latter has two roots and a short crown, which is obliquely truncate posteriorly. The second premolar has a larger crown than the third. It has no anterior or posterior basal lobes, but the posterior face is concave. The third premolar has a distinct though short posterior heel. The fourth premolar has the posterior heel with a medan cutting edge; there is a small anterior basal lobe. There is no internal lobe, nor the rudiment of it, as seen in *Didelphodeus absarokæ*. The first true molar is smaller than the second and third, which are nearly equal. The heels of all the true molars have a raised border, which incloses a basin; the inner wall is notched at its junction with the base of the principal cusp. The two inner cusps are of about equal elevation, and are much less elevated than the external one.

The enamel of all the teeth in both jaws is smooth.

The symphysis mandibuli forms but a slight bevel of the general surface, is not strongly sutural, and does not extend posterior to the second premolar.

The ulna is rather robust. The section of the shaft is triangular, the base upwards representing the section of a plane on which the radius rests. This plane turns to a slope of 45° inwards near the distal part of the shaft, where it is at first bounded by sharp superior and interior ridges. The inferior edge is obtuse, and the inner and outer sides are concave, especially the external. The distal extremity of the tibia is expanded and its malleolar face is oblique, and not vertical, as in true carnivora. A compressed process projects posteriorly in the plane of the posterior face, and is separated from the posterointernal angular border by a shallow groove. No groove or process on the external face, as is seen in *Canis*.

Measurements.

		M.
Length of palate		.073
Width of palate at last molar teeth		.029
Width of palate at first premolar teeth		.0115
Width of posterior nares		.0117
Width of nasal bones at middle		.0116
Width of frontal between orbits		.0385
Diameters of base of canine	anteroposterior	.0095
	transverse	.0060
Length of premolar series		.033
Length of base of first premolar		.0068
Diameters fourth premolar	fore and aft	.008
	transverse	.007
Length of true molars		.019
Diameter second true molar	fore and aft	.008
	transverse	.0105

	M.
Length of mandible with condyle	.109
Elevation at coronoid process	.052
Elevation of condyle	.027
Elevation at last molar	.021
Elevation at first premolar	.013
Thickness at first true molar	.008
Length of molars from canine	.059
Depth of ulna at middle	.010
Anteroposterior width of distal end of fibula, with process	.012

The *Stypolophus whitiæ* is dedicated to Frances Emily White, M. D., the professor of physiology in the Woman's Medical College of Pennsylvania.

I insert here what should have been noted under the head of the suborder *Creodonta;* the successional modifications of the superior molar teeth to be seen in the various genera of this group. In *Mioclænus* and in some of the teeth of *Mesonyx* the extremes are to be observed, viz: In the former two well-separated conic external tubercles, and in the latter but one. In *Deltatherium* and *Peratherium* these tubercles are flattened externally, and directed inwards in true *Didelphoid* fashion. In *Didelphodus* and *Stypolophus* they are close together and more conic, but the angle extending from the posterior cusp foreshadows a sectorial blade. In *Pterodon* and *Oxyæna* this blade is realized, and the two cusps are flattened and nearly fused, producing a type of sectorial peculiar to the family of the *Oxyænidæ*.

Stypolophus aculeatus Cope.

Report Capt. G. M. Wheeler, U. S. Geog. Geol. Surv. W. of 100th Mer., iv, pt. ii, p. 112. *Tricodon aculeatus* Cope, Proceed. Amer. Philos. Soc., 1872, p. 460. Paleontological Bulletin No. 1, p. 1, July 29, 1872.

Plate XXIV, figs. 6–7; XXVII, figs 1–2.

This species was first described from a premolar and a true molar, teeth of the inferior series. I now give a description of a considerable part of the dentition of both jaws. I am enabled to do this through the kindness of Professor Guyot of Princeton College, who placed at my disposal a specimen acquired by him for the beautiful museum under his direction.

The Princeton specimen agrees precisely with my type in its corresponding parts. It indicates the largest species of the genus, with the possible exception of *S. hians*, and one of about the size of the red fox, or a little larger.

The last superior molar, although narrower than the others, has a triangular outline, the posterior angle being nearly right. Its anterior face is straight. The crown bears three cusps, of which the interior is the most robust, the median the smallest, and the exterior the most acute. The latter is bounded by a low cutting edge on each side, extending from the base to the anterior and posterior external angles. The penultimate superior molar is subrectangular in outline, the right angle being external and anterior, and the internal being obtusely rounded. The two external median characteristic cusps are lenticular in section, and acute. At the inner and marginal base of each is a low tubercle, the posterior one rudimental. These inclose a basin with the larger inner cusp; round the base of the latter is a distinct cingulum. The antepenultimate molar is lost. The preceding one has a single large conical cusp and a posterior low cutting heel on the external side; also a low basal tubercle in front. The inner lobe of the crown is rather large, and is contracted so as not to originate from the entire inner side of the external portion. It supports an inner cusp only, and has no basal cingulum.

Measurements of superior molars.

	M.
Length of posterior four molars	.041
Length of the anterior, on base	.010
Width of the anterior, on base	.009
Length of penultimate, on base	.009
Width of penultimate, on base	.011
Length of last molar, on base	.004
Width of last molar, on base	.009

The mandibular ramus is rather compressed. The masseteric fossa is well marked and is bounded by a strong ridge in front, but has no distinct border below. The front of the base of the coronoid process is concave, and its inner border is the most prominent. A short space separates the last molar from it. The second true molar is the largest, and the first and third are of about the same size, and not much smaller than the second. The last differs from the first in having the heel narrower in transverse diameter. The heels are all basins with the external wall somewhat within the base of the crown. The anterior part of the crown is much elevated above the heel, and consists of the usual three cusps, whose base forms a right-angled triangle, of which the shearing portion forms the hypothenuse. The last premolar is large, rather longer than the first true molar. Its crown consists of a large conic median cusp of wide lenticular section, behind which is a heel with obtuse cutting edge, and an internal basal cingulum. There is a well-marked anterior basal tubercle, and a rudiment of a posterior lobe of the median cusp. No lateral cingula. Of the other premolars it can only be said that the base of the third is as large as that of the fourth.

Measurements of inferior molars.

	M.
Length of last four molars on base	.0355
Length of true molars on base	.025
Length of first true molar on base	.0083
Width of first true molar on base	.0050
Length of second true molar on base	.009
Width of second true molar on base	.0058
Length of third true molar on base	.008
Length of heel of third	.0035

MIACIS Cope.

Paleontological Bulletin, No. 3, p. 2, Aug. 7, 1872. Proceed. Amer. Philos. Soc., 1872, 470. *Uintacyon* Leidy, nomen nudum, Proc. Acad. Nat. Sci. Phila., 1872, p. 277 (December, not published until 1873). Report U. S. Geol. Surv. Terrs., i, p. 118, nomen nudum.

This genus was proposed for a species which was represented at the time by a portion of a mandibular ramus, which had supported the last three molars. The portions of the latter preserved were stated to resemble corresponding parts of *Canidæ*, with approximations to those of *Stypolophus*. Subsequently Dr. Leidy described the mandibular ramus, containing most of the teeth, of a larger species; and a fragment of the lower jaw of a still larger species. From the former of these specimens I derive the greater part of the following diagnosis. I premise with the statement that there are in this specimen five premolar teeth, the third of which is apparently three-rooted, and stands partially transversely to the axis of the jaw. I suspect this tooth to be an abnormal production, and do not propose to include it in the generic diagnosis. I have pointed out a similar example in the inferior dentition of the *Coryphodon latidens*.*

Dentition below; I. ? C. 1; Pm.; M. 3; first premolar one-rooted; first true molar with a broad heel, one edge of which is submedian and a little elevated above the other. Last two molars tubercular, the second with conic tubercles in front and a short heel posteriorly.

This genus appears to be the canine representative among the flesh-eaters of the Eocene, as *Oxyæna* is the feline. There is no more reason for suspecting it of Marsupial affinities, as is suggested by Leidy, than in the case of any others of the *Creodonta*. The fact that the well-preserved inferior border of the ramus of the *M. edax* is not inflected, is evidence to the contrary.

The five species of *Miacis* differ in their measurements as follows: That of *M. vorax* is derived from Leidy; those of *M. edax* from the typical specimen, which Professor Leidy kindly lent me.

a Last inferior molar with two roots.

Length of inferior molar series, M. .042; of true molars, .020; depth of ramus at M. II, .016 .... *M. canavus.*

Length of inferior molars, .037; of true molars, .018; depth of ramus at M. II, .013 .... *M. brevirostris.*

a a Last inferior molar with one root.

Length of last three molars on base, .013; depth of ramus at M. II, .008. *M. parvivorus.*

Length of last three molars on base, .014; depth of ramus at M. II, .010. *M. edax.*

Length of last three molars, .017 .... *M. vorax.*

MIACIS CANAVUS Cope.

Bulletin U. S. Geol. Surv. Terrs., vi, p. 189, February 26, 1881.

Established on the mandibular rami of two individuals, which display the roots and some of the crowns of all the teeth exclusive of the incisors.

The root of the canine indicates that the crown is of large size and compressed at the base. The first premolar is one-rooted, and is separated from the second by a short diastema. The second has two well-distinguished roots, which are separated from those of the third by a diastema like that in front of them. Posterior to this there are no diastemata. The

*See Report Captain Wheeler's Expl. Surv. W. of 100th Mer., iv, pt. ii, p. 215.

second root of the fourth premolar is much larger than the anterior. The sectorial, though the largest tooth, is of but moderate dimensions; its heel supports two posterior tubercles. The first tubercular is a little shorter. It presents the three anterior tubercles of the sectorial, but they are obtuse and placed close together. The heel is well developed, and its external border is elevated into a ridge, which extends obliquely inwards and forwards.

The second tubercular is a very small tooth, but has two roots, the posterior of which is posterior to the anterior border of the ascending ramus.

According to Leidy's measurements, this species is about the size of his *M. vorax* of the Bridger formation. That species has, like the two others of that horizon, a second tubercular tooth with only one root.

Measurements.

	M.
Length of dental line posterior to canines	.0440
Length of premolar series	.0250
Length of base of fourth premolar	.0065
Length of base of sectorial	.0085
Length of base of first tubercular	.0060
Length of base of second tubercular	.0040
Depth of ramus at second premolar	.0150
Depth of ramus at second true molar	.0100

From the Wind River beds of Wyoming, J. L. Wortman.

MIACIS BREVIROSTRIS Cope.

Bulletin U. S. Geol. Surv. Terrs., v, p. 190, February 26, 1881.

This species differs from those of the Bridger epoch in the same way that *M. canavus* does, *i. e.*, in the biadicate last inferior molar. Its dimensions are intermediate between those of *M. edax* and *M. vorax*, hence a little smaller than those of the *M. canavus*. This difference is partially seen in the shortening of the premolar series of teeth They are closer together than in the *M. canavus*, and the roots are larger. The sectorial tooth is shorter. The fourth premolar has a low anterior basal cingulum; the posterior part of the crown is robust. The first tubercular molar is wide, and consists of a basin-shaped heel and a short anterior portion which is more elevated. The latter consists of two cusps, which are connected by an anteriorly convex ledge, but there is no third anterior tubercle as in *M. parvivorus*. The ramus is quite robust, and the basis of the canine tooth is unusually large. Mental foramina are below the anterior parts of the second and fourth premolars, respectively. Last inferior molar small.

Measurements.

	M.
Length of molar series	.0380
Length of premolars	.0200
Length of base of fourth premolar	.0060
Length of base of sectorial	.0072
Length of base of first tubercular	.0048
Length of base of second tubercular	.0042
Depth of ramus at second premolar	.0140
Depth of ramus at second true molar	.0140

Wind River beds, J. L. Wortman.

MIACIS PARVIVORUS Cope.

Paleontological Bulletin, No. 3, p. 2, Aug. 7, 1872. Proceed. Amer. Philos. Soc., 1872, p. 470. *Viverravus parvivorus* Cope, Pal. Bull., No. 12, p. 3. An. Rep. U. S. Geol. Surv. Terrs., 1872 (1873), p. 560. Plate xxiv, fig. 12.

Established on a portion of the right ramus mandibuli, containing portions of three molars, the penultimate being perfect. As in *Canidæ*, the molars diminish in size posteriorly, the last being single-rooted, the penultimate being two-rooted. The structure of that tooth is approximately that of *Stypolophus*, *i. e.*, with three trihedral cusps in front and a heel behind, but the cusps are of equal height, and their point of union not raised above the surface of the heel. This is a valley bounded by a sharp margin, which is incurved to the outer cusp, leaving a vertical groove on the outer side, as in *Stypolophus* sp. This species is further characterized by the single-rooted small tubercular posterior molar, which is also present in *M. edax* and *M. vorax*. The antepenultimate molar is much larger than the penultimate. The crown of the latter is laterally expanded, and bears a cingulum at the base anteroexternally. Enamel smooth.

	M.
Depth of ramus at penultimate molar	.0080
Length of crown of penultimate molar	.0040
Elevation of crown of penultimate molar	.0025
Width of crown of penultimate molar	.0033

Found on Black's Fork of Green River.

DIDYMICTIS Cope.

System. Cat. Vert. Eocene New Mexico, U. S. Geog. Surv. W. of 100th M., 1875, p. 11. Rept. U. S. Geog. Surv. W. 100th M., vol. iv, part 2, p. 123.

Inferior molars six, consisting of four premolars and two true molars. True molars, a posterior tubercular, and an anterior tubercular-sectorial, *i. e.*, with three elevated cusps and a posterior heel. Premolars with a lobe behind the principal cusp. The canine teeth are directed forwards, and are very close together, so that it is doubtful whether there were any incisors. An ungual phalange of the typical species is strongly compressed.

The humerus in this genus is distally expanded transversely, and the margin is pierced by the humeral artery. The astragalus exhibits two entire trochlear faces; the wider external and directed interosuperiorly, the inner presenting superointeriorly. They are separated by an obtuse longitudinal angle, and are little or not at all concave transversely. The form is depressed. The head supports a single transverse convex facet for the navicular, and, with the neck, is as long as the trochlear portion.

In this genus the sectorial tooth of the lower jaw is of a very primitive type, resembling especially inferior molars of marsupials of carnivorous habits. This is seen in the close approximation of the anterior cusp to the two immediately succeeding it, and in its relatively small elevation in comparison with the external cusp. The latter is much elevated in this genus. The heel of the same tooth is low; its length is in direct relation to the size of the species; that is, it is relatively shortest in the smallest species. The rudimental sectorial cusps of the tubercular tooth in *D. haydenianus* show that but few changes of form are necessary to connect the inferior dentition of this genus with that of *Oxyæna.*

The longest known species of this genus is the *D. protenus*, from the Wasatch formation of New Mexico. Three additional species were afterwards discovered by Mr. J. L. Wortman in the Wind River country of Wyoming. The species range from the size of a mink to that of a coyote. Their characters are as follows, as derived from the mandibular teeth:

I. Inferior tubercular molar with the three anterior cusps well developed.
Length true molars .0125; last molar narrow........................ *D. haydenianus.*
II. Inferior tubercular with rudimental cusps.
* Inferior tubercular molar oval in outline, with a heel.
Length true molars .025; last three premolars .035; last molar short...... *D. altidens.*
Length true molars .019—.020; last three premolars .036; last molar elongate.. .... *D. protenus.*
Length true molars .016—.018; last three premolars .028—.030; last molar narrow. *D. leptomylus.*

Length true molars .010; last three premolars .0135; last molar narrow.. *D. dawkinsianus.*
** Inferior tubercular molar short, subquadrate in outline.
Length true molars .011; depth of ramus at sectorial .010 .... *D. massetericus.*
Length true molars .018; depth of ramus at sectorial .017 .... *D. curtidens.*

DIDYMICTIS HAYDENIANUS Cope.

Proceed. Amer. Philos. Soc., 1882, p. 464. Paleontological Bulletin No. 35, p. 464, 1882, Nov. 11.

Plate XXIII c, figs. 12–13.

This Creodont is represented by parts of the maxillary and mandibular bones of the left side, the former supporting the four and the latter supporting the three last molars. The arrangement of the superior molars is much as in *D. protenus*, the fourth premolar being a true sectorial. The third premolar has no internal lobe, although the section of the base of the crown is narrowly triangular. It has anterior and posterior basal lobes, and a posterior lobe on the cutting edge. In the sectorial the median lobe is a good deal more produced than the posterior, though the two form together the usual blade. The anterior basal lobe is distinct, and the internal is larger and is conic. The first true molar has the anterior external base of the crown produced. Its two external cusps are conic and distinct. The internal part of the crown is rounded and supports a conic internal tubercle, which is separated from the external cones by two small concentric tubercles. The second true molar is considerably smaller, and is transverse, its external border being very oblique. It has an acute internal lobe.

The character of the species is well marked in the inferior true molars. The first has the form seen in other species of *Didymictis*. The heel is large, and with a median basin between lateral cutting edges. The two anterior inner cusps are of equal elevation and are near together; the external is much larger. The last molar is elongate, but reduced in size. Its anterior three cusps, rudimental in other species, are here elevated, forming the triangular mass seen in the first true molar. They are not so elevated, however, as in that tooth, and thus not so much developed as in *Oxyæna*, *Stypolophus*, etc. The fourth premolar has a median cutting edge on the short heel.

Measurements.

		M.
Length last four superior molars		.022
Length Pm. iii		.0065
Length Pm. iv		.0085
Width Pm. iv		.0050
Diameters M. i	anteroposterior	.0055
	transverse	.0088
	oblique external	.0072
Diameters M. ii	anteroposterior	.0027
	transverse	.0055
Diameters inferior M. i	anteroposterior	.007
	transverse	.005
Diameters inferior M. ii	anteroposterior	.0055
	transverse	.003
Depth of ramus at M. ii (squeezed)		.010

The peculiar characters of the last inferior molar distinguish this species from its congeners. The last superior molar is relatively smaller than in the *D. protenus*. In size this species is equal to the *D. dawkinsianus*, and is smaller than the *D. leptomylus*. As already remarked, the inferior dentition approaches that of *Oxyæna*. A slightly-increased development of the anterior cusps of the last (second) inferior true molar would give two inferior tubercular sectorial molars, as in that genus. The superior dentition is, however, totally different, for there is no approach in the first true molar to the sectorial type characteristic of the *Oxyænidæ*. It is dedicated to the distinguished geologist, Dr. F. V. Hayden.

Puerco epoch of New Mexico; D. Baldwin.

DIDYMICTIS ALTIDENS Cope.

Bulletin U. S. Geol. Surv. Terrs., vi, 1881, Feb. 26, 190. American Naturalist, 1880, Oct. 746.

Plate XXV a, figs. 13-14.

This species is represented in my collection by the jaws of four individuals in a fragmentary condition. One of these supported the last five inferior molars; in others the inferior molars are separate from the jaws.

The tubercular molar is relatively small in this species, not exceeding the average size of that of *D. protenus*, while the sectorial is considerably larger. The anteroposterior diameter of the fourth premolar is equal to that of the sectorial, and that of the third premolar is a little less. The premolars are closely placed, the first not seen.

The fourth premolar is not much widened posteriorly, and the lobe of the posterior edge is well marked. The cuspidate part of the sectorial is about equal in anteroposterior diameter to the length of the heel. The external cusp is much higher than the two interior, and the latter are equal in elevation. They are all obtuse, as are the continous edges which represent the blade of the sectorial tooth of a higher carnivore. The length of the heel from the base of the internal tubercle is equal to the height of the latter. It carries a ridge-like tubercle, which extends from its posterior external border forwards and inwards. Between its inner side and the internal rim of the heel is an oblique concavity. The tubercular in its details is a reduced copy of the sectorial, the anterior cusps being represented by low tubercles and occupying relatively very little space. The posterior oblique tubercle and concavity are there. The crown differs from that of the sectorial in having the external basal cingulum stronger; there is no internal basal cingulum on either tooth. No internal, and a trace of external cingulum on the fourth premolar.

The mandibular ramus is compressed and deep. The masseteric fossa is well defined below as well as anteriorly. There is a well defined area of insertion (? for the internal pterygoid muscle) on the inner side of the base of the coronoid process.

Measurements.

No. 1.

		M.
Length of bases of last five inferior molars		.0620
Diameters of fourth premolar	anteroposterior	.0150
	transverse	.0065
Diameters first true molar	anteroposterior	.0145
	transverse	.0080
Diameters second true molar	anteroposterior	.0088
	transverse	.0062
Depth ramus at second true molar		.0250

No. 2.

	M.
Anteroposterior length of sectorial	.0150
Length of heel	.006
Elevation of external side of crown anteriorly	.015
Width at same point	.009
Length of crown of tubercular	.009
Elevation anteriorly	.005
Width of same	.006

Wind River beds, Wyoming. J. L. Wortman.

Didymictis leptomylus Cope.

American Naturalist, 1880, Dec., p. 908; Bulletin U. S. Geol. Surv. Terrs., 1881, vi, p. 191.

Plate XXV a; fig. 12.

The specimens which I refer at present to this species belong to two varieties, which may perhaps be specifically distinct; but this cannot be demonstrated at present They differ in dimensions only. Thus the true molars of the type, which comes from the Wind River beds, measure M. .016 in length. Five specimens from the Big Horn basin agree in having this dimension .018. The entire inferior molar series is only a little shorter than that of the smaller variety of the *D. protenus* from New Mexico. (See my report to Captain Wheeler, Plate XXXIX.) The species is characterized by the narrow and relatively elongate form of the tubercular molar. Its heel is considerably produced behind the posterior oblique ridge, which is not the case in the *D. altidens*. Its anterior part has the three low cusps well defined and close together, and behind them is the oblique longitudinal cutting edge. The middle of the posterior margin rises into a tubercle. The external cusp of the tubercular-sectorial is much elevated. The heel has a strong external cutting edge and internal ledge, which reaches the posterior border, and is not quite so long as the internal tubercle is high. The cusps are rather obtuse, especially the internal pair, which are of equal height. The representative of the blade is not very sharp. There are no basal cingula on these teeth.

	M.
Length of tubercular-sectorial	.009
Width of same	.005
Length of tubercular	.007
Width of same in front	.004

Big Horn basin, Wyoming. J. L. Wortman.

DIDYMICTIS DAWKINSIANUS COPE.

Bulletin U. S. Geol. Surv. Terrs., vi, 1881, Feb. 26, p. 191.
Plate XXV a; fig. 11.

This flesh-eater is represented by more or less imperfect mandibular rami of ten individuals. The most complete of these lacks only the portions posterior to the coronoid process, and those anterior to the first premolar, and supports all the teeth excepting the first and second premolars. The premolars are all two-rooted excepting the first. The base of the fourth premolar is considerably longer than that of the third. Both of these teeth have a short posterior heel, and above it a cutting lobe. The fourth has a well-marked anterior basal tubercle. The heel of the sectorial is relatively short, and the anterior portion of the tooth elevated. The anterior and inner cusps are high, and about equal, but the external cusp is much higher. The external border of the heel is more elevated than the inner. The tubercular molar is elongate, and has a small triangular anterior portion somewhat elevated, in slight resemblance to the sectorial tooth. This portion consists of two opposite cusps and a lower one in front of the anterior inner, which connects with the external by an anterior ledge. The posterior portion has a tubercle on the external side, besides a posterior elevation. The ramus is rather slender, and the masseteric fossa is bounded by a prominent ridge in front, but fades out below.

The measurements show this to be the smallest species of the genus, being much less than *D. leptomylus*.

Measurements.

	M.
Length of dental series, including first premolar	.0265
Length of premolar series	.0168
Length of base of fourth premolar	.0055
Length of base of sectorial	.0053
Width of base of sectorial at middle	.0035
Elevation of sectorial	.0055
Length of first true molar	.0044
Width of first true molar in front	.0028
Elevation of first true molar in front	.0025
Depth of ramus at second premolar	.0064
Depth of ramus at tubercular molar	.0070

This species is dedicated to my friend Prof. W. Boyd Dawkins, the distinguished geologist and paleontologist, of Manchester, England.

Five specimens were obtained from the Wind River basin, and five from that of the Big Horn, Wyoming Territory.

DIDYMICTIS PROTENUS COPE.

Plate XXV d, figs. 4, 5.

System. Cat. Vert. Eocene New Mexico, U. S. Geog. Survs. W. of 100 M., 1875, p. 11. Report upon U. S. Geog. Survs. of W. 100 M. In charge of First Lieut. Geo. M. Wheeler, Corps of Engineers, U. S. Army, under the direction of Brig. Gen. A. A. Humphreys, Chief of Engineers U. S. Army; Part II, Vol. IV, Paleontology, 1877, p. 123. Washington. Paleontological Bulletin No. 34, p. 159, Feb. 20, 1882.

Jaws, more or less complete, of six individuals from the Big Horn basin, are referable to this species. They agree closely in measurements and belong to the larger variety of the species figured on Plate XXXIX of the report to Captain Wheeler.

A left maxillary bone containing the last four molars furnishes the best characters for this part of both genus and species yet obtained. The third premolar has no interior lobe, but the inner base is more convex than the external, and has a low cingulum. There is a short posterior heel, and a shorter anterior basal tubercle. The fourth premolar has three external lobes, and an internal conic lobe which is opposite to the space between the anterior and middle external lobes. The anterior external lobe is the smallest and is subconical. The middle lobe is a flattened cone with a three-sided base. The third lobe is a blade directed outwards as well as backwards. Its free edge forms, with the posterior edge of the middle lobe, a sectorial blade divided by a median fissure. The first true molar is triangular, with the longest side anterior and the external and posterior sides equal. The external side is very oblique, subtending an obtuse angle with the posterior ide. It is also concave medially, and the anterior lobe projects outwards and forwards. The two external cusps are small, conic, well separated, and situated much inside of the external border. The internal cusp is large, and is separated from the externals by a small triangular tubercle on each border of the crown. Each of the latter descends into a cingulum which extends outwards. A strong cingulum surrounds the internal base of the crown, disappearing at the intermediate tubercles mentioned. The second true molar is much smaller than the first, and the details of its structure are the same. The anterior exterior angle is not so much produced.

The fourth premolar is as effective a sectorial tooth as that of the species of *Galecynus* and *Temnocyon*, and has the anterior basal tubercle which is wanting to that genus but present in *Aelurodon*.

Measurements.

		M.
Length of last four molars		.040
Length of third premolar		.011
Width of third premolar		.006
Diameters sectorial	anteroposterior	.015
	transverse	.010
Diameters first molar	anteroposterior	.008
	transverse	.015
Diameters second molar	anteroposterior	.005
	transverse	.009

Didymictis masstericus Cope.

Paleontological Bulletin No. 34, p. 160, 1882, Feb. 20. Proceed. Amer. Philos. Soc., 1881, p. 159, Feb. 1882.

Plate XXIV e; fig. 11.

This species is intermediate in size between the *D. leptomylus* and the *D. dawkinsianus*, and is characterized by the peculiar form of its tubercular molar, and the deeply excavated masseteric fossa. It appears to have been a rare species, as only one mandibular ramus was found by Mr. Wortman. This is broken off in front of the fourth premolar, and supports the last three molar teeth.

The tubercular molar is subquadrate in form, and consists of three low tubercles in front, and a wide heel behind, which has an elevated posterior border. The tubercular-sectorial has a short and narrow heel. Its anterior cusps are not very acute, and the two internal are equal, and a good deal shorter than the external. The fourth premolar is relatively shorter than in any other species of the genus, and the posterior marginal lobe is a mere thickening of the edge of the heel. There is a low anterior basal tubercle. The enamel is smooth.

The ramus is compressed and not deep. The angle is prominent, and is not inflected; it does not extend so far posteriorly as the posterior border of the condyle. The inferior border of the masseteric fossa is an angular line, without abrupt excavation, but the face of the fossa descends rapidly. The anterior border of the fossa is abrupt and is formed by the usual sub-vertical ridge.

Measurements.

	M.
Length between Pm. iv and condyle, inclusive	.0520
Length of posterior three molars	.0170
Length of tubercular-sectorial	.0070
Elevation of tubercular-sectorial	.0070
Depth of ramus at sectorial	.0100

From the Big-Horn River, Wyoming.

DIDYMICTIS CURTIDENS Cope.

Paleontological Bulletin No. 34, p. 160, Feb. 20, 1882. Proceed. Amer. Philos. Soc., 1882, Dec. p. 160. Plate XXIV d; fig. 10.

As in the case of *D. massetericus* the present species is represented by a single fragmentary mandibular ramus. This supports a sectorial tooth of the size and form of that of the *D. protenus*, and is thus much larger than that of the species just named. This tooth is placed nearer to the base of the coronoid process than is seen in any other species, and only leaves space for a short tubercular tooth. This is lost from the specimen, but the alveolus shows pretty clearly its small dimensions. The base of the fourth premolar remains, and it is evident that this tooth was like that of *D. protenus* in form and proportions. The base of the posterior marginal lobe is present. The ramus is deeper and larger than in the *D. massetericus*.

Measurements.

	M.
Length of bases of last three molars	.0285
Length of bases of fourth premolar	.0120
Length of bases of sectorial on base	.012
Width of bases in front	.008
Depth of ramus at sectorial	.017

Big Horn basin. J. L. Wortman.

OXYÆNA Cope.

Report on Vertebrate Fossils obtained by the Wheeler Survey in New Mexico, 1874, p. 11 (extracted from Report of Lieut. Wheeler to Chief of Engineers). System. Cat. Vert. Eocene New Mexico, 1875, p. 9. Report U. S. Geog. Survs. W. of 100th Mer., p. 95.

Dental formula: I. $\frac{3}{?0}$; C. $\frac{1}{1}$; Pm. $\frac{4}{4}$; M. $\frac{2}{2}$. Two small median superior incisors and a very large external one separated by a diastema from the canine. The latter is large, and is followed with little interval by the first premolar. The two last premolars and all the molars of the superior series with an internal heel; the last molar transverse; third and fourth upper premolars with an anterior cone and posterior cutting-lobe; the first true molar with two anterior acute cones, the posterior forming a sectorial edge with the posterior lobe; last superior molar with a single trenchant edge.

In the mandibular dentition, the canine teeth are directed forward and upward without intervening incisors. First premolar one-rooted; second and third consisting of an anterior elevated cone and posterior heel, which is elevated and trenchant in the middle. The fourth premolar is nearly similar, with the posterior tubercle sharp-edged. The two true molars with an anterior elevated portion and small, low heel; the former consisting of three acute tubercles, of which the largest or exterior forms with the anterior a sectorial blade oblique to the axis of the mandibular bone.

The exterior portion of the posterior transverse superior molar is a transverse blade, interior to which is one or probably two subtriangular cusps. The blade shuts down in contact with the plane posterior face of the united middle cusps of the last inferior molar, and the cusp shuts down on the inner side of the heel of the same, where the surface is often seen to be worn obliquely by it. The elevated cusps of the last inferior molar close into a deep fossa of the maxillary bone; the blades of the external and anterior cusps shearing against the inner side of the posterior median cusp and posterior blade of the penultimate superior molar. The inner heel of the latter opposes transversely the posterior heel of the penultimate inferior molar, shearing somewhat with the posterior border of the united median cusps. The external and anterior cusps of the penultimate inferior molar, with their external shear, fit within the median cusp and posterior blade of the antepenultimate superior molar, and are received into a corresponding pit of the maxillary bone, which is not so deep as the posterior fossa. The surface of the maxillary between this tooth and the last premolar is only slightly concave. Thus, in this genus, and the arrangement is similar in *Stypolophus*, each inferior tubercular-sectorial tooth makes two shears with two corresponding superior molars, viz, a posterior-transverse with the superior molar behind it, and an external-oblique with the superior molar corresponding to it. This does not occur in any recent *Carnivora*, and is a more complex, although much less powerful, arrangement than they possess.

The skull in this genus is robust. In the *O. forcipata* there is an elevated sagittal crest, and the superior walls of the cranium are massive. The crest divides on the posterior part of the frontal region, and disappears. The zygomata are short and deep, and laterally expanded. The malar bone rises in a strong postorbital process, partially inclosing the orbits, as in the Cats. The angle of the mandible is not inflected in the least degree.

The scapula has a well-developed coracoid hook. The spine rises abruptly from near the glenoid fossa. The tuberosities of the humerus are not very prominent, and are separated by a rather wide bicipital groove. The deltoid crest is continuous with the edge of the greater tuberosity and is quite prominent. At the distal extremity there is an epitrochlear foramen. The condyle has the internal flange and external cylinder of carnivorous mammalia, the cylinder with a deep notch on the posterior inferior face, the inner border of the notch continuing into a flange on the posterior side. The epicondyles are not so much expanded as in the species of *Stypolophus* and other forms of *Creodonta*. The head of the radius is a regular transverse oval. The only irregularity is a slight concavity of the superior border. The face is gently concave, with a point directed proximad on the superior border. The carpal extremity of the radius is triangular. The surrounding tuberosities are distinct. The carpal extremity of the ulna is somewhat like the head of a rib in its obliquity and its distal and lateral tuberosities. But few bones of the fore foot are preserved. The most important is the cuneiform, which differs much from that of the *Carnivora*, but resembles that of *Esthonyx*. It is flat and not oblique. The unciform facet is concave, and about as large as each of the two superior facets. The latter are transverse and subequal, and are little concave, and are separated by an obtuse ridge. The bone resembles the cuneiform of *Ursus* more than that of *Procyon*, and still less those of *Canis* and *Felis*. It differs from *Ursus* in its less obliquity and its external production into a tuberosity. The proximate ends of the first, second, and fifth metacarpals are a good deal like those of *Ursus*, but the trapezial facet of the pollex is more concave in the transverse direction than in that genus.

In a fragmentary skeleton of probably *O. morsitans* a portion of the ilium is preserved. It exhibits a tuberosity above the acetabulum which represents the "anterior inferior spinous process" of human anatomy, and is larger than in the existing genera *Ursus*, *Canis*, and *Felis*. The ischium is wide and flat, and its posterior external thin edge is prominent proximad to the spine. The latter is an unimportant angle a considerable distance beyond the line of the acetabulum. In *Didelphys* and *Sarcophilus* it is wanting, while in *Phascolarctos* and nearly all forms of *Carnivora* it is near the posterior line of the acetabulum. The only exception I find is in the *Viverridæ*, where a *Herpestes* has it in much the same position as in *Oxyæna*. The superior border is, however, not expanded.

The middle of the shaft of the femur is wanting in all our specimens of this genus. The proximal portion of that of *O. morsitans*, is wide and flat, and has a large great trochanter about equal in elevation to the head, which does not inclose a deep or large fossa. The fossa for the ligamentum teres is at the fundus of a deep emargination of the rim of the head. The distal part of the femur is flattened as in *Amblyctonus*, and the patellar groove is not elevated as in *Stypolophus viverrinus*, but wide, although less so than in the Bears. The head of the tibia displays a spine and median groove, but the crest is not prominent.

The distal end of the tibia exhibits the ungrooved astragalar surface of the other *Oxyænidæ*, with abruptly projecting internal malleolus. Its border is less regular than in other genera described. The outer extremity is narrowed, and gives rise to a longitudinal external ridge of the lower part of the shaft, and there is a tuberosity on the posterior and one on the inner side of the lower extremity. The posterior as well as the anterior astragalar border is angulate at the base of the malleolar process. The tendinous grooves are shallow.

The astragalus is like that of *Sarcophilus* and different from that of *Didelphys* and *Phascolarctos* in the absence of the oblique fibular facet, which is here vertical and lateral. The trochlea is slightly concave above, and the malleolar facet does not present so oblique a face as in *Didymictis*. It differs from the marsupial genera, and resembles the carnivorous in its large neck and head. The proximal part of the calcaneum displays the usual two astragalar facets well separated. It is remarkable for the obliquity of the facet for the cuboid, which presents upward as well as forward (when in the supine position). The calcaneum is wide, especially in its postero-inferior face, and the posterior free portion is narrow and oblique, indicating a plantigrade habit. Its flatness exceeds that in *Ursus arctos*, and the expanse of the anterior portion is similar to that genus, while greater than in *Canis* and *Felis*. The obliquity of the cuboid facet is not seen in either of the recent genera named. The navicular is shallow, cup-shaped, and has three distal facets and an internal tuberosity. The cuboid is a very characteristic bone, and is unlike that of any other genus known to me. The proximal or calcaneal face is very oblique to the long axis of the bone, presenting outwards when the axis is placed antero-posteriorly. It is, however, evident that the long axis diverges from that of the foot, outwards. A more truly proximal facet is the rather wide one for the astragalus, which makes a right angle with that for the calcaneum. Owing to the divergence of this bone from the others, the ectocuneiform articulated with it as much as with the navicular, an arrangement seen in *Didelphys*. It is possible that the hinder foot may have been divided somewhat as in some of the lemurs, the two external digits antagonizing the three internal. Cuneiforms lost.

The metatarsals preserved include the I, II, III, and V of one foot, and the I, III, and IV of the other. They resemble much those of *Ursus*. The first has no lateral facets for II, and its facet is not more concave than in *Ursus;* hence it was not probably opposable. The II is the only one with concave transverse section; that of the others is convex in both directions. They underlap each other from the external inwards, as in various carnivora. The V presents a considerable proximal free process outwards. Numerous phalanges have been obtained. They are depressed, with their distal articular facets slightly emarginate. None of them present the triangular section characteristic of many recent *Carnivora*. Their proportions are not different from those seen in the *Ursus arctos*. A claw is moderately compressed, and terminates abruptly and obtusely. The extremity is deeply fissured, and each of the two apices is rugose.

A few vertebræ of this genus have been preserved The relative proportions of the cervicals are unknown. The two venous foramina in the floor of the neural canal of the dorsals are very large. The caudals are long and stout.

Restoration.—The Oxyænas had the characteristic peculiarities of the Creodonta and of the carnivorous Marsupials in their general proportions. The head was relatively larger, and the limbs were smaller than in true *Carnivora*. The feet were plantigrade, and had five toes anteriorly and posteriorly. The hind foot was either divided so that the external two toes opposed the internal three or the entire foot was directed outwards from the line of the calcaneum. In the latter case the hallux may have been opposable, as in the opossum, but in a much less degree. The tail was long and stout.

Species of this genus were abundant during the Wasatch epoch in New Mexico and Wyoming, and probably over the entire continent. They have not yet been reported from higher Eocene beds, not even occurring in the Wind River. A small species is found in the Puerco A species (*O. galliæ*) has been recently detected in the Eocene of France by M. H. Filhol.

This genus resembles *Pterodon*, as described and figured by Gervais, in the dentition of the maxillary bone; but the teeth of the lower jaw are totally distinct in character, approaching more nearly those of the *Palæonyctis* of De Blainville. According to Gervais, the inferior molars of *Pterodon* are like those of *Hyænodon*, without interior tubercle, and the inner lobes of the superior molars are not so large as in *Oxyæna*. The latter differs from *Palæonyctis* in the character of the antepenultimate lower molar, which in *Oxyæna* is characterized by the presence of a median blade, but in *Palæonyctis* by a heel supporting (in the typical species) two tubercles.

Oxyæna forcipata Cope.

Report Vert. Foss. New Mexico, 1874, p. 12. Report Capt. G. M. Wheeler, U. S. G. G. Expl. Surv. W. of 100th Mer., iv, ii, p. 105, 1877; pl. xxxv, xxxvi, xxxvii.

Plate XXIV b, figs. 11-15; XXIV c, figs. 1-18.

This formidable animal was abundant in Northern Wyoming during the Wasatch epoch. At least ten individuals are represented in the collection from the Big Horn basin. The following are the dimensions of the mandibles of the five best preserved.

	1	2	3	4	5
Length of dental series	.103	?	.100	.100	.107
Length of premolar series	.042	.045	.044	.051	.054
Depth of ramus at M. iii	.042	.039	.037	.042	.047

The measurement .035 for the length of the premolars given in my report to Captain Wheeler, *loc. cit.*, refers to the anterior three teeth, which were originally supposed to be the only premolars.

The specimen above noted as No. 2 presents a good many parts of the skeleton, from which I derive the following characters:

	M.
Length of centrum of a dorsal vertebra	.021
Length of centrum of a lumbar vertebra	.025
Length of centrum of a caudal vertebra	.027
Length of centrum of a caudal vertebra	.030

The dorsals measured are depressed artificially, but their length is not apparently altered. The caudals are nearly perfect. Their neural arch is complete, but is relatively shorter on the shorter centrum than on the longer one; on the latter it has no spine. There are two transverse processes on each side separated by a notch. On the shorter vertebra they are nearer together, and the posterior is the larger. The inferior surface is regularly convex medially; at the extremities it presents two tuberosities, of which the anterior are the most prominent.

The radio-carpal articular facet is short transversely. Surrounding it are four tuberosities. One of these forms the internal angle of the bone; the others are near the external end, one superior and one inferior; the fourth is at the superior side of the ulnar facet.

Measurements of fore limb.

		M.
Diameter of head of humerus from bicipital groove		.023
Transverse width of condyles of humerus distally		.025
Anteroposterior of humerus at middle		.013
Anteroposterior of humerus at external rim		.021
Diameters of head of radius	vertical	.013
	transverse	.020
Diameters distal extremity of radius	vertical	.019
	transverse	.024
Width of carpal facet of radius		.016
Length of tuberosity of distal end of ulna		.008
Width of lateral tuberosity of distal end of ulna		.016
Diameters cuneiform bone	anteroposterior	.0125
	transverse	.018
Proximal width of first metacarpal (total)		.011
Proximal width of second metacarpal (total)		0055
Proximal depth of second metacarpal (total)		.011
Proximal width of fifth metacarpal (total)		.011

The head of the tibia is characterized by the failure of the internal femoral facet to reach the posterior border. It thus leaves a free ledge.

The anterior face of the distal extremity of the tibia is slightly concave. The malleolar process is large and truncate, and is not grooved, but rises into a low, wide tuberosity at the base. The fibular face is oblique to the anterior face, inclining to an angle of 40°. Its interior extremity is continued to a rather acute process, which is separated by an open notch from the base of the malleolus. This carries the tendons of the *flexor longus digitorum* and *tibialis posticus* muscles. The head of the fibula projects considerably external to its tibial facet, the resulting section being trapezoidal, approaching triangular. The posterior face of the head is gently concave and is surmounted by a low free rim. A ridge extending forwards from this gives the proximal end of the head a roof-shaped form.

The posterior groove of the astragalus is wide. Its internal bounding angle is prominent, forming an obliquely descending tuberosity. The external bounding tuberosity is not so prominent. The anterior angle bounding the external calcaneal condyle is not more prominent, differing thus distinctly from that of *Mesonyx ossifragus*. The lateral trochlear angles of the astragalus are obtuse, especially the internal. The inner base of this bone has an open median notch. When the calcaneum is in position it is evident that the animal walked partly on its external side. This is bounded externally in front of the external condyle by a horizontal crest. The cuboid facet only covers the external two-thirds of the distal extremity of the calcaneum. The internal third, however, retreats rapidly posteriorly inwards. The very oblique calcaneal face of the cuboid, already described, is deeply notched externally by the proximal part of the groove for the *flexor digitorum* tendon. This groove forms a quarter of a circle, passing downwards, outwards, and backwards. The distal face of the cuboid is concave and is undivided. The entocuneiform facet of the navicular is situate more than its width away from the internal margin of the bone. The mesocuneiform is wide. The ectocuneiform is narrower than the latter, and bevels the external extremity of the navicular, thus looking towards the cuboid.

Measurements of posterior limb.

		M.
Diameters of proximal end of tibia	fore and aft, with crest	.035
	transverse	.031
Diameters of distal end of tibia	fore and aft	.018
	transverse	.024
Fore and aft diameter of proximal end of fibula		.017
Total length of astragalus		.032
Diameters of trochlea	anteroposterior externally	.0155
	vertical externally	.0145
	transverse	.017

	M.
Width of head of astragalus	.016
Depth of head of astragalus (greatest)	.009
Length of calcaneum externally	.045
Length of fore part of calcaneum	.018
Width of fore part at sustentaculum	.021
Width of cuboid facet of sustentaculum	.013
Depth of cuboid facet of sustentaculum	.0008
Length of cuboid bone (greatest)	.016
Length between calcaneal and metatarsal faces of cuboid on external side	.007
Diameters astragalar face of cuboid { anteroposterior	.010
{ transverse	.0055
Diameters metatarsal face of cuboid { anteroposterior	.0095
{ transverse	.0105
Diameters of navicular { anteroposterior	.013
{ transverse	.017
Diameters metatarsus i { longitudinal	.030
{ proximal { fore and aft	.013
{ transverse	.0105
Diameters metatarsus ii { longitudinal	.040
{ proximal { fore and aft	.012
{ transverse	.007
Diameters metatarsus iii { longitudinal	.044
{ proximal { fore and aft	.012
{ transverse	.007
Diameters metatarsus iv proximally { fore and aft	.0128
{ transverse	.009
Diameters metatarsus v proximally { fore and aft	.010
{ transverse	.012
Diameters ungual phalange { longitudinal	.013
{ proximal { vertical	.006
{ transverse	.006

REMARKS.—From the above measurements, which are confirmed by more than one other skeleton, it can be seen that there is in this species a remarkable disproportion between the size of the skull and that of the limbs. While the dimensions of the jaws are like those of the jaguar, those of the limbs do not exceed those of the cheetah; while the digits are not only much shorter, as those of a plantigrade animal, but are more slender. The ungual phalange preserved shows that the claws had no prehensile power, and were not effective as weapons or for digging. This is a further indication that the species of *Oxyæna* were aquatic in their habits.

PROTOPSALIS Cope.

American Naturalist, 1880, p. 745; Bulletin U. S. Geol. Surv. Terrs., vi, 1881, p. 193.

Inferior molars: one like those of *Oxyæna*, *i. e.*, with large heel and internal cusp; another, probably the last, larger, without internal tubercle, and with a rudimental heel, thus resembling the inferior sectorial of various existing *Carnivora*. A median dorsal vertebra distinctly opisthocœlous.

Femur with a weak third trochanter. The proximal extremity of fourth metatarsal of the right side furnishes instructive characters. The external side is deeply excavated below the cuboid facet, to receive a correspondingly prominent interlocking tuberosity of the fifth metatarsal. The excavation is not divided by a longitudinal groove, as in the cats, but its surface extends continuously from front to rear. On the inner side of the fourth there is a subvertical facet for the third metatarsal, which is bounded posteriorly by the usual deep vertical ligamentous groove.

The form of the true sectorial tooth, together with that of the metacarpal, approximate this genus to the *Felidæ* more closely than to any other family of existing *Carnivora*. The resemblance seen in the sectorial is, however, probably delusive, as it is not the same tooth as the sectorial of the *Carnivora*. The resemblance in the metacarpal is real, as the characters are unlike those of *Canidæ* or *Hyænidæ*.

It is probable that this genus should be placed in the *Oxyænidæ* between *Pterodon* and *Oxyæna*. But one species is yet known.

PROTOPSALIS TIGRINUS Cope.

American Naturalist, 1880, p. 745; Bulletin U. S. Geol. Surv. Terrs., vi, 1881, p. 193.
Plate xxv *b*, figs. 1–7.

Size about that of the tiger or jaguar, exceeding that of any other flesh-eater of the Eocene period. The heel of the smaller tubercular-sectorial is not large, and has a plano-concave superior surface. The principal cusp is much elevated, while the internal cusp is small. The sectorial differs from that of a *Hyæna* in having the posterior cusp more and the anterior cusp less elevated; the heel is only a strong posterior cingulum, which is continued as a narrow line along the inner base of the tooth. A rough cutting ridge forms the posterior inner angle of the principal cusp. There is a wide longitudinal groove of the inner face of the inferior canine, whose enamel surface is impressed-punctate. The inner side of the crown is so worn as to lead to the belief that the external incisor is of large size.

The inferior border of the mandibular ramus rises below the last molar tooth. The masseteric fossa shallows gradually below, so that its inferior outline is not well defined. The dental foramen is of large size. The articular faces of a median dorsal are a little wider than deep, and the width is not equal to the length of the centrum. The outlines of the latter are gently concave, and it is not keeled below or on the sides. Surface smooth. The femur is preserved, lacking the distal extremity inclusive of the rotular groove. It is about as long as that of the jaguar, and is moderately slender. The head extends rather further proximally than the great trochanter, and is defined by a distinct neck. The fossa ligamenti teris is a wide posterior emargination of its edge. The great trochanter is recurved on its external border posteriorly, but the trochanteric fossa is open proximally and fades out anteriorly and below. The little trochanter is large and has a long base. The third trochanter is a thickened angular concavity of the external border opposite a point a short distance below the little trochanter. The feet were evidently large; the proximal extremity of the fourth metatarsal is about equal in dimensions to that of a lion.

Measurements.

		M.
Diameters of crown of sectorial	anteroposterior	.025
	transverse	.014
	vertical	.022
Length of heel of tubercular-sectorial		.006
Width of same		.006
Vertical diameter of base of crown of canine		.022
Depth of mandible at last molar		.044
Length of femur (condyles inferential)		.310
Diameter of shaft at middle		.034
Diameters of proximal extremity of fourth metatarsal	anteroposterior	.021
	transverse	.013
Length of centrum of middle dorsal vertebra		.027
Diameters anterior articular face of middle dorsal vertebra	transverse	.022
	vertical	.018

MIOCLÆNUS Cope.

Paleontological Bulletin No. 33, p. 489, Sept. 30, 1881; Ibid., No. 34, p. 187, Feb. 20, 1882; Amer. Nat. 1881, p. 830, Sept. 22; Proceed. Amer. Philos. Soc., 1881, p. 489; *Loc. cit.* 1883, p. 547.

Dental formula; I. ?; C. $\frac{1}{1}$; Pm. $\frac{?}{4}$; M. $\frac{3}{3}$. First and second superior premolars without internal lobe; fourth with one external cusp, and a more or less developed internal heel or cingulum; all the inferior premolars without internal cusp. True molars of superior series, with but one internal tubercle, connected by a low ridge with two intermediate tubercles and two external tubercles. Inferior molars tubercular, the third with a fifth lobe or heel.

More or less of the dentition of nine species of this genus is known. The only one of which any part of the skeleton is known is the *M. ferox*. The typical species is the *M. turgidus*.

The bones of the *Mioclænus ferox* enable me to refer the genus approximately to its proper position in the system. Although we do not possess the corresponding parts of the *Mioclænus turgidus*, the type of the genus, it is probable, if not certain, that they agree in generic characters. The agreement in dentition extends to all the principal technical points, though the specific differences are marked.

The skeleton is that of a creodont. The peculiar involution of the zygapophyses of the posterior vertebræ, is seen in *Mesonyx* and in some Artiodactyles. The unequal phlanges are compressed claws, and the metapodial bones have protuberant condyles. The astragalus has a simple head with convex surface, and the trochlea is a shallow open groove.

The tubercular dentition refers this genus to the *Arctocyonidæ*.* With this family it is accordingly placed provisionally. It differs from the known fossil genera in the single tubercle of the internal part of the crown of the superior molars.

The species *M. brachystomus* and *M. etsagicus* of the Wasatch epoch have been removed from this genus. I have shown that the former is an Artiodactyle. Now, in technical points, the dentition of those species is identical with that of *Pantolestes* Cope, as well as with *Mioclænus*. Although the skeleton of the type of *Pantolestes*, *P. longicaudus* of the Bridger Beds, is yet unknown, it is safe to suppose that it does not differ from that of the

* For the dentition of this family see Lemoine, Annales, Sci. Nat., 1878, July.

M brachystomus. I therefore refer the two species first mentioned to *Pantolestes* and place that genus in the Artiodactyle sub-order.

The species of *Mioclænus* can be only as yet compared in their dentition. The characters thus derived are the following:

I. Second inferior true molar without any internal cusp on the heel, but a low ridge.
Last true molars much smaller than the others; length of inferior true molars, .018 .... *M. turgidus.*
Last true molars not reduced; second superior, .015 by .013; largest .... *M. ferox.*
Last true molars not reduced; length of true molars, .012, least .... *M. minimus.*

II. Second inferior true molar with internal posterior cusp.
Last inferior molar reduced; inferior premolars robust; length of true molars, .016 .... *M. baldwini.*
Last inferior molar not reduced; inferior true molars, .020; inferior premolars compressed conic; superior true molars, .026 .... *M. subtrigonus.*

III. Second inferior true molar unknown.
Larger; second superior molar, .012 by .010 .... *M. corrugatus.*
Large; lower true molars, .023, last little reduced; second superior, .011 by .008. *M. protogonioides.*
Small; superior true molars about .018; second larger than first, which is much larger than fourth premolar .... *M. bucculentus.*
Large; cusps of inferior molars obtuse; inferior pm. iii, .008, its heel short and small .... *M. mandibularis.*

Mioclænus turgidus Cope.

American Naturalist, 1881, p. 830, Sept. 22; Paleontol. Bull. No. 33, p. 489, Sept. 30, 1881; Proceed. Amer. Philos. Soc., 1881, p. 489.

Plate XXV e, figs. 19–20; LVII f, figs. 3–4.

The remains of this species have been more frequently obtained than those of others of the genus, eight individuals being represented in my collection. The characters displayed by the typical specimen are as follows: There are no cingula on the second, third, or fourth premolars. The last two are wider than long, and the external face is a little flattened. The tubercles in the third and fourth are conic; the external has a small one at the anterior base and a rudiment at the posterior base, and there is a low one on the posterior side at the middle. The second true molar is wider than the first. The tubercles are all round in section. Besides those already mentioned, there is a rudiment of a posterior inner on the first, which is represented by a cingulum on the second. The latter has basal cingula all around except on the inner side; the same are visible on the first true molar in a rudimental condition. Enamel nearly smooth.

The inferior molars are of robust proportions. Their sizes are, commencing with the largest: Pm. iv; M. ii; M. i; M. iii. The last molar is only half as large as the penultimate. It has two anterior and an external lateral tubercle and a heel. On the penultimate molar there are two anterior tubercles with a trace of anterior inner; also a broad flat heel, with a low tubercle on the external side. The constitution of the first true molar is identical. The fourth premolar has a rudimental heel, consisting of a low tubercle only. The principal cusp is conic, and is over the middle of the transverse diameter, and a little behind the middle of the anteroposterior diameter. No cingula. Enamel nearly smooth.

Measurements.

Maxillary bone.

		M.
Length of base of P-m. iv, M. i, and M. ii		.0175
Diameters base P-m. iv	anteroposterior	.0055
	transverse	.0065
Diameters base M. i	anteroposterior	.0060
	transverse	.0070
Diameters base M. ii	anteroposterior	.0060
	transverse	.0095

Mandible.

		M.
Length of bases of last four molars		.0250
Diameters P-m. iv	anteroposterior	.0070
	transverse	.0055
Diameters M. i	anteroposterior	.0060
	transverse	.0060
Diameters M. iii	anteroposterior	.0055
	transverse	.0043
Depth of ramus at M. i		.0115
Thickness of ramus at M. i		.0085

Another specimen includes the last four superior molars. The third true molar is even smaller than the corresponding inferior tooth would lead one to suppose, the grinding face having about one-fifth the superficial area of that of the second superior molar. The fourth premolar has a little greater transverse diameter than the first true molar. In another specimen, which includes part of the skull with some superior molars, the third superior molar is a little larger than in the last mentioned, displaying one external and one internal tubercle In this specimen the second premolar has a sub-triangular base with broadly rounded angles, and the crown is simple, with a conic apex. In the present species the characters of *Mio-*

clænus are best seen in the subconical tubercles of the premolars, particularly that of the heel of the fourth inferior premolar. In the other species this heel is more of a crest and is connected with the principal cusp by a low ridge.

All the specimens of the *Mioclænus turgidus* are from the Puerco beds of New Mexico, where they were found by Mr. D. Baldwin.

MIOCLÆNUS MINIMUS Cope.

Paleontological Bulletin No. 35, 1882, p. 468.

Plate XXV e; figs. 22–24.

This is one of the least mammalia of the Puerco fauna, exceeding by a little the *Hyopsodus acolytus*. It is represented by parts of two mandibles, which display all the true molars. The premolars are strictly those of *Mioclænus*.

The two anterior cusps of the true molars are higher than the heel, and they are united together to a point above the level of the heel. The section of both those of the M. ii is round; that of the external one of the first is crescentic; of the inner cusp, round. The heel is wide, and supports a cusp at the posterior external angle. It is bounded posteriorly and on the inner side by a raised ridge, which gives with the cusp, on wearing, a comma-shaped surface. A transverse ridge closely appressed to the anterior cusps connects them anteriorly. In one of the specimens there is a cingulum on the external side of the second inferior molar; on the other specimen it is wanting. Enamel smooth.

The mandibular ramus is rather deep and compressed, and displays an external ridge on the anterior border of the coronoid, which is not continued downwards.

Measurements (No. 2).

		M.
Length of base of true molars		.0125
Diameter M. ii	anteroposterior	.0040
	transverse	.0035
Depth of ramus at M. ii		.0073

From the Puerco beds of New Mexico. D. Baldwin.

MIOCLÆNUS BALDWINI Cope.

American Naturalist, 1882, p. 853. (October, published Sept. 28.)
Plate XXV f; fig. 16.

Represented by a right mandibular ramus which supports the last four molars, and contains the alveoli of the second and third premolars as well.

The specimen shows that the premolars are large, the third the largest, and the second and fourth of equal length, and as long as the first true molar. The fourth premolar is oval in section and its heel is well developed, and supports a median cusp. The internal posterior cusp of the true molars is well developed. The second true molar has a well-developed anterior inner cusp, which is wanting in *Hemithlæus opisthacus*. The true molars grow successively narrower posteriorly, so that the last molar is relatively smaller than in *H. opisthacus*. The ramus becomes shallow anteriorly. It is also compressed throughout. The masseteric fossa is not marked, and the posterior part of the ramus is not incurved. The base of the coronoid process rises, so as to elevate the heel of the third inferior molar.

Measurements.

	M.
Length of last six inferior molars	.035
Length of last four inferior molars	.022
Length of P-m. iv	.0057
Length of M. i	.0053
Length of M. iii	.0055
Depth of ramus at M. ii	.0100
Depth of ramus at Pm. ii	.0065

From the Puerco beds; discovered by Mr. D. Baldwin, to whom I have much pleasure in dedicating the species.

MIOCLÆNUS FEROX Cope.

Proceedings American Philosophical Society, 1883, p. 547.
Plate XXIV f; fig. 6 et seq.

This species is represented by four specimens. One of these includes various separate teeth and a considerable portion of the skeleton; a second includes loose teeth and a smaller number of bones of the skeleton; and the third consists of a part of a mandibular ramus, which contains the three true molars. These indicate the largest species of the genus yet known, the first individual above mentioned being about the size of a wolf.

The canines are well developed, and have a robust root. The crown is rather slender and is very acute. It is rounded in front, but has an acute angle posteriorly. It is not grooved, and the enamel is smooth. The single-rooted first superior premolar is situated close to the canine, and behind it is a short diastema. I have the probable first true molar or fourth premolar. The external cusps are rather small, and are well separated from each other. The inner outline of the crown is rather broadly rounded. The internal tubercle is connected on wearing, with an anterior transverse crest which terminates near the inner base of the anterior external cusp, in an intermediate tubercle. There is a posterior intermediate tubercle. There is a cingulum all round the crown excepting at the posterior intermediate tubercle. The second (? first) true molar is like the one just described, but has relatively greater anteroposterior width. In this tooth the cingulum extends all the way round the crown.

There are but two inferior molars of this individual preserved, the second and third true. The former of these has a parallelogrammic outline with rounded angles. There are two posterior, and two anterior tubercles; an anterior transverse ledge; and a narrow external and posterior cingulum, the latter rising into the internal posterior tubercle. The latter is a mere angle and is much smaller than the external posterior, which has a wide crescentic section. Of the anterior tubercles the interior is much the larger, and has a circular worn base. The third true molar is triangular in outline. Its crown includes two anterior and an external median tubercle. The inner and posterior parts of the crown form a wide shelf, with the internal edge denticulate. A weak external cingulum.

Measurements of Teeth.

		M.
Diameters base of crown of incisor	anteroposterior	.0045
	transverse	.004
Diameters base crown of canine	anteroposterior	.0130
	transverse	.0095
Diameters crown, superior M. i	anteroposterior	.0095
	transverse	.0120
Diameters M. ? ii	anteroposterior	.0110
	transverse	.0110
Diameters of inferior M. ii	anteroposterior	.0120
	transverse	.0105
Diameters of inferior M. iii	anteroposterior	.0125
	transverse	.0090

The second individual includes part of the superior walls of the skull. The fragment displays a high sagittal crest, which is fissured in front so as to keep the temporal ridges apart to near its anterior apex. The brain surfaces show small, smooth, flat hemispheres, separated by a constriction from the wide and large olfactory lobes. The navicular bone shows three well-defined distal facets, indicating probably five digits in the pes. The teeth of this specimen include a posterior superior molar, and an inferior third or fourth premolar, with other teeth. The premolar is like that of a creodont. Its principal cusp is a simple cone. To this is added a short wide heel, whose superior surface is in two parts, a higher and a lower, divided by a median ridge. A low anterior basal lobe, and a weak external cingulum.

The third specimen belonged to an individual a little smaller than the other two. It includes the first inferior true molar, a tooth lost from the others. Its form is somewhat narrowed anteriorly, where it has two low, but well separated anterior inner tubercles, which form a V with the external anterior.

Specimen No. 1 is accompanied by fragments of vertebræ and limbs. The former are principally from the lumbar region, but fragments of the atlas remain. This vertebra is of moderate length, and the cotylus is somewhat oblique. The vertebrarterial canal is rather elongate, and its anterior groove-like continuation in front of the diapophysis is not deeply excavated. The lumbar vertebræ are remarkable in the characters of their zygapophyses. These display subcylindric surfaces of the posterior pair, which indicates that the anterior ones are involuted, as in the specialized Artiodactyles and Perissodactyles of the later geological ages. Such a structure does not exist among carnivora, nor in any mammals of the Lower Eocene, to my knowledge, excepting some creodonta. I do not find it in *Didelphys* nor *Phascolarctos*, but it exists in a moderately developed degree in *Sarcophilus*. It is, however, entirely similar to the arrangement in ***Mesonyx obtusidens***, which see. The articular surface forms more than half of a cylinder, and its superior portion is bounded within by an anteroposterior open groove. The surface within this is not revolute, as in *Bos* and *Sus*, but the articular surface disappears, as in *Cervus*. Eight such postzygapophyses are preserved, all disconnected from their centra. Two of them are united together. There are two other separated zygapophyses of smaller size, which have but slightly convex surfaces. One is probably a prezygapophysis of a dorsal vertebra. No centrum is preserved.

Of the anterior limb there is a probable distal half of a radius. It is of peculiar form, and resembles that of *Sarcophilus ursinus* more than any other species accessible to me. One peculiarity consists in the outward look of its carpal surface, which makes an angle of about 45° with the long axis of the shaft. The obliquity in *S. ursinus* is less. The external border of the shaft in *M. ferox* is, however, straight, and terminates in a depressed tuberosity. Beyond this, the border extends obliquely outwards to the carpal face, which it reaches at a right angle. The internal border of the shaft is gradually curved outwards to the external border of the carpal face. Its edge is obtuse, while the external one is more acute for a short distance, and rises to the anterior (superior) plane of the shaft. The carpal face is spherically subtriangular with rounded angles. It displays two slightly distinguished facets, one of which is superior, and the other is larger and surrounds it, except on the superior side. The internal marginal projection, or "styloid process," is not so prominent as in *S. ursinus*, and is a roughened raised margin. Joining it on the inferior edge of the carpal face is another rough projection of the margin. Immediately opposite this, on the superior edge of the carpal face, is a rough tuberosity, which incloses a small rough fossa, between itself and the styloid process. Internal to it is a shallow groove for an extensor tendon of the manus; then a low short ridge, and internal to that a wide shallow depression for other extensors. The carpal face differs greatly from those of *Sarcophilus* and *Didelphys* in having the inner portion wider than the outer, instead of the reverse, and in having no distinct styloid process. It indicates that the manus was turned outwards much more decidedly than in those genera. I have described a bone very similar to this one in the *Conoryctes comma*, as the extremity of the tibia (p.), which reference is probably erroneous.

Of carpal bones the only recognizable one is the unciform. Its proximal articular surface rises with a strong convexity entad, and descends to an edge ectad. The metacarpal surface is concave in anteroposterior section, forming a wide shallow groove, extending in the direction of the width of the foot. Its two metacarpal areas are not distinguished. The entire first and second metacarpals, with the heads of the third and fourth, are preserved. They considerably resemble those of *Sarcophilus ursinus.* The distal articulations are injured in both, but both display a sharp trochlear keel posteriorly, which on the second extends nearly to the superior face of the articulation. The condyle is subround, and is constricted laterally, and at the base above. The second metacarpal is short and robust, shorter than in *Sarcophilus ursinus.* The first is also robust, but is relatively longer, as it is three-quarters the length of the second. Its head is expanded, especially posteriorly, and the large trapezial face is subtriangular, with round apex directed inwards as well as forward. The posterior face of the head is notched ectad to the middle. On the external side of the head there is a vertical facet with convex distal outline, for contact with the second metacarpal. The head of the latter is narrow, and is concave between the sides. The concavity is bounded posteriorly by a raised edge. The anterior part of the proximal facet is decurved. The shaft is deep proximally, but on the distal half is wider than deep. The lateral distal fossæ are remarkably deep and narrow, the condyle very much contracted. The head of the supposed third matacarpal is as wide as the second anteriorly, but narrows to the posterior third, and then contracts abruptly to a narrow apex. The supposed external side of the head is perfectly straight, and is continuous with the side of the shaft without interruption. The entad side displays no facet, but has a depression below the head which adapts itself very well to the head of the first metacarpal. In fact, if the metacarpals just named second and third, exchange places, so that second is placed third and third second, the metacarpal series fits far better. The fourth fits the so-called second much better than the so-called third. This may therefore be the true order, although that first used agrees better with the carpus of *Sarcophilus.* The head of the so-called third is slightly convex anteroposteriorly, and is oblique laterally, descending a little to the inner side. The fourth metacarpal is wider anteriorly than either the second or third The inner edge is straight, while the outer is concave, the head being narrower before than behind. It has a lateral facet on each side; the inner plane, the external concave in the vertical as well as in the anteroposterior direction It thus approaches the form of a metatarsal, but is not so strongly excavated, nor is the head notched on either side. The unciform face is convex anteroposteriorly and plane transversely.

The *femur* is broken up so that I cannot restore it. The head of the *tibia* is gone, but a considerable part of the astragalar-face is preserved. This is transverse to the long axis of the tibia. It is narrowed anteroposteriorly next the fibular facet. Malleolus lost. The shaft is robust, and does not expand distally for articulation with the astragalus. Three centimeters proximal to the distal end, the external side throws out a low, rough, ridge-like tuberosity. Above the middle, the crest turns outwards, leaving the internal face convex. There is a broken patella, which has one facet much wider than the other.

The *astragalus* has the trochlear portion a little oblique. That is, the internal crest is a little lower than the external, and the inner face is a little sloping. The latter is impressed by a fossa above the posterior part of the sustentacular facet, which runs out on the neck. The trochlea has a shallow groove which is nearer the external than the internal crest, and which passes entirely round the posterior aspect to the plane of the inferior face of the astragalus. The groove for the flexor tendon is thus entirely inclosed and issues on the inferior face at the posterior extremity of the groove which separates the sustentacular from the condylar facets. The external crest of the trochlea is less prominent posteriorly than the internal, thus reversing the relations of the superior part. The internal ridge becomes quite robust, but does not flatten out and project sub-horizontally as in *Oxyæna forcipata*. The fibular face is vertical; neither its anterior nor posterior angles are produced. The neck is somewhat contracted (the internal side is injured). The head is a transverse oval, strongly convex vertically, moderately so horizontally, and without flattening. A *mesocuneïform* (or possibly *ectocuneïform*) bone is wedged-shaped in horizontal section, without posterior tuberosity, and its anterior face is a slightly oblique square. The narrower facet is oblique in the transverse sense.

The *metatarsals* are represented, excepting the first and second. The only complete one is the fifth. The heads of the third and fourth are much like those of *Oxyæna forciputa*, and of about the same size. Their anterior width is equal, and in both the external side is more oblique than the internal. Both have a notch at the middle of the internal side, but they differ in that the third has an open notch on the external side which is wanting to the fourth. The lateral excavations of the external sides are deep and rather large, and thin out the anterior external edge. The lateral facets are correspondingly large on the fourth and fifth; on the third metatarsal it is small, and a mere decurvature of the proximal surface. That of the fourth is longer proximo-distally than transversely. That of the fifth is about as long as wide, and presents more anteriorly; or, to express it more accurately, the shaft and head present more outwardly than those of the fourth. The proximal, or cuboid facet is narrow anteroposteriorly, and is curved, the external side being concave. On the external side just distal to this facet the head of the bone expands into a large outward-looking tuberosity, which is separated from the posterior tuberosity by a strong notch. Between it and the head proper, on the anterior face, is a large fossa. The entire form is something like that of the proximal extremity of a femur with head, neck, great trochanter and trochanteric fossa. A somewhat similar form is seen in the corresponding bone of *Oxyæna forcipata*. The shaft of the fifth metatarsal is one-fifth longer than that of the second metacarpal (? 3d) above described. Its direction is straight, but it is somewhat curved anteroposteriorly. Its section is subtriangular, the apex external. The condyle is narrowed and subglobular above, and spreads laterally behind, the external expansion being wide and more oblique. The keel is prominent, and is only visible from above (in front) as an angle. The distal extremities of some other metatarsals differ in being flatter at the epicondyles and concave between them on the posterior face. The condyles are more symmetrical, and are bounded above on the anterior face by a profound transverse groove. Several *phalanges* are preserved, including part of an unguis. They are all depressed, and with well-marked articular surfaces, of which the distal are well grooved, and the proximal notched below. The lateral areas of insertion of the tendons of the flexors are well marked on the edges of the posterior faces. An ungual phalange is much compressed at the base. The basal table is well marked, and has a free lateral edge. The nutritive foramen enters above the posterior extremity of this edge. No trace of basal sheath.

Measurements of No. 1.

Measurement		M.
Length of atlas at anterior vertebrarterial foramen		.0165
Expanse of postzygapophyses of a lumbar vertebra		.0230
Diameter radius at middle of shaft		.0100
Greatest distal width of radius		.0220
Diameters carpal surface	vertical	.0140
	transverse	.0185
Diameters of unciform	vertical (interiorly)	.0130
	anteroposterior (greatest)	.0140
	transverse (in front)	.0150
Diameters head metacarpal i	anteroposterior	.0130
	transverse	.0120
Length of metacarpal i		.0310
Width metacarpal i at epicondyles		.0110
Diameters head metacarpal ii	anteroposterior	.0110
	transverse	.0070
Length of metacarpal ii (or iii)		.0400
Width of metacarpal at epicondyles		.0120
Diameters head of M. iii (or ii)	anteroposterior	.0125
	transverse	.0075
Diameters head of M. iv	anteroposterior	.0120
	transverse (at middle)	.0070
Width of patella near middle		.0190
Diameters of tibia .07 M. from astragalus	anteroposterior	.0185
	transverse	.0130
Anteroposterior width of astragalar face		.0200
Total length of astragalus		.0310
Diameters of the trochlea	length on groove	.0210
	width above	.0160
	elevation externally	.0130
Greatest width of astragulus below		.0225
Length anterior to internal crest of trochlea		.0100
Diameters head of metatarsal iii	anteroposterior	.0130
	transverse (in front)	.0110
Diameters head of metatarsal iv	anteroposterior	.0140
	transverse	.0105
Diameters head M. v	without tuberosity, anteroposterior	.01[illegible]0
	without tuberosity, transverse, with lateral facet	.0080
	without tuberosity, transverse, without lateral facet	.0040
	transverse over all	.0170
Length Mt. v		.0460
Width Mt. v at epicondyles		.0120
Width Mt. v at condyle above		.0065
Width of M. iii or iv at epicondyles		.0120
Width of proximal end of phalange		.012
Length of smaller phalange (1st series)		.0230
Proximal diameter of smaller phalange	vertical	.0070
	transverse	.0110
Ungual phalange, vertical diameter of cotylus		.0090

The specimen which has been partially described in the preceding pages as No. 2 has many pieces which are identical with those preserved in specimen No. 1. Among these may be mentioned the glenoid cavities of the squamosal bone. These display, besides the large postglenoid process, a well-developed preglenoid ridge, as in *Arctocyonoidæ*, *Oxyænidæ*, and *Mesonychidæ*. A large distal caudal vertebra of elongate form indicates a long tail. An articular extremity of a flat bone is intermediate in form between the proximal end of the marsupial bone of *Didelphys* and that of *Sarcophilus*. Its principal and transverse articular surface is transversely convex, as in the latter (*S. ursinus*), but the lesser articular face is separated from it by an even shorter concave interspace than in the opossum. It has almost exactly the form of that of the latter animal. It is a short, flat cone, with two faces presenting on the same side; the one, part of the concavity mentioned, the other, flat and presenting away from it. This piece has a slight resemblance to the very peculiar head of the fibula in the opossum, but is not like that of *Sarcophilus ursinus*. I, however, think it much more probably the proximal extremity of a marsupial bone.

A supposed *cuneïform* is subtransverse in position, and resembles in general those of *Oxyæna* and *Esthonyx*. It has the two large transverse proximal facets, the anterior one-quarter wider than the posterior. The distal facet (trapeziotrapezoidal) is simple. The *navicular* is much like that of *Oxyæna forcipata*, but is more robust. Its external tuberosity is flattened anteroposteriorly, and is produced proximally. The three distal facets are well marked, the median a little wider than the external, while the internal is subround, convex, and sublateral in position. The *entocuneïform* is a flat bone, with cup-shaped facet for the navicular and narrow facet for the first metatarsus. This facet is transverse transversely and concave anteroposteriorly. It shows (1) that there is a pollex; (2) that it is probably small; and (3) that it was not opposable to the other digits, as is the case in the opossum; (4) it does not show whether the pollex has an unguis or not.

Measurements No. 2.

		M.
Transverse width condyle of mandible		.0230
Anteroposterior width condyle of mandible (at middle)		.0103
Diameters head of *os marsupii*	transverse	.0220
	anteroposterior	.0068
Diameters cuneïform	vertical	.0075
	anteroposterior	.0115

Diameters navicular	vertical in front	.0085
	transverse	.0180
	anteroposterior (middle)	.0110
Diameters entocuneïform	vertical at middle	.0100
	anteroposterior (middle)	.0140
	transverse distally	.0060

Two other bones of specimen No. 2 I cannot positively determine. The first resembles somewhat the trapezium of *Sacrophilus ursinus*, and still more that of *Didelphys*. I will figure it, as a description without identification will be incomprehensible. The next bone is of very anomalous form. It may be the magnum, which is the only unrecognized bone of importance remaining, or it may be a large intermedium. It has no resemblance to the magnum of any mammal known to me. It was evidently wedged between several bones, as it has eight articular facets. Two are on one side; the largest (convex and oval) is on one edge; three are on one end, and two, the least marked, are on the other flat side, opposite to the first.

Restoration.—We can now read the nature of the primitive mammal *Mioclænus ferox*, in so far as the materials above discussed permit. It was a powerful flesh-eater, and probably an eater of other things than flesh. It had a long tail and well-developed limbs. It had five toes all around, and the great or first toe was not opposable to the others, and may have been rudimental. The feet were plantigrade and the claws prehensile. The fore feet were well turned outwards. There were in all probability marsupial bones, but whether there was a pouch or not cannot be determined. These points, in connection with the absence of inflection of the angle of the lower jaw, render it probable that the nearest living ally of the *Mioclænus ferox* is the *Thylacynus cynocephalus* of Tasmania. The presence of a patella distinguishes it from Marsupials in general. Its dentition, glenoid cavity of the skull, and other characters, place it near the *Arctocyonidæ*. Should the forms included in that family be found to possess marsupial bones, they must probably be removed from the *Creodonta* and placed in the *Marsupialia*.

This species is about the size of a sheep. The bones are stated by Mr. Baldwin, who discovered it, to be derived from the red beds in the upper part of the Puerco series.

MIOCLAENUS SUBTRIGONUS Cope.

Paleontological Bulletin, No. 33, p. 491. Proceed. Amer. Philos. Soc., 1881, p. 491; 1883, p. 555. Plate XXIV f, fig. 4; LVII f, fig. 5.

This species was originally represented by a portion of a cranium anterior to the orbits and lacking the extremity of the muzzle, distorted by pressure. It exhibits nearly all of the molar teeth. The species differs from *M. turgidus* in the greater acuteness of all its cusps, and in the equilateral form of the fourth premolar. It is too large to belong to the *M. minimus*, which is represented by mandibles only; and too small to be the *M. mandibularis*, whose maxillary dentition is unknown.

The inner borders of the molar teeth are shorter than the outer, especially in the last two molars. The last true molar is smaller than either of the others. The cusps are all subconical, but the internal is connected with the intermediate by ridges, which give it a triangular section. The latter form a V, homologous with that in *Anisonchus*, but not so distinct, and the intermediate tubercles are not lost in its branches as in that genus. The posterior inner lobe of that and other genera is represented by a thickening of the cingulum. This cingulum extends entirely round the P-m. iv and M. i, and M. ii; the M. iii is injured. The sides of the base of the base of the P-m. iv are slightly concave. The enamel of all the molars is wrinkled.

Measurements.

		M.
Length of bases of last five molars		.0285
Diameters of base of P-m. iv	anteroposterior	.0060
	transverse	.0050
Diameters of base of M. i	anteroposterior	.0060
	transverse	.0060
Diameters of base of M. ii	anteroposterior	.0060
	transverse	.0075
Diameters of base of M. iii	anteroposterior	.0040
	transverse	.0060

I now give the characters of the inferior molar series, which have been found, by Mr. Baldwin, associated with the true superior molars. Of the latter, it may be remarked that the second true molar is not so much longer than the first as in *M. bucculentus*, although the difference in size is very evident. The third is smaller than the first, and ovoid in outline, while the first and second are subquadrate. The external cusps are conic and widely separated, and the intermediate areas are distinct. There is a cingulum all round the crown of the last two, and round that of the first, except at the inner side, and at the anteroexternal angle.

The last three inferior premolars are higher than long at the base, and are compressed, and the apex acute. The posterior edge of the third and fourth is truncate, and simple. Each has a posterior cingulum which forms a narrow heel on the fourth. No other cingula. Of the true molars only the second is wanting. The form of these is like those of the *M. ferox*, with the cusps more prominent. The first only has trace of the anterior V; in the others, the two anterior tubercles are opposite and connected by a short anterior ledge. The heel of the first consists of a basin bounded by three tubercles, of which the external is pyramidal and largest. The median posterior is small. The heel of the third is narrow and prominent, and the internal lateral tubercle is represented by a short raised edge. The enamel of all the molars is wrinkled, and the inner side of the premolars is grooved with the height of the crown. A weak external cingulum on M. iii.

Measurements.

		M.
Length of last three superior molars		.0265
Diameters of M. i	anteroposterior	.0060
	transverse	.0060
Diameters of M. ii	anteroposterior	.0062
	transverse	.0072
Diameters of M. iii	anteroposterior	.0047
	transverse	.0060
Length of last inferior molars		.0340
Length of last three premolars		.0140
Length of P-m. iv		.0050
Elevation of P-m. iv		.0050
Diameters of M. i	anteroposterior	.0057
	transverse	.0042
Diameters of M. iii	anteroposterior	.0070
	transverse	.0035

Rather larger than the pine weasel, *Mustela americana.*

MIOCLÆNUS MANDIBULARIS Cope.

American Naturalist, 1881, p. 831, Sept. 22, 1881. Paleontological Bulletin, No. 33, p. —. Plate LVII f; fig. 7.

The typical specimen of this species is represented by two fragments of the left mandibular ramus which were apparently found together, but which do not fit, owing to the loss of an intermediate fragment, and accumulation of hard ferruginous stone. That they belong to the same ramus is probable but not certain. I have therefore not referred it to its position in the genus as determined by the second inferior molar tooth, although that tooth is preserved

The second or third inferior premolar is a rather large tooth formed somewhat after the pattern of the corresponding one of the species of *Haploconus*. It is compressed, has an elevated confluent anterior lobe, a large median lobe, and a low, short heel. The posterior face of the median lobe is truncate, and is bounded by two edges, of which the internal is continuous with the inner edge of the heel. The latter has a weak median keel which rises to a point of the posterior margin. Surface smooth, no cingula. The second inferior true molar has a large, wide heel, whose external side supports a large tubercle. The posterior border is raised, and the size continues round the inner side, supporting two small lobes, one posterior, the other internal. The anterior cusps are large and closely approximated, and there is a small anterior inner cusp. This is a little inside of the middle line and is connected with the external anterior cusp by a ledge. No cingula.

Measurements.

		M.
Length of Pm. iii		.0080
Width Pm. iii at middle of base		.0038
Diameter M. ii.	anteroposterior	.0060
	transverse	.0060

The structure of the second inferior molar places this species between the *M. turgidus* and the *M. brachystomus*. It is as large as the former, but had larger jaws and muzzle, judging by the size of the premolar tooth.

From the Puerco Beds of New Mexico, D. Baldwin.

MIOCLÆNUS PROTOGONIOIDES Cope.

American Naturalist, 1882, p. 833, (October, published Sept. 28).
Plate XXV f, fig. 17.

The third in size of the genus, represented by the superior true molars. It is an exaggerated form of the *M. subtrigonus*. The internal angle of the V, as well as the intermediate tubercles at the ends of its limbs, are distinct. Cingula extending entirely around the crown, the posterior with a small tubercle on the M. ii as in *A. subtrigonus;* none on M. iii, which is .75 the area of the M. ii. Diameters M. ii, anteroposterior, .008; transverse, .010. Diameters M. iii, anteroposterior, .007; transverse, .009.

From the Lowest Puerco of New Mexico, D. Baldwin.

Mioclænus bucculentus Cope.

Proceedings American Phiosophical Society, 1883, p. 555.
Plate XXIV g; fig. 2.

A part of the right maxillary bone which supports three molars indicates this species. The molars are Pm. iv, M. i and M. ii. This series is characterized by the remarkably small size of the fourth premolar, and large size of the second true molar. The first true molar is intermediate.

The fourth premolar consists of an external cone and much smaller internal one. There is a weak posterior basal cingulum. The reduced size of the internal cone suggests the probability that the third premolar has no internal cusp, and that there may be but three premolars. In the latter case the species must be distinguished from *Mioclænus.*

The first and second true molars have conic well separated external cusps, and a single pyramidal internal cusp. The intermediate tubercles are distinct. There is a posterior cingulum which terminates interiorly in a flat prominence. There is an anterior cingulum and a strong external one, which form a prominence at the anterior external angle of the crown. Enamel wrinkled.

Measurements of Superior Molars.

		M.
Length of bases of Pm. iv, M. i and ii		.0180
Diameters Pm. iv	anteroposterior	.0040
	transverse	.0046
Diameters of M. i	anteroposterior	.0060
	transverse	.0065
Diameter of M. ii	anteroposterior	.0070
	transverse	.0085

From the Puerco region of New Mexico, D. Baldwin.

Mioclænus corrugatus Cope.

Paleontological Bulletin No. 36, p. 560, 1883.
Plate XXIV g; fig. 1.

This species is known from a right maxillary bone which contains the last four molar teeth, with parts of pelvis and other bones of one individual. This species is intermediate in size between the *M. protogonioides* and *M. ferox* (see p. 325). The superior molars are more nearly quadrate than in the other species of the genus, owing to the better development of the posterior internal tubercle, which is, however, as in the others, a mere thickening of the posterior cingulum. It is wanting from the last superior molar.

The cusps on the true molars are, as in the *M. ferox*, small, and not large and closely placed as in *M. protogonioides*. The intermediate ones are nearly obsolete. The crowns are all entirely surrounded by a cingulum. The entire enamel surfaces wrinkled so as to be rugose, although the teeth are those of an adult and well used. The second superior molar is larger than the first, exceeding it in the transverse rather than the fore-and-aft diameter. The third is the smallest, and is of oval form with obliquely truncate external face. It is less reduced than in the *M. turgidus*.

The fourth premolar consists of a strong compressed-conic cusp with three basal cusps of small size, viz., an anterior, a posterior, and an internal. The last is the larger, though small; is formed like a heel, and is connected with the others by a cingulum. No external cingulum.

Measurements.

		M.
Length of last four molars		.036
Diameters P-m. iv	anteroposterior	.010
	transverse	.008
Diameters M. i	anteroposterior	.010
	transverse	.010
Diameters M. iii	anteroposterior	.008
	transverse	.011

From the Upper Puerco beds; D. Baldwin.

ACHÆNODON Cope.

Paleontological Bulletin, No. 17, p. 2, Oct. 25, 1873. U. S. Geol. Surv. Terrs. Ann. Rep. for 1873, p. 457.

Known from the skull only. Dentition: I. $\frac{?}{3}$; C. $\frac{1}{1}$; Pm. $\frac{3}{4}$; M. $\frac{3}{3}$ Canines well developed; dental series without diastema First superior premolar two-rooted; crown of second (third) compressed, simple. Crown of third (fourth) with one external and one internal principal tubercles. Superior molars quadritubercular. First inferior premolar one-rooted. Crowns of third and fourth compressed, simple. First and second true molars quadritubercular; of third true molar with five lobes, the posterior one forming a heel. Symphysis not coössified.

This genus agrees with other *Arctocyonidæ* in the presence of a preglenoid crest. Its relationships to the other genera of the family appear to be quite close. It is not yet absolutely certain whether there are four or three inferior premolars. In case the latter is the correct number, the first premolar is two-rooted.

It is probable that this genus, in common with the other *Arctocyonidæ*, stands in ancestral relation to the existing families of the arctoid carnivora. It is, in the restriction of its premolar series, rather less primitive than *Arctocyon*, and prepares the way for the *Procyonidæ*, the least specialized of the true *Carnivora*. From this family, by a modification of the fourth premolar towards a carnassial form, we derive the extinct genus *Hyœnarctos* Falc. of the upper Miocene, and the existing *Aeluropoda* M. Edw. From the latter the bears have doubtless been derived by a process of divergence from the general carnivorous line, in a special direction of their own.

Although the Arctoids are well represented in America, an Eocene ancestor has hitherto been a desideratum. It is now happily supplied.

But one species is known from North America. It is a large and formidable animal.

ACHÆNODON INSOLENS Cope.

Paleontological Bulletin, No. 17, p. 2, Oct. 25, 1873. Ann. Rep. U. S. Geol. Surv. Terrs. 1873, p. 457, Jour. Acad., Phila., 1874., Fig. 5, p. 10, March.

Plates LVII and LVII *a*.

The only specimens of this species which I possess consist of mandibular rami. The Princeton Exploring Expedition of 1877 discovered a good skull with lower jaw, at the same locality, and the director of the museum of Princeton College, Professor Guyot, kindly permitted me to examine it. I am under obligations to him and to Professors Scott and Osborne for the means of ascertaining the characters of the cranium and superior molar teeth above given. Perhaps before the present article appears, their memoir on this species will have been published. I confine myself to the general characters given, and direct my readers for fuller details to their monograph.

Judging from the size of the size of the skull, this species was about as large as a lion. It had a muzzle of about the length characteristic of that animal, and a huge sagittal crest. The eyes were remarkably small.

The rami are robust and rather shallow. The external face is but little convex below the last premolar, but projects much beyond the alveolar border opposite the last two molars. The anterior border of the masseteric fossa is quite prominent, but the fossa shallows out below. The inferior outline of the ramus is gently convex. The symphysis is rather long and rises at an angle of 45°. Its posterior termination is opposite the middle of the third premolar. There are two mental foramina, one under the second and one under the fourth premolars. The incisor teeth present forwards.

The tubercles of the true molars form two pairs, the third with a large fifth lobe. These paired lobes are more or less united at the base, while the pairs themselves are well separated from each other. The anterior pair is a little more elevated than the posterior pair. Last premolar with longer basis than first molar; its posterior heel tubercularly plicate. The crown of the penultimate premolar is a slightly compressed simple cone with elongate base, but little shorter than that of the first molar. Molars with smooth enamel; an anterior cingulum on the second and third. A small posterior median tubercle on the second molar, and a short external cingulum from the base of the posterior cone forwards, on the third. Canines very large, sub-erect, enamel smooth.

Length of molar series	.180
Diameter of canine tooth	.033
Length of premolars	.093
Length of premolar No. 3	.035
Length of molar No. 1	.024
Length of molar No. 2	.027
Width of molar No. 1	.022
Length of molar No. 3	.041
Width of molar No. 3	.024
Depth of ramus at molar No. 2	.073

The type specimen was found by Mr. Samuel Smith at the mammoth Buttes near the head of South Bitter Creek. The formation is the Washakie basin of the Bridger. The *Achænodon insolens* shares with the *Protopsalis tigrinus* and *Mesonyx ossifragus* the distinction of being the largest of the Eocene flesh-eaters. It was a formidable beast, and a worthy associate of the large *Palæosyops vallidens* and huge *Dinocerata* of the same period and region.

DISSACUS Cope.

American Naturalist, 1881, p. 1019 (November 29).

This genus is only known from upper and lower jaws. These possess a dentition much like that of *Mesonyx*, which differs, however, from it in one essential respect. The apices or cusps of the last two molars are double, not simple as is the case in *Mesonyx*. This constitutes an approach to *Sarcothraustes*, where there are three apices to the main cusp. Two species are known, both from the beds of the Puerco Epoch.

DISSACUS NAVAJOVIUS Cope.

American Naturalist, 1881, p. 1019. *Mesonyx navajovius* Cope, Proceed. Amer. Philos. Soc. 1881, p. 484 (September 30).

Plate XXV c; fig. 1.

Smaller than the two known species of *Mesonyx*, and with the crowns of the molars more compressed and the blades of the heels of the inferior series more acute. Molars seven, the first one-rooted. Last molar with a cutting heel like the others, and with the penultimate, with a rudimental anterior inner cusp. All the molars with an anterior basal tubercle except the first, second, and third. No basal cingula. Principal cusp elevated and compressed, as in the premolars of *Oxyæna*. Enamel minutely rugose. Mandibular rami and inferior canine teeth compressed, the angle of the latter not inflected.

Measurements.

	M.
Length of inferior molar series	.078
Length of premolar series	.046
Length of base iv premolar	.010
Pm. iv, elevation of cusp	.008
Length of ii true molar	.012
Elevation of true molar	.010
Width of heel of true molar	.005
Depth of ramus at M. ii	.020
Diameter of base of crown of canine	.009

This species was a little larger than the red fox. Its remains were found by Mr. D. Baldwin in Northwestern New Mexico.

DISSACUS CARNIFEX Cope.

American Naturalist, 1882, p. 834 (October 5).
Plate XXIV g; figs. 3–4.

This creodont differs from its only congener in its greater size, and in the presence of an anterior basal lobe on the third inferior premolar. This is wanting in *D. navajovius*. As compared with the latter the six inferior molars are as long as its seven, and the mandibular ramus is much deeper. Like it the Pm. iv and the true molars have an anterior basal tubercle; and the last two true molars have an internal supplementary cusp. After the *Sarcothraustes antiquus*, the largest flesh-eater of the Puerco, and about equal in dimensions to the coyote.

Measurements.

	M.
Length of last six molars	.075
Length of true molars	.038
Length of Pm. iv	.0125
Length of M. ii	.0135
Length of M. iii	.0130
Depth of ramus at M. ii	.029

Northwestern New Mexico; D. Baldwin.

SARCOTHRAUSTES Cope.

Paleontological Bulletin, No. 34, p. 193, February 20th, 1882; Proceed. Am. Phil. Soc., 1881, p. 193.

We have in evidence of the characters of this genus the last two superior molars, the last one lacking the crown; and parts of both mandibular rami, which exhibit teeth as far posteriorly as the first true molar inclusive; all belonging to one individual. A part of a skeleton of a second individual, which includes a fragment of lower jaw, belongs probably to this species.

Sarcothraustes resembles both *Amblyctonus* and *Mesonyx*, but it is probably to the latter genus that it is allied. The last superior molar is transverse, much as in *Oxyæna*. The crown of the penultimate is subtriangular and transverse. It has two external subconic cusps and a single internal lobe, whose section on wearing is a V, each branch of the face extending to the base of the corresponding external tubercle. There are three small inferior incisors, and a large canine. There are probably only three inferior premolars, the first one rooted. The crown of the second has no heel The crown of the third has a short wide heel. The crown of the first true molar consists of an anterior elevated cone and a posterior heel. The latter is wide, having a posterior transverse, as well as a longitudinal median keel. The fragments of the supposed second individual include two large glenoid cavities with strong preglenoid crests, as in *Mesonyx*.

As compared with *Mesonyx*, this genus differs in the V-shaped crest of the penultimate superior molar; in *Mesonyx* it is represented by a simple cone. The last superior molar of *Mesonyx* is triangular and not transverse, but the composition of the crown of that tooth in *Sarcothraustes* must be known before the value of this character can be ascertained. If the view that *Sarcothraustes* has but three inferior premolars be correct, this character distinguishes it from *Mesonyx*, as do also the transversely expanded heels of the molars.

SARCOTHRAUSTES ANTIQUUS Cope.

- Loc. sup. cit.
Plate XXIV *e*; figs. 19–22.

The penultimate superior molar has a strong posterior cingulum which commences within the line of the internal bases of the external cusps, and rises into considerable importance behind the internal cusp. There is also an anterior cingulum which does not rise internally, and which is continuous with a strong external basal cingulum. The latter passes round the posterior base of the posterior cone, and runs into the posterior branch of the internal V. The posterior cone is smaller than the anterior cone, and its apex is well separated from the latter. The appearance of this tooth is something like that of a carnivorous marsupial.

The symphysis mandibuli slopes obliquely forwards, and is united by coarse suture. The ramus is stout and deep, as compared with the size of the molar teeth. The roots of the teeth are relatively large, especially those of the first two premolars. The crown of the canine is lost. The first premolar points forwards, nearly parallel with the canine, and divergent from the second premolar. The crown of the second premolar is small and subconic, and has a rudimental heel, and no anterior basal tubercle. The first true molar resembles considerably that of *Mesonyx*. There is a small anterior basal tubercle on the inner side of the principal cusp. The expansion of the heel is transverse only, there being no longitudinal lateral edges or tubercles. The enamel is obsoletely, rather coarsely wrinkled. There are two rather large mental foramina; the posterior below the anterior root of the first true molar, and the anterior below the posterior root of the second premolar.

Measurements.

		M.
Diameters of superior M. ii.	anteroposterior externally	.015
	transverse	.024
Anteroposterior diameter of base of M. iii		.0095
Anteroposterior diameter base of crown of inferior canine		.020
Length of bases of three inferior premolars		.038
Diameters inferior M. i.	anteroposterior	.019
	transverse	.0095
	vertical	.0110
Depth of ramus at Pm. iii		.0520
Width " "		.022

MESONYX Cope.

Paleontological Bulletin No. 1, p. 1, July 29, 1872. Proceed. Amer. Philos. Soc., 1872, p. 460. *Ibid.*, 1873, p. 198. The Flat-Clawed Carnivora of the Eocene of Wyoming, April 19, 1873, p. 1. ? *Synoplotherium* Cope. Pal. Bull. No. 6, p. 1, Aug. 20, 1872. Proceed. Amer. Philos. Soc., 1872, p. 483. Flat-Clawed Carnivora, etc., p. 5, plates 1 and 2. Proceed. Amer. Philos. Soc., 1873, p. 203. Annual Report U. S. Geog. Geol. Surv. Terrs., Hayden, 1872 (1873), pp. 550, 554, plates 5, 6.

This genus is known from the dentition of the lower jaw and part of that of the upper, and from many parts of the skeleton. The superior molar teeth are yet unknown. Of the known parts of the skeleton the anterior limbs are derived from the *M. lanius*, while the vertebræ and posterior limbs are those of the *M. obtusidens*. As I have hitherto placed these species in different genera, I must explain my reasons for uniting them in *Mesonyx*. The two species were obtained from different localities, the *M. lanius* from the Washakie, and the *M. obtusidens* from the Bridger basins. It is not certain that these beds belong to exactly the same horizon. Through the kindness of Professor Guyot, I have been able to examine the lower jaw and teeth, with some other portions, of a species related to *M. lanius*, if not the same, from the Bridger basin. This specimen contains the molar teeth characteristic of *M. obtusidens;* the corresponding ones being worn down by attrition in the type specimen of *M. lanius*.

Dentition: I. $\frac{3}{?}$; C. $\frac{1}{1}$; Pm. $\frac{4}{4}$; M. $\frac{?}{3}$. The space between the inferior canines is so narrow that it is probable that the incisors were wanting or reduced in number. The external superior incisor conical and larger than the others. Superior canines vertical; inferior canines subhorizontal. First inferior premolar one-rooted. Premolars from the third similar to the molars, consisting of a posterior median blade, an anteromedian conic cusp, and an anterior basal tubercle. The true molars diminish in size posteriorly, but their composition is identical with that of the first, so far as determinable.

The *cranium* of the *M. lanius* is fractured above. There remain the squamosal and periotic bones, occipital condyles, malar and part of maxillary, both premaxillaries and the greater part of both mandibular rami. The squamosal process of the zygoma is produced inferiorly far below the auditory meatus, even further than in the bears. Its proximal portion includes, on the lower face, a strong glenoid groove at right angles to the axis of the cranium, with its anterior margins acute and prominent. This is the well-defined glenoid cavity of the feline type. The zygoma has a wide curvature, indicating a powerful temporal muscle. The posterior angle of the malar extends well posteriorly. Its anterior portion projects, forming a longitudinal rib; there is no produced postorbital process. The tympanic bone is produced upwards and outwards and forms a tube with everted lips. The opisthotic (mastoid) separates it entirely from the exoccipital, and overlaps the posterior half of the tube by a laminar expansion. A pit in this bone near the *meatus externus* represents the insertion of the stylohyal ligament. There is no bulla, the tympanic chamber being small, with thick walls and without any trace of septa. The character of this region resembles that seen in the bears more than that of any other carnivorous type.

The *premaxillaries* are vertico-oblique in position, presenting the nareal opening directly forwards as in cats, but with a still less prominent alveolar border. The horizontal part of this border is indeed very short, including but two small incisors. It then rises vertically, and turns obliquely backwards to the maxillary, inclosing a deep sinus with the canine tooth. From the anterior side of this sinus the larger external incisor issues, with its root extensively exposed externally. A rib ascends from the front of its alveolus to the anterior or nareal margin of the bone. The triturating surfaces of the incisors are directed backwards, and the alveolar edge is thickened in front of them with a tuberosity. The teeth are much worn so that the forms of the crowns cannot be determined, but at .25 inch beyond the alveoli they are compressed, the large outer tooth with a longitudinal angle in front.

The *cranium* of the specimen of *M. obtusidens* is fragmentary. The malar bone of the right side is similar in position and form to that of the *Canidæ*, especially in the presence of a weak angle only, to mark the posterior border of the orbit. It has a much less expanded union with the maxillary than in these animals, and is proximally shallower, thicker, and more prominent. Its posterior portion is more plate-like.

The cervical *vertebræ* of the *M. obtusidens* are damaged. The dorsals are strikingly smaller than the lumbars, being less than half their bulk. They are opisthocœlous with shallow cups, and the centra are quite concave laterally and inferiorly. The centra of the lumbars are more truncate, with a trace of the opisthocœlous structure, and are quite depressed in form. The median part of the series is more elongate than in the corresponding vertebræ of the genus *Canis*. They exhibit an obtuse median longitudinal angle, on each side of which, at a little distance, a nutritious artery entered by a foramen. The zygapophyses of the posterior lumbars have interlocking articulations, the posterior with a convex exterior articular face, the anterior with a concave interior one. The sacrum is not completely preserved; three coössified centra remain. These are more elongate, and the diapophyses have less expansion than in *Felis*, *Hyæna*, *Canis*, or *Ursus*. They are much flattened, and the middle one has two slight median longitudinal angles. The caudal vertebræ indicate a long tail, with stout base. Its proximal vertebræ are depressed, and with broad, anteriorly directed diapophyses The more distal vertebræ have subcylindric centra; the terminal ones are very small.

The bones of the *fore limbs* of the *M. lanius* are stout in their proportions. The *humerus* has a well-marked rugose line for muscular insertion on its posterior face, but no prominent angle. Distally the inner and outer condylar tuberosities are almost wanting, and there is neither external aliform ridge nor internal arterial foramen. The olecranar and coronoid fossæ are confluent, forming a very large supracondylar foramen. The condyles are moderately constricted medially, and there is a well-marked submedian rib separated from the outer condyle by a constriction. The latter is continued as an acute ridge on the outer side of the olecranar fossa. The inner condyle is the more prominent, and its outer margin is a sharp elevated crest. The *ulna* has a very prominent superior process, continuing the cotylus upwards. The coronoid process, on the other hand, is rather low. The radial cotylus is flat and broad. The distal end is not preserved. The *radius* has a more transverse head than *Canis* or *Felis*, and has three articular planes, the inner being a wide oblique truncation of the edge. The shaft is angulate below, and becomes a little deeper than wide near the distal end. The extremity is lost. The *carpal* bones are probably all present. The fore foot was found in place so that the relations of the bones are known with certainty. (Plate xxix *a*, fig. 1.) The scaphoid and lunar are distinct. The former exhibits proximally the inner tuberosity, then a slight concavity, and then the convexity for the radius, and then it is obliquely truncated so as to give a general rhomboid outline. Beneath there are but two facets, the inner for the magnum the deepest, and divided lengthwise by the truncation of the bone. The larger facet fits correctly the 0.0. trapezium and trapezoides. The lunar preserved, lacks the posterior extremity. It has a short anterior or external face, and a very convex proximal one, with a subquadrate cross-section at its greatest convexity, which is near its middle Below it presents the usual two facets, the one more concave than the other, and soon cutting off the latter, meeting the internal facet behind it. The upper face is convex. The cuneiform is large and concave lengthwise above, for the narrow extremity of the ulna. Below, it has a large concave facet for the unciform. The pisiform is of unusual size, and is as stout as the largest metacarpus, and nearly half as long as the outer (5th) metacarpal. It articulates with a thick V-shaped facet of the cuneiform. Its extremity is obtuse and expanded. The trapezium is large and is attached to its metacarpus laterally, sending a process downwards posteriorly. It supports a narrow articular surface for a rudimental first metacarpus, which is not preserved, but which could not have been larger than that of the spotted hyena. The trapezoid is smaller and of a triangular outline, with the base forwards. The magnum is a rather small bone, articulating as usual with the metatarsals 2 and 3. It is depressed in front. The unciform is a large bone with a considerable external anterior surface. Two-thirds of its upper surface is in contact with the cuneiform, the remaining part projecting upwards with convex face to unite with the lunare. Below, it supports metatarsals 4 and 5.

There were probably four digits of the fore foot, the pollex being very rudimental. The proportions are stouter than in the dogs and hyenas, but not so much so as in the bears. The proximal extremities of the *metacarpals* interlock as in the hyenas, and much more than in the dogs. The external side of each is excavated to receive an oblique facet of the one adjoining, as in *Protopsalis;* but there is no abrupt prominence as in the cats. The *phalanges* have a length similar to that seen in some bears, but the metatarsals are more elongate. The lengths of the latter are, fifth shortest, then second, third, and fourth. Their condyles are broad, with median keel behind, and shallow supracondylar fossa in front. The first phalanges are about one-third the length of the metacarpals; the second of digit No. 2 broad and stout, and half as long as the phalange of the first row. An ungual phalange has a singular form, so that the claw might be supposed to have a subungulate character. It is flat, considerably broader than high and with expanded and obtuse extremity. The articular extremity is depressed and transverse concave in vertical, convex in transverse section. The anterior three-fifths of the superior middle line is occupied by a deep gaping fissure, which separates the extremity into two points. The inferior face is entirely flat, there being no tendinous tuberosity. The sides are grooved, and give entrance each to a large arterial foramen proximally. These claws resemble remotely those of seals, and differ remarkably from those of existing terrestrial *Carnivora*.

The glenoid cavity of the *scapula* of *M. obtusidens* is shallow; the coracoid process is a short hook, separated by a strong groove from the edge of the cup. The spine is well developed. In the character of the coracoid, this genus resembles *Felis* more than *Canis* or *Ursus*. The ulna exhibits little trace of articular face for the radius, less than in *Felis* or *Canis*. Its humeral glenoid face is more convex transversely in its anterior or vertical portion, than in those genera, and a little more than in *Ursus*.

Of the *hinder limb* of the *M. lanius*, the only characteristic pieces remaining are the navicular, cuboid, and an external cuneiform bone. The cuboid is rather stout, with a slightly concave proximal facet and two distal ones, one of them smaller and sublateral. The navicular is wide and flat, and with a strongly concave astragaline facet. Below, it presents two deep oblique concave facets for the cuneiforms, with a small sublateral one on the outer side. The facets of the cuboid indicate that the fourth digit is well-developed, but the presence of the hallux cannot be positively ascertained.

In the hind limb of the *M. obtusidens* the *femur* resembles that of true *Carnivora* in all essentials. The rotular groove is narrow and elevated, the inner margin a little higher. The condyles are rather narrow, the inner with less transverse and anteroposterior extent, and separated by a wide and deep fossa. The patella is narrow, thick, and truncate at one end. The proximal end of the *tibia* exhibits a very prominent and well elevated crest or spine, which bounds a deeply excavated fossa. The articular faces are separated by a deep notch behind; the external is a little the larger and is produced into a point outwards and backwards; it lacks the notch of the anteroexterior margin so distinct in *Canis*, but possesses an emargination at the outer base of the crest, homologous with it. The general form is, however, more like that of *Canis* than of *Felis*, and least like that of *Ursus*. The distal extremity of the tibia presents Carnivorous characters. The two trochlear fossæ are deeply impressed, the outer wall of the external one being formed by the fibula only. The anterior marginal crest is more elevated than the posterior, and presents an overlapping articular face between the fossæ for a corresponding tuberosity of the neck of the astragalus. The inner malleolus is entirely without the groove for the tendon of the *tibialis posticus* muscle, and therefore different from many of the digitigrade *Carnivora*. It has an ovate truncate surface. On the anterior face opposite the inner trochlear groove is a rather small but deep fossa.

The *astragalus* has an elongate oblique neck and a navicular extremity slightly expanded inwards. The trochlear ridges are well elevated, and not very oblique to the true vertical plane, being much as in the dog. The distal extremity is quite different from *Felis*, *Hyæna*, *Canis*, and *Ursus* in having a rather narrow convex facet next the cuboid bone extending from front to rear, and in having the navicular facet pulley-like or slightly concave in transverse section, while it is strongly convex anteroposteriorly. This is part of the peculiarity presented by the hind foot in this genus. Behind the navicular facet, on the superior face, is a tuberosity which stops the flexure of the foot by contact with the tibia; a trace of it is seen in the dog. The calcaneum has the compressed form of the digitigrades, but the broader internal, and convex external astragaline facets resemble much more those in the bears. The sustentacular facet looks as much forwards as upwards. The cuboid facet is a frustum of a triangle with the apex directed inwards and downwards.

The *metapodial* bones are rather elongate, and flattened so as to be transverse in position. A second metatarsal is more flattened than corresponding bones of *Canis* and *Felis*. Its cuneiform facet is somewhat concave transversely. It has two facets for the third metatarsal, as in *Hyæna;* that is, they are a good deal more distinct than in *Felis*, where they are more distinct than in *Canis*. The distal condyles are furnished with a posterior and inferior carina, which is wanting above; the articular face is wide above as in *Canis*, and is bounded by a transverse fossa as in digitigrade genera. The phalanges of the first series are elongate and curved as in *Felis*, being relatively longer than in *Ursus*. Phalanges of the other series are quite short. The ungues are short and flattened, their inferior surface is nearly plane, and the superior but little convex. A shallow groove divides the upper face longitudinally to the extremity. The margin below is acute to a slightly contracted neck. There is no indication of collar for reception of the horny sheath, except perhaps a slight area of fracture on each side, and there is no projecting tuberosity below for insertion of flexor tendon. The middle of the proximal part of the unguis is a raised plane, and on each side of it, at the neck, two arterial foramina enter. There is a small foramen in the groove, and several smaller ones near the margin. These ungues resemble somewhat those of some tortoises. They were found with the other phalanges, with which they agree in size and articulation, and no doubt belong to the same animal. It is evident that they differ in character from those of most existing *Carnivora*. The penultimate phalanges agree with them in the depressed form of their proximal articular faces, wanting entirely the triangular form so characteristic of *Carnivora*, especially of the cats and dogs. The short flat shaft of the same is almost equally peculiar.

It is clear that there were only four anterior digits in *M. lanius*, and but four posterior ones in *M. obtusidens*.

Affinities. These have been already considered in their general bearings. The genus is the type of a distinct family, which must be placed nearest to the *Amblyctonidæ*. From this family the *Mesonychidæ* differ in the complete trochlear articulation of the ankle-joint. At first sight *Mesonyx* appears to have some similarity to *Hyænodon* in its dentition, but close examination shows that the resemblance is rather to *Amblyctonus*. There is no true sectorial blade on any tooth of *Mesonyx*, the long heel furnishing the only cutting edge, as in the premolars of several genera. As in *Hyænodon* there is no internal tubercle, it is true, but the median tubercle is a cone, and the anterior is rudimental, so that there is no sectorial structure. The structure of the feet in *Hyænodon* being yet unknown, it is not possible to state the relations seen in these parts. The hind feet are, as already pointed out, entirely different from those of other *Creodonta*, to which group the fore feet refer the genus. The flat claws are a unique peculiarity, and suggest affinity to the seals, and an aquatic habit. The teeth, moreover, show a tendency in the same direction, in the simplicity of their crowns. The structure of the ankle forbids the supposition that these animals were exclusively aquatic, as it is of the type of the most perfect terrestrial animals. The reduced number of digits—four both anteriorly and posteriorly—as in the *Hyænidæ*, is also opposed to any suggestion of aquatic habits.

Species. Two are certainly known, and a third may be in our collections. The former are distinguished as follows, by the dentition of the lower jaw:

Smaller in all dimensions, except in the first true molar, which is equal that of *M. lanius* .... *M. obtusidens.*
Larger by one-half, except in the first true molar, which equals that of the other species .... *M. lanius*

The specimen in the Princeton museum, already mentioned, is of similar form and proportions to the *M. lanius*, excepting that the last two inferior true molars are only four-fifths as long as those of the *S. lanius.*

History. I have been unable to find any reference to this genus other than those included in the citations at the head of this article.

MESONYX OBTUSIDENS Cope.

Proceedings American Philosophical Society, 1872, 480 (July 29), et loc. sup. cit.

Plate xxvi, figs. 3–12; xxvii, figs. 1–24.

Of the typical specimen of this species, there are preserved portions of the skull with teeth, chiefly mandibular; numerous vertebræ from all parts of the column; parts of scapula, ulna, and fore feet, portions of pelvis, femora, tibiæ, tarsals, metatarsals, and phalanges.

There are numerous *teeth* preserved, but separate from the skull and mostly mandibular. The inferior canine is stout, especially in the root, which is a flat oval in section. The crown is but little curved, slightly compressed, and without edge or groove. The premolars graduate into the molars, so that the line of distinction is not easily drawn. The first premolar has a single root; the crown is slightly conic, with a small tubercle at the base behind. This tubercle increases in size on the premolars 2 and 3, and becomes on the true molars a longitudinal cutting edge extending along the axis of the crown, not much elevated above a wide base. It occupies half the length of the crown in the larger molars, and is preceded by an elevated conic cusp. In front of the base of this, a small conic tubercle projects forwards, which appeared as a rudiment on the third premolar. The number of mandibular teeth would appear to be, Pm. 4, M. 3. No portions certainly referable to the superior molars were found.

There are no cingula on the teeth, and the enamel is perfectly smooth. The appearance of the crowns as well as of the bones indicates an adult animal.

The measurements are as follows:

	M.
Length of malar bone	.073
Depth of malar bone in front	.016
Depth of malar bone at postorbital angle	.023
Depth of malar bone at middle of orbit	.015
Thickness of malar bone at middle of orbit	.013
Length of crown of canine tooth (worn)	.020
Diameter of base fore and aft	.013
Diameter of premolar (1) fore and aft	.006
Length of crown of premolar	.006
Length of base of premolar (2)	.011
Height of crown of premolar	.009
Length of crown of true molar	.018
Width of crown of true molar	.008
Height of cutting edge	.005

The internal face of the trochlear portion of the astragalus is nearly plane, not oblique, as in most of the *Creodonta*. The external side is a little more inclined, especially opposite the anterior extremity of the trochlea, where the inferior portion is produced into a horizontal process. The head is not expanded transversely, the internal tuberosity being very low, and the infero-external side concave. The sustentacular facet is small and isolated. There are no horizontal processes of the distal extremity of the calcaneum as are commonly seen in Creodont genera. The inferior face of the calcaneum is concave in cross-section for its distal half. The second metatarsal has its cuneiform surface convex anteroposteriorly and concave transversely. There is a small proximal facet looking inwards at the front of the inner side of the principal facet, apparently for the entocuneiform bone. There is a well marked ligamentous insertion on the anterior face of the shaft next to the facets. The carina of the posterior side of the condyle of several of the metapodial bones is continued for a short distance on the posterior face of the shaft:

	M.
Transverse diameter of glenoid cavity of scapula	.025
Transverse diameter of ulnar cavity for humerus	.014
Length of centrum dorsal vertebra	.019
Diameter of centrum, transverse	.014
Diameter of centrum, vertical	.014
Length of centrum of a median lumbar	.030
Diameter of centrum, transverse	.025
Diameter of centrum, vertical	.016
Diameter of centrum vertical, first sacral	.014
Diameter of centrum transverse, first sacral	.026
Expanse of sacrum	.046
Length of two sacral vertebræ	.049
Length of proximal caudal	.022
Expanse of diapophyses caudal	.036
Diameter of centrum caudal, vertical	.009
Diameter of centrum caudal, transverse	.015
Diameter of centrum distal caudal, vertical	.007
Diameter of centrum distal caudal, transverse	.007
Chord of femoral trochlea and condyles	.038
Width of trochlear groove	.013
Width of condyles	.029
Width of tibia, proximally	.038
Diameter of tibia, anteroposteriorly	.039
Diameter of shaft, .050 M. from end	.017
Diameter of distal extremity, transversely	.026
Diameter of distal extremity, anteroposteriorly	.018
Length of patella	.025
Width of patella	.015
Length of astragalus	.030
Width of astragalus above	.016
Width of astragalus distally	.017
Width of astragalus, neck	.012
Width of cuboid facet of calcaneum	.016
Depth of cuboid facet of calcaneum	.011
Width of a second metatarsal (shaft),	.012
Depth of a second metatarsal (head)	.014
Width of a second metatarsal distal end	.010
Length of proximal phalange	.0290
Width proximally	.0100
Width proximally of a penultimate phalange	.0085
Length proximally of a penultimate phalange	.0110
Length of ungual phalange	.0150
Width medially	.0065
Width proximally	.0070

Besides the inferior size, this species apparently differs from the *M. lanius* in the form of the ungual phalanges. Those preserved are much narrower than the single one of the *M. lanius* that is known. It is, however, not certain that these phalanges were alike on all the digits.

This species was as large as a wolf. While the proportions of the limbs were not very different, the form was rather more slender behind. The orbit was smaller, and the cheek bone more prominent than in those animals. The long tail added to the general resemblance to the dogs. The narrow navicular facet of the astragalus renders it probable that the inner toe is wanting or rudimental, and that there are four digits on the hind foot. The claws are flat, and altogether without prehensile use, but rather adapted for aquatic life.

I obtained the bones above described on a bluff of Cottonwood Creek, near Fort Bridger, Wyoming, during my expedition of 1872. All the pieces were found in close juxtaposition, and without admixture of those of any other animal.

MESONYX LANIUS Cope.

Synoplotherium lanius Cope. Paleontological Bulletin No. 6, p. 1, August 20, 1872. Proceed. Amer. Philos. Soc., 1872, p. 483. *Ibid.*, 1873, p. 207. Annual Report U. S. Geol. Surv. Terrs., 1872, p. 557, pls. 5–6.

Plate xxvii, fig. 25; Plates xxviii, xxix, figs. 1–6; Plate xxix *a*, fig. 1.

Besides the typical and, so far, the only specimen I have obtained of this species, there is a second probably referable to it in the museum of Princeton College. For the opportunity of examining the latter I am indebted to the kindness of Professor Guyot. My own specimen is represented by a large part of the skull with nearly complete dentition, the superior molars loose; lumbar and caudal vertebræ; large portions of both fore limbs, including the bones of the feet; smaller portions of the hind limbs and feet.

The *mandibular rami* are quite elongate, and indicate a cranium near the size of that of the brown bear (*Ursus arctos*). Their form is slender, and they have a long, rather narrow, symphysis, which projects obliquely forwards. The angle is not preserved. The mental foramen is large and issues just behind the canine teeth.

The *dentition* is I. $\frac{3}{?0}$; C. $\frac{1}{1}$; M. $\frac{?7}{7}$. The canine is of very large size, especially the part protruded beyond the alveolus. The crown is stout at the base, but is soon compressed and obliquely truncated by the attrition of the inferior canine on its inner face. Two superior molars preserved are three-rooted, and the section of the crown is more or less equally trilobate. The number in the maxillary bone is estimated at seven, the number found in the ramus of the mandible. There are six two-rooted molars below, and probably one single-rooted premolar, though this is indicated by an alveolus only. The molars are rather narrow anteroposteriorly, and are not very different in size, except that the penultimate is a little longer, and the last a little shorter than the others. There was evidently a longitudinal cutting-edge behind, and some other shorter process on the front of the crown; the edge is preserved on the last tooth and resembles that of *M. obtusidens*, so that I have little doubt that the remainder of the tooth was, as in that genus, a conic tubercle. This opinion, based on my imperfect specimen, is shown to be correct by the Princeton specimen. Here the teeth are as in *M. obtusidens*. The most remarkable feature of the genus is seen in the inferior canines. These are very large teeth, and are directed immediately forwards, as in the case of the cutting-teeth of rodents. They work with their extremities against the retrorse crowns of the two external incisors above, and laterally against the superior canine. They are separated by a space about equal to the diameter of one of them. In this space I find no alveoli nor roots of teeth; the outer alveolar wall extends far beyond the inner. The latter terminates opposite the middle of the superior canine. It may be that there are no inferior incisors.

Some of the vertebræ display stout triangular neural spines; on the lumbars the posterior zygapophyses are embraced laterally by the grooved correspondents of the succeeding vertebra. Some of the caudal vertebræ are long, slender, and without neural arch, indicating that this genus, like *M. obtusidens*, had a long, slender tail.

Measurements.

	M.
Length of glenoid cavity	.045
Width of glenoid cavity	.025
Diameter of zygomatic fossa	.058
Width of opisthotic inside foramen stylohyoideum	.014
Diameter of meatus auditorius externus	.012
Diameter of cavum tympani	.009

	M.
Length of ramus mandibuli preserved	.228
Length of series of seven molar teeth	.131
Length of last molar, crown	.0155
Width of last molar, crown	.0080
Length of penultimate, crown	.0215
Width of penultimate, crown	.010
Length of exposed part of inferior canine	.024
Length of exposed part of superior canine	.032
Length of exposed part of outer upper incisor	.023
Diameter of triturating-surface inferior canine	.028
Diameter of triturating-surface inferior canine, transverse	.0166
Diameter of superior canine, anteroposterior	.024
Diameter of the two inner incisors	.010
Diameter of exterior incisor, oblique	.010
Diameter of symphysis mandibuli	.044
Diameter of nareal orifice	.040
Depth of nareal orifice	.031
Depth of mandibular ramus at M. 6	.049
Thickness below of mandibular ramus at M. 6	.014
Length of a superior molar, crown	.020
Diameter of condyle of humerus	.047
Diameter of shaft of humerus, compressed	.0410
Diameter of condyles of humerus	.0415
Diameter of condyles of humerus, anteroposterior	.032
Diameter of head of radius, transverse	.0282
Diameter of head of radius, vertical	.0162
Diameter of shaft of radius	.016
Diameter of cotylus of ulna, long	.030
Depth of ulna at coronoid process	.034
Length of carpus and digit 2 without unguis	.112
Length of two phalanges without unguis	.037
Length of metacarpal without unguis	.061
Length of metacarpal No. 3	.974
Length of metacarpal No. 4	.070
Length of metacarpal No. 5	.053
Length of scaphoid, transversely	.023
Length of cuneiform, transversely	.027
Length of pisiform, transversely	.027
Width of pisiform, distally	.016
Length of unciform, transversely	.020
Width of unciform, anteroposteriorly	.013
Width of trapezoid, anteroposteriorly	.0155
Width of trapezium, anteroposteriorly	.0114
Length of trapezium, vertically	.016
Width of scaphoid, anteroposteriorly	.015
Width of navicular, anteroposteriorly	.0155
Length of navicular, transversely	.0255
Length of ungual phalange	.016
Width of ungual phalange	.010
Diameter of centrum of lumbar vertebra	.029
Diameter of centrum of caudal vertebra	.009

Restoration. The fore feet are like those of both dogs and bears. The very prominent postglenoid ridge, and the narrow tympanic chamber are decided points of resemblance to the bears, but the *cavum tympani* is even less expanded than in those animals. The characters of dentition are more like those of the *Hyænodontidæ* and *Amblyctonidæ* than any other group, and even the remarkable incisor-like inferior canines are approximated by the anteriorly directed canines of *Hyænodon leptorhynchus* Laiz. et Par.

The *Mesonyx lanius* was considerably larger than the *M. obtusidens*, equaling the black bear (*Ursus americanus*) in size. It had a large head, with a long, rather narrow, and truncate muzzle. The limbs were relatively smaller, not exceeding those of the black bear in length and thickness. The tail was long and slender as in the cats, while the claws were broad and flat as in the beaver.

Habits. The molar, canine, and incisor teeth of my specimen, as well as that of the Princeton fossil, are much worn by use. This is especially true of the canines of both, while the crowns of the molars of the Bitter Creek specimen are almost entirely worn away. The same peculiarity is to be observed in the specimens of the allied *Amblyctonus sinosus*, which I obtained in New Mexico.* It is probable that these species chewed hard substances. The peculiar approach of the lower canines is a special modification for peculiar habits, which I suspect to have been the devouring of the turtles which so abounded on land and in the waters of the same period. The slender symphysis could most readily be introduced into the shell, while the lateral pressure of the upper canines with the lower, would be well adapted for breaking the bony covering of those reptiles. The breaking of these shells in the attempt to masticate their contents would produce the unusual wear of the teeth observed.

History. I originally placed this species in a genus distinct from the *M. obtusidens* on the ground of a supposed difference in the number of molar teeth. The Princeton specimen renders it extremely probable that the two species are congeneric.

The dental series is uninterrupted from the canine if, as I believe, there is an alveolus for a simple premolar behind it. This I overlooked

*See Report of Lieut. G. M. Wheeler, Expl. Surv. W. of 100th Mer., 1 v., pt. ii, Pl. xxxiii, figs. 1–3 and 11.

when first describing the species, and hence gave the molars as 6 instead of 7.

The typical specimen was found by myself on a terrace of the Mammoth Buttes, near South Bitter Creek, Wyoming, in the beds of the Washakie basin. A portion of the bones had fallen a few feet from a remaining mass of the softer bed, where I soon found the rest of the specimen in place. The skull and anterior foot were taken out from close juxtaposition.

MESONYX OSSIFRAGUS Cope.

Plates XXVIII a, Fig. 1; XXVIII b; XXVIII c; XXVIII d; ? XXIV e, figs. 14-19.

American Naturalist, 1881, p. 1019, December. Proceedings Am. Philos. Soc., December, 1881, p. 165. *Pachyæna ossifraga* Cope, Report Vert. Fossils New Mexico, U. S. Geog. Surv. W. of 100th Mer., p. 13, 1874. Id., Ann. Report Chief of Engineers, 1874. Report Lieut. Wheeler, p. 125. Report Capt. Wheeler, U. S. G. G. Surv. W. of 100th Mer. iv, ii, p. 94, 1877, pl. xxxix, fig. 10.

I was so fortunate as to receive from Mr. Wortman the greater part of a skull of this species, together with some bones of the limbs, belonging to one individual. These were mingled in great confusion with the bones of two individuals of *Phenacodus*, which I was able to distinguish through the fortunate possession of a complete skeleton of the *P. primævus.* Besides this individual, Mr. Wortman obtained jaws and some of the bones of three individuals from the Big Horn basin.

M. ossifragus was the largest Creodont of the Eocene, equaling the largest grizzly bear in the size of its skull. In a cranium with lower jaw and almost complete dentition, the length to the premaxillary border from the postglenoid crest is M. .365; the largest *Ursus horribilis* in my collection gives .270 for the same length. This specimen has the dental formula I. $\frac{3}{2}$; C. $\frac{1}{1}$; P-m. $\frac{4}{4}$; M. $\frac{3}{3}$. The claws have the flattened form which I discovered in *M. lanius*, and the proximal phalanges have much the shape of those of a Perissodactyle. The astraglus has much the character of the animals of that order, and has the distal facets as I originally detected them in the *M. obtusidens* The form of this bone is shorter and wider than in the latter species.

The skull already mentioned lacks the brain case and basicranial axis, embracing the muzzle, zygomata, pterygoid region, and lower jaw, with nearly complete dentition. The muzzle is contracted and is rather short; the zygomata are widely expanded. The premaxillary extends well posteriorly along the nasal, but it does not probably reach the frontal bone, which is lost from the specimen. The maxillary is contracted behind the canines. The molars arise from it in a ridge which commences above the fourth premolar. The free part of the bone presents an angle downwards just beyond the maxillary, and posterior to this point has a thin inferior border without the bevel indicating the insertion of the masseter muscle usual in carnivora. The maxillary projects posteriorly in a free angle separated from the base of the pterygoid process. The posterior extremity of the molar is a little anterior to the glenoid cavity, and has a horizontal internal expansion The superior border of the bone has a very slight postorbital angle. The glenoid cavity is wide and deep. Both the preglenoid and postglenoid crests are large, and are most elevated externally. The meatus auditorius externus is small, and is closed below by the posttympanic process without visible tympanic bone. The posttympanic is not coössified with the postglenoid process, but is in contact with it. Just interior to the meatus the posterior face of the former becomes somewhat tuberous. It has a free superior border, which for a short distance forms the posterior border of the zygomatic fossa. I do not see postglenoid or postzygomatic foramina.

The mandible is distinguished for its long and slender rami and symphysis. The inferior border is gently convex, and the symphyseal portion is in the line of the remaining part of the rami. The length of the symphysis is unusual, being one-third of the total to the base of the condyle. Its inferior face is distinctly separated from the lateral face by an obtuse angle. The condyle projects much beyond the angle, and is quite large. Its face looks upwards and backwards. The angle is shallow and little prominent, its posterior border extending obliquely forwards. Its thin inferior edge is directed somewhat obliquely inwards, though not distinctly inflected. The base of the coronoid process is very wide, equaling the length of the inferior molar series, omitting the first premolar. Its anterior border slopes obliquely backwards, and is obtusely rounded to the summit. The latter is not elevated and is very obtuse. The posterior border descends obliquely and then nearly vertically to the base of the neck of the condyle. The masseteric fossa is not defined either anteriorly or inferiorly.

The superior incisors are slender, and the crowns are very short and acuminate. The external incisor, though the largest, is not so large absolutely or relatively as in the smaller species *Mesonyx lanius* The precanine diastema large, equaling that posterior to the first superior premolar. The canines are very large and have an oval section at the base lying anteroposteriorly. The crown is destitute of ridges, and the enamel is perfectly smooth. The first premolar has one root with an oval section. The two roots of the second are large in comparison with the crown. The latter has no anterior basal lobe, has a simple cone, and a heel with cutting median edge. The cone has no edges or ridges. The third premolar has the same form, with the addition of a rudimental anterior basal tubercle. The heel has a wide base. The fourth premolar differs from the third in having a large internal conic lobe and an anterior basal tubercle. The posterior edge of the principal or external cone carries a small lobe; its anterior face is rounded. On the first and second true molars this small lobe becomes a second external cone smaller than the anterior external and than the internal cone, which are about equal. Both these teeth have an anterior and posterior basal lobes. The second true molar differs from the first, in that the posterior basal lobe is smaller, and that the external cingulum into which it continues is wider than that of the first true molar. It also extends with an interruption to the anterior basal lobe of the second molar, but is rudimental on the anterior external part of the first. The outline of the bases of both of these teeth is trifoliate, the anterior external lobe a little the smallest. The posterior molar is more nearly triangular, the external part of the crown having less anteroposterior extent. It supports one external cone with a small posterior basal lobe, and a posterior, external, and anterior cingulum. The internal cone is well developed, but is not so large as the external. None of the molars have an internal cingulum.

The crowns of the inferior incisors are lost, but their bases are small. The canines are large, and have an anteroposterior oval section without angles or grooves. Enamel smooth. The first premolar is one-rooted, and is directed very obliquely forwards. The second and third premolars are like the corresponding superior teeth, except that the heel of the third is not so wide, and the median edge is more prominent. The remaining four molars are alike. They consist of a principal conic lobe with a small anterior basal lobe and a large posterior heel. The cone has a low median anterior edge, and more prominent median posterior edge. The heel has an obtuse cutting edge, which rises in a compressed trihedral form, as it is truncated behind as well as flattened at the sides. No cingula on the inferior molars. Enamel obsoletely rugulose.

Measurements of cranium.

		M
Length from premaxillary border to postglenoid		.365
Length from premaxillary border to end of last molar		.193
Length of dental series, including canine and last molar		.166
Width of premaxillary teeth on bases		.045
Width between bases of canine teeth		.051
Width at convexities of zygomata		.282
Width of glenoid cavity	transversely	.057
	fore and aft	.034
Depth of zygoma posteriorly		.077
Depth of molar at middle		.048
Diameters of base of canine	anteroposterior	.025
	transverse	.020
Length of series of true molars		.057
Length of base of P-m. ii		.0125
Diameters P-m. iv	anteroposterior	.019
	transverse	.015
Diameters M. i	anteroposterior	.0215
	transverse	.018
Diameters M. ii	anteroposterior	.0205
	transverse	.0205
Diameters M. iii	anteroposterior	.015
	transverse	.017
Length of mandibular ramus		.362
Length of mandibular ramus to posterior base of coronoid		.322
Length of mandibular ramus to anterior base of coronoid		.190
Length of mandibular ramus to end of symphysis		.112
Depth of ramus at P-m. i		.047
Depth of ramus at M. iii		.068
Depth of ramus at coronoid process		.122
Depth of ramus at neck of condyle		.025
Diameters P-m. iii	vertical	.014
	anteroposterior	.017
	transverse behind	.009
Diameters P-m. iv	vertical	.015
	anteroposterior	.020
	transverse	.010
Diameters M. iii	vertical	.015
	anteroposterior	.022
	transverse	.011

Two posterior dorsal vertebræ display a number of peculiarities. They are moderately elongate, and moderately depressed. They have a strong median inferior angular ridge and concave sides. The articular faces are oblique and slightly opisthocoelous. There is a longitudinal lateral ridge, and a second lateral ridge above it which is continued into a short anapophysis. The external edge of the prezygapophysis is also produced backwards as a ridge, but terminates abruptly at the middle of the side of the neural arch. The postzygapophysis presents a free angle outwards. The base of the neural spine is compressed, and extends over the whole neural arch.

The peculiarity of these vertebræ as compared with the corresponding ones of *Canis* and *Feblis*, is the absence of the metapophyses, and the consequent horizontal spread of the postzygapophyses. The inferior keels and oblique opisthocoelous centra are not found in those genera. They are in fact something like cervicals, but the absence of vertebrarterial canal and presence of anapophysis, forbids such reference.

The fore-limbs of this individual are represented by both humeri, ulnæ and radii, and by the lunar, cuneiform and magnum, with four or five metacarpals of one side. Both humeri are a little distorted; one is abnormally shortened, and the other is elongated proximally. The peculiarity of this bone is its shortness, as compared with the bones of the fore-arm and of the posterior leg. The great tuberosity is prominent, and the deltoid crest extends far down on the shaft, terminating only a little above the radial fossa. The external epicondyle is not prominent, and is marked by shallow fossæ. The ridge which forms the external edge of the posterior face of the distal part of the humerus is prominent, but disappears before reaching the epicondyle. The internal epicondyle is very prominent, and rises into a ridge which bridges over the supracondylar foramen. The olecranar fossa is wide and deep. The condyle consists of the internal flange and the external cylinder. The former is not very prominent nor acute; the latter is rather short transversely, and is a little convex in the transverse section. Its anterior face is shorter than its posterior face, and the latter has a low external raised border. This humerus resembles considerably that of the *M. lanius*, from which the internal epicondyle and foramen have been lost.

I have also probably restored the length of the shaft (Plate XIX, Fg. 1) so as to be too long.

The radius and ulna are rather stout and are moderately decurved. They present many peculiarities. The inner edge of the ulna is raised so as to be in contact with the radius throughout its length. The external edge of the shaft of the ulna is also elevated nearly its entire length, thus inclosing a wide deep groove with the external edge. This external ridge is the origin of the supinator brevis muscle, and of the extensores pollicis in *mammalia*, and would appear to indicate unusual power of supination of the hand, and of extension of the thumb. But the form of the head of the radius forbids the idea that that bone could be rotated so as to supinate the hand to much extent. The olecranon is long, and is deep near the coronoid process, and contracts towards its extremity. The coronoid process is elevated and unusually wide, the sides extending upwards so as to be nearly parallel, forming a truncate instead of the usual acuminate summit. The facet band for the radius is slightly concave. The humeral face rises above it on a process of the inner side. Immediately anterior to it a narrow and rather deep fossa extends along the inner superior edge of the ulna, opening onto the general surface of the shaft within a short distance. The interno-inferior face of the shaft of the ulna is convex. The distal extremity is acuminate; its inferior face is flat and is bounded by a ridge with a tuberosity externally. A convexity and then a concavity of the internal surface adapt it to the radius. The carpal extremity projects beyond that of the radius, and is quite narrow in both directions. It has not the double-rib head-like form of that of *Oxyæna.*

The radius is rather stout. The head is transverse and deeper at the internal than the external side. The humeral surface is double; one part is concave and occupies the middle of the head; the other is convex, and is turned outwards to correspond with the internal flange of the humeral condyle. Its prominence is continued from it, presenting inwards, and ceases distally abruptly in a semicircular edge. This strong recurvature of the humeral surface is characteristic of the species. The distal extremity is rather large. Its superior surface presents the wide open groove for the extensor tendons, which is bounded on the outer side by an obtuse ridge-like tuberosity. The external face above the ulna is slightly concave, and is separated from the ulnar face by an angular ridge. The internal face has the same relation to the inferior face. The scapho-lunar facet is undivided, and has a wide crescent shape, the concavity above and the angles rounded.

As in the other species, the lunar bone is distinct from the scaphoid. The cuneiform is produced outwards into a narrow tuberosity. The magnum is small, and has a subquadrate anterior face. It rises above, as usual, and has a short posterior tuberosity. The metacarpals are not long, and the median ones are rather robust. The posterior keel of the distal extremity is obtuse and somewhat oblique, and separates parts of the condyle of unequal width and prominence. The supracondylar transverse groove is not deep, and the superior face of the condyle is rather flat. The head of a metacarpal has but one proximal facet, which is convex anteroposteriorly. There is a band-like lateral facet at right angles to the proximal, below which the bone is excavated abruptly. A phalange is wide and depressed, appropriately to the flat claws. Its distal condyle is not recurved above.

Measurements of anterior limb.

		M.
Length of humerus (partly inferential)		.165
Transverse diameter of head (partly inferential)		.041
Width at epicondyles		.055
Width of condyles	anteriorly	.023
	posteriorly	.025
Anteroposterior width condyles	internally	.031
	externally	.029
Length of ulna		.282
Length of olecranon		.070
Depth of olecranon at ex'remity of olecranon		.026
Depth of olecranon at coronoid process		.051
Depth of olecranon at head of radius		.026
Depth of olecranon at middle of shaft		.019
Width of cuneiform facet of ulna		.013
Length of radius		.202
Diameters of head of radius	vertical	.0185
	transverse	.035
Width of shaft at middle		.020
Width at distal end		.039
Width of carpal facet		.028
Depth of carpal facet		.015
Depth of face of lunar		.015
Width of face of lunar		.013
Depth of face of cuneiform		.011
Width of face of cuneiform		.026

		M.
Length of a metacarpal		.069
Width of a metacarpal distally		.019
Width of condyles		.015
Diameters of a head of a metacarpal	fore and aft	.017
	transverse	.013
Length of a phalange		.024
Diameters of a phalange proximally	vertical at middle	.012
	transverse	.017

The posterior limb is quite elongate. The tibia has the length of that of the *Ursus americanus*, while the femur is a little shorter than the femur of the same species. In *Mesonyx ossifragus* it is a little longer than the tibia; in *Ursus americanus* the difference in the proportions of the two bones is greater. The femur has more prominent greater and lesser trochanters than either *Ursus americanus*, *Canis lupus*, or *Uncia concolor*. The great trochanter projects beyond the line of the head, and its extremity is compressed anteroposteriorly. The trochanteric fossa is large, and is more open than in either of the three Carnivora above mentioned. The lesser trochanter is compressed and large. The third trochanter commences a little below the line of the inferior edge of the little trochanter. It has a long base, but is obtuse and little prominent. The fossa ligamenti teris is much larger than in either of the species above mentioned, and extends to the neck of the femur. The shaft of the femur is rather stout, and has a large medullary cavity, which in the specimen is filled with calcareous spar The walls are not thicker than in some *Dinosauria.* The linea asper is impressed, and vanishes inferiorly. The rotular groove is wide, and its bounding crests are rather high, and are subequal. The proportions are about as in *Uncia concolor*, being less elevated than in *Canis lupus*, and more so than in *Ursus americanus.* The edges are reflected above the rotular surface, a peculiarity not seen in either of the species named. The fossa at the external base of the rotular face is present as in *Ursus* and *Canis.* The condyles are subequal, are regularly convex, and are not much produced backwards.

Measurements of femur.

	M.
Length from summit of great trochanter	.315
Width of head	.077
Anteroposterior diameter of head	.032
Width at third trochanter	.048

Width below third trochanter	.032
Width above condyles	.061
Width of condyles	.058
Width of rotular groove	.030
Depth of inner condyle with rotular crest	.058

The tibia has a nearly straight shaft, which is rather slender below. Its section below the head is triangular, the base being posterior. That of the inferior front of the shaft is also triangular, the base of the triangle being the inner side. The crest is quite prominent, not flat, as in *Ursus*, but not quite so strong as in *Uncia concolor*. It is replaced by a gentle convexity just below the middle of the shaft. The inner femoral cotylus does not overhang the inner side of the head. The latter has a wide, low, longitudinal ridge posterior to the middle, which distinguishes concavities anterior and posterior to it. The external femoral face is decurved posteriorly, and rises into a spine posterior to the middle of the boundary between it and the internal face. There is a transverse depression at the summit of the spine. The latter has a superior and an inferior tuberosity. The internal malleolus is produced. Its internal face carries a groove for the tendons; part of the surface is damaged, so that more than one cannot be determined. The trochlear face is divided into two fossæ, which are not so deep as those of *M. obtusidens*. The fibular proximal facet is not distinct.

Measurements of tibia.

		M.
Total length		.275
Diameters of head	anteroposterior (total)	.065
	transverse	.062
Diameters just below middle of shaft	fore and aft	.022
	transverse	.023
Diameters of distal end	anteroposterior	.028
	transverse	.039

The only tarsal bones preserved are the astragalus and calcaneum. These were found nearly in place, adherent to the distal extremity of the tibia. They are about the size of those of the black bear, and larger than those of the *M. obtusidens*. The astragalus has the same peculiarity as that of the *M. obtusidens*, in the distinct band-like facet of the external side of the distal extremity for articulation with the cuboid bone; a peculiarity unknown elsewhere among Creodonta and among Carnivora. The width of this surface is about one-half that of the navicular surface, and it is of uniform width, extending obliquely to the middle line. The navicular face is deeper than wide, and is convex anteroposteriorly and concave transversely. Its internal part is reverted to the inferior side, no doubt to accommodate the posterior process of the navicular. The general appearance of this double articulation is much like that of a perissodactyle ungulate. The trochlea is wider and not so convex nor so deeply grooved as in *M. obtusidens*. It is hour-glass shaped, and the external face has considerably greater extent than the internal. The former is vertical except the anterior part, which descends as an angular process into a fossa in front of the calcaneal condyle, which is flared outwards. An articular band marks the inner face of the trochlea continuous with the general surface. The postero-internal angle is produced, as in *Oxyæna*, into a flat subhorizontal rounded process, which overhangs the sustentaculum. The neck of the astragalus is much shorter than in *M. obtusidens*. The posterior outline is widely and deeply notched.

The calcaneum is long and rather narrow. The sustentaculum and the superior condyle are both small. Below the latter, on the external face, is a strong longitudinal crest continuous to the cuboid face. This crest in *Oxyæna* and in carnivora stops at or before the vertical line of the anterior end of the condyle. Beneath it the side of the calcaneum has a deep longitudinal concavity. The free extremity is truncate. The cuboid surface is large, but imperfection of the specimen obscures its external boundary. It is flat and not very oblique to the axis of the bone.

Measurements of tarsus.

		M.
Length of astragalus on inner side		.044
Length of internal arc of trochlea		.024
Length of external		.031
Depth of inner side at trochlea		.023
Length of neck of astragalus		.015
Anteroposterior diameter of distal end		.022
Diameters navicular facet	oblique	.024
	transverse	.016
Diameters cuboid facet	oblique	.023
	transverse	.007
Length of calcaneum		.077
Depth at free extremity		.027
Depth at condyle		.034
Depth at cuboid facet		.026
Width at free end		.022
Width at sustentaculum		.041
Depth (vertical) of cuboid facet		.023

Portions of a second individual of *Mesonyx* were found by Mr. Wortman in the Big Horn region. Of this I must observe that the head of the radius has not the anterior flare of the inner side of the *M. ossifragus*, but it resembles more nearly the corresponding part of the *M. lanius*, though it presents some minor differences. The superior concavity is strong. The axis vertebra has a rather long odontoid, with circular section and confluent articular surfaces. The centrum has a strong hypapophysial heel, which is laterally expanded posteriorly. The posterior articular face is moderately oblique, and is plane excepting a central depression. The vertebrarterial canal is wide.

Comparison of this axis with those of *Carnivora* and *Marsupialia* yields the following results: It differs from all the principal genera of the former order, and from *Didelphys* and *Phascolarctos* of the latter, in the nearly round form and downward extension of the atlantal facets. The form is approached by that of the *Sarcophilus ursinus*, but is not so pronounced in the latter. The strong hypapophysis is approached by *Felidæ* only among *Carnivora*, and by *Didelphys* among *Marsupialia*. The size of the axis is about that of the *Crocuta maculata*.

Two ungual phalanges which accompany this specimen have much the form of those of *M. lanius*. The proximal part is claw-like, but the distal part narrow hoof-like. The extremity is deeply fissured; the sides are acute and flat, and there is a median inferior table, which widens posteriorly. It is separated by a groove from the lateral edge, which is deeply impressed posteriorly. The acute lateral edges are spongy.

Measurements of No. 2.

		M.
Length of axis		.065
Length of axis to base of odontoid		.048
Diameters of atlantal facet	vertical	.022
	horizontal	.021
Width of centrum behind atlantal facet		.032
Diameters of posterior face centrum	vertical	.024
	transverse	.030
Diameters of head of radius	vertical	.017
	transverse	.029
Length of phalange		-
Diameters distal end phalange	vertical	.0075
	transverse	.013
Length of ungual phalange		.021
Diameters proximally	vertical	.009
	transverse	.010
Width of expanse		.0105

Portions of several other individuals were found by Mr. Wortman in the bad lands of the Big Horn Basin.

RESTORATION.—From the preceding investigation we can form a general idea of the form and proportions of the *Mesonyx ossifragus*. We can depict an animal as large as a large-sized American black bear, with a long stout tail, and a wide head as large as that of a grizzly bear. The fore limbs are so much shorter than the hind limbs that the animal customarily sat on its haunches when on land. In walking, its high rump and low withers would give it somewhat the figure of a huge rabbit. Its neck was about as long as that of an average dog. Its tread was plantigrade, and its claws like those of various rodents, intermediate between hoofs and claws. The animal, to judge from its otter-like humerus, was a good swimmer, although there is nothing specially adapted for aquatic life in the other bones of its limbs. Its teeth, on the other hand, are of the simple construction of the mammals which have a diet largely composed of fishes. We cannot but consider this animal as one of the most singular which the Eocene period possessed. In size it was not exceeded by any other flesh-eater of the period, but was equaled by the *Protopsalis tigrinus*.

CHIROPTERA.

Species of this order were first detected in the Eocene formations of France by Cuvier, who named a species from the gypsum (Upper Eocene) *Vespertilio parisiensis*. In North America, Professor Marsh has recorded them from the Middle Eocene (Bridger), but whether they belong to existing generic forms or not is yet unknown. The oldest North American species is described below. It is from the Wind River region, which represents the Lower Bridger.

VESPERUGO Keys & Blas.

Wirbelthiere Europas, 1810, p. 45.

I. $\frac{2}{3}$; C. $\frac{1}{1}$; Pm. $\frac{2}{2-3}$; M. $\frac{3}{3}$. First and second superior true molars with two external Vs, and an internal heel, which supports a more or less elevated cusp. Inferior molars like those of *Didelphys*, with an anterior triangle of three cusps and a cuspidate heel.

VESPERUGO ANEMOPHILUS Cope.

Bulletin U. S. Geol. Surv. Terrs. vi, 188¹, p. 184; American Naturalist, 1880, p. 745.

Represented by the anterior part of a skull without lower jaw. Dentition: I. ?; C. 1; Pm. 2; M. 3. Posterior molar narrow; its posterior external V rudimental; first and second molars subequal. Fourth premolar elevated and acute, with an external basal cingulum; second premolar simple, acute. Profile steeply elevated behind orbital region, less steep in front of it; zygomas wide.

Measurements.

	M.
Length from interorbital region to above canine alveolus in front	.010
Interorbital width	.005
Width of zygomas	.012
Width between outsides of last molar teeth	.010
Length of molar series	.008
Length of true molars	.004

Found by J. L. Wortman.

TAXEOPODA.

Cope, American Naturalist, 1882, p. 523, June (May 20).

Ungulate; carpus with the bones of the second row directly succeeding those of the first. The lunar bone is supported by the magnum, and little or not at all by the unciform, and the scaphoid is supported by the trapezoides and not by the magnum. In the same way the bones of the second series of the tarsus do not alternate with those of the first series. The astragalus articulates exclusively with the navicular, and the calcaneum with the cuboid.

This comprehensive division is, so far as present knowledge extends, well distinguished from the best known orders of ungulates, the *Amblypoda*, the *Perissodactyla*, and the *Artiodactyla*.

The ungulata are here understood to be the hoofed placental mammalia with enamel-covered teeth, as distinguished from the ungulate or clawed, and the mutilate or flipper limbed, and the edentate or enamelless groups. The exact circumscription and definition has not been attempted, though probably the brain furnishes an additional basis of it in the absence of the crucial and presence of other fissures, etc. Suffice it to say that it is on the whole a rather homogeneous body of mammalia, especially distinguished as to its economy by the absence of forms accustomed to an insectivorous and carnivorous diet, and embracing the great majority of the herbivorous types of the world.

The internal relations of this vast division are readily determined by reference to the characters of the teeth and feet, as well as other less important points. I have always insisted that the place of first importance should be given to the feet, and the discovery of various extinct types has justified this view. The predominant significance of this part of the skeleton was first appreciated by Owen, who defined the orders *Perissodactyla* and *Artiodactyla*. Professor Gill[1] has also used these characters to a certain extent, but without giving them the exclusive weight that appears to me to belong to them. Other authors have either passed them by unnoticed, or have correlated them or subordinated them to other characters, in a way which has left the question of true affinity, and therefore of phylogeny, in a very unsatisfactory condition. Much light having been thrown on these points by recent discoveries in paleontology, the results as they appear to me are here given.

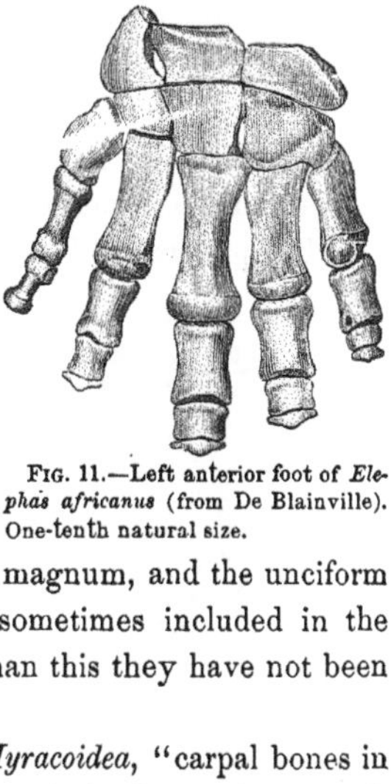

Carpus.—It is well known that in the *Perissodactyla* and *Artiodactyla* the bones of the two rows of the carpus alternate with each other; that the lunar, for instance rests on the unciform, and to a varying degree on the magnum, and that the scaphoides rests on the magnum and to some degree on the trapezoides and trapezium. It is also known that in the *Proboscidea* another state of affairs exists; *i. e.*, that the bones of the two rows do not alternate, but that the scaphoides, lunar, and cuneiform, rest directly on the trapezium and trapezoides, the magnum, and the unciform respectively. The preceding characters are sometimes included in the definitions of the respective orders. Further than this they have not been used in a systematic sense.

FIG. 11.—Left anterior foot of *Elephas africanus* (from De Blainville). One-tenth natural size.

Professor Gill says of the carpus of the *Hyracoidea*, "carpal bones in two interlocking rows; cuneiform extending inwards (and articulating with

[1] Arrangement of the families of Mammals prepared for the Smithsonian Institution. Miscellaneous Collections, 230. Nov., 1872.

magnum); * * * unciform and lunar separated by the interposition of the cuneiform and magnum." Professor Flower[1] gives a figure which justifies these statements, but neither the one nor the other agrees with my specimens. In the mantis of a *Hyrax capensis* (from Verreaux Paris) I find the following condition of the carpus. The bones of the two series are articulated consecutively, and not alternately; they do not interlock, but inasmuch as the magnum is a little narrower than the lunar, the latter is just in contact (anteriorly) with the trapezoides (centrale) on the one side, and the unciform on the other. My specimen agrees with Cuvier's figure of *Hyrax capensis* in all respects. It is probable that Professor Flower has figured some other species under that name, which besides its peculiarities, is of smaller size.

FIG. 12.—Left anterior foot of *Phenacodus primævus*, one-third natural size (original).

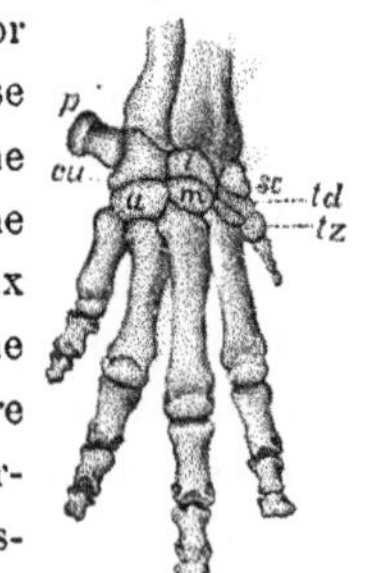

FIG. 13.—Right anterior foot of *Hyrax capensis* (from Cuvier). *Sc.* scaphoid bone; *l.* lunar; *cu.* cuneiform; *p.* pisiform; *tz.* trapezium; *td.* trapezoides; *m.* magnum; *u.* unciform. Natural size.

In April, 1875,[2] I described the manus of *Coryphodon* (Bathmodon), showing that the lunar was supported below by the magnum and by parts of the trapezoides and unciform. This carpus has the character of that of *Hyrax capensis*, with the two last named articulations more extensive. This was the first description of the carpus of the *Amblypoda*. In February, 1876,[3] Professor Marsh described the carpus of *Uintatherium* (*Dinoceras*), in which he stated that the bones "form interlocking series." He however states that "the magnum is supported by the lunar and not at all by the scaphoid," a state of things which does not belong

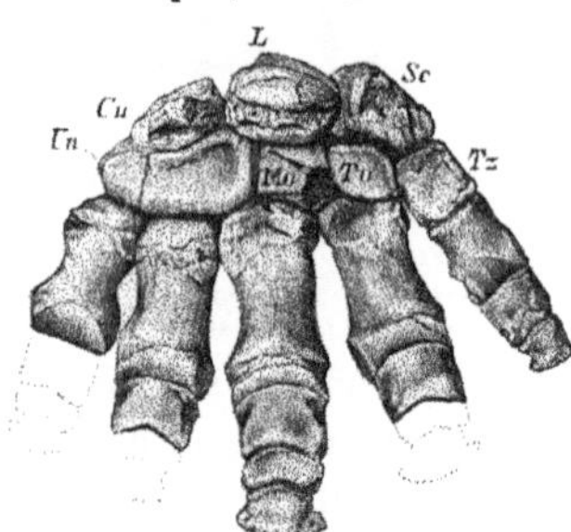

FIG. 14.—Right manus of *Coryphodon*, one third natural size. Original, from Report of Capt. G. M. Wheeler, Expl. W. of 100 Mer., vol. iv, 1877.

[1] Osteology of the Mammalia, p. 266: fig. 92.

[2] Systematic Catalogue of the vertebrata of the Eocene of New Mexico, p. 24 (U. S. Geol. Survey W. of 100th Mer.).

[3] Amer. Journal Sci. Arts, xi, p. 167; pl. vi, fig. 2.

to the interlocking carpus. The trapezoïdes does not join the lunar, but the unciform does so, as in *Coryphodon*. Professor Marsh's figure as to the articulations of the magnum does not agree with his description, as it makes that bone articulate with the scaphoid. The second description is however correct, and the carpus is identical with that of *Coryphodon*.

In the *American Naturalist*, June, 1882,[1] I have shown that the carpus of the *Condylarthra* is essentially like that of the *Hyracoidea*.

Tarsus.—In the tarsus of the *Perissodactyla* and *Artiodactyla* it is well understood that the cuboid extends inwards so as to articulate with the astragalus, giving the latter a double distal facet. It is also well known that the astragalus of the *Proboscidea* has but a single distal articulation, that with the navicular. It is, however, true that the cuboid is extended inwards, but that it articulates with the distal extremity of the navicular instead of that of the astragalus. It was shown by Cuvier that the astragalus of the *Hyracoidea* articulates with the navicular only, and that the cuboid is not extended inwards so as to overlap the latter. In 1873[2] Marsh stated that the astragalus of the *Amblypoda* articulates with both cuboid and navicular. Finally I discovered in 1881[3] that the astragalus of the *Condylarthra* articulates with the navicular only, and that the cuboid articulates with the calcaneum only. In the tarsus then there are four types of articulation, which are represented by the *Condylarthra*, the *Proboscidea*, the *Amblypoda*, and the *Artiodactyla* respectively.

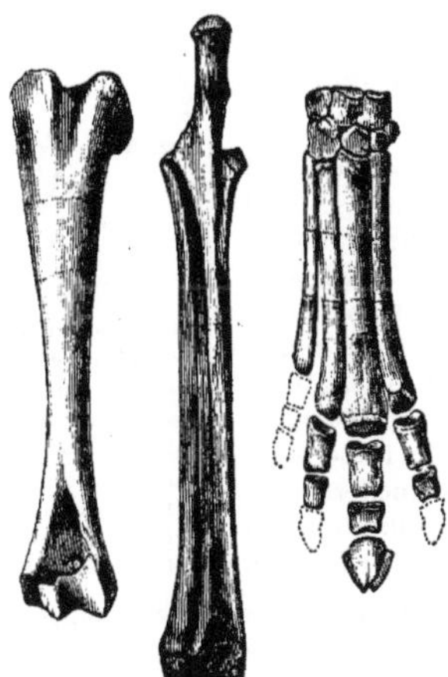

FIG. 15.—Fore leg and foot of *Hyracotherium venticolum* (original). Two-thirds natural size.

Orders.—From the preceding considerations we derive the following definitions of the primary divisions of the Ungulata, which should be called orders. In the first place I find the diversity in the structure of the carpus to be greater in the relations of the magnum and scaphoides, than in the relations between the unciform and the lunar. In other words the trape-

[1] Page 522. [2] American Journal Science and Art, January, 1873. [3] American Naturalist, 1881, p. 1017.

zoides and magnum are more variable in their proportions than is the cuneiform. This is directly due to the fact that the reduction of the inner two digits is more usual than the reduction of the external two. I therefore view the relations of these bones as more characteristic. In the tarsus the really variable bone is the cuboid. It is by its extension inwards that the additional facet of the astragalus is produced. Its relations will therefore be considered rather than that of the astragalus in framing the following definitions:

Order I. Scaphoides supported by trapezoides and not by magnum, which supports lunar. Cuboid articulating proximally with calcaneum only. *Taxeopoda.*

Order II. Scaphoides supported by trapezoides, and not by magnum, which supports lunar. Cuboid extended inwards and articulating with the distal face of the navicular.... *Proboscidea.*

Order III. Scaphoides supported by trapezoides and not by magnum, which with unciform supports the lunar. Cuboid extended inwards and articulating with astragalus.... ..*Amblypoda.*

Order IV. Scaphoides supported by magnum, which with the unciform also supports the lunar. Cuboid extended inwards so as to articulate with the astragalus. *Diplarthra.*

Fig. 16.—Left posterior foot of *Phenacodus primævus*, one-third natural size (original).

Fig. 17.—Right posterior foot of *Hyrax capensis* (from Cuvier). *Ca.* calcaneum; *a.* astragalus; *n.* navicular; *cu.* cuboid; *ecc.* ectocuneiform; *mc.* mesocuneiform; *enc.* entocuneiform. Nat. size.

The sub-orders are defined as follows:

I. Taxeopoda.

There are two, perhaps three sub-orders of the *Taxeopoda*, the *Hyracoidea*, the *Condylarthra*, and perhaps the *Toxodontia*.[1] The *Toxodontia* are however not sufficiently known for final reference.[2] The sub-orders are defined as follows:

A post-glenoid process; no fibular facet of calcaneum, but an interlocking articulation between fibula and astragalus; ungual phalanges truncate.......... *Hyracoidea.*

A post-glenoid process; no fibular facets on either calcaneum or astragalus; a third trochanter of the femur; ungual phalanges acuminate...... ..... *Condylarthra.*

[1] See my remarks on Toxodon, Proceedings Amer. Philosoph. Society, 1881, p. 402.

[2] The considerable resemblance between the dentition of *Toxodon* and *Hyrax* must not be overlooked.

II. Proboscidea.

There may be two sub-orders of this order, the *Proboscidea* and the *Toxodontia.* I do not know the carpus of *Toxodon*, but if it does not differ more from that of the elephants than the tarsus does, it is not entitled to subordinal distinction from the *Proboscidea.* The sub-order of *Proboscidea* is defined as follows:

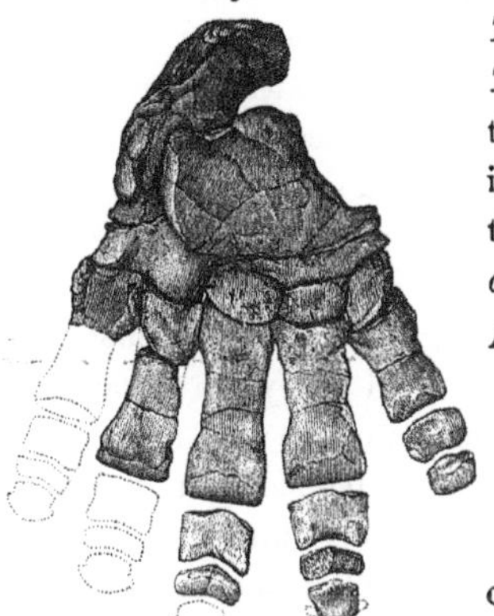

Fig. 18.—Posterior foot of *Coryphodon* (original). From Capt. G. M. Wheeler's Report Surv. W. of 100 Mer., iv. One-third natural size.

A fibular articulation of the calcaneum; no post-glenoid process; no third trochanter of femur *Proboscidea.*

III. Amblypoda.

The sub-orders of this order, as I pointed out in 1873, are two, defined as follows:

Superior incisor teeth; no ali-sphenoid canal; a third trochanter of femur........*Pantodonta.*
No superior incisors, nor ali-sphenoid canal, nor third trochanter of femur ....*Dinocerata.*

The difference between the *Proboscidea* and the *Amblypoda* consists chiefly in that the navicular of the latter is shortened externally so as to permit the cuboid to articulate with the astragalus. The cuboid has the same form in both. The peculiar character of the navicular gives the astragalus a different form.

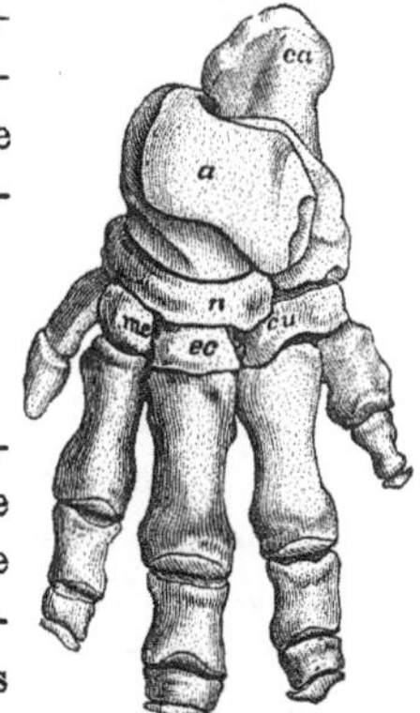

Fig. 19.—Left pes of *Elephas indicus* one-tenth natural size; after Cuvier.

IV. Diplarthra.

This order is called by some authors the Ungulata, but that name is also used in the larger sense in which it is here employed. This appears to be its legitimate application, as the name should, if possible, be used for hoofed Mammalia in general, as its meaning implies. The two well-known sub-orders are the following:

Astragalus truncate distally; number of toes odd, the median the largest. *Perissodactyla.*
Astragalus with a distal ginglymus; number of toes even, the two median largest. *Artiodactyla.*

Phylogeny.—The serial arrangement of the bones of the carpus and tarsus seen in the *Taxeopoda*, is probably the primitive one, and we may expect numerous accessions to the order on further exploration of the early Eocene epochs. The modification seen in the more modern orders of *Perissodactyla* and *Artiodactyla*, may be regarded as a rotation to the inner side, of the bones of the second carpal row, on those of the first. This rotation is probably nearly coincident with the loss of the pollex, as it throws the weight one digit outwards, that is on the third and fourth digits, rendering the first functionally useless to a foot constructed solely for sustaining a weight in motion. The alternation of the two rows of carpals clearly gives greater strength to the foot than their serial arrangement, and this may probably account for the survival of the type possessing it, and the extinction of nearly all the species of the type which does not possess it. Here is applied again the principle first observed by Kowalevsky in the proximal metapodial articulations. This author shows that the types in which the metapodials articulate with two carpal or tarsal bones, have survived, while those in which the articulation is made with a single carpal or tarsal have become extinct. The double articulation is, of course, mechanically the more secure against dislocation or fracture.

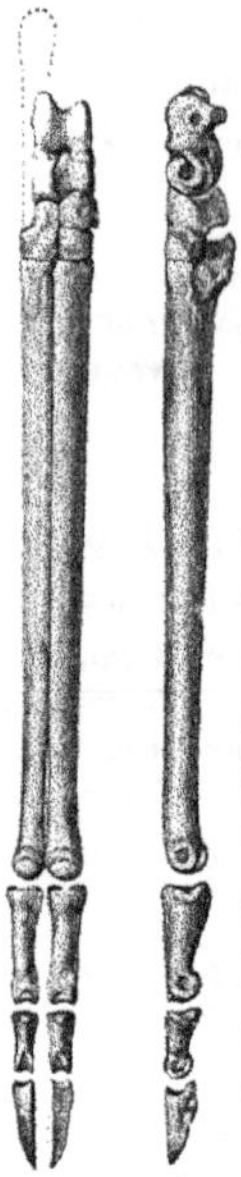

FIG. 20.—Hind foot of *Poebrotherium labiatum* (original). About one-third natural size.

As regards the inner part of the manus I know of no genus which presents a type of carpus intermediate between that of the *Taxeopoda* and *Amblypoda* on the one hand, and the *Diplarthra* on the other. Such will, however, probably be discovered. But the earliest *Perissodactyla*, as for instance *Hyracotherium*, *Hyrachyus*, and *Triplopus*, possess the carpus of the later forms, *Rhinocerus* and *Tapirus*. The order *Amblypoda* occupies an interesting position between the *Taxeopoda* and *Diplarthra*, for while it has the carpus of the primitive type, it has the tarsus of the later order. The bones of the tarsus alternate, thus showing a decided advance on the *Taxeopoda*. This order is then less primitive than the latter, although in the form of its astragalus it no doubt retains some primitive peculiarities which none of the known *Taxeopoda* possess. I refer to the absence of the trochlea, a character which will yet be discovered in the *Taxeopoda*, I have no doubt.

The *Taxeopoda* approach remarkably near the *Bunotheria*, and the unguiculate and ungulate orders are brought into the closest approximation in these representatives. In fact I know of nothing to distinguish the *Condylarthra* from the *Mesodonta*, but the ungulate and unguiculate characters of the two divisions. In the *Creodonta* this distinction is reduced to very small proportions, since the claws of *Mesonyx* are almost hoofs. Some of the genera allied to *Periptychus*, present resemblances to the *Creodonta* in their dentition also.

The facts already adduced throw much light on the genealogy of the Ungulate Mammalia. The entire series has not yet been discovered, but we can with great probability supply the missing links. In 1874 I pointed[1] out the existence of a yet undiscovered type of Ungulata, which was ancestral to the *Amblypoda*, *Proboscidea*, *Perissodactyla*, and *Artiodactyla*, indicating it by a star only in a genealogical table. This form was discovered in 1881, seven years later, in the *Condylarthra*. It was not until later[2] that I assumed that the *Diplarthra* are descendants of the *Amblypoda*, although not of either of the known orders, but of a theoretical division with bunodont teeth.[3] That such a group has existed is rendered extremely probable in view of the existence of the bunodont *Proboscidea* and *Condylarthra*. That the *Taxeopoda* was the ancestor of this hypothetical group as well as of the *Proboscidea*, is extremely probable. But here again neither of the sub-orders of this group represent exactly the ancestors of the known *Amblypoda*, which have an especially primitive form of the astragalus not found in the former. In the absence of an ankle-joint, the *Pantodonta* are more primitive than any other division of the Ungulata, and their ancestors are not likely to have been more specialized than they. It is probable that a third sub-order of *Taxeopoda* has existed which had no trochlea of the astragalus, which I call provisionally by the name of *Platyarthra*.

The preceding paragraphs were written in May of the year 1882.

[1] Homologies and Origin of Teeth, etc., Journal Academy Nat. Science, Philada., 1874, p. 20.
[2] Report U. S. Geol. Survey W. of 100th Mer., p. 282, 1877.
[3] This hypothetical sub-order is called in the above scheme *Amblypoda hyodonta*.

On my return home, September 1st, after an absence of three months, I found that various parts of the skeleton of *Periptychus*[1] have reached my museum. On examination, I find that the astragalus of that genus fulfils the anticipation above expressed. *It is without trochlea*, and nearly resembles that of *Elephas*. As it agrees nearly with that of *Phenacodus* in other respects, I only separate it as a family from the *Phenacodontidæ*. One other type remains to be discovered which shall connect the *Periptychidæ* and the hypothetical *Hyodonta*, and that is a Taxeopod without a head to the astragalus—unless, indeed, the "*Hyodonta*" should prove to have such a head. I think the latter the less probable hypothesis, and hence retain the term *Platyarthra* for the hypothetical Taxeopod without trochlea or head of the astragalus.

These relations may be rendered clearer by the following diagram:

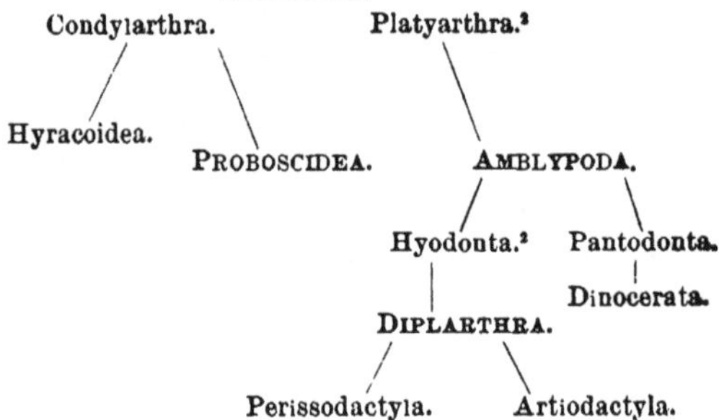

The preceding classification was first published in the Proceedings of the American Philosophical Society, October, 1882, after having been read at the Montreal meeting of the American Association for the Advancement of Science, August 29, of the same year.

CONDYLARTHRA.

Cope, American Naturalist, 1881, p. 1018, November 29.

In a paper on the "homologies and origin of the molar teeth of the Mammalia Educabilia, published in March, 1874,[3] I ventured the generali-

[1] See American Naturalist, October, 1882.
[2] Hypothetical.
[3] Journal of the Academy of Natural Sciences, Philadelphia.

zation that the primitive types of the Ungulata would be discovered to be characterized by the possession of five-toed plantigrade feet, and tubercular teeth. No Perissodactyle or Artiodactyle mammal was known at that time to possess such feet, nor was any Perissodactyle known to possess tubercular teeth. Shortly after advancing the above hypothesis, I discovered the foot structure of *Coryphodon*, which is five-toed and plantigrade, but the teeth are not of the tubercular type. For this and allied genera, I defined a new order, the *Amblypoda*, and I have published the confident anticipation that genera would be discovered which should possess tubercular (bunodont) teeth. This prediction has not yet been realized. The discovery of the *Condylarthra* went far towards satisfying the generalization first mentioned, and indicates that the realization of the prophecy respecting the *Amblypoda* is only a question of time.

In 1873[1] I described, from teeth alone, a genus under the name of *Phenacodus*, and although a good many specimens of the dentition have come into my possession since that date, I have never been able to assign the genus its true position in the mammalian class. The teeth resemble those of suilline Ungulates, but I have never had sufficient evidence to permit its reference to that group. Allied genera recently discovered by me have been stated to have a hog-like dentition, but that their position could not be determined until the structure of the feet shall have been ascertained.[2]

In his recent explorations in the Wasatch Eocene of Wyoming, Mr. J. L. Wortman was fortunate enough to discover nearly entire skeletons of *Phenacodus primævus* and *P. vortmani*, which present all the characters essential to a full determination of the place of *Phenacodus* in the system. The unexpected result is, that this genus must be placed in a special group of an order which includes also the *Proboscidea*.[3]

The astragalus in this sub-order is absolutely undistinguishable from that of the flesh-eating groups *Creodonta* and *Carnivora*. The humerus also presents a character of the unguiculate orders, in possessing an epicondylar

[1] Palæontological Bulletin No. 17, Oct. 1873, p. 3; also, Report G. M. Wheeler, U. S. Engineers Expl. W. 100 Mer., iv, p. 174, 1877.

[2] Proceed. Amer. Philosoph. Society, 1881, p. 495.

[3] American Naturalist, June, 1882 (May 17).

foramen, which is elsewhere unknown among ungulates. The humeral condyles have the generalized character of the same type of the *Amblypoda*, and of the lower *Perissodactyla*, in lacking an intertrochlear crest.[1] The *Condylarthra* may then be further defined as follows:[2] Astragalus with one uniformly convex distal articular face; humerus with epicondylar foramen. This sub-order has as yet been only found in the lowest horizons of the Eocene period, the Puerco and Wasatch, and only on the North American Continent. Appropriately to this position in time, its structure indicates that it is the most primitive type of the *Ungulata*. A number of genera and species belong to it, and these fall into three families, which are defined as below. They conform to the definitions of the order in possessing the normal mammalian type of dentition, without specialization, and a third trochanter of the femur. The approximation to the *Hyracoidea* is greater than that of any other group of the *Taxeopoda*. That order agrees with the *Condylarthra* in the simple articular extremity of the astragalus, which is, however, less convex; but it has a very peculiar articulation with the anterior face of the extremity of the fibula, seen in no other group of ungulates. In the manus of the *Hyracoidea* the lunar bone is very peculiar, not being divided below into two facets, as in most other ungulates, and generally extending to the carpals of the trapezoides series (the intercalare), as well as to the unciform. In this point the *Hyracoidea* come nearer to the *Amblypoda*. In *Hyrax* there is also no epicondylar foramen. The three families of *Condylarthra* are defined as follows:

Dentition bunodont; toes 5—5; astragalus without trochlea; neck very short; premolars very simple above and below .... *Periptychidæ*.
Dentition bunodont; toes 5—5; astragalus with trochlea; neck longer; premolar teeth different from the molars above and below .... *Phenacodontidæ*.
Dentition lophodont, with crescents and deep valleys; premolars partly like molars below; neck longer? .... *Meniscotheriidæ*

The bunodont dentition, with very simple premolars, flat astragalus, and five toes on all the feet, give the *Periptychidæ* the lowest place in the sub-order and order, as the most generalized type known. The *Meniscotheriidæ* have a quite specialized dentition, and until I learned its Condylar-

[1] American Naturalist, April, 1882, p. 334.
[2] Amer. Naturalist, 1881, p. 1017, Nov. 29.

throus character, I was at a loss to account for the presence of such perfection in so old a type. The number of the toes is yet unknown. The family appears to have had no descendents, and is a good illustration of Dr. Kowalevsky's views as to the persistence of the "adaptive" over the "non-adaptive" types of articulation. Kowalevsky observed that the types of Ungulata, which have the carpo-metacarpal and tarso-metatarsal articulations simple and not alternating, have become extinct. In those which persisted, the metapodials articulate with two bones of the carpal or tarsal series. The same rule has generally applied in the Ungulates to the distal astragalar articulation. The orders with the double articulation have left descendents, while the *Condylarthra* with the single articulation have disappeared without leaving a trace. The *Proboscidea*, which have the same simple distal articulation, still remain, however, to show an exception to the generalization. They have, however, an alternation in the second series of the posterior foot not present in the *Taxeopoda*. The relations of the genera of these three families are as follows:

Periptychidae.

In *Periptychus* only are the posterior feet known. The carpus is yet unknown. The successional modifications are seen in the addition of cusps to the inner sides of the premolars of both jaws, and the true molars of the upper jaws. In *Periptychus* we have the largest number of dental cusps and lobes, and in *Anisonchus* the next. With that genus the inferior premolars lose their inner ledges, and the true molars their anterior internal lobes. The latter are still further reduced in *Hemithlæus*, and the former in *Haploconus*. It is possible that *Conoryctes* belongs to this family. In all of my specimens of this genus the faces of the molars are so worn that I cannot see the pattern, and the ungues are unknown. There is, however, a general agreement in the known parts of the skull and skeleton. If it enters here it will be characterized by the perfect V's of the inferior molars. The following are the characters of the genera:

I. Intermediate tubercles present; inferior premolars with internal lobes.

Superior molars with two internal cusps besides apex of V; superior premolars with internal lobes .... *Periptychus*.

II. Intermediate tubercles wanting; inferior premolars without internal lobes.

Superior molars with posterior internal cusp only, besides internal V; last two superior premolars with internal lobes .... *Anisonchus.*

Superior molars with internal V only, no other internal lobes; last two superior premolars with internal cusps .... *Hemithlæus.*

Superior molars with posterior internal cusp only, besides apex of V; fourth superior premolar only with internal lobe .... *Haploconus.*

PHENACODONTIDÆ.

The genera of this family display a uniformity in the structure of the true molar teeth not seen in the *Periptychidæ.* Their range of grade is seen in the premolars, especially those of the superior series. Thus in *Protogonia,* all of those teeth have but a single external lobe. In *Phenacodus,* the fourth has two external lobes. In *Diacodexis,* the second, third, and fourth teeth have two external lobes. The premolars are unknown in *Anacodon.* While *Protogonia* is primitive in its superior premolars, its inferior true molars come nearer to developing V's than any other genus of the family. The definitions are as follows:

Last superior premolars with but one external cusp; inferior molars with V's. *Protogonia.*

Last superior premolars with two external cusps; inferior molars with well-developed cusps .... *Phenacodus.*

Inferior molars with flat grinding faces; no cusps .... *Anacodon.*

Second, third, and fourth superior premolars with two external cusps; those of inferior molars well developed .... *Diacodexis.*

MENISCOTHERIIDÆ.

This family includes the single genus *Meniscotherium.*

Superior molars with intermediate tubercles, the anterior crescentic, the posterior oblique, forming a crest with the posterior inner; anterior inner conic. Inferior molars and last premolar with two V's; other inferior premolars without internal lobes; fourth superior premolar with two external lobes .... *Meniscotherium.*

The geological distribution of the nine genera of these families is as follows:

	Puerco.	Wasatch.
Periptychidæ.		
Haploconus	4	....
Hemithlæus	2	....
Anisonchus	3	....
Periptychus	3	....
Phenacodontidæ.		
Protogonia	2	....
Anacodon	....	1
Phenacodus	2	6
Diacodexis	....	1
Meniscotheriidæ.		
Meniscotherium	....	3
	17	11

PERIPTYCHUS Cope.

American Naturalist, 1881, p. 337 (March); *Catathlæus*, Cope, American Naturalist, 1881, 829, Sept. 22. Proceeds. American Philosoph. Soc., 1881, p. 487.

Dental formula: I. $\frac{2 \text{ or } 3}{3}$; C. $\frac{1}{1}$; Pm. $\frac{4}{4}$; M. $\frac{3}{3}$. Premolars of superior series consisting of external conic cusp, and an internal crescentic crest, which is like a developed cingulum surrounding the cusp. Crowns of true molars supporting seven tubercles, as follows: two external; one principal median internal, which has accessory cusps, one anterior and one posterior to it; lastly, two, an anterior and a posterior intermediate tubercles. Diastemata in both series small. Inferior incisors small; canines of moderate size. Inferior premolars consisting of one principal external cusp, and an internal cingulum. This rises medially into an accessory cusp, and extends posteriorly into a narrow heel, and anteriorly into a small anterior basal tubercle. The inferior true molars sustain four principal cusps opposite in pairs, with accessory median ones on the anterior and posterior borders. The posterior median lobe is so developed in the last inferior molar as to constitute a fifth lobe or heel.

The angle of the lower jaw is not reflected nor inflected. A part of the condyle with adjacent regions shows several features. The coronoid process rises near the condyle, and in precisely the same plane as the angle of the jaw. Both leave the condyle near its inner extremity. The articular face of the latter looks upwards and backwards about equally, and is rather flat.

A fragmentary skull shows a postglenoid crest, and the robust post-tympanic and paroccipital processes united, and leaving the meatus auditorious externus widely open below. The os petrosum is small and not inflated. The foramen ovale is not separated from the meatus auditorius below. There is a postglenoid foramen, and a supraglenoid foramen. There is also a well-marked mastoid foramen. The mastoid bone is extensively exposed. The cranial walls are thick.

Such part of the cast of the brain as appears, gives the following points: The hemispheres are very narrow, and rather elongate, and are separated by a long flat crus from the olfactory lobes. The latter are very large and nearly as wide as the hemispheres.

The posterior three premolars are preceded by temporary teeth in both jaws. Of these the anterior is protruded at about the same time as the first true molars, and is the last one shed, remaining until after the last true molar is fully protruded. The last milk premolar differs from the corresponding permanent one in its greater elongation. The extension is posterior, in the form of a heel with three tubercles, of which the median is very small, the crown resembling a permanent true molar, except that the anterior portion is a little more elongate and compressed. The anterior basal lobe is a mere elevation of the cingulum, as in the permanent premolar, but the internal cusp is more distinct than in the latter. The penultimate milk premolar is more like the corresponding permanent tooth, but is a little more flattened and elongate, and the heel is not tubercular. The first milk molar is a little more compressed than the corresponding permanent tooth, and the edge of the heel is not tubercular. Otherwise they are similar. It was on a specimen supporting the last two milk molars, with the first true molar so injured that its true character could not be ascertained, that the *Periptychus carinidens* was established.

The base of the diapophysis of the atlas is perforated as in the succeeding cervicals. The lateral perforating foramen is entirely isolated. The axial facets are separated. The axis is depressed and moderately short. The odontoid is depressed, and has an oval section. The cervical vertebræ are much shorter than in *Phenacodus*, being deeper than long, and wider than deep. They are very slightly opisthocælous. The caudal vertebræ are quite robust, indicating a powerful tail. Dorsals not found.

The tuberosities of the humerus are small in proportion to the size of the head. The condyle is much like that of a creodont, with internal flange and external cylinder, without intertrochlear crest or ridge The internal epicondyle is large, and is pierced above by an epitrochlear foramen. The olecranon is compressed. The head of the radius has a flat articulation with the ulna. Its outline is a transverse oval, narrowed at the external extremity. The scapula has a well developed coracoid hook, and the spine rises abruptly from the neck.

In the *P. rhabdodon* the femur is not materially larger than the humerus. The great and little trochanter are well developed, and the third trochanter is situated low down, as in *Phenacodus*, and not opposite the little trochanter as in *Creodonta*.

The posterior foot has five digits. The astragalus is much like that of the *Proboscidea* in form. The head is moderately long, and is depressed. Its distal extremity is regularly convex from side to side. The trochlea is horizontal, and is not grooved medially, but is very slightly concave. Fibular face vertical; malleolar face slightly oblique and occupied by a deep central fossa. The head is not as convex as in *Phenacodus*, but is more recurved on both sides. On the external side it is so far recurved as to be continuous (in *P. rhabdodon*) with the sustentacular facet, and a part of this face is probably in contact with the cuboid, as in many creodonta, but which cannot be said therefore to overlap the astragalus, as in the *Amblypoda*. If this facet were distal, and in the plane of the navicular facet, it would be necessary to refer this genus to that order.

The calcaneum is robust. Its astragalar facets, especially the external, are rather flat. The cuboid facet is large, and is supplemented by an external tuberosity. The sustentaculum is well developed. Free portion lost in my specimens. The cuboid is robust, and not flat, extending well beyond the navicular. Its distal extremity does not display distinct facets.

The navicular much resembles that of *Phenacodus*. It is proximally concave; its inferior aspect has three facets, of which the internal is largely posterior. The ectocuneiform has an elongate posterior tuberosity. It has also a proximal facet of the external side which corresponds with one of the navicular, for the cuboid. Its distal face is concave.

Portions of two posterior feet preserved, display five metatarsals, and several phalanges. The distal carina of the former is posterior and weak. The latter are rather narrow for an ungulate, but are not elongate, and are rather depressed; the distal ones are more robust, and are rather more narrowed distally than usual in ungulata, and the neck of a broken ungual phalange of an external digit is nearly round in section. The third digit is longest, and the first, shortest; it is not very short, and is quite slender. Sesamoid bones are probably present The posterior foot is that of a plantigrade animal.

I have obtained a cast of the top and sides of the cerebral hemispheres, and the proximal portion of the olfactory lobes, from a skull of a *Periptychus* in which the teeth are preserved, and prove the species to be the *P. rhabdodon*. I describe it in detail under that species, but state here that the olfactory lobes are enormous, and the hemispheres small and very flat. The *mesencephalon is entirely exposed.*

The position of this genus and its immediate allies *Anisonchus*, *Haploconus*, and *Hemithlæus*, is not yet positively determined. But three references are possible, viz, to the *Taxeopoda Condylarthra*, the *Bunotheria Creodonta*, and the *Marsupialia*. As no undoubted marsupial characters have been found, discussion of their affinities to that order is deferred. Nevertheless it must be remembered that there are *no osteological characters common to all Marsupialia*, and that the undoubted characteristics of that order *can be found in the soft parts only*. The determination of extinct *Marsupialia* will, on this account, always be difficult. The sculpture of the premolar teeth is not unlike that seen in the fourth premolar of *Ptilodus*. The character of the condyles of the humerus is, however, totally unlike that of *Catopsalis* and *Meniscoëssus*. The dentition is against reference to the *Creodonta*, excepting as regards the family of *Arctocyonidæ*. The only suggestion of such affinity is found in the form of the astragalus, and in the narrow external phalanges. The astragalus of all the *Condylarthra* is undistinguishable from that of the *Creodonta*. It is almost certain that the middle ungual phalanges in *Periptychus* are much wider than the lateral ones, so that they are probably truly ungulate. The reference to the *Condylarthra* is indicated by the large inferior third trochanter of the femur; also by the vertical aspect of the mandibular condyle, with the identity of plane of the angular and coronoid plates. In the *Creodonta*, where this part is known, the coronoid process and the angular plate are, as in *Carnivora*, different. I except from this *Mesonyx*, where they are in the same plane.

There are three species of *Periptychus*, two of which differ principally in dimensions.

PERIPTYCHUS RHABDODON Cope.

American Naturalist, Sept. (Oct.), 1882. *Catathlærhus rhabdodon* Cope. American Naturalist, 1881, October, p. 830 (Sept. 22). Paleontological Bulletin, No. 33, p. 487, 1881. Proceed. Amer. Philo. Soc., 1881, p. 487.

Plates XXIII f, XXIII g, figs. 1–11; LVII f, figs. 1–2.

This species was evidently very abundant during the Puerco epoch, portions of fifty individuals having come into my possession. These consist mostly of fragments of the jaws, superior and inferior. The most important of these specimens includes most of the dentition of the mandible and the posterior part of that of the maxillary bone; fragments of the skull; a number of vertebræ; considerable parts of both limbs; part of the posterior foot. A second specimen includes both jaws, parts of scapula and ulna, cervical vertebræ, femur and part of posterior foot. A third specimen includes part of the lower jaw with condyle and teeth; some vertebræ, and the astragalus. Another includes one mandibular ramus with symphysis. There are many maxillary bones with teeth, but no premaxillaries.

All the teeth of this species (incisors unknown) are characterized by a remarkable sculpture of sharply-defined grooves and ridges. The ridges extend from the bases of the crowns to the apices of the cusps of the premolars and molars, and on the external sides of the teeth are straight. As they converge some of the ridges cease. On the interior faces of the crowns the grooves are less profound, and the ridges are more irregular in direction and less closely placed. On the inner sides the sculpture may disappear with age, but never on the external side.

In their unworn state the internal cingulum-crest of the premolars is coarsely serrate. They extend round the external cone to the posterior base, and to the outer base anteriorly, so as to appear as a wide oblique margin projecting anterior to the cone, on an external view. The external cone is compressed at the apex. The first superior premolar is unknown. No external cingula on the premolars, but a distinct one on the external base of the true molars. The latter have no internal cingulum, but there is one on both the anterior and posterior bases of the crowns which extend from the anterior and posterior internal tubercles, respectively. The two tubercles last mentioned project farther inwards than the median internal tubercle, and are distinguished from it on the inner face of the crown by grooves. All the tubercles, including the external, have a subround section. The intermediate tubercles are the smallest, the other four being subequal. The premolars are remarkably large as compared with the true molars; the disproportion being greater in this than in any mammal known to me, except the *Tetraconodon* of Falconer.

The crowns of the inferior premolars are convex on the external side, and flat on the inner. The principal and external cone is compressed at the apex in the anterior three premolars. The internal cusp is flat, and appressed to the external, and its apex is joined to the side of the latter. On trituration the two speedily become confluent, while the anterior and posterior basal lobes remain distinct, forming two lesser areas. The first inferior premolar is a simple cone flattened on the inner side, and with a low posterior heel. Of the four principal tubercles of the inferior true molars the anterior external is the largest, and extends a little farther anteriorly than the anterior internal. The anterior tubercle is on the inner side of the middle. On the last molar, the heel may have one or two accessory tubercles.

The ramus is moderately robust and compressed. The inferior outline extends to the incisive border by a gentle slope, and rises below the last molar teeth. The masseteric fossa is well marked in front, but has no distinct posterior or inferior boundary. The superior true molars are placed well posteriorly, or rather the orbit is anterior, for its anterior border is above the middle of the last premolar. The foramen infraorbitale is above the middle of the third superior premolar. The anterior part of the malar bone is quite prominent, overhanging the maxillary, and bounding a shallow fossa which is open below. A crushed skull of the large variety displays a strong sagittal crest, arising from gradually converging rather obscure temporal ridges. Another skull has the parietal region much depressed, which is not altogether due to pressure, as the sagittal crest is partly intact. There is a strong supraoccipital crest, which is bilobed, the convex lateral portions being separated by a deep median notch. Between their bases the occipital face of the skull is concave. The paroccipital process is separated from the occipital condyle by a strong notch. It does not extend quite so far as the rounded, robust mastoid or posttympanic process, to which it is closely attached. At its inner base there is a rather large ? stylomastoid foramen. The postglenoid process is peculiar, being an angular ridge terminating interiorly in a low angular tuberosity. Its posterior side, or that of the base of the zygoma, is flat, and slopes forwards and downwards. The glenoid cavity is nearly flat, and has no preglenoid crest, on the inner part at least; the rest of the border being destroyed in my specimens. The surface is cut off within by the groove, which, entering the roof of the meatus auditorius, represents the foramen ovale. On the internal side of this groove there is a ridge, which, at its middle, swells into a tuberosity. The meatus auditorius is widely open. The postglenoid and supraglenoid foramina are rather small, and the latter is exactly above the postglenoid ridge. The mastoid foramen is on a line with the suprameatal ridge. The superior part of the mastoid bone just behind the posttemporal crest is pierced by another smaller foramen.

The lengths of the last five inferior molars in this species vary from M. .065 to .055; and the inferior true molars from .035 to .033. The following measurements are taken from two individuals:

Measurements.

No. 1.

	M.
Length of inferior dental series, less incisors	.090
Length of inferior premolar series	.046
Length of inferior true molar series	.034
Length of base Pm. i	.008

		M.
Length of base Pm. iii		.013
Width of base Pm. iii		.0095
Length of base Pm. iv		.013
Width of base Pm. iv		.010
Diameters M. i	anteroposterior	.011
	transverse	.009
Diameters M. iii	anteroposterior	.012
	transverse	.008
Depth ramus at Pm. ii		.0225
Depth ramus at M. iii		.0295
Depth of ramus posterior to M. iii		.052
Length of bases of two superior incisors		.012
Diameter of base of crown superior canine		.007
Length of superior true molars		.028
Diameters of superior molar i	anteroposterior	.0096
	transverse	.011
Diameters of superior molar iii	anteroposterior	.009
	transverse	.010

No. 2.

Length superior premolar series		.049
Diameters Pm. i	anteroposterior	.009
	transverse	.0095
Diameters Pm. iii	anteroposterior	.013
	transverse	.014
Elevation of occipital crest laterally above foramen magnum		.044
Anteroposterior diameter of paroccipito-posttympanic process		.017
Anteroposterior diameter of meatus auditorius		.009
Anteroposterior diameter of glenoid cavity at postglenoid process		.015

The cervical vertebræ, as already remarked, are about as long (or short) as those of the species of *Elephas*, but are more depressed. The axis is less shortened, as its centrum has much the proportions of that of *Rhinocerus*. The atlas is shorter than in the latter genus. Unlike most genera of mammalia, in *Periptychus*, the vertebrarterial canal pierces the base of the vertically compressed diaparapophysis precisely as in *Elephas*. Its position is a little above opposite the fundus of the odontoid foramen, and nearly in the position it occupies in the *Elephas africanus*. The condylar cotyli have full convex borders, and their articular surface is without constriction. Axial facets flat and subround. The axis is preserved in Nos. 1 and 3. The atlantal facets are distinct from each other, are subround, and slope at an angle of 45° with the middle line. Vertebrarterial canal complete (neural arch and extremity of transverse process lost). The floor of the neural canal is wide and flat, and is pierced by two small foramina posterior to the middle. Inferior face flat, with a faint trace of angular keel. The centra of the succeeding cervicals, although slightly convex in front in vertical section, are concave in transverse section. The transverse diameter exceeds the vertical in the third vertebra more than in the sixth, where it is still, however, greater than the transverse. The inferior faces of all the centra are smooth and a little convex, and a little oblique to the vertical diameter, indicating the anterior elevation of the neck. Posterior to the third, the centra become a little concave transversely on the posterior side, but some of the epiphyses are lost. The bases of the neurapophyses are nearly round. On the fourth cervical the postzygapophysis looks almost entirely outwards. The diapophyses are round at the vertebrarterial canal, while the parapophyses are flat at the same position.

The cervical vertebræ just described are from No. 3. Two proximal caudals accompany the same. The latter are robust, and have plane articular extremities. They have also complete neural arches, one with, the other without, spine. The bases of the diapophyses spread anteriorly into ridges, which nearly reach the anterior edge of the centrum. Inferior face convex in the middle transversely, concave longitudinally, the convexity spreading posteriorly into two angles for base of supposed chevron bone. No. 4 exhibits a more distal caudal. It is large and has elongate proportions. This specimen evidently had a long tail. Unfortunately no dorsal vertebræ have been preserved.

Measurements of caudal vertebræ.

No. 3.

		M.
Diameter of centrum behind	vertical	.013
	transverse	.016
Elevation with neural spine		.019
Length of second centrum		.019
Diameter in front	vertical	.012
	transverse	.013
Total elevation		.016
Diameter centrum behind	vertical	.0135
	transverse	.016

No. 4.

	M.
Length of centrum of caudal vertebra	.029
Width of centrum at extremity	.025
Width of centrum at middle	.012
Depth of centrum at middle	.010

The glenoid cavity of the scapula is rather narrow. Its posterior border is regularly rounded, while the lateral borders converge rapidly to an apex which rises on the external base of the coracoid process. The internal base of the latter is excavated into a groove. The posterior face of the neck of the scapula is moderately wide, and is bounded by a groove near its internal edge. The spine rises near this edge, and far from the thin anterior edge. Posterior to it the posterior edge is a little recurved outwards.

Measurements.

No. 1.

		M.
Diameters articular face glenoid cavity	anteroposterior	.027
	transverse	.018
Length of coracoid beyond face		.008
Width of neck		.026
Distance from glenoid cavity to spine		.011

The humerus is a robust bone with large head and condyles, and shaft contracted below the middle. The tuberosities are small, like those of the *Proboscidia*, and not produced as in the higher ungulates. The bicipital ridge is large, with a flat back and recurved edge, which is quite oblique, ending below in an angular projection, which marks the middle of the anterior face of the shaft. Below this point the section of the shaft is triangular, the posterior side being the longest, and bounded externally by the external epicondylar ridge. This ridge does not develop a prominent external epicondyle. The internal epicondyle is, on the contrary, very prominent. It projects abruptly from the middle of the internal condyle, and has a truncate narrow external edge, which is a little oblique to the axis of the shaft. Its superior edge rises to the shaft, and forms the bridge over a large transverse oval foramen epicondyloideum. This opens distally on the anterior face of the humerus opposite the superior part of the coronoid fossa, and above and internal to the internal flange of the condyle. The coronoid fossa is deeper than the olecranar, and the two are separated by a thin septum. The condyles have a great transverse extent compared with their anteroposterior; a character more marked than in *Phenacodus*. The internal flange is moderately prominent, and the internal roller is separated from the trochlear groove by a convexity of the surface. The trochlea is wide, and a little wider than the roller anteriorly, and it expands a little and has raised edges posteriorly. No roller posteriorly. The internal border of the trochlea posteriorly is separated by a deep subround fossa.

Measurements of humerus.

No. 1.

			M.
Total length with flange			.135
Length to distal end of bicipital ridge			.076
Transverse diameter of head with greater tuberosity (extended by pressure)			.039
Diameters of shaft at inferior extremity of bicipital ridge	anteroposterior		.022
	transverse		0.15
Diameters of humerus below do.	anteroposterior		.014
	transverse		.018
Width at epicondyles			.055
Diameter condyles anteriorly	tranverse		.033
	anteroposterior	roller	.015
		flange	.018
Length of end of epicondyle			.022

The ulna is stout and compressed throughout its length; its vertical diameter not diminishing much distally. The olecranon is compressed so as to be in a vertical plane; the superior edge being acute, and the inferior rounded, and becoming wider near the extremity, where it rises obliquely inwards. The external side of the humeral cotylus is a little, the inner side more, convex. The flanges of the humeral cotylus of the ulna are peculiar. The posterior is only developed on the internal side of the shaft, and does not exist on the inner side. This is the condition in *Coryphodon*, but the external is present in *Phenacodus* and *Hyrax*. It is wanting in *Carnivora*, but the internal is smaller in them than in *Periptychus*. The radial contact is entirely flat, and there is no anterior flange except on the external side. It is here horizontally extended, but not vertically, as is the case in *Carnivora*. There is a longitudinal groove on the external side of the shaft near the superior margin, which soon terminates. A much wider groove, with defined inferior edge, commences below the radial facet, on the inner side of the shaft, and extends along the latter, becoming wide and shallow. The radius has the proximal half of its shaft quite slender, more so than in *Phenacodus primævus* or *wortmani*, and there is no bicipital tuberosity. The head expands abruptly from the shaft, and the humeral face is transverse. It is not much recurved at the internal and wider extremity, while the external or narrower extremity is obtuse and rounded. The anterior foot is unfortunately unknown.

Measurements of ulna and radius.

No. 1.

		M.
Length of olecranon from coronoid		.030
Depth of olecranon at middle		.022
Depth of olecranon at coronoid		.028
Depth of olecranon at radial facet		.018
Depth of olecranon in front of radial facet		.016
Width of humeral cotylus at middle		.0135
Diameters head of radius	transverse	.025
	vertical	.014
Diameters shaft of radius	transverse	.011
	vertical	.008

In the other specimen, already designated No. 3, from which the cervical vertebræ have been described, a femur is preserved. This is a robust bone, with shaft flattened behind, and large extremities. The fossa ligamenti-teris is large. The great trochanter does not extend beyond the head, and the exterior edge is thick and strongly recurved, inclosing a small transversely deep trochanteric fossa. The little trochanter is a prominent acute convex edge. The rotular groove is moderately wide and moderately prominent. Its lateral ridges are equal. The femoral condyles are subequal and stand well apart. Their surface is not cut off by notches from the rotular.

Measurement of femur.

No. 3.

		M.
Total length on inner side		.147
Proximal width at head		.046
Transverse diameter of head		.025
Width at little trochanter		.031
Diameters shaft below third trochanter	transverse	.020
	anteroposterior	.014
Distal width of femur		.043
Depth at inner side of rotular groove		.030
Width between condyles at middle		.007

Specimen No. 4 exhibits a perfectly preserved third trochanter, which is broken in Nos. 1 and 3. It has one concave side and a truncate extremity slightly recurved to the concave side.

In Nos. 1 and 3, tibiæ which have lost their epiphyses are preserved. The crest is prominent, and continues to below the middle of the shaft. At the latter point it displays the peculiarity of becoming more prominent than above it, and is acute and recurved outwards, as is sometimes seen in the extremity of the bicipital crest of the humerus. Just below it the section of the shaft is triangular, with subequal sides, the posterior flat. The distal extremity is triangular, the posterior side longer than the others. This bone is long for the length of the femur and humerus.

Measurements of tibia without epiphyses.

Nos. 1 and 3.

		M.
Total length (derived from two tibiæ)		.135
Diameter at middle of shaft	anteroposterior	.020
	transverse	.011
Diameter at distal end	anteroposterior	.022
	transverse	.027

The calcaneum is depressed. The border of the sustentaculum is truncate and not acute. The cuboid facet is transverse, wide, diamond-shaped, with opposite angles, the external angle supplemented by a prominent flattened tuberosity. The inferior surface has a wide external groove. Free process lost. The external inferior cotylus of the astragalus is very little concave, conformably to the form of the calcaneum. The angle at its externo-anterior termination is not produced downwards, as in *Phenacodus vortmani* and *P. primævus*; still less than in *Mesonyx ossifragus*. The postero-internal angle is produced beyond the trochlea, but not so much as in *Oxyæna;* its connection with the head is by a flat horizontal ridge. The head is extended laterally in both directions, so as to be wider than the neck. The median inferior or sustentacular facet is separated by deep grooves from the other facets, but is connected with the inferior recurvature of the head. The longitudinal median groove terminates posteriorly in a deep foramen penetrating upwards, and issuing in a posterior notch of the trochlear surface. This foramen is caused by the closing of the usual tendinous notch, and is also seen in the genus *Bathmodon*. The trochlea is strongly convex anteroposteriorly, though nearly flat transversely. There is a depressed fossa beneath its antero-external angle.

As already remarked, this astragalus resembles somewhat that of *Oxyæna*. It is preserved in three or four individuals. The navicular is rather shallow, and is wider transversely than anteroposteriorly. Its median facet is the largest, and rises highest in front. The cuboid, which is also preserved in No. 1, has a square anterior face. Its proximal face is convex anteroposteriorly, and the distal face is concave anteroposteriorly to a less degree. There is a short lateral facet on the inner side for the navicular. The external tendinous tuberosity is prominent, and extends across the inferior face of the bone. The front of the ectocuneiform is slightly oblique; it is half as high again as wide, and has a low tuberosity near the internal proximal corner. Posterior tuberosity large. The metatarsals are short. The third and fourth are of equal lengths; the second is a little and the first much shorter. The proximal extremity of the fourth is externally concave for the head of the fifth. There are no deep transverse grooves above the distal condyles These are rather narrow, and have a short posterior heel. The distal faces of the phalanges are infero-posterior, but on the phalangines they are somewhat recurved. The lateral ligamentous fossæ are moderate. The phalanges are slenderer than is usual in ungulate mammalia, and resemble those of some carnivora. They are more depressed than in these animals, and resemble most those of *Mesonyx*. The distal extremities of the metapodials, though narrow, display no tendency to the convex form seen in carnivora.

It is evident from the proportions of the posterior feet in the two specimens preserved, that these members are relatively small in the *Periptychus rhabdodon*. It is also evident that they were wide and flat and plantigrade; more so than in the species of *Phenacodus*.

Measurements of posterior foot.

No. 1.

		M.
Width of calcaneum at sustentaculum		.026
Diameters sustentacular facet	anteroposterior	.013
	transverse	.013
Distal width of calcaneum		.034
Diameters cuboid facet	vertical	.013
	transverse	.017
Astragalus; greatest	length	.030
	width	.028
Trochlea of astragalus; greatest	length	.019
	width	.023
	height, externally	.014
Head of astragulus, greatest	width	.022
	height	.012
Length of third metatarsal		.050
Diameter of third metatarsal	proximally anteroposteriorly	.015
	medially transversely	.009
	at distal fossæ transversely	.0135
Length of fourth metatarsal		.046

	M.
Anteroposterior width of head of fourth metatarsal	.0135
Width of first metatarsal at middle	.007
Length of cuboid in front	.019
Width of cuboid in front	.0165
Diameters distal facet { anteroposterior	.010
Diameters distal facet { transverse	.014

No. 4 (large form).

Diameter shaft humerus at middle	.016
Depth head of radius	.013
Width calcaneum at sustentaculum	.026
Width navicular	.018
Length navicular	.007
Length ectocuneiform	.012
Width ectocuneiform	.009
Depth ectocuneiform	.019
Length of a short metatarsal (? iv)	.030
Depth of the same metatarsal proximally	.015
Width of the same metatarsal distally	.012
Length of phalange of M. i or ii	.019
Width of phalange distally	.008
Length phalangine of M. i or ii	.011
Width proximally	.009
Width distally	.007
Length astragalus of a second specimen	.027
Width of trochlea	.019
Elevation externally	.013

Brain.—The cast of the middle part of the brain-case already mentioned (p. 388) presents interesting characters. The cerebral hemispheres are very flat, and are only differentiated from the olfactory lobes by a moderate contraction and depression, which forms the peduncle of the latter. Only the proximal part of the olfactory lobe is preserved, but this expands so as to be only a little narrower than the hemispheres. The peduncle has a ridge on the median line, and a shallow fossa on each side of it. The lateral outlines of the hemispheres diverge, and the widest part is posterior. There is no indication of sylvian fissure. The transverse section of the hemispheres would be a flat arch but for the presence of a longitudinal oval protuberance on each of them, which do not quite touch the median line, and which have definite boundaries. If their limits determine the size of the cerebral hemispheres, then the latter are wider than long, but they probably pass gradually into the mesencephalon behind them. These bodies remind one of the corpora olivæformia, and may represent the superior or median frontal convolutions. They are probably, however, not to be homologized with any convolution, representing rather the cerebral vault of the lateral ventricle. Posterior to them the flat surface descends gently without indication of corpora quadrigemina or other irregularity, and at a distance about equal to the length of the oval bodies, begins to rise gently. The cranium is broken here, and no cast of the cerebellum was obtained.

I may remark that the cranium from which this cast is taken is not crushed, and that it consists of parts of the parietal and squamosal bones only. The latter remain as far as the incurvature to the pterygoid processes in front of the glenoid cavity.

Measurements of brain.

	M.
Length from posterior rise to base of olfactory lobes	.037
Length of oval bodies of hemispheres	.018
Width of proximal part of olfactory lobes	.027
Width of olfactory peduncles	.021
Length from olfactory lobes to oval bodies of hemispheres	.005
Diameter of hemispheres at posterior part of oval bodies	.038
Depth from sagittal crest to olfactory lobes	.024

Restoration.—This remarkable animal was about the size of the collared peccary, though the skull was perhaps a little larger. It must have had a peculiar appearance, and unlike that of any known mammal. The long legs with plantigrade feet must have given it the form of a bear, but its very short neck is only paralleled by that of the elephant. While the shorter legs forbid near resemblance to that animal, and the shape of the head is very different, yet the resemblances in the figure cannot be overlooked. It had a long tail, stout at the base. It was a smaller animal than the *Phenacodus primævus*, but the head was of near the same size. The dental system does not furnish any weapons of offense or defense, and none are known from any other part of the skeleton. Its habits were omnivorous, judging again from dental characters. It is the most abundant mammal of the Puerco, and to this time the largest discovered.

The large variety already mentioned is less abundant than the typical form, three individuals only having been sent by Mr. Baldwin. The specimens are frequently weathered from the rock matrix so as to be in beautiful condition.

PERIPTYCHUS CARINIDENS Cope.

American Naturalist, 1881, March, p. 337. Paleontological Bulletin No. 33, p. 484, Sept. 1881. Plate XXV *a*, fig. 16; XXIII*d*, figs. 14–15; XXIV*g*, fig. 5.

This species is, with my present knowledge, only distinguished from the *P. rhabdodon* by its inferior size and its longer and narrower last inferior molar. This difference is seen in both adults and young of corresponding ages, and chiefly in the true molar teeth. While the length of those of the lower jaw varies in the *P. rhabdodon* from 33 to 35 mililmeters, in the *P. carinidens* it only reaches 29 millimeters, or with the posterior two premolars 50 millimeters. The details of the adult dentition do not differ from those of the *P. rhabdodon*, except that the last inferior molar is narrowed and produced posteriorly.

The permanent dentition of this species is represented on Plate XXIII*d*, figs. 14–15; and the milk dentition on Plate XXV*a*, fig. 16. As the species was established on the latter specimen, I give a description of it.

The second and third milk molars support a principal median cusp, a broad heel, and a prominent anterior cingulum. The heel is more or less divided into tubercles; the anterior cingulum is on the inner side, and represents the anterior cusp of a sectorial tooth. On the inner side of the principal cusp a cingulum rises, forming a flat internal tubercle. The heel of the second supports three tubercles, of which the external is the largest. The anterior cingulum supports a small cusp and then rises to the internal tubercle, which is compressed. The sides of all the cusps are marked with distinct, well-separated vertical ridges. Each extremity of the internal cusp is connected with the principal cusp by a ridge. The first true molar has few cusps. Those of the heel are scarcely distinct, and form a border which rises prominently into the flat internal tubercle, which forms a narrow longitudinal blade. The anterior cingulum has no cusp, and does not rise into the inner tubercle. The principal cusp has a strong entering groove next the inner tubercle.

In these deciduous teeth the compressed semisectorial character seen in the first inferior true molar of *Haploconus* is carried back to the second true molar. The sectorial character is increased by the heel-like character of the posterior front of both first and second true molars, which lacks the cusps characteristic of *Phenacodontidæ* and ungulate mammals. These teeth therefore resemble premolars rather than true molars. Before I was acquainted with *Haploconus*, I provisionally referred this genus to the *Creodonta;* the resemblance to *Haploconus* is, however, unmistakable. The ridging of the enamel is also present in the milk dentition.

Measurements of milk dentition.

	M.
Length of crown of first molar	.0115
Width of crown of first molar	.006
Elevation of crown of first molar	.006
Length of second molar	.011
Width of second molar	.007
Elevation of second molar	.0065
Depth of ramus at second molar	.020

The species was obtained by Mr. D. Baldwin from "below all the Wasatch Sandstones," in the Puerco.

PERIPTYCHUS DITRIGONUS Cope.

Plate XXIII *g*; fig. 12.

This rare species is known from a right mandibular ramus which exhibits part of the symphyseal suture, with the alveoli of the molar teeth except the first. The only well-preserved crown is that of the second true molar.

The second true molar presents very peculiar characters, and the mandibular ramus is shallower and thicker than in the two other species of *Periptychus.* The former has a wide external cingulum which is not present in the other species, and there are only six cusps instead of seven. These are peculiarly arranged. The anterior three are much as in *P. rhabdodon*, the anterior inner being not quite so far internal as the posterior inner, close to it and as large as the anterior external. The posterior three are a posterior inner and posterior median as in *P. rhabdodon*, and a peculiarly placed posterior external. This is not opposite the posterior inner, but is anterior to such a position, and intermediate between the latter point and the one occupied by the median tubercle in *P. rhabdodon.* It is as large as the anterior external tubercle. All these tubercles are conical and not connected by angles or ridges. The position of the posterior external cusp leaves the cingulum wide posteriorly, and near its edge develops some small tubercles. There are also some small tubercles at other points on the edge of the crown, but no other cingula. The enamel is not regularly ridged as in *P. rahbdodon*, but has a rather coarse obsolete wrinkling.

Measurements.

	M.
Length from Pm. ii to M. ii inclusive	.052
Diameters of M. ii { anteroposterior	.011
Diameters of M. ii { transverse	.010
Depth of ramus at M. ii	.022
Width of ramus at M. ii	.016
Depth of ramus at Pm. ii	.019

From the Puerco formation of New Mexico; D. Baldwin, discoverer.

HEMITHLÆUS Cope.

American Naturalist, 1882, p. 832. (Sept. 28.)

Known from dentition only. Dental formula: I. $\frac{?}{?}$; C. $\frac{1}{1}$; P-m. $\frac{4}{4}$; M. $\frac{3}{3}$. Superior true molars with two external cusps and an internal V. No intermediate tubercles, nor internal tubercles or cusps, other than the apex of the V. Third and fourth premolars with one external and one internal cusp. Inferior true molars with two anterior and two posterior cusps, with some supplementary tubercles. First inferior premolar one-rooted; second with two roots and a simple crown. Crowns of second and third unknown.

This genus has the premolar teeth of the species of *Anisonchus* with the inner lobe conical. Its principal peculiarity consists in the absence of the posterior internal cusp, which is such a prominent feature in *Anisonchus* and *Haploconus*.

Two species are known, both from the Puerco formation.

HEMITHLÆUS KOWALEVSKIANUS Cope.

American Naturalist, 1882, p. 832. (September 28.)
Plate XXV *f*; figs. 6–7.

This animal is represented in my collection by parts of the crania of two individuals. One of these includes a right maxillary bone with bases of six molars, and crowns of four, the last being absent; and a left mandibular ramus with bases of six molars, and crowns of four, the Nos. i, ii, v, and vi. The second specimen includes a right maxillary with the posterior four molars preserved, and the right ramus with the true molars i and ii preserved.

In size this species yields to the *Anisonchus coniferus* and the *Haploconus entoconus*, but is larger than the *A. sectorius*. The third and fourth superior premolars are of about equal size, and larger than the true molars, but of much more simple construction. They have anterior and posterior cingula (weak on the third), but none on the external or internal bases. The bases of the superior true molars are wider than long, but are not so much extended inwards as in the *Anisonchus coniferus*. Their external cones have short acute apices with circular section. They are partly confluent at the base. The ridges which form the limbs of the V originate, one at the internal base of each, and unite at a more acute angle than in *Anisonchus sectorius* or *Haploconus lineatus*. There is an external cingulum, weak on the last superior molar, and an anterior and a posterior one, but no internal one. The anterior and posterior cingula are equally developed, and they rise with a curve to the edge of the V, reaching it a little exterior to the apex. The last superior true molar is smaller than the others, but has similar proportions.

The crown of the first inferior premolar is much smaller than that of the second, and has a lenticular section. That of the second is large and robust, and has a stout lenticular section. The first and second true molars have a small but distinct anterior median tubercle, and a rather larger median posterior tubercle. The sections of the lateral posterior tubercles are lenticular. The anterior pair soon become confluent on wearing, as their bases are closer together than those of the posterior pair. No cingula.

The enamel is smooth in all the specimens.

In the second individual the malar bone is deep and flat on its external face, and a slight convexity of its superior border indicates the posterior edge of the orbit. The posterior edge of the last superior molar is opposite the middle of this enlargement, while the notch between the posterior border of the maxillary bone and the malar enters a little in advance of this point.

Measurements.

No. 1.

		M.
Length of maxillary bone posterior to canine, to M. iii exclusive		033
Length of bases of molars ii, iii, iv, and v		.020
Diameters of Pm. iii	anteroposterior	.005
	transverse	.007
Diameters of Pm. iv	anteroposterior	.005
	transverse	.007
Length of bases of anterior six inferior molars		.027

		M.
Length of four premolars		.017
Length of base of crown of Pm. ii		.0045
Diameters of inferior M. ii	anteroposterior	.0046
	transverse	.004

No. 2.

Length of superior true molars		.013
Diameters M. ii	anteroposterior	.0045
	transverse	.007
Diameters M. iii	anteroposterior	.004
	transverse	.006
Diameters Pm. iv	anteroposterior	.005
	transverse	.007

This species was found by Mr. D. Baldwin in the lowest beds of the Puerco formation in Northwestern New Mexico. I dedicate it to Dr. Waldemar Kowalevsky of Moscow, one of the most able of the European paleontologists.

HEMITHLÆUS OPISTHACUS Cope.

Proced. Amer. Philos. Soc. 1882, p. 467. *Mioclaenus opisthacus* Cope, American Naturalist 1882, p. 833 (September 20).

Plate XXV*f*; figs. 8–9.

This species is known from fragments of four mandibles, and a broken last superior molar tooth, found together by Mr. Baldwin. The mandibles belong to one species, and there is nothing to cast doubt on the reference of the superior tooth to the same. This tooth refers the species to the genus *Hemithlæus*.

This species resembles its congener in the abrupt diminution in size of the premolars anterior to the third. The contrast between the second and third is greater than in the *H. kowalevskianus*. The anteroposterior extent of the former is little more than half that of the latter. The third and fourth premolars are large and oval in horizontal section, and quite similar in size and form. The heel of each is small, and has a median elevation, and that of the fourth is a little the larger. The fourth premolar is not larger than the first true molar. The true molars become narrower posteriorly, and the first is as large as or larger than the second. The anterior two cusps of the molars are more elevated than the posterior; they soon unite on attrition. There is no interior median cusp, but a narrow ledge in front of the anterior cusps, on the second and third molars; the first is worn on all the specimens. The posterior two cusps are well developed, and soon connect on attrition, and the external sends a long point forwards and outwards. The third molar has an elongate heel, and internal and external median cusps, of which the latter is crescentic in section. The ramus is compressed, and becomes shallow anteriorly.

The superior molar has the characters of those of *H. kowalevskianus*, the external wall being lost.

Measurements.

		M.
Length of posterior six inferior molars (No. 1)		.036
Length of base of Pm. ii (No. 2)		.0036
Diameters Pm. iii	anteroposterior (No. 1)	.007
	transverse (No. 1)	.004
Diameters Pm. iv	anteroposterior (No. 3)	.0068
	transverse (No. 3)	.0045
Diameters M. ii	anteroposterior (No. 1)	.006
	transverse (No. 1)	.005
Diameters M. iii	anteroposterior (No. 1)	.0066
	transverse (No. 1)	.004
Depth of ramus at Pm. iii (No. 1)		.008
Depth of ramus at M. ii (No. 1)		.011

Besides the difference in the relations of the inferior premolars above cited, this species differs from the *H. kowalevskianus* in its superior size.

The jaws described were found by Mr. D. Baldwin in the Puerco beds on the Rio San Juan, New Mexico.

ANISONCHUS Cope.

Paleontological Bulletin, No. 33, p. 488, Sept. 30, 1881. Proceedings American Philosophical Society, 1881, p. 488.

This genus is only known from dental characters, and only the molar teeth have been preserved. These are Pm. $\frac{?4}{4}$; M. $\frac{3}{3}$. The first superior premolar is unknown. The third and fourth consist of an external conic cusp, and an internal elevated crest or lobe, as in *Periptychus*. Molars supporting two external tubercles, an internal V, and a posterior internal cusp cut off from the internal V. The limbs of the V represent the intermediate tubercles of *Periptychus* and other genera, and the apex of the V represents the internal median tubercle of that genus. The posterior internal cusp is separated from it by a vertical groove on the inner face of the crown, and is continuous with a posterior cingulum of the crown.

The inferior premolars have no internal cusp or crest, but the P-m. iii and iv have a heel. The true molars have anterior and posterior intermediate tubercles. The last true molar has a heel. The premolars in both jaws are large.

The difference between this genus and *Periptychus* is well marked in the superior true molars; the two genera are otherwise much alike. The inferior premolars of *Anisonchus* differ in the absence of the internal cusp.

There are three well-marked species of this genus, which differ as follows:

Internal lobes of superior premolars conical; width of base of second true molar, .010; large .... *A. coniferus.*
Internal lobes of third superior premolar conic; of fourth, flattened and concentric; width of M. ii, .007; length, .004; small .... *A. gillianus.*
Internal lobes of the superior premolars flattened and continuous with cingula; width of M. ii, .006; length, .0046; medium .... *A. sectorius.*

The type, and first discovered species, is the *A. sectorius*, which is also the most abundantly represented in my collection.

ANISONCHUS CONIFERUS Cope.

American Naturalist, 1882, p. — (Sept. 28).
Plate XXIV*g*; fig. 6.

This is the largest species of the genus, and the largest of the *Periptychidæ* after the species of *Periptychus*. My knowledge of it is based on fragments of two skulls which exhibit the superior molars following the P-m. i, and the crowns of the inferior P-m. ii, and M. ii and iii. A calcaneum of appropriate size accompanies the jaw-fragments.

The second superior premolar is a robust tooth with a subtriangular base, and simple conical external cusp, and a small internal basal cusp. A trace of anterior, but no other cingulum. The third premolar is more robust, and probably has a larger internal cusp, but this is worn off by mastication. There are faint anterior and posterior cingula not connected with this cusp; none externally or internally. The fourth superior premolar has a much greater transverse extent. Its external cusp is simple, and without accessories, while the internal is large and is connected by its edge with a well-marked anterior cingulum. A weak posterior cingulum; none external. Crown worn. The true molars are distinguished by their transverse extent. The accessory cusp, posterior in *A. sectorius* and in *Haploconus lineatus*, is nearly median on the second and third true molars, though continuous with the posterior cingulum. Median V shortened; external cusps conic. True molars with an external, as well as anterior and posterior, cingula. No internal cingulum. Enamel smooth.

Fragments of the mandibles show that the masseteric foramen is not impressed anteriorly, but the anterior ascending coronoid ridge is strong. The first inferior premolar had a single robust root. The second is two-rooted. Its crown is robust; and has a small transverse rudiment of a heel. The second true molar is rather abbreviated. It has two elevated cusps anteriorly, which are confluent at the base, and are supplemented by a small tubercle in the middle of their anterior base, between them. The posterior part of the crown supports three cusps, of which the internal and posterior are small, and the external is an angle from which a ridge extends anteriorly and inwards. The posterior inferior molar is longer, having a heel. The anterior cusps are confluent at the base, and there is a small basal tubercle in front of the inner lobe. There is a narrow cingulum at the internal base of the internal anterior cusp, but at no other part of the crown.

The infraorbital foramen is above the middle of the second superior premolar. There is a mental foramen below the first inferior premolar.

Measurements.

No. 1.

		M.
Length of bases from superior Pm. iii to M. ii, inclusive		.023
Diameters Pm. iii	anteroposterior	.0065
	transverse	.0075
Diameters Pm. iv	anteroposterior	.0065
	transverse	.0095
Diameters M. i	anteroposterior	.0045
	transverse	.0090
Depth of ramus mandibuli at M. ii		.0130

No. 2.

		M.
Diameters Pm. ii	anteroposterior	.0050
	transverse	.0060
Diameters M. iii	anteroposterior	.0037
	transvarse	.0076
Length of base of inferior Pm. ii		.0050
Depth of ramus at inferior Pm. ii		.0100
Diameters inferior M. ii	anteroposterior	.0045
	transverse	.0045
Length of base iii		

Both of the specimens of this species were found in the lowest Puerco beds of New Mexico by Mr. D. Baldwin.

ANISONCHUS GILLIANUS Cope.

Proceed. Amer. Philos. Soc., 1882, p. 467; *Haploconusgillianus* Cope, American Naturalist, 1882, p. 686 Plate XXV *f*; figs. 10–11.

This animal is the smallest of the family of the *Periptychidæ* which is yet known. There are parts of five individuals in my collection which include the dentition of both jaws exclusive of the most anterior teeth. Two of these consist of fragments of the lower jaw only. Two others include parts of both jaws, and one includes only the right maxillary bone with teeth.

The typical specimen displays the second and third superior premolars and first two true molars, with the second and fourth inferior premolars, and last two true molars. The fourth superior premolar is lost from this specimen, but I exposed it in a second one, after removing the decidious tooth which preceded it. The second superior premolar has but two roots, the anterior and interior being fused. The section of the base of the crown is a spherical triangle with the apex anterior. It has a low cingulum except at the external base. The apex of the crown is compressed so as be a fore and aft edge. The third superior premolar is similar as to its external cusp, which is larger than that of the second. The internal cusp is three-quarters the height of the external, and the apex is compressed so as to be antero-posterior. The diameter of its base is only half that of the external cusp. There is a cingulum which is weak externally and wanting internally. The second true molar has greater transverse extent than the first or fourth. The external cusps are slightly convex externally. The V is not produced inwards, while the posterior internal cusp does stand well inwards, its base being especially prominent. It is posterior to the middle line, though entirely interior to the apex of the V. Besides the posterior cingulum, the strong anterior cingulum reaches nearly to its anterior base. The third true molar is equal to the first.

In the inferior series the second premolar has a small anterior basal tubercle, a robust acute median cusp, and a small heel whose outline is a circular ridge. The fourth premolar is like the second, but all the parts are larger. The true molars have the crown contracted upwards from the base. The cusps are rather elevated, the anterior the most so. The anterior median is nearly as high as the anterior inner, and like it, is connected with the anterior external by a thin ridge, the two forming a V. The posterior cusps are connected by a continuous edge, in which the posterior median is barely distinguishable. On the last molar, however, it forms a large and prominent fifth lobe. The cingula are very rudimental. Enamel smooth.

In the second specimen the crown of the fourth premolar was exposed after removing the temporary molar from above it. The internal crest is like that of the corresponding tooth in the *A. sectorius*, continuing into a wide cingular ledge at the anterior base of the external cusp. The superior true molars of this specimen have the general form of those of the type, and differ from those of the *A. sectorius*. They have, however, a stronger external cingulum than those of the latter. This character is seen to be still more marked in the third cranial specimen, where the cingulum is more produced at the interior and posterior angles of the crown. In the specimen the external faces of the external cusps are quite flat, and on the second true molar their apices are inclined inwards. Specimen No. 2 is intermediate between 1 and 3 in this respect. It is quite possible that Nos. 2 and 3 belong to a species distinct from No. 1, and if this is the case they must be regarded as the types of *Anisonchus gillianus*, as the first description of the species was first principally drawn from them. The inferior first true molar of No. 2 differs from that of No. 1 in the smaller size of the anterior inner and anterior median cusps, and in the larger size of the posterior median. Another inferior molar is intermediate between the two.

The temporary fourth premolar of both jaws is preserved in No. 2, and of the superior series in No. 3. That of the superior series resembles a true molar more than it does its successor. In fact, it is identical with the true molars, excepting in the smaller size of the internal cusp, which is only a little posterior. The external cusps differ from each other; the anterior is erect; the inner is a little inclined inwards, forming a V. Its appearance reminds one of the permanent first true molar in *Coryphodon*. The last inferior milk molar only differs from a true molar in the smaller size of the anterior inner and anterior median cusps, and the greater anterior prolongation of the anterior lobe. It differs from its successor much as that of *Triïsodon quivirensis* does its successor, and resembles the permanent fourth premolar of such types (*Didelphys*, e. g.) in which that tooth resembles the true molars. This specimen shows the anterior cusp of the third and fourth permanent premolars.

Measurements.

No. 1.

		M.
Length of last six superior molars (Pm. iv estimated)		.023
Length of true molars		.0112
Width of base of Pm. ii		.0038
Diameters Pm. iii	anteroposterior	.0048
	transverse	.0052
Diameters M. ii	anteroposterior	.0040
	transverse	.0065
Diameters base of M. iii	anteroposterior	.0036
	transverse	.0072
Length of last four inferior molars (M. i estimated)		.0175
Length of base Pm. ii		.0040
Diameters Pm. iv	anteroposterior	.0050
	transverse	.0038
Diameters M. ii	anteroposterior	.0047
	transverse	.0034
Diameters M. iii	anteroposterior	.0050
	transverse	.0035
Depth ramus at M. ii		.0090

No. 2.

Diameters superior deciduous M. iv	anteroposterior	.0040
	transverse	.0040
Diameters permanent M. i	anteroposterior	.0040
	transverse	.0060
Length base inferior deciduous M. iv		.0041
Length base inferior permanent M. i		.0044

This species is dedicated to my friend, the distinguished zoölogist, Prof. Theodore Gill, of Washington. Its horizon is the lower Puerco beds, where it accompanies the *Haploconus entoconus*. D. Baldwin.

ANISONCHUS SECTORIUS Cope.

Mioclænus sectorius Cope, American Naturalist, 1881, p. 831 (September 22). *Anisonchus sectorius* Cope, Paleontological Bulletin No. 33, p. 488, September, 1881. Proceed. Amer. Philos. Soc., 1881, 488.

Plate XXV *c;* figs. 5, 6, and 8.

This species is known from the maxillary bones with teeth of five individuals; two accompanied by mandibles with teeth, and by a number of separate mandibular rami. The third and fourth superior premolars cover a larger base than either of the true molars. The external cusp has a base extended anteroposteriorly, but the apex is conical, and there are no basal tubercles. The inner cusp has a crescentic base, as in *Periptychus*, but the apex is narrowed and compressed conic. The external tubercles of the true molars are subconic, and do not develop any external ridges. They are connected by the crescentic slightly angular crest, or V, whose apex forms the inner anterior boundary of the crown. This crest is not divided into parts homologous with the intermediate tubercles. The crowns of the M. i, ii, and iii are surrounded by a basal cingulum, which in the M. i develops a tubercle at the anterior external angle. The posterior inner lobe is more posterior in this species than in any of the others, and has a V-shaped apex. It projects further inwards than the anterior inner lobe. It is represented by a mere tubercle of the cingulum in *Mioclænus*. No internal or external cingulum on Pm. iv. Enamel nearly smooth.

The ramus of the mandible is rather slender anteriorly. The Pm. iv is robust, and the cusp is above the middle of the base of the crown. Median anterior tubercle small, but distinct. The heel is short and narrow, and has a raised border, connected with the base of the main cusp. The cusps of the true molars are elevated and conic, the anterior the highest, and others subequal. The base of the posterior pair is a little narrower than that of the anterior pair. There is no central tubercle as in *Periptychus rhabdodon*, and no basal cingulum on any tooth.

Measurements.

		M.
Length of three superior molars		.0160
Diameters superior Pm. iv	anteroposterior	.0055
	transverse	.0070
Diameters superior M. i	anteroposterior	.0052
	transverse	.0060
Length of inferior molar series		.0610
Length of inferior true molar series		.0160
Diameters inferior Pm. iv	anteroposterior	.0060
	transverse	.0040
Diameters inferior M. ii	anteroposterior	.0050
	transverse	.0040
Depth ramus at M. ii		.0090

A number of minor points will distinguish this species from those included among the *Mesodonta*, and also from those of *Pantolestes*, which it much resembles. The molar teeth are narrower behind, and the fourth premolar is larger.

From the Puerco beds of Northwestern New Mexico. Discovered by D. Baldwin.

HAPLOCONUS Cope.

American Naturalist, 1882, p. 417 (April 25).

Dental formula: I. ?; C. $\frac{1}{1}$; P-m. $\frac{4}{4}$; M. $\frac{3}{3}$; no diastemata. Canines well developed; superior first premolar one-rooted; third superior premolar a flattened cone, without accessory crest or cusp; fourth superior premolar with crown of an external cusp, and an internal crest, as in the genera just preceding. Superior true molars with crowns consisting of two external cusps, a median V directed inwards, and a distinct posterior internal cusp. No other cusps of superior molars. Inferior premolars without internal tubercle or cusp; the first one-rooted. True molars with four tubercles in pairs, and a posterior median tubercle. Anterior median tubercle present or absent. Third inferior true molar with a heel. Angle of mandible not inflected or reflected.

The skeleton of this genus is unknown. The true molars of the superior series are those of *Anisonchus*, but the third premolar is entirely different. Some of the species are distinguished by the absence of the anterior median tubercles of the inferior true molars, while others possess it. I cannot divide them into two genera on this account, as the tubercle in question is sometimes very small, and in some cases indistinguishable. Besides this, the species differ like those of *Anisonchus*, in the form of the internal tubercle of the superior fourth premolar. In the one group it is a cone; in the other a crest, or strong cingulum rising into a cone. This will also probably prove to be an evanescent character in some species not yet discovered. An intermediate form is seen in the fourth superior premolar of the *Anisonchus coniferus;* and the two forms are displayed in a less typical condition in the two species of *Protogonia*.

The four species are all known from their mandibular rami, and the superior molar series of two of them is known. The distinctive characters of the former are as follows:

I. No anterior median tubercle of inferior true molars.

a. Fourth inferior premolar robust.

Length of last four molars M. 018; of true molars ii and iii, .0082........*H. angustus.*

Length of last four molars, .022; of true molars ii and iii, .010; third and fourth premolars equal..*H. lineatus.*

aa. Fourth inferior premolar compressed and sharp.

Length of last four molars, .020; third premolar shorter than fourth......*H. xiphodon.*

II. Anterior median tubercle present.

Anterior median cusp smaller; last four molars, .024; true molars ii and iii, .0115; largest...*H. entoconus.*

The characteristics of the superior molars of two of the species are as follows:

I. Internal lobe of fourth superior premolar crest-like and concentric.

True molars less transverse; length .012; premolars striate ....*H. lineatus.*

II. Internal lobes of fourth premolar conic.

True molars more transverse; length, .0145; premolars smooth ....*H. entoconus.*

All the species are from the Puerco Eocene.

Haploconus angustus Cope.

Mioclænus angustus Cope. American Naturalist, 1881, 831, Sept. 22. Paleontological Bulletin No. 33, p. 491.

Plate LVII*f*; fig. 6.

The least species of the genus, with the teeth about the size of *Hyopsodus paulus* Leidy, but with more robust jaw. The molar teeth diminish in size regularly posteriorly from the P-m. iv. They all have three subequal posterior cusps which are less elevated than the anterior ones. The median is enlarged into a heel on the last tooth. The anterior are opposite, and the external is larger than the internal. There is no anterior internal. The external wears into an anteroposterior narrow grinding surface, which looks like a combination with an anterior median. The latter is, however, not separate on the least worn molars. The anterior outer cusp increases in size anteriorly, and is the large cusp of the P-m. iv. It sends a branch backwards on the inner side of the crown which forms the edge of the narrow concave heel. There are no cingula except a short one on the anterior corners of the base of the crown of the P-m. iv. Enamel obscurely wrinkled.

Measurements.

		M.
Length of posterior four molars		.0180
Diameters of P-m. iv	anteroposterior	.0050
	transverse	.0035
Diameters of M. i	anteroposterior	.0050
	transverse	.0035
Diameters of M. ii	anteroposterior	.0040
	transverse	.0032
Diameters of M. iii	anteroposterior	.0045
	transverse	.0030
Depth of ramus at M. i		.0110
Thickness of ramus at M. i		.0060

From the Puerco beds of Northwestern New Mexico, one specimen found by D. Baldwin.

Hoploconus lineatus Cope.

American Naturalist 1882 (May), p.

Plate XXV *e*; figs. 1–4.

This species appears to have been more abundant than the other species, Mr. Baldwin having sent me parts of fifteen individuals. Two of these include parts of crania, the best preserved lacking the posterior half of the brain-case and the extremity of the premaxillary only. It is somewhat damaged anterior to the orbits, so that some of the foramina cannot be seen.

In this species the predominant size of the premolars, seen in all the species of this group, is restricted to the fourth in the superior series, and the third and fourth below. Alveolæ of two of the superior incisors are preserved. The canines have a vertical position. The second premolar is two-rooted. The third premolar has the posterior root wider than long, but not divided into two. The fourth premolar has three roots. The crown of the third is compressed so that the apex wears into a narrow anteroposterior oval. The internal crest of the fourth premolar is large, and is extensively visible back of the external cusp when viewed from the side. The external cusps of the true molars are somewhat flattened, the posterior more so than the anterior. The internal V is more open, and thus less prominent inwards than in the *Anisonchus sectorius.* The posterior internal cusp is more prominent and acute than in that species, and it is continuous with a posterior cingulum which extends to the base of the posterior external cusp. An anterior cingulum which only extends to the base of the angle of the V. A weak external cingulum on the second and third true molars; a slight one on the antero-external part of the P-m. iv, and a weak one on the anterior half of the P-m. iii. Enamel smooth on the true molars; weakly striate on the external side of the P-m. iii and iv, and more strongly striate on the inner side of the P-m. iv.

The mandibular ramus is compressed and rather deep. The condyle is well above the level of the crowns of the molars, and its articular face on its inner half (external unknown) looks upwards. The coronoid process is large, and rises close anterior to the condyle, and is in the plane of the angle. The latter is mostly in a vertical plane. Its inferior border is decurved, and then recurved to an apex, which projects a little posterior to the line of the condyle, and is a little incurved. The masseteric fossa is defined anteriorly to the middle of the depth of the horizontal ramus, but below that point definition is wanting.

The inferior canine is rather abruptly recurved. There is no diastema behind it. The crown of the second premolar is rather elevated, is compressed, has a posterior acute heel, and a vertical anterior outline without basal tubercle. The third premolar has the same characters, except that the heel is truncate behind, and has an inclosed fissure above. The fourth premolar is more robust, and has a wider heel whose basin is open on the external side. There is no anterior basal tubercle, but a low cingulum extending from the inner base of the principal cusp. As in the other premolars this cusp is compressed, and it has a free postero-external edge. The second true molar is a little smaller than the first or third. The cusps of the anterior pair are closely connected, and the anterior ridge of the internal one descends to the base, and forms an anterior cingulum which turns back to the anterior external base of the external cusp. The internal anterior cusp also has a sharp posterior edge which joins a corresponding edge of the external posterior cusp. All of the cusps are rather acute. A cingulum descends from the posterior median cusp and extends to the posterior base of the anterior external cusp. It is most prominent on the last molar, where the posterior cusp forms a prominent heel. The edge-like compression of the anterior inner cusp is most marked on the first true molar, and least marked on the third.

The portion of skull preserved shows a moderately elongate muzzle, somewhat compressed between the canine alveoli and the bases of the malar bones. The orbits are rather small and are open posteriorly. There is an obtuse postorbital angle of the frontal bone, and sharp temporal angles originate from these and converge posteriorly, and probably form a sagittal crest; but this part is broken off. The infraorbital foramen is above the posterior part of the third superior premolar. The frontal bone is convex downwards to the postfrontal angles, but is concave medially.

Measurements.

No. 1.

		M.
Length of posterior five superior molars		.0226
Diameters third premolar	anteroposterior	.0045
	transverse	.0040
Diameters fourth premolar	anteroposterior	.0050
	transverse	.0055
Diameters first true molar	anteroposterior	.0045
	transverse	.0058
Diameters third true molar	anteroposterior	.0035
	transverse	.0056

No. 2.

		M.
Length of posterior five superior molars		.0220
Length from canine to M. iii inclusive		.0340
Long diameter base of crown superior canine		.0050
Width between postorbital angles		.0260
Length base of inferior P-m. ii		.0045
Length base of inferior P-m. iii		.0070
Width base of inferior P-m. iv		.0040
Diameters M. i	anteroposterior	.0050
	transverse	.0030
Diameters M. ii	anteroposterior	.0050
	transverse	.0038
Length M. iii		.0052
Depth ramus at P-m. iii		.0100
Depth ramus at M iii		.0145

No. 3.

	M.
Length of ramus from canine to edge of angle	.0760
Depth ramus at M. iii	.0155
Elevation of condyle above base of ramus	.0300
Anteroposterior width base of coronoid	.0220

The compression and enlargement of the external anterior cusp of the inferior true molars is a peculiar feature of this species. It is perhaps homologous with the crest which extends from this cusp to the anterior median in the *Anisonchus gillianus*. It suggests a more or less carnivorous habit for the species.

From the upper Puerco beds of Northwestern New Mexico.

This animal is a little larger than the *Anisonchus sectorius*, and about equal in the size of its skull to the gray fox.

Haploconus xiphodon Cope.

Proceed. Amer. Philos. Soc., 1882, p. 466.
Plate XXV e; figs. 5–6.

This species is represented by a mandibular ramus and perhaps by three rami. The one on which the species rests contains five molars, the middle one of the series broken, so that its form cannot be positively ascertained. It is probable that it is the first true molar, so that the animal exhibits the last true molar not entirely protruded, and is therefore nearly adult. But there are some reasons for suspecting the animal to be young. Thus the last inferior molar does not exhibit more of a heel than the second usually does, and the supposed third premolar is smaller than that tooth is in the other species, having nearly the proportions of the second premolar. The teeth present may then be supposed to be the molars from the second to the sixth inclusive. But opposed to this view is the fact that the supposed third premolar has more the structure of that tooth in details than that of the second, and the specimens accompanying, which have the temporary dentition apparently of the same species, present premolar teeth of a very different character. In any case the present specimen represents a third species of the genus, and I describe it at present as an adult.

The third premolar has a simple, compressed crown, about as high as the length of its base, and without anterior basal tubercle. It has a narrow triangular posterior face which is concave, and truncated by a cingulum below; no heel proper, nor lateral cingula. The fourth premolar is an elongate tooth consisting of a compressed principal median lobe, an anterior lobe connate with it, and a heel. The latter has elevated posterior and interior borders. A rudiment of an exterior border is seen in a narrow ridge on the external side of the posterior face of the principal lobe of the tooth. The sides of the premolars present rather distant ridges, as in *Periptychus carinidens*. The second true molar has two anterior and three posterior tubercles; the latter close together, pointed, and of about equal size. Of the anterior tubercles the external is much the larger and more elevated. It is compressed, and has a curved subacute anterior edge, which extends much in front of the internal tubercle. There is no anterior inner tubercle, nor are there any cingula. The enamel of the sides of the crown presents a few vertical ridges. The last inferior molar only differs from the second in the greater size of the median posterior lobe, which is nevertheless smaller than in the two other species of *Haploconus*.

There is a mental foramen below the posterior edge of the alveolus of the second inferior premolar.

Measurements.

	M.
Length of last five inferior molars	.0250
Length of third premolar	.0050
Length of fourth premolar	.0066
Length of second true molar	.0050
Width of second true molar	.0032
Length of third true molar	.0050
Depth of ramus at Pm. iii	.0095
Depth of ramus at M. iii	.0130

The two rami with the temporary premolars exhibit the last true molar. inclosed in the jaw. The third and fourth premolars are much like the fourth premolar of the specimen above described, but the fourth is a little more robust than that of the latter, which is very much like the third of the deciduous series. The space occupied by the supposed first premolar of the type specimen is too short for the fourth premolar of the deciduous series, otherwise it might be supposed to have occupied that position. The two true molars resemble those of the type, excepting that the last one does not extend so far into the base of the coronoid process, and its posterior lobe is smaller, in accordance with its position as No. two in the series.

The specimens were procured by Mr. D. Baldwin in the Puerco beds of New Mexico.

HAPLOCONUS ENTOCONUS Cope.

American Naturalist, 1882, p. 686.

Plate XXV*f*; figs. 4-5.

This is the largest species of *Haploconus*, and represents a group in it slightly different from that which is typical. Mr. Baldwin has sent me fragments of the skulls and jaws of seven individuals, so that its molar dentition is entirely known.

The premolars are more robust than the molars in this species. The third of the superior series has a subtriangular base, with broadly rounded angles. The crown is absolutely simple, the apex having a nearly round section. There are traces of anterior and posterior cingula. The base of the fourth superior premolar is more extended transversely, to sustain the conical internal cusp. The external cusp is but little longer than wide. Traces of anterior and posterior cingula, which are not connected with the internal lobe. The external cusps of the superior true molars are not flattened on the external side, and have a subround section. The internal V is not very prominent inwards, and is uninterrupted. The internal cusp rising from the posterior cingulum is large, and is a little posterior to median in position, and its base projects well inwards. The last molar is a little smaller than the others, and the external face is oblique. All the true molars have an external cingulum, which extends into the anterior limb of the V in front, and posteriorly rises to the posterior external cusp. The posterior limb of the V descends to a posterior cingulum. The internal cusp descends to both anterior and posterior cingula, of which the latter is the stronger.

The alveolus of the inferior canine indicates a large tooth directed at an angle of 45° forwards. The first premolar has a single stout root. The second premolar is two-rooted, and has a subconic crown with a rudimental heel behind. The third and fourth premolars are similar to it, differing in their increasing size, and the transverse extent of the small heels. None of them have anterior tubercle or cingulum, as in the other species. In all of them the heel of the fourth inferior premolar tooth is longer than in *H. entoconus.*

The inferior true molars are of robust form, subequal in size, and smaller than the last two premolars. They have three posterior, three anterior, and no central tubercles, and of these the median posterior and anterior are the smallest, except on the third molar, where the posterior forms the robust heel. The lateral posterior are the apices of externally directed angles of the summit of the crown, and are less elevated than the principal anterior. These are opposite and are fused at the base. The external is more elevated than the internal. The true molars all have a trace of anterior cingulum, and a trace externally between the cusps. The only internal cingulum surrounds the base of the anterior tubercle, on the second and third true molars. The enamel of all the teeth is smooth, except a faint striation on the superior premolars.

The symphysis mandihuli extends posteriorly to below the anterior margin of the third premolar. The inferior border of the ramus is greatly convex. The border of the masseteric fossa is only distinct on the anterior edge of the coronoid process. There is a small meatal foramen below the first premolar. The infraorbital foramen is above the posterior part of the third premolar. The malar bone does not extend anterior to the base of the zygomatic process of the maxillary, excepting in a narrow prolongation forming the external edge of the inferior border of the orbit. This prolongation rises on the anterior border of the orbit to a point considerably above the line of the infraorbital foramen.

Measurements.

No. 1 (*superior molars*).

	M.
Length of bases of last five superior molars	.026
Length of superior true molars	.014
Diameters of P-m. iii { anteroposterior	.0065
Diameters of P-m. iii { transverse	.007
Diameters of P-m. iv { anteroposterior	.0055
Diameters of P-m. iv { transverse	.008
Diameters M. i { anteroposterior	.005
Diameters M. i { transverse	.0085
Diameters M. iii { anteropostorior	.0035
Diameters M. iii { transverse	.008

No. 2 (*both jaws*).

	M.
Length of last five superior molars	.0275
Length of superior true molars	.014
Length of last five inferior molars	.029
Length of inferior true molars	.0155
Diameters M. i { anteroposterior	.0055
Diameters M. i { transverse	.045
Measurements M. iii { anteroposterior	.0055
Measurements M. iii { transverse	.0038
Depth of ramus at M. ii	.0146

No 3 (*Inferior Molars*).

	M.
Length of inferior molar series	.0410
Length of inferior true molars	.0175
Length of base of P-m. ii	.0054
Length of base of P-m. iii	.0068
Width of base of P-m. iii	.0050
Length of base of P-m. iv	.0068
Width of base of P-m. iv	.0050

This species has only been found as yet in the lower part of the Puerco formation, by Mr. Baldwin; as the *H. lineatus* has been only obtained from the upper Puerco.

PROTOGONIA Cope.

Paleontological Bulletin, No. 33, p. 492, Sept. 30, 1881. Proceed. Amer. Philos. Soc., 1881, p. 492.

Fourth superior premolar with one external and one internal lobe. True molars with two external, two internal, and two intermediate lobes, both the latter connected with the anterior internal by a ridge. Supposed inferior true molars with two V's with weak anterior branches; last true molar with heel. Fourth inferior premolar with internal cusp.

In the superior true molars the anterior transverse crest of the *Hyracotherium* is represented, but not the posterior. This is replaced by a low ridge running across the course it pursues in *Hyracotherium*. The posterior median tubercle is also not found in the latter genus. *Protogonia* differs from *Limnohyus* in the subconic character of the external lobes of the superior molars. If the tubercles, excepting the posterior inner, should be converted into crescents, the genus *Meniscotherium* would be produced. It probably represents the ancestral type of the *Meniscotheriidæ*. The simple premolars give it a position nearer the *Periptychidæ* than that occupied by *Phenacodus*. An approach to it is made by the *Phenacodus puercensis* of the same geological horizon. Two species of *Protogonia* are known to me.

PROTOGONIA PLICIFERA Cope.

American Naturalist, 1882, p. 833 (Sept. 28).
Plate XXV *f*; figs. 2–3.

This species, although not the first described, is most expressive of the characters of the genus. The external lobe of the fourth superior premolar is absolutely simple, as in most Artiodactyla, while in the *P. subquadrata* there is a minute rudiment of the second or posterior external cusp, which is well developed in *Phenacodus*. A single individual is all that is known of the species. This is represented by a maxillary bone, which contains the fourth premolar, and anterior two true molars, and the accompanying mandibular ramus, which supports the three corresponding inferior teeth.

This species is smaller than the *P. subquadrata*, and differs from it especially in the form of the internal cusp of the fourth superior true molar. In this species it is the apex of a V, whose limbs form the anterior and posterior cingula. In *P. subquadrata* it is a simple cone, disconnected from the cingula. Between it and the internal side of the base of the external cusp is a rudiment of the anterior intermediate tubercle. No posterior intermediate tubercle. The external face of the external lobe of the crown is flat. The anterior and posterior cingula terminate externally at the angles of the crown, the anterior rising into a prominent angle. Internally they extend to near the internal border of the crown, but do not pass round it The external cusps of the true molars are conical, as are the smaller intermediate tubercles. The anterior internal tubercle is the apex of a V, and is larger than the posterior internal. The latter is well developed, and is a cone. A distinct cingulum extends round the crowns except on the inner side. It rises to an angle at the anterior external corner of the crown.

The first and second inferior true molars support six cusps, three posterior and three anterior. The three posterior are arranged round the posterior raised edge of the crown. The three anterior are two internal and one external. The last is the largest, and the anterior internal is the smallest, and is close to the posterior internal. The latter is a little behind opposite to the anterior external, and is connected with the posterior external by a low oblique ridge, thus completing a V. The three anterior tubercles are connected by ridges, and form on wearing, a narrow transverse U. No cingula. The fourth inferior premolar has a short, wide heel, with reverted edge and a large principal cusp. On the inner posterior side of the latter is the internal cusp, closely connected; and at the anterior base is a prominent tubercle, inside the longitudinal axis of the principal cusp. No cingula. Last true molar unknown.

Measurements.

		M.
Length of superior P-m. iv with M.'s i and ii		.0215
Diameters P-m. iv	anteroposterior	.0060
	transverse	.0070
Diameters of M. ii	anteroposterior	.007
	transverse	.0095
Length of inferior P-m. iv, with M.'s i and ii		.023
Diameters P-m. iv	anteroposterior	.007
	transverse	.0046
Diameters of M. i	anteroposterior	.0075
	transverse	.006

Puerco formation of New Mexico, discovered by D. Baldwin.

PROTOGONIA SUBQUADRATA Cope.

Paleontological Bulletin No. 33, p. 493, 1881. Proc. Amer. Philos. Soc., 1881, p. 492.

Plate LVII*f*; figs. 11-12.

Probably two specimens; one supporting three superior molars; the other including damaged superior molars, and the last two inferior molars. The pertinence of the latter specimen to this species is doubtful. The animal was about the size of the red fox. The external cusp of the fourth superior premolar is flattened externally, and has a small lobe on its posterior edge. The inner tubercle is conic and is separated by a tubercle from the anterior base of the external. True molars without external ridges. The external cusps of the true molars are lenticular in section. The posterior inner cusp is in nearly the same anteroposterior line with the anterior, its section about equaling that of the intermediate cusps. The first and second molars have an external, an anterior and posterior, but no internal, basal cingula. The enamel is somewhat wrinkled where not worn.

The heel of the last inferior true molar is elevated, and its worn surface forms the extended posterior branch of the posterior **V**. The posterior edge of the penultimate molar is elevated and curved forwards on the inner side of the crown. The anterior cusp forming the angle of the **V** of this tooth is higher than the posterior angular cusp, but the anterior limb descends rapidly as in *Coryphodon*. A weak antero-external, and postero-external cingula. Enamel wrinkled where not worn.

Measurements.

No. 1.		M.
Length of bases of three superior molars		.025
Diameters of superior P-m. iv	anteroposterior	.0066
	transverse	.0086
Diameters of superior M. i	anteroposterior	.0085
	transverse	.011
Diameters of superior M. ii	anteroposterior	.009
	transverse	.011

No. 2.		
Length of bases of last two inferior molars		.0225
Diameters of last inferior molar	anteroposterior	.0114
	transverse	.0066

		M.
Diameters of inferior M. ii	anteroposterior	.0112
	transverse	.0080
Depth of ramus at M. ii		.0240
Thickness of ramus at M. ii		.0110

From the Puerco beds of Northwestern New Mexico. Discovered by Dr. Baldwin.

ANACODON Cope.

Paleontological Bulletin, No. 34, p. 181, Feb. 20, 1882. Proceed. Amer. Philos. Soc., 1881, p. 181.

Char. gen. Known only from mandibles supporting molar teeth. Probably family *Phenacodontidæ*. Last inferior molar with heel. Crowns of molars without distinct cusps, but with a superior surface consisting of two low transverse ridges separated by a shallow valley. Unworn grinding surface with shallow wrinkles. Perhaps only three premolars.

The only species known is from the Wasatch epoch.

ANACODON URSIDENS Cope.

Loc. sup. cit., p. 182.

Plate XXV *e*; fig. 11.

Broken mandibular rami of two individuals constitute the basis of my knowledge of this species. It is of the size of the *Phenacodus primævus*. The last inferior molar is wedge-shaped with the very obtuse apex posterior. It displays two slight transverse elevations anteriorly, which represent the usual cusps. Grinding surface generally nearly flat. The posterior half of the crown of the penultimate molar is flat, and is separated from the anterior half by a transverse groove. Its surface is marked by shallow branching grooves.

The molar preceding this one in the broken specimen is probably the first. It is possible from its slightly worn condition that it is the fourth premolar, but the form is that of a true molar. The surface of the crown is marked by shallow grooves not very closely placed. The three premolar teeth in advance of this tooth are broken off. Their bases are narrow.

There are no basal cingula on the molars.

Measurements.

		M.
Length of posterior true molars		.033
Diameters of M. iii	anteroposterior	.015
	transverse	.010
Diameters of M. ? i	anteroposterior	.015
	transverse	.011
Depth of ramus at M. ii		.030

The characters of the teeth of this species are something like that of some of the *Palæochæri* of the Miocene, and resemble more those seen in some of the bears. From the Big Horn River; J. L. Wortman.

PHENACODUS Cope.

Paleontological Bulletin, No. 17, p. 3, 1873. Report on Vertebrate Fossils of New Mexico, U. S. Geog. Survs. W. of 100th meridian, 1874, p. 10. Ibidem, p. 122. Report U. S. Geol. Survs. W. of 100th meridian, IV, Part II, p. 173. American Naturalist, 1881, p. 1017. Paleontological Bulletin, No. 34, p. 177, 1881.

The characters of this genus are derived from the typical species, the *P. primævus*, and from the *P. vortmani*, both of which are represented by almost perfect specimens.

The skull is distinguished by the anterior abbreviation of the nasal bones, and the consequent partly superior aspect of the anterior nostrils. Posteriorly these bones are carried farther than usual, extending between the orbits; in *P. primævus* constituting a slightly nearer approach to the living genus *Tapirus* than is usual in other Eocene genera now known. The premaxillaries are slender, and do not unite in front. Posteriorly they reach to the nasals, but not to the frontals. There are no postorbital processes either of the frontal or malar bones. Sagittal crest present. The palate is excavated beyond the posterior border of the last superior molar, and the pyramidal processes of the palatine bone are not separated from the maxillaries by a notch. Postglenoid processes prominent, no trace of preglenoid ridge. Posttympanic process short, widely separated from the postglenoid in front by the meatus auditorius, and from the paroccipital behind by the robust bases of both processes. Petrous bones lost, but probably small.

Frontal bone deeply and widely excavated anteriorly to receive the nasal bones. Parietals extending forwards to their usual position posterior to the orbits. The mastoid bone well exposed between the occipital and squamosal. Alisphenoid joining the parietal. Lachrymal bone larger, extending anteriorly and posteriorly to the anterior border of the orbit.

Incisive and infraorbital foramina present. Lachrymal foramen not exterior to the orbit. An alisphenoid canal, into which the foramen rotundum opens. Foramen ovale separated from the foramen lacerum anterius. A postglenoid foramen. Carotid foramen grooving the side of the basioccipital bone forwards and upwards. The condition of the foramina lacera uncertain, owing to the loss of the petrous bones Foramen condyloideum distinct. Postparietal and supraglenoid foramina present.

The mandibular rami are not coössified at the symphysis. The condyle is elevated, standing a considerable distance above the line of the molar teeth, and its articular surface looks mostly upwards. On the inner half it is decurved so that the posterior half looks posteriorly. The coronoid process is elevated, and the angular region is in the same vertical plane.

The dental formula is: I. $\frac{3}{3}$; C. $\frac{1}{1}$; P-m. $\frac{4}{4}$; M. $\frac{3}{3}$. The incisors have transversely expanded sharp-edged crowns. The canines are not very large. The anterior premolars are a little spaced, but there are no distinct diastemata, except in *P. primævus*, a short one behind the first inferior premolar. The first premolar in both jaws is one-rooted and has a simple crown. The third and fourth superior premolars have internal lobes and two external cusps. The true molars have two internal, two intermediate, and two external tubercles. The external faces of the latter are convex. Anterior cingular cusp rudimental. The third and fourth inferior premolars have an internal median cusp, and the fourth is not like the first true molar. The true molars support four cusps opposite in pairs, and an odd one between and behind the posterior two. This tubercle is developed into a heel on the last molar.

The numbers of the vertebræ are C, 7; D, 14–15; L, 6 or 7; S, 3 or 5. The numbers are certainly known in *P. vortmani*, and also in *P. primævus*, excepting in the case of the dorsals. Of these there are thirteen preserved in the specimen of *P. primævus;* one is probably lost from the front of the series, and one, probably the thirteenth, from the posterior part of the column. The number of the caudal vertebræ is not known, but they were numerous.

The transverse process of the axis is short and flat, and not narrow as in *Tapirus*, and its base is perforated posteriorly by the vertebrarterial canal. The latter issues on its inferior face, and perforates the neural arch transversely just above the superior border of the cotylus for the occipital condyle. The axis has a cylindric odontoid, and atlantal facets well separated. Its neural spine is well developed. A vertebrarterial canal. The articular faces of the succeeding cervicals are oblique and slightly opisthocœlous. The transverse processes are narrow, are directed posteriorly, and have no descending processes except on the sixth centrum. Seventh cervical without vertebrarterial canal. The second dorsal has rather elongate diapophyses without metapophyses; the neural spine is narrow and rather elevated. The metapophyses are quite elevated on the posterior part of the series, and the centra become somewhat opisthocœlous and depressed. No anopaphyses on either dorsals or lumbars. The latter have well-developed flat diapophyses, and the centra are keeled below. The postgygapophyses are not revolute, but they become oblique, so that the external parts of the surfaces are nearly vertical. The centra of the sacrum are rather elongate, and are of equal length. The diapophyses of the first sacral are short and deep, and present no articular facets for the diapophyses of the last lumbar. The iliac articular surface extends over the first and part of the length of the second sacral vertebræ. The rest of the sacrum has united diapophyses continuous with it. The caudal vertebræ were probably numerous, the proximal have strong transverse processes, and the median are robust in the *P. primævus*. Some of them have chevron bones.

In the ribs the capitular and tubercular facets are well developed. The shafts are flattened excepting at the posterior part of the series. The episternum is elongated and compressed, and it has two costal articular facets near its middle. The anterior sternal segments are narrow; the posterior are wider and flattened. No clavicle. The scapula is widened anteriorly as in the ursine *Carnivora*, the *Amblypoda*, and the *Proboscidea*, and has a short coracoid process. The spine rises abruptly from the neck, and turns forward in a strong acromion. The tuberosities of the humerus are well developed as in tapiroids. The condyle has no intertrochlear ridge, but the cylindrical part is swollen, as in *Oreodon*, *Anoplotherium*, and *Mesonyx*. The epitrochlear foramen is a peculiar feature already mentioned. The head of the radius is transversely oval, without inferior interlocking angle. The distal extremity is truncate, and the scaphoid and lunar facets are not distinctly separated. The distal extremity of the ulna is a compressed tuberosity which is produced beyond the radius, to which it is oblique.

The scaphoid is rather larger than the lunar, and has a considerable posterior tuberosity. Its radial facet is transverse and anterior, and the median part of the bone is concave. The inferior face is one large facet for the trapezoides. The facet for the trapezium is small and lateral. The facet for the magnum is small and is entirely lateral. The radial face of the lunar is larger, and the rest of the superior side of the bone slopes downwards and is less concave. The inferior face is occupied in front almost exclusively by the facet for the magnum. A narrow facet for the unciform forms a narrow band along the external part of the inferior face. The cuneiform has considerable transverse extent. Much of its superior face is occupied by a concave facet for the ulna; the external part is contracted, and terminates acutely. The inferior face is nearly flat. The trapezoides is a shallow disciform bone, entirely supporting the scaphoid, and not joining the lunar. The magnum has a resemblance to that of *Coryphodon*. It has an anterior face, a posterior tuberosity, and an elevated smaller superior tuberosity. It is narrow, and the inferior facet is simple and concave. The unciform is a good deal like that of *Tapirus*, but has a small posterior tuberosity. Viewed from before it is subtriangular, the cuneiform face being the hypothenuse. There is no distinct facet for the lunar articulation, as is seen in *Coryphodon* and *Rhinocerus*. It has two inferior facets. The trapezium is lost from the specimens of both species.

The distal keels of the metapodial bones are distinct, but are quite short, and entirely posterior.

The pelvis is of normal proportions. The ilium is not much expanded, but its posterior superior border is thinned out. The anterior border is decurved in front. The peduncle is distinct and rather short, and has a triangular section, the narrower face being anterior and only apparent near the acetabulum. A trace of an anterior inner spine. Posterior edge at acetabulum not expanded. Scissura acetabuli open. The ischia are shorter than the ilia, and the shaft is flat, and the tuber not enlarged. There is an angular spine. The base of the pubis is compressed.

The head and great trochanter of the femur are distinct, and the latter is as elevated as the former. It is excavated by a deep trochanteric fossa. The little and third trochanters are prominent, the latter near the middle of the shaft. The condyles are well separated. The rotular groove is not short or wide, is well excavated, and the internal ridge is the most produced. The tibia is not elongate, and has a large head. The crest is truncate obliquely above. The distal extremity is small and robust, and the malleolus is prominent. The astragalar grooves are not oblique. The fibula is distinct throughout, and is rather stout. Its extremities are not much expanded.

The posterior foot was not plantigrade, but proportioned much as in *Tapirus*, adding a digit on each side to the three possessed by that genus. These digits are arranged in the arc of a circle, so as not to give unusual width to the foot. The third is longer than the second and fourth, and the fifth is longer than the first. The astragalus has much the form of that of a carnivorous mammal. The lateral crests are well developed, and of unequal elevation, the external being the most elongated. The median groove is wide and deep. The neck is distinct, and is rather long, and is bent slightly inwards. The head is transversely oval, and its articular surface is uniformly convex in all directions. The calcaneum is of average length, and has a compressed shaft. It has no horizontal extension at the astragalar condyle, and the sustentaculum is robust and short. The cuboid facet is oblique towards the astragalus, with which it is about equally produced. The navicular is a simple, short, saucer-shaped bone, much like that of the carnivora. The ectocuneïform is robust, and has a prominent posterior tuberosity. The mesocuneïform is shorter, and is transverse. The entocuneïform is longer and narrower than the others. The cuboid is a robust bone, with a large posterior tuberosity. Its proximal extremity has but one articular face, which is convex anteroposteriorly. The phalanges are not shortened, and the ungues are well developed, flat, and obtusely angular.

A cast of the cranial cavity gives the following as the general characters of the brain. The cerebral hemispheres are remarkably small, each one being less by one quarter than the cerebellum. They are separated from the latter and from the large olfactory lobes by strong constrictions. The posterior one is occupied by a thick tentorium. In like manner a wide groove for a robust falx separates the hemispheres above. A notch represents the Sylvian fissure, and the lobus hippocampi is quite large. The vermis of the cerebellum is quite distinct, and the lateral lobes are large. They are impressed laterally by the petrous bones, as in various ruminants. The anterior columns of the medulla are not visible. There are traces of the convolutions on the hemispheres.

I refer nine species to this genus. Of these only two, *P. primævus* and *P. vortmani*, are sufficiently well known to render the generic reference certain. The others are mostly known from teeth, and it is highly probable that most of them belong to the genus. They are all from either the Puerco or Wasatch formations—three from the former and six from the latter.

Species which are known from mandibular teeth may be distinguished as follows:

Large; true molars .041; Pm. iv .014; depth of ramus at M. ii, .027.....*P. primævus*.
Medium; true molars .027; depth at M. ii, .017; last molar smaller .. ..*P. vortmani*.
Like the last, but the superior true molars without median external marginal tubercle. *P. puercensis*.
Length of true molars .0258; some of the premolars with flat expanded heels. *P. calceolatus*.
Smaller; true molars .022; depth at M. ii, .013; last molar elongate..*P. macropternus*.
Smaller; last four molars .027; Pm. iv, .007; depth at M. ii, .013; last molar with short heel..*P. brachypternus*.
Smallest; true molars .017; depth at M. ii, .012; heel long; cusps elevated. *P. zuniensis*.

Besides these, the *P. nuniënus* is founded on some bones without teeth, and the *P. hemiconus* on superior molar teeth only. The *P. sulcatus* of the Report of Capt. G. M. Wheeler, Vol. IV (1877), probably belongs to some other genus.

PHENACODUS NUNIËNUS Cope.

Plate LVII b; figs. 1–2.

The largest species is represented by humeri and parts of a scapulæ of two individuals, and probably by other bones, which were found mingled together with the bones of a *P. primævus* and a *Mesonyx ossifragus* in great confusion. These bones can be compared with those of *P. primævus*, and several differences besides the superior dimensions appear.

The glenoid cavity of the scapula is longer anteroposteriorly and narrower transversely than in *P. primævus*. The proximal part of the spine rises nearly half way between the anterior and posterior borders of the neck; in *P. primævus* it is much nearer the posterior edge of the neck. The head of the humerus is more nearly rounded and less extended obliquely forwards and outwards than in *P. primævus*. The tuberosities are less prominent, and the greater has a more obtuse extremity in the *P. nuniënus*. The subscapularis facet on the inner side of the lesser tuberosity is more distinctly defined, forming a well-marked groove running upwards and forwards. The deltoid crest extends down the shaft, and is represented by an angle as far as the fossa for the head of the radius. The epicondyles are not prominent; the inner is the most so, and is marked by the arterial foramen. The bridge inclosing this is quite narrow. The condyles have a thin-edged flange on each side posteriorly. Anteriorly the external disappears, and the internal becomes beveled so as not to be acute.

Measurements.

	M.
Total length of humerus	.245
Width at proximal extremity	.066
Anteroposterior diameter of head	.044
Width of humerus distally	.065
Width of condyles in front	.040
Width of condyles posteriorly	.027
Median diameter of condyles	.022
Diameters proximal end scapula { anteroposterior	.052
Diameters proximal end scapula { transverse	.031
Length of glenoid cavity	.040
Width of neck	.049

Several posterior feet accompany the bones, some of which are clearly those of *P. primævus*. The calcanea are all of the same size. Some have larger cuneiform bones, and the navicular less produced posteriorly, while

From the Big Horn region, Wyoming. J. L. Wortman.

the metatarsals have the same length. These may belong to the *P. nuniënus*.

PHENACODUS PRIMÆVUS Cope.

Paleontological Bulletin No. 17, p. 3, October 25, 1873. Annual Report U. S. Geological Survey Territories, 1873, p. 458 (1874). Report on the Vertebrate Fossils of New Mexico. U. S. Geogr. Geol. Surveys W. of 100th Mer., Capt. G. M. Wheeler, 1874, p. 10. Report U. S. G. G. Surv., W. of 100th Mer., G. M. Wheeler, iv, pt. ii, p. 174.

Plates LVII b, LVII c, LVII d, LVII e, LVII h, LVIII; fig. 11.

The remains of this representative species are locally distributed, and it is only recently that they have been obtained in any abundance. The first one was dug by myself from a bank of the Wasatch Eocene marl on Bear River, Wyoming, in association with *Coryphodon latipes* and *Hyracotherium index*. The next specimens were obtained in only moderate abundance from the Wasatch deposit of Northwestern New Mexico. It is in the Big Horn Basin, Northern Wyoming, that the bones have been found abundantly by Mr. Wortman, who also probably obtained it in the Wind River region. Among the Big Horn specimens are several parts of skeletons, and one almost entire one. This last specimen is the basis of the following description, and with it the others will be compared. A restoration of the species can be made from it. From the measurements, it is evident that the animal was intermediate in size between a sheep and a tapir.

Cranium.—The skull is rather elongate, with the anterior border of the orbit a little anterior to the middle of its length. The premaxillary region is tapering, and the occipital region is not wide, but the interorbital space is wide and flat. The sagittal crest, though elevated, is short, commencing a little anterior to a line connecting the anterior borders of the glenoid cavities. The zygomatic arches are rather straight, and are but little expanded posteriorly. The lateral occipital crests are well marked. The anterior prolongation of the skull is confined to the jaws, the facial surface being short through the anterior abbreviation of the nasal bones.

The premaxillaries are produced and narrowed anteriorly, and their inferior and interior surfaces are flat and are connected by a convex exterosuperior surface. Above the space between the third incisor and the canine teeth, the superior surface is excavated so as to be longitudinally concave. It then rises into its vertical laminar ascending process which bounds the nareal opening, and articulates with the anterior part of the nasal bone. The middle and inferior parts of this lamina have a smaller vertical diameter than the part adjacent to the nasal bone. The maxillary is a little concave above the middle premolar teeth. Its superoposterior angle is above the anterior border of the orbit. The superciliary border of the orbit is thick, and its posterior termination is marked by a very obtuse angle with a roughened surface. The anterior temporal ridge is not produced, but is a strong roughened angle. It becomes a ridge before joining its fellow, and includes with it a narrow deep groove, which is even continued on the anterior part of the sagittal crest. Although the brain cavity is small, the walls of the temporal fossæ are gently convex in the squamosal region. The orbital border of the malar bone is acute, while the inferior face is truncate, with a narrow bevel between it and the external face.

The alveolar face of the maxillary bone terminates in a free point above the processus triangularis of the palatine. The anterior border of the zygomatic fossa falls a little anterior to the posterior edge of the last superior molar. The pterygoid alæ are broken away, but their bases are quite narrow, and they extend posteriorly to the line of the anterior border of the glenoid cavity. The basioccipital bone is rather wide, and the part immediately anterior to the foramen magnum is concave in longitudinal section, with a weak median keel. Anterior to this point it is not keeled, and the only interruption to the regularity of the surface is seen in a low angular tubercle on each side, at the position of the occipito-sphenoid suture. Basisphenoid plane. The postglenoid process is quite prominent, and is but little extended transversely. A shallow groove passes round within the internal extremity of its base. The meatus auditorius is wide, and the posttympanic process is a short obtuse knob. The paroccipital process is short and compressed so as to be tranverse. The space between it and the posttympanic is traversed by a low ridge, which has a nearly anteroposterior direction. The occipital condyles are excavated on their inferoanterior border, so as to produce a prolongation along the inferior edge of the foramen magnum to near the middle line. The palatal edge of the posterior nares is reverted. The lachrymal bone is large, forming the anterior third of the superciliary border, and the anterior half of the roof of the orbit. The anterior edge of each nasal bone is wide, and is notched. The width continues posteriorly to above the anterior part of the orbit, where it increases. Above the middle of the orbit the nasals begin to narrow, and they terminate a little behind the line of the postorbital angles, with a twice angular outline.

The incisive foramina are narrow and extend to the third incisor inclusive; they are separated by a very narrow septum. The anterior infraorbital foramen is large, and issues above the middle of the first true molar. The posterior part of the squamoso-parietal suture has a raised squamosal border, and two or three postparietal foramina enter at this point. There are two or three strong postsquamosal foramina, and a small supraglenoid, much like that of the species of *Anchitherium*. The foramen magnum is a transverse oval. Its superior border is notched medially. The foramina rotundum and ovale are not large and of subequal size. The former is only a little nearer to the latter than it is to the carotid groove (or foramen). The jugular foramen is smaller than either. The postglenoid foramen is situated at the posterior side of the internal base of the postglenoid process. It is about the size of the *F. rotundum*. The *F. condyloideum* enters the foramen magnum at the middle of its side, and is small, like the *F. jugulare*.

Mandible.—The horizontal ramus is slender, and the inferior outline is slightly convex below the true molars. It is slightly concave below the front of the coronoid, but descends again to the full and rounded angle. The outline is then not like that of *Coryphodon*. The symphysis has a very gradual rise. The coronoid process is at its anterior base broadly truncate. The flattening narrows as it ascends and becomes oblique at the summit, presenting outwards. The anterior margin is here strongly curved, even descending a little to the posterior apex. The third concave posterior border descends to very near the anterior border of the condyle. The internal extremity of the condyle is more robust than the external, and overhangs the flat internal face of the angle. The external extremity is the summit of a short vertical ridge, which disappears below. The masseteric fossa is only defined anteriorly, and there is no distinct internal pterygoid fossa. The condyle is elevated as far above the tooth-line as it is below the apex of the coronoid process. There are two prominent mental foramina, one below the posterior border of the first premolar and one below the middle of the fourth premolar. There are some less important foramina in the symphyseal part of the jaw. The symphysis extends to below the posterior border of the second premolar.

Dentition.—The superior incisors are lost. The alveoli are nearly in contact, and are round except the third, which is oval. The precanine diastema is as long as the third incisive alveolus. The canine alveolus has nearly twice as wide a diameter as that of the third incisor. The section of the crown is subtriangular, the anterior face wearing flat. The first premolar is situated near to the canine, and the single alveolus is a little larger than that of an incisor. The crown is lost. The second premolar is two-rooted. The crown is longer than wide or high, and consists of a single angular cusp. The external face is flat, and the internal face is convex. The base of the crown widens posteriorly, but does not produce a heel, as its surface is everywhere a steep continuation of the inner face. No anterior basal tubercle. The base of the third premolar is triangular in form, the base being posterior. The external elevation has a triangular outline, but the posterior edge is interrupted by a secondary cusp less elevated than the principal one. The internal cusp is opposite this posterior one, and is not quite so elevated. Its base projects well inwards. No cingulum except at the anterior base. The fourth premolar has a subquadrate outline, the internal angles being rounded off. The two external cusps are subequal in elevation, and they are only distinguished externally by a short groove, which does not extend to the base of the crown. The general external surface is nearly flat. There is an anterior intermediate tubercle, but no posterior one. No transverse crests.

The outlines of the true molars are obliquely subquadrate, the last molar being the most oblique. Their external tubercles are round at the base, but somewhat flattened at the apex. The internal tubercles are rounded on the inner, and flattened on the outer side. The intermediate tubercles are distinct from both. There is a narrow cingulum interrupted at points. It rises into a weak anterior external angle, and a strong tubercle at the notch which separates the external tubercles. On the last molar the posterior external tubercle is reduced in size, and the posterior cingulum is wide and tuberculate. It rises into a large posterior internal tubercle. The external cingulum and tubercle are weak. The enamel is slightly obsoletely wrinkled except where worn.

There is no precanine diastema in the lower jaw. The middle incisors project almost horizontally forwards, and the alveoli are oval in outline. The crown of the canine is a little recurved. It is rather short, and of triangular section. The sides are a little convex, and the anterior angle not acute. No ridges or grooves. First premolar one-rooted, separated by a short space from the canine, and separated by a longer space, equal to its own long diameter, from the second premolar. Crown lost. The crown of the second premolar is triangular, and is in profile an equilateral triangle, with the base longer than the sides. There is no heel, but the crown is wider behind and has a vertical fossa on the inner side of the middle line. The inner edge of this fossa terminates above in a rudiment of an internal cusp. Similarly there is in the unworn tooth a rudiment of an anterior cusp on the superior part of the anterior edge; probably not a constant character. No cingula. The third premolar has a form like the second, but it is larger and is wider behind. The posterior face has two grooves, the internal the shorter, and terminated above by a well-marked internal cusp. The external posterior bounding ridge is somewhat tubercular. No heel nor cingulum; a rudiment of anterior basal cusp. The fourth premolar tooth is a larger tooth than the third, and is as long as the first true molar. There is a large, but low anterior basal tubercle, and the internal anterior cusp is larger than the external, and a little posterior to it. The only posterior cusp, the external, is the size of the anterior basal tubercle. Internal edge of heel not elevated; no posterior median tubercle.

In the true inferior molars, the internal anterior tubercle is a little larger than the external. The anterior basal ledge is distinct, but there is no other cingulum. The only median tubercle is posterior, and is well developed, though smaller than the others. A low ridge connects the posterior or external tubercle with the base of the anterior inner. The last true molar is a little smaller than the second, and is more narrowed behind than they. The posterior median tubercle is as large as the posterior laterals, and is in contact with their bases. In some specimens the interior extremity of the anterior ledge develops a low tubercle, and in most a similar low tubercle is found on the posterior edge of the anterior inner cusp. On specimens where these tubercles are unusually distinct, I proposed the species *Phenacodus trilobatus*. Present information does not substantiate its distinctness. The enamel of the inferior molars has a very obsolete rugosity. In the description of this species in the Vol. IV Wheeler U. S. Geological Survey, p. 175 and Plate XLV, fig. 2, I was guilty of a transposition of some of the separate superior molars, on which it was based. Thus the tooth regarded as first superior molar is really the last, and the teeth should form a series from the left to the right, instead of from the right to the left.

Measurements of skull and teeth.

		M.
Length of cranium from apex of premaxillary to superior border of foramen magnum inclusive (measured below)		.237
Length from apex of premaxillary to vertical line from free end of nasal bone		.068
Length from premaxillary to anterior edge of orbit		.115
Length from premaxillary to palatal notch		.122
Length from premaxillary to postglenoid process		.191
Long diameter of meatus auditorius externus		.0095
Width of palate between canines		.027
Width of palate between last molars		.032
Width between postglenoid processes		.050
Width between carotid grooves		.020
Width of occiput at foramen magnum		.062
Width at anterior parietal region (least)		.037
Width of interorbital region		.070
Greatest width of anterior nares		.039
Depth at front of orbit		.058
Length of premaxillary to front of canine (axial)		.025
Long diameter of canine alveolus		.010
Length of premolar series		.048
Length of base of second premolar		.011
Diameters third premolar	anteroposterior	.012
	transverse	.010
Diameters fourth premolar	anteroposterior	.012
	transverse	.012
Diameters second true molar	anteroposterior	.013
	transverse	.015
Diameters third true molar	anteroposterior	.011
	transverse	.013
Length mandibular ramus		.195
Length symphysis (in oblique axis)		.048
Elevation of condyle of mandible		.071
Elevation of coronoid process		.096
Depth at posterior border of symphysis		.028

		M.
Depth at posterior border of third premolar		.027
Depth at third true molar		.029
Length of inferior dental series		.123
Length of base of third premolar		.012
Diameters fourth premolar	anteroposterior	.014
	transverse	.009
Diameters first true molar	anteroposterior	.013
	transverse	.011
Diameters third true molar	anteroposterior	.014
	transverse	.011
Width of mandibular condyle		.0225

Besides the general features already described under the head of the genus, the *brain* of the *Phenacodus primævus* displays the following features: The olfactory lobes are as wide as long, and they diverge, having two external sides. In section they are triangular, presenting an angle downwards. The hemispheres are depressed, and widen posteriorly. They are well separated from each other and from the cerebellum; so much so that it is quite probable that the corpora quadrigemina are exposed. Their outlines are however not distinguishable on the flat surface which connects the hemispheres posteriorly. No further indication of sylvian fissure can be seen in the cast, beyond an entering angle defining the lobus hippocampi anteriorly. The latter is prominent externally and less so downwards. There are distinct indications of convolutions. There are three on each side above the sylvian convolution, and a fourth extends from the sylvian upwards and posteriorly below the posterior part of the third or external convolution. The sulci separating the convolutions are very shallow. The internal and external convolutions unite anteriorly, passing round the extremity of the median convolution. The space between this gyrus and the base of the olfactory lobe is only three millimeters.

The cerebellum is longer than a single hemisphere. Its superior surface is somewhat flattened and descends forwards; the lateral boundary of this face is a projecting edge which rises behind to an angle of the vermis. The posterior face is shorter than the superior, and is vertical. It is separated by a space from a very prominent lateral convolution, while the region of the flocculus is concave from the internal form of the ascending portion of the petrous bone. This concavity is open anteriorly. The base of the fifth pair of nerves is below its apex, and that of the sixth below the inferior extremity of the laterral convolution. Section of the medulla oblongata a transverse oval; its inferior face, and that of the pons varolii smooth. A deep fossa just anterior to the bases of the optic nerves.

Measurements of brain.

	M.
Length from vermis to olfactory lobes inclusive	.070
Length of olfactory lobes from above	.015
Length of hemisphere from above	.030
Length of cerebellum from above	.024
Depth of olfactory lobe	.010
Depth of hemisphere	.023
Depth of cerebellum and medulla	.026
Depth of medulla at vermis	.015
Width of olfactory lobes at middle	.030
Width of hemispheres in front	.024
Width of hemispheres behind	.044
Width of cerebellum	.036
Width of medulla at vermis	.020

The *cervical vertebræ* are considerably smaller than the dorsals. They are not elongate nor yet shortened, but have much the proportions of those of the *Tapirus terrestris*, excepting the axis, which is relatively longer and narrower. The expanse of the atlas is less, owing to the shortness of the diapophyses, and it is also smaller relatively to the other cervicals. The fore edge of the diapophysis is regularly convex, and extends inwards anteriorly, having only a notch between it and the body. The neural arch is rather wide anteroposteriorly, and has an acute edge posteriorly. The arterial groove, parallel to the edge of the condylar cotylus, is well marked. The axial facets are rather wide, and are well separated from each other below. The neurapophyses of the axis are rather narrow fore and aft, so that the anterior part of the neural arch is without support, and projects as a trihedral wedge to above the base of the odontoid process. The bridge inclosing the vertebrarterial canal is also short, and the paradiapophysis sends a short narrow process backwards. The atlantal facets are somewhat convex, and are directed upwards and backwards, and have a moderately flared posterior edge. The inferior edge at its inner extremity is swollen into a transverse tuberosity which does not extend to the middle line. The latter has a low obtuse keel anteriorly, which rises posteriorly to a triangular flat hypapophysial tuberosity, with a roughened surface. The posterior articular face is oblique and quite concave. The postzygapophyses are well developed, although rather small. Their articular surface is inclined at an angle of 45° outwards and upwards, and 15° backwards and upwards. The neural spine is quite narrow anteroposteriorly and rises at an angle of 45°, overhanging the third vertebra. Its anterior edge is acute to the anterior extremity of the arch, and the posterior face is rounded, with a narrow median keel.

The third to the sixth cervicals inclusive are preserved engaged in the matrix. Their centra are rather smaller than those of the *Tapirus terrestris*, and are not relatively so robust. Their articular faces have about the same degree of obliquity as in that species, and they have a pentagonal outline deeper than wide. The middle line below is a prominent keel, which expands posteriorly into a triangular hypapophysial table, which is truncate posteriorly. This table is smaller on the fourth vertebra than on the third, but on the fifth it is larger than on the third. On the sixth it is about as large as that on the third, but is less distinct in its outlines, while on the seventh its median line is prominent, and the lateral boundaries almost obsolete. On the third and fifth centra its lateral borders are especially prominent. The sides of all the centra are concave. The paradiapophyses are narrow in vertical diameter; that of the third is acuminate posteriorly, the apex extending to the line of the posterior margin of the middle line below. Its inferior edge is only convex. On the fifth this convexity becomes an angle, and the diapophysis begins to project. On the sixth the parapophysial free border is prominent downwards and forwards as well as backwards, and the diapophysis is quite prominent. On the seventh the diapophysis only remains. The floor of the neural canal is strongly ribbed on the median line. The zygapophyses of one side are connected by a prominent ridge, and the general surface of the neural arch above is flatter than in *Tapirus terrestris;* the lateral concavities are more evident on the sixth than on the third vertebra, and the roof is more oblique on each side. Neural spines small. Zygapophysial facets all rather oblique, more so on the seventh than the third vertebra.

Measurements of cervical vertebræ.

		M.
Diameters diaparapophysis atlas	anteroposterior	.025
	transverse from foramen	.014
Diameters axial facet atlas	vertical	.016
	transverse at middle	.014

		M.
Diameters axis	transverse at atlantal facets	.035
	transverse posteriorly	.018
	anteroposterior without odontoid	.036
	vertical posteriorly	.017
Length of neural spine axis	from anterior edge neural arch	.060
	from posterior edge neural arch	.032
Length of centrum of third cervical (below)		.023
Length of centrum of fourth cervical (below)		.022
Length of centrum of fifth cervical (below)		.022
Length of centrum of sixth cervical (below)		.022
Length of centrum of seventh cervical (below)		.022
Diameters third cervical centrum in front	vertical	.0185
	transverse	.015
Diameters seventh cervical centrum in front	vertical	.0195
	transverse	.0165
Greatest width of third cervical at postzygapophyses		.035

The vertebrarterial canal is relatively of smaller size than in the *Tapirus terrestris*, appropriately to the smaller brain.

The anterior *dorsal vertebræ* are partly concealed in the matrix. They have rather elongate neural spines, narrow in both diameters, and triangular in section. The diapophyses are elongate, but shorten posteriorly, so that on the eleventh and twelfth they project very little, especially at the inferior side. They there sustain a large tubercular articular facet, which has the long diameter directed upwards and forwards. At the superior base the prominent metapophysis stands erect, the entering angle between the two being excavated into a strong fossa. It is distinct from the external edge of the prezygapophysis, and its summit is truncate, and extended to a little horizontal apex, both anteriorly and posteriorly. The centra of the eleventh and twelfth dorsals are depressed, and their inferior face is regularly convex in transverse section. The lateral margins are expanded posteriorly, terminating in the cup-shaped capitular articular facets. This lateral prominence is not seen in *Tapirus terrestris* nor in *Hyracotherium venticolum*. The opisthocœlism is well marked, but not so strong as in the cervicals. The anterior articular face of the eleventh is flat above, obliquely truncate at the sides, and wider and convex below. On the fourteenth dorsal the median inferior prominence assumes the form of a rib. On the fifteenth it has become a prominent keel. On this vertebra the neural spine has a wide anteroposterior diameter and its posterior edge slopes forwards. On the fourteenth there is a small tuberosity on the superior border of the capitular facet, and the lateral edge of the centrum is uninterrupted from this facet to the posterior one.

The anterior *lumbar vertebræ* are not depressed, this feature only appearing in the centrum of the sixth and last. All the centra are keeled below, and all except the last very prominently keeled. The diapophysis is well developed on the first lumbar, and its base is rather robust and its middle is anterior to the middle of the centrum. In the succeeding vertebræ its base is more depressed and more extended posteriorly. The diapophysis is directed anteriorly excepting on the last centrum, where it is transverse. The neural spines are directed forwards, and the base of that of the sixth lumbar is shorter and less compressed than that of the others. On the first lumbar there is a narrow metapophysis on the external side of the prezygapophysis. On the second, third, and fifth it is less elevated, and is continued as a lateral crest, half way to the postzygapophysis. It is broken from the fourth. On the sixth it is less elevated, and does not extend beyond the line of the middle of the base of the neural spine. On the first lumbar the articular faces of the postzygapophyses are nearly flat and rise outwards at an angle of 45°. On the second they are more convex and are a little produced upwards. They are separated from each other by a deep notch, which is continued as a groove along the basal posterior part of the neural spine. As compared with those of *Tapirus terrestris* the lumbar vertebræ are more elongate and much more strongly keeled. Their proportions are more like those of *Hyracotherium venticolum*, but they are less depressed and more strongly keeled. The postzygapophyses are totally different from those of the *Tapirus* and are more like those of *Hyracotherium*, but are, so far as preserved, less revolute.

The sacrum is rather narrow, and each of the posterior three centra of the four of which it is composed is concave in all its inferior longitudinal lines. The first is obtusely prominent on the middle line below, and gently concave between the latter and the descending distal edges of the diapophysis. The centra are coössified, but the lines of junction are prominent. They are slightly flattened at the middle line, producing an angular section, or even a slightly double concave one, at the junction of the third and fourth, where there are three longitudinal ridges. The continuous diapophysial plate is most contracted at the junction of the second and third centra. Beyond this point the edge narrows, becomes more prominent laterally, becomes thin-edged, and then narrowly truncate again. The neural spines are compressed, and are well developed on all the centra, that of the second having the largest anteroposterior diameter. The articular face of the last centrum is a deep transverse oval.

The *caudal vertebræ* are represented by eleven centra and parts of centra. They indicate a long robust tail; probably rather more robust than that of the lion, but not so prolonged. As in most *Mammalia* with long tail, the proximal centra are short and robust, and they successively increase in length to past the middle of the series. The distal centra are rather smaller than those of the lion or hunting leopard. On the anterior centra the diapophyses are robust and elongate, and are directed posteriorly. Their posterior base is at the posterior articular face. They are triangular in section, owing to the presence of a ridge which, on the superior aspect, extends diagonally backwards. As the diapophysis becomes reduced to a hook on the posterior extremity of the centrum, a prominence arises in the same line at a point marking the anterior third of the length. As the diapophysis diminishes, this process increases in length, but both processes are finally reduced to mere tubercles. The articular faces are slightly opisthocœlous They are round in the anterior part of the series, and become more depressed posteriorly. This is chiefly due to a flattening of the inferior side. When the neurapophyses fail to meet above, they remain as crests and thin ridges, on the posterior extremity only of the centrum, while the prezygapophyses remain as prominent widely separated processes at the anterior extremity of the centrum. The inferior face of the anterior centra is ribbed below; the rib widens at the extremities, its angles at the posterior edge probably giving attachment to the chevron bones. This flattened rib is finally represented at the extremities of the centrum only, and disappears first at the posterior extremity.

Measurements of vertebral column.

	M.
Length of cervical series	.180
Length of dorsal series	.681
Length of sacral series	.140

			M.
Diameters of eleventh dorsal	anterior	vertical	.015
		transverse	.026
	posterior transverse		.040
	longitudinal		.028
Diameters of second lumbar	anterior	vertical	.025
		transverse	.025
	posterior transverse		.030
	longitudinal		.034
Transverse extent of postzygapophyses			.026
Anteroposterior width of diapophyses at base			.023
Diameters centrum of first sacral	anterior	vertical	.021
		transverse	.038
	anteroposterior		.030
Expanse of diapophyses of first sacral			.068
Diameters fourth sacral	posterior	vertical	.019
		transverse	.022
	anteroposterior		.035
Diameters first caudal	posterior	vertical	.021
		transverse	.021
	anteroposterior		.025
Length of caudal with one strong diapophysis			.029
Length of diapophysis of do			.014
Length of caudal with two short diapophyses, the posterior longer			.034
Length of diapophysis, do			.007
Diameters caudal with two equal diapophyses	anteroposterior		.037
	anterior vertical		.028
	anterior transverse		.019
Length of diapophysis			.005
Length centrum without neural arch			.038
Length distal centrum			.030
Posterior diameters of do.	vertical		.011
	transverse		.014

The neck of the *scapula* is not much contracted, but the anterior border is turned abruptly forwards, before assuming its anterior convexity. The apex is rounded, and not acuminate, as in the amblypodous genera known; on the contrary, the distal posterior curve is strongly convex. From this point to the neck, the posterior edge is nearly straight. The spine is elongated. The proximal part of its edge is broken, but at the middle it widens to a narrow roughened surface, and thence gradually descends, and disappears three centimêters from the edge. The glenoid is a wide oval with an anterior acumination on a short acuminate tuberosity. The coracoid is an abruptly recurved hook with obtuse apex vertically compressed, looking inwards. Its base is separated from the edges of the glenoid cavity by a groove. The inferior face of the scapula is without noteworthy ridges or grooves. I find no indication of clavicle. The outline of the scapula is unlike that of any Perissodactyle, but the head is more like that of *Hyracotherium venticolum* than any other member of that order accessible to me. The coracoid is less incurved in *Hyrachyus agrestis*, and is entirely flat in *Tapirus terrestris*.

Measurements of scapula.

	M.
Length measured along spine	.179
Width, .030^{m} from proximal end of spine	.077
Semi-diameters of glenoid cavity { anteroposterior	.036
Semi-diameters of glenoid cavity { transverse	.029
Width of neck	.036
Length of coracoid from internal base	.008

The *sternum* is represented by six segments with a possible seventh concealed by the fore-arm which lies across it. Its characters are unlike those of the *Perissodactyla*, and resemble rather those of some of the *Conivora*. It has some resemblance also to that of *Hyrax*. The presternum is much compressed, and two-fifths of its length projects in front of the facets for the first rib, as a compressed prow. The compression is less marked at the posterior extremity. The second segment is in section a vertical parallelogram. The fourth segment is an oblong flattened plate, and the fifth is a nearly square flattened plate with the angles truncated. The sixth is wider than long, with the posterior angles much more broadly truncate than the anterior, and the posterior border excavated. The seventh is much narrower than the sixth, and is flat below. Its distal extremity is lost. It is probably the xiphisternum, so that the number of segments is the same as in *Hyrax* and the *Ruminantia*, and one more than *Tapirus* and the hogs. It differs from that of Hyrax in the ossification of the presternum, and the greater width of the mesosternal segments. The latter are therefore of a totally different form from the corresponding segments in *Perissodactyla*, including *Hyrachyus*, but similar segments are seen in *Coryphodon*.

Measurements of sternum.

	M.
Length of presternum	.085
Width at rib-facet	.026
Diameters second segment { anterior { vertical	.018
Diameters second segment { anterior { transverse	.016
Diameters second segment { anteroposterior	.035
Length fourth segment	.029
Width fourth segment behind	.031
Diameters fifth segment { transverse (least)	.022
Diameters fifth segment { anteroposterior	.026
Diameters sixth segment { transverse (least)	.026
Diameters sixth segment { anteroposterior	.020
Width of second segment medially	.014

The *humerus* is rather robust. Its muscular ridges are well developed. The deltoid extends two-thirds the length, and the external epicondylar ridge rises as it disappears, and is most prominent near its origin. The head is an anteroposterior oval supplemented by the very large greater tuberosity, and the small but prominent lesser one. The edge of the former is raised, and near its posterior end overhangs a distinct large round facet for the teres minor tendon. The ridge for the insertion of the teres major rises to the inferior overhanging edge of the head a little posterior to this facet. The subscapularis facet of the lesser tuberosity is large and transversely lenticular in shape. It is bounded below by a strong angular ridge. There is a low ridge in the fundus of the wide bicipital groove. The inner side of the shaft for some distance above the condyle is flat. An inch above the condyle its section is triangular. The epicondyles are not prominent; the internal is a small tuberosity, which is continued into the narrow and short bridge that incloses the epitrochlear foramen.

The condyles have greater transverse extent than in most *Diplarthra*, resembling rather that of Proboscideans and *Carnivora*. The internal flange is prominent, and the external cylinder is quite convex in longitudinal direction, as in *Oredon*, but less prominently.

The *Phenacodus primævus* displays its primitive and *Proboscidian* character in the relations of the bones of the cubitus. Instead of having the ulna more slender than the radius, as in *Hycracotherium* and the modern *Perissodactyla*, or equal to it in diameter, as in *Hyrachyus*, the shaft of the ulna is more robust than that of the radius. The olecranum is not acuminate, but is obliquely truncate, and is flattened below. The middle of the shaft is compressed; its internal side is nearly flat, or a little concave, while a ridge marks the middle of the external side. This ridge becomes an angle on the distal third of the bone, and almost disappears. Above this ridge there is a wide, rather shallow, groove, which is bounded above by the angular external edge of a narrow superior truncation on which the radius rests. At the distal extremity the external face has a shallow median groove. On the inner side the tuberosity which is applied to the radius is three centimeters from the extremity. The carpal facet on the latter is a narrow oval, convex in both directions.

The section of the middle of the shaft of the radius is an oval placed obliquely with the superior extremity truncate. This truncation narrows and becomes the narrow convex internal edge proximally. Distally it expands into the inner face of the superior arched surface. The head is a transverse parallelogrammic oval, with the internal extremity flared forwards at an angle of 45°, and narrowed. The middle line of the distal two-fifths, below, is a prominent angular ridge, which disappears proximad to the extremity. The latter is enlarged, and contracts to the carpal facets, especially from above and below. The lunar facet is wider anteroposteriorly than the scaphoid, and has an indication of distinction from it. The cuneiform surface of the ulna is placed at an open angle to them, and is narrower than either. The shaft of the radius is curved, and the superior outline is convex.

Measurements of anterior limb.

			M.
Length of humerus			.173
Diameter of proximal end	anteroposterior (greatest)		.061
	transverse	at bicipital groove	.032
		at lesser tuberosity	.042
Anteroposterior diameter of shaft at middle			.027
Width at epicondyles			.044
Width of condyle			.035
Anteroposterior diameter condyle	externally		.023
	internally		.030
Length of ulna			.204
Depth at extremity of olecranon			.026
Depth at coronoid process			.038
Depth at middle of shaft			.021
Diameters of distal tuberosity	vertical		.015
	transverse		.022
Diameter of carpal facet	vertical		.009
	transverse		.016
Length of radius			.145
Diameters of head of radius	vertical		.015
	transverse		.030
Diameters at middle shaft	vertical		.016
	transverse		.015
Diameters of distal enlargement	vertical		.025
	transverse		.030
Diameters of carpal facet	vertical (lunar)		.013
	transverse		.024

The *carpus* is entirely preserved in my specimen. The general features have already been described under the head of the genus *Phenacodus*. The radial surface of the scaphoid and lunar is very convex in the anteroposterior section, descending well on the anterior face, and leaving the anterior external face wider than deep at the middle. The external face of the cuneiform is four times as wide as deep. The posterior tuberosity of the scaphoid is acuminate; that of the lunar is shortly truncate, and that of the cuneiform vertically compressed and narrowly rounded. The concave ulnar facet of the last bone covers three-fifths the length. There is but one inferior facet of the scaphoid, which is a little wider than deep, and has a median angle directed posteriorly. On the external side posteriorly there is a vertical oval facet for the magnum. The inferior face of the lunar has three facets; one anterior for the magnum; one internal, which widens posteriorly, for the magnum; and one external, narrow claviform, narrowing posteriorly, for the unciform. The unciform face of the cuneiform is two-thirds its transverse, and all of its anteroposterior width. The trapezium is a flat bone, with slight contact with the scaphoid, and extending distal to the trapezoides. It is wide anteroposteriorly and has a narrow distal facet. The trapezoides have a wider anterior face than the magnum. Its shape is discoid, and its anteroposterior diameter exceeds the transverse but little. There is no posterior tuberosity. The trapezial facet is accompanied by a notch of the external superior edge. The inferior edge of the anterior face is convex. The magnum is the smallest bone of the carpus (except perhaps the trapezium). It is compressed, and has the usual elevated median superior tuberosity. The anterior face is subquadrate. There is but one inferior facet, which is concave anteroposteriorly. This is bounded on the inner side of the bone by a vertical facet half as wide as the inferior, for the external proximal edge of the second metacarpal. The facet for the unciform on the external side extends from the entire anterior face upwards and backwards to the superior tuberosity. The posterior tuberosity extends very little posterior to the inferior facet. The external or anterior face of the unciform is a right-angled triangle with the perpendicular side inwards, and equal to the horizontal, which is a little convex downwards. The internal side has two facets, one superior oblique for the magnum, the other anterior and triangular for the edge of the third metacarpal. The inferior facets are concave anteroposteriorly, and that for the fifth metacarpal is two-thirds the width of that for the fourth.

Measurements of carpus.

		M.
Transverse diameter of proximal row		.046
Diameters scaphoid	anteroposterior	.023
	longitudinal	.013
	transverse in front	.013
Diameters lunar	anteroposterior	.018
	longitudinal	.015
	transverse in front	.014
Diameters cuneiform	anteroposterior at middle	.012
	longitudinal	.010
	transverse	.027
Diameters of trapezoides	proximal, anteroposterior	.014
	proximal, transverse	.013
	longitudinal	.008
Diameters magnum	vertical, in front	.010
	vertical, at middle	.015
	anteroposterior	.020
	transverse in front	.011
Diameters unciform	vertical (inner side)	.015
	transverse	.017
	anteroposterior	.016

The *metacarpal bones* are all preserved, including that of the pollex. The latter is however broken, leaving a slight uncertainty as to its exact length. The proximal facets of these bones differ from those of the metatarsals in their being more curved anteroposteriorly, or less flat. They are a little shorter and less robust than the corresponding metatarsals. The external or fifth is the same size as the fifth metatarsal, but is differently formed proximally. The second and third metacarpals have external lateral bevels or facets for the next carpal. The second metatarsal is the only one of its series that has such a facet. The proportions of the three principal metacarpals are very similar to those of the corresponding metatarsals. That is, the second and fourth have the same distal production, reaching the ligamentous fossa of the median metacarpal, but the second is a little the longer owing to its more proximal origin. In all three each proximal extremity extends a little over that of the one adjoining it on the external side, the external face being concave below the lateral facet. The condyle of each is flat anteriorly, and the groove that cuts it off from the anterior face of the shaft is very shallow. The distal surface is very little convex. The keel is quite short, and rather prominent. Both anterior and posterior faces of the shaft are flat. The form of the distal articulations of the metacarpals is that of an ungulate rather than that of a clawed mammal. Those of

Mesonyx, which a good deal resemble them, have the unguiculate character, although not very pronounced. The fifth metacarpal is just half the length of the third. In its proportions it reverses those of the corresponding metatarsal, for the proximal extremity is only one-half the width of the distal. In the metatarsal the proximal is the wider extremity. The proximal facet of the metacarpal is sublenticular, and descends anteriorly, the anterior apex being at the inner side. Opposite the latter, on the external side of the shaft, is a low tuberosity. The distal extremity is a little oblique, but the posterior extremity of the heel is nearly median (though it commences at the inner side), and the lateral distal faces are in the same plane. The internal side of the proximal extremity of the second metacarpal is roughened with two longitudinal fossæ, bounded by bands of the surface, of which the anterior is flattened.

Measurements of metacarpus.

			M.
Diameters of second	longitudinal		.065
	median transverse		.014
	distal	anteroposterior	.014
		transverse	.015
Diameters of third	longitudinal		.072
	proximal	anteroposterior	.017
		transverse	.012
	median transverse		.014
	distal	anteroposterior	.013
		transverse	.016
Diameters of fourth	longitudinal		.060
	proximal	anteroposterior	.015
		transverse	.011
	median transverse		.012
	distal	anteroposterior	.013
		transverse	.013
Diameters of fifth	longitudinal		.036
	proximal	anteroposterior	.013
		transverse	.007
	median transverse		.010
	distal	anteroposterior	.010
		transverse	.011

The proximal and distal measurements given above are those of the articular surfaces.

There is nothing exceptional about the *phalanges*. They are all depressed, those of the first and fifth digits the least so. The distal articular facets of the first phalanges are not much emarginate nor reverted, and those of the second are more reverted, and not in the least emarginate.

The latter are about half as long as the former. Of the ungual phalanges that of the third or median digit is flat and has a broadly rounded free outline, being a little longer than wide. It is not quite symmetrical, the apex being a little exterior to the median line. The ungues of the second and fourth digits are more narrowed and unsymmetrical, although their apices are obtuse. Those of the first and third are still narrower, and are relatively deeper at the proximal extremity. None of the ungual phalanges are fissured at their extremity, but the second and fourth, and probably the third, have a foramen within the margin on the middle line.

Measurements of phalanges.

I.—*Of first series.*

				M.
First digit	proximal diameters	vertical		.009
		transverse		.012
	length on middle line			.015
	distal diameter	of facet	vertical	.007
			transverse	.008
		greatest width		.010
Second digit	length on middle line			.024
	width distally			.011
Third digit	length at middle			.027
	distal diameter	vertical		.007
		transverse		.0145
Fourth digit	length at middle			.021
	proximal width (total)			.016
	distal	width (greatest)		.013
		depth		.007
Fifth digit	proximal diameter	vertical		.0085
		transverse		.0125
	length at middle			.015
	distal diameters	vertical		.007
		transverse		.010

II.—*Of second series.*

			M.
Second digit	length on middle above		.013
	distal diameters	vertical	.007
		transverse	.0115
Third digit	length		.014
	width of distal facet		.012
Fourth digit	length above		.013
	distal diameters	vertical	.0075
		transverse	.0115

III.—*Of ungual phalanges.*

			M.
First digit	length		.014
	width	of neck	.0065
		greatest	.008
Second digit	length		.019
	width	at neck	.010
		greatest	.014

			M.
Third digit	length		.019
	width	at neck	.014
		greatest	.016
Fourth digit	length		.018
	width	at neck	.0115
		greatest	.013
Fifth digit	length		.013
	width	at neck	.007
		greatest	.008

The *ilia* are flat, and not very wide. The superior edge expands rather abruptly at the neck, and then follows a curve nearly parallel with that of the anterior border. The external face of the expansion is concave. The anterior apex of the crest is subacute and decurved. It is robust and subtriangular in section at base. Just above the position of an anterior inferior spine, and near the edge of the acetabulum, is a deeply impressed subtriangular fossa. The obturator foramen is very large. The anterior edge of the ilium is much thickened and for most of its length is one-third the transverse diameter of the plate The plate is abruptly thinner above this line. The internal face of the innominatum is concave behind the acetabulum, and the posterior edge is thickened and reverted at this point. The inner face of the ischium is flat; the plane of its distal portion is nearly at right angles to that of the plate of the ilium. The position of the tuber ischii is but little thickened, and is very little curved outwards. It is less prominent distally than the inner part of the posterior border. The entire ischium is rather thin. The external face of its shaft is convex near the obturator foramen.

Measurements of pelvis.

		M.
Total length through acetabulum		.280
Length of ilium to acetabulum		.140
Width of plate of ilium at middle		.060
Width of peduncle	interiorly	.027
	anteroexteriorly	.019
	posteroexteriorly	.026
Diameter pubis at base	anteroposteriorly	.026
	anterointernally	.009
Length of obturator foramen		.070
Width ischium at spine		.036
Width of ischium at middle of obturator foramen		.026
Width ischim from tuber to symphysis		.069
Diameters acetabulum	vertical	.038
	anteroposterior	.038

The *femur* is a rather stout bone, rather more robust than that of the *Tapirus terrestris*. That is, it is shorter, but of equal diameter, and with condyles nearly equal in proportions. The head has a peculiar shape, which is also found in the *Phenacodus vortmani;* it is oval instead of round; the long diameter crossing that of the trochanters at a very wide angle outwards and backwards. The posterior face is most flattened, and here the fossa ligamenti teris forms a wide notch in the edge. The great trochanter is large, and projects as far as the head, from which it is separated by a deep notch. Its superior edge is transverse, terminating in an anteriorly directed apex, and inclosing with the prominent posterior edge a deep trochanteric fossa. Summit of great trochanter beveled obliquely outwards. The little trochanter is prominent. It commences opposite the lower fundus of the trochanteric fossa, and terminates opposite the superior edge of the third trochanter. The latter is above the middle of the length of the femur, that point marking the inferior edge of its truncate extremity. It is well-developed, and its truncate edge has a lenticular outline. The borders of the latter are prominent, so that the anterior and posterior sides are moderately concave. The internal side of the shaft is rather broadly rounded. The external edge is angular. The rotular crests are prominent and thick, and the interior is the more prominent and prolonged upwards. It is not, however, swollen as in some recent Perissodactyles. At the posterior base of the external condyle is a shallow fossa. A much wider and deeper one lies above the inner condyle and the inner half of the intercondylar fossa. The latter is wide, equaling the internal condyle in width posteriorly, and is a little narrower than the external condyle. The greater width of the latter is caused by the prominence of its external border, which encloses with it a groove, which is very shallow except at its superior termination at the superior quarter of the face. Just above and external to this part of the groove is a smaller fossa. Other than this and a shallow fossa of the internal side, the lateral or epicondylar fossæ are not marked, nor are there any between the condyle and the rotular ridge, as in *Hyracotherium venticolum.*

The *patella* is a short, wide, thick bone, terminating in a rounded angle below. Above it is truncate, with a transverse groove at the middle.

Measurements of femur.

		M.
Total length from head		.222
Length to line of middle of third trochanter		.097
Length to internal rotular crest		.185
Diameters of head	long	.035
	short, at middle	.028
Total width proximally		.071
Total width at third trochanter		.047
Total width below third trochanter		.031
Width of rotular groove, with crests		.025
Greatest width of condyles		.058
Depth of condyle at inner rotular crest		.068
Diameters of patella	longitudinal	.032
	transverse	.028
	anteroposterior above	.019

The *tibia* is about as long as the femur, and, like it, is a rather robust bone. The crest is thick and not prominent, and at its proximal part is broadly truncate. The halves of the spine are well separated from each other. The external cotylus is a wide anteroposterior oval, slightly concave. The interior femoral face is of somewhat similar form, but is much decurved posteriorly, so as to be convex anteroposteriorly. The popliteal notch is wide and regularly concave. The external lateral fossa is well marked only proximally; the posterior edge of this side is obtuse. The posterior edge of the external side is more acute above, but disappears below. The section of the shaft is triangular, but the obtuse anterior angle becomes the anterior edge of the malleolus. The latter bounds most of the internal astragalar groove, sinking posteriorly. The tibia projects behind the groove mentioned, so that the posterior face is not flat in the transverse direction like the anterior face. The transverse diameter of the astragalar articulation coincides with that of the head of the tibia, and the astragalar grooves are at right angles to it.

The *fibula* is moderately slender, with rather large extremities. The proximal surface is an anteroposterior oval, concave to fit a posterior convexity of the tibial facet. The distal extremity is truncate, and has a nearly vertical internal face. Its anterior face is about as wide as its exterior, and the angle they form has a low tuberosity. No tendinous grooves.

Measurements of tibia and fibula.

		M.
Total length of tibia		.210
Diameters of head	anteroposterior at internal cotylus	.048
	anteroposterior at popliteal notch	.034
	transverse	.055
Anteroposterior width of shaft at distal third		.020
Diameter distal end tibia	anteroposterior (greatest)	.033
	transverse	.032
Length fibula		.186
Proximal diameter	anteroposterior	.028
	transverse	.011
Diameter of shaft		.011
Distal diameter	anteroposterior	.025
	transverse	.019

As compared with the tibiæ of *Hyracotherium venticolum* and *Hyrachyus agrestis*, that of the *Phenacodus primævus* displays its simplicity in the absence of the notch of the anterior external border of the head and of the tuberosity which, especially in the *H. agrestis*, bounds it in front. The distal extremity agrees with that of the *Hyracotherium* in being in the same plane as the head, but it differs from both the species in the large flat malleolus. In the *Perissodactyla* named, it has a small anteroposterior extent, and its extremity is widely beveled in a fashion not seen in the *Phenacodus*.

The *posterior foot* is one-half longer than the anterior foot. The calcaneum and astragalus are about the size of those of *Hyrachyus agrestis*, but the median toes are a little shorter. The trochlea of the astragalus is about the size of that of the last-named species, but the groove is not so deep nor so wide. The head is nearly as wide as the trochlea, and its face is in front of the inner two-thirds of the latter, and extends in front of the malleolus of the tibia. Its anterior and external outlines viewed from above form a regular curve, a segment of a wide ellipse, to the line of the front of the trochlea, where the outline contracts abruptly. The navicular articular face is wider than deep, and is turned somewhat under. Superior face of head flat. There is no deeply-impressed fossa on either side of the trochlea of the astragalus as in *Hyrachyus*, etc. The calcaneum has much the shape of that of *Hyrachyus agrestis*, but is more robust. The sustentaculum is much more massive, having the vertical sides and a flat inferior face of equal width. The cuboid facet is transversely diamond-shaped, nearly as deep as wide, and without notch near the astragalus. There is a quite prominent external ridge which descends forwards from the base of the astragalar condyle to the external angle of the cuboid facet, which is wanting in the *Hyrachyus.* The posterior extremity is truncate, excepting a tuberosity of the inner edge, which is acuminate.

The navicular is wider than deep, and has a notch in its posterior border. There is a short tuberosity on the postero-external corner. The externo-posterior angle is somewhat produced, but it does not extend to the posterior side as in *H. agrestis.* The cuboid is a robust bone longer than wide. Its proximal face is wider than deep, is straight and oblique downwards and outwards transversely, and moderately convex anteroposteriorly. The tuberosity looks downwards, and extends across the entire posterior face, including a deep transverse groove. The ectocuneiform is much the largest of the cuneiforms; its is twice as long as the mesocuneiform, and a little wider. Its outline in front would be square but that a concavity exists on the distal half of the external face. The anterior face of the mesocuneiform is a transverse parallelogram. The entocuneiform is a flat bone, as long as the ectocuneiform, and with a narrow anteroposterior articulation with the navicular. Its external face is straight longitudinally and convex anteroposteriorly. It is a little wider anteroposteriorly below than above. Its inner face is half in contact with the mesocuneiform and half with the second metatarsal.

The proportions of the *metatarsals* are a good deal like those of the metacarpals. The length of the first is two-fifths that of the third, and that of the fifth is one-half that of the third. The second has a lateral facet for the ectocuneiform; the others have a single proximal facet. The second and fourth reach to the distal ligamentous fossæ of the third. The middle three are rather robust, especially distally. The distal condyles are segments of plane transverse cylinders, interrupted posteriorly by the short obtuse keel. A shallow transverse groove bounds the condyles proximad in front, being better marked than in the anterior foot. The proximal extremity of the first metatarsal is compressed, and the distal extremity is compressed and oblique. The proximal extremity of the fifth has two faces, the internal, naturally, for the fourth. The external is a tuberosity which nearly touches the posterior tuberosity of the cuboid.

The phalanges are absolutely longer and narrower than those of *Hyrachyus agrestis;* those of the second and fourth digits are also flatter. The first phalange of the hallux is subround proximally, and its form is somewhat irregular. The superior face is concave; there is a low tuberosity on the external side below the middle, and the side next the second is flat distally; the distal facet is deeper than wide, and is not recurved above. The proximal facets of the second and fourth phalanges are less transverse than that of the median phalange; *i. e.,* they are narrower and more convex above. Their distal facets are not emarginate, and are but little recurved. The second phalanges are nearly square, or as wide as long. The proximal superior edge is a little produced; the distal facets are not emarginate, and are very little recurved. The ungual phalanges are even more obtuse than those of the anterior foot, especially those of the first, third, and fifth digits. As in the case of the other phalanges, the median is the most depressed, and they become less depressed to the first and fifth. The ungues of the second and fourth digits are rather more acuminate than those of the first and fifth; while that of the fourth is broadly rounded. They all have a median foramen just within the border. The neck is but little constricted, and the proximal angles are not produced. The inferior sides of the ungual phalanges have a flat proximal area rather larger than a semicircle, between which and the margin is a concave space of a very wide crescentic form.

Measurements of posterior foot.

	M.
Total length from heel	.162
Length of tarsus	.106
Length of median metatarsus	.073
Length of median digit	.066

I.—*Tarsus.*

	M.
Length of calcaneum	.082
Depth at heel	.022
Width at sustentaculum	.035
Depth at astragalar condyle	.031
Diameters cuboid facet { vertical	.021
Diameters cuboid facet { transverse	.024
Length of astragalus on inner crest	.042
Length of astragalus on outer crest	.033
Width of trochlea, including crests	.025
External elevation of trochlea	.022
Diameters navicular face { vertical	.016
Diameters navicular face { transverse	.024

	M.
Diameters navicular { longitudinal	.011
Diameters navicular { proximal { anteroposterior	.024
Diameters navicular { proximal { transverse	.026
Diameters cuboid { proximal { anteroposterior	.017
Diameters cuboid { proximal { transverse	.018
Diameters cuboid { longitudinal median	.024
Diameters cuboid { median { anteroposterior	.018
Diameters cuboid { median { transverse	.025
Diameters ectocuneïform { anteroposterior	.030
Diameters ectocuneïform { vertical	.016
Diameters ectocuneïform { transverse	.016
Anterior diameters mesocuneïform { vertical	.009
Anterior diameters mesocuneïform { transverse	.011
Diameters ectocuneïform { proximal { anteroposterior	.011
Diameters ectocuneïform { proximal { transverse	.006
Diameters ectocuneïform { longitudinal	.016

II.—*Metatarsus.*

Length first	.028
Anteroposterior diameter { proximally	.013
Anteroposterior diameter { distally	.012
Length of second	.068
Proximal diameters { anteroposterior	.019
Proximal diameters { transverse	.015
Distal diameters { anteroposterior	.014
Distal diameters { transverse	.018
Width of head of third	.018
Greatest width above condyle	.021
Width at middle of shaft	.014
Distal diameters { anteroposterior	.015
Distal diameters { transverse	.019
Length of fourth	.066
Proximal diameter, transverse	.015
Median diameter, transverse	.013
Anteroposterior diameter of condyle	.014
Length of fifth	.038
Proximal diameters { anteroposterior	.014
Proximal diameters { transverse	.013
Anteroposterior diameter at middle	.009
Distal diameters { anteroposterior	.013
Distal diameters { transverse	.010

III.—*Phalanges, first row.*

Length of first	.014
Proximal diameters 1st { vertical	.009
Proximal diameters 1st { transverse	.010
Distal diameters 1st { vertical	.006
Distal diameters 1st { transverse	.006
Length of second	.025
Proximal diameters 2d { vertical	.012
Proximal diameters 2d { transverse	.018
Length of third	.028
Proximal diameters { vertical	.012
Proximal diameters { transverse	.022

	M.
Length of fourth	.023
Proximal diameters { vertical	.011
Proximal diameters { transverse	.018

IV.—*Second phalanges.*

Length on third digit	.017
Proximal width	.017
Distal width	.016
Distal depth	.007
Length on fourth digit	.015
Proximal diameters { vertical	.010
Proximal diameters { transverse	.015

V.—*Ungual phalanges.*

Length of first	.014
Proximal { depth	.008
Proximal { width	.009
Length of second	.022
Proximal { depth	.010
Proximal { width	.014
Width at shoulder	.015
Length of third	.024
Proximal { depth	.010
Proximal { width	.018
Width at shoulder	.019
Length of fourth	.021
Proximal { depth	.009
Proximal { width	.013
Width at shoulder	.0145
Length of fifth	.013
Proximal { width	.009
Proximal { depth	.008
Width at shoulder	.009

Restoration.—The foregoing measurements show that this species was as large as a big-horn; that its body was rather longer than in that animal, and its legs shorter and more robust. It was in fact proportioned more as in the common American tapir, but was of smaller size. The middle three toes of both feet reached the ground, while the first and fifth projected laterally and posteriorly, like the dew-claws of the hogs. The tail was longer and heavier than that of any of the living hoofed mammals, resembling in its proportions that of the wolf. The eyes were small and the muzzle long, but was singularly soft above near the extremity. Whether this soft part was pierced by valvular nostrils, as in the hippopotamus, or was produced into a short proboscis, as in the saiga or in the tapir, cannot be certainly ascertained, but there are indications of the insertion of important cartilages, if not muscles, on the superior faces of the premaxillary bones.

The animal was probably omnivorous in its diet. It was not furnished with any weapons of offense or defense pertaining to the osseous system, so that it must have sought refuge in flight. The well-developed muscular insertions of its limbs, and the digitigrade character of its step, indicate that it may have had considerable speed.

Distribution.—The bones of this species have been found wherever the beds of the Wasatch Epoch occur, but most abundantly in Northern Wyoming. From the Wind River Valley Mr. Wortman brought two specimens, and ten from the Big-Horn Basin.

Synonyms.—Two names have been given to supposed species which may be identical with this one. These are *P. omnivorus* and *P. trilobatus.* The latter is considered under the head of the dentition. The former was founded on an inferior molar from New Mexico. It remains for future examination to ascertain its true position.

PHENACODUS HEMICONUS Cope.

Paleontological Bulletin, No. 34, p. 179, Feb. 20, 1882. Proceed. Amer. Philos. Soc., 1881, Dec. (1882), p. 179.

Plate XXV *e;* fig. 16.

Represented by the posterior two superior molars of an individual intermediate in size between the *P. primævus* and *P. puercensis.* The posterior molar is peculiar in the very rudimental character of the posterior internal lobe, which is reduced to a mere wart on the cingulum. The posterior external tubercle is also rudimental, not exceeding the posterior inner in dimensions. The anterior tubercles, including the intermediate, are well developed, the internal exceeding the external. The cingulum is wide and crenate, and is only wanting on the external base of the crown. The penultimate molar does not differ so much from that of *P. primævus,* but the two internal cones are not so deeply separated at their base. The tubercles are all but little worn, and are conical in form, the external flattened on the external faces. Enamel wrinkled.

Measurements.

		M.
Diameters of M. ii	anteroposterior	.009
	transverse	.012
Diameters of M. iii	anteroposterior	.010
	transverse	.013

The size of this species precludes the possibility of its identity with any of the other species described here.

The Big-Horn Basin, Wyoming; J. L. Wortman.

PHENACODUS VORTMANI Cope.

Bulletin U. S. Geol. Surv. Terrs., F. V. Hayden, vi, 1881, p. 199; Paleontological Bulletin, No. 34, 1882, 179, Feb. 20. *Hyracotherium vortmani* Cope, American Naturalist, 1880, p. 747. *Phenacodus apternus* Cope, Paleontological Bulletin, No. 34, pp. 179, 180.

Plates LVII *g*, LVII *h*, LVIII; figs. 8–10.

This was a rather abundant species in Wyoming during the Wasatch Epoch. The best specimen in my possession is a nearly perfect skeleton found by Mr. J. L. Wortman near the Big-Horn River in Northern Wyoming. This I now describe.

Cranium.—Pressure has distorted the skull so that the superior surface is oblique to the sides, and the frontal region is a little crushed in. By viewing the two sides, however, a fair idea of the general form may be attained. The premaxillaries are not so produced as in the *P. primævus*, and they unite on the middle line at their extremity in a Gothic arch. The free portion of the nasal bones is short, is somewhat contracted, and is obtuse at the extremity. The profile is probably nearly straight. The orbit is of moderate size, and extends as far forwards as above the first superior true molar. The facial plate of the maxillary bone is uniformly gently concave. The sagittal crest is well developed, of medium height, and commences at a point one-third way between the posterior border of the orbit and the supraoccipital crest. The latter is bilobate, moderately notched on the middle line, and the alæ have not much transverse extent. It continues into the zygoma as the posterior temporal crest, and sends a distinct branch downwards to near the extremity of the posttympano-paroccipital process. There is a small lachrymal tuberosity on the anterior edge of the orbit. The supraorbital border is rather thick, and terminates in a distinct short angular postorbital process, which is directed downwards. The zygomatic arch is not stout nor convex The orbital edge of the molar is a little flared outwards, and it supports no trace of postorbital angle. The postglenoid process is prominent, is rather broad, with a convex inferior margin. The meatus auditorius is widely open below. The posttympanic and paramastoid processes are united into a short tuberosity with narrowed apex. The occipital crest is curved backwards, and the surface between the external end of the occipital condyle and the lateral occipital crest is concave.

The premaxillary bone exposes an external face of nearly equal width, from the second incisor to its superior extremity at the anterior lateral edge of the nasal bone. Posterior to this point the premaxillary bone does not extend. The nasal bones expand posteriorly so as to occupy the superior facial surface just anterior to the line of the anterior border of the orbits. How far posteriorly they extend cannot be certainly determined in the fossil, but what looks like the fronto-nasal suture crosses the face just in front of the anterior border of the orbits. If this be the posterior boundary of the nasals they do not have such a posterior extension as in *P. primævus*, although they are expanded in the same manner. The fronto-parietal suture apparently crosses the superior surface just posterior to the line connecting the postorbital processes. The anterior part of the malar bone is a band of equal width extending along the border of the orbit to the lachrymal bone. The latter extends a little on the superior orbital border, and about equally far below and above the lachrymal tuberosity for a moderate width on the facial surface. The squamoso-malar suture is straight, and the two elements of the zygoma terminate each in an acute angle.

The foramen infraorbitale opens above the middle of the fourth premolar tooth. The lachrymal foramen is within the orbit. Postglenoid foramen rather large, issuing in a groove of the postglenoid process. A postparietal, and a single mastoid foramen. A rather large foramen just within the position of a supraglenoid foramen, above the postglenoid process, and entering the squamosal bone posteriorly and inwards. Foramen magnum incomplete in the specimen.

Mandible—The mandibular ramus is not deep. Its inferior outline is gently convex from below the last molar tooth to the symphysis, which rises very gradually to the alveolar border. The inferior edge below the coronoid process is concave, and the border of the angle is broadly convex, turning gently forwards before ascending to the condyle. The external angle of the latter is prominent, and sends downwards a low ridge which becomes the anterior edge of a beveled border of the convex angle of the jaw, which ceases on reaching the inferior surface. The face of the condyle is transverse and narrow, and presents exclusively superiorly. The base of the coronoid process is narrow fore and aft. The masseteric fossa is well marked, excavating the ascending ramus above the line of the teeth; below this it is only defined for a short distance in front by a downward continuation of the anterior ridge. Symphysis not coössified. There is a single mental foramen, which is below the first inferior premolar.

Dentition.—The first and second incisor teeth are very similar, and different in the form of the crown from the third. The crowns are expanded anteroposteriorly, so as to touch each other, although the roots are separated. But there is a space nearly equal to the lengths of the edges of both crowns between those of the first incisors of opposite sides. Wearing makes the edges of the first and second acute. The third incisor is equal in size to the others, but its crown is canine-like, and with recurved apex. A short space separates this tooth from the canine, and the alveolar edge is arched upwards to receive the crown of the inferior canine. The canine is not large. Its crown has a spherical triangular section, the posterior edge not acute.

The first premolar leaves a single rather large oval alveolus, which occupies the space between the canine and the second premolar. The latter has a simple crown, which is a triangle in profile. Its base is wider posteriorly than anteriorly, but there are no accessory tubercles. The anterior and posterior edges are subacute. The base of the crown of the second superior premolar is nearly a right-angled triangle, the posterior and exterior edges contributing to the right angle. The external face of the crown is flat and triangular in shape, a slight convexity indicating the basis of a small posterior external tubercle worn off by mastication. The internal cusp is also worn off. Its base is entirely posterior. The base of the fourth premolar more nearly approaches an isosceles spherical triangle, the posterior border being a little more convex than the anterior. The external cusps have subequal bases, which have a round section on attrition. The internal cusp has a semicircular section on wear. There is space for a posterior intermediate tubercle, but if it were present, it is worn off. Anterior and posterior, but no external nor internal cingula. The anterior cingulum rises into an angle at the antero-exterior corner of the crown, which is not very prominent.

Of the true molars the first and second are about equal and the third is considerably smaller. The external face of the second is more oblique posteriorly than that of the first. The third is an oval with the narrower end external. The external cusps are subequal and subround in section in the first and second molars, and the internal cusps are also subequal and subround. The intermediate tubercles are present except on the posterior half of the second molar, where it is wanting, agreeably to the more contracted space at that point. Both these teeth have cingula all round the crown excepting on the internal side. The external cingular tubercle is obtuse. In the third true molar the two posterior tubercles are much reduced in size, and there is no intermediate tubercle between them. There is one between the anterior pair, which are of usual proportions to the crown. The anterior and exterior cingula are the only obvious ones on this tooth.

The inferior incisors are directed very obliquely forwards, and have the crowns compressed and wider than the roots. The crown of the third is not certainly known. The crown of the inferior canine is directed straight upwards, is about the size of that of the upper jaw, and has a subround section at the base. The molar series is continued from it without diastema. The first premolar is a good deal smaller than the second, has one root, and a crown higher than long with its base rising forwards and its apex acute. The second premolar has a simple compressed conic crown with apex above the anterior root. The third has a similar form viewed externally, but is larger, and has a posterior face, which in its present worn state is a plane narrow triangle, with the base downwards. It is probable that in the unworn state of this tooth the internal tubercle was rudimental or wanting, and the heel very short. The fourth premolar has a large heel, and well developed internal and anterior tubercles. The ridges connecting the external posterior tubercle and the external median with the internal median and anterior, respectively, form two V's like the inferior molar of *Rhinocerus.* No internal posterior tubercle, so that the fourth premolar is unlike the first true molar.

The first and second inferior true molars are subequal, while the third is smaller. The former support three posterior tubercles and two anterior opposite each other. The median posterior is smaller than the others, and does not give the posterior edge of the crown an angulate outline. The section of the interior tubercles is round; that of the exterior is crescentic, the anterior horn extending into a low ridge which, in the case of the posterior tubercles, reaches the anterior inner tubercle. In the first true molar it extends round the anterior base of the crown as a wide ledge; in the second and third molars this ledge is much less distinct. The median posterior tubercle of the third true molar forms a prominent heel, but its prominence is less obvious because of the proximity of the inner median tubercle which is not opposite the external median, but posterior to it. No cingula on the external or posterior side of the teeth; inner side concealed by matrix.

Measurements of skull.

	M.
Length of cranium from apex of premaxillary bone to superior border of foramen magnum, inclusive; measured below	.145
Length from apex of premaxillary to a vertical line from end of nasal bone	.0185
Length from end of premaxillary to anterior edge of orbit	.064
Length from end of premaxillary to postglenoid process	.116
Long diameter of meatus auditorius externus	.006
Width of palate between canines	.015
Width of palate between postglenoid processes, exclusive	.027
Width of occiput at foramen magnum	.038
Width (least) at anterior parietal region	.016
Width at postorbital processes	.036
Width of interorbital space	.035
Greatest width of anterior nares	.013
Greatest depth at front of orbit (without molar)	.036
Greatest width at zygomata	.055
Length of mandibular ramus	.110
Depth of mandible at condyle	.042
Depth at heel of third true molar	.020
Depth at second true molar	.017
Depth at first premolar	.011
Length on palate from premaxillary border to line of canine alveolus	.019
Length of canine alveolus	.0065
Length of molar series	.053
Length of premolar series	.030
Length of base of second premolar	.0065
Diameters superior P-m. iii { anteroposterior	.007
Diameters superior P-m. iii { transverse	.006

		M.
Diameters superior P-m. iv	anteroposterior	.0075
	transverse	.008
Diameters M. ii	anteroposterior	.008
	transverse	.010
Diameters M. iii	anteroposterior	.006
	transverse	.009
Length of dental line of mandible		.076
Length of base P-m. ii		.0065
Diameters P-m. iii	anteroposterior	.008
	transverse	.004
Diameters P-m. iv	anteroposterior	.009
	transverse	.0055
Diameters M. i	anteroposterior	.009
	transverse	.0066
Diameters M. iii	anteroposterior	.010
	transverse	.006
Elevation of crown of inferior canine		.008

Vertebræ.—The vertebral column is, probably, excepting the caudal series, complete. The dorsals are disarranged anteriorly, and the median lumbars are out of line, otherwise the centra are in place. There are seven cervicals, fourteen dorsals, seven lumbars, and three sacrals. Eleven caudals are preserved.

The vertebrarterial canal is present on six of the cervical vertebræ. Their centra are moderately elongate, and a little opisthocœlous. The articular faces are oblique to the long axis, indicating the elevation at which the head was held above the body. They are all obtusely keeled below, and the keel divides, forming a narrower or wider flat or concave triangular area with tuberous angles. The atlas is the shortest of the cervicals, and its neural arch is twice as wide anteroposteriorly as the inferior bar, and is flat except a double tuberosity on the anterior border. The latter, viewed from above, is quite concave. The diaparapophyses are short, but wide anteroposteriorly. They are more extended transversely in their posterior half; their border contracts anteriorly, and the edge is a little wider. The vertebrarterial canal pierces it above and within its posterior border, and issues on the inferior face a little anterior to the middle. The artery notched the anterior base of the diapophysis and then pierced the neural arch on the superior face equidistant between the diapophysis and the edge of the cotylus. The facets for the axis are distinct below, and have a slight angle on their internal edge. There is a small tuberculum atlantis with smooth truncate face. Centrum smooth below.

The axis is of moderate elongation, and has a very oblique, slightly concave, posterior articular face. Its form is that of a deep semicircle, with a truncate protuberance below. The spine is produced obliquely upwards and backwards, and is low in front. The postzygapophyses are large. The diaparapophyses unite into a short, acuminate backwards directed process, which extends to below the middle of the postzygapophyses. The atlantal facets are well turned upwards, and are hence deeper than wide. The odontoid process is cylindric, and its inferior articular surface is distinct from that of the atlantal surfaces. The inferior keel of the centrum is weak in front, but projects behind into the protuberance already described.

In four succeeding cervicals the articular faces of the centra are subround. In the third, fourth, and fifth the transverse process extends but little posterior to the posterior articular face, and the inferior edge does not descend to the level of the hypapophysial edge. On the sixth vertebra the diapophysial portion is directed upwards from the vertebrarterial canal, while the parapophysial portion extends almost vertically downwards, so that its horizontal inferior edge falls considerably below the hypapophysial line. On this vertebra also the neural spine first assumes important dimensions. The seventh cervical is mostly concealed by matrix. The zygapophyses in all the cervicals posterior to the axis are connected by a prominent but very obtuse ridge, and the roof of the postzygapophysis is thickened. The neural arches are quite flat on top.

The modifications seen in the forms of the dorsal vertebra, as we trace them from before backwards, are the following: In the first vertebra the anterior articular is a little wider than deep; the relative width increases until on the posterior centra it much exceeds the depth. The usual change in the prezygapophyses is seen. On the first they are like those of the cervicals, horizontal and looking upwards and inwards; on the third they have already become sessile convexities, looking upwards and forwards, and it is not until the eleventh vertebra that they begin to resume the horizontal up-looking character. On the first dorsal the diapophysis is robust and prominent, and has a large concave capitular facet. It is less prominent on the sixth, and has a subround extremity, and on the ninth it is surmounted by a small metapophysis. On the twelfth the diapophysis is oval in section, and the metapophysis forms the external border of the prezygapophysis. On the thirteenth the diapophysis is short and flat, and the metapophysis is elevated. On the fourteenth the diapophysis is flatter and the metapophysis is an elongate process directed upwards and forwards. The centrum of the fifth dorsal is uniformly rounded below in transverse section, and a little concave anteroposteriorly. The dorsals become a little concave laterally as we follow them backwards, until the fourteenth becomes keeled below by the moderate excavation of the sides. The neural spines are quite elongate and slender on the anterior part of the column, and have the usual posterior slope upwards. On the twelfth the spine is first vertical, wide, and truncate at the extremity. On the thirteenth and fourteenth the spines have the posterior edge a little longer and more oblique than the anterior.

The lumbar vertebræ continue the modifications of form already commenced in the dorsals. The centra become larger and longer up to the seventh and last, which is a little shorter than the first lumbar. From the first to the last the articular faces of the centra increase in relative depth, the sides become more concave, and the hypopophysial keel more distinct, excepting on the last. Here the middle line below is more rounded. The posterior articular face of the last lumbar is not flattened as is frequently the case, but is but little wider than deep. The diapophyses are flattened and rather thin, and they originate from the anterior two thirds of the side of the centrum. The metapophyses are much elevated on all of the lumbars excepting on the last, where they form only a prominent keel. The neural spine is preserved on the fourth lumbar. Its base is longer than on the dorsal vertebræ, is directed forwards as well as upwards, and the extremity is rounded, not truncate. The neural spine of the seventh lumbar is more nearly vertical and its sides are parallel.

The three sacral vertebræ are thoroughly coössified. The diapophyses form a narrow uninterrupted border, which is decurved and expanded on the anterior vertebræ. The centra are rounded below. They diminish much in diameter posterior to the anterior or lumbar articular face. The distal articular extremity of the third and last sacral is only half the linear diameter of that of the last lumbar, and is subround, and somewhat oblique, the inferior border being the more prominent.

Of the eleven caudal vertebræ preserved, five, and perhaps six, are continuous from the sacrum. These diminish rapidly in bulk, retaining uniform length. The articular faces are subround; the anterior have them rather oblique, appropriately to the descending direction of the tail at the base. The diapophyses are present on the six, but very small on the sixth. They are depressed, acuminate, and turned backwards. The neural spines of the first and second are elongate but not elevated. They are squarely truncate. The inferior face of the centrum of the first caudal is round transversely, and concave anteroposteriorly. On the second the middle line below is narrowed medially and expanded posteriorly. In the third this feature is more marked, the posterior expansion being concave medially. On the fourth and fifth there is a strong median rib, which divides into two posterior ribs, giving small surfaces for chevron bones. On the sixth the median inferior rib is gone. Distal to this point the centra increase in length. The neural canal is reduced to a groove between two widely spreading tuberosities in front, and two closely approximated ones at the posterior extremity. The diapophyses are represented by a small process at each extremity of the centrum, of which the posterior disappears first. On the inferior face, the anterior extremity is more protuberant downwards than the posterior, and the median and chevron ridges are gradually effaced.

In comparing this column with that of the *P. primævus*, the principal differences are seen in the numbers of some of the series. In that animal the dorsals are probably fifteen, the lumbars are six, and the sacrals four. In the present species they are, dorsals fourteen; lumbars seven; sacrals three. The difference in the number of the sacrals is more important than that of the lumbars and dorsals It is only necessary to shift one from the lumbar to the dorsal series in the *P. primævus*, to have an agreement with the *P. vortmani*. The centra of the cervical vertebræ are a little shorter in the *P. vortmani* than in the *P. primævus*, and the caudal diapophyses are more depressed and not so robust. The metapophyses are already distinct on the ninth dorsal vertebra in the *P. primævus*, while they first appear on the twelfth in the *P. vortmani*.

Measurements of Vertebræ.

I.—*Cervicals.*

	M.
Total width of atlas	.044
Diameters diaparapophysis of atlas, anteroposterior	.015
Diameters diaparapophysis of atlas, transverse from foramen	.008
Diameters axial facet of atlas, vertical	.0105
Diameters axial facet of atlas, transverse at middle	.009
Diameters of axis, transverse, at atlantal facets	.023
Diameters of axis, transverse, posteriorly	.012
Diameters of axis, anteroposterior without odontoid	.022
Diameters of axis, vertical posteriorly	.0105
Length centra below, of third cervical	.013
Length centra below, of fourth cervical	.013
Length centra below, of fifth cervical	.013
Length centra below, of sixth cervical	.013
Diameters third cervical, anteriorly, vertical	.009
Diameters third cervical, anteriorly, transverse	.010
Diameters third cervical, posteriorly, vertical	.0104
Diameters third cervical, posteriorly, transverse	.012
Vertical diameter of neural canal of third cervical	.0088
Expanse of postzygapophyses of third cervical	.0216
Expanse of diaparapophyses of third cervical	.0295
Total elevation posteriorly to base neural spine	.021
Diameters centrum sixth cervical in front, vertical	.0086
Diameters centrum sixth cervical in front, transverse	.010
Total elevation of sixth with neural spine	.0274
Expanse of prezygapophyses	.026
Expanse of diapophyses	.029
Length of parapophyses from canal	.0105

II.—*Dorsals.*

	M.
Diameter centrum first dorsal in front, vertical	.0085
Diameter centrum first dorsal in front, transverse	.0115
Diameter neural canal of first dorsal in front, vertical	.008
Diameter neural canal of first dorsal in front, transverse	.010
Expanse of prezygapophyses of first dorsal in front	.0225
Expanse of diapophyses of first dorsal in front	.0345
Elevation neural spine third dorsal from postzygapophysis	.043
Length centrum fifth dorsal	.014
Width of posterior face of centrum fifth dorsal	.015
Diameters eighth dorsal, anteriorly, vertical	.0077
Diameters eighth dorsal, anteriorly, transverse	.0115
Diameters eighth dorsal, anteroposterior	.015
Expanse of diapophysis of eighth dorsal	.026
Length neural spine do. from postzygapophysis	.033
Diameters thirteenth dorsal, anteriorly, vertical	.010
Diameters thirteenth dorsal, anteriorly, transverse	.0135
Diameters thirteenth dorsal, anteroposterior	.020
Total elevation with neural spine	.0345
Diameters of neural canal, vertical	.008
Diameters of neural canal, transverse	.007
Elevation of neural spine posteriorly	.013
Anteroposterior diameter of spine at summit	.010

III.—*Lumbars and sacrum.*

			M.
Length of centrum of	first lumbar		.020
	fourth lumbar		.024
	sixth lumbar		.025
	seventh lumbar		.019
Diameters centrum sixth, in front,	vertical		.014
	transverse		.014
Depth centrum seventh, posteriorly			.0135
Total elevation of fourth lumbar			.051
Elevation of neural spine of fourth lumbar from base			.024
Length of sacrum			.062
Diameters distal end sacral centrum	vertical		.0095
	transverse		.011

IV.—*Caudals.*

Diameters first caudal	posterior	vertical	.010
		transverse	.009
	anteroposterior		.017
Diameters third caudal	posterior	vertical	.008
		transverse	.010
	anteroposterior		.014
Expanse of postzygapophyses do			.012
Diameters sixth or seventh centrum	posteriorly	vertical	.008
		transverse	.008
	anteroposterior		.018
Diameter centrum with two equal diapophyses	anterior	vertical	.0085
		transverse	.008
	anteroposterior		.020
Length centrum without neural arch			.022
Length of a distal centrum			.019

V.—*Totals.*

Length of cervical series	.101
Length of dorsal series	.236
Length of lumbar series	.159

Ribs.—These are flattened proximally, but cylindric for the greater part of the length. They are flattened again distally. The head is round and the tubercle oval, and looking forward, as well as upwards.

Sternum.—The præsternum is elongate, and the elongation is posterior to the articulation for the first pair of ribs, which is half as long again as the portion anterior to that articulation. At the articulation the posterior portion is depressed so as to be wider than deep. The posterior extremity is subequally quadrate. The costal articular facets are pedunculate and directed posteriorly. The anterior margins of the peduncles extend half way to the extremity, which is narrowly compressed. The inferior keel of this portion is not deep, and disappears posteriorly on the depressed base of the posterior portion of the shaft. Three succeeding segments of the sternum are preserved, which increase in width posteriorly. The last (fourth) is longer than wide, and flat, with truncate extremities. As compared with the sternum of the *Phenacodus primævus*, the praesternum of the *P. vortmani*, is much less compressed posterior to the costal articulations, and the peduncles of the latter are moré prominent. The fourth sternal segment in the *P. primævus* is shorter and wider. I find no trace of clavicle in this species.

Measurements of ribs and sternum.

			M.
Widths of fourth rib	transverse	from head to tubercle	.006
		of tubercle	.0055
		of shaft near tubercle	.0065
	anteroposterior at proximal fourth		.0025
Length of 12th or 13th rib			.109
Diameter of 12th or 13th rib at middle			.0045
Length of presternum			.0515
Length to posterior base of rib facet			.023
Depth at anterior base of rib facet			.009
Depth at posterior extremity			.008
Width at posterior extremity			.008
Diameter second or third segment	transverse	anterior	.0065
		posterior	.011
	anteroposterior		.022
Diameters fourth or fifth segment	transverse	anterior	.0015
		posterior	.013
	anteroposterior		.0195

The *scapula* is like that of *P. primævus* reduced in size. The oval glenoid cavity with anterior acumination separated by a shallow groove from a short subconic posteriorly turned coracoid process, is the same in both species. Both the scapulæ are somewhat mutilated, so that while the protuberance of the anterior edge above the neck is evident, it is not possible to know whether it is as much extended as in *P. primævus*. One portion lost in that species is present in the specimen of *P. vortmani;* the acromial process. As in the *Amblypoda* and *Proboscidea*, the spine of the scapula is recurved, forming a prominent shelf along its proximal portion; while the proximal extremity is extended forwards and downwards in a considerable acromion. As its extremity is broken off the length cannot be ascertained. This scapula resembles that of the *Dinocerata* and *Proboscidea*, with an acromion added. This gives it very much the character of that of some rodents, as *Sciurus*.

Measurements of scapula.

		M.
Total length		.106
Diameters of glenoid cavity	anteroposterior	.019
	transverse	.012
Width of neck		.020
Elevation or spine at acromion		.011
Length of coracoid from internal base		.004

The *humerus* is relatively somewhat more slender than that of *P. primævus.* In its details it is much like that of that species, but differs in the following points: The head is rounded in outline, or not compressed as it is in some degree in *P. primævus.* The subscapularis fossa is much less extended anteroposteriorly, and is more oblique. The bicipital and epitrochlear ridges are less prominent, and the epitrochlear bridge is rather longer. The three characters do not differ from those of *P. primævus.* The tuberosities are well developed, though not produced as in most *Diplarthra.* The proximal external edge of the greater tuberosity forms a convex ridge. The anterior boundary of the smooth surface of the head becomes a ridge on the inner half, and ends in a low angular tuberosity in the fundus of the bicipital groove. The *fossa teris minoris* is large and impressed. The *crista teris major* is weak proximally, but becomes prominent at the proximal third of the shaft. It is continued downwards on the front of the shaft as the rather obtuse angular anterior edge. It is less prominent than in *P. primævus* and *Periptychus rhabdodon*, and less evidently continuous with the deltoid crest than in many ungulates. The epicondyles are not very prominent, but the internal is the most so. The epicondylar foramen is rather larger than in *P. primævus.* The condyles are not much extended transversely. There is considerable trochlear surface on the distal external side of the flange, and terminates in a fossa posteriorly. The roller is quite convex, and the posterior trochlear groove is wide, deep, and somewhat oblique.

The ulna is stouter than the radius to near the distal extremities, where the long diameters are equal. The carpal extremity of the ulna is of course the most contracted. The ulna is however not so robust that the radius appears to cross it as in the *Proboscidea*, but it is compressed so as to be deep. The inferior edge is an angle, except at the extremity of the olecranon, and has a gently convex profile to about the middle of the length when it rises and for the remainder of the length is slightly concave. Between the humeral cotylus and this rise the shaft is triangular in section. Distad to this point is a compressed oval, since the external ridge descends to become part of the inferior border, forming an obtuse rib across the external side of the shaft. The olecranon is rather long and compressed, with superior acute edge and concave internal side. It is about as deep as the deepest part of the shaft, and is thickened below and distally. The humeral cotylus is flared behind on the external side only, and below most on the internal side. Below the internal flare to the middle of the length the side of the shaft is gently concave. The distal extremity has a flattened triangular section, and terminates in a small rounded tuberosity.

The radius is not much expanded at the head, and a horizontal section of its humeral surface is sigmoid, to suit the humerus. The external angle is obtuse and short, and turned slightly posteriorly. The shaft at its middle is a little flattened horizontally; more distally it is cylindric. The distal extremity is stout and not depressed. The superior tendinous groove is wide and shallow, while the supero-external groove is better defined by a short prominent longitudinal ridge below it. The articular surface is not so much contracted as in the *P. primævus*, and the facets are relatively wider anteroposteriorly. The scaphoid and lunar facets are faintly distinguished from each other, and in each the anteroposterior exceeds the transverse diameter. A small flat internal tuberosity and a similar posterior (inferior) one. External tuberosity larger and more prominent.

Measurements of fore limb.

			M.
Length of humerus			.113
Diameter of proximal end	anteroposterior (greatest)		.032
	transverse	at bicipital groove	.0155
		at lesser tuberosity	.021
Anteroposterior diameter of shaft at middle			.013
Width at epicondyles			.028
Width of condyles distally			.019
Anteroposterior diameter of condyle	externally		.012
	internally		.015
Length of ulna			.131
Depth of ulna at	extremity of olecranon		.012
	coronoid process		.019
	middle of shaft		.010
Diameters of distal tuberosity	vertical		.005
	transverse		.004

		M.
Diameter of carpal facet	vertical	.004
	transverse	.008
Length of radius		.090
Diameters of head of radius	vertical	.008
	transverse	.015
Diameter at middle of shaft	vertical	.006
	transverse	.009
Diameters of distal enlargement	vertical	.011
	transverse	.015
Diameters of carpal facet	vertical (at lunar)	.0075
	transverse	.011

There is a similarity between *Phenacodus* and *Hyrax* in the proportions of the ulna and radius. The carpal articulation of the ulna is, however, quite different in the two. In *Hyrax* it is larger and more transverse, as in *Elephas*. It is also larger in *Hyrachyus*. In fact, it is in *Phenacodus* much like that of a carnivore.

A part of the left anterior foot of this specimen is preserved. It is smaller than the posterior foot, the metacarpals and phalanges having smaller lengths and diameters. The thumb is as well developed as in the *P. primævus*, so that the presence of five toes in the manus is demonstrated. But one bone of the carpus is preserved, the trapezoides. It differs considerably in its form from that of the *P. primævus*, having a nearly quadrate anterior face, instead of a transverse one. It is considerably deeper anteroposteriorly than wide, and has not the navicular-like shape of that of the *P. primævus*. On this account I compared it with the magnum, but in this genus that bone has the usual posterosuperior convexity and posterior tuberosity. Both superior and inferior faces are concave anteroposteriorly. The posterior extremity is narrowed, and is recurved obtusely above, in an angle below.

The head of the second metacarpal has the internal third defined as the facet of the trapezium. The free edge of this facet overhangs a little the first metacarpal in front, the latter fitting an oblique groove of the second. To the third metacarpal, the second presents a crescentic facet, which is bounded below by a fossa. The shaft is not elongate, and the section is oval, the edge next the pollex being thicker than that next the third metacarpal. Its distal extremity is about the size of that of the third, and larger than that of the fourth. Its keel is distinct and exclusively posterior, and is directed somewhat towards the third metacarpal. The first metacarpal is small. Its distal extremity is cut off on the side next the M. ii, so that the keel is near the inner edge. The internal edge is oblique. The first phalange is rather elongate. The ungual has a round section near the base, and the lateral edge is but little developed. The first phalange of the second digit does not differ much from the corresponding piece of the posterior foot except in its inferior size.

The distal extremity of the third metacarpal is somewhat contracted, and does not have the flatness of that of the *P. primævus*. Its posterior keel is thick, and directed a little towards the thumb. The first phalange of the third digit differs from that of the pes in its less depressed form and slightly reduced length. The actual proximal depth is a little greater, while both extremities are distinctly narrower. Its distal ligamentous fossæ are well marked. The distal extremity of the fourth metacarpal is truncate on the inner side and oblique on the exterior; the superior or anterior part of the condyle being as narrow as in a carnivore, but not bounded proximally by the fossa usually seen in that order. Keel distinct.

The extremity of the fourth metacarpal is smaller than that of the second. Its external epicondyle is more prominent than the internal. The penultimate phalange is a little narrower than the corresponding one of the pes.

Measurements of manus.

			M.
Diameters of trapezoides	anteroposterior		.009
	anterior	vertical	.006
		transverse	.0053
Distal transverse diameter of M. i			.006
Length of 1st phalange of pollex			.008
Length of M. ii			.0352
Distal diameters M. ii	anteroposterior		.006
	transverse		.0095
Length of 1st phalange of M. ii			.015
Distal diameters M. iii	anteroposterior		.007
	transverse		.0082
Length of 1st phalange of M. iii			.018
Proximal diameters of 1st phalange of M. iii	anteroposterior		.007
	transverse		.010
Distal diameters of 1st phalange of M. iii	anteroposterior		.005
	transverse		.0072
Distal width of M. iv			.008

The *pelvis* has an intermediate type of form. It has no resemblance to that of a marsupial or a rodent, but resembles rather that of a carnivore, as *Canis* and *Ursus*. It is in fact intermediate between that of those genera and *Hyrax*, the ilium more resembling that of the latter. It has no tendency to the triradiate type of the *Diplarthra*, nor of the expanded form seen in the *Amblypoda* and *Proboscidea*. The infero-anterior edge of the ilium is longitudinally concave, though not so much so as in *P. primævus*, since the anterior angle is not so much produced. Most of the thin edge of the crest is preserved, and it forms a regular curve of moderate convexity, largely parallel to the infero-anterior border. It contracts gradually to the apex, which is a short truncate thickened border. Posteroinferiorly the posterior edge suddenly contracts to the wide neck. The infero-anterior edge is a flat edge of moderate width, which gradually expands at the neck to two-fifths the width of the external face. There is a rudimental anterior-inferior spine.

The superior edge of the acetabulum marks the middle of the length of the innominate bone, as it does in *P. primævus*. In *Hyrax capensis* it marks two-fifths the length from the tuberosity of the ischium. The latter is in *Phenacodus vortmani* strongly convex and not thickened. The spine of the ischium is a well-marked angle half way between the anterior edge of the acetabulum and the *tuber ischii*. The internal superior angle of the innominatum is obtuse opposite the acetabulum, and not expanded as in many *Diplarthra*. The pubes are lost excepting the symphyseal portion. This is considerably thickened, and the longitudinal median keel is distinct. This is crossed by an angle parallel to the acute anterior border of the pubes which is caused by a bevel of the antero-inferior face. The ischiopubic symphysis is narrow, indicating, with the form of the ischium, a large obturator foramen.

Measurements of pelvis.

		M.
Total length through acetabulum		.157
Length of ilium to acetabulum		.079
Width of plate of ilium at middle		.030
Width of peduncle	interiorly	.014
	anteroexteriorly	.010
	posteroexteriorly	.019
Diameter of pubis at base	anteroposteriorly	.010
	anterointernally	.0043
Length of obturator foramen		.038
Width of ischium at spine		.019
Width ischium at middle of obturator foramen		.015
Diameters of acetabulum	vertical	.016
	anteroposterior	.020

The *posterior limb* is more slender than that of the *P. primævus*. The tibia is a little longer than the femur, which equals the length of the innominate bone to the posterior edge of the obturator foramen. In *P. primævus* the tibia is a little shorter than the femur, and is six-sevenths the length of the innominate bone to the posterior edge of the obturator foramen.

The *femur* has mnch the character of that of *P. primævus*, including the transversely oval form of the head, and the large third trochanter. The long axis of the proximal extremity makes an angle of 45° with the long axis of the distal extremity, but the long axis of the head is parallel with that of the distal extremity. It therefore does not coincide with the long axis of the proximal extremity. The latter is at right angles to the long axis of the proximal extremity of the great trochanter. The latter is obliquely truncate, so that its external angle is as high as the head, and the internal angle is longer than the head. The latter sends down a strong ridge which bounds the large trochanteric fossa posteroexternally. The small trochanter is prominent. The rotular groove is rather narrow and elevated, and the internal ridge is a little more prominent on its convexity than the external. The internal condyle is narrower than the external. The latter has a fossa just external to the edge of its anterior portion. At its posterior base the face of the shaft has an external shallow fossa, and an internal rib. The former is bounded externally by a faint ridge which extends towards the great trochanter. Other than this there is no trace of *linea aspera*.

The *patella* is short and robust It is truncate above, and acuminate below. Anteriorly it is convex with two proximal lateral facets.

The *tibia* is rather slender. The crest is moderately prominent, and continues as the obtuse angular front of the shaft. Its superior extremity is truncate downwards. The spine is not prominent. The cotyli of the head are not very unequal; the internal longer fore and aft, the external longer transversely. The posterior face below the head is concave, with equal lateral bounding ridges. The internal ridge is the more acute, and continues as the angular internal edge of the shaft to near the distal extremity. The internal edge of the head is acute; the external edge is truncate in front of the fibular facet. The long axis of the distal extremity makes an angle of 35° with the long transverse axis of the head. The astragalar grooves are on the contrary parallel to the long anteroposterior axis of the head, and hence oblique to the long axis of the distal extremity. They are deep and with rounded fundus. The posterior apex of their dividing ridge is equally produced with the anterior angle, which is just exterior to the end of the dividing ridge. The internal malleolus is truncate below, and a little more produced than the two angles in question. The inner side of the internal malleolus is nearly flat, and but shallowly concave.

The head of the *fibula* is expanded anteroposteriorly. The shaft is slender and subround in section. The distal extremity is less expanded than the proximal, and its transverse and anteroposterior widths are equal. It has two distal faces; one oblique for the astragalus, the other transverse and free. The tendinous groove of its external face is only marked by a short prominent ridge on its anterior border.

Measurements of posterior limb.

		M.
Total length femur from head		.133
Length to line of middle of third trochanter		.049
Length to internal rotular crest		.106
Diameters of head	long	.018
	short at middle	.013
Total width femur	proximally	.036
	at third trochanter	.024
	below third trochanter	.013
Greatest width of condyles		.028
Depth of condyle at inner rotular crest		.038
Diameters of patella	longitudinal	.020
	transverse	.015
	anteroposterior above	.011
Total length of tibia		.143
Diameters of the head	anteroposterior at internal cotylus	.034
	anteroposterior at popliteal notch	.025
	transverse	.030
Anteroposterior diameter of shaft at distal third		.010
Diameter of distal end of tibia	anteroposterior (greatest)	.018
	transverse	.018
Length of fibula		.128
Proximal diameter anteroposteriorly		.014
Diameter of middle of shaft		.005
Distal diameter	anteroposterior	.0115
	transverse	.010

Besides its greater slenderness, the tibia of this species differs from that of *P. primævus* in the greater obliquity of the astragalar grooves. In the latter species, the internal malleolus is more protuberant, and not truncate, as in the *P. vortmani;* and the anterior face of the distal end is flatter.

The *posterior* foot is almost entirely preserved. The astragalus is a diminutive of that of *P. primævus*, having the trochlear groove profound and the neck elongate. It differs from it in having the trochlear ridges less unequal, and the head not so flat and laterally extended. The depth of the navicular face is two-thirds the transverse length, while in *P. primævus* it is about half the same. Its convexity is flattened in the *P. vortmani*, but entirely convex in *P. primævus*. Its surface is strongly recurved on the internal side as in that species. The median fossa of the internal face of the trochlear portion is distinct. The inferior anterior angle of the external side of the same is flared outwards, as in *P. primævus.*

The *calcaneum*, like the bones of the leg, is rather more slender than that of *P. primævus*. It resembles that element of the latter in most respects, but the cuboid face does not present the internal angle, but is truncate next the astragalus. It is quite oblique, both in the transverse and the vertical direction. The cuboid is like that of *P. primævus*, but is less robust. The navicular also resembles that of the same animal. The part of its distal face which is applied to the ectocuneiform is convex, to fit the proximal concavity of that bone. Its posterior down-looking tuberosity is well developed. The ectocuneiform and mesocuneiform are relatively more elongate than in *P. primævus*. The latter is quite flat in *P. primævus*, but its length in *P. vortmani* equals its width at the second metacarpal. On the other hand the entocuneiform is shorter than that of *P. primævus*, and its inferior border is compressed, and little or not at all truncate. The metatarsal facet is, on the contrary, quite strong in the *P. primævus.*

The metatarsals are all rather more slender than those of *P. primævus.* The fourth is as long as the calcaneum in *P. vortmani*, but is shorter than that element in *P. primævus*. When in place it is not quite as much produced as the second in *P. vortmani;* in *P. primævus* it reaches the same point on the third. The second is produced proximally more than the others, and hence has a lateral facet for the ectocuneiform. The relative lengths of the toes are, beginning with the shortest, I, V, IV, II, III, as in *P. primævus*. The fifth is half as long as the third; it has two proximal articular facets; one lateral, the other proximal for the cuboid. The first has a single narrow proximal facet. The distal articulations are not so stout as in *P. primævus*, but the distal posterior keels are well developed. The ligamentous fossæ are well defined.

The first metatarsal is somewhat flattened, and its shaft is divergent from that of the second. The proximal extremity of the shaft is obliquely truncate where it is in contact with the concave side of that of the second. The inner part of the extremity is an anteroposterior tuberosity. The head of the second metatarsal is also somewhat obliquely directed inwards, and there is a low tuberosity on the external edge of the shaft just below it, which fits a fossa of the applied edge of the third metatarsal. The distal extremity is less flattened than that of the first, but is narrower than in most three-toed ungulates.

The phalanges of the proximal set are rather elongate, and are flat. Their distal extremities are narrower than the proximal and are depressed, the articular faces not being recurved to the superior surface. Distal ligamentous fossæ strong. The phalanges of the second set are longer than wide, and less robust than those of *P. primævus*. Their distal articulations are more recurved. Four ungual phalanges are preserved in a fair condition. They are rather narrow, and display the inferior proximal subtriangular table without lateral groove. Extremity notched; inferior surface concave. That of the hallux is short and subconic.

Measurements of posterior foot.

	M.
Total length of foot	.148
Length of tarsus	.058
Length of median digit	.043
Tarsus.	
Length of calcaneum	.045
Depth at heel	.012
Width at sustentaculum	.019
Depth at astragalar condyle	.009
Diameters of cuboid facet { vertical	.009
Diameters of cuboid facet { transverse	.010
Length of astragalus on inner crest (total)	.023
Length of astragalus on outer crest (diameter trochlea)	.0125
Width of trochlea, including crests	.011
External elevation of trochlea	.011
Diameters of navicular face { vertical	.009
Diameters of navicular face { transverse	.012
Diameters of navicular bone { longitudinal	.006
Diameters of navicular bone { proximal { anteroposterior	.014
Diameters of navicular bone { proximal { transverse	.017
Diameters of cuboid { proximal { anteroposterior	.010
Diameters of cuboid { proximal { transverse	.011
Diameters of cuboid { longitudinal (median)	.014
Diameters of cuboid { median { anteroposterior	.012
Diameters of cuboid { median { transverse	.010

		M.
Diameters of entocuneiform	anteroposterior	.010
	vertical	.011
	transverse	.007
Anterior vertical diameter of mesocuneiform		.0055
Diameters of entocuneiform	proximal { anteroposterior	.006
	proximal { transverse	.003
	longitudinal	.008

Metatarsus.

Length of first		.022
Anteroposterior diameters	proximally	.007
	distally	.006
Length of second		.047
Proximal transverse diameters		.008
Distal diameters	anteroposterior	.008
	transverse	.0085
Length of third		.048
Proximal transverse diameters		.009
Width at middle of shaft		.007
Width at above condyle (greatest)		.0095
Distal diameters	anteroposterior	.008
	transverse	.009
Length of fourth		.044
Proximal diameters	anteroposterior	.010
	transverse	.006
Median width		.005
Distal diameters	anteroposterior	.007
	transverse	.008
Length of fifth		.025
Proximal diameters	anteroposterior	.006
	transverse	.0055
Anteroposterior diameter at middle		.004
Distal diameters	anteroposterior	.006
	transverse	.006

Phalanges of fore or hind foot.

First row.

Length of first (or fifth) digital		.010
Distal diameters of first (or fifth) digital	vertical	.003
	transverse	.004
Length of second digital		.018
Proximal diameters of second digital	vertical	.006
	transverse	.0095
Distal diameters of second digital	vertical	.0045
	transverse	.007
Length of third digital		.020
Proximal diameters of third digital	vertical	.007
	transverse	.011
Distal diameters of third digital	vertical	.005
	transverse	.008
Length of fourth digital		.015
Width of fourth digital	proximal	.009
	distal	.006

Second series.

		M.
Length of third digital		.012
Proximal diameters	vertical	.006
	transverse	.0085
Distal diameters	vertical	.004
	transverse	.008

Ungual phalanges.

Length		.014
Proximal diameters	vertical	.0055
	transverse	.007

It is, unfortunately, not possible to determine whether the phalanges above measured belong to the anterior or posterior foot, owing to the fact that they were associated with a collection of the bones of both in one portion of the slab containing them. The bones of the posterior foot as far as the phalanges, were continuous, while those of the metacarpus were wanting. It is, therefore, probable that these phalanges belong to the posterior foot.

Besides the specimen above described, Mr. Wortman brought portions of four mandibles from the Wind River bad lands, and twelve from those of the Big Horn.

Restoration.—The limbs of this species are rather elongate for an eocene mammal, and the anterior limbs are distinctly shorter than the posterior. The size of the animal is that of a bull-dog, but the head is smaller, and the neck rather shorter, and not nearly so robust. The limbs have about the same proportions to the body as those of a bull-dog, but the anterior ones are shorter. The proportions of the parts of the limbs, and of the fore and hind limbs to each other, excepting the feet, are much as in the collared peccary. The feet of the latter animal are longer than in *Phenacodus vortmani.*

We can thus imagine the *Phenacodus vortmani* as an animal of the comparatively slender build of the bull-dog, with a neck and head proportioned more as in the raccoon, and with the rump more elevated than the withers, as in the peccary. The feet resembled those of a tapir or rhinoceros, but had a pair of short toes on each side, which did not reach the ground. To this add a tail much like a dog's in proportions, and the picture is complete. The diet of this animal was omnivorous, with a smaller proportion of animal food than the hogs, for instance, use. The food is more likely to have resembled that of the quadrumana. What means of defense this species had is not easily surmised, as the canine teeth and hoofs are not large.

I have named this species after Mr. Jacob L. Wortman, of Junction City, Oregon, whose explorations in the West have been more than usually productive of important results.

PHENACODUS CALCEOLATUS Cope.

Proceeds. Amer. Philos. Soc., 1883, p. 559.

Plate XXIV *g;* fig. 7.

This species is founded on fragments of the skull and limbs, with teeth, of a single individual. The teeth consist of two superior and four inferior molars of one side, and a smaller number of those of the opposite side.

The teeth are of the size of those of the *Phenacodus puercensis*, and, like that species, there is no median external cingular cusp of the superior molars. In these teeth the external basal cingulum is weak, but there is a strong anterior cingulum, distinct from any of the cusps. No internal cingulum. External cusps conical, well separated; intermediate cusps rather large; internal cusps rather large, close together, but deeply separated. The last superior molar is reduced in size. It has well-developed anterior and posterior cingula, a weak external and no internal cingula. The intermediate tubercles are rather large, and there is one large internal tubercle.

The heel of the last inferior molar is short, wide, and rounded. The posterior tubercle is but little behind opposite the posterior internal tubercle. The latter is separated from the anterior inner by a deep fissure, while the opposite side of the crown is occupied by a large median external cusp, which has a semicircular section. The large anterior cusps are confluent on wearing. No anterior cingulum in the worn crown. The crowns of the first and second true molars of the specimen are rather worn. They show that the posterior median tubercle is very indistinct and probably absent The bases of the smaller inner cusps are round, and on wearing unite with the larger external cusps. Of the latter the posterior is the larger. Anterior cingulum rudimental or wanting. No lateral or posterior cingula. The principal peculiarity of the lower dentition of this species and the one from which it is named, is the form of the third or fourth (probably third) premolars, both of which are preserved. They have a compressed apex, which descends steeply to the anterior base, without basal or lateral tubercle. The base of the crown spreads out laterally behind, and is broadly rounded at the posterior margin, so as to resemble the toe of a wide and moccasined foot. It is depressed, the surface rising to the apex from a flat base.

Measurements.

		M.
Diameters of second superior molar	anteroposterior	.0080
	transverse	.0100
Diameters of last superior molar	anteroposterior	.0067
	transverse	.0085
Length of inferior true molars		.0258
Diameters of M. ii	anteroposterior	.009
	transverse	.008
Diameters of M. iii	anteroposterior	.0085
	transverse	.0068
Diameters of the Pm. iii	anteroposterior	.008
	transverse	.005

About the size of the *P. puercensis.*

PHENACODUS PUERCENSIS Cope.

Proceeds. Amer. Philos. Soc., 1881, p. 492. Paleontological Bulletin No. 33, p. 492, Sept. 30, 1881.

Plate XXV *e;* figs. 12–13. LVII *f;* figs. 8–9.

In the Paleontological Bulletin No. 34 I combined this species with the *P. vortmani,* which it greatly resembles in size and proportions. It is an abundant species in the Puerco formation of New Mexico, and I have recently secured numerous jaws with teeth. Among these there are six fragments of maxillary bones with molars. I find, on comparing these with the true *P. vortmani* of the Wasatch beds, that they uniformly differ in the absence of the external tubercle which rises from the cingulum opposite the space between the principal external cusps. In this respect they resemble the *Protogonia subquadrata.* The difference is important, and I anticipate that other characters will be found confirmatory of the distinction of the species here maintained.

Typical specimens have the following characters:

Last superior molar smallest; first and second true molars with six tubercles, two external, two median, and two internal. A strong basal cingulum, except on inner side. Inferior true molars, besides the usual five tubercles, furnished with an anterior ledge with a tubercle at its interior extremity. A weak external basal cingulum. A little larger than the *P. vortmani.* Length of superior true molars, M. .021; length of base of crown of M. iii, .006; do. of M. i, .008; width of do., .008; length of base of crown of inferior M. iii, .0085; width of do. in front, .006; depth of ramus at M. i, .019.

There is some variation in the form of the smallest superior molar, some being nearly round and others suboval in outline. Two mandbular rami with molars represent smaller individuals. They measure as follows: Length of bases of last four molars, M. .033; do. of true molars, .025; length of crown of inferior M. iii, .0082. Depth of ramus at M. ii, .0145.

Besides the absence of the external median tubercle, this species differs from the *P. vortmani* in two other points. The posterior external cusp of the fourth superior premolar is smaller, showing in this point also an approximation to the genus *Protogonia.* The second point is the presence of the posterior intermediate tubercle on the second superior true molar. It is wanting in the *P. vortmani.*

The temporary fourth premolar is in both jaws a more complex tooth than the corresponding permanent tooth, as usual among ungulate mammalia. The fourth superior milk molar is a little longer than wide, and the anterior external cusp is compressed. In front of this the cingulum throws up an angle. The internal cusps are large, and the anterior intermediate is, opposite the anterior external, and anterior to the anterior internal. It results that, on wearing, the anterior, internal, and intermediate, form a curved surface. There are posterior and anterior cingula, but no external or internal.

The fourth inferior milk molar is extended chiefly anteriorly, where the worn anterior cusp makes a crescent with the anterior external, become in this tooth a median external. The interior median is opposite the exterior. The heel has three cusps, two large lateral and a small median. The external posterior sends a low ridge to the interior median. The second and third premolars also have milk predecessors.

Numerous specimens were found by Mr. Baldwin in the Puerco beds of New Mexico.

PHENACODUS MACROPTERNUS Cope.

Paleontological Bulletin, No. 34, pp. 179–180, Feb. 20, 1882.

Plate XXV *e*; fig. 15.

This species is apparently rare, being represented by only one mandibular ramus, which supports the posterior three molars, and a possible second ramus with molars iv and v. The first and second true molars are much like those of *P. vortmani*, but the third is relatively larger, and has an especially elongate heel. In *P. vortmani* the last molar is constricted, and narrower than the penultimate. In *P. macropternus* there is a weak external, and no internal cingulum. The tubercles of the last two molars are quite regularly conical, while the external pair of the first molar wear into crescents. Smaller than the *P. vortmani*.

Measurements.

		M.
Diameters M. i	anteroposterior	.0062
	transverse	.0050
Diameters M. iii	anteroposterior	.0095
	transverse	.0065
Depth of ramus at M. ii		.0130

From the Wasatch of the Big Horn, Wyoming; J. L. Wortman.

PHENACODUS BRACHYPTERNUS Cope.

Paleontological Bulletin, No. 34, p. 180, Feb. 20, 1882.

Plate XXV *e*; fig. 14.

Three mandibular rami are the only specimens of this species found by Mr. Wortman in the Big Horn region. They all display the fourth premolar, which has the characters of this genus, as distinguished from *Mioclænus*. The species is materially smaller than the *P. vortmani*, and its last inferior molar is intermediate between those of the latter and the *P. apternus*, in form. Both the internal and external intermediate tubercles are very full, and give the tooth posterior width. The posterior or fifth tubercle is large, and gives the posterior outline of the crown a trifoliate form. The posterior median tubercles of the M. ii and i are well marked. The molars gradually increase in size forwards, and the fourth premolar is longer than any of them, and rather narrow. Its form is more elongate than that of any other species, as it has a well-developed anterior lobe. It is apparently different in form from the deciduous molar of *P. primævus* and *puercensis*, and as the third true molar is in place and worn, the tooth is doubtless one of the permanent series. When the superior molars are discovered it may be found that the species belongs to the genus *Diacodexis.* The third premolar has an elongate principal cusp, whose anterior part is broken off in my specimens. Its heel is short and wide. The heel of the Pm. III is short and wide. On the true molars a weak external cingulum. Enamel slightly wrinkled.

Measurements.

		M.
Length of posterior four molars		.0277
Length of crown of Pm. iv		.0076
Diameters M. iii	anteroposterior	.0070
	transverse	.0046
Depth of ramus at Pm. ii		.0100
Depth of ramus at M. ii		.0128

From the Big-Horn Wasatch formation; J. L. Wortman.

PHENACODUS ZUNIENSIS Cope.

Paleontological Bulletin, No. 33, p. 492. Sept. 30, 1881. Proceeds. Amer. Philo. Soc., 1881, p. 492. Plate, LVII*f*; fig. 10.

This is the smallest species of the genus yet known. It is represented by many mandibular rami, the best preserved of which supports the last four molar teeth.

In the last true molar the anterior external cusp is larger than any of the others. There is a minute anterior inner. The external lateral is the apex of a crescentoid crest; the corresponding inner one is smaller, and is part of the raised border, which culminates in a small median tubercle posteriorly. The first and second true molars are narrowed in front, and there is no distinct anterior ledge, only a minute anterior inner tubercle. The external cingulum is more distinct and the enamel is wrinkled. The fourth premolar has a short base, and the inner cusp is much smaller than the principal one; it has a wide heel and an anterior basal tubercle. Length of true molars, M. .018; of last true molar .006; of first true molar .006; width of do. .004; depth of ramus at do. 011.

The anterior border of the masseteric fossa is well marked to the middle of the depth of the ramus. The fossa is elsewhere undefined.

From the Puerco epoch of Northwestern New Mexico; D. Baldwin.

DIACODEXIS Cope.

American Naturalist, 1882, p. 1029. (Dec. 2 (?).)

Inferior and superior molars tubercular; the superior with intermediate tubercles, but no external cingular tubercle. Principal external cusps subconic; the anterior internal cusp continuing into the anterior intermediate on trituration. There are four superior premolars, of which the posterior three have two external cusps. The third and fourth have well-developed internal lobes. The inferior premolars and molars are unknown, except the last true molar. This has distinct cusps, not joined by crests, and a prominent heel or fifth lobe.

It is yet to be positively determined whether this genus belongs among the *Condylarthra* or with *Pantolestes* in the suilline *Artiodactyla.* It somewhat resembles *Dichobune,* but differs in the absence of the intermediate crescentic lobe of the superior molars, and the development of a transverse crest on wearing. These teeth are in fact like those of the *Perissodactyla,* and this fact, together with the bunodont type of the true molars, induces me to place the genus provisionally in the *Phenacodontidæ.*

But one species is known, which is from the Wasatch formation.

DIACODEXIS LATICUNEUS Cope.

American Naturalist, June 1882, p. 1029. (Dec. 2(?).) *Phenacodus laticuneus*, Paleontological Bulletin No. 34, 1882, p. 181.

Plate XXV*e;* figs. 17–18.

This is a small species, and is represented by six superior molars and the last inferior molar in a fragment of the lower jaw. The latter tooth exhibits peculiar characters already mentioned. The superior molars differ from those known to belong to the *Phenacodus primævus* and *P. puercensis* in having a vertical fissure of the inner side which separates the bases of the two internal tubercles. This gives them some resemblance to the superior molars of the species of *Anisonchus*, but the important difference remains in the separation of the anterior inner tubercle from the intermediate tubercles. The three are confluent into a V in the genus last mentioned.

The external cusps of the superior molars are rather acute, and lenticular in section, their external sides forming a convex rib There is no rib between the external sides. There is a strong anterior cingulum, which terminates externally in a low angular cusp. There is no cingulum on any other part of the crown. The second, third, and fourth premolars have two external cusps, and much resemble the corresponding teeth in *Hyracotherium*. The second is longer than wide, and has an internal ledge; the third is as wide as long, and has a wide internal ledge; the fourth is wider than long, and has an internal and two intermediate cusps, and an anterior and posterior cingulum. They all have a weak external cingulum, of which a trace exists in the true molars.

The last inferior molar has a double anterior inner cusp, as in some *Mesodonta*, and the external anterior cusp is robust. All the cusps are conical and with round section, and their bases are close together. The outline of the base of the crown is almost an isosceles triangle with rather wide base in front.

Measurements.

		M.
Length of last six superior molars		.0350
Length of true molars		.0160
Diameters of M. ii	anteroposterior	.0055
	transverse	.0080
Long diameter base of Pm. ii		.0050
Diameters Pm. iii	anteroposterior	.0060
	transverse	.0060

From the Wasatch beds of the Big Horn River, Wyoming; J. L. Wortman.

MENISCOTHERIUM Cope.

Report on Vertebrate Fossils from New Mexico; U. S. Geog. Geol. Survey W. of 100th Meridian, 1874, p. 8. Report U. S. Geogr. Geolog. Survey W. of 100th Meridian, 1877, Vol. IV, Pt. II, p. 251. American Naturalist, 1882, p. 334.

As already indicated, with this genus we enter a new family of the Condylarthra, and one which superadds to its general structure a considerable specialization of the molar teeth. The present genus is the only one of the family yet known, and it is represented by but three species. With present knowledge it must be asserted that the range of this genus was limited both in time and space. The remains of the species have been derived from the Wasatch beds of New Mexico, and from a horizon from near its base, which overlies the Puerco. The genus has not been found in any other of the areas of the Wasatch deposits, but it occurs abundantly in its locality, so that it is possible to e erm ne its general structure.

Dental formula, I. $\frac{?3}{3}$; C. $\frac{1}{1}$; Pm. $\frac{4}{4}$; M. $\frac{3}{3}$; without diastema. Incisors and canines in both jaws small. Fourth superior premolar unlike first true molar. Fourth inferior premolar like first inferior true molar. Third inferior true molar without third lobe or heel. Superior incisors with small oval crowns. Inferior incisors with short depressed crowns. First and second superior premolars without internal lobes. First, second, and third inferior premolars without internal lobes. True molars supporting two external Vs, which are separated by a vertical ridge. Two internal lobes; the anterior conic, the posterior crescentic, the anterior horn of the crescent connecting with a central intermediate lobe, so as to form an oblique cross-crest on wearing An anterior intermediate lobe forming a crescent concentric with the anterior V. Last superior molar without the posterior internal V. Fourth superior premolar with two external Vs, two intermediate lobes, and an internal cone. Third superior premolar with external wall and internal cone.

First inferior premolar one-rooted; third, two-rooted, with compressed crown. Fourth premolar and true molars consisting of two Vs, whose adjacent limbs join each other. Inferior canine incisor-like.

Orbits widely open posteriorly. Palate excavated between the molars posteriorly. A postglenoid process. Auricular meatus widely open below. Posttympanic and paroccipital processes united. Occipital and sagittal crests. Foramen ovale scarcely separated from *f. lacerum* anterius. Mandibular symphysis coössified. Mandibular condyle uplooking. Brain not so small as in *Phenacodus*.

The cervical vertebræ are, like those of *Phenacodus*, of medium length, and slightly opistheocœlous. They, with the lumbars, have an inferior keel. The disparity in size between the anterior dorsals and lumbars is marked. The odontoid process is cylindric. Metapophyses of lumbars well marked. No anapophyses. The neural canal is rather large.

The head of the scapula has a curved coracoid process. The spine rises abruptly from the neck. The humerus is much like that of *Phenaco-*

dus or a carnivore. The tuberosities are not produced, and there is no external epicondyle. The internal is large, and sends upwards the bridge that incloses the epicondylar foramen.

The pelvis is not well preserved in the specimens. It shows a rather narrow, triangular neck, a well-marked anterior inferior spine, and an open acetabular groove. The femur shows a *fossa ligamenti teris*, and the third trochanter on the middle of the shaft, and well developed. The tibia has an anterior crest, and no notch on the antero-external edge of the head. The internal malleolus is a prominent tuberosity, and the astragalar face is scarcely grooved, and oblique, as in the *Creodonta.* The distal extremity of the fibula articulates with the sides of the astragalus, but not with the calcaneum. The calcaneum much resembles that of *Phenacodus.* It is elongate, and the astragalar facet is not longitudinal, but very oblique. The astragalus has trochlear keels of unequal height, and a shallow groove between them, much less marked than in *Phenacodus,* but not so flat as in *Periptychus.* The neck is elongate, and the distal articular surface is convex in every direction.

The number of toes in *Meniscotherium* is unknown. Metapodial bones preserved are rather narrow, leading to the supposition that the digits are similar to those of *Phenacodus* and *Periptychus.* The posterior keels of the distal extremities of the metapodials are distinct.

Three species of this genus are known to me. They may be distinguished as follows:

Length of last four molars, M. .034 .. .. *M. terrærubræ.*
Length of last four molars, M. .029 .... *M. chamense.*
Length of last four molars, M. .024 .... *M. tapiacitis.*

The superior molars of this genus have some resemblance to those of *Hyopotamus*, but the inferior molars are different, and of the *Anchitherium* type. The temporary last molar differs from its successor, the permanent fourth premolar, in its more complex character. It is in fact identical in structure with the true molars. The second and third milk molars differ from their successors in their greater anteroposterior extent. The third has an internal lobe opposite the posterior half of the external wall.

MENISCOTHERIUM TERRÆRUBRÆ Cope.

Paleontological Bulletin No. 33, p. 493, Sept. 30, 1881. Proceed. Amer. Philos. Soc., 1881, p. 493. Plate XXV*f*, figs. 12–14: XXV *g*.

I have parts of skeletons of many individuals of this animal, but each one in a very imperfect condition. It is practicable from these to learn the osteology of the species, but in using the measurements it must be borne in mind that there is some range of variation in the different individuals.

A specimen with both jaws crushed in a limestone nodule displays two superior incisors. They are short, and a little expanded laterally. The second has median angle and an external oblique edge. On the third the apex is at the internal angle, and the edge is all external and oblique. Surface smooth.

Measurements of incisors.

	M.
Length of crown of superior I. i	.004
Width of crown of superior I	.004
Length of crown of superior I. ii	.005
Width of crown of superior I. ii	.004
Length of inferior true molars of same	.025

In a second specimen the superior molar series is most complete.

The dimensions of the superior molars increase to the penultimate, while the external and posterior sides of the last molar are contracted, reducing its size. The external faces of the external Vs of the true molars are considerably impressed; those of the premolars are nearly flat.

The second premolar is two-rooted, and has a compressed crown, without either heel or cingulum, except a thickening of the posterior base. The base of the crown is triangular. The external plate of the third premolar is simple, and is connected with the internal cusp by a cingulum on the posterior base of the crown. The crown is transverse, and the inner tubercle rather small. The fourth premolar is much larger than the third. Its external plate is divided into two apices, which are not impressed. Their external faces are separated by a faint ridge and are divided medially by a faint ridge. The anterior external angle is rather prominent. The anterior and posterior cingulum extend to and round the inner base of the interior tubercle. Within the anterior external apex is a well-developed intermediate crest parallel to it; and there is a corresponding crest within the posterior external apex. This one turns inwards at its posterior extremity, which is on the posterior cingulum.

The anterior angle or horn of each external crescent of the true molars is very prominent. They are sections of short vertical ridges, which unite near the base of the crown, giving abruptness to the impression of the external surface of the anterior lobe. The middle of each face has a faint median ridge. The two molars have an anterior basal cingulum, but no posterior or internal, excepting a trace between the bases of the internal lobes. The anterior intermediate crescent is quite parallel with the external; the anterior internal tubercle has a slightly V-shaped section. The posterior inner tubercle is quite confluent with an oblique intermediate crest, as in *M. chamense.* In the last true molar, as there is only one internal tubercle, this crest is short, terminating at the posterior border. The last true molar is like the last premolar, except in its two impressed external crescents. Enamel smooth.

Measurements of superior molars.

		M.
Length of superior molars, less Pm. i		.046
Length of true molar series		.028
Length of base of Pm. ii		.005
Diameters of base Pm. iii	anteroposterior	.006
	transverse	.007
Diameters of base Pm. iv	anteroposterior	.008
	transverse	.010
Diameters of base M. ii	anteroposterior	.011
	transverse	.013

There are some variations in the characters of the fourth superior premolar. In the specimen above described, which is the one on which the species was based, the external face of the crown is uninterrupted, and there is a weak cingulum round its internal base. In another specimen there is a narrow vertical ridge distinguishing the two lobes on the external face, as is found in the true molars, but less developed, and the internal cingulum is strong In another specimen with the external ridge, there is no internal cingulum.

A characteristic of this species is the great contraction of the premolar series. This condition in an imperfect specimen led me to suspect in first describing the species that the fourth inferior premolar was only the third. The inferior incisors, canines, and premolars form an uninterrupted series. The incisors are subequal, and have flat crowns, projected obliquely forwards. The crown of the canine is but little larger than that of the third incisor. It is not canine-like, but has a compressed crown, which is inclined forwards, and has therefore an oblique base. The superior edge consists of a long posterior and short anterior portions. The crowns of the first and second premolars are compressed conic, convex in front, and straight behind, the second with a minute heel. The root of the latter has a figure 8 section; its two elements not being separated. The third premolar is a large tooth, and the crown consists of an anterior compressed obtuse lobe, and a wide heel, with a median cutting edge. There is a small anterior basal lobe, and the interior side may be vertically grooved behind it, as in various *Artiodactyla*, but not necessarily. The fourth premolar is abruptly very much larger than the third, and is as long as the first true molar, and a little narrower anteriorly. In details of structure it is exactly like the true molars.

The internal extremities of the posterior branches of the Vs composing the inferior molars rise into an acute cusp. The anterior branches of the Vs, on the other hand, are curved backwards, partly closing the open mouths of the Vs. The resulting pattern on wearing is a good deal like that of the *Anchitherium*. The anterior limb of the anterior V is a little more produced in the fourth premolar. The last true molar is like the others. No cingula.

Measurements of inferior teeth.

		M.
Length of dental series, including canine		.055
Length of true molar series		.027
Length of alveolus of canine		.0025
Height of crown of canine		.006
Length of crown of third premolar		.0058
Diameters fourth premolar	anteroposterior	.009
	transverse	.006
Diameters second true premolar	anteroposterior	.0092
	transverse	.0062

Several fragmentary crania are preserved. The best of these lacks the muzzle and the zygomata. This one is remarkable for fullness of the frontal region, the profile descending from it both anteriorly and posteriorly. The brain-case is not narrowed, but the muzzle contracts rapidly. The orbits are large and the malar bones prominent. Their posterior border is indicated by postorbital angles only; there are no zygomatic angles bounding them below. The former rise into strong temporal ridges, which rapidly converge and unite to form a low, straight, and narrow sagittal crest. The parietal bones over the hemispheres are convex in all directions, and posterior to them are a little concave, where the surface rises posteriorly into the supraoccipital crest. This projects upwards, and, being concave in the middle, has a bilobed outline, though not so strongly as the *Periptychus rhabdodon*. The occiput is concave on each side above the position of the condyles. The lateral crest continues into the posttympanico-paroccipital crest, which is incurved on the inferior base of the skull as far as the occipital condyle. The basioccipital has a faint median keel. The basisphenoid is flat. Between the postglenoid and the foramen ovale there is a small concave face, which extends from the anterior base of the zygoma to the external meatus on the inferior surface. On its interior edge is the base of the styloid process. In front of the latter its border is cut by the large foramen ovale, which is not cut off from the *foramen lacerum*. The palate is doubly excavated posteriorly to opposite the middle of the anterior external lobe of the third superior molar.

The infraorbital foramen is large, and issues above the middle of the fourth superior premolar. There are three postparietal foramina, serially arranged, the anterior on the squamosal suture. There are three suprasqamosals, also linearly placed, the last near the parietal suture. There are two inferior squamosals near the base of the zygomatic process, one above the other. The nasal bones widen posteriorly, but not so much as in the species of *Phenacodus*. Their posterior outlines are acuminate. The malar bone extends to the lachrymal. The lachrymal foramen is not external to the orbit. Palate rather concave. Frontal region slightly concave in transverse section.

The mandible is contracted forwards. The symphysis is narrowed, flat, and at the incisor teeth nearly horizontal. The rami are compressed and rather robust. The coronoid processes are elevated, and have nearly vertical anterior edges. The angles are produced posterior to the line of the condyles and are broadly rounded. The basal part is compressed and thin, but the posterior edge is thickened and projects inwards opposite the line of the molars. The condyles are close to the posterior base of the coronoids, and their superior articular face extends to the coronoid. The external expansion is semicircular in outline, and extends to the base of the coronoid anteriorly. The internal expansion is triangular, with the apex decurved and continuous with the posterior edge of the condyle. The articular surface descends on the latter, occupying the interior half of the condyle. There is no ridge or tuberosity below the condyle posteriorly as in many perissodactyles. The masseteric fossa is shallow, and only defined by the low anterior ridge. This ceases rather abruptly at the middle of the depth of the ramus below the last molar. The internal pterygoid fossa is not marked. The inner edge of the base of the coronoid is rather prominent. The dental foramen enters a little below the tooth line. There is one mental foramen, which is below the first premolar.

Measurements of skull.

	M.
Length from canine to foramen magnum, inclusive	.089
Width of skull at base of zygoma	.063
Width of palate between fourth premolars	.017
Length from occipital crest to beginning of sagittal crest	.039
Length from occiput to line of anterior border of orbits	.0635
Width at pos'orbital angles	.012
Width at auricular meatus	.034
Width of nasal bones behind (both)	.020
Width of sphenoid at foramen lacerum	.010
Width of foramen magnum	.0085
Depth of skull (without molars) at front of orbit	.034
Length of mandibular ramus (total)	.099
Length of ramus to base of coronoid	.066
Depth of mandible at condyle	.048
Depth of mandible at coronoid	.053
Depth of mandible at last molar	.017
Depth of mandible at fourth premolar	.015
Depth of mandible at canine	.006

The lower jaw above described belongs to the same individual as the skull, but lacks the condyles. These I derived from a perfect lower, which has the good series of teeth which furnished the description of them already given. For vertebræ recourse must be had to still other specimens.

The axis and third cervical *vertebræ* are so much like those of *Phenacodus vortmani* that the description of them will apply to these. They are somewhat smaller. The inferior keel terminates posteriorly in a triangular table with concave middle in both. The articular surfaces are very oblique, indicating the elevation at which the head was held. The neural canal is large.

Another specimen, or set of specimens of equal size, found together, furnish three cervical, seven dorsal, seven lumbar, the sacrum, and three caudal vertebræ. These are accompanied by mandibular rami of three individuals of identical dimensions. The first peculiarity observable in these vertebræ is the more than usually increased size of the lumbars. Were not the intermediate grades present, the dorsals and lumbars might be supposed to belong to different animals. The centra of the dorsals are depressed and a very little opisthocœl; their inferior surface is regularly rounded transversely, and a little concave anteroposteriorly. The anterior lumbars are similar, but larger. The posterior lumbars are not elongate, as in *Phenacodus* and *Hyracotherium*, but have the same proportions as the dorsals. They are depressed, and are strongly keeled below. The sacrum includes only three vertebræ, whose centra are much depressed. The diapophyses posterior to the first, form a continuous ledge interrupted only by the intervertebral foramina. The proximal caudal vertebral centra are as stout as those of the dorsal vertebræ and indicate a tail, stout at the base.

In the following measurements all the vertebræ measured are from the lot above described.

Measurements of vertebræ.

			M.
Length of axis without odontoid			.016
Expanse of axis at atlantal facets			.019
Diameters posterior face centrum	vertical		.007
	transverse		.0095
Diameters centrum of a cervical	posterior	vertical	.008
		transverse	.010
	anteroposterior		.009
Height of neural arch			.007
Expanse of postzygapophyses			.019
Diameters of an anterior dorsal	posteriorly	vertical	.0063
		transverse	.0085
	anteroposterior		.009
Diameters of a posterior dorsal	posteriorly	vertical	.006
		transverse	.0095
	anteroposterior		.010
Diameters of an anterior lumbar	posteriorly	vertical	.006
		transverse	.012
	anteroposterior		.011
Diameters median lumbar	posteriorly	vertical	.008
		transverse	.015
	anteroposterior		.012

			M.
Diameters posterior lumbar	posteriorly	vertical	.008
		transverse	.015
	anteroposterior		.015
Height of neural canal			.006
Expanse of metapophyses			.018
Length of sacrum			.038
Diameters anteriorly	vertical		.006
	transverse		.013
Diameters of first caudal	posteriorly	vertical	.0065
		transverse	.010
	anteroposterior		.010

The glenoid cavity of the *scapula* is an oval, acuminate anteriorly to an apex projecting downwards. The coracoid is prominent, and strongly incurved. Length of cavity, .009; width, .006. From a different individual from the one last described.

The *humerus* is much like that of *Phenacodus vortmani*, differing in the smaller size, less prominent tuberosities, and slightly different condyles. The shaft is flattened anteroposteriorly below the head, but there is no bicipital ridge, and the groove is restricted to the notch between the tuberosities. The latter project little, the greater not at all anteriorly, but it rises proximally into a convex crest. The teres minor fossa is flat, and the teres major ridge is distinct and extends on the proximal two-fifths of the length. The subscapularis fossa is small and flat, and is bounded by an angular ridge which rises upwards and outwards, forming the prominence of the lesser tuberosity. The external epicondyle is wanting. The internal is prominent and compressed, and sends a bridge over the epitrochlear foramen. The latter is as large as in the Creodonta, larger than in the species of *Phenacodus*, but not so large as in *Periptychus rhabdodon*. The crest of the external distal region is acute, but not very prominent. The shaft above the coronoid fossa is triangular, the anterior angle obtuse, the external acute. The condyle has an internal flange, and external to that a prominent rim not seen in *Phenacodus* nor *Periptychus*. It is represented by a short cylinder in *Diplarthra* generally, and the edge of it is turned up in *Hyracotherium venticolum*. In the *M. terrærubræ* it is a rim springing directly from the contraction of the roller, on the anterior side only. The convexity of the roller is marked, but too wide to be an intertrochlear crest, such as is seen in *Hyracotherium* and other *Diplarthra*. As compared with the *Phenacodus vortmani*, this humerus has a generally weaker structure.

Measurements of humerus.

			M.
Length of humerus			.092
Diameters proximally	anteroposterior (greatest)		.024
	transverse	at bicipital groove	.015
		at lesser tuberosity	.019
Diameters of shaft at middle	anteroposterior		.012
	transverse		.007
Width at epicondyles			.023
Width of condyles distally			.015
Anteroposterior diameter of condyles	externally		.010
	internally		.013
	at constriction		.008

The above described humerus is from the individual which furnished the vertebræ. It includes also femora, tibiæ, and a calcaneum, which I now describe.

The *femur* is rather robust. The heads are unfortunately lost from the specimen under consideration. The trochanters are evidently prominent. The third trochanter commences above the inferior termination of the lesser trochanter, and the base extends much below it. The patellar trochlea is prominent; the external condyle is a little wider than the internal, and less prominent. There is a fossa at its posterior base, while the inner has a shallow one on the middle of its side.

The *tibia* is a characteristic bone. It resembles considerably that of *Phenacodus*, but differs in many points. First, the crest is not obliquely truncate next the head, but is fully rounded. Second, it extends farther down the shaft, terminating rather abruptly a little above the middle of the latter. Third, the ridge which is its continuation, twists abruptly to the internal edge of the shaft, leaving the anterior face flat, instead of continuing down the middle of the front of the shaft, and turning a little outwards below. Fourth, the early disappearance of the external posterior angle, and its reappearance on the distal two-fifths of the length, and termination in an angle of the astragalar surface. Fifth, the disappearance of the posterior external angle on the middle of the length of the shaft. The posterior face of the shaft on its distal half is roughened for muscular insertion. The roughened area narrows upwards, and terminates in a narrow apex. Sixth, the obliquely truncate internal malleolus, which is acuminate as in many Creodonts. The long axis of the astragalar face is slightly oblique to the transverse long axis of the head, tending backwards towards the inner side. In this respect it is identical whith *Phenacodus vortmani.* In *Hyracotherium venticolum* the two axes are parallel. The tibia is much smaller than that of *Phenacodus vortmani*, and with the femur, prove that the posterior limb is little if at all longer than the anterior limb, as is the case in that species. The patella is an elongate oval, less robust than that of *Phenacodus vortmani.* It is rounded above and narrowed below.

Measurements of hind leg.

		M.
Diameter of head of femur No. 1		.0115
Length femur No. 2 from base of little trochanter, distad		.059
Length femur No. 2 from inferior base of third trochanter, distad		.049
Width of condyles posteriorly		.023
Elevation of rotular groove at external condyle		.024
Width of rotular groove with crests		.0095
Length of patella		.016
Proximal diameter patella	anteroposterior	.007
	transverse	.0085
Length of tibia		.097
Diameters of head	anteroposterior	.020
	transverse	.023
Diameters of shaft at middle	anteroposterior	.008
	transverse	.0065
Diameters distal end	anteroposterior	.0095
	transverse	.012

The *calcaneum* is remarkably elongate, and is but little expanded anteriorly. The free portion is compressed. The astragalar condyle has a peculiar position. It is placed diagonally across the middle superior ridge of the free portion, and presents its articular surface chiefly forwards. The sustentaculum is deep, short, and narrowed to a knob. The cuboid facet is concave vertically, and very oblique transversely, retreating on the inner side. Its form is subquadrate, with the sides oblique, and the superior and exterior angles produced and rounded. The latter marks the extremity of the external distal ridge, which extends from the calcaneal condyle.

Numerous *astragali* are preserved, but none accompany the specimen which has furnished most of the skeleton above described. This bone in this species looks much like that of a species of *Stypolophus.* The inner trochlear crest is much less elevated than the outer, and is obtusely rounded. The inner side of this part of the astragalus is oblique, and has a central fossa and a prominent shelf-like angle below it. The external face is vertical; it is separated by a groove from the concave calcaneal facet below. The trochlear groove is shallow, and extends to the inferior face posteriorly, after a posterior interruption by the ligamentous fossa. The sustentacular facet is a long oval, isolated all round. The neck is long and diverges from the axis of the trochlea. The navicular face is oblique to the base of the astragalus, in the same direction as the trochlea, but more so. It extends a little on the internal side to a point. On the external side the articular surface does not extend so far posteriorly, and is broadly rounded. In common with the entire astragalus, it is depressed.

Measurements of tarsus.

	M.
Length of calcaneum	.0215
Depth of calcaneum at middle	.009
Depth of calcaneum at astragalar condyle	.0116
Width of calcaneum at sustentaculum	.014
Diameters cuboid facet { vertical	.009
Diameters cuboid facet { transverse	.009
Length of astragalus (total)	.020
Length of trochlea on external side	.0115
Height of trochlea on external side	.0085
Width of trochlea at middle above	.0085
Width at trochlea below	.013
Diameters head { vertical	.007
Diameters head { transverse	.009

The phalanges preserved are flat, much like those of *Phenacodus*. Ungues not obtained.

Restoration.—This species was about the size of a fox, but with a very different physiognomy. The profile was curved, the muzzle short, and the eyes large. The body was not so slender as in *Phenacodus* or a fox, having the more robust proportions of a raccoon. The fore and hind legs were rather short, and of equal length, so that the rump was not elevated as in the peccary. There was a large tail. The species is one-third (linear) larger than the *Hyrax capensis*. It was probably a vegetarian.

The *Meniscotherium terrærubræ* differs from the *M. chamense* in two features. The first is its superior size. The second is the flattened form of the external faces of the true molars and the absence of the convexity of the external bases of the crown.

There is some variation in size. A few specimens apparently belong to a smaller race, whose last four superior molars measure M. .032.

All the specimens of this species in my possession were obtained by D. Baldwin in the Wasatch beds of New Mexico.

MENISCOTHERIUM TAPIACITIS Cobe.

Proceed. Amer. Philos. Soc., 1882, p. 470.

Plate XXV*f*, fig. 15.

The species now to be described is a good deal smaller than the *M. chamense*, and *a fortiori* than the *M. terrærubræ*. It is known to me from the nearly entire rami of a single mandible. These support the last five molars of one side or the other, and alveoli of two others and of the canine tooth.

Two characters besides the small size are observable in this jaw: First, the symphysis has not the shallow convex inferior outline in transverse section, but is on the contrary angular, having subvertical sides separated from a convex middle by a rounded angle. The symphysis is thus deeper than in *M. terrærubræ*. Second, the crown of the third inferior molar tooth has the form of that of the second of the *M. terrærubræ*. It is anteroposteriorly short, and has no keel nor anterior basal lobe; its section is lenticular, and profile subconic. In *M. terrærubræ* this tooth is elongate; with well developed keel and anterior lobe. The alveolus of the canine is relatively larger than that of the *M. terrærubræ*. The coronoid process does not rise so close to the last molar tooth, nor so steeply, as in the latter species. The posterior recurvature of the internal extremity of the anterior limb of the posterior V of the true molars is but little marked.

Measurements.

		M.
Length of true molars on base		.018
Diameters M. ii	anteroposterior	.006
	transverse	.0044
Diameters M. iii	anteroposterior	.0065
	transverse	.0038

	M.
Diameters Pm. iii { vertical	.0045
Diameters Pm. iii { anteroposterior	.004
Width of inferior face of symphysis	.008
Depth of ramus at Pm. iii	.009
Depth of ramus at M. iii	.0103

This species was obtained by Mr. D. Baldwin from beds of probably lowest Wasatch age, in New Mexico.

AMBLYPODA.

Cope System. Cat. Vert. Eocene New Mexico, U. S. Geog. Survs. W. of 100th M., 1875, p. 28. Report do. iv, pt. ii, p. 179.

Mammalia with small (? smooth) cerebral hemispheres, which leave the olfactory lobes and the cerebellum exposed. The feet short and plantigrade, with numerous (in the known genera five) digits, terminating in flat hoof-bearing ungual phalanges. The seven bones of the carpus distinct; the unciform articulating with the lunar as well as with the cuneiform. The astragalus flat, without trochlear groove, and attached to the tibia with little freedom of movement; its distal extremity divided into two facets, one for the navicular, and more or less of the other for the cuboid bone. Molars invested with enamel, with wide crowns and crests. A post-glenoid process.

The above characteristics are the only ones which can, in the author's estimation, be admitted into the ordinal category, for although the animals embraced in the *Amblypoda* present many other peculiarities, they are such as may readily vary within the limits of an order, and in fact do so in the families of many of the orders known to us. The above definition displays a double set of affinities, viz: those indicated by the structure of the feet, and those expressed by the type of the brain. The former exhibit a close resemblance to the feet of the *Proboscidea*, the approach being greatest in the hind foot. The principal difference in this extremity, is seen in the extension of the navicular articulation over the entire distal end of the astragalus in the *Proboscidea*, while in the *Amblypoda*, the navicular is shortened, thus permitting the cuboid to come in contact with the external part of the distal extremity of the astragalus. The cuboid is alike in the two orders, having considerable transverse extent, and supporting the external two metacarpals on its distal face. This lengthening of the navicular is a specialty of the *Proboscidea* among hoofed Mammals, the shorter form being characteristic of the lower types of both *Perissodactyla* and *Artiodactyla*, where the astragalus has two distal articulations. In the *Perissodactyla*, the extent of the navicular increases until the highest genus, the Horse, is reached, where it almost covers the entire end of the astragalus; but, in the *Artiodactyla*, the extension of the cuboid over the astragalus does not diminish. The nearest approach to the distal articulation of the astragalus of the *Amblypoda*, outside of the order, is seen in the Miocene Perissodactyle genus *Symborodon*. Here the cuboid and navicular facets are flat, and are separated by an oblique line, so as to be similarly incapable of hinge-like movement. The resemblance to the lowest *Artiodactyla* (*e. g.*, *Oreodon*, *Hippopotamus*) is very remote, for there the two facets are parallel, offering a ginglymus to the articulating bones.

The difference between the fore foot of the *Amblypoda* and that of the *Proboscidea* consists in the alternating position of the elements of the two carpal rows. This is also a character of the two other living orders of hoofed Mammals, and maintains itself with great persistency in both of them. It is essentially a primitive character, the alternating position being usual in the cold-blooded *Vertebrata*, and is the persistence of the oblique relation of the original divergent branching rays, to which digits have been traced. In the *Proboscidea* and *Hyracoidea*, the elements of the two rows assume an opposite and longitudinal relation. The structure of the fore foot in the *Amblypoda* appears to be about equally related to that of the *Proboscidea*, the *Perissodactyla*, and the *Artiodactyla*.

In the cubito-carpal articulation, the resemblance is again to the *Proboscidea* in the relatively large proportion of it belonging to the ulna, and the consequent lateral position of the latter bone. In this respect, it differs much more from the other two living orders of hoofed Mammals, although here again the lower forms of both resemble the *Amblypoda* more than do the higher forms. As is well known, both of the hoofed orders display a constantly diminishing extent of the ulno-carpal articulation, and increase of the radio-carpal, until, in the Horse and Ox, the ulna becomes a mere splint attached to the radius.

The relationships indicated by the brain are to the lissencephalous orders *Chiroptera*, *Insectivora*, and *Edentata*. As an ungulate order, the *Amblypoda* are distinguished from the first two, were other characters wanting. We may here notice, however, some curious resemblances between the forms of the teeth and lower jaw of *Coryphodon* and *Bathyopsis* and some *Insectivora*, and the still more curious resemblance between the tibio-tarsal articulation in the order and that of the cotemporary Creodont allies of the *Insectivora*. Comparison with the Ungulate forms of *Edentata* only is necessary, and from these the enamel sheathing of the teeth separates the *Amblypoda* at once. The small size of the brain doubtless relates these animals to the other Eocene *Ungulata* described by Lartet, still more nearly than to existing *Lissencephala*. In the small size and smoothness of the hemispheres, and relatively large development of the optic and olfactory lobes, the brain of the *Amblypoda* more nearly resembles that of the *Creodonta* than that of any division of recent animals. The resemblance between the brains of *Amblypoda* and those of the Carnivorous *Arctocyon* (*fide* Gervais) is so great as to testify to a similar degree of cerebral development in both the clawed and hoofed types of Eocene *Mammalia*.

As a *résumé* of the relations of the *Amblypoda*, it may be said that they are the most generalized order of hoofed *Mammalia*, being intermediate, in the structure of their limbs and feet, between the *Proboscidea*, the *Perissodactyla*, and the *Artiodactyla*. This fact, together with the small size of the brain, places them in antecedent relation to the latter, in a systematic sense, connecting them with the lower *Mammalia* with small and smooth brains, still in existence; and, in a phylogenetic sense, since they preceded the other orders in time, they stand in the relation of ancestors. It is doubtless true that the *Amblypoda* were the ancestors of all living Ungulates, although no genus of the latter can yet be traced to any known genus of the former, such genera remaining for future discovery.* Standing in this antecedent relation, comparison with other classes of *Vertebrata* is in place. The proportionate size of the brain is, in the *Dinocerata*, as has been discovered by Marsh, more like that characteristic of many

*A discussion of these and other general relations of the *Amblypoda* may be found in a paper read by me before the American Association for the Advancement of Science, August, 1875, and published in the Penn Monthly Magazine, December, 1875.

Reptiles than of Mammals, and I have shown that the immovable tibio-tarsal articulation is a Reptilian feature as well. These are, however, but hints of a relationship doubtless very remote.

Before proceeding to a more detailed consideration of the genera of this order, I give the distinguishing characteristics of the two suborders into which they naturally fall:

I. A third trochanter of the femur, and fossa for the round ligament; no alisphenoid canal; superior incisors present............ *Pantodonta.*
II. No third trochanter nor fossa for the round ligament; an alisphenoid canal; no superior incisors.................................. *Dinocerata.*

The differences presented by these suborders are thus very decided, but they agree in some important points, not necessarily of ordinal value. Thus the *foramen ovale* is distinct from the *foramen lacerum anterius*, and the *meatus auditorius* is not closed inferiorly. In the first point, they agree with *Symborodon* and *Rhinocerus* more than with any *Proboscidea* or *Artiodactyla*. In the latter respect, they agree with the Tapirs and *Rhinoceridæ*, but not with other Ungulates. The cervical vertebræ are short, and not united by ball-and-socket joint, and are intermediate in character between those of *Proboscidea* and other Ungulates. In both suborders, the scapula is acuminate at its superior border and expanded behind, as in *Proboscidea*, while the abrupt origin of its spine is a character of *Proboscidea, Artiodactyla*, and many other Mammals, but not of *Perissodactyla*. In the rudimental spine and crest of the tibia, we have again especially Proboscidian resemblances, which are confirmed by the shape of the ilium. This bone expands immediately from the acetabulum into a broad plate, which has a continuous convex crest, and is altogether different from the pedunculate ilium of the Rhinoceros and Hippopotamus.

As regards the points in which the suborders differ, it may be observed that the *Pantodonta* in their dentition and femur resemble the *Perissodactyla* more than do the *Dinocerata*, while the absence of alisphenoid canal in *Coryphodon* is a suilline character, and the only one which I find in the group. In the form of the femur, the *Dinocerata* resemble closely the *Proboscidea*, but in the presence of the alisphenoid canal they agree with both *Perissodactyla* and *Proboscidea*. It is not unlikely that, in future, genera will be found which connect both these orders more nearly with primitive types of *Artiodactyla*, but as yet we are not acquainted with them.

The order *Amblypoda* was first defined by the writer in the Systematic Catalogue of the Vertebrata of the Eocene of New Mexico, published, in April, 1875. The two suborders *Pantodonta* and *Dinocerata* were originally defined by the writer in "The Short-footed Ungulata of Wyoming", published March, 1873, in the following language:

"No incisors; nasal bones elongate; astragalus articulating with both navicular and cuboid; no third trochanter .... *Dinocerata.*

"Dentition complete, *i. e.*, incisors present; ? nasal bones; astragalus articulating with both navicular and cuboid; a rudimental third trochanter .... *Pantodonta.*"

The name *Dinocerata* was then proposed as a correction of "*Dinocerea,*" originally introduced by Professor Marsh* for the animals which it includes, under the belief that it constituted a distinct order of *Mammalia*, which, however, he did not characterize. Shortly afterward (January, 1873†), I gave the first general synopsis of the characters of the species of the group then contained in my collection, in which they resemble the *Proboscidea*, as follows: "1. The shortness of the free portion of the nasal bones; 2. The malar bone is rod-like, and forms the middle element of the zygomatic arch; 3. The cervical vertebræ are exceedingly short and transverse; 4. The femur is without third trochanter; 5. Its condyles are contracted, and the narrow intercondylar fissure is prolonged far forward; 6. The spine of the tibia is wanting, and the glenoid cavities separated by a longitudinal keel; 7. The astragalus is not hourglass-shaped above, but with a uniform face; 8. The phalanges are short and stout, and represent several toes." To these may be added two external characters, which directly result from the osteological, viz: 9. The possession of a proboscis; this is proven by the extreme shortness and stoutness of the free portion of the nasal bones, by the very short cervical vertebræ, and by the fact that the nasal and premaxillary bones are deeply excavated at their extremities, with surrounding osseous eminences for the origin of the muscles of the trunk; 10. The extension of the femur below the body, so that the leg was extended with

*Amer. Jour. Sci. and Arts, 1872, October, 1872 (separata September 27).

† The extra copies of this paper, which contained all except the character number "1", were published January 16.

the knee below and free from the body, as in Elephants, Monkeys, and Men. Other characters common to the *Proboscidea* and some other Ungulates are: 11. The scapula acuminate above the spine, and with a very short coracoid; 12. Broad truncate occiput, with widely separated temporal fossæ: 13. The greatly expanded iliac bones."

These characters were adduced in support of the view that these animals should be referred to the *Proboscidea.* Although I have subsequently referred them to a new and special order, the above characters express the affinities which I claimed for the group, although several of them are found not to be common to all the species. Thus the characters of the malar bone and cervical vertebræ are not common to all of the *Dinocerata*, while these, with the characters of the femur, are not found in the *Pantodonta.* It was not, however, until a few weeks afterwards that I discovered the near affinity between these suborders. As regards the possession of a proboscis, there is every reason to believe that some of the species possessed one, though it may have been short as in the Tapir, while it is possible that in others it was wanting, or not more developed than in the Hog.

The first attempt to define the *Dinocerata* as an order of *Mammalia* was made by Prof. O. C. Marsh, of New Haven, in a paper published a short time subsequently* to my essay quoted above. The characters which he brought forward, and which had mostly already appeared in the descriptions of species published by him and by myself, are the following: "1. The absence of upper incisors; 2. The presence of canines; 3. The presence of horns; 4. The absence of large air-cavities in the skull; 5. The malar bone forms the anterior portion of the zygomatic arch; 6. The presence of large post-glenoid processes; 7. The large perforated lachrymal forming the anterior portion of the orbit; 8. The small and horizontal nareal orifice; 9. The greatly elongated nasal bones; 10. The premaxillaries do not meet the frontals; 11. The lateral and posterior cranial crests; 12. The very small molar teeth and their vertical replacement; 13. The small lower jaw; 14. The articulation of the astragalus with the navicular and cuboid bones; 15. The absence of a true proboscis."

* The extra copies of this paper bear date January 28, 1873.

This heterogeneous list of characters could not define any natural group, as many of them are of not more than generic or family value.* Several of the most important are not shared by the genus *Coryphodon*, a form at that time unknown to Professor Marsh, but which clearly belongs to the same order of *Mammalia*. My conclusion has been that the *Dinocerata* do not alone constitute an order of *Mammalia*, but that they form a division of an order which includes also *Coryphodon*, and doubtless many other still unknown types, whose position is, as I first stated, between the *Proboscidia* and the *Perissodactyla*, but which has no affinities with the *Artiodactyla*, as has been asserted.

Full descriptions of the species and genera of this order first appeared in my essay, "On the Short-footed Ungulata of Wyoming", above quoted (published March 14, 1873). I there described the existence of five toes in the pes of the genus *Eobasileus*, and the co-ordinal relations of *Coryphodon* (*Bathmodon*). In a note published by Professor Marsh, October, 1873, that author asserts that the *Dinocerata* have "but four toes in the pes"; but in a paper on *Uintatherium* (*Dinoceras*), which subsequently appeared, he admits that that genus has five toes in the pes (Am. Jour. Sci. and Arts, Feb., 1876). We owe to later observations of Professor Marsh two of the most important points in the structure of the *Dinocerata*, viz, the superficial structure of the brain, and the arrangement of the bones of the carpus. He shows (*l. c.*, July, 1874, and February, 1876) that the cerebral hemispheres are so small as not to cover any part of the olfactory lobes and the cerebellum; that their surface was nearly smooth, and that their combined diameter was less than that of some parts of the neural canal of the vertebral column. The brain is relatively the smallest among known *Mammalia*, and resembles strongly that of the Creodont *Arctocyon* of the French Eocene, figured by Professor Gervais, in the "Archives du Muséum", 1870. I subsequently showed† that the brain of *Coryphodon* presents similar characters.

The structure of the carpus of *Uintatherium*, described by Marsh (*l. c.*, February, 1876), is essentially identical with that of *Coryphodon*, which I

* As I pointed out in an article in the American Naturalist, May, 1873.

† My description of the brain of *Coryphodon* appeared in the Proceedings of the American Philosophical Society for March, 1877, and was published April 25th. Professor Marsh's description appeared in the July number of the American Journal of Science and Arts for July, 1877, which appeared near the middle of June.

described in the Systematic Catalogue of the Vertebrata of the Eocene of New Mexico (April, 1875).

The *Pantodonta* are confined, so far as discoveries extend at present, to the Lower Eocene or Wasatch beds, in the Rocky Mountain region, while the *Dinocerata* are confined to the higher or Bridger Eocene strata. The former suborder includes three genera, *Bathmodon* Cope, *Coryphodon* Owen, and *Metalophodon* Cope; the *Dinocerata* four, *Bathyopsis* Cope, *Uintathe-rium* Leidy, *Eobasileus* Cope, *Loxolophodon* Cope.

I have anticipated* the discovery of *Amblypoda* with tubercular teeth No such have yet been found, but the probability of such discovery remains as strong, in my opinion, as at the time I ventured to suggest it.

PANTODONTA.

As already pointed out, the structure of the limbs and feet in this suborder is as in the order generally, and the scapula has the same form in general. The symphysis mandibuli is furnished with teeth, and forms a long solid spout. The astragalus has a very peculiar form, being even

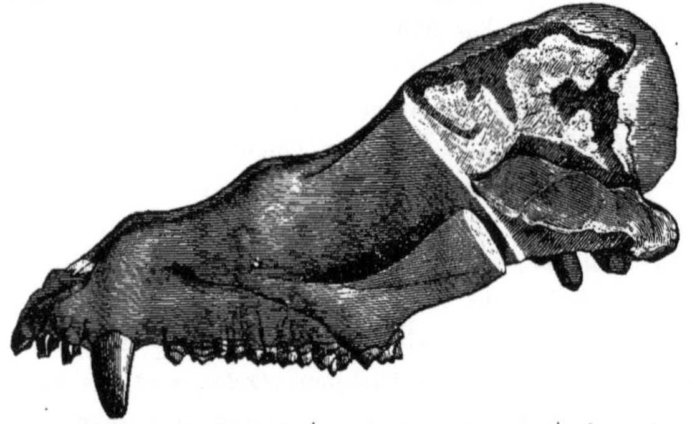

FIG. 21. Skull of *Coryphodon elephantopus*, displaying brain-cavity, $\frac{3}{8}$ nat. size.

more exceptional than in *Uintatherium*. The superior articular surface is flat or concave in the middle. It is turned inward in front of the articular face for the inner malleolus, terminating in a long point. The cuboid articular face is quite small and sublateral, and sessile like the navicular.

* Jour. Acad. Nat. Sci. 1874, March, "On Homologies of Molar Teeth of Mammalia Educabilia", p. 20.

The fibular facet is vertical and is extensive, and the internal lateral is well marked.

On the other hand the coracoid process is produced into a curved hook, and is thus more largely developed than in the *Proboscidians* or *Perissodactyles*. The neck is longer than in the *Dinocerata*. But one family of this suborder is yet known, the *Coryphodontidæ*. Its characters and those of the genus *Coryphodon* were fully described by myself in the final report of the United States Explorations and Surveys West of the One hundredth Meridian, by Captain Wheeler, iv, pt. ii, p. 187–206, and to this work I refer for full details

History.—The first piece recognized as belonging to a distinct genus, under the name of *Coryphodon*, was described by Professor Owen in 1846. This was a fragment of the mandible, supporting the last inferior molar. In describing it, Owen noticed the peculiar form of the posterior part of the ramus. He also referred to some superior molars, one of which is figured by Cuvier (Ossemens Fossiles, pl. 77, fig. 6) as probably belonging to the same genus. These were alluded to by the French anatomist as the *Lophiodon de Soissons* and the *Lophiodon de Lyonnais*. Owen named the species observed by him *Coryphodon eocænus*, and it is believed that the teeth described by Cuvier belong to another species, to which De Blainville gave the name *Coryphodon anthracoideus*. Little, however, was known of this form until ten years later, when Professor Hébert, of Paris, obtained some additional material from Meudon near Paris, and other localities. In the Annales des Sciences Naturelles for 1856, he gave a full account of the characters of the dentition, and described the femur. He explained correctly the homologies of the dental structure, and added a species, *C. oweni*, which is of smaller size than the *C. eocænus*.

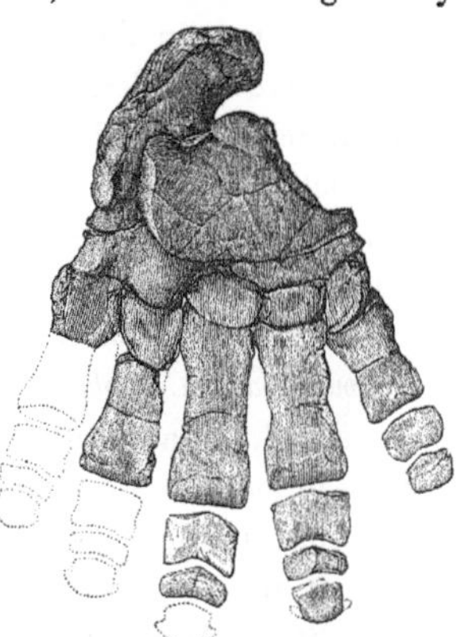

FIG. 22. Right posterior foot of a species of Coryphodon half natural size. From Report Expl. W. of 100th Meridian, WHEELER, IV, pt. ii.

The first American species was made known by the writer in February, 1872, under the name of *Bathmodon radians*, and the description was at that time confined to the superior molars, the femur, and the humerus. In February, 1873, in "The Short-footed Ungulata of the Eocene of Wyoming", I described the characters of the scapula and astragalus and its connections, which furnished reasons for removing the genus from the *Perissodactyla* and placing it in the *Proboscidia*, under a subordinal division, which was named the *Pantodonta*. The same course was pursued in the Annual Report of the United States Geological Survey of the Territories for 1872 (1873), where a second American species, *B. latipes*, was added. The next additions to our knowledge of the osteology of *Coryphodon* are due to the exploration of the Eocene beds of New Mexico, conducted by the writer, in 1874, in connection with the United States Geographical Surveys West of the 100th Meridian. On November 28, 1874, in extracts from the Report of the Chief of Engineers, published in advance, I gave the general characters of the skull, and of the hind foot, determining the number of toes in the latter. The temporary dentition and three new species were also described. In some remarks before the Academy of Natural Sciences of Philadelphia, made March 9, 1875, I pointed out the near relation of these animals to the species of *Coryphodon*. In the "Systematic Catalogue of Vertebrata of the Eocene of Wyoming", April, 1875, the structure of the fore foot, including the number of digits, was pointed out, and the group *Pantodonta* removed, with the *Dinocerata*, to a new order, the *Amblypoda; Coryphodon* was referred to the *Pantodonta*, and two species were added. I announced the generic identity of the American and European forms in a communication before the Academy of Philadelphia, made April, 1876 (published April 26), pointing out the subordinate characters of the respective types.* My final report to Captain Wheeler, United States Army (1877), completed the description of the osteology, so far as accessible to me; and I may add that this is so far complete that there only remain unknown the number of the ribs and dorsal and lumbar vertebræ, and the structure of the hyoid apparatus.

Genera.—There are five genera of *Coryphodontidæ: Coryphodon, Bath-*

*A short time after this, Prof. O. C. Marsh asserted the identity of *Coryphodon* and *Bathmodon* in the Amer. Journ. Sci. and Arts.

modon, *Ectacodon*, *Manteodon*, and *Metalophodon*. The last is easily distinguished from the first by its dental characters, but *Bathmodon* rests on a peculiar foot structure, and its dental characters are unknown. As the feet of *Metalophodon* are not known, it is possible that it is identical with *Bathmodon*. Several species also have been referred to *Coryphodon*, of which the foot structure has not been ascertained, so that it is possible that some of them are Bathmodontes. The solution of these questions will depend on future discoveries; meanwhile I leave most of the species in *Coryphodon*. In determining the last-named genus I have had the advantage of access to the specimens contained in the Museums of London and Paris, a privilege which I owe to Professors Owen, Gervais, and Hébert. To the last-named gentleman I am under especial obligations, as the specimens of *Coryphodon eocœnus* contained in the Museum of the Sorbonne are much the most perfect in Europe.

I give the characters of the genera below:

I. Last superior molar with two interior cusps.
All the superior molars with a well-marked external posterior V.......... *Manteodon*.
II. Last superior molar with but one inner cusp or angle.
a. Last superior molar with posterior external cusp.
Anterior two molars with posterior external V.............................. *Ectacodon*.
aa. Last superior molar without external posterior cusp.
†Anterior two molars with posterior external V.
Astragalus transverse, with internal hook.............................. *Coryphodon*.
Astragalus subquadrate, without internal hook .... *Bathmodon*.
††First superior molar only, with posterior external V .... *Metalophodon*.

MANTEODON Cope.

Proceed. Amer. Philoso. Soc. 1881, Feb., 1882, p. 166. American Naturalist, 1882, p. 73 (Dec. 30, 1881).

The generic characters have been already pointed out in the key above given. They are a little more like those of the superior molar teeth of such *Perissodactyla* as *Limnohyus* and near allies, than those seen in the typical *Coryphodon*. The posterior transverse crest of that genus is here represented by a complete V, but the anterior lobe of that crest which represents the anterior V of the Perissodactyle is only a lobe, as in *Coryphodon*. The tooth in fact is much like the penultimate molar of the latter genus.

The two internal cusps are unique in the family The additional one is a growth of the inner extremity of the posterior cingulum, and is separated from the anterior inner cusp by a deep and wide notch. It is opposite to the posterior V, as the anterior inner cusp is opposite the anterior rudimental V. The premolar and incisor teeth are similar to those of *Coryphodon.* The skeleton is unknown.

MANTEODON SUBQUADRATUS Cope.

Loci supra citati.

Plate XLIV*a*; figs. 1–5.

The characters of this pantodont are learned from a series of teeth which were found together by Mr. Wortman free from admixture of others. They are not worn, excepting by moderate use by the animal when living.

The last superior molar is not of the oval form belonging to the species of *Coryphodon,* but is quadrate, with the internal side shorter, and with rounded lateral angles. The first anterior cingulum, which represents the anterior basal cingulum of the *Lophiodontidæ,* is as elevated as in the species of *Coryphodon.* Externally it rises in a protuberance with sharp edge, which curves posteriorly and disappears on the external side of the crown. The inner extremity terminates abruptly, forming the anterior interior tubercle. The anterior external lobe is rather flat, and is not conical nor elevated above the anterior cingular lobe. It is not deeply separated from the latter, nor from the posterior V; its edge is rough. The posterior V projects well inwards, and is rather narrow. Its posterior border extends as far outwards as the point of junction of its anterior border with the anterior external lobe, and terminates in a slight elevation of its border. The base of the crown extends external to the base of the V, and forms a strong posterior external protuberance. This causes the outline of the external base to be concave. This side of the crown has several small protuberances and rugosities. The posterior basal cingulum extends as far externally as the posterior V, and terminates internally in the posterior internal cusp. The second or basal anterior cingulum is well developed. There are no external nor internal cingula. The surface of the enamel is strongly and closely rugose where not worn.

The posterior inferior molar exhibits a transverse posterior crest, without any tubercle or ridge in the mouth of the posterior V-shaped valley. There is a strong posterior cingulum, amounting to a narrow heel. As in the case of the superior molar, the enamel where not worn is closely and strongly wrinkled. The first superior premolar is characterized by the very small development of its internal lobe, which is only a strong basal cingulum. The crown proper has a sub-triangular outline, and the external face is flat and not concave. No external cingulum; enamel wrinkled. An external incisor has a large transversely extended crown, without cingula. A low rib on the median line of the inner side. Enamel wrinkled. In this and in another incisor the base of the crown is considerably expanded laterally.

Measurements.

		M.
Diameters of crown M. iii, sup.	anteroposterior	.035
	transverse	.041
	vertical	.020
Width of M. iii inferior, posteriorly		.022
Diameters of Pm. i, sup.	anteroposterior	.018
	transverse	.014
Diameter base crown I, ii		.024
Length crown i, ii		.019
Width base crown i, iii		.026

The crowns of the molars of this species are about as large as those of an ox. From the basin of the Big Horn, Northern Wyoming.

ECTACODON Cope.

American Naturalist, 1882, 73 (Dec. 30, 1881). Proceed. Amer. Philos. Soc., 1881, Dec., p. 167, Feb. 1882.

In *Ectacodon* the last superior molar has more of the elements of a posterior external V than in *Coryphodon*, but not so much as in *Manteodon*. The posterior transverse crest, it is true, has no oblique posterior ridge joining it, to form with it more or less of a V. But the external posterior angle of the crown supports a cusp homologous with the vertical rib found at the basal or external angles of the Vs in *Palæosyops* and allied genera, and indicating the outlines of a V which lacks its posterior side, in a manner not seen in *Coryphodon*. The penultimate and ante-penultimate superior molars are like those of the latter genus. Skeleton unknown. I have a single species of this genus.

ECTACODON CINCTUS Cope.

Paleontological Bulletin, No. 34, p. 167, Feb. 20, 1882. Proceed. Amer. Philo. Soc., 1881, Dec., pl. 167 (1882). Plate XLIV *a*; fig. 6.

Six superior molars of one skull represent this species. They belong to a large animal, one about the size of the *Manteodon subquadratus*. The last superior molar has a characteristic outline. It is not oval as in the species of *Coryphodon*, nor quadrate as in *Manteodon* sp., but sub-parallelogrammic. The transverse diameter exceeds the anteroposterior, and the anterior and posterior sides are parallel. The external outline is slightly oblique and slightly notched in the middle. The internal border is regularly rounded. The basal or second cingulum extends entirely round the tooth from the posterior external cusp, round the inner base to the anterior external base of the crown, being absent only from the external base. The first cingula, both anterior and posterior, are well developed, as in the species of *Coryphodon*, and unite in the prominent internal angle. The posterior first cingulum joins the posterior basal cingulum at the middle of its length. The anterior first cingulum extends to the anterior external part of the crown, and then turns downwards and posteriorly, and terminates at the middle of the external base. The posterior crest is not transverse, but quite oblique, sloping at an angle of 45° with the axis of the jaw. The part of the crest which represents the posterior V is a good deal larger than the part representing the anterior V, and is closely joined with it. The latter is well separated from the anterior first cingular ridge and its anterior exterior elevated portion. The enamel of this tooth is finely wrinkled, and is more readily worn smooth than in the *Manteodon subquadratus*.

The penultimate superior molar has the posterior V well developed, and its posterior basal or external angle is marked by a tubercle homologous with that which is so prominent on the last molar. The anterior V is a conic tubercle closely joined with the posterior V, and well separated from the anterior first cingular lobe. The basal cingula are well developed, but do not meet on the inner base of the crown. The first or superior cingula meet as usual in an interior angle, but there is a contraction of the anterior crest just before reaching this angle. The first true molar is smaller than the second and has the same general structure. Here, however, the anterior first cingulum is more prominent near the internal angle than the posterior. The characters of the premolars do not differ from the corresponding ones of species of *Coryphodon.* The enamel is delicately wrinkled. The first superior premolar is not preserved.

Measurements.

		M.
Diameters of crown of M. iii	anteroposterior	.034
	transverse	.043
	vertical	.015
Diameters M. i	anteroposterior	.028
	transverse	.033
	vertical	.012
Diameters Pm. iii	anteroposterior	.023
	transverse	.030

It is probable that this species was about the size of an ox.

CORYPHODON Owen.

History British Fossil Mammals, 1846, p. 299, fig. 103, 104. Bron Lethæa Geognostica, 1856, p. 842. Hébert, Annales des Sciences Naturalles, 1856, p. 87. Cope, Synopsis Extinct Vertebrata Eocene of New Mexico, 1875, p. 28. Marsh Amer. Journ. Sci., 1877, 81. Cope, Report U. S. Geol. and Geog. Expl. Survs. W. 100th Meridian, iv, pt. ii, p. 187, part 1877. Proceed. Amer. Philos. Soc., 1881, p. 166.

Dental formula: I. $\frac{3}{3}$; C. $\frac{1}{1}$; P-m. $\frac{4}{4}$; M. $\frac{3}{3}$. Canines large, no diastema. Superior premolars different from the molars, consisting of an external and an internal V, the external the larger. The first and second superior true molars have two external Vs, the posterior well developed, and the anterior represented by a subconic cusp; a single internal cusp with V-shaped section, the anterior branch extending as a high transverse crest to the external border of the crown, forming the anterior triturating surface. The third true molar differs from the others in having the external Vs united into a crest, which is placed transversely, forming the posterior border of the crown. The inferior premolars differ from the true molars. Their crowns form a V, with the apex outward, and a short heel behind it with a more or less median keel, and increasing in transverse width in the

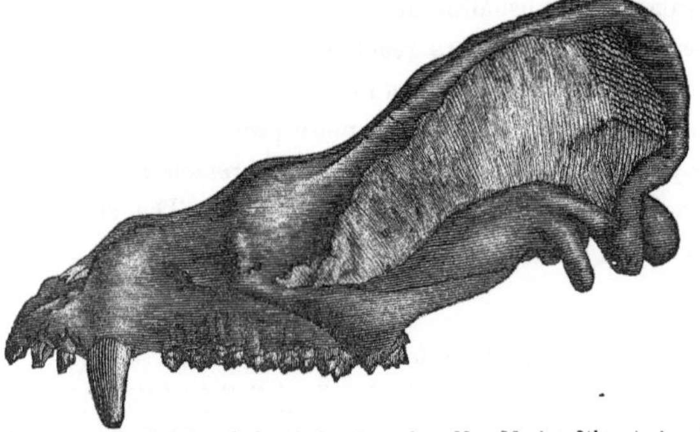

FIG. 23.—Profile view of skull of *Coryphodon elephantopus*, from New Mexico, $\frac{2}{9}$th nat. size. From Report of Lieut. G. M. Wheeler to Chief of Engineers, 1875, pl. v.

posterior teeth. The true molars support two transverse crests, each of which may send obliquely inward from its external extremity a low ridge. The anterior of these ridges is the best developed, and is always present.

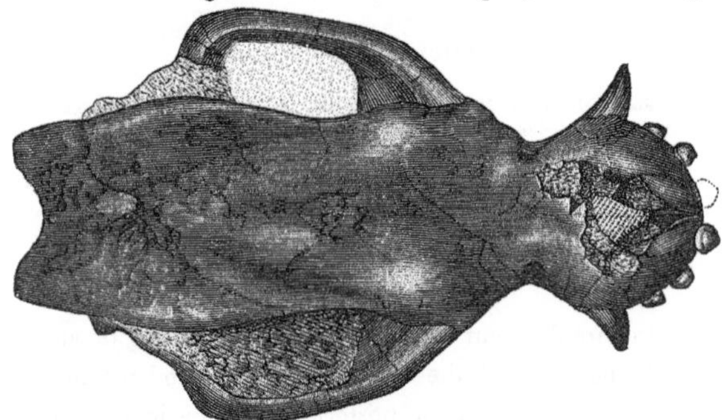

FIG. 24.—Superior surface of skull of *Coryphodon elephantopus*, above figured, $\frac{2}{9}$th nat. size. From final report of Lieut. G. M. Wheeler, vol. iv, pt. ii, 1877.

Frontal and parietal regions flat, bounded by separated temporal ridges. Feet short, plantigrade. Digits 5–5, with short, wide ungues. Neck and tail of medium length.

More or less perfect specimens of one hundred and fifty individuals of the *Coryphodontidæ* were obtained during my exploration in Northwestern New Mexico, under the organization of the survey of Capt. G. M. Wheeler. Portions of many others have since been obtained by Mr. Wortman and myself in Wyoming. Those specimens which include astragali are mostly referable to this genus, but a few from both regions belong to *Bathmodon.* It has, however, generally happened that specimens which include

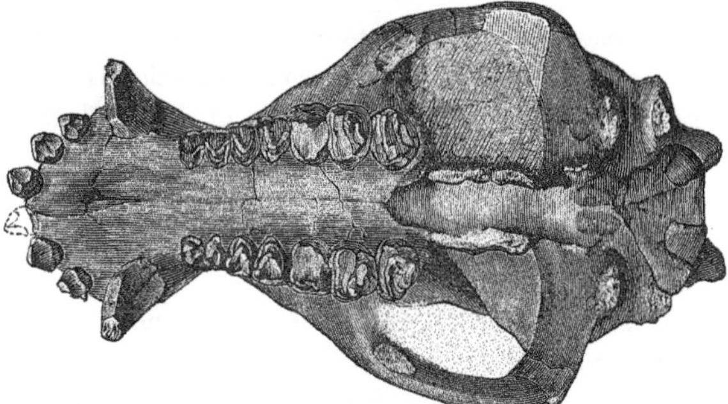

FIG. 25.—Skull of *Coryphodon elephantopus*, from below, ⅛th natural size. From Report Expl. Surv. West of 100th Mer., G. M. Wheeler, vol. iv.

teeth do not possess astragali, and *vice versa*, so that I have been compelled to disregard the generic position of the species for the time being, and have relied chiefly on the mandibular teeth for the characters, owing to the more irregular occurence of other elements; and the tubercles of the teeth are not always constant. The following key expresses the relations of these forms, most of which, perhaps all, are good species:

I. The last inferior molar with three posterior cusps, the internal sometimes represented by a ridge; or the posterior inferior molars with an accessory cusp or tubercle on the inner side between the crests (*Coryphodon*, Owen):

An internal tubercle; last upper molar with an anterior cross crest and anterior external crest closely connected; size largest .... *C. anax.*

An internal conic cusp; posterior crest oblique; heel very small; size medium .... *C. cuspidatus.*

An internal crest; posterior crest oblique; heel small; size medium .... *C. obliquus.*

An internal tubercle; posterior crest little oblique; heel large; size large .... *C. lobatus.*

II. Posterior inferior molars without internal accessory tubercle.

a. Posterior inferior molars with small or no heel:

Large; posterior superior molar oval, with distinct straight posterior crest; inferior molars elongate; symphysis mandibuli produced and narrower; premaxillary elongate .... *C. latipes.*

Medium; inferior molars nearly as wide as long; premaxillary short .... *C. latidens.*

aa. Posterior inferior molars with prominent or wide heel:

Medium; posterior superior molar with posterior angle, and angulate posterior crest; inferior molars elongate; symphysis mandibuli broad and short; premaxillary elongate; tusk trihedral *C. elephantopus.*

Smaller; premaxillary bone short; tusk trihedral .... *C. simus.*
Medium; premaxillary elongate; tusk compressed and grooved .... *C. molestus.*
Large; last superior molar oval, with angulate posterior crest; its anterior lobe connected with anterior cingular crest .... *C. repandus.*

III. Last inferior molar with but one posterior cusp from which a curved crest extends round the posterior border of the crown.

Superior true molars narrow; external incisors sharply angulate on external face........ *C. curvicristis.*

IV. Posterior inferior molar unknown.

Posterior superior molar oval; posterior crest straight; internal crest fissured (? normally); a complete internal cingulum .... *C. marginatus.*

The *C. eocænus* Owen, belongs to Section I, and the *C. anthracoideus* Blv., and *C. owenii* Héb., from France, belong to Section II.

Professor Marsh has published a figure of the skull of *Coryphodon* (Amer. Journ. Sci., 1877, Pl. iv, fig. 1), which represents the temporal ridges as converging towards a sagittal crest. This is not found in the *C. eocænus*, nor in any of the American species whose crania are known to me (*C. elephantopus, C. molestus*).

The first known American specimens of the family were discovered by Dr. Hayden, on Bear River, near Evanston, Wyoming. They represented two species, *Bathmodon radians* and *Coryphodon latipes*. I subsequently visited the same locality and obtained numerous specimens, which further demonstrate the fact that two species were entombed there, and indicate that a third species was associated with them. The *C. cuspidatus* has been obtained in New Mexico and Wyoming; and the *C. obliquus, C. lobatus, C. latidens, C. elephantopus, C. simus*, and *C. molestus* have only been found in New Mexico.

Restoration.—The general appearance of the Coryphodons, as determined by the skeleton, probably resembled the Bears more than any living animals, with the important exception that in their feet they were much like the Elephants. To the general proportions of the Bears must be added a tail of medium length. Whether they were covered with hair or not is, of course, uncertain; of their nearest living allies, the Elephants, some were hairy, and others naked. The top of the head was doubtless naked posteriorly, and in old animals may have been only covered by a thin epidermis, as in the Crocodiles, thus presenting a rough, impenetrable front to antagonists.

The movements of the Coryphodons, doubtless, resembled those of the Elephant, in its shuffling and ambling gait, and may have been even more awkward, from the inflexibility of the ankle. But, in compensation for the probable lack of speed, these animals were most formidably armed with tusks. These weapons, particularly those of the upper jaw, are more robust than those of the *Carnivora*, and generally more elongate, and attrition preserved rather than diminished their acuteness. The size of the species varied from that of a Tapir to that of an Ox.

There is no evidence that these animals possessed a proboscis, as was probably the case with some of the *Dinocerata*.

We must suppose that the Coryphodons were vegetable feeders, but not restricted to any particular class of food. They were doubtless, to a large extent, like the hogs, omnivorous.

Position.—The genus *Coryphodon* is characteristic of the Lower Eocene formation in North America and Europe. In the former country it is confined to the Wasatch and Wind River epochs; in Europe, to the Suessonian and, perhaps, the Thanetian of France and England. It is absent from the Upper Eocene of both countries.

Coryphodon cuspidatus Cope.

Report U. S. Geol. Survs. W. of 100th Meridian, iv, pt. ii, p. 206, pl. xlvi, fig. 1.

This species is well characterized by the strong additional cusp found on the inner side of the crown of the last inferior molar between the interior extremities of the transverse crests. To give it more place, the posterior crest is more oblique than in other species. The accessory cusp is represented by a crest in the *C. eocœnus*, and the New Mexican *C. obliquus*.

The typical specimen, including only a fractured last molar with a supporting fragment of the jaw, was found by myself in New Mexico. A second specimen was found by my assistant, Mr. J. L. Wortman, in the bad lands of the Big Horn Valley, Wyoming. It includes, however, only some jaw fragments with three inferior premolars, and two entire and two broken inferior true molars. These indicate a species of smaller size than the *Bathmodon radians*, and agree with the typical specimen above mentioned. The molars preserved, excepting the posterior, do not differ materially from those of other species, excepting the *C. latidens*, which is peculiar.

Measurements.

		M.
Dimensions of inferior Pm. IV	anteroposterior	.020
	transverse	.017
Dimensions of inferior M. II	anteroposterior	.025
	transverse	.020
Width of crown of inferior M. III, behind		.023
Width of crown at crest		.019
Width of crown at tubercle		.024

CORYPHODON LATIPES Cope.

Bathmodon latipes Cope, Annual Report U. S. Geol. Surv. Terrs., F. V. Hayden, 1872 (1873), p. 588. Proceed. Amer. Philos. Soc., 1873, p. 70.

Pl. xxix *b*, figs. 4–5; xlviii, figs. 7–14.

The remains of *Coryphodontidæ* obtained by Dr. Hayden, near Evanston, Wyoming, in 1871, included parts of two distinct species, a large and a small, which I described as *Bathmodon radians* and *Coryphodon latipes*, respectively. The bones were mixed together when they came into my possession, but in most instances it is not difficult to distinguish them. An expedition which I made to the same locality, in 1873, enabled me to procure many additional bones of both species, and to confirm the distinctions already observed. It remains, however, difficult to assign a few of the bones to their proper species, and an astragalus indicates the possible existence of a third one.

The smaller specimens obtained by Dr. Hayden, include atlas, axis, dorsal and lumbar vertebræ, scapula, humerus, phalange, femur, astragali, &c.

The transverse process of the atlas is stouter and less flattened at the base than in *B. radians*. The axis is but little oblique, and has a low, obtuse hypapophysis below. Its form is much as in the larger species, being rather elongate, but shorter than in *Rhinocerus* and other Perissodactyles. The dorsals and lumbars are short and plane; the former are obtusely, the latter acutely, keeled below.

The glenoid portion of the scapula differs notably from that of the *B. radians*. Besides being smaller, it is narrower, its outline contracting regularly from the posterior border to the coracoid process, instead of curving outward to it as in *B. radians*. The external face at the base of the coracoid is nearly plane in *C. latipes*, but is strongly concave in *B. radians*. The astragalus is considerably broader than long, the apex turned outward in front of the inner malleolus, being especially produced. The tibial face is concave transversely, convex anteroposteriorly at the front, and plane behind. There is a posterior submarginal foramen, which is not bridged over in one specimen, producing a deep notch. The navicular facet has considerable transverse extent, and the anterior side of the bone is more transverse than in *B. radians*. The calcaneal facets are diagonally opposite to each other; the outer is subround, the inner anterior narrow and transverse. It differs in the two specimens, the perforating foramen not being bridged over in the one (the type) with the similar posterior interruption described above. This may be due to fracture. The only ungual phalange has the articular face not quite sessile on the transverse rugose free extremity.

Measurements.

		M.
Diameter of diapophysis atlas (fore and aft)		0.036
Diameter of axial facet of atlas		.034
Diameter of centrum of dorsal	fore and aft	.040
	vertical (total)	.043
	transverse	.059
Diameter of neural arch of same	vertical	.018
	transverse	.032
Diameter of centrum of lumbar	anteroposterior	.041
	vertical (total)	.050
	transverse	.063
Diameter of head of femur		.060
Length of astragalus fore and aft		.050
Width of astragalus		.065
Length of navicular facet		.049
Width of navicular facet		.020
Width of cuboid facet		.018
Length of cuboid facet		.016
Width (fore and aft) of anterior calcaneal facet		.018
Length of posterior calcaneal facet		.022
Length of fibular facet (axial)		.041
Length of terminal phalange		.012
Width of terminal phalange, proximally		.015
Width distally		.030
Diameter of glenoid cavity of scapula	transverse	.054
	anteroposterior	.080

The specimens procured by myself were taken from the bone bed on October 5 and 6, 1873. Those taken on the former day are of a ferruginous color, and include right scapula, ulna and radius, sacrum, right ilium, and left femur. Those obtained on the following day are of a black color, and include left humerus, left tibia without distal extremity, patella, and right femur. With each lot is a cervical vertebra, but I am not sure whether these belong to this species or to the *B. radians*. These bones, or at least the femora, represent two individuals. On October 8, I obtained part of another scapula of this species. I did not procure any astragalus.

Their uniformly inferior size distinguishes these bones from those of the *B. radians*, which I procured at the same place on October 5. The difference in the form of the glenoid extremity of the scapula, pointed out above, is maintained by these specimens. The important peculiarity in the form of the proximal extremity of the humerus remains to be noticed. In the *B. radians*, obtained by Dr. Hayden and myself, the greater tuberosity is hugely developed, its proximal extremity forming a subhorizontal table nearly as wide as the articular head of the humerus at its middle. The surfaces of the head and tuberosity are separated by a low curved crest which descends to the external border of the bicipital groove. In the *C. latipes* the great tuberosity, though large, is not nearly so wide proximally, its transverse diameter being scarcely half that of the articular face, on a transverse line passing just posterior to the bicipital groove. This difference is partly due to a more external position of the ridge separating the articular from the rough surfaces. The teres and pectoralis facets are large and impressed in the *C. latipes;* the latter being bounded below by a sudden retreat of the surface. The lesser tuberosity is not large, and consists of the anterior prominence of the inferior border of the pectoralis facet. The deltoid ridge is not distinct proximally, but becomes very prominent below the middle of the shaft, where it is twisted forwards as it descends. It is more prominent than in the humerus I figured in Captain Wheeler's report, Pl. LXII, fig. 1, from New Mexico. The internal epicondyle is prominent, while the external is wanting. The external twisted marginal crest is prominent and soon descends into the shaft.

Measurements of humerus.

			M.
Total length			.328
Proximal diameters	transverse in one line	of greater tuberosity	.030
		of articular surface	.067
	anteroposterior	lesser tuberosity	.092
		at bicipital notch	.075
Distal diameters	transverse	total	.120
		of condyles at middle	.071
	anteroposterior external condyle		.058

The spine of the scapula is very prominent, and its free edge is wide, so that both its anterior and posterior faces are concave. The interior face of the scapula is concave or curved backwards.

Measurements of scapula.

		M.
Proximal longitudinal width	with coracoid	.084
	of glenoid cavity	.063
Proximal transverse width		.056

The radius is a robust bone with large extremities, of which the distal is the larger. The carpal articular surface is, however, smaller than the humeral, the greater width being caused by the large external tuberosity. The head is a transverse oval, as in the New Mexican species. The shaft is oval in section, and is flattened in the plane of the distal extremity except close to the head, where it rises a little inwards.

Measurements of radius.

		M.
Length		.220
Diameters of head	transverse	.068
	vertical	.043
Width of carpal articular face		.044

The *sacrum* includes four vertebræ, and as the last one is broken, the presence of a fifth cannot be asserted. The intervertebral sutures are visible. The centra and neural arches are depressed; the distal ones very flat. The neural spines are represented by a low keel. The transverse processes are expanded, and are thoroughly coössified distally, inclosing large intervertebral foramina, the anterior being the largest.

Measurements of sacrum.

		M.
Length of fragment		.165
Width at first intervertebral foramen		.168
Anteroposterior diameter of intervertebral foramina	first	.029
	third	.018
Diameters of first sacral centrum	vertical	.032
	transverse	.058

The *femur* is but little shorter than that of *B. radians*, but is more slender in all its parts, notably in the condyles. The fossa ligamenti teris is well developed, and the head is more prominent than the great trochanter. The little trochanter has considerable longitudinal extent, and the third trochanter is only moderately prominent. It is situated at the middle of the shaft. The posterior face of the shaft is everywhere flat. The patella is an elongate oval and has not the tendinous prolongation of the one figured in Wheeler's report IV, Pl. LVIII, fig. 4.

Measurements of femur.

		M.
Total length		.376
Proximal width		.109
Transverse diameter of head		.056
Transverse diameter of great trochanter		.053
Width of shaft at little trochanter		.080
Width of shaft at third trochanter		.066
Width of shaft below third trochanter		.055
Distal diameters	transverse, below	.092
	anteroposterior at external trochlear ridge	.083

The *tibia* is characterized by the remarkable slenderness of its shaft as compared with the diameters of its head. Above and at the middle the shaft is compressed, so that its long axis is nearly anteroposterior. The crest is truncate by the wide insertion of the *quadriceps extensor* muscle, and incloses scarcely any fossa on the external face of the bone. The cotyli are separated by a wide open emargination behind, and a narrow groove in the position of the spine. The interior is one-fourth wider anteroposteriorly than the external.

Measurements.

		M.
Proximal diameter of tibia	transverse	.095
	anteroposterior	.076
Diameters of external cotylus	transverse	.047
	anteroposterior	.051
Diameters of shaft at middle	transverse	.028
	anteroposterior	.0[illegible]6

The teeth discovered with these specimens are superior molars, premolars, incisors, and canines. No inferior true molars were obtained, so that the relationships of the *C. latipes* to the species discovered in New Mexico remains uncertain When the characteristic teeth are obtained, it may be found to be identical with some of the latter. I look for future investigations in Wyoming, now under way, to decide this question.

From the Wasatch beds on Bear River, near Evanston, Wyoming.

I refer seven individuals from the Big-Horn Wasatch Basin provisionally to this species. Three of these are represented only by superior teeth, &c., and in four the last inferior molar is preserved. Of the latter, three have an angle, sometimes almost a crest, descending from the posterior inner tubercle, as in *C. obliquus*, but the specimens are all of superior size to that species, some of them very much exceeding it. It is also possible that this ridge is not a constant character. This species has the dentition which I have referred to the *Bathmodon radians*, but no astragalus of that species occurs in the Big-Horn collection. I refer it to the *C. latipes*, although the teeth of the typical specimen have not yet been identified. I hope soon to be able to decide this question.

CORYPHODON ELEPHANTOPUS Cope.

Report of Vert. Foss. New Mexico, U. S. Geogr. Surveys W. of 100th Mer., 1874, p. 10. Id. Annual Rep. U. S. Geol. Surv. W. of 100th Mer., 1875, p. 95, plates v, vi. Report do. iv, pt. ii, 1877, p. 217. Plates L–LIV.

Plate XXIX *c*.

Portions of the dentition of both jaws, including the last molar teeth of two individuals, prove that this species inhabited Wyoming in the early Eocene period. One of the individuals, represented only by the last molars of both jaws, is a little smaller than the typical specimen of which an entire cranium is figured in Captain Wheeler's report (4to, 1877, Pl. LI–III), while a second specimen, which includes the entire superior molar series, is a little larger than the same.

This species is characterized by the obliquity of the edge of the posterior crest of the posterior superior molar backwards away from a transverse line; and by the slope of the external side of this crest. In other words, the inner half of the posterior crest nearly forms a V, like that of the penultimate molar. The posterior edge of the V is present, running outwards from the inner end of the posterior crest, which thus becomes the apex of the V. The *C. elephantopus* thus most nearly approaches the genus *Manteodon* of all the species. To accommodate the obliquity of the crest the posterior outline of the last upper molar is strongly angulate, giving a subtriangular outline. The heel of the last inferior molar is insignificant

Specimens were collected by J. L. Wortman in the Big-Horn region, Wyoming.

CORYPHODON SIMUS Cope.

Report Vert. Foss. New Mexico, U. S. Geogr. Surveys W. of 100th Mer., 1874, p. 8. Report U. S. Geogr. Surveys W. of 100th Mer., 1877, iv, pt. ii, p. 225. Pl. LV.

A broken mandible and maxillary bone, with several teeth, represent this small species in the Big-Horn collection. Maxillary teeth of similar proportions were found by Russell S. Hill in the Eocene of Powder River, Wyoming.

CORYPHODON REPANDUS Cope.

Paleontological Bulletin No. 34, p. 171, Feb. —, 1882. Proceed. Amer. Philos. Soc., 1881, p. 171 (1882).

Plate XLIV *e*; figs. 1–4.

This large species is known from the posterior portions of the dentition of both jaws, with an entire symphysis.

The last superior molars are intermediate in outline between the regular oval of the *C. latipes* and the subtriangular form of the *C. elephantopus*. The peculiarities of the species are seen in the posterior crest. The two lobes of which this is composed do not form a continuous line as in *C. latipes* and *C. simus*, but form an angle with each other as in *C. anax*. The anterior lobe is compressed, and its long axis is nearly that of the jaw; the second lobe leaves it at a right angle, but curves backwards as it extends inwards, giving a concave exteroposterior border. There is no ridge descending outwards from the inner extremity of the crest, to form a V, as in *C. elephantopus*. But the posterior basal cingulum extends to the external side of the tooth, which is not the case in any other species known to me excepting the *C. marginatus*. The anterior cusp is closely joined to the external elevation of the anterior first cingulum as in *C. anax;* a character which separates it from all other species. A strong trace of a cingulum passes round the inner base of the crown. No external cingulum. The first true molar does not differ materially from that of other species. It is considerably smaller than the last. The apex of the premaxillary bone with the second incisor and alveolus of the first is preserved. The bone is rather short. The crown of the incisor is regularly convex externally, and is not expanded at the base. There is a strong internal cingulum.

A fragment of the lower jaw supports the last two molars. The internal angle of the last one is unfortunately broken. The posterior crest is, however, perfectly transverse, which is not the case with the species with three posterior tubercles. The preserved part of the posterior border shows a distinct, rather narrow heel. The anterior Vs are well developed, and there are no lateral cingula. The symphysis is flattened out by pressure. The inferior canine is large. It is sub-triangular at base and has an anterior basal angular projection.

Measurements.

		M.
Diameters of superior M. iii	transverse	.046
	longitudinal	.037
Diameters of superior M. i	transverse	.036
	longitudinal	.032
Diameters crown i. 2	vertical	.018
	transverse	.018
Diameters inferior M. iii	transverse	.028
	anteroposterior	.040
	vertical in front (restored)	.024
Length of symphysis		.107
Depth of ramus at M. iii		.056

The superior molars of this species might readily be taken for an undersized individual of *C. anax*, but the last inferior molar is of a different type, and refers the species to a different section of the genus.

The Big-Horn Basin; J. L. Wortman.

CORYPHODON CURVICRISTIS Cope.

Paleontological Bulletin No. 34, p. 172, Feb., 1882. Proceed. Amer. Philos. Soc., 1881, Dec. 172, (1882).

Plate XLIV c.

The fragments which represent this species belong to one individual. They include a considerable part of both mandibular rami with numerous molar teeth, and most of the inferior incisors loose. Also the second superior molar, some superior premolars, the canine, and three or four incisors, two of them in place in an incomplete premaxillary bone. None of the bones of the skeleton were obtained, so far as known.

The ramus of the mandible is both robust and deep. Its inferior border does not rise posteriorly so much as in some species, as *e. g.*, *C. latidens*, and the angle is well below the horizontal line of the dental alveoli. The dental foramen is just about in this line. The inferior premolars and molars do not differ from those of several other species, but the last molar has several peculiarities. The external cusp is the only one of the posterior pair which is present. It gives origin to two crests, both of them curved. The posterior represents the usual posterior transverse crest, but is gently convex backwards, and turns forwards on the inner side of the crown, only terminating at the external base of the anterior cross crest. The other curved crest is low, although higher than in most species, and extends to the middle of the base of the anterior cross crest. There is a distinct heel which is elevated at the middle and disappears gradually at each end, not being abruptly incurved as in *C. anax*. The anterior part of this tooth is as peculiar as the posterior. The external cusp gives origin to three crests, two of them the usual limbs of the anterior V; while a third descends to the anterior border a little exterior to its middle. It incloses a deep groove with the anterior ridge of the anterior V. This arrangement is not seen in any other species.

The inferior canine is robust, and has its anterior angle prominent, but not alate. The crowns of the inferior incisors are regularly convex exteriorly, and have no cingula. They are regularly graded in dimensions.

The superior molar preserved is probably the penultimate. Its anterior portion is broken. The posterior external V is narrower than usual for a second molar, and resembles somewhat that of the last superior molar of the *Manteodon subquadratus*. A slight contact face on the posterior cingulum shows that this tooth is not the last molar. The said cingulum extends to the external base of the V; in rising to the internal cusp it forms a sigmoid curve. The cingulum below this, on the inner base of the crown, is rudimental. The superior canine has a long and robust crown, with a triangular section to the apex. The posterior face is a little wider than the other two, which are equal. The anterior is slightly concave in cross-section, and the posterior slightly convex transversely, although concave longitudinally. There is a weak ridge nearly parallel to and near the postero-external angle, and traces of others on the postero-external face of the crown in front of this one. The antero-internal angle is swollen at the base.

The superior incisors present characteristic features. The ridge of the external face, which is weakly developed in some of the species, and is wanting in others, is here represented by a strong longitudinal angle, which extends from the base of the crown to its apex, dividing the external face into two distinct planes. This character is most marked on the external incisor, where the planes are sub-equal, and concave. On the second the anterior plane is smaller, and on the first it is a good deal smaller. These incisors have a weak internal cingulum, but no external one.

Measurements.

		M.
Length of ramus from Pm. iv, inclusive		.257
Length of inferior true molars		.098
Diameters of M. i inferior	anteroposterior	.0275
	transverse	.020
Diameters of M. iii inferior	anteroposterior	.036
	transverse	.029
Depth of ramus at M. iii		.075
Diameters of M. ii superior	anteroposterior	.0315
	transverse	.039
Diameters of crown of superior canine	longitudinal	.094
	anteroposterior	.022
	transverse	.034
Diameters of crown of I. iii	vertical	.022
	transverse	.024

The numerous characteristic marks show that this species is one of the most distinct of the genus. It is also one of the largest, being second only to the *C. anax*.

From the Wasatch formation of the Big Horn; J. L. Wortman.

CORYPHODON MARGINATUS Cope.

Paleontological Bulletin No. 34, p. 176, Feb., 1882. Proceed. Amer. Philos. Soc., 1881, Dec., p. 176, (1882.)

Plate XLIV*e*; fig. 5.

This is one of the smaller species, having nearly the dimensions of the

C. molestus. It is only represented by the superior canine, first inferior premolar, and last superior molar of one individual found together by Mr. Wortman. Their size, mineral condition, and degree of wear render it probable that all belong to one individual.

The superior molar is of the oval type, without posterior shoulder. The posterior crest is therefore straight, and parallel with the anterior crest. Its inner extremity does not display the least tendency to form a V, as is seen in *C. elephantopus.* Its exterior extremity is widely separated from the external prominence of the anterior crest (cingulum). The latter displays, at its inner extremity, the peculiarity of a deep fissure of the anterior side, which nearly divides the crest, and partially isolates the internal tubercle. Adjacent to the fissure its crest is tuberculate. The posterior upper cingulum descends from the inner cusp to the basal cingulum. The basal cingulum is well developed on the anterior and interior sides of the crown, and on the posterior as far as the base of the inner cusp of the posterior crest, where it gradually fades out. Enamel wrinkled.

The superior canine is remarkable for its small size. The posterior face is a little the widest, and its bounding edges are sharp, but not expanded. There are no prominent ridges of the enamel. The anterior face is moderately wide. The first inferior premolar presents no peculiarities.

Measurements.

		M.
Diameters of M. iii superior	anteroposterior	.028
	transverse	.038
	vertical	.019
Diameters of Pm. i inferior	anteroposterior	.015
	transverse posterior	.009
Diameters of C. superior	anteroposterior	.014
	transverse posterior	.018

The superior molar is but little worn, and shows that the animal was just adult. The canine is more worn than the molar.

There are several characters which mark this species as distinct from those previously known. It is the only member of the genus which has a complete internal cingulum. The fissure of the anterior crest, if normal, is peculiar to this species. The superior canine is disproportionately small.

Big-Horn Wasatch, Northern Wyoming.

Coryphodon anax Cope.

Plate XLIV *e*, fig. 6; XLVIII *a*.

Mr. Wortman sends me a number of teeth of probably two individuals, which exceed in size those of any species of *Coryphodon* yet known, and differ in certain details of form from all of them. The specimens consist of incisors, premolars, and molars of both jaws of one animal, and an inferior canine, which from its separate wrapping, I suppose to have been derived from a different locality.

The incisors and premolars have the form usual in species of the genus, differing only in their large size. A well-preserved superior true molar is probably the third. It has the form usual in the genus, but exhibits two peculiarities. The posterior transverse crest is divided more deeply than usual by a deep notch which enters it from the transverse valley. The external portion is the shorter, and exhibits the peculiarity of being connected with the external part of the anterior transverse crest. It is as closely connected with this crest as it is with the internal portion of the posterior crest. The external connection does not exist in the other species of the genus, where the two crests are separated at their outer extremities by a deep valley. The posterior basal cingulum is obsolete, while the anterior is well developed. The enamel of this tooth, where not worn, is wrinkled.

The posterior part of the last inferior molar is characteristic. The posterior transverse crest is short and very oblique, its inner extremity striking the posterior margin near the middle. Here it is elevated into a cusp, which rises above the surrounding parts in a characteristic manner. There is no ledge round its posterior base, but the border expands outwards at the base of the true crest. The additional inner marginal tubercle is low and compressed, as in *C. lobatus.* A second inferior true molar is normal, with well-developed anterior marginal ridge. The inferior canine mentioned is of large proportions, exceeding by one-half, the dimensions of the inferior canine of *C. lobatus.* Its crown is curved outwards, and has a basal alate expansion of its internal ridge.

Measurements: Diameters of last superior molar, anteroposterior, .039 M.; transverse, .051 M. Diameters of second inferior true molar, anteroposterior, .039 M.; transverse, .028 M. Length of inferior canine, .160 M. Length of crown of canine, .090 M. Diameters of base of crown of canine, ver-

Among the large Coryphodontes from the Big Horn Basin two sizes may be distinguished, a larger and a smaller. The latter usually have the posterior crest of the posterior inferior true molar transverse, and there is no interior crest or tubercle in front of it. Those I have referred to under the head of *C. latipes*, and the subject of Plate XLVI represents them, excepting only the spaces between the premolar teeth. But some of them display a more or less distinct trace of the internal crest, and in one specimen it is present on one side and not on the other. The larger specimens always possess the internal tubercle, and I refer these to the *C. anax*. The tubercle and posterior crest vary somewhat in proportions. In two specimens the latter is very short and oblique, its internal extremity having the position of the heel of a five-lobed molar; in three specimens it is longer and more normal. This is the character of the *Coryphodon lobatus*. In such specimens the internal lobe is nearer the extremity of the transverse crest, and the posterior border of the tooth is not angulate backwards. The specimens of this kind are as large as those typical of *C. anax*, and perhaps should be regarded as belonging to the *C. lobatus*. Owing to the want of superior molars, I cannot ascertain the full characters of the latter species.

A partially complete skeleton of a true *C. anax* includes the following bones: most of the mandible with several, including the last, molars; a few vertebræ, including axis; glenoid cavity of scapula; humerus and ulnæ complete, with parts of radius, parts of carpus; femur and tibia complete, parts of fibula, carpus and phalanges.

The ramus of the lower jaw is considerably deeper than that of the *C. latipes* (cfr. Pl. XLVI), and its inferior border rises backwards just below the posterior border of the last inferior molar. The symphysis extends to below the posterior border of the third inferior premolar, and the gutter for the tongue is quite deep. The surface spreads out to the bases of the canines and incisors, losing its concavity at the median border. The mental foramen is large and is situated below the first premolar on one side, and is double, and is below the second premolar on the other. The second true molar has the usual characteristic form of the genus, including a narrow posterior basal cingulum. The last molar is like that of other species (excepting the *C. curvicristis*) anteriorly, but the posterior crest forms a rectangle with the angle directed posteriorly. The distance from the lobe into which this angle rises to each of the intermediate tubercles is about equal. There is no heel, but a short, narrow, outward-looking cingulum extending from the external side of the posterior lobe, represents it. No other cingula; enamel finely wrinkled where not worn. A superior incisor has the crown little expanded laterally and quite thick anteroposteriorly. It has a well-marked obtuse external angle, and internal basal cingulum. The external faces are not flat or concave as in *C. curvicristis*. The superior canine has an immense root, and a three-sided crown. The external and posterior faces are gently convex in section, the former gently curved downwards, and neither have any ridges.

Measurements of jaws.

	M.
Depth of ramus mandibuli at M. iii	.092
Length of bases of inferior true molars	.100
Diameters of M. ii { anteroposterior	.032
Diameters of M. ii { transverse	.024
Diameters of M. iii { anteroposterior	.043
Diameters of M. iii { transverse	.0275
Diameters of base of crown of inferior canine { vertical	.031
Diameters of base of crown of inferior canine { transverse	.027
Width between bases of canines	.095
Width between bases of Pm. ii	.044
Length of superior canine, apex supplied	.214
Length of crown of canine, apex supplied	.072
Transverse diameter of base of crown, apex supplied	.033

The centrum of the axis is considerably wider than long, and is not keeled below. The dentate process has a round section. The superior or ligamentous facet is separated from the inferior or atlantal face by a groove all round. Three additional cervicals, a dorsal and a lumbar vertebræ, were found thirty feet down the ravine from the jaws and limb bones, according to Mr. Wortman. As no other vertebræ were found with the latter, and no other parts of a skeleton were associated with these vertebræ, all are viewed as pertaining to the same animal. They, however, lack epiphyses. As the last inferior molars, though fully protruded, are not worn on the posterior lobes, the association of these vertebræ is not impossible. The cervicals are quite short, and the centra are somewhat oblique, and the articular faces wider than deep. The posterior is deeper than the anterior. The inferior face is neither keeled nor strongly angular, and its posterior border is produced so as to overlap the succeeding centrum to its middle.

The dorsal centra are short and contracted at the middle, and without keel or hypapophysis. The capitular fossæ are strong on both faces of the bases of the diapophysis. The centrum of the lumbar is relatively a little longer, and has a median thickened ridge below.

Measurements of vertebræ.

		M.
Length of axis with dentatus		.105
Diameters of centrum axis	anteroposterior	.65
	transverse	.102
Diameters of anterior face C. iii	vertical	.047
	transverse	.076
Diameters of centrum dorsal	vertical	.035
	transverse	.066
	anteroposterior	.053
Length of centrum lumbar		.037

The glenoid cavity of the scapula is rather narrow, and its anterior region is narrowed even more than in the *C. latipes*, and is produced into an acute tuberosity. This is close to but distinct from the coroacoid, which is broken off from the specimen. The spine rises abruptly from a thick base from near the edge of the glenoid. The proximal extremity of the humerus is a little crushed, but shows that the proximal extremity of the great trochanter is half as wide as the articular face measured on a transverse line. Its proportions are thus intermediate between those in the *C. latipes* and the *Bathmodon radians* (see page 276). The shape of the proximal extremity is subtriangular. The inferior portion of the deltoid crest is prominent, and twists inwards, continuing by an angular ridge to the internal edge of the condyle. The internal epicondyle is prominent, both outwards and backwards. The internal condyle is also prominent and twists backwards in a crest which disappears on the inferior half of the posterior face of the shaft. The condyle is rather short; the inner flange is prominent, and the inner cylinder has its external part beveled quite steeply. The olecranar fossa is mostly external to the middle line.

Measurements of scapula and humerus.

			M.
Diameters at glenoid cavity	anteroposterior		.105
	transverse		.063
Width of neck of scapula			.095
Diameters of head of humerus	transverse	with tuberosity	.142
		articular surface	.093
	anteroposterior		
Diameter shaft at middle			.075

Diameters distally	total transverse		.128
	condyles transverse		.082
	anteroposterior	flange	.072
		cylinder	.058

When the olecranon is placed with its base on a level surface the radial facet of the ulna slopes inwards at an angle of 30°. It is nearly plane. The olecranon narrows to an apex by the gradual descent of the superior border, the steeper rise of the inferior surface, and the oblique truncation of the inner side. The shaft of the ulna is rather short, and is compressed, the long axis being directed downwards and inwards at an angle of 45°. Its inner side is divided equally by a low straight interosseous ridge, and is bordered below at the distal extremity by a horizontal projecting ridge, continuous with the inferior surface. The external face is plane, and the inferior edge nearly straight. The carpal face is compressed in the direction of the shaft, and is directed obliquely inwards, coming from the distal extremity. A comparison of this ulna with that of a specimen here referred to the *C. latipes*, shows that while the latter has nearly the same vertical diameters, it is considerably shorter both in the shaft and the olecranon. The head of the radius of the *C. anax* is transversely expanded, and the internal border is reflected posteriorly or proximally, and the external border distally. There is a flat tuberosity on the external part of the superior border of the cup, and no notches anywhere.

Measurements of ulna and radius.

		M.
Length of ulna		.350
Length of olecranon		.075
Length of olecranon, plus glenoid		.155
Depth of middle of olecranon		.058
Depth at coronoid process		.083
Depth at middle of shaft		.050
Depth of carpal facet		.046
Diameters of head of radius	vertical	.043
	transverse	.071

The pisiform bone is robust and is considerably expanded distally. The unciform has the usual form. The three inferior facets are distinct, and the inner one is very oblique, connecting by equal angles the median distal facet with that for the os magnum The posterior aspect presents a conical claw-like tuberosity. The lunar face is not separated from the cuneiform face by an angle.

Measurements of carpus.

			M.
Length of pisiform			.054
Width of shaft			.028
Width of distal extremity			.048
Diameters of unciform	proximally	fore and aft	.032
		transverse	.058
	vertical anteriorly		.028

A segment of the sternum is subquadrate and flat, as in the *C. latidens*, but thicker. The length is M. .058; width at middle, .044.

The femur is rather elongate, with the shaft less robust than in the *Bathmodon radians* (see Plate XLVII). The head projects beyond the great trochanter, and has a deep *fossa ligamenti teris* near the posterior border. The little trochanter is a prominent crest of two inches in length on the internal edge of the bone. The third trochanter is broken off. Its base is at the middle of the shaft. The rotular groove is rather wide, and the lateral borders are equal, and not much elongated. It extends well proximally. The epicondyles are insignificant; the external is impressed with three fossæ, one on the edge of the condyle, and the middle one is the largest. The internal condyle has the widest sweep, and looks obliquely towards the external condyle. The inter-condylar fissure is rather narrow and deep.

Measurements of femur.

	M.
Length	.477
Width proximally, with great trochanter	.138
Transverse diameter of head	.073
Width of shaft below third trochanter	.064
Width at epicondyles	.105
Anteroposterior diameter inner condyle	.077
Anteroposterior diameter outer condyle	.064

The tibia is much shorter than the femur, and is rather more slender than in some species of the genus. It presents the generic characters of flat crest and ungrooved distal articular face. The most slender part of the shaft is below the middle. The inner femoral cotylus has a considerably greater anteroposterior diameter, and there is a free edge behind it which increases the projection beyond external cotylus. There are various longitudinal rugosities on the internal face of the proximal third of the bone, and the wide face of the crest is grooved striate. The malleolar process is not large, is acuminate, and turns backwards. The posterior border has an angle near the base of the malleolus, but there is no groove for the flexor tendons of the foot and digits. The patella is flat, and does not exhibit the distal prolongation seen in some specimens; *e. g.*, the one figured in my report to Captain Wheeler, Vol. IV of his general report, Plate LVIII, fig. 4. The proximal end of the fibula is unusually large; it has a nearly quadrangular section, and the tibial facet covers about half the oblique proximal face. The calcaneum is quite robust. It presents the usual characters. The cuboid facet is small and more than usually oblique inwards. A tuberosity continues the inferior face between this facet and the base of the sustentaculum, and is separated from the former by a fossa. The sustentaculum is narrow, and its astragalar facet oblique downwards and forwards. The external astragalar facet is flat; external to it a large smooth surface receives the extremity of the fibula. Below this is a deep longitudinal fossa, and below this a strong horizontal crest, which extends to the cuboid facet distally and to the middle of the heel posteriorly. The posterior extremity of the calcaneum has a nearly quadrate outline.

Measurements of posterior foot.

	M.
Length of tibia on front to spine	.300
Width of head	.105
Anteroposterior diameter external cotylus	.075
Anteroposterior diameter internal cotylus	.053
Least diameter of shaft	.042
Diameters distal end tibia { anteroposterior	.044
Diameters distal end tibia { transverse	.084
Length of patella	.094
Transverse width head of fibula	.052
Length of calcaneum	.106
Width calcaneum at sustentaculum	.081
Width calcaneum at free extremity	.048
Depth calcaneum at free extremity	.042
Depth calcaneum external facet	.045
Depth calcaneum cuboid facet	.029
Width of cuboid facet	.037

Not having an astragalus or posterior foot of the *C. anax*, it is not yet possible for me to state whether this species is a *Bathmodon* or a true *Coryphodon*. The large *Bathmodon pachypus* is distinguished from the *C. anax* by various characters which are mentioned under the head of that species.

The *Coryphodon anax* has been only found so far, in the Wasatch beds of the Big-Horn Basin of Wyoming. J. L. Wortman discovered it.

It was originally described in the Paleontological Bulletin No. 34, p. 168, and is figured in the present work in Plates XLIV *a*, figs. 7–12; XLIV *b*; XLIV *e*, fig. 6; XLIV *f*.

BATHMODON Cope.

Proceed. Amer. Philos. Soc., 1872, p. 417, Feb. 16, 1872. Annual Report U. S. Geol. Surv. Terrs., F. V. Hayden, 1872, p. 586 (1873). Systematic Catalogue Vertebrata Eocene New Mexico, 1875, p. 24. American Naturalist, 1882, January, p. 73. Paleontological Bulletin No. 34, p. 165, Feb., 1882.

This genus reposes as yet on the structure of the posterior foot, which is however not completely known. The astragalus is less transversely extended than in *Coryphodon*, and the internal apex is recurved, and carries on its interior extremity an articular facet not found in that genus. It looks inwards, and stands below the middle of the inner side of the tibial face of the bone. It is not certain what element articulates with this facet. It may be a proximal extension of the proximal extremity of the entocuneiform. This is questionable in view of the square truncation of the facet and its entirely inward presentation. It may be a supernumerary bone forming a spur.

So far as known the other characters of the genus are as in *Coryphodon.* The skull has not been found. Some bones found with the type specimen of *B. radians* have the following characters:

Some anterior sternal segments are cylindric; in one, the articulations for the hæmapophysis project laterally, giving the piece a T-shaped form. The atlas has a flat diaparapophysis, presenting its edges fore and aft; the arterial canal traverses it obliquely. The coracoid is double, having a tuberosity on the edge of the glenoid cavity, and a prominent hook just outside of it. The lumbar vertebræ are quite short. The cuneiform bone is narrow pyriform, with two triangular facets on one side, the smaller being sublateral, and one twisted over the other. The ungual phalanges are very short, somewhat flattened, and with the terminal portion transverse and rugose as in some toes of *Elephas.*

BATHMODON RADIANS Cope.

Proceed. Amer. Philos. Soc., 1872, Feb. 16, pp. 417, 418. Hayden's Geol. Surv. of Montana, 1871, p. 350. Annual Report U. S. Geol. Surv. Terrs., 1872 (1873), p. 587. *Coryphodon radians* Cope, Report U. S. Geol. and Geog. Expl. Surv. W. 100th Meridian, G. M. Wheeler, iv, part II, p. 206. ? *Bathmodon semicinctus* Cope, *l. c.*, p. 418. U. S. Geol. Surv. Terrs., 1872, 1873, 588.

Plates XLV, XLVI, XLVII, XLVIII; figs. 1–6.

The original specimens on which this species was recorded, were obtained by Dr. Hayden near Evanston, Wyoming. They include parts of two individuals represented by maxillary teeth, scapulæ, humerus, femur, tibiæ, fibulæ, astragalus, and cuneiform bones. I subsequently obtained from the same person who exhumed these bones, both rami of the lower jaw of probably one of these individual specimens. I visited the locality in 1873, and with many specimens of bones of *Coryphodon latipes*, I obtained a humerus and a femur, agreeing with those of the *Bathmodon radians* obtained by Dr. Hayden.

This species is larger than the *Coryphodon latipes*, and is one of the largest species of the family known to me. It is distinguished from the latter not only by the form of the astragalus already mentioned, but by marked peculiarities in the scapula, humerus, and fibula. Those distinguishing the first two bones, have been noted under the head of the *C. latipes.*

The inferior premolars are all two-rooted, and form an uninterrupted series. The basis of the malar part of the zygomatic arch originates opposite the adjacent parts of the penultimate and last molars. The premaxillary bone is massive, and with but little area for attachment with its fellow in front. The incisor teeth are large, with subcylindric roots, and their alveoli are well separated. The crowns are expanded transversely, with convex cutting edge.

The premaxillary bone is elongate, flat, and with a sloping superior face, which rises gently inwards. The bases of the incisors stand obliquely outwards. The inferior surface is flat, and the basis of the broken palatal spine is rather small. An incisor tooth has a transversely diamond-shaped crown, slightly twice concave on the inner faces, strongly convex on the outer, with a faint external cingulum near the external angles. Enamel obsoletely striate.

A posterior superior premolar has a cingulum on the inner obtuse apex. The crest of the inner crescent, descending on each side of the apex of the outer, forms a cingulum-like ledge at its base as far as the angle formed by the descent of the apex of the outer crescent. The outline of the crown of this tooth, viewed from above, is narrow cordate, with obtuse apex. The convexity of the outer crescent inwards is very strong, and the base of the crown is externally two-lobed. Enamel striate rugose. In a more anterior premolar (with three roots) there is no internal cingulum, and the crest of the inner crescent is not carried to the external basis of the tooth, and is entirely wanting on the posterior face of the tooth. The external crescent is more vertical and less concave. Outline of crown subtriangular.

The transverse diameters of all the superior molars exceed their longitudinal. In the penultimate, which may serve as a type, this superior or outer plane of the inner crescent ridge extends along about .66 of the posterior of the outer crescent. In the last molar the surface is very wide on the posterior and inner side of the external crescent; it then contracts, and expands again on the posterior side, its outer bounding crest reaching to the external margin of the crown.

Besides these points, the molars possess a strong cingulum along the posterior base of the crown, which unites with the surface near the inner protuberance of the latter in the penultimate; in the last molar it reappears, forming a short lobe on the posterior face. The enamel where not worn is slightly rugose.

In the mandible the incisors radiate around the narrow extremity of the trough-like symphysis, and have transversely expanded crowns. The canine is inclined forwards, and forms part of the same series. Its crown is triangular in section, the outer face convex. In the males it was enormously enlarged, as indicated by a symphysis in my possession. The anterior premolar approached the canine. The premolars have an external chevron directed inward, whose exterosuperior surface of enamel is acute cordate. Besides this is a little longitudinal ridge, which represents another chevron of the true molars. On the first of the latter, both chevrons are developed, the posterior the least, both with their anterior ridge boundaries lowered; they sink entirely on the last two molars, which become thus two-crested, as in those of some Tapiroids, and the premolars of *Dinotherium*. The last inferior molar has a narrow ledge in place of a heel, which sends a narrow ridge upwards and outwards. There is no trace of tubercle or crest in the interval between the external ends of the cross-crests. The oblique anterior connecting ridges are low in the inferior molars of this species. The first and second molars have posterior cingula relatively as large as that of the last molar. The premolars are separated by interspaces; a character I have not seen in any species of *Coryphodon.*

The transverse process of the *atlas* is rounded distally, and is about as long as wide; the surface for the axis is directed obliquely inward.

The apex of the *scapula* is a massive flattened acumination with truncate extremity. The spine is elevated and truncate next the glenoid cavity, which is a wide oval, much produced at the coracoid margin.

In the *humerus*, the hook of the greater tuberosity is developed, but is not much elevated above the proximal plane. It originates from an external expansion of the greater tuberosity, which bears a shallow concavity separated from the head by a low curved subtransverse ridge. As pointed out under *Coryphodon latipes*, this expansion of the great tuberosity is very large, its width equaling that of the articular surface of the head.

An almost perfect femur of *B. radians* is preserved. The third trochanter is not very prominent The little trochanter is little developed; the great trochanter is large but does not equal the head. The latter is subglobular, and the ligamentous fossa extends to its rim. The distal rotular surface is prominent, the inner edge more so than the outer. Its articular surface is broadly continuous with those of the condyles; a slight emargination of the outlines only marking the usual constriction on each side. In this it resembles *Cervidæ* and some *Antilopidæ.* The inner condyloid surface is cut off by the emargination in *Toxodon* and *Bos bubalus;* the emarginations are deep, but do not cut off either, in *Equus, Camelopardalis*, and three species of *Bos;* while they are so deep as to cut off both in *Rhinoceros*, 5 sp., *Hippopotamus, Bos brachycerus, B. sondaicus*, and in *Catoblepas.*

The *fibula* has the inner sharp edge prolonged to the proximal end: the form of the latter is much as in *Eobasileus.* The *astragalus* is slightly concave in both directions on the trochlear face, most so anteroposteriorly. The anterior outline of the same is strongly and obliquely convex, and the surface is produced sideways into a lateroanterior apex. The inner malleolar border is thus very concave; the outer is gently convex with a long fibular facet. The posterior margin is concave, the inner tuberosity prominent. The *navicular* facet is as broad as long, and nearly sessile, being probably separated by a groove from the tibial. The cuboid facet is subround, small, and sublateral. The calcaneal are situated diagonally opposite each other. The anterointernal is twice as large as the other, and is transverse and truncate internally by a facet near the apex, at right angles. The other calcaneal facet is subround.

Measurements.

	M.
Longitudinal diameter of last superior molar	.035
Transverse diameter of last superior molar	.0455
Longitudinal diameter of penultimate	.032
Transverse diameter of penultimate	.039
Longitudinal diameter of posterior premolar	.024
Transverse diameter of posterior premolar	.034
Longitudinal diameter of anterior premolar	.0215
Transverse diameter of posterior premolar	.0265
Length of premaxillary bone	.082
Transverse width of posterior suture	.628
Width of premaxillary at middle suture	.043
Length of basis of last two inferior premolars	.057
Transverse diameter of edge of mandible at first premolar	.017
Length of ramus mandibuli to anterior margin of coronoid process	.310
Length of premolars and molars	.218
Length of last molar crown	.040
Width of last molar crown	.030
Width of last premolar crown	.018
Length of last premolar	.025
Width of symphysis at canines	.045
Diameter of canines ♂	.028
Diameter of canines ♀	.023
Length of exposed portion of incisor 2	.026
Width of crown of incisor 2	.0245
Length of diapophysis atlas	.047
Width of diapophysis atlas	.056
Width of facet for axis	.053
Width of glenoid cavity of scapula (straight)	.086
Length of coracoid from inner basis	.045
Diameter of condyles of femur	.104
Diameter of head and great trochanter	.130
Diameter of head alone	.062
Diameter of shaft with third trochanter	.076
Supposed length of femur (16.75 inches)	.415
Transverse diameter of head of tibia	.092
Anteroposterior diameter, internal	.061
Anteroposterior diameter, external	.045
Length of proximal articulation fibula	.027
Length of distal articulation fibula	.042
Diameter of shaft of fibula	.042
Total length of astragalus (fore and aft)	.072
Total width of astragalus	.065
Length of navicular facet	.045
Width of navicular facet	.034

	M.
Width of cuboid facet	.025
Length of cuboid facet	.023
Length of anterior calcaneal facet	.040
Width of anterior calcaneal facet	.024
Length of posterior calcaneal facet	.021
Length of fibular (axial)	.043

From the Wasatch Lower Eocene at Evanston, Wyoming. Numerous specimens of teeth of both jaws obtained by myself in New Mexico, and by Mr. Wortman in the Wind River and Big Horn regions of Wyoming, agree closely with those found with the bones of this species. In lack of diagnostic bones it is not possible to identify them positively with the *B. radians*. A humerus having much the character of that of this species is figured on Plate LIV, fig. 5, of my Report to Capt. G. M. Wheeler.

BATHMODON PACHYPUS Cope.

Proceedings Academy Philadelphia, 1882, p. 294.
Plates XLIV c; XLIV d; XLIV e; figs. 7-13.

The largest species of the *Pantodonta* yet known is represented by portions of two individuals. Neither of these includes any teeth, but the characteristic bones of the skeleton have been preserved. They indicate an animal of the dimensions of the middle-sized species of *Uintatherium*, but with shorter legs. All the specimens were found by Mr. J. L. Wortman in the Wasatch beds of the Big Horn Basin of Wyoming. It is possible that some of the large teeth of the collection made by Mr. Wortman belong to this animal, but there is as yet no means of proving the connection.

One of the specimens includes a large part of the humerus, with head of radius; pelvis entire; femur entire; and a good many carpal and tarsal bones, metapodials, and phanlanges. The only cranial fragment is a mandibular condyle. The bones of the fore limb and foot do not differ from those of the specimen above described, except in their superior size, with the following exceptions: The outline of the proximal extremity of the great trochanter is a semicircle rather than a triangle. There is little indication of the tuberosity of the superior border of the head of the radius. The proximal end of the pisiform is preserved. It is more expanded transversely than the distal extremity, as I have represented in figs. 2 and 5, Plate LIX, of my report to Captain Wheeler.

Measurements of fore limb.

			M.
Transverse diameter of humerus	proximal end		.154
	distal end		.151
Width of head of radius			.082
Length of pisiform			.064
Width of pisiform proximally			.055
Diameters unciform	transverse		.065
	anteroposterior		.051
	depth in front (greatest)		.032
Diameters magnum	anteroposterior		.054
	transverse in front		.027
	vertical	in front	.019
		medially	.039

The width of the condyle of the mandible is M. .102.

As no complete pelvis of a *Coryphodon* has been hitherto procured, I describe the present specimen in detail. Its expanded ilia present the ungulate type, and the wide peduncles resemble those of the elephant. The plate is, however, not so expanded anteroposteriorly as in the genus *Elephas*, and the crest is more convex. If the crest of the ilium of the camel were considerably more convex, the bone would resemble that of this species of *Bathmodon*. It is less dilated anteroposteriorly than in *Rhinocerus* sp., and has a stouter peduncle than in that genus, *Equus*, *Tapirus*, or *Sus*. The symphysis is shorter than in any of the genera above mentioned. This is partly due to the fact that the posterior parts of the ischia are not so wide anteroposteriorly as in any of them, the nearest approach being seen in *Rhinocerus*. The obturator foramen is larger than in any of these genera, excepting *Rhinocerus*, and there is no *tuber ischii* at all comparable to what is seen in all those genera. The pubes are slender, and are not so transverse as in those genera, but slope posteriorly. The notch of the acetabulum is opposite the shaft of the ischium, and not opposite the obturator foramen, as in all the genera named. The sacral articulation is rather long and marks a bevel of the edge of the plate of the ilium.

Measurements of pelvis.

	M.
Axial length	.572
Axial length of ilium	.340
Width of plate of ilium	.340
Width of peduncle of ilium	.112
Length of pubis	.140
Long diameter of pubis at middle	.030
Length of ischium	.125
Long diameter of ischium at middle	.065
Length of ischio-pubic symphysis	.090
Length of obturator foramen	.106
Width of obturator foramen	.070

There is no important difference between the femur and that described under *C. anax*, excepting the superior size. The little trochanter is not quite so prominent in the larger bone.

Measurements of femur.

	M.
Length	.500
Width proximally with great trochanter	.165
Transverse diameter of head	.085
Width at third trochanter	.105
Width of shaft below third trochanter	.077
Width at epicondyles	.138
Anteroposterior diameter inner condyle	.092
Anteroposterior diameter outer condyle	.074

The astragalus has the characters of that of *B. radians*, with an approach to some of those of *Coryphodontes*. The internal hook is well developed, and is recurved so that its posteriorly-directed apex is separated by a deep notch from the posterior median tuberosity. In *C. latipes* and other species the apex is directed inwards and not backwards. This is a well-marked character, and one in which it resembles the *Bathmodon radians;* but in that species the internal hook is still shorter, though its tibial face is not carried so far posteriorly. The apical facet, mentioned as characteristic of the genus *Bathmodon* has the same relative size and proportion as in *B. radians*, but its surface is not cut off from the anterior part of the apex by a deep groove, as in that species. A comparison of the inferior side with that of the astragalus of *B. radians*, displays the following differences: The external facet in the latter is cut off from the posterior tuberosity by a deep groove, which is continued from a deep notch in the posterior border of the bone. In *B. pachypus* the notch is very shallow, and the groove is not connected with it, the external facet being continuous with that on the inferior side of the posterior tuberosity. The notch is more apparent on the superior than on the inferior side of the astragalus in the *B. pachypus*. In the latter, the anterior inferior facet is separated by a wide space from the supernumerary internal facet; in *B. radians* they are almost in contact. The anterior facet is absolutely smaller in the *B. pachypus*, and is more largely extended on the posterior tuberosity than in *B. radians*. Both the navicular and cuboid facets have greater transverse extent in the *B. pachypus*, and the latter extends to behind the middle of the former, which it scarcely does in *B. radians*.

Measurements of astragalus.

		M.
Diameters, greatest	anteroposterior	.080
	transverse	.094
Diameters navicular facet	vertical	.037
	transverse	.069
Diameters cuboid facet	vertical	.022
	transverse	.051
Depth of fibular facet		.025
Transverse width of anterior inner astragalar facet		.032

The calcaneum differs decidedly from that of *Coryphodon anax*. While its general dimensions are greater, the length is the same. The section of the posterior extremity is a transvere parallelogram; in the *C. anax* it is square. The sustentaculum has two facets, a superior narrower, and an anterior; the *C. anax* has the anterior only. In the former the posterior face of the sustentaculum rises to the posterior edge of the facet; in the *B. pachypus* it has a superior surface, which is separated from the superior facet by a groove. The cuboid facet is transverse and extends inwards, overlapping the sustentacular; in *C. anax* it is very oblique, and does not reach the sustentacular. The median inferior tuberosity of the distal extremity is present in both species, as is the inferior marginal crest of the distal half of the internal side. In both species this ridge shows two oblique grooves, which perhaps accommodated the tendons of the *tibialis posticus* and *flexor digitorum* muscles. In this order there is no groove of the cuboid bone to accommodate them.

Measurements of calcaneum.

		M.
Total length		.107
Diameter of free end	vertical	.048
	transverse	.061
Width at sustentaculum		.095
Width of cuboid facet		.049
Greatest depth of facet		.026

The calcaneal and astragalar facets of the cuboid are about equal, and make a strong angle with each other; the calcaneal is the more oblique to the distal face. The distal extremity is subquadrate, the anterior three-fifths being occupied by the two metatarsal facets, which are little distinct from each other. The external is less than half as large as the internal.

Measurements of cuboid.

		M.
Measurements	vertical in front	.031
	anteroposterior below	.050
	transverse below	.050
Length of a metopodial bone		.080
Diameters proximally	anteroposterior	.040
	transverse	.038

			M.
Diameters distally	anteroposterior		.028
	transverse		.040
Diameters of a phalange	proximally	anteroposterior	.037
		transverse	.024
	longitudinally (lateral)		.022

A second individual, apparently, of this species is represented by an entire humerus and radius. The articular extremities of the humerus and radius agree closely with those of the specimen first described, differing only in being a little larger.

The humerus differs from that of *Coryphodon anax* much as the calcaneum of the first described specimen does from the corresponding bone of the latter species. It is nearly identical in length, but considerably more robust in every part. In the size of its articulations it equals some of the large species of *Uintatherium*, but the shaft is relatively longer in the latter. The radius is also remarkable for its very robust form and the large size of its articular extremities. The head resembles that of the specimen first described in wanting the prominent tuberosity of the superior border found in *C. anax.* There is, however, a convexity in the corresponding position in the *B. pachypus.* The inferior surface next to the ulnar band is rough, with longitudinal ridges and grooves, but there is no defined fossa. The long axis of the distal extremity is oblique to that of the head, the external face looking outwards and upwards. The carpal face presents several peculiarities in which it differs much from the corresponding part of a *Coryphodon* figured on Plate LXIV, fig. 6, of my final Report to Captain Wheeler. The carpal facets only occupy a part of the extremity of the bone, which is much expanded outwards. On this obliquely truncate club-shaped extremity the long axis of the facets extends backwards and outwards, their margin only coinciding with that of the former at the point of contact with the ulna. The facets for the scaphoid and lunar are distinct, the former being plane, the latter concave. The scaphoid, moreover, has only about one-fifth the area of the lunar, a relation quite different from that which prevails among mammals generally. The extremity has a ligamentous fossa between the scaphoid facet and the internal border of the dimensions of the latter. The ulnar face is not distinct. There is an angular ridge just above the external border of the head, and another on the inner side, defining it from the inferior side just above the head. There

Measurements.

		M.
Length of shaft of humerus		.400
Width at middle		.085
Width at epicondyles		.163
Width of condyles		.115
Diameter of head anteroposteriorly		.105
Diameter heads of radius	vertical	.055
	transverse	.088
Length of radius		.260
Greatest width of distal end radius		.090
Diameters scaphoid facet	anteroposterior	.025
	transverse	.020
Diameters lunar facet	anteroposterior	.051
	transverse	.052
Long (transverse) diameter of shaft at middle		.048

From the preceding it is evident that this species is distinguished for its robust form, as indicated by the proportions of its legs and feet. It is the heaviest species of the *Coryphodontidæ* yet known.

METALOPHODON Cope.

Proceed. Amer. Philos. Soc., 1872, p. 542, Sept. 20. Annual Report U. S. Geol. Surv. Terrs., 1872 (1873), p. 589.

This genus differs from *Coryphodon* in the structure of the true molar teeth; the formula is the same as in that genus: I. 3; C. 1; Pm. 4; M. 3. All the premolars have a single external V, extended well inwards, and a single internal cusp, whose anterior and posterior edges are continued into cingula which extend along the anterior and posterior bases of the external V. The first true molar, like that of *Coryphodon*, has two external V's, but the anterior is represented by a subconic cusp, while the posterior is large, and is well produced inwards to its apex. The anterior basal cingulum of the premolars has become an elevated transverse crest of the crown, terminating in the crescent-edged single inner tubercle. The second true molar, instead of resembling the first, as is the case in *Coryphodon*, is like the third or last true molar in its structure. In both molars the posterior external V has become a stout transverse crest, connected with the subconic representative of the anterior V, the two forming a cross-crest nearly parallel to the anterior cingular cross-crest. They are separated from the latter by wide interruptions at both extremities. The canine in the only known species, is compressed, and has a wide groove extending along its posterior edge.

The existence of the posterior transverse crest on the second true molar constitutes a point of nearer resemblance to the *Eobasileidæ* than is found in the other known *Coryphodontidæ*.

METALOPHODON ARMATUS Cope.

Proceed. Amer. Philos. Soc., 1872, p. 542. Annual Report U. S. Geol. Surv. Terrs., 1872 (1873), p. 589.

Plate xlix, figs. 1-7.

This, the only known species of the genus, was found by myself near Black Butte, Wyoming, in the Wasatch formation. The typical specimen consists of a number of teeth of both jaws which I took from two decayed skulls which were partly exposed in a bank of argillaceous material. These skulls were lying close together, and belonged to an adult and young animal, respectively, as indicated by the relative wear of the teeth. The adult series consists of three incisors, two superior premolars, and three superior true molars, with three inferior premolars. The young animal is represented by one incisor, the superior and inferior canines, three superior premolars, and two superior true molars; also by one inferior premolar and one inferior true molar. The second true molar of the younger specimen is unfortunately wanting, so that it cannot be demonstrated that it belongs to this species, but the other teeth, especially the first and third true molars, correspond so closely as to induce me to suspect that the two animals were members of the same species, and probably of the same herd.

I first describe the teeth of the adult:

The *incisors* are well developed, those of the premaxillary subequal in size. The crown has a convex cutting edge and flat inner face. The outer face is convex. In some the inner face is more concave, and is bounded by a weak cingulum next the root.

The superior *premolars* present a single external crescent of acuminate outline, and a smaller, more transverse one, within. A cingulum bounds the crown fore and aft, but is wanting at both external and internal bases of the crown. The posterior sends a branch to the apex of the internal lobe. In the fourth the crescent is more open and the crown less transverse.

The first true *molar* presents an increase in transverse extent of the posterior external crescent, and the anterior one is a flattened cone. In the posterior two, the anterior ridge curves round at the apex, but is separated by a considerable interruption from the posterior. The latter is shortened, and terminates externally in a conic tubercle, which approaches the recurved outer extremity of the anterior ridge. In the last molar the posterior ridge is shorter, nearly straight, terminating abruptly at each extremity. On all the true molars there is a strong anterior basal cingulum below the anterior transverse ridge. In the first there is also a strong posterior basal cingulum; but this is wanting from the second and third. A short curved cingulum passes round the base of the internal extremity of the posterior crest, but none passes around the interior or exterior base of the anterior crest.

The *inferior premolars* are represented by the first, third, and fourth. The first has but one root. The crown is compressed, and the posterior heel is very short, consisting of a median keel and two cingula The posterior external basal angle of the V is very prominent in this species in the posterior premolars, as in *Coryphodon*. It is much more elevated than the posterior basal angle. The heel has considerable transverse extent in the fourth premolar, the internal cingulum expanding into a table. The cingula do not extend to the internal or external bases of the crown.

The teeth of the younger animal present the following characteristics:

The *canines* are damaged, but were of large size, amounting in the upper jaw to a tusk. The superior is compressed, with acute edges. The inner face is gently convex, the outer more strongly so, with an acute ridge on its anterior convexity inclosing an open groove with the interior cutting edge. The surface of the dentine, when exposed, has a transversely wrinkled character, but no trace of engine-turning in the fractures. The inferior canine is subequilaterally triangular in section at the base. The posterior face is slightly convex, and is separated by a very prominent angle from the external face. The latter is concave next this crest-like angle, and is convex anteriorly.

The first superior true molar is similar to that of the adult specimen, and is unworn. The large posterior external V is nearly flat on its externo-superior face. The cingula do not pass round the internal base of the crown. The enamel is finely wrinkled; it is thrown into vertical ridges on the inner convex extremity of the crown. Similar ridges are prominent in the corresponding part of the crown of the last superior molar. The last inferior molar has a distinct but rather narrow posterior heel, and there is no accessory tubercle or accessory ridge as in the typical species of *Coryphodon.*

The size of this species is about that of the *Bathmodon radians.*

Measurements of the teeth.

No. 1.

	M.
Total length of superior incisor	.057
Length of crown (inner face) superior incisor	.015
Width of crown (oblique) superior incisor	.020
Width of crown (oblique) inferior incisor	.023
Length of crown (inner face) superior incisor	.018
Width of posterior molar	.039
Length of posterior molar	.028
Elevation of posterior crest of molar	.016
Width of anterior true molar	.035
Width of premolar	.028
Length of premolar	.0215
Length of premolar (first)	.016
Width of premolar (first)	.008
Length of premolar (inferior)	.024
Width of premolar (inferior)	.020

No. 2.

Width of superior canine .030 from apex		.020
Width of inferior canine at external base		.028
Diameters of first superior molar	anteroposterior	.033
	transverse	.037
Width of last lower molar		.023
Length of last lower molar		.037

METALOPHODON TESTIS Cope.

Paleontological Bulletin, 34, Feb., 1882, p. 175. Proceed. Amer. Philo. Soc., 1881, Dec., 175.

Plate XLIV *a*; fig. 13.

The genus *Metalophodon* was described by me in 1872. Up to December, 1881, it had remained without further illustration of importance, as no good specimens of it had been obtained by any of my expeditions up to that year. The material then obtained consists of the entire superior molar series of the right side, and the superior molars of the left side, in beautiful preservation. These display the characters on which the genus was proposed, *i. e.*, the conversion of the posterior external V of the second true molar into a transverse crest similar to that of the last true molar. It follows that the first true molar is the only one which exhibits this V. It also follows that in this genus the peculiarities of the dentition of *Coryphodontidæ* are carried further than in *Coryphodon*, where two molars display the V, and one the crest; or than in *Manteodon*, where all three have a V, and none the crest. The genera then stand in the order of evolution, *Manteodon, Coryphodon, Metalphodon.*

The first superior premolar has lost its crown. The other premolars do not display any marked peculiarities. The internal cusps are well developed, and are most prominent posterior to the line of the apex of the exterior crest. They connect with the posterior cingulum by a broad ledge, but do not connect with the anterior cingulum. The two cingula nearly connect round the inner base of the crown on the third premolar.

The first true molar is well worn. The base of the posterior external V can be seen, and the anterior and posterior cingula. There is no internal cingulum. The second true molar is the largest of the teeth. It is subtriangular in outline, its external side forming, with the posterior, a right angle. Its general character is much like that of the *Coryphodontes*, but it presents the remarkable exception which constitutes the character of the genus *Metalophodon*. The posterior crest does not include a V, but is straight, and consists of the same elements as the posterior crest of the third true molars, but differently proportioned. The part representing the anterior V is a cone, much shorter than the part corresponding to the posterior V. As there is a postero-exterior angle of the crown there is an oblique surface rising to this part of the crest, which represents the external face of the V. There is also a small tubercle at the angle, where a similar one is found in the corresponding tooth of *Ectacodon cinctus.* Altogether this tooth is like the posterior molar of *Coryphodon elephantopus*, with a more prominent postero-external angle added. The anterior and posterior basal cingula are well developed, the latter being strong interiorly to the point where it sends a branch upwards to the internal cusp. There is no internal cingulum.

The last superior molar is a transverse oval, more regular than usual in the species of *Coryphodon*, since the diameters of the internal and external portions are about equal. The characters of the posterior crest differ from those seen in the genus named in that the internal portion is much smaller than the external, having a small conic apex, distinct from that of the exterior portion. Its postero-external face is nearly vertical, and it diverges a little posterior to parallel with the anterior crest. The latter (the first cingulum) is elevated, and is widely separated externally from the posterior crest, to whose base it descends on the external extremity of the crown. The basal cingulum is present all round the crown except at the base of the posterior crest, and externally. It is narrow on the inner extremity of the crown. It sends upwards a strong branch to the apex of the internal cusp. The enamel of all the molars is strongly wrinkled, but is worn smooth wherever rubbed.

Measurements.

		M.
Length of superior molar series		.179
Length of premolar series		.085
Diameters P-m. ii	anteroposterior	.019
	transverse	.025
Diameters M. i	anteroposterior	.029
	transverse	.032
Diameters M. ii	anteroposterior	.036
	transverse	.042
Diameters M. iii	anteroposterior	.0285
	transverse	.041
	vertical	.015

The *Metalophodon testis* differs from the *M. armatus* in the more triangular form of its penultimate superior molar. Its form is quite different from that of the last molar, while in *M. armatus* the two teeth resemble each other closely. The species are of about the same size. The individual from which the above description is taken is rather aged.

DINOCERATA.

The characters of this suborder have been already pointed out and discussed. The differences from the *Pantodonta* are well marked, but the resemblances are such as to render it impossible to refer the *Dinocerata* to a different order. Their strong resemblances to the *Proboscidia* are generally admitted, but the few characters which distinguish them are of the first importance. These are, first, the very small size of the brain, especially of the cerebral hemispheres, and, second, the double distal articulation of the astragalus, where the facet for the cuboid bone is nearly as large as that for the navicular.

Within the above definition there is room for much variation, which, however, the known genera do not display. They agree in various points of minor importance Thus there is no sagittal crest of the skull, the temporal ridges being lateral, and there is a great supraoccipital crest. These crests are more or less furnished with osseous processes or horns. One of these consists in part of the maxillary bone, and stands in front of or over the eye. The nostrils are well roofed over by the nasal bones. There is always a diastema behind the canine tooth in both jaws. There is less difference between the premolar and molar teeth in the known genera than in the *Pantodonta*, and they are all constructed on the same pattern. Thus in the upper jaw the crowns of the molars support two oblique cross-crests, which unite to form a V with the apex inwards. There is no internal cusp or tubercle. The inferior molars consist essentially of an outer V and a heel; the true molars differ in having the heel a little larger, and more recurved on its posterior border, but it does not rise into a transverse crest, as in the *Coryphodontidæ*. Messrs Spier, Scott, and Osborne show that the inferior incisors in *Loxolophodon* are compressed and two-lobed.

In *Loxolophodon* the malar bone forms the middle element of the zygomatic arch, sending a narrow strip only forward to the neighborhood of the lachrymal. In *Uintatherium*, according to Marsh, its extension towards the side of the face is rather greater, much as in some Perissodactyla.

The known genera agree with the typical *Proboscidia* in the posterior expansion of the scapula, and its apical acumination, in the short cervical vertebræ, in the flat carpal bones, in the absence of pit for round ligament of the femur, in the flattened great trochanter, contracted condyles, and fissure-like intercondylar fossa of the same bone: in the longitudinal crest of the tibia separating glenoid articular faces which are on a transverse line. Also in the short calcaneum, which is wider than long and tubercular on the inferior face; in the five digits, the acetabulum not separated by a peduncle from the iliac plates, and the lack of angular production of the latter beyond the sacrum.

Genera.—Owing to the imperfect character of the material which I have had the opportunity of examining, it is not possible to state the number of these with absolute certainty. There are certainly three of these, and probably four. So far as present knowledge goes, they pertain to one family, which I have called the *Eobasiliidæ.* The three genera mentioned differ in the forms of the mandible; the fourth has certain cervical vertebræ of a peculiar form, but the form of the mandible is unknown. I can only contrast the genera as follows:

A Mandible unknown.
  Certain cervical vertebræ short and flat, as in *Proboscidia*.......... *Eobasileus.*
AA Symphysis of mandible with four teeth on each side.
  a Mandible without inferior expansion.
    Cervical vertebræ not very short.......................... *Loxolophodon.*
  γ Mandible expanded below, its entire length.
    Cervical vertebræ unknown.............................. *Bathyopsis.*
AAA Symphysis of mandible with three or two teeth on each side.
    Mandible with very narrow symphysis .... *Uintatherium.*

History, &c.—I originally* referred the *Eobasileidæ* to the *Proboscidia* on account of the structure of the limbs, and subsequently stated a number of reasons for this conclusion at a meeting of the Academy Natural Sciences of Philadelphia, held January 14, 1873 (published January 16.) In the present paper numerous confirmatory characters are added. The *Bathmodontidæ* I at first referred to the *Perissodactyla.*

Professor Marsh, in describing a species of this group, *Titanotherium* (?) *anceps* (July, 1871), compares it with Perissodactyle species, and in describing the tibia says that it, "at its proximal end, has the femoral surfaces contiguous, with no prominent elevation between them, resembling in this respect some of the *Proboscidia.*" A few days before the publication of my conclusions, in a foot-note (July 22, 1872) he altered the name *Titanotherium* to *Mastodon,* thus indicating that the species is a proboscidian. Shortly after (American Journal of Science and Arts, September 27) he altered his view, constructing a supposed new order, "*Dinocerea,*" for their reception.

EOBASILEUS Cope.

Proceed. Amer. Philos. Soc., 1872, p. 485. Palæontological Bulletin, No. 6, p. 2, Aug. 20, 1872. Annual Report U. S. Geol. Surv. Terrs., 1872 (1873), p. 575.

This genus was established on a species which is represented by a considerable part of the skeleton, but without cranium or teeth; hence most

* Paleontological Bulletin No. 6, August 20, 1872, p. 2, reprinted in the Proceedings of the American Philosophical Society of like date.

of its characters remain unknown. The very short cervical vertebra which belongs to it serves to distinguish it from other genera. (See Plate XXX, Fig. 3.) A second specimen (*E. furcatus*) found near the first may belong to it; it includes a fragmentary cranium, but unfortunately no cervical vertebræ. Its introduction into this genus is therefore purely arbitrary.

The typical species is of large proportions, only second in size to the *Loxolophodon cornutus.*

EOBASILEUS PRESSICORNIS Cope.

Loxolophodon pressicornis Cope. Proceed. Amer. Philos. Soc., 1872, p. 580 (published August 19). Loc. cit., p. 488 (August 22). *Eobasileus cornutus* Cope, l. c., p. 485 (August 20), not *Loxolophodon cornutus* Cope, l. c., August 19. *Eobasileus pressicornis* Cope, Annual Report U. S. Geol. Surv. Terrs., 1872 (1873), p. 575, portion.

Plate xxx, figs. 1-5; Plate xxxi.

The typical specimen embraces cervical, dorsal, and lumbar vertebræ, ulna, both femora and tibiæ, astragalus, navicular, &c., and large parts of the scapulæ and pelvis.

The *scapula*, in its proximal portions, differs in little from that of *Loxolophodon cornutus*, besides its inferior size. The coracoid is a compressed tubercle inclosing a groove with the glenoid cavity.

	M.
Diameter of glenoid cavity (longitudinal)	.168
Diameter of glenoid cavity (transverse)	.098

The *os pubis* displays a strong pectineal rugosity, commencing near the acetabulum.

	M.
Long diameter of acetabulum	.143
Length of common public suture	.108
Diameter of pubis, near acetabulum	.052

The *femur* is nearly as long as that of *Loxolophodon cornutus*, but is more slender, and has a relatively smaller head. It is flattened fore and aft, and the great trochanter is much expanded and with a shallow concavity on the posterior face. There is a marked concavity on the posterior face of the shaft above the condyles. There is a rudiment of the little trochanter. The *tibia* is scarcely three-fourths the length of the femur, and has a rather contracted shaft, which is in section rounded triangular, one angle presenting forwards. There is no spine except a rudiment in the swollen upper portion of the anterior ridge. The articular surfaces are together rather narrowly transverse. They are separated by a keel which is undivided posteriorly; anteriorly, the contiguous margins of the cotyli separate. The long axis of the inner of these is directed anteroposteriorly outwards in front; of the other, similar but much more transverse. It overhangs the shaft outward and backward, and supports beneath the subround down-looking fibular articular surface. The distal articular surface is distinguished from allied species by the downward prominence of the malleolar process, the anteroposterior width, and the greater extent of the fibular articular face. The face is slightly concave anteroposteriorly, and openly sigmoidal transversely.

Measurements of leg.

	M.
Length with astragalus in place	1.200
Femur, length	.750
Femur, diameter of ball	.118
Femur, width at great trochanter	.220
Femur, width at middle shaft	.091
Femur, depth at middle shaft	.060
Tibia, length	.470
Tibia, width of proximal surfaces (transverse)	.147
Tibia, width of proximal surfaces (anteroposterior)	.070
Tibia, transverse diameter of shaft	.061
Tibia, anteroposterior shaft	.065
Tibia, anteroposterior shaft, distal articulation	.092
Tibia, transverse shaft, distal articulation	121
Fibula, length	.430
Fibula, transverse width at middle	.032
Fibula, width of proximal articular face	.042
Fibula, width of malleolar articular face (transverse)	.052
Fibula, width of malleolar articular face (longitudinal)	.044

A section of the fibula, near the proximal end, is subtriangular; a short distance below, subcircular; on the distal two-thirds it is flat, with the thinner edge convex inward.

The *astragalus* is a flat bone, with its entire superior face occupied by the tibial articular surface. This is as broad as long, and very little convex. It is broader in front than behind; the outer margin is concave, the inner slightly convex. The posterior margin projects most on the outer side, and it is divided by a pit-like cavity, which sends a groove to the inner margin. The outer malleolar surface is an anteroposterior oval; the inner, a concavity, beyond which the inferior portion of the bone projects. The inferior face is divided by a prominent transverse angle, between sub-anterior and subposterior faces. The latter receives the calcaneum on two oval surfaces, which are joined behind by a narrow strip. The navicular face is subrhomboid, the cuboid one-third as large, and triangular, with a round base outward. The margin of the former scarcely projects beyond the superior face.

Measurements of astragalus.

	M.
Total width	.128
Total length	.107
Width of tibial face in front	.090
Length of tibial face, externally	.088
Length of internal malleolar face	.045
Length of outer calcaneal, malleolar face, anteroposteriorly	.050
Length of navicular facet	.085
Width of navicular facet (anteroposterior)	.060
Length of cuboid facet	.065
Width of cuboid facet (anteroposterior)	.035

The centrum of *a cervical vertebra*, which lacks epiphyses, is very short, and the articular face is a wide, transverse oval. Both are slightly concave, and the axis being slightly oblique, the anterior is the more elevated. The surface of the latter is quite rugose, except on the margins. The cervical canal is wide, and the neurapophyses and parapophyses narrow. Inferior surface regularly convex.

Measurements of cervical vertebra.

	M.
Length of centrum	.044
Length of basis of neurapophysis	.040
Length of anterior articular face	.102
Depth of anterior articular face	.086
Width of neural canal at base	.060

The centra of the dorsal vertebræ are too much distorted by pressure for description.

This amblypod was a huge animal, little less elevated than the *Loxolophodon cornutus.* Its limbs were more slender in their proportions. It is in this species that I find much evidence in favor of the presence of a proboscis of greater or less length. Should several of the other cervical vertebræ have been as short as the one preserved, it is evident that the animal could not possibly have reached the ground with a muzzle so elevated as the long legs clearly indicate. The reader is referred to Plates XXX and XXXI for the evidence in pictorial form.

The bones above described were discovered by the writer in an amphitheater of the bad lands of the Washakie Basin, known as the Mammoth Buttes, in Southwestern Wyoming. They were in greater or less part exposed, lying on a table-like mass of soft Eocene sandstone. A description of this remarkable locality is given in the Penn Monthly Magazine for August, 1872.

EOBASILEUS FURCATUS Cope.

Loxolophodon bifurcatus Cope, in extra copies on Proboscidians of the Eocene of Wyoming, August 19, 1872.* *Loxolophodon furcatus* in the same, Proceed. Amer. Philos. Soc., 1872, p. 580, August 20. L. c. 488, August 22. *Eobasileus furcatus* Cope, Annual Report U. S. Geol. Surv. Terrs., 1872 (1873), p. 580.

Plates xxxii, xxxiii.

This species was originally described from a posterior horn, which was obtained near the locality which furnished the typical specimen of *E. pressicornis*. It was found in an old camp separate from the other specimens. The trail from this camp passed the front of the bad-land bluffs, and where it reached the foot of the latter I found projecting from the rock parts of a skull and skeleton, which I suspect to be the animal to which the horn belonged. It is very probable that the horn was picked up at this point, although, of course, there is no direct evidence to that effect. It is represented in Fig. 5, Plate XXXIII, and presents the following characters:

The basis is very narrow and lenticular; a short distance above it the outer side is convex. The anterior and posterior extensions of the base differ; the one is thinner, the other more massive and with a shallow groove above its commencement. The latter is posterior. The compressed apex of the horn-core sends down a rib outwardly to the anterior base and one inwardly, which disappears on the convex base. The general form is spatulate, with the apex expanded obliquely across the lateral crest of the skull, and regularly rounded in superior outline. Its anterior face is flat, the posterior convex; its surface is grooved by very small blood vessels.

As compared with the posterior horn-core of *Loxolophodon cornutus*, there is every difference. That is continuous with one margin of the crest, this erect above it; that has a round base, this a lenticular one. It is more

*See Proceed. Amer. Philos. Soc., 1872, p. 515, where this name is recorded.

like that of *Uintatherium mirabile*, which I only know from Marsh's figure, but is abundantly distinct. It is much more elongate, especially above the (?) posterior part of the crest, and is flattened, and not triangular in section as in that species.

Measurements of horn-core.

	M.
The total length above crest (5.5 inches)	.135
The total length above anterior base (7¼ inches)	.180
Width across apex (in front)	.095
Thickness across apex (in front)	.028
Thickness at base	.040

The supposed remainder of the specimen (Plates XXXII, XXXIII, Figs. 1–5) includes various parts of the cranium, without teeth; portions of the atlas, femur, and fore and hind feet.

The fore part of the nasal bone is preserved. The apex is rather acuminate, and its profile forms a descending curve continuous with the superior plane. The tuberosities are mainly lateral, and cause an abrupt expansion of the lateral outlines. Their obtuse external face is longitudinally oval, and descends slightly backward. The inferior lateral marginal ridge is contracted, and incloses a concave median space. The tuberosity sinks to the level of the median suture. The posterior part of the nasal rises to the apex of the middle horn-core, forming its inner face. The postero-superior angle of the premaxillary reaches to near the base of the horn, and is not drawn out to a narrow apex as in *L. cornutus*. The horn is compressed anteroposteriorly at the base; at the apex obliquely inward and forward. The outer face is concave on the lower half, the inner, convex. The posterior face is concave and the anterior convex, when viewed from the side in profile.

Measurements of front.

	M.
Width of both nasal bones at tuberosity	.124
Width of both at base of distal cone	.060
Depth of suture at front of tuberosity	.030
Length of suture, from premaxillary to horn-core	.035
Length of horn-core (in front), (6 inches)	.150
Diameter (externally)	.080
Diameter of apex	.048

The occipital region is furnished with an enormous transverse crest which extends upward and backward. Its margin is gently convex, and its superoanterior face concave. The posterior face is narrowed by the inferior crest-like margins of the temporal fossæ, which extend from the squamosal part of the zygoma and gradually contract, terminating abruptly in a low knob where it joins the transverse crest. The posterior face between the former, is divided into two planes by a low vertical ridge, which terminates some distance below the summit. The transverse crest is continued in a curve forward on each side as the superior margin of the temporal fossa. These are very stout, but are broken off near the position of the horns.

Measurements of occiput.

	M.
Elevation from *foramen magnum*	0.180
Width between inferior temporal crests	.250
Width of condyles with foramen	.180
Elevation above internal sinuses at angles	.180

The mastoid tuberosity is short and stout; the mastoid foramen is large and not piercing a crest. The ex-occipital suture is obliterated. The V-shaped crest behind the meatus in *Loxolophodon cornutus* is little marked here. The surface of the bone has various muscular impressions. The basi-occipital exhibits a low median crest dividing lateral concavities; transverse width at condyles .077 M. The fragments of teeth are too uncharacteristic for specific description. Numerous cranial fragments accompany the above, but have not yet been properly placed.

The *atlas* is broken; its cotyloid cavities are rather shallow, and the diapophyses small. Its anteroposterior diameter below at the middle line, is .070; at base of diapophysis .070. The condyles of the *femur* present the characters of the group. There is a deep vertical groove on the inner side just above the condyle. The latter approach each other closely on each side of the intercondylar fossa and are flattened on the superior posterior margins. Width across extremities M. .150.

The *unciform* bone is in form a little less than a quarter of a circle, and the external (anterior) depth is one-half its transverse length. Its superior surface is slightly convex. It displays, as in living proboscidians, four inferior facets, thus proving the existence of a fifth toe to the hind foot. The external facet is deeply concave, and contains a pit. It is oblique, and unites with the superior face by an acute angle. The internal inferior facet probably supported the small inner toe by its metatarsus directly. The other two are more nearly on one plane, and are deeper than wide.

Measurements of unciform.

	M.
Depth in front	.048
Width (transverse)	.097
Width of external facet	.035
Width of second facet	.026
Width of third facet	.047
Width of internal facet	.023

On account of the absence of corresponding parts, it is not possible to be certain whether the individual above described belongs to the species *Eobasileus pressicornis* or not. It is possible that the bones belong to the same individual as that described under the latter name, although the localities where they were found are about one hundred yards apart. At about the same distance from both the above specimens I found a humerus, which I describe briefly. The distal articular face is very oblique to the transverse axis, but is about equally developed on opposite sides of the shaft. The condyles are unequal, have parallel axes, and are hour-glass shaped, with a shallow concavity. The supra-condylar fossæ of opposite sides are not very large nor deep.

Measurements of humerus and ulna.

	M.
Transverse diameter, distally	.175
Anteroposterior diameter of inner condyle	.104
Anteroposterior diameter of outer condyle	.125
Transverse diameter, olecranon	.110

The portion of *ulna* just measured belongs to a young individual found near to the one above described as *Eobasileus furcatus*. It is represented on Plate XXIX, Fig. 7.

Relations.—Besides the difference in the development of the anterior nasal tuberosities, which might be sexual only, this species differs from *L. cornutus* in the simple naso-maxillary horn-cores, which want the interior tuberosity of that species, and in the fact that they are composed exclusively on their inner sides of the nasal bones to the apex, the maxillaries forming the outer face. *E. pressicornis* has also a much wider and less massive parieto-occipital basin, with lighter horn-cores. Minor differences have been already mentioned.

Restoration.—As in all the species of *Uintatherium* in which the horns are known, these appendages stood in front of the orbits, it is probable that such was the case in the *Eobasileus furcatus* also. The muzzle is materially shorter and more contracted, and the true apex of the muzzle was not overhung by the great cornices seen in *Loxolophodon cornutus*. The occipital and parietal crests are much more extended in this species than in the *L cornutus*, so that in life the snout and muzzle had not such a preponderance of proportion as in that species. All the species of this genus were rather rhinocerotic in the proportions of the head, although the horns and tusks produced a very different physiognomy. The extremities of the nasal bones are strongly pitted and exostosed, and this, taken in connection with the elevation of the head, renders it probable that this species possessed a proboscis of less or greater length.

History.—This species was originally described by the writer in a short paper, which was published and distributed August 19, 1872, under the generic name *Loxolophodon*. I shortly afterward referred it to the new genus *Eobasileus*.

LOXOLOPHODON Cope.

Proceed. Amer. Philos. Soc., 1872, p. 580, extra copies published August 19th, and p. 488 (August 22d). Annual Report U. S. Geol. Surv. Terrs., 1872 (1873), p. 565. ? *Tinoceras*, Marsh, American Journal of Science and Arts, 1872 (October), published September 21st.

The cranium in this genus is elongated and compressed. The muzzle is posteriorly roof-shaped, but is anteriorly concave and flattened out into a bilobed protuberance which rises above the extremity of the nasal bone. This extremity is subconic and short and decurved. A second pair of horn-cores stands above the orbits, each one composed externally of the maxillary bone, and internally of an upward extension of the posterior part of the nasal. Behind this horn the superior margin of the temporal fossa sinks, but rises again at its posterior portion, ascending above the level of the middle of the parietal bones. This portion of the skull is injured in my only specimen. The occipital rises in a wall upwards from the foramen magnum, and supports, probably, a little in front of the junction with the superior and inferior ridges bounding the temporal fossa, a third horn-core on each side. The base of this core is as stout as that above the orbit, and subround in section. The temporal fossa has its principal extent posterior to the zygomatic arch, and is in form like a trough, the inferior edge being recurved from the squamosal process to the summit of the occipital crest. It is narrow within the zygomatic arch, which is short, inclosing a space whose length is less than one-fourth that of the cranium.

The occipital bone extends but a short distance on each side of the condyles, and is separated from the mastoid by an irregular suture, which is pierced by a large mastoid foramen. On the inferior face, near to each condyle, and one-third the distance from its inner extremity, is a posterior condyloid foramen, isolated by a narrow bar from the extremity of the *foramen lacerum posterius.* The paroccipital process is represented by a small tuberosity, and the mastoid by a rather larger one, some distance anterior to it.

The *meatus auditorius* opens upward just below the external ridge of the temporal fossa, and at a little distance behind the post-glenoid process. Its canal contracts rapidly, and extends upward and backward toward the labyrinth. It is separated from the *foramen lacerum* by but a thin wall, and if there was an expansion of the *cavum tympani* it must have been exceedingly small, owing to the close approximation of the mastoid to the basi-occipital and sphenoid at this point. The labyrinth is lodged in a petrous mass opposite the occipito-mastoid suture, and the canals are small.

The *basioccipital* contracts anteriorly, and with the sphenoid forms an uninterrupted boundary of the *foramen lacerum* This terminates opposite to the posterior boundary of the external meatus, and gives rise to a wide, shallow groove, which extends anteriorly between the pterygoid ala and the post-glenoid process, and, turning outward round the latter, grooves it. Opposite to the post-glenoid process and just posterior to the end of the pterygoid, a foramen enters, which, though remarkably small, is the *foramen ovale.* Almost continuous with it is a canal which pierces the base of the pterygoid longitudinally, and issues in an excavation of its external face near the sphenoid. This is the alisphenoid canal.

There appears to be no *foramen postglenoidale*, but there is one in the position of the stylomastoid on the inferior side of the mastoid bone and posterior to the meatal groove.

The *pterygoids* are remarkable for their great length, inclosing, as they do, with the palatine process, a deep, narrow, trench-like fossa, which measures almost the entire length of the zygomatic fossæ. Processes of the sphenoid contribute to these walls (which are thus double), and the sphenoid roof is strongly concave. The *alisphenoid* is elongate antero-posteriorly, and is principally in contact superiorly with the frontal; anteriorly it has a short suture with the lachrymal. I cannot determine whether the orbito-sphenoid is distinct. Almost its entire length is traversed by a shallow groove, which terminates in a small *foramen opticum*, opposite to a point marking the posterior third of the zygomatic fossa. The *foramen rotundum* issues as usual between the alisphenoid and the pterygoid, within the alisphenoid canal, which exceeds it in diameter.

The *lachrymal* is a large bone, of a triangular outline, the shorter side being inferior. It is entirely on the inner face of the orbit, and, as in the elephant, separates the frontal and maxillary by its superior prolongation. Its inferior border is slightly notched in front by the large *foramen infraorbitale posterius*, and the anterior is deeply emarginate, passing behind the small *f. lachrymale.*

The *palate* is remarkable for its length and narrowness. Its roof is chiefly composed of the maxillaries, but a very short portion is formed by the palatine plates of the *o.· o. palatina.* These are produced into a median point behind between the nares, and exteriorly form the inner wall of the postnareal trough for a considerable distance. The *maxillaries* also form the outer wall for a short distance, being produced in a contracted form behind the molar teeth. The two bones inclose a small foramen in this prolongation, and a larger one on the anterior suture of the palatine, the *foramen palatinum.* The palate is deeply concave anteriorly. There is an elongate foramen close to the alveolus of the first premolar, extending anterior to it. The premaxillaries are longitudinal and separated anteriorly, for two-fifths of their length, by a large *foramen incisivum*, which they do not inclose in front. They extend on the side of the muzzle into an acute angle upward and backward, and are prolonged forward above the exterior nares, which the suture reaches by an abrupt descent. The maxillary supports the malar on a posteriorly directed process which reaches to the end of the anterior third of the arch below, half that distance on the side, and is bordered by a narrow strip of the malar on the inner side, as far as the anterior boundary of the orbit.

The *dentition* is I. 0; C. 1; Pm. 3; M. 3. The canine is a tusk of compressed form, with anterior and posterior cutting edges, and a strong posterior curvature. Its fang is embraced one-third by the premaxillary bone, and is inclosed in a rib-like swelling of the sides of the cranium, which extends upward and backward. The premolars have transverse cordate surfaces of attrition. These have probably resulted from the wearing down of a chevron of two crests converging inward, in some with an inner tubercle. On the molars this crescent is represented by a V, with the apex inward; on the last, the inner tubercle is at one side (the posterior) of the apex.

Name.—I first applied the name *Loxolophodon*, with a diagnostic description, to this genus, in a short paper published August 19, 1872, as above cited. The *L. cornutus* was there cited as the first species, and is here retained as the type. I again described it more fully in a paper published August 22d, citing *Eobasileus* (August 20th) as a synonym, in which I was in error, as indicated in the present book. The same nomenclature was employed in a paper read before the American Association for the Advancement of Science, held at Dubuque, commencing August 23, 1872.

Prior to the issue of the paper of August 22, I had (February 16, 1872) provisionally applied the name *Loxolophodon* to the species there called *Bathmodon semicinctus* Cope, without generic character. With further material it appears that the *Bathmodon semicinctus* is very near to the *B. radians*, so that the name *Loxolophodon* was cancelled in this connection, and was used again for the present genus without interference, especially as it was first published as a *nomen nudum*.

Professor Marsh, in the American Journal of Science and Arts, 1872 (September 21),* applied the name *Tinoceras* to a species (*T. grandis*) which he has led us to infer belongs to this genus, but did not specify the generic characters. He had previously applied it without description to the *Uintatherium anceps*, August 24 (and 19,† in an erratum, where *Mastodon anceps*

* I did not receive this, and most of the other papers of Professor Marsh on this fauna, till early in December, 1872.

† These papers were not received by me till early in December, 1872.

is altered into *Tinoceras anceps*). As no characters whatever were assigned to it on either of these occasions, it had no value in zoological nomenclature.

Three species of this genus are known to me, the *L. cornutus* and the *L. galeatus*, which are represented in my collection, and the *L. spierianus*, which is founded on a skull preserved in the museum of Princeton University, New Jersey. The last named was as large an animal as the two others, and had a very elongate skull with weak horns and narrow, high occiput. Its median horns are situated well anterior to the orbit, and its zygomatic fossa is remarkably small. It was discovered by the Princeton scientific exploring party at the same locality that produced the other species, viz., the Mammoth Buttes of Western Wyoming.

The three species may be distinguished as follows:

Median horns triangular in section, with internal tuberosity, and above orbits; occiput narrow .... *L. cornutus.*
Median horns subquadrate in section without internal tuberosity; occiput and nasal tubercles wide .... *L. galeatus.*
Median horns subround and without tuberosity, in front of orbits; occiput and nasal tubercles narrow .... *L. spierianus.*

Messrs. Scott, Spier, and Osborne, of Princeton, have published a description* of the lower jaw and teeth of a species of *Loxolophodon*, which they identify with the *L. cornutus*, which was derived from the locality and horizon of the species above mentioned. They show that the descending flange of *Uintatherium* and *Bathyopsis* is only represented by a convex ridge on each side of the symphysis. They point out the characters of the dentition, which are remarkable. The molars much resemble those of *Bathyopsis*. The canines and incisors are alike in form, and in a continuous series. The crowns are compressed so as to be extended anteroposteriorly, and are deeply emarginate, so as to be bilobed, the lobes with subacute edges. This form of incisors is unique, resembling only remotely the large median incisors of certain *Insectivora*. Resemblance to mammals of the same type may be traced in the molar teeth.

* American Journal of Science and Arts, xvii, 1879, p. 304. See review of this paper in American Naturalist, 1879, p. 334, where some corrections are made by the writer.

LOXOLOPHODON CORNUTUS Cope.

Loxolophodon cornutus Cope. Proceed. Amer. Philos. Soc., 1872, p. 580 (August 19); l. c., 1872, p. 488 (August 22). *Eobasileus cornutus* Cope, Amer. Nat., 1872, p. 774. Annual Report U. S. Geol. Surv. Terrs., F. V. Hayden, 1872 (1873) p. 568, Plates I–IV. ?? *Tinoceras grandis* Marsh, Amer. Journ. of Sci. and Arts, October, 1872 (published September 21), *fide* Marsh.

Plates xxxvii to xlii.

Established on the remains of a single individual, which consist of a nearly perfect cranium, the right scapula complete, several vertebræ, including the sacral, the first or second rib, the pelvis complete, and the entire right femur; also probably the proximal end of a radius.

The species is remarkable for the narrow form of the *cranium*, its width at the middle being one-fourth its length. A little in front of the middle are situated the horn-cores. These diverge, the upper portion having an outward curvature. The base of each is triangular, with obtuse angles in section, and the inner angle is the section of a rib-like projection which commences half way up the horncore and extends across the middle line to its fellow. Above its rather abrupt termination the core is transversely compressed, with oval obtuse apex. The core measures M. .240 (9.5 inches) from its base in front, M. .108 (4.25 inches) in width at the base behind, and .077 (3 inches) in diameter at the apex. A slight swelling of the sides of the muzzle descends obliquely forward from the base of each horn, which enlarges below into a prominent rib, which incloses the alveolus of the canine tusk. In front of the horns the muzzle is roof-like; anteriorly it flattens out, and swells a little above the posterior end of the nasal meatus. In front of this it expands again, and rises gently to the extremity of the bilobed nasal shovel, which overhangs the premaxillaries, the nasal meatus, and the greater part of the apex of the nasal bones. The latter is short and with a wide base, and resembles two lateral cones flattened together, their extremities obliquely truncate outward and excavated. In the nasal bones of the *Eobasileus furcatus* Cope, the shovel is represented by a tubercle only on the side of a continuous surface. The composition of the upper surface of the cranium is somewhat difficult to determine, owing to the injured state of the posterior part. If we regard the bone which bounds the lachrymal behind and above, as frontal, as I did in originally describing the species, it gives an extraordinary extent to the nasals, for the common suture of these bones extends V-shaped backward, to a point opposite to the middle of the zygomatic arches. It gives to the nasals an extent equal to that of the frontals and parietals combined. They not only support the anterior shovels, but form the inner half of the median horn-cores, rising as high as the tuberosity above described. The immense length of the snout in *Loxolophodon* looks as though the nasal bones had extended themselves forward, so as to ossify the basal portions of an elephantine proboscis.

The frontals descend behind the horns, with a very obtuse or rounded continuation, to the inner side of the fossa, and without any superciliary margin. They form with the posterior part of the nasals a shallow median basin. The suture with the parietals is very indistinct, but if I have truly discovered it, it forms another posteriorly directed chevron, and leaves but a narrow superciliary portion of the frontals. Above the postglenoid processes the parietals rise again to the transverse occipital crest, but to what height is uncertain. At the mastoid region, the cranium widens a little, and is excavated at the sides by the temporal fossæ. Near where the lateral and supraoccipital crests join the inferior ridge-like border of the temporal fossa, which position is occupied by a knob in *E. furcatus,* is a strong horn-core with subcylindric base. It stands obliquely backward toward the junction of the posterior squamosal and transverse crests, and is connected to these by an oblique ridge, one side of which is marked with irregular, short, longitudinal rugosities. At the base of these elevations are three sinuses. This portion was found close to the skull, but separated from it, and the loss of intervening fragments prevents actual fit.

The occiput is preserved for four inches above the condyles; it doubtless displayed a posteriorly sloping transverse crest as in *E. furcatus.* The paroccipital and mastoid tuberosities are narrowed and extend obliquely downward and forward. The lower part of the exoccipital suture runs along a ridge, and there is a tuberosity in front of the mastoid foramen. An irregular Λ-shaped crest extends upward with the apex at the inferior temporal crest, and its anterior limb forms part of the posterior boundary of the *meatus auditorius.* The inferior temporal crest is directed outward below, but forward above.

The narrowness of the cranium is readily seen on comparing the postglenoid processes. These are strong, and have considerable transverse extent, and are separated by a space only a little greater than the transverse diameter of each. A strong groove passes along the inner base of the postglenoid process to the *foramen lacerum*. It is overhung by a prominent horizontal ridge of the postglenoid. The zygomatic arches are compressed posteriorly, with crest-like superior ridge, but rounded above anteriorly. There is not the least trace of posterior boundary of the orbit. The squamosal process overlaps the malar bone extensively, terminating in a point, which ends obtusely. The malar is supported in front by a maxillary process, which is united with it by a zigzag suture on the outer face, and a squamosal one within and below. The *foramen infraorbitale exterius* is large, and issues a short distance in front of the orbit, not so near it as in the elephants. From this point to the ridge inclosing the canine alveolus the side of the maxillary bone is deeply concave, and the palatal surface correspondingly contracted. The bone is continued upward and outward as the external part and apex of the middle horn-cores. Anteriorly it is bounded by the premaxillary to a point as far anterior to the base of the horn as the width of the latter; behind that point it is in contact with the nasals. The premaxillary is prolonged upward and backward into a narrow tongue. Its inferior portion is convex above on each side, concave below, with projecting alveolar borders, which are flat and slightly concave fore and aft. The extremity of each is rugose below, supports a prominent tubercle medially and a smaller one at the superior angle.

The exterior nares are not separated by osseous septum. Their lateral border is marked on the inferior surface of the nasal and premaxillary roof by a curved ridge or crest, which converges forward and bounds the interior concavity of the roof. This gave support to muscular or ligamentous attachments. The posterior angle of the nares is abruptly excavated, and with thickened walls. The palate is remarkably narrow, and is most deeply excavated between the alveoli of the tusks, or at the maxillo-premaxillary suture. From near this point to the palatine suture a low but sharp crest extends along the middle line. The width of the palate at the diastema is one-ninth of its length. The diastema is more than half the length of the molar series. The pterygoid process of the palate has two convergent grooves on its inferior surface. The alisphenoid canal is larger, exceeding in diameter both the *foramen rotundum* and *f. ovale.*

The *vomer* does not unite with the superior crest of the palatine bones. The sphenoid flattens out behind the postnareal trough and is co-ossified with the basioccipital. The latter is marked by two oval surfaces at the place of suture, with a slight prominence between. No lower jaw was found with this specimen, but from the contraction of the parts opposed to it, it was evidently very small.

The *teeth* are remarkable for the extent of the exposure of their slender roots, as well as their very small size as compared with the size of the animal. The tusk is slightly turned outward at the tip and the inner face is worn by attrition with some opposing tooth for one-third its length on the posterior third. The superior margin of the enamel on this side is chevron-shaped, the apex being only one-third the length from the extremity. It extends further upward in front and on the outer side, but is worn in an oval patch at the apex of the chevron of this side by contact with the inferior teeth, as above described. The enamel is smooth behind, rugose in front. The crown contracts regularly to a flattened obtuse apex. The fang is hollow for about half its length. The enamel of the molars is nearly smooth. Each one has a strong cingulum fore and aft, which is discontinued on the inner and outer faces except in Pm. 1 and M. 2. In the former it is continued on the outer side at the base of the concavity of the exterior face; on the latter, it is continued round the inner side. The grinding surface of the Pm. 1 is tripodal, and probably composed of a worn crescent and inner tubercle. The others are transverse arrow-shaped; the worn M. I. indicates mature age. M. II. is larger than M III. Its oblique crests have evidently been worn from before. All the molars have three roots, but the posterior pair are united for part of their length in all.

Cranial measurements.

	M.
Length from end nasals to end occipital condyle (3 feet 1.5 inches)	.930
Width just behind end nasal shovels	.192
Width in front of horns	.132
Width at base of horns in front	.205
Width behind horns at apex of frontal suture	.185

	M.
Width above posterior edge *meatus auditorius*	.310
Width between apices middle horn-cores	.370
Width of basis of supraoccipital horn-core	.100
Width including zygomatic arches (greatest)	.320
Length of nasal bones to ridge between horn-cores	.410
Length of nasal bones to frontal suture	.540
Length of zygomatic fossa above	.230
Length from angle nares to end of shovel	.205
Length from angle nares to end of premaxillary	.155
Length from end of premaxillary to basis of canine	.120
Length from end of premaxillary to basis of Pm. 1	.276
Length of molar series	.185
Length of palate	.450
Length of pterygo-palatine crest	.200
Length of sphenoid axis	.185
Length of basioccipital (with condyles)	.128
Width between tips of premaxillaries	.070
Width at canine alveoli	.185
Width between canine alveoli	.080
Width at diastema	.050
Width between last molars	.070
Width between pterygo-palatine crests	.065
Width of post-glenoid process	.095
Width between post-glenoid processes	.095
Width of basioccipital at front	.073
Width of basioccipital at condyles	.200
Width of space for tympanic chamber	.034
Length of tusk on curve (12.7 inches)	.320
Diameter at middle (anteroposterior)	.050
Diameter at base (anteroposterior)	.063
Diameter at middle (transverse)	.030
Diameter of crown of Pm. 1 transverse	.022
Diameter of crown of Pm. 1 anteroposterior	.024
Diameter of crown of M. 2 anteroposterior	.035
Diameter of crown of M. 2 transverse	.034
Diameter of crown of M. 3 transverse	.043
Diameter of crown of M. 3 anteroposterior	.045
Elevation of shovel above base of apex of nasal	.060

The measurements may require some correction in respect to the supra-orbital width, where the cranial walls have suffered from compression. The frontal of one side has been pushed so as to overlap that of the other by about an inch.

The *scapula* is of a sub-triangular form, the front being vertical, the apex directed backward and an angle upward. The posterior expansion is considerable, as in the elephants, while the superior angle is acuminate and much produced and massive. The spine is much elevated, bounding a deep supraspinous fossa. It is truncate proximally, descending to near the border of the glenoid cavity. Its extremity is dilated in alate fashion, equally fore and aft, and not posteriorly only as in the elephants. The glenoid cavity is flattened so as to be longitudinal, and the coracoid is a rudimental tuberosity.

Measurements of scapula.

	M.
Total length (25.25 inches)	.640
Total width	.480
Length of apex from spine	.140
Elevation of spine proximally	.125
Length of glenoid cavity	.185
Width of glenoid cavity	.110

The interior side of the scapula is strongly convex by the development of two longitudinal ribs, one corresponding to each fossa, but concave in longitudinal section.

The proximal end of the *radius* exhibits two facets oblique to each other, the larger concave and transverse, the other oblique downward. Transverse width M. 0.130; vertical .070. The extremity of a humerus not found with this individual, to which the radius applies pretty well, has a very oblique trochlear face, and measures seven inches across the condyles. It, however, belongs to a smaller species.

The *femur* is entire. Like that of other species of the group it is much expanded proximally and deep distally, with the shaft contracted and somewhat flattened in the plane of the great trochanter. The latter is in one plane, with its external margin turned a little backwards. The head is part of a globe, and is a little more elevated than the trochanter, and separated from its apex by a shallow concavity. There is no little trochanter. The rotular face is not elevated nor wide, and with lateral borders subequally developed. The anteroposterior axis of the condyles is somewhat oblique to a line at right angles to the proximal end. This is because the interior condyle is the longer; its articular face is continuous with the rotular, with a marginal notch; the outer condyle is continuous with continuous outer margin. Strong ridges revolve from above the condyles to the posterior face of the shaft, the inner near the condyle. The outer runs parallel to the main axis as a low external ala, and backwards three inches above the condyle. The face between them is concave.

Measurements of femur.

	M.
Total length (31.75 inches)	.747
Total proximal width	.255
Diameter of ball	.136
Transverse diameter at middle of shaft	.096
Anteroposterior diameter at middle of shaft	.074
Anteroposterior diameter of condyles posteriorly	.150
Transverse diameter of condyles posteriorly	.160
Transverse diameter of condyles distally	.145

The *pelvis* has a large transverse expansion. The iliac plates are ovoid in outline, with the apex outward and downward. The margins are rather thin, excepting the internal above the acetabulum. These are massive, and with a longitudinal excavation. They terminate in a deep oblique excavation for the diapophyses of the sacrum. The external margin rises compressed from just above the acetabulum. The latter is large for the size of the ilia, and its margins rise to a slight elevation beneath the exterior margins of the latter. The *incisura acetabuli* is obclavate and nearly symmetrical. The *os ischium* is compressed and deeper than the pubis. It possesses a tuberosity on the posterior inferior margin. The *obturator foramen* is small, and is a vertical oval. The *pubis* is rather slender and short. Its section at base is subtriangular; beyond, it becomes more compressed, and is spirally twisted on itself through a part of a circle. Its anterior margin near the symphysis is strongly rugose for the origin of the *pectineus* muscle; which rugosity extends into a band on the outside of its proximal portion.

Measurements of pelvis.

	M.
Long diameter of ilium	.605
Transverse diameter at acetabulum	.430
Length of inferior free margin at acetabulum	.250
Long diameter at acetabulum	.150
Shorter diameter at acetabulum	.130
Shorter diameter at obturator foramen	.070
Width of ischium at tuberosity	.140
Length of ischium at tuberosity	.110
Diameter of pubis at obturator foramen	.062
Expanse of ilia laid on a flat surface and with sacrum in place (4.2 feet)	1.280

The general character of the pelvis is more like that of the elephant than that of any Perissodactyle. It agrees with the former and differs from that of the rhinoceros in the shortness of the pedestals of the ilia or rather in the sessile position of the latter on the acetabula; also in the absence of production of the iliac crests in advance of and above the sacrum. It is also elephantine in the shortness of the inferior elements of the pelvis.

Of *vertebræ* there are preserved a dorsal, two lumbar, and some sacral. The first is very short and transverse. It is so injured that I can only give some measurements. The base of the transverse neurapophysis is a flat oval; both capitular articular surfaces are deep. The anterior lumbar is longer, but still short; its articular faces are slightly concave. The neural arch is wide and supports the diapophysis. The sides of the centrum are concave and pierced by foramina, and there is a strong rugose hypapophysis. The section at the middle is subtriangular. I have three sacral vertebræ, which are separated by very distinct sutures. They diminish very rapidly in size, and the centra become flattened-transverse. It is doubtful whether there was a fourth vertebra, and the tail must have been short and slender. The articular face of the first is a transverse, rather broad ellipse, and twice the diameter of the third distally. The diapophysis of the second is much the stoutest. It unites with the subvertical plate-like diapophysis of the first as well as with that of the third. It is concave above, and terminates distally in a massive L-shaped surface of articulation with the ilium. The foramina inclosed by the diapophyses are quite large. The inferior face of the first sacral centrum is slightly concave with a hypapophysial tuberosity in front; it is strongly concave in the second.

Measurements of vertebræ.

	M.
Anteroposterior diameter of dorsal	.044
Diameter at bottom neural arch of dorsal	.040
Length of base of neurapophysis	.041
Diameter of centrum lumbar (vertical)	.090
Diameter of centrum lumbar (transverse)	.110
Diameter of centrum lumbar (anteroposterior)	.080
Length of three sacral vertebræ	.226
Transverse extent of sacrum (15 inches)	.380
Diameter of first vertebra at free end (transverse) (4.6 inches)	.122
Diameter of first vertebra at free end (vertical)	.093
Diameter of last vertebra at free end (vertical)	.021
Diameter of last vertebra at free end (transverse)	.065
Total expanse of heads of rib	.106
Diameter of capitular face (vertical)	.048
Diameter of tubercular (vertical)	.030
Width of rib just below head	.050

Restoration.—We may ascribe to the *Loxolophodon cornutus* form and proportions of body similar to those of the elephant. The limbs, however, were somewhat shorter, as the femur is stouter for its length than in the *E. indicus*. It was intermediate in this respect between the latter species and the species of *Rhinocerus*. The tibia is relatively still shorter. The tail was quite small. The neck was a little longer than in the elephants, but much less than in the rhinoceroses; the occipital crest gave attachments to the *ligamentum nuchæ* and muscles of the neck, which must needs have been powerful to support the long muzzle with its osseous prominences, and to handle with effect the terrible laniary tusks. The head must have been supported somewhat obliquely downward, presenting the horns somewhat forward as well as upward. The third or posterior pair of horns towered above the middle ones, extending vertically with a divergence, when the head was at rest. The posterior and middle pair of horns were no doubt covered by integument in some shape, but whether dermal or corneous is uncertain. Their penetrating foramina are smaller than in the *Bovidæ*. The cores have remotely the form of those of the *Antilocapra americana*, whence I suspect that the horns had an inner process, or angle as in the prong-horn at present inhabiting the same region. The nasal shovels may have supported a pair of flat divergent dermal tuberosities, but this is uncertain; they are not very rugose.

The elevation of the animal at the rump was about six feet, distributed as follows, allowance being made for the obliquity of the foot.

	Inches.
Foot	4.50
Tibia	20.50
Femur	31.75
Pelvis	16.00
	72.75

The anterior limbs were stouter than the posterior, judging from the proportions in various species, and were no doubt longer if of the Proboscidian character. This would give us the hypothetical elevation at the withers:

	Inches.
Leg	61.00
Scapula (actual)	21.00
Neural spines (extremities)	7.00
Or 7 feet 5 inches	89.00

These measurements are made from the plantar and palmar surfaces, allowance being made for the pads.

The obliquity of the anteroposterior axis of the anterior dorsal vertebra indicates that the head was posteriorly elevated above the axis of the dorsal vertebræ. Owing to the lack of cervical vertebræ, the length of the neck cannot be determined. It may have been short, as in the *Eobasileus furcatus*, or longer as in the species of *Uintatherium*. In the former case it is entirely clear that the muzzle of this animal could not have reached the ground by several feet, and that, as occurs in the similar cases of the tapirs and elephants, there was a proboscis to supply that necessity. The indications derived from the bones of the muzzle point to the attachment of a proboscis or a heavy upper lip, and the ridges of the symphysis of the lower jaw, pointed out by the Princeton paleontologists, suggest a strong prehensile lower lip. The fact that the *foramen infraorbitale* of the *Loxolophodon* is less than in the elephants, in no wise militates against the possession of a proboscis, for it is still smaller in the tapir, which has one, and larger in many rodents which are without it. The numerous rugosities of the posttympanic and mastoid regions indicate the insertions of strong muscles. Some of these may have adductors of large external ears.

The inferior incisor teeth have no adaptation for cutting off vegetation. The mental foramen is small, but the small nutrient artery thus indicated is not adverse to belief in a prehensile under lip to make up for the uselessness of the teeth. The projecting nasal regions would prevent short lips from touching the ground. The posterior position of the molar teeth indicates use for a long, slender tongue.

This species was probably quite as large as the Indian elephant, for the individual described is not adult, as indicated by the freedom of the epiphyses of the lumbar vertebræ, and fragments of others in my possession indicate considerably larger size.

Habits.—The very weak dentition indicates soft food, no doubt of a vegetable character, of what particular kind it is not easy to divine. The long canines were no doubt for defense chiefly, and may have been useful in pulling and cutting vines and branches of the forest. The horns furnished formidable weapons of defense. That the anterior nasal pair were

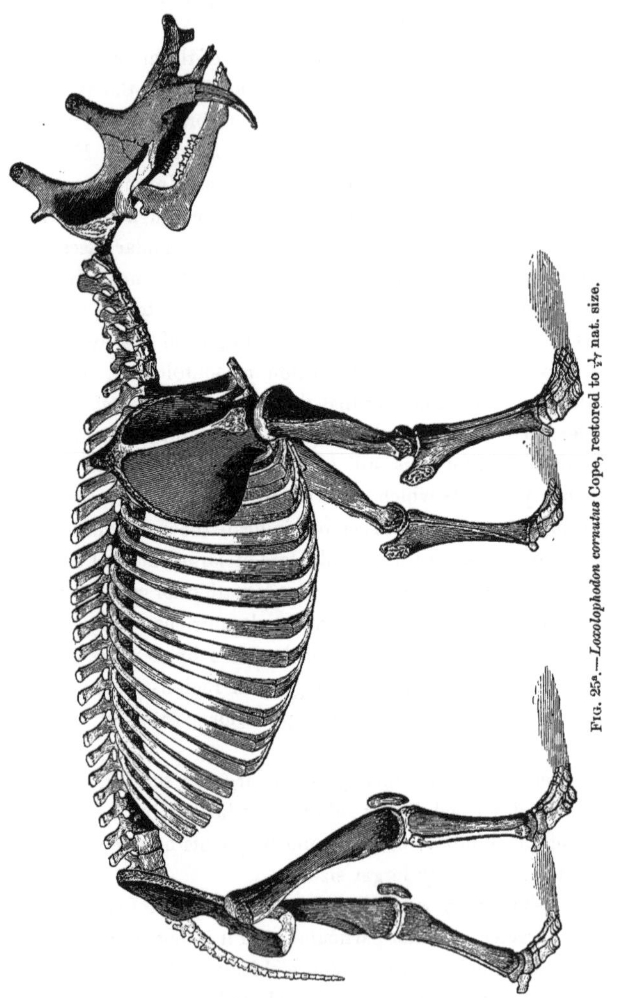

FIG. 25^a.—*Loxolophodon cornutus* Cope, restored to $\frac{1}{27}$ nat. size.

not used for rooting in the earth is evident from the elevation of the head, which would render this impossible.

This huge animal must have been of defective vision, for the orbits have no distinctive outline, and the eyes were so overhung by the horns and cranial walls as to have been able to see but little upward. The muzzle and cranial crests have obstructed the view both forward and backward, so that this beast probably resembled the rhinoceros in the ease with which it might have been avoided when in pursuit.

Synonymy.—According to Marsh, he described this species September 21, 1872, under the name of *Tinoceras grandis*, which thus becomes a synonym of *Loxolophodon cornutus.* As the name *Tinoceras* had never been described prior to that date, although applied to the *Titanotherium* (?) *anceps* Marsh, previously without description, this name becomes a synonym of *Loxolophodon.*

Locality.—The remains of the *Loxolophodon cornutus* were found together by the writer in August, 1872, in a ravine of the bad lands of Wyoming. The greater part of the cranium and the femur were excavated from the base of a cliff of perhaps 250 feet in height, on the side of a ravine elevated about 1,000 feet, in the Mammoth Buttes, on South Bitter Creek. As the basin of Bitter Creek is 7,500 feet above the sea, the fossil was taken from an elevation of 8,500 feet. The horizon is the Washakie Group of the Eocene, of Hayden.

An account of the remarkable locality in which I found the specimen was published in the Penn Monthly Magazine for August, 1873.

LOXOLOPHODON GALEATUS Cope.

Eobasileus galeatus Cope, Annual Report U. S. Geol. Surv. Terrs., 1873 (1874), p. 456, pl. i. Paleontological Bulletin No. 17, p. 1, October 25, 1873.

Plates xliii, xliv.

Represented by the upper portion of a cranium of an individual of the size of the *Loxolophodon cornutus.* The maxillary, palatal and basicranial regions and teeth are wanting, and there is no lower jaw.

The species possesses a greatly elevated occipital crest, whose superior border presents a median angle upward. A short distance in front of it, and connected by a very stout, lateral ridge, there arises, on each side, a large, erect horn-core. The base is very massive, subquadrate in section, and flattened in front. Posteriorly, it presents a very shallow groove, which is bounded on the outside by a low ridge. The shaft expands gradually, and is proportionally flattened from behind forward. The posterior face is flat; the anterior gently convex. The extremity is transverse-convex, and pitted for cartilaginous or corneous attachment. These horns stand on the parietal bones The frontals extend to their bases, and send a laminar expansion backward to the margins of the lateral and posterior crests, covering the parietal in the fundus of the basin, which the former inclose.

The median horns are very stout, and are connected with the posterior by an acute supratemporal ridge. Their inner face is composed to near the apex, of the nasal bones. Where they terminate, the apex contracts, and is composed of a cylindric production from the maxillary. The section of these cores at the middle is subquadrangular, and longitudinally oval at the base.

The extremity of the nasal bones is small and contracted, and is extensively overhung by the cornice-like, flat cores above them. Thus the end of the snout has a bilobate outline when viewed from above.

The occipital face is concave in vertical section, and presents a V-shaped depression, with the angle downward, and a low ridge on the middle line to the transverse superior border.

Measurements.

	M.
Width of the foramen magnum and occipital condyles	.2100
Elevation of occiput (8 inches)	.2500
Width of basin between lateral crests	.3250
Height of posterior horn-core (7 inches)	.2300
Width of base of posterior horn-core anteroposteriorly	.1300
Width of base of posterior horn-core transversely	.0900
Width at summit	.1230
Height of median horn-core	.1750
Diameter of base anteroposteriorly	.1000
Diameter of base transversely	.0800
Diameter of summit	.0650
Projection of nasal cornices beyond apex	.0630
Length of posterior molar, crown	.0450
Width of posterior molar, crown	.0550

This species is equal in size to the largest known from the Bridger formation. It differs from *Loxolophodon cornutus* in the posteriorly-truncate base of the posterior horn-cores, the quadrate instead of triangular section of the median cores, their greater stoutness, their lack of internal angle, and the extent of their inner face covered by the nasal bones. It more nearly resembles the *E. furcatus* Cope, and may possibly prove to represent an old male of that animal. There is, however, a considerable disparity in their sizes; and the horns of *L. galeatus* differ in the greater stoutness, having twice the diameter, with little greater height. They differ also in form, in the abrupt contraction just below the apex. The cornice-like cores of the nasal bones represent the tubercles of the *E. pressicornis*. The posterior horns differ in many ways from those of the *E. furcatus*, and are probably sufficient to indicate a different species.

From the bad lands of South Bitter Creek, or the Mammoth Buttes, of Southwestern Wyoming; the Washakie Basin.

UINTATHERIUM Leidy.

Proceed. Acad. Nat. Sci. Phila., 1872, p. 169 (published August 1). *Uintamastix* Leidy, loc. cit.: ? *Dinoceras* Marsh, Amer. Journ. Sci. Arts, 1872, October, 1872 (published September 27).

This genus resembles *Eobasileus* in its general proportions, but differs in its more elongate cervical verterbræ. The centra of these are flat at both extremities, but have not such a marked elephantine abbreviation as seen in the genus named. This enabled the head to approach the ground more nearly, and as the limbs were shorter in some of the species, they no doubt modified the length of the muzzle and lips.

Several names have been applied to this genus. Professor Leidy's name, here employed, bears date early in August. Under date of September 27, Professor Marsh proposed the name *Dinoceras* (Amer. Journ. Sci. Arts, 1872) for the *U. mirabile*, but did not give his reasons for separating it from his *inoceras* (the names of the two bear an objectionable resemblance), or those published by Dr. Leidy or myself. As it is propably synonymous with *Uintatherium*, I include it here, as is done by Dr. Leidy.

I am acquainted, by autopsy, with two species of this genus. None of them are so large as the *Eobasileus pressicornis*. *U. robustum* Leidy, is smaller, and the *U. lacustre* Marsh, smaller still. *U. mirabile* (*Dinoceras*)

Marsh, is about the size of the *U. robustum*, but differs from it in the absence of a tubercle on the last molar. I therefore retain three species, as follows: *Uintatherium robustum* Leidy, *U. mirabile* Marsh, *U. lacustre* Marsh. For convenience I compare these species with those of *Eobasileus.*

The naso-maxillary horn-cores have been seen in *E. furcatus* and *U. mirabile*, and the nasal tubercles in the same. The posterior horn-cores are known in the *U. mirabile.* The posterior and lateral crests of the cranium inclose a basin-shaped concavity above in all these species. It has been observed in all but *E. pressicornis.* The dentition is similar to that in *Loxolophodon*, *i. e.*, I. 0; C. 1; Pm. 3; M. 3. The first premolar in *U. lacustre* has an internal cone and outer concave crest. The worn surfaces of the other teeth in that species, *U. robustum* and *U. mirabile*, are narrow ovate, with a deep exterior emargination. The true molars support two crests, which converge inward and unite with a small tubercle behind the apex in *U. lacustre* and *U. robustum.* The tusk is long, compressed, and double-edged, as in *Loxolophodon.* The last inferior molar in *U. robustum* possesses three transverse crests, the posterior two parallel, and obliquely directed inward toward the axis of the anterior, which is the highest.

In a specimen of one of the smaller species the *ulna* widens considerably distally, being nearly as wide as the much-expanded olecranon. The latter is large, flattened, and subtransverse, and presents a sharp ridge internally. On the inner side of the distal part of the articular face for the humerus is a tubercle, from which a short, wide groove runs out on the inner face of the bone. The head of the radius is a little exterior to the middle line, and the shaft crosses the ulna in an open, shallow groove, to the inner side.

The *cuboid* is flat, and displays two proximal and two distal articular facets in an undetermined species described below. The astragalus of the same species is subbifurcate posteriorly, and has internally an extensive oblique malleolar fossa. The *calcaneum* is short and massive, with two superior and one small anterior articular facet.

The species, including *Eobasileus furcatus*, may be thus distinguished:

1. Large species (occipital condyles extending over about M. 0.170.)
   Naso-maxillary horns long; posterior horn-cores flat, elevated...... *E. furcatus.*

2. Species of intermediate size.
   Molars smaller, with an additional tubercle on the last............ *U. robustum.*
   Last molar without additional tubercle; maxillary horn-cores low, triangular; posterior horn-cores short, triangular in section (Marsh) ... . *U. mirabile.*
3. Smallest species (occipital condyles extending over about M. 0.95).
   Molar teeth larger, the last with a posterior expansion .. .. *U. lacustre.*

Uintatherium robustum, Leidy.

Proceed. Acad. Nat. Sci., Phila., 1872, p. 169, August. Report U. S. Geol. Surv. Terrs., i, pp. 96, 333, pl. xxv, figs. 6–12; pl. xxvi, figs. 1–3; pl. xxvii, figs. 30–34. *Uintamastix atrox*, Leidy, *l. c.*

I have been able to examine, through the kindness of Professor Leidy, the type of his description, and find it to belong to a smaller species than any of those above described. The lateral-parietal and supra-occipital crests are well developed, and the latter extends obliquely backward. Several peculiarities are to be observed in the dentition. Thus there is a great inequality in the height of the transverse crests of the posterior upper molar, the anterior, or the arched one, rising to a high cusp at its outer extremity. A small tubercle exists on the side of the inner angle of the grinding surface in the penultimate molar. The same angle is much elevated in an anterior molar. The canine is wider distally than in *L. cornutus* and less recurved. The mastoid process is quite prominent. The humerus has a prominent internal condyloid ridge and tuberosity, and the condyles are not very oblique. The inner posterior lobe of the tibial face of the astragalus is quite well defined. There is no median ligamentous pit on the trochlear face.

Measurements (from Leidy).

	Inches.
Depth of lower jaw at last molar	3.25
Length of humerus, about	21.00
Diameter at condyles	7.50

I obtained a fragmentary lower jaw of this species from Henry's Fork of Green River in 1873. One ramus contains the roots of six posterior molars; the other sustains large portions of the crowns of the last two. The latter agree with the corresponding teeth as figured and described by Leidy. I have represented them on Plate XXXVI, Fig. 2. The symphysis of this mandible is preserved, and is represented on Fig. 1 of the same plate. It is of singular form, being solidly coössified and very much compressed. Its inferior middle line is an obtuse keel, which rises anteriorly with a regular curve. Its extremity is broken off in the specimen, but the deep alveoli of two incisors remain on each side. The form of these shows that the roots of the incisor teeth are much compressed. A small mental foramen issues below the middle of the second or posterior one.

This symphysis is a good deal more compressed than that of *Loxolophodon cornutus*, as described by Messrs. Scott, Spier, and Osborne, but has the same general character.

I have given the following more detailed description of this specimen in the Proceedings of the Philadelphia Academy for 1882, p. 295. I had previously mentioned it in the Annual Report of the Hayden Survey for 1872, p. 565.

The presence of but two teeth on each side of the symphysis is an important peculiarity, since the other species so far as described, are said to have four teeth on each side of the symphysis, viz: three incisors and one canine. Those present in the present species I suppose to be incisors. The molar teeth are so much like those of *Uintatherium robustum*, that I believe the specimen to belong to that species.

Symphysis very much compressed, so that the incisor teeth of opposite sides are close together; its inferior outline curved upwards to the alveolar edge, in an obtuse keel. Base of flange for superior canine distinct, commencing below the posterior edge of the posterior alveolus, and immediately preceded by a mental foramen. Middle line of symphysis rugose. Ramus at last molar robust, owing to the prominence of the inferior part of the anterior masseteric ridge. In adaptation to the oblique position of the head, the inferior molars are oblique to the long axis of the ramus, sloping upwards and backwards, with exposed anterior roots. The molars increase in size posteriorly, and the last one is abruptly larger than the penultimate. Their structure is as in *U. robustum, i. e.*, with an obliquely transverse high crest in front, and a low posterior transverse edge of the heel, and a short oblique crest between the two. The last named is short, and is directed obliquely outwards and forwards towards the external extremity of the anterior crest, but disappears before reaching it. The internal extremity of this and the low posterior crest, with the external extremity of the anterior crest, rise into cusps. At the middle of the anterior base of the anterior transverse crest there is a tubercle, which represents the anterior limb of the anterior V in *Coryphodon.* The crowns of the premolars are broken away in the specimen.

The alveoli of the incisors are flat, and are directed forwards at an angle of only 20° from the horizontal until near their orifices, where the angle is greater. The roots of the incisors are thus curved upwards and forwards. There is but little space between the anterior alveolus and the anterior angle of the symphysis.

Measurements.

		M.
Length from anterior edge of symphysis to anterior base of canine flange		.074
Width of symphysis below at bases of lateral flanges		.032
Depth of symphysis between bases of lateral flanges		.040
Width of symphysis above between posterior incisors		.017
Length of bases of posterior five molars		.148
Length of bases of true molars		.110
Diameters crown, M. ii	anteroposterior	.031
	transverse in front	.020
Diameters crown, M. iii	anteroposterior	.035
	transverse in front	.025
Width of ramus at posterior edge of M. iii		.040

Although the crowns are somewhat worn, the enamel is wrinkled intermediately between coarse and fine.

It may be remarked here that it is by no means certain that the other species here included under the head of *Uintatherium* have the number of incisors ascribed to the *U. robustum.*

Uintatherium lacustre, Marsh.

?*Dinoceras lacustris*, Marsh, *l. c.*, October, 1872 (published September 27, 1872).

Plate xxxvi, figs. 3–8.

I have the occipital and parts of the parietal and squamasal bones, with some teeth of a specimen which I found together, and which I refer doubtfully to this species. Parts of the same cranium were sent to Professor Marsh by his collector, who left in the rock the fragments which I afterward procured. Professor Marsh's description of *Dinoceras lacustris* followed soon after, and his measurements agree with those of the specimens obtained by me.

The species is distinguished from its congeners, apart from its smaller size, by the large size of teeth. These are nearly as large as those of *Loxo-*

lophodon cornutus, and considerable larger than those of *U. robustum* and *U. mirabile*. The occipital condyles are not larger than those of the elk, *Cervus canadensis*. Their anterior inferior angles are separated by a triangular fossa of the basioccipital bone. This bone is expanded posteriorly, but narrows forwards, and is prominent and rib-like where it joins the basisphenoid. There is a short paroccipital process, which is anterior to the external part of the occipital condyle, its internal border rising from the condyloid foramen. Its anterior base, together with the other parts of the exöccipital bone, is closely adherent to the posstympanic process, which forms with it a large transverse process behind the auricular meatus. The posttympanic element in this process is much the larger, and projects further externally, where it presents a narrow edge. In this respect it differs much from the *Loxolophodon cornutus* and *Eobasileus furcatus*, where the external face of this process is wide and otherwise characteristic (See plates XXXIII and XXXVII). The base of the postglenoid process is not so extended transversely as that of the posttympanic, and is well separated from it. Its outline is that of one extremity of an ellipse. The glenoid surface has a considerable transverse extent in consequence of the sudden expansion of the zygoma behind. Its external extremity is marked by a low tuberosity. The auricular groove is bounded below inwards from the internal base of the postglenoid process by an external extension of the os-petrosum. It is fragile and bears on its inferior face a short process, perhaps the stylohyal. The pterygoid process of the sphenoid originates opposite the anterior border of the posterior alisphenoid foramen. The latter is large. Its superior border is continued as a ridge backwards and a little outwards, and overhangs the foramen ovale. This foramen is smaller than the alisphenoid, and is a narrow oval, looking outwards and forwards. It is situated .015 M. external to the anterior apex of the foramen lacerum anterius, and a short distance within the small internal glenoid tuberosity above mentioned. The condyloid foramen is round and large, and is situated near the anterior border of the basioccipital, where the posterior border of the foramen lacerum posterius is excavated for the jugular vein. There is a small and thin-walled otic bulla. There are no postglenoid nor supraglenoid foramina.

The posterior molar has a wide floor extending from the posterior or straight transverse crest, to the cingulum This crest is low, and has a low tubercle near its apex behind The other molars have strong fore and aft cingula, but none at the ends. The worn surfaces are first V-shaped, later arrow-shaped. The first premolar has a curved outer crest and an inner conic tubercle.

Measurements.

	M.
Diameter of occipital foramen and condyles	.092
From exterior end condyle to external border of mastoid	.058
From exterior end condyle to post-glenoid process	.053
Width of skull at glenoid surface between post-alisphenoid foramina	.292
Width of skull at glenoid surfaces between post-alisphenoid foramina	.040
Width of basioccipital anteriorly	.015
Transverse diameter of last upper molar	.045
Transverse diameter of third premolar	.029
Length of molar series	.163

Found by the writer in the Bridger formation of South Bitter Creek, Wyoming, in the Mammoth Buttes of the Washakie Basin.

? UINTATHERIUM sp.

Uintatherium sp., Cope, Annual Report U. S. Geol. Surv. Terrs., 1872 (1873), p. 582.

Plates xxxiv, xxxv.

A part of a large skeleton of a species of the *Dinocerata* was found by myself lying on a plateau of the Mammoth Buttes, Wyoming. In the absence of any part of the cranium, I have been unable to indentify it either as to genus or species. A cervical vertebra preserved, is so much more robust than that of *Eobasileus pressicornis*, that I originally referred the species to *Uintatherium*, in accordance with the descriptions of the corresponding vertebræ of the genus given by Marsh. It is the belief of Messrs. Scott, Spier, and Osborne that the cervical vertebræ of *Loxolophodon* are not as much abbreviated as in *Eobasileus*, so it becomes possible that the present animal is a species of the former genus. As the Princeton paleontologists have, however, described some species of supposed *Uintatherium* from the same neighborhood, (*U. leidianum* and *U. princeps*), the present specimen may belong to one of these. The description and figures are given to elucidate the characters of the suborder.

The species is one of medium size in the order, but was absolutely larger than the Indian rhinoceros.

The bones consist of several vertebræ, some carpal bones, and the entire hind limb of the left side except the toes and the cuneiform and navicular bones.

The odontoid process is very stout, with a descending trihedral apex Length M. .078; diameter at base, .048. A dorsal vertebra, with a single (anterior) capitular articular face, is quite concave in front

	M.
Diameter anteroposteriorly	.057
Diameter vertically	.094

A cervical vertebra has the proportions of the dorsal as to its centrum, thus differing materially from species previously described. The articular surfaces are slightly concave.

	M.
Length (anteroposteriorly)	.065
Diameter, vertical	.087
Diameter, transverse	.100

The *femur* resembles that of the other species already described, but is remarkable for the relatively small size of the head. While the lengths of the bone are not very different, and the expanse of the great trochanter about the same, the head of *L. cornutus* is large, the present one is very much smaller, and that of *E. pressicornis* intermediate. There is a rudimental third trochanter, and the condyles are as large as, and similar to, those of *E. pressicornis*. The external marginal condylar ridge is quite short. The shaft is broken and some small pieces lost; it is now 26 inches long, but was no doubt longer when complete.

Measurements of femur.

	M.
Expanse of great trochanter	.230
Diameter of head	.109
Diameter of shaft at middle	.093
Diameter above condyles	.152
Diameter at extremity of condyles	.139
Diameter (vertical) of inner condyle	.125

The *tibia* is perfectly preserved. It is short and stout, and with massive extremities. The outer basal part of the spine remains, and is prominent. The cotyli are not oblique; the inner is sub-round, the outer transverse, widening outwardly; their long axes are at right angles to each other. The spine is a low ridge of contact of the cotyli. The superior fibular face is a transverse oval; the inferior much smaller than in *E. pressicornis*. This shaft is contracted, and flattened behind and on the inner side. The distal extremity is transverse, less truncated for the fibula than in *E. pressicornis*, less convex behind, and with a less prominent external malleolus. The point dividing the astragalus behind is more prominent.

Measurements of tibia.

	M.
Total length	.398
Diameter of head, longitudinal	.080
Diameter of head, transverse	.138
Diameter of shaft, transverse	.063
Diameter of shaft, anteroposterior	.060
Diameter of distal articulation, anteroposterior	.077
Diameter of distal articulation, transverse	.113
Diameter of distal extremity, fore and aft	.093
Diameter of distal extremity, transverse	.125

These measurements show that this bone is considerably shorter than in *E. pressicornis*, though of equal distal diameter. In both species the measurements considerably exceed those given by Marsh for his *Titanotherium* (?) *anceps*. The form of the articular extremities differs from both in being more narrowed and transverse

The *fibula* is larger proximally and smaller distally than in *E. pressicornis*. Diameter proximal articular face .039; of the distal .045.

The *astragalus* is similar in size and form to that of *E. pressicornis*, but differs in two points. The posterior margin is deeply incised for the ligamentous insertion, and the outer lobe is clearly cut to this fossa, on the inner side. There is a pit for a ligament on the convexity of the inner part of the middle of the tibial articular face. A third difference is seen on the inferior face The inner calcaneal facet is longer and narrower, and is margined on the inner side by a large fossa parallel to its axis, which is wanting in the other species. The *calcaneum* is short and wide; its only anterior articulation is with the cuboid, and is small. The heel is deeper than long, and is obliquely truncate downward and inward.

Measurements of calcaneum.

	M.
Length	.105
Width	.092
Depth in front	.056
Length of heel	.047
Depth of heel	.055
Length of cuboid facet	.038

The *cuboid* is a flat subtriangular bone with two unequal articular faces below.

	M.
Length	.064
Width	.070
Depth	.031
Length of cuneiform (anteroposterior)	.040
Depth of cuneiform	.017

The *humerus* of a third specimen may or may not belong to this species. It was found in another locality. Its condyles are much less oblique than in that one described under *E. pressicornis*, and the olecranar fossa is shallower. It belongs to a larger animal; see

Measurements of humerus.

	M.
Transverse diameter distally (7.75 inches)	.195
Transverse diameter, inner condyle	.125

Remarks.—The remains were discovered by the writer in the Bridger Bad Lands, on South Fork of Bitter Creek, Wyoming.

BATHYOPSIS Cope.

Amer. Naturalist, 1881, p. 75. Bull. U. S. Geol. Surv. Terrs., 1881, p. 194.

The characters of this genus can only be given as seen in the mandible, the only part of the skeleton in my possession. Dentition: I. 3; C. 1; Pm. 4; M. 3. Incisors, canine, and first premolar forming an uninterrupted series, which is separated by a diastema from the molar series. The molar and premolar teeth are constructed on an identical pattern, presenting slight modifications from front to rear. This consists of an anterior elevated transverse crest, and a posterior heel, with raised posterior border; between these is situated on the external side an elevated cusp, which sends a low ridge inward and forward. The inner extremity of the anterior crest is cusp-like, and is accompanied by a second internal anterior cusp immediately posterior to it. The mandibular ramus has great vertical depth, its inferior border being convex downward throughout its entire length. Symphysis coössified.

The above characters indicate a new genus of considerable interest. Its form differs from that of the two genera where it is known, viz, *Uintatherium* and *Loxolophodon*, in the much greater development of the inferior expansion. In *Loxolophodon* it has been shown by Messrs. Speir and

Osborne to be represented by a mere convexity. In *Uintatherium*, Marsh has discovered it to be confined to the anterior part of the jaw, as in the saber-tooth cats. In *Bathyopsis* it extends to the entire length of the ramus, giving an outline in profile much like that of *Megatherium*. The anterior extremity of the symphysis projects beyond the line of the anterior border of the inferior expansion.

The characters of the inferior molars in this and other genera of *Dinocerata* are very peculiar. In *Bathyopsis* they are constructed on the plan of those of insectivorous marsupial and placental mammals, so as to lead to the suspicion that its food consisted of crustacea, or insects of large size, or possibly of thin-shelled mollusca.

Bathyopsis fissidens Cope.

Loci supra citati.

Plate XXIX *b*, figs. 1–3.

The lower jaw is not much larger than that of the Malayan tapir, but as this part of the skull is disproportionately small in this order, it is probable that this species is of considerably larger size than the one mentioned. The symphysis mandibuli is quite narrow, and its superior excavation is deep. It extends as far posteriorly as the middle of the diastema. It has considerable vertical thickness. The anterior edges of the lateral expansions are truncate, and present an obtuse angle outward, which forms the anterior boundaries of the slight concavity of the lateral face. The middle of the expansion below the first premolar tooth is slightly convex. This wall incloses a large internal expansion of the dental canal, which issues in a large mental foramen. This foramen is situated near the middle of the vertical diameter of the expansion, and below the anterior part of the diastema. It looks downward and forward. The external face of the posterior part of the ramus is nearly plane The inner face is vertical to a line which corresponds with the inferior border in *Coryphodon*, and then slopes obliquely outward to the inferior margin. The base of the coronoid process rises vertically from the line of the alveolar border, and its external edge forms an anterior border for the masseteric fossa. The inferior border of the fossa is not defined. The inferior border of the ramus is decurved posteriorly, and projects inward considerably beyond the plane of the jaw.

The premolars differ from the molars in having all their diameters, excepting the vertical, reduced. The fourth premolar only differs from the first true molar in the less elevation of the posterior border of the heel, and in a little smaller transverse diameter. The external part of the heel of the last molar rises into an obtuse cusp; the remainder of the border is tubercular The heels of the other true molars end in simple recurved transverse edges. On the premolars their posterior extremities are not recurved. The anterior face of the anterior cross-crest of all the molars is concave, and on the second premolar it looks obliquely inward. The posterior or second anterior inner cusp is obsolete on the second premolar. The enamel on all of these teeth is, excepting where worn, rather finely wrinkled. The first premolar is not preserved, but its alveolus indicates that it is one-rooted and rather robust. The sizes of the alveoli of the other anterior teeth are arranged in the following order, commencing with the largest: C.; I. 2; I. 3. The alveolus of the canine is compressed, and has more than twice the anteroposterior diameter of the largest incisor. The alveoli of the first and third are subround; that of the second is somewhat compressed.

Measurements.

		M.
Length from the middle of the second incisive alveolus to the extremity of the last molar		.1950
Length of the series of consecutive molars		.1170
Length of diastema		.0240
Diameter of alveolus of Pm. I		.0090
Diameter of alveolus of canine		.0250
Length of premolars		.0460
Diameters of Pm. II	vertical	.0130
	anteroposterior	.0130
	transverse	.0090
Diameters of Pm. IV	vertical	.0120
	anteroposterior	.0105
	transverse	.0120
Diameters of M. I	vertical	.0100
	anteroposterior	.0155
	transverse	.0110
Diameters of M. III	vertical	.0176
	anteroposterior	.0260
	transverse	.0180

The appearance of the ridges of the anterior part of the jaw of the *Bathyopsis fissidens*, together with the remarkably large dental canal and mental foramen, strongly suggest that the animal possessed a large and perhaps prehensile lower lip.

From the Wind River, Wyoming. J. L. Wortman.

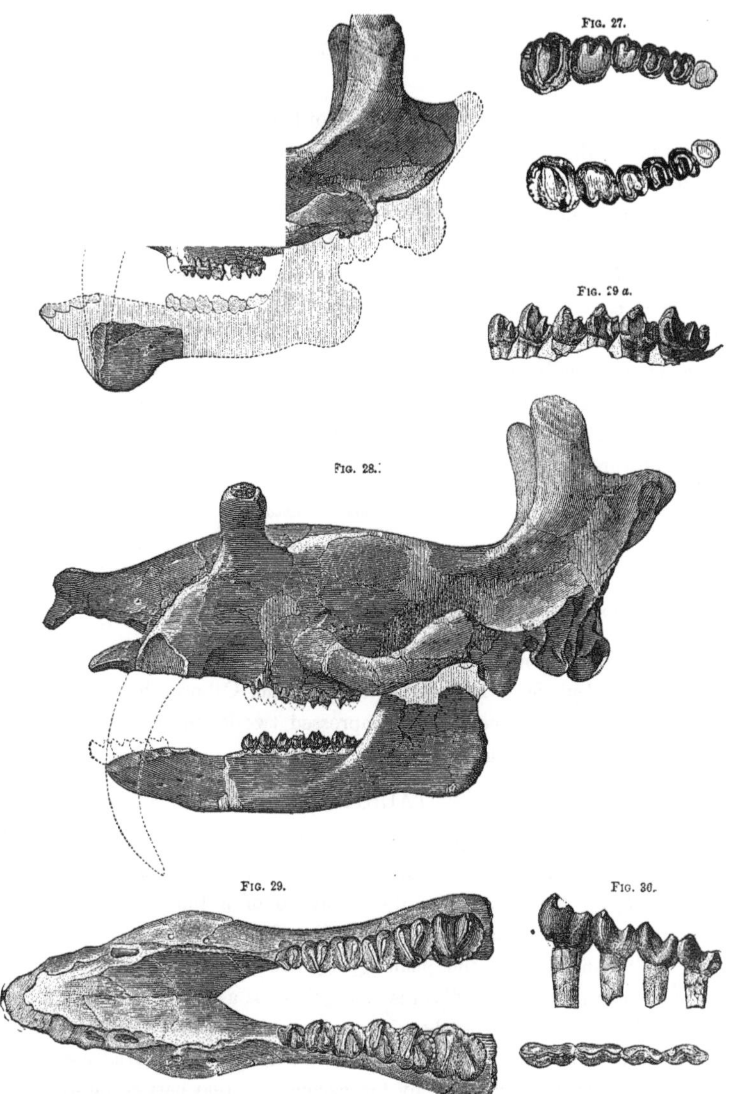

Figures copied from Osborne's Memoir on *Loxolophodon* and *Uintatherium*. Figs. 26 and 28, one-eighth natural size; figs. 27, 29, and 30, one-fourth natural size. Fig. 26, *Uintatherium leidianum*. Fig. 27, superior molar teeth of do. from below. Fig. 28, *Loxolophodon spierianum*. Fig. 29, mandible of supposed *Loxolophodon* from above; *a*, inferior molars, external side. Fig. 30, inferior incisors of supposed *Loxolophodon*, external and superior views.

A considerable part of the dentition of the mandible of this species was found in the Big-Horn bad lands. This includes an incisor tooth, which is quite characteristic, and renders it probable that the anterior parts of the jaws differ considerably from those of other *Uinatheriidæ*. The root is subround.. The crown resembles a good deal that of the species of *Coryphodontidæ*. It is higher than wide and has a subacute apex. One edge of the crown is convex, and the other concave. The external face is concave in both directions, and has no ridges nor cingulum. The inner face is concave longitudinally and convex transversely. The convexity is median and has a longitudinal concavity on each side of it. No internal cingulum except a trace at the base of the concave edge. The edges are obtuse even when unworn, and the enamel is obsoletely rugulose.

Measurements of incisor.

		M.
Diameters of crown	anteroposterior	.012
	transverse	.020
	vertical	.020
Diameters of root	anteroposterior	.012
	transverse	.014

This incisor (see Plate LVIII *a*, Fig. 7) is very different from the kind seen in *Loxolophodon*. Messrs. Scott, Spier, and Osborne have shown that genus to have these teeth with compressed two-lobed crowns, a type unknown elsewhere among *Mammalia*.

TALIGRADA.

Since the preceding pages defining the suborders of the Amblypoda were printed, I have learned and described the characters of the genus *Pantolambda*[1]. These indicate the existence of a third suborder of the Amblypoda, which I have called the Taligrada.

In the first place, the phalanges (including ungual) show that the genus is ungulate. Secondly, the astragalus has a large distal facet for the cuboid bone. This proves that the genus cannot be referred to the Taxeopod order. The question as to whether it belongs to the *Amblypoda* or the *Diplarthra* would be decided by the carpus, but that part is unfortunately not preserved, and I have to rely on empirical indications for a provisional

[1] American Naturalist, 1883, April, p. 406 (March 15).

determination. Apart from the astragalus, the characters are those of the *Condylarthra* rather than of the *Perissodactyla*, and it is therefore to be supposed that the carpus has also the characters of that order. This would place the genus in the *Pantodonta*, which has the carpus nearly that of the *Taxeopoda*, and the tarsus of the *Diplarthra*. The points of resemblance to the *Condylarthra* are the following: The ilium is narrow. The humerus has an epitrochlear canal. The superior molar teeth have but one internal lobe. The resemblances to the *Pantodonta* are these: The cervical vertebræ are plane and short. The femur has a third trochanter. The premaxillary bone is dentigerous. The astragalar trochlea is as in the *Periptychidæ*, and the *Proboscidia;* that is, without groove, and slightly convex anteroposteriorly, thus differing from that of the *Pantodonta*. The dentition is especially like that of the *Amblypoda* in general, and that of the superior series is unlike anything known in the *Diplarthra*.

I propose to place this genus in the *Amblypoda* for the present, next to the *Pantodonta*, but it cannot enter that sub-order on account of the form of its astragalus. These sub-orders of *Amblypoda* will be compared as follows:

Astragalus with a head distinct from trochlea, with distal articular facets .. *Taligrada*.
Astragalus without head; distal facets subinferior .... *Pantodonta*.

In the sub-order *Taligrada*, the single family *Pantolambdidæ* presents the following characters:

Superior and inferior molars with the cusps developed into Vs. Postglenoid process present; postympanic and paroccipital not distinct. All the vertebræ with plain articulations. Humeral condyles without intertrochlear ridge. Femur with third trochanter. Digits of posterior foot probably five. Metapodial keels small and posterior.

PANTOLAMBDA Cope.

American Naturalist, 1882, p. 418. Proceedings American Philosophical Society, 1883, p. 558.

A considerable part of the skeleton of a species of this genus having been sent me by Mr. D. Baldwin, I am able to throw much light on the affinities of this curious genus.

Canine teeth distinct; dental series continuous. Superior molars all triangular; that is, with a single internal cusp. External cusps of premolars unknown; of molars two. Internal cusp V-shaped, sending its horns externally as cingula to the anterior and posterior bases of the external side of the crown, without intermediate tubercles. Inferior true molars with a crown of two Vs, the anterior the more elevated. Premolars consisting of one open V, with a short crest on a short heel, as in *Coryphodon.*

Dental formula: I $\frac{?3}{3}$; C. $\frac{1}{1}$; P-m. ? $\frac{4}{4}$; M. $\frac{3}{3}$; the last inferior with a heel. A strong sagittal crest. Auricular meatus widely open below. Large postparietal, postsquamosal and mastoid foramina.

Cervical vertebræ rather short; other vertebræ moderate, the lumbars not elongate. A large tail. Humerus with a large internal epicondyle. Femur with all the trochanters large. Ilium with the anterior inferior spine well developed. Metacarpals short, plantigrade. Phalanges of second series flat, and of subquadrate outline. The astragalus has a wide head, but no neck, as it is not separated from the trochlear portion by a constriction. It is as wide as the trochlear portion, but about one-third of its length extends within the line of the malleolar face of the trochlear portion. The navicular face is flat, that of the cuboid bone is convex vertically, and one half as long horizontally as the navicular, and only half as deep. These two facets are continuous with the sustentacular below. Interior to all of these, on the internal tuberosity of the head, is a sub-round facet looking inwards, like that characteristic of the genus *Bathmodon*, but relatively larger. A continuous facet is seen on the adjacent edge of the navicular. The use of these facets is unknown.

The brain case indicates small and nearly smooth hemispheres, extending with little contraction into a rather large cerebellum. The olfactory lobes are produced anteriorly at the extremity of a rather long isthmus

If we consider the dentition alone, *Pantolambda* is the ancestor of the *Coryphodontidæ.* The history of the feet requires further elucidation.

In describing this genus and species, I remarked, *loc. cit.*, that they were "founded on a mandibular ramus, which supports the first true molar, and the last two premolars. The characters of these teeth remarkably resemble those of *Coryphodon.* * * * * It will be for additional material to demonstrate whether this genus belongs to the *Amblypoda* or *Perissodactyla.*"

From what has preceded it is quite evident that the ensemble of characters is predominantly that of the former order.

It is now apparent that the type of superior molar tooth which predominated during the Puerco epoch was triangular; that is, with two external and one internal tubercles. Thus of forty-one species of Mammalia of which the superior molars are known, all but four have three tubercles of the crown, and in this thirty-eight triangular ones we may include those of three species of *Periptychus*, which have a small supplementary lobe on each side of the median principal inner tubercle.

This fact is important as indicating the mode of development of the various types of superior molar teeth, on which we have not heretofore had clear light. In the first place, this type of molar exists to-day only in the insectivorous and carnivorous Marsupialia; in the Insectivora, and the tubercular molars of such Carnivora as possess them (excepting the plantigrades). In the Ungulates the principal traces of it are to be found in the molars of the *Coryphodontidæ* of the Wasatch, and *Dinocerata* of the Bridger Eocenes. In later epochs it is chiefly seen only in the last superior molar.

It is also evident that the quadritubercular molar is derived from the tritubercular by the addition of a lobe of the inner part of a cingulum of the posterior base of the crown. Transitional states are seen in some of the *Periptychidæ (Anisonchus)* and in the sectorials of the *Procyonidæ.*

PANTOLAMBDA BATHMODON Cope.

American Naturalist, 1882, p. 418.

Plate XXIX *d*.

The bases of the inferior Pm. iii and iv are subquadrate, the inner side rounded, that of the iv relatively the wider. On the iii the median keel constitutes the heel; on the iv, the keel is in the center of a wide heel. No cingula. The first true molar has an anterior cingulum, but no external one. The enamel is wrinkled where not worn. Diameters of Pm. iii, anteroposterior, .009; transverse, .007; of Pm. iv, anteroposterior, .009; transverse, .0085. Width of first true molar in front, .0083. Apparently about the size of a sheep.

The second and more perfect specimen of this species includes a considerable part of the skull, with maxillary bone and teeth, the latter bone not continuous with the others. There are thirteen vertebræ from all parts of the column; more or less of two humeri, neither with the head, both ilia; the proximal part of the femur; the astragalus, navicular, and several metatarsals and phalanges.

The posterior part of the *skull* is very opossum-like. It is long and narrow, and the sagittal and lateral occipital crests are much produced. Above the foramen magnum the occipital bone is deeply concave. The lateral occipital crest divides at its inferior portion into two ridges, which run nearly parallel and inclose between them a shallow lenticular fossa. The posterior ridge terminates in a short oval tuberosity, the paramastoid process, which bounds the auricular meatus posteriorly. The anterior terminates higher up, at the superior anterior angle of the tuberosity, without expansion, on a level with the superior edge of the meatus, thus not producing a post-tympanic process. The internal extremity of the postglenoid process is produced into a narrow ridge which extends inwards and then forwards, bounding an oval fossa arranged anteroposteriorly. Its internal border is indicated by a narrow tuberosity, between which and the ridge before mentioned is a longitudinal groove. The anterior temporal ridge is an obtuse angle continuing the half of the sagittal crest to the postorbital angle. The muzzle is narrowed upwards, and the nasal bones, though not narrowed, are together, obtusely roof-shaped. The facial portion of the premaxillary bone is remarkably wide anteroposteriorly, the width nearly equaling the depth in front of the canine tooth. The malar bone is very prominent, extending laterally much beyond the maxillary above the molar teeth. It presents no postorbital angle, but has on its external face a wide groove. This is overhung by a strong ridge anteriorly, which descends anteriorly and ceases above the anterior part of the second true molar. The groove runs out below, and becomes distally two planes, a lateral and an inferior separated by a prominent angle. The zygomatic fossa terminates anteriorly in a sinus which is opposite to the anterior edge of the third premolar. The posterior border of the palate is a little concave near the internal extremity of the last molar, and then curves posteriorly. The median portion is lost.

The mandibular rami are partly preserved, together with the part of the symphysis between the canine teeth. The symphysis is coössified, has a nearly vertical direction, and is flat transversely in front. The rami are compressed, and somewhat thickened below. The inferior outline is concave below the coronoid process (which is lost), and the angle is very prominent. It is broadly rounded in profile, and extends a considerable distance posterior to the base of the condylar process. Its external border is somewhat revolute; the inferior border is very slightly incurved. The masseteric fossa has an anterior border, and is marked off from the angle posteriorly, but it fades out below.

The parieto-squamosal suture is rather elevated in position above the base of the zygoma, but descends posteriorly to reach the occipital below the mastoid foramen. The nasal bones encroach considerably on the frontals posteriorly, reaching to opposite the postorbital angle. In this this animal resembles some of the species of *Phenacodus*. The nasal, naso-maxillary, and maxillo-premaxillary sutures are distinct and even.

There are two post-squamosal foramina, and five postparietal, with two or three on the squamoso-parietal suture. The condylar foramen is distinct from the *foramen lacerum posterius*. It is probable that there is a postglenoid foramen. The relations of the foramen ovale are not certain. There are several small foramina in the front of the symphysis.

The superior *true molars* only are preserved with parts of the alveoli for the fourth premolar. The first and second molars are similar, but the second external V of the third is much reduced, owing to the oblique truncation of the postero-external side of the crown. It is broken off in the specimen. The external Vs of the first and second molars are much flattened inwards, and have slightly concave external faces. There is no ridge or tubercle, but a weak angle only, to distinguish their external planes. The anterior is a little larger than the posterior, on account of the extension of the anterior external angle. The latter has a slight anterior hook or recurvature. The internal V is rather acute, and is a little anterior to a transverse line dividing the crown equally. There are anterior and posterior basal cingula which do not meet at the internal extremity of the crown (this portion is damaged on the second molar), and which disappear at the line of the apices of the external Vs. There is a weak cingulum at the external base of the external Vs. The enamel is delicately wrinkled, or smooth where subject to attrition. The alveoli of the canine and third superior incisor show that they are large teeth.

Of inferior teeth only parts of the crowns of the M. i and M. iii are preserved. The limbs of the anterior Vs are elevated and sharp-edged, and the posterior terminates in something like a cusp internally. The third true molar has a well-developed median angle on its external side. The heel is long. The inferior canine alveoli indicate flattened roots and a crown directed upwards. The incisive alveoli form with the canines a continuous series, and the median is a little larger than the first and third. Crowns lost.

Measurements of skull.

	M.
Length from supraoccipital crest to posterior end of nasal bones	0.1035
Length of sagittal crest	.087
Width between superior edges of squamosals	.025
Anteroposterior extent of glenoid surface	.017
Anteroposterior extent of meatus auditorius	.010
Length of cavity containing cerebrum and cerebellum	.050
Long diameter of occipital condyle	.018
Width of both nasal bones at middle	.0185
Length of facial process of premaxillary anteroposteriorly	.015
Elevation of orbit above molar teeth	.019
Production of angle of mandible beyond posterior base of condyle	.014
Depth of produced portion of angle	.027
Length from angle to anterior border of masseteric fossa	.050
Depth of ramus at posterior edge of last molar	.0285
Depth at first true molar	.0275
Thickness of ramus below first molar	.0115
Width of symphysis above between the canines	.018
Length of superior true molars	.0285
Diameters M. i { anteroposterior	.011
Diameters M. i { transverse	.017
Diameters M. ii { anteroposterior	.011
Diameters M. ii { transverse	.017
Transverse diameter of M. iii	.0165
Length of inferior true molars on base	.032
Diameters M. i { anteroposterior	.0105
Diameters M. i { transverse	.008
Diameters *of base* of M. iii { anteroposterior	.014
Diameters *of base* of M. iii { transverse	.007
Anteroposterior diameter of superior canine alveolus, not less than	.013

The *vertebræ* preserved include two cervicals, six dorsals, two lumbars, and two caudals. The centra increase regularly in size to the lumbars. Their articular faces are distinctly biconcave, except those of a median distal caudal, where their convexity exceeds the median concavity.

The cervicals are rather short, having proportions like those of *Coryphodon* and various carnivorous animals. The centra are wider than deep, and the posterior concavity is greater posteriorly than anteriorly, and is surrounded by the thickened edge of the centrum. Below the parapophysis the inferior face is concave, and each fossa diverges outwards, leaving the median plane wider behind than before. In the vertebræ preserved there is no hypapophysial keel. In a sixth or seventh cervical there is no vertebraterial canal, unless it perforate the "transverse process" at a distance from the centrum. A strong groove from the neural canal excavates the superior side of the parapophysis and turns anteriorly to where it is broken off. A capitular rib-fossa is not distinct, hence I am not sure that this centrum is the seventh.

In the first three dorsals preserved, which represent the more important part of the thoracic region, the posterior articular face is more concave than the anterior. In the posterior of these, the anterior costal facet is larger and more distinct than the posterior. In the two others the posterior is large, and the anterior is not to be seen, and is probably wanting, although the centra are somewhat injured at that point. These three dorsals are of different dimensions, showing a rapid increase posteriorly. None of them have inferior keels or fossæ, but the first has a median angle. In the third the breadth is greater in relation to the length than in the first; and the second is intermediate. In the remaining three dorsals the increased width is retained, and the length increases. The centra become more depressed, and more excavated below the diapophyses. The middle line becomes more prominent in the fifth and sixth. The bases of the diapophyses are narrow and depressed. The costal facets are anterior, and present laterally.

The best preserved lumbar is elongate and depressed, and the anterior face is quite concave. It is not keeled below, and there is a considerable nutritive foramen on each side of the middle line, which is connected with the inferior base of the diapophysis by a shallow vertical wide groove. An anterior dorsal with robust and flattened diapophysis is longer than wide, and equally moderately concave at both extremities, which are subround. The neural spine and zygapophyses are well developed. There are considerable spaces for the attachment of a chevron bone, which are continuous with the rim of the articular face of the centrum. The inferior surface is concave on each side of a medium median surface, which is concave anteroposteriorly. A more elongate caudal has two diapophyses on each side separated by a sinus, the anterior the smaller. The middle line both above and below is angulate medially and flattened at the extremities. Anteriorly above there is a small tuberosity directed upwards and forwards on each side, the remains of the neurapophyses.

Measurements of vertebræ.

		M.
Diameters of C. v or vi	anterposterior	.014
	transverse in front	.017
	vertical in front	.011
Diameters C. vi or vii	anteroposterior	.0126
	vertical in front	.0113
	transverse in front	.017
Diameters of an anterior dorsal	anteroposterior	,017
	vertical in front	.013
	transverse in front	.0156
Diameters of a posterior dorsal	anteroposterior	.019
	vertical in front	.0125
	transverse in front	.0175
Diameters of a more posterior dorsal	anteroposterior	.022
	vertical in front	.014
	transverse in front	.0225
Diameters of a lumbar	anteroposterior	.026
	vertical in front	.0125
	transverse in front	.022
Diameters of an anterior caudal	anteroposterior	.0225
	vertical in front	.0147
	transverse in front	.017
Diameters of distal caudal	anteroposterior	.024
	vertical in front	.013
	transverse in front	.0125

Zygapophyses of four vertebræ are preserved. They are flat except in one case they are a little and in another more convex. They are not subcylindric as in *Mioclænus*, *Mesonyx*, &c.

A supposed *manubrium sterni* is represented by the anterior T-shaped portion. It is broadly truncate in front, and the superior straight transverse edge is recurved. The body of the bone which extends posteriorly has a vertical oval section, the inferior angle narrowed and subacute. It is broken off. The superior point of the truncate anterior face looks as though divided between two flat articular faces, perhaps for clavicles. Immediately behind the lateral superior angles of the front are two small deep fossæ, each perhaps for a first hæmapophysis. The position of the bone is, however, uncertain.

Measurements.

		M.
Diameters of front	vertical	.016
	transverse above	.022
Diameters of shaft	vertical	.0105
	transverse	.008

The *humerus* is robust and of moderate length. The bicipital or deltoid crest is very prominent, and becomes more so distally where it projects in a narrow twisted tuberosity and then descends abruptly to the posterior side of the shaft. There is a narrow oval strong muscular insertion on the posterior edge of the inner side of the shaft proximad of the end of the deltoid crest. From this point distad, the posterior face is flat. Three and one-half centimeters distad the external edge expands into a laminar crest which soon presents a straight edge, which terminates in the very obtuse and short external epicondyle. Above the olecranar fossa the posterior face is slightly concave. So long as any trace of the deltoid crest remains, the shaft has a triangular section. The internal epicondyle is very prominent and is truncate. The bridge over the epicondylar foramen is well developed. The condyles are not as much extended transversely as in *Periptichus rhabdodon*, but is more so than in *Phenacodus primævus*, and is a good deal as in *Coryphodon*. The flange is moderately prominent, and the cylinder quite convex anteriorly, descending to the short internal extremity. The angle between the two surfaces is posteriorly converted into a deep narrow groove, which posteriorly separates the external condyle from a low flange of the inner edge of the condyle. The latter is as usual concave posteriorly in the transverse direction, and has strong lateral and posterior outlines, which are formed like the boundaries of one side with two angles of a parallelogram. Between the internal angle and the base of the epicondyle is quite a fossa.

Measurements of humerus.

		M.
Length from distal end of bicipital crest to edge of flange of condyle		.064
Diameters at end bicipital crest	anteroposterior	.028
	transverse	.016
Diameters at proximal angle distal external crest	anteroposterior	.015
	transverse	.0265
Width at epicondyles		.041
Width of condyles	anteriorly	.025
	posteriorly	.016
Depth of condyle	at flange	.018
	at concavity of roller	.0125

A *metacarpal bone*, probably the second, is quite short, and the shaft is contracted, and the distal end expanded. The proximal extremity has the facet slightly concave, and with the posterior edge broadly rounded, and a little raised, and with a narrow lateral facet on each side. The one of these slopes outwards below; the other inwards to an open excavated surface. The shaft is slightly convex in front, both in length and width. The distal condyle is flat and truncate, and is not bounded by a groove proximally. The keel is distinct but not strong. The condyle is truncated by a round ligamentous fossa on each side, and a wide transverse one posteriorly proximad. There is a fossa of the anterior face just below the proximal facet; and there is an open groove on each side proximally, below the lateral facets, which are not cut by them.

Some fragments of metapodials and some phalanges are preserved which may belong to the manus or pes. The phalanges are those of the middle place. Their outlines are nearly square, the narrowing distally destroying its regularity. They are quite depressed, and the proximal cotylus looks partly upwards. The lateral distal ligamentous fossæ are round, and impressed. The lateral border proximad to them is a narrow ridge. The distal condyle is not produced posteriorly, but extends posteriorly and is cut off by a rather shallow fossa. An ungual phalange is quite flat, but not much expanded. The distal extremity is deeply fissured above. The inferior table fades out distally; its proximal lateral angles are bounded by fossæ, two of which are, perhaps, foramina. There is a short free angle on each side of the proximal cotylus. The superior surface is longitudinally roughened on each side of the median groove.

Measurements of foot.

			M.
Length of metacarpus ii			.032
Proximal diameters metacarpus ii	anteroposterior		.011
	transverse in front		.008
Distal diameters metacarpus ii	anteroposterior		.008
	transverse	condyle	.011
		epicondyles	.013
Length of phalange of second row			.010
Diameters proximally	transverse		.0105
	vertical		.0055
Diameters distally	transverse		.007
	vertical		.004
Length of ungual phalange			.016
Width of ungual phalange (greatest)			.0095
Width of ungual phalange (least)			.0085
Width of ungual phalange, proximal			.0105

Of the *pelvis* the ilia are the most important pieces preserved. This bone is elongate and of triangular section. The external side is the widest, and is of nearly uniform diameter from one end to the other. Its plane joins that of the inferior side at an angle of forty degrees. Its superior border is gently convex, and is decurved at the apex or crest proper. Its surface is concave at the middle, but flat at the apex. The inferior face is concave anteroposteriorly with the curve of the bone, while transversely it is a little convex. The interior face is apparently adapted to a long sacrum, which extends to opposite the middle of the anterior inferior spine. The anterior inferior spine is a prominent tuberosity. The ilium cannot be said to have a distinct peduncle.

Measurements of ilium.

		M.
Length from acetabulum (apex restored)		.092
Width of external face	above anteroinferior spine	.025
	near anterior extremity	.029
Width of inferior face	above spine	.022
	near apex	.014
Width of internal face	at spine	.015
	at curve of crest	.025

The only fragment of the *femur* is broken off at the superior part of the third trochanter, and lacks the head and neck. The trochanteric fossa is long and deep. The anterior face of the bone at the little trochanter is quadrately convex, and that trochanter is prominent, and the apex is recurved.

Measurements of femur.

		M.
Length from little to great trochanter, inclusive		.060
Diameters just below little trochanter	anteroposterior	.0165
	transverse	.022

The *astragalus* displays a very slight concavity of the trochlea in the transverse direction, while the anteroposterior convexity is more pronounced. Although the head is not distinguished by a neck, there is between the tibial and the navicular facets, a transverse, oval fossa. The external face of the trochlea is nearly vertical, the anterior inferior angle alone diverging a little outwards, and still less downwards. Both anterior and posterior sides of this face are vertical. The former is somewhat concave. The inner side of the trochlea is oblique, and its posterior extremity is broadly rounded. The internal distal facet, peculiar to this genus and to *Bathmodon*, is vertical, and is more than half a circle. The posterior flexor tendinous groove is roofed over, so as to be a foramen; its superior opening is large and transverse and looks posteriorly.

The navicular is a short and flat bone. Its internal extremity is narrow in front, and obliquely truncate. The external side is squarely truncate for contact with the cuboid. The posterior tuberosity is small and rounded, and is at the postero-external angle. The ectocuneïform facet is flat and trapezoidal in form, is wider transversely than that of the mesocuneïform, and does not extend so far posteriorly. It is separated from the latter by a small round fossa posteriorly. The mesocuneïform facet is convex anteroposteriorly, while the entocuneïform is smaller and subtriangular.

Measurements of tarsus.

		M.
Length of astragalus at internal side		.025
Length of astragalus at external side		.016
Total width at trochlea		.027
Total width of trochlea		.015
Total width of head		.020
Total width of navicular facet		.014
Depth of navicular facet at middle		.013
Depth of cuboid facet at middle		.008
Depth of external face in front		.012
Diameters of navicular	anteroposterior	.0165
	transverse	.017
Diameters of ectocuneïform facet of do.	anteroposterior	.0085
	transverse	.008
Vertical depth of cuboid facet of do		.005

DIPLARTHRA.

Cope; Proceeds. Amer. Philo. Soc., 1882, p. 444. Paleontological Bulletin, No. 35, p. 444 (Nov. 11, 1882).

This order includes the two suborders of Perissodactyla and Artiodactyla, as already explained (page 378). It is the Ungulata of some authors, *e. g.*, Professor Gill, but I have used this term in a wider sense in imitation of other writers.

The two suborders which compose the order Diplarthra are distinguished in their entirety by but few characters; but these are universal, and easily observed. The constitution of the carpus and tarsus is essentially the same in both, but the tarsus of the Artiodactyla is more perfectly adapted for rapid movement than that of the Perissodactyla.[1] To the former order belong the species having the swiftest and most graceful movement. Besides the inferior ginglymus of the astragalus, the Artiodactyla are characterized by equal development of the third and fourth digits instead of the preponderating development of the third. But in some members of the order, the third and fourth digits are not entirely equal; the posterior foot of *Eurytherium* has three toes. In the Perissodactyla the third and fourth anterior toes of *Menodus* are nearly equal. In some members of the latter order (*Hyracotherium*) the distal facets of the astragalus are somewhat convex, so approaching the Artiodactyla. On these and similar grounds, I think these two groups are scarcely to be distinguished as orders, and it is highly probable that future paleontological discoveries will approximate their early representatives still more closely.

The Perissodactyla abounded in Eocene time, and have diminished in numbers to the present period, when but few species remain. They appeared in the Wasatch Eocene, in North America, since none have yet been discovered in the Puerco epoch. The Artiodactyla are not certainly known from the Puerco, and are rare in all the later Eocene beds. They begin to be abundant in the Miocene, and culminated in the Pliocene and recent epochs.

I have been aided in the investigation of this order by the work of Kowalevsky, printed in 1873, "Monographie der Gattung Anthracotherium Cuv. und Versuch einer natürlichen Classification der fossilen Hufthiere."

[1] For a discussion of the question of the origin of the characters of these orders, see my paper "On the effect of impacts and strains on the feet of Mammalia," American Naturalist, 1881, p. 542.

PERISSODACTYLA.

This, the first suborder of the Diplarthrous Mammalia naturally occupies a position between the *Amblypoda* and the *Artiodactyla.* Its lower forms are more specialized in the structure of the feet than the *Amblypoda,* while its highest types do not reach the perfection of structure seen in the *Artiodactyla.* This is particularly indicated by the form of the astragalus, which has but one, the tibial trochlea, and never displays the distal one characteristic of the cloven-footed families. The *Perissodactyla* occupy, as regards their dentition, a position parallel with the *Artiodactyla.* They are always superior in dental complication to the *Proboscidia* and the suilline *Artiodactyla,* but only one series, that of the horses, reaches the complexity of molars general in the *Ruminantia.* The dentition of the mass of the *Perissodactyla* might be described as intermediate between that of the *Proboscidia* and the lowest selenodont *Artiodactyla.*

The families of this order form a closely connected series, and the division of them into three divisions, the "Pachydermata," "Solipeda," and *Perissodactyla,* has no warrant in nature. Especially unnatural is the conjunction of the genera included under the first name, with the *Proboscidia* and certain suilline *Artiodactyla,* in a single order, as was proposed by Cuvier. The modifications of dentition from the simple type seen in *Menodus,* to the most complex, as in *Equus,* are close and consecutive. So, also, the gradual diminution in the number of digits from 4–4 to 1–1 can be traced through all the intervening stages.

The following definitions of families are applicable in the present stage of knowledge. Those of all but three were published in the Bulletin of the United States Geological Survey of the Territories, 1879, p. 228. A modification in the diagnoses of the families *Chalicotheriidæ* and *Palæotheriidæ* was introduced into a subsequent memoir on the classification of the order in the Proceedings of the American Philosophical Society for 1881:

I. Anterior exterior crescent of superior molars shortened, not distinguished from the posterior by external ridge; inferior molars with cross-crests; premolars different from molars.

1. Toes 4–3 .... *Lophiodontidæ.*
2. Toes 3–3 .... *Triplopodidæ.*

II. Exterior crescents of superior molars as in I; inferior molars with cross-crests; superior molars and premolars alike, with cross-crests.
3. Mastoid bone forming part of the external wall of the skull...... *Hyracodontidæ.*
4. Mastoid bone excluded from the walls of the skull by the contact of the occipital and squamosal .... *Rhinoceridæ.*
III. Exterior crescentoid crests of superior molars subequal, distinct; inferior molars with cross-crests.
5. Superior molars and premolars alike and with cross-crests; toes 4–3........ *Tapiridæ.*
IV. The external crescentoid crests of the superior molars subequal, separated by an external ridge; inferior molars with crescents.
A. Superior premolars different from molars; with only one internal cusp.
6. Toes 4–3; a vertebrarterial canal................................ *Chalicotheriidæ.*
7. Toes 3–3; no vertebrarterial canal .... *Macrauchenidæ.*
AA. Premolars like molars, with two internal lobes above.
8. Toes with digits, 4–4.. *Menodontidæ.*
9. Toes with digits, 3–3.. *Palæotheriidæ.*
10. Toes with digits, 1–1.. *Equidæ.*

The genera included in these families are the following. The table shows their geological distribution:

	Eocene.		Miocene.				
	Lower.	Upper.	Lower.	Middle.	Upper.	Pliocene.	Recent.
Lophiodontidæ.							
Systemodon Cope	2						
Hyracotherium Ow	12						
Pliolophus Ow	4	2					
Heptodon Cope	4						
Helaletes Marsh		3					
Lophiodon Cuv	2	11					
Hyrachyus Leidy		9					
Colonoceras Marsh		1					
Triplopidæ.							
Triplopus Cope		2					
Hyracodontidæ.							
Hyracodon Leidy			2				
Rhinocerontidæ.							
Orthocynodon S. and O		1					
Aceratherium Kaup			3				
Cœnopus Cope			2				
Diceratherium Marsh				3			
Zalabis Cope							
Aphelops Cope					4		

	Eocene.		Miocene.			Pliocene.	Recent.
	Lower.	Upper.	Lower.	Middle.	Upper.		
Rhinocerontidæ—Continued.							
Ceratorhinus Gray				1	2		2
Rhinocerus Linn					4		2
Peraceras Cope					2		
Atelodus Pom					2	1	2
Cœlodonta Bronn						3	
Elasmotherium Fischer						2	
Tapiridæ.							
Tapirus Linn						6	5
Elasmognathus Gill							1
Chalicotheriidæ.							
Ectocium Cope	1						
Leurocephalus S. S. and O		1					
Palæosyops Leidy	1	5					
Limnohyus Leidy		3					
Lambdotherium Cope	2	1					
Pachynolophus Pom		2	2				
Chalicotherium Kaup		3					
Nestoritherium Kaup				?1			
Macraucheniidæ.							
Macrauchenia Ow						2	
Menodontidæ.							
Diplacodon Marsh		1					
Menodus Pom			2				
Symborodon Cope			6				
Dæodon Cope				1			
Palæotheriidæ.							
Anchilophus Gerv		2					
Paloplotherium Ow		6					
Palæotherium Cuv		3					
Anchitherium Kaup			4	4			
Anchippus Leidy				1	1		
Hippotherium Kaup					8	1	
Protohippus Leidy					5	1	
Equidæ.							
Hippidium Owen					2	3	
Equus Linn						5	7

Total number of well determined species, one hundred and ninety-two.

From the preceding table it can be readily seen that this order was abundantly represented during the Eocene period, and that the recent species are comparatively few. It may also be observed that certain families predominated during certain periods. Thus the prevalent *Perissodactyla* of the Eocene are *Lophiodontidæ* and *Chalicotheriidæ;* those of the Miocene are *Rhinocerontidæ* and *Palæotheriidæ.* The *Tapiridæ* and *Equidæ* characterize the latest Tertiary epochs. A genealogical tree of the order may be constructed as follows:

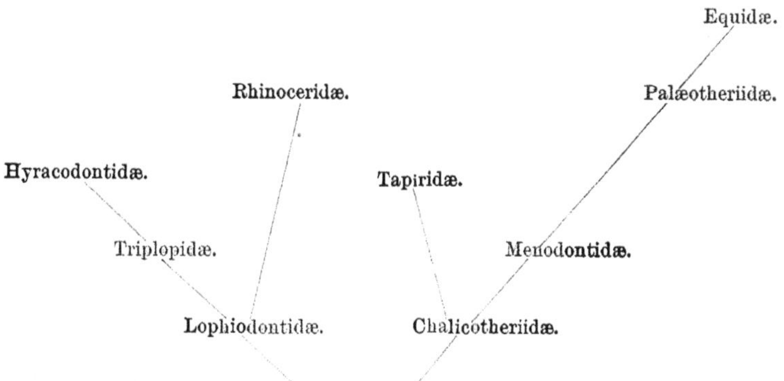

The types of the *Lophiodontidæ* and *Chalicotheriidæ* differ only in the two points of the separation or non-separation of the exterior crescents of the superior molars, as already pointed out. That no great modification of known forms (as *Ectocium* in the *Chalicotheriidæ*, and *Hyracotherium* in the *Lophiodontidæ*) would be necessary to obliterate this difference, is quite clear. The parent types of the order, which present the most generalized dentition, *Systemodon* and *Ectocium*, were cotemporaries of Lower Eocene age.

LOPHIODONTIDÆ.

This family embraces a larger number of known species than any of the others of the order. With one exception all the species belong to the

Eocene period. They range from the size of a rabbit to that of an ox. They resembled most, among living animals, the tapirs.

The genera are characterized as follows:

I. External lobes of superior molars separated and little flattened; lobes of inferior molars scarcely united (*Hyracotheriinæ*).

A. No diastemata.

a. Fourth inferior premolar unlike the first true molar.

Last inferior molar with five lobes; superior premolars four .... *Systemodon*.

AA. A diastema behind the first premolar and canine in both jaws.

a. Last inferior premolar different from first true molar.

Last inferior molar with heel; cross-crests of superior molars interrupted . . *Hyracotherium*.

aa. Last inferior premolar like first true molar.

True molars as in *Hyracotherium* .... *Pliolophus*.

II. External lobes of superior molars flat, not well distinguished (*Lophiodontinæ*).

"*A*. No diastema in lower jaw.

Last inferior molar with third lobe .... *Helaletes*."

AA. Lower jaw with diastema.

*No diastema behind first premolar.

a. No inferior premolars like the true molars.

Superior molars 7.

Last inferior molar with heel .... *Heptodon*.

Superior molars 6.

Last inferior molar with heel .... *Lophiodon*.

Last lower molar without heel, no horns...... *Hyrachyus*.

Last lower molar?; "an attachment for a dermal horn on each nasal bone"..... .... *Colonoceras*.

FIG. 31. Part of right maxillary bone of *Heptodon singularis* Cope; from the Wasatch beds of New Mexico, from Captain Wheeler's report, iv, ii, pl. lxvi.

The geographical range of these genera is as follows:

North America only .... *Systemodon*, *Heptodon*, *Helaletes*, *Colonoceras*.

North America and Europe . . *Hyrachyus*, *Hyracotherium*, *Pliolophus*.

Europe only .... *Lophiodon*.

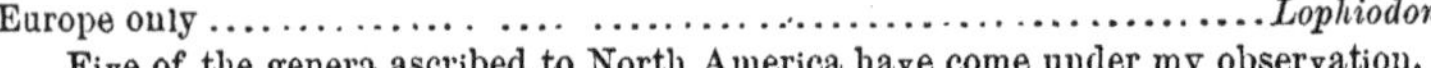

Five of the genera ascribed to North America have come under my observation.

SYSTEMODON Cope.

American Naturalist, 1881, p. 1018. Paleontological Bulletin No. 34, p. 183.

Dental formula: I. $\frac{?}{3}$; C. $\frac{1}{1}$; P-m. $\frac{4}{4}$; M. $\frac{3}{3}$; series in both jaws continuous, or without distinct diastemata. Third and fourth superior premolars with two external and one internal lobe. True molars with continuous transverse crests; external cusps distinct, with regular lenticular section; anterior external cingular cusp well developed. Inferior molars with uninterrupted cross-crests; last with heel. Fourth premolar with but one posterior cusp. Skeleton unknown.

From specimens of the *Hyracotherium tapirinum* m. brought from the Big Horn region I learn that the dental system is different from that characterizing the species of *Hyracotherium*. There is no diastema posterior to the superior canine, while in the latter genus there are two. Anterior to the superior canine there is a considerable one in the *Hyracotherium*. This part is not preserved in any of the specimens of *S. tapirinum*. The characters mentioned have induced me to separate the latter as a type of a distinct genus, *Systemodon*. The position of this genus is at the base of the *Lophiodontidæ*, as regards its dental characters, since the continuous dentition is a primitive condition as compared with the presence of the diastemata seen in *Hyracotherium*. An examination of the figures and descriptions given by Dr. Lemoine of his *Pachynolophus gaudryi* found by him in the neighborhood of Reims, shows that it belongs to the genus *Hyracotherium*. It is therefore distinct from either of the species of *Systemodon*, and is to be compared with the *H. craspedotum* of the Wind River country, with which it agrees in size.

Two species of the genus are known, both from the Wasatch formation. They differ as follows:

Largest; superior molars .032; premolars .030; internal lobe of second superior large. *S. tapirinus.*

Smaller; superior molars .030; premolars .025; internal lobe of second superior small. *S. semihians.*

SYSTEMODON TAPIRINUS Cope.

Paleontological Bulletin No. 34, p. 183, 1881. American Naturalist, 1881, p. 1018. *Hyracotherium tapirinum* Cope. Systematic Catalogue of the Eocene Vertebrata of New Mexico, 1875, p. 20. Report U. S. Geol. Surv. W. of 100th Mer., Capt. G. M. Wheeler, iv. ii. p. 263, Pl. lxvi. figs. 12–16.

Plate LVI; figs. 1–2.

This species was originally discovered in New Mexico by myself, but it was Mr. Wortman who found its locality of greatest abundance, the Big Horn basin of Northern Wyoming.

The *Systemodon tapirinus* may be readily distinguished among the allied forms by its superior dimensions, and by the large size of its premolar teeth. The latter character implies an elongate muzzle, a feature also indicated by the elongate and narrowed symphysis of the lower jaw. The orbit is rather small, and is bounded posteriorly above by a short, sharp decurved postorbital process. The anterior temporal ridge is an angle which extends diagonally to meet the one of the opposite side, probably to form a sagittal crest, which is, however, lost from the specimens in my possession.

The superior canine is lost, but the alveolus has nearly a vertical direction. The first premolar is two-rooted and is situated close to it. The sections of the crown, both vertical and horizontal, are triangular; the anterior edge is acute. The second premolar has a double external cusp, and low anterior and posterior basal lobes. The internal lobe is prominent but not elevated. It is connected with the external angles by a narrow anterior cingulum, and a wide posterior one. No external cingulum. The third superior premolar has much the form of the second, but the internal lobe is more prominent inwards and more elevated. No intermediate tubercles The anterior outline of the base is concave, partly owing to the large size of the anterior external cingular cusps, and the posterior is convex. The fourth premolar is more transverse owing to the greater development of the internal lobe. The anterior border of the base is not concave. There are a strong anterior and a strong posterior basal cingulum, but no external nor internal cingula. The internal cusp is connected with the external lobes by two obtuse cross-crests. No intermediate tubercles, nor external cingulum. The true molars only indicate the primitive intermediate tubercles by a slight thickening of the transverse crest, and this is scarcely perceptible on the last molar. The last-named does not differ in size and proportions from the second molar. All three have a well-developed anterior external cingular cusp, and an anterior and posterior cingulum. There is a short one between the internal cusps, and an external one at the base of the posterior cusp. On the first two molars there is a rudimental one at the base of the anterior cusp also. The enamel of all these teeth, where not worn, is finely wrinkled.

The first premolar, canine, and incisors of the lower jaw form an uninterrupted series. The incisors are closely packed, and have narrow roots and transverse crowns. They are directed at an angle of 45° forwards and upwards. The canine is much larger than the incisors and first premolar, and the base of the crown has a round section. The first premolar has one root with a round section. There is variation in different individuals as to its size relative to the canine, as it is sometimes larger and the canine sometimes smaller. The crown is lenticular in section. It has a posterior basal cingulum with a little apex. The remaining three premolars have bases of about equal anteroposterior extent. The second has a simple compressed and acute crown, while the two others have an internal lateral tubercle. None have an anterior basal tubercle, but the third and fourth have an anterior lobe below the apex. The heels of all three are wide behind, and that of the second is short. The heels of the third and fourth are longer, and have a submedian keel and no tubercles. No cingula on the premolars.

The transverse crests of the inferior true molars are slightly oblique, and their extremities are a little more elevated than the middles, the internal extremity the most so. They all have a well-marked transverse anterior ledge, and the first and second have a posterior cingulum which is thickened at the middle. There is a trace of an external cingulum, but none of an internal one. The heel of the third molar is very large and has an elevated border. Enamel slightly wrinkled where not worn.

The symphysis mandibuli is coössified, and there is a slight diastema behind the first premolar. The other teeth are in a continuous series. At the diastema there is a concavity in the side of the superior part of the ramus, narrowing the alveolar edge. There is a mental foramen below the anterior edge of the second premolar, and another below the anterior edge of the canine, which looks downwards. The masseteric fossa is defined on the side of the ramus by an obtuse angulation of the surface, and anteriorly by a rounded ridge. The inferior edge of the ramus is narrower.

Measurements.

		M.
Length of superior molar series		.0630
Length of premolar series		.0300
Length of base of Pm. i		.0050
Diameters crown of Pm. iii	transverse	.0094
	anteroposterior	.0090
Diameters crown of Pm. iv	transverse	.0110
	anteroposterior	.0090
Diameters M. ii	anteroposterior	.0110
	transverse	.0130

		M.
Length of inferior dental series from I. i to M. iii, inclusive		.0920
Length of inferior molar series		.0780
Diameters base inferior canine	anteroposterior	.0075
	transverse	.0080
Length last three premolars		.0265
Length base Pm. ii		.0090
Length base Pm. iii		.0087
Length inferior true molar series		.0320
Diameters M. ii	anteroposterior	.0106
	transverse	.0076
Diameters M. iii	anteroposterior	.0153
	transverse	.0076
Length of symphysis (oblique)		.0360
Depth of ramus at diastema		.0200
Depth of ramus at Pm. iii		.0230
Depth of ramus at M. ii		.0280
Vertical diameter of orbit (? distorted)		.0285

Jaws and teeth of more than twenty individuals of this species were brought from the Wasatch beds of the Big Horn River by Mr. J. L. Wortman. They vary somewhat in size, some being a little smaller than the individual above described. The latter was as large as a fully grown sheep.

SYSTEMODON SEMIHIANS Cope.

Palæontological Bulletin No. 34, p. 184, Feb. 20, 1882.

Plate LVI; figs. 3–4.

This species was also abundant in the Big-Horn region, jaws and teeth of sixteen individuals having been obtained. Its dimensions are a little smaller than those of the *S. tapirinus*, especially as to the premolar teeth. There is also a short postcanine diastema, which is not seen in the *S. tapirinus*.

The proportions of the maxillary series are represented by a left maxillary and premaxillary bone, with all the teeth in place, but the crowns lost from the first premolar anteriorly. The crowns of the true molars are somewhat worn, so I confine the description of these to the premolars. The third and fourth have considerable transverse extent, the latter being wider than long. The second has scarcely any internal tubercle, but only a low postero-internal heel. The internal tubercle of this tooth is large in *S. tapirinus*. The crown has two cusps, the posterior lower. The last two premolars have two external cusps close together. They have also an anterior external cingular lobe, as in the true molars. There is a posterior external basal lobe in the third premolar, but none or a rudiment on the fourth. No internal cingulum on the premolars. The superior true molars, although worn, show a prominent anterior external basal lobe, and no complete internal cingulum. The base of the crown of the first premolar is narrow anteroposteriorly, and it has two roots as in *S. tapirinus*. It is in close contact with the second premolar, and is separated from the base of the canine by a space a little less than its own anteroposterior diameter, and less than the diameter of the canine. The base of the crown of the latter shows that it is not a large tooth, and has a wide lenticular section. The base of the external incisor is rather large, and is compressed.

Measurements of superior teeth.

		M.
Total length of superior series		.0720
Total length of molar series		.0310
Total length of premolar series		.0250
Diameters base of canine	anteroposterior	.0055
	transverse	.0040
Length of base of Pm. i		.0040
Diameters Pm. iii	anteroposterior	.0070
	transverse	.0078
Diameters Pm. iv	anteroposterior	.0070
	transverse	.0090
Diameters M. iii	anteroposterior	.0100
	transverse	.0125

Some superior molars in better condition than those last described exhibit the following characters: The intermediate tubercles are fused with the internal, forming a continuous cross crest, but their apices are distinguishable. The external cusps are subconical and are well separated. The anterior and posterior cingula are strong, the external is weaker, and it is wanting from the posterior part of the internal base of the crown.

A portion of a mandibular ramus, supporting six molars, presents the following characters: The teeth are a little smaller than those of *S. tapirinus*, the reduction being especially visible in the premolars. The cones of the crowns are more distinctly separated by notches than in that species, and are quite distinctly conic. The anterior ledge of the true molars is distinct, and there is a median posterior tubercle of the first two, which is represented by the wide crenate-edged heel of the third true molar. The anterior-internal cusps of the last two molars are double or bilobed; that of the first is last. The anterior cones of the fourth premolar are subequal, and the posterior external cone is elevated. There is a trace of the posterior internal. There is also an anterior ledge. The heel of the third premolar rises to a median blade and posterior cusp. The anterior cusp is elevated and compressed, and supports a small internal lateral cusp. The base of the crown of the third premolar is elongate. All the teeth are rather compressed, and there is only a trace of an external cingulum.

The ramus is compressed and moderately deep. The dental foramen is large, and its superior border is on a level with the posterior base of the crown of the third true molar. Its inferior base is in line with the base of the crown of the second true molar.

Measurements of mandible.

		M.
Length of last six molars		.0530
Length of true molars		.0310
Diameters third premolar	anteroposterior	.0065
	transverse	.0040
	vertical	.0052
Diameters second true molar	anteroposterior	.0092
	transverse	.0060
	vertical	.0062
Diameters third true molars	anteroposterior	.0120
	transverse	.0060
Depth of ramus at Pm. iii		.0170
Depth of ramus at front of M. iii		.0220

The nearest ally of this species outside of the genus *Systemodon* is probably the *Hyracotherium craspedotum* Cope. This species was brought from the Wind River bad lands, and does not occur in the Big-Horn collection. It is about the size of the *S. semihians*, but is a true *Hyracotherium*, with a diastema behind the first premolar. The strong cingulum which characterizes it is not found in the *S. semihians*, and the inferior molars are wider and more robust.

HYRACOTHERIUM Owen.

Transactions of the Geological Society of London, 1841, pp. 203–208. British Fossil Mammals, pp. 419–423. Gervais Paleontologie Française, p. —. Cope, Report U. S. Geol. and Geog. Expl. Surv. W. of 100th Mer. Capt. G. M. Wheeler, 1877, iv, pt. ii, p. 258.

Dental formula, I. $\frac{3}{3}$; C. $\frac{1}{1}$; Pm. $\frac{4}{4}$; M. $\frac{3}{3}$. Three diastemata in the upper jaw, *i. e.*, behind the third incisor, the canine, and the first premolar, respectively; two in the lower jaw, behind the canine and first premolar, respectively. Superior incisors with flat, transverse crowns. Superior first premolar without inner lobe. All the other molars with two external tubercles in the form of compressed cones. The true molars and last premolar have an additional external tubercle in front of the anterior cone, which is an elevation of the basal cingulum. It is variously developed on the second and third premolars. Second premolar with the inner lobe like a wide cingulum; the third and fourth premolars have one internal tubercle, and one or two smaller ones between it and the external tubercles, the anterior of which forms, on wearing, with the inner tubercle, a single transverse band of dentine. True superior molars with two internal and two intermediate tubercles, which form on wearing, oblique transverse bands. The anterior one of these passes anteriorly to the base of the anterior external cusp.

The first and second inferior premolars have no internal tubercles. The third and fourth have them on the anterior part of the crown, but not on the posterior. The true molars have two pairs of opposite tubercles, the third and last adding a posterior fifth lobe or heel. The internal tubercles of the true molars have a circular section; the external ones are V-shaped in section. The posterior limb of the V is directed toward the internal posterior tubercle; the anterior limb extends diagonally toward the other internal tubercle. The limbs of the anterior V have corresponding directions, the anterior terminating at the anterior border of the crown of the tooth.

The species of *Hyracotherium* differ in the relative developments of the intermediate tubercles of the superior molars, of the distinctness of the tubercles of the inferior molars, of the lengths of the diastemata, and of the development of the cingula.

In this genus the nasal bones are well developed, and join the premaxillaries to form the boundaries of the external nares. There is a strong postorbital process of the os-frontis, which does not inclose the orbit posteriorly by reaching the zygomatic arch. The latter presents no postorbital process upward. The postglenoid process is rather short and obtuse. The meatus auditorius externus has no osseous floor external to the petrous chamber. The paroccipital process is well developed, and is distinct from the smaller posttympanic. There is a sagittal crest. The hamular processes are well developed.

The infraorbital foramen is of medium size The incisive foramina are large. There is an alisphenoid canal. The foramen anterior to its anterior opening I take to be the rotundum, and in front of the latter is a larger foramen, probably the *f. orbitosphenoidale,* but as I cannot find the optic foramen anterior to it, it may include the latter also. It is very probable that the *foramen ovale* is not distinguished from the *f. lacerum anterius.* The *f. postglenoideum* is distinct, grooving the posterior base of the postglenoid process. There is no supraglenoid foramen, but there are two or three postsquamosals. The *f. condyloideum* is quite distinct from the *f. lacerum posterius.* I cannot be sure that there is a postparietal foramen; two near the proper location in *H. venticolum* appearing to be in the squamosal bone. The mastoid, if present, is not enlarged into a fissure, as in various recent genera.

The symphysis of the mandible is coössified at maturity. The condyle is well elevated and the coronoid process is produced.

The cranial characters above enumerated are derived from skulls of *H. venticolum* and *H. craspedotum.* The details are generally much like those of *Anchitherium* (*anceps*), omitting of course the generic and family characters, seen in the double-crested premolars in continuous series In the absence of the supraglenoid foramen we see an approximation to types of *Perissodactyles* other than those of the equine line. The following generic characters in the skeleton are derived from the *H. venticolum.*

Twenty vertebræ of one individual of *H. venticolum* are preserved; of these, three are cervicals, twelve are dorsals, and six are lumbars. The axis has the cylindric odontoid of other primitive types, and the atlantal facets are well separated. There are no transverse or vertical foramina, but the vertebrarterial canal is present. The paradiapophysis is acuminate, and is directed backward. The spine is elevated, and is extended both anteriorly and posteriorly. The third or fourth cervical is rather elongate, and has the vertebrarterial canal. The transverse process is extended both forward and backward, and each extremity is acuminate. The articular faces of the centrum are weakly opisthocœlous, and are very oblique. Neural spinal low. The centra of the dorsals are weakly opisthocœlous. On these vertebræ the interspinous foramina are never cut off from the interspinous intervals, as is the case in *Tapirus*. Except on the first two or three dorsals, the prezygapophyses are flat or convex surfaces directed downward and forward, and not recurved posteriorly or laterally, and separated on the median line by a notch. The neural spines are high, and the diapophyses are short. On the posterior dorsals anterior metapophyses are developed, and the tubercular facets are nearly sessile on their external sides. The metapophyses are separated by a groove from the prezygapophyses. The former send a ridge downward and backward on the last dorsals. On the lumbars the prezygapophyses embrace the postzygapophyses laterally, but do not embrace them in the manner characteristic of the *Ruminantia*. A diapophysial ridge appears and becomes prominent, its base extending the entire length of the centrum.

The axis of this genus is that of a primitive ungulate of any type, and would even pass for that of a carnivore. The later cervicals are characteristically ungulate, but the further subdivision is not indicated. The remaining parts of the column show decided indications of equine rather than tapiroid affinity, in two points. These are: first, the absence of isolated interspinous foramina; second, the narrow form and more revolute articular surfaces of the postzygapophyses.

The scapula has a well-developed incurved coracoid process, resembling in this respect *Coryphodon* and *Anchitherium*, and differing from recent *Perissodactyla*.

Both tuberosities of the humerus are well developed, and inclose a simple bicipital groove, as in *Tapirus* and *Anchitherium*, having no trace of the additional groove seen in the later equine types. The condyles are characterized by the presence of a prominent intertrochlear crest, and by the fact that the condylar surface external to it, does not extend all the way on the posterior aspect. There are no epicondyles. The ulna and radius are distinct. The former is slender, most so distally; the radius maintains its proportions throughout. Its head is a transverse oval, without interlocking angle below. The carpal surface of the ulna is simple, that is, without dividing ridges.

There are eight distinct carpal bones, those of the second series alternating with those of the first, and also with the metacarpals. The first bone of the second row, probably the trapezium, was found displaced to a position behind the others. It carries no facet for a first metacarpal. There are four well-developed metacarpals, of which the V is shorter than the II. They carry the usual keel on the posterior distal face. The phalanges are rather short, and the unguis of digit III is well expanded.

The femur resembles in general the one figured by Owen as that of *Pliolophus vulpiceps*,* and that of *P. vintanus*† figured by myself. The great trochanter projects far above the head, and there is very little neck The *fossa ligamenti teris* is large, as is the trochanteric fossa. The third trochanter is large, but less produced than in *Tapirus* and *Equus*. The rotular crests are about equally elevated, thus resembling *Tapirus* more than other recent *Perissodactyla*. The tibia is long and slender. It has a prominent crest, which is cut off from the external border of the head by a deep notch. The crest is rather more prominent than in recent genera. There are both proximal and distal faces of attachment for the fibula, but I do not find the bone among those preserved. The inner malleolus is short, and its external face is obliquely beveled in front. It carries no facets, and is marked by a wide open vertical groove behind the middle. The trochlear grooves are deep and quite oblique The patella is robust.

The calcaneum is elongate and compressed. The astragalar trochlea is deeply grooved and has subvertical sides. The head is short but not sessile, and carries two distal facets. The external is quite narrow and fits the cuboid bone. The latter is longer than deep and deeper than wide. The peroneal hook is present. The navicular is relatively deeper than in the horse, and is closely united to the ecto- and mesocuneiform. The entocuneiform is a large flat and oval bone, on the posterior side of the tarsus. The mesocuneiform is the smallest. There are only three metatarsal bones, and no rudiment of a fourth. Each one is applied to the entire distal surface of its corresponding cuneiform bone, without oblique articulation with an adjoining one. The distal articular surface of the median metatarsal has

* Quarterly Journ. Geol. Soc., London, xiv, p. 54.

† Report U. S. Geol. Geog. Expl. Surv. W. of 100th Mer., G. M. Wheeler, iv, pl. lxv.

the usual posterior keel. The corresponding extremities of the lateral metatarsals are unsymmetrical, the adjacent part of the condyle being more prominent, and the free portion shortened. In this respect they differ materially from those of *Hyrachyus*, where the lateral metatarsals are nearly symmetrical, having the same character as the median one.

The following comparisons may be made between the limbs of *Hyracotherium* and other *Perissodactyla*. The coracoid process is larger than in either *Triplopus* or *Hyrachyus*. The proximal part of the humerus is a good deal like that of both those genera. In the distal extremity the external condyle is more prominent outwards than in either. The disproportion in diameters of the ulna and radius is greater than in *Hyrachyus*, but not so great as in *Triplopus*. In the carpus *Hyracotherium* is more like *Hyrachyus* than *Triplopus*, in the great inequality between the anterior faces of the trapezoides and magnum, from which it results that the metacarpus II has considerable contact with the latter. In *Triplopus* there is very little such contact, as the trapezoides presents as large an external face as the magnum. The unciform is also wider in the two genera with four metacarpals. There is considerable resemblance between the bones of the posterior limb of *Hyracotherium* and *Hyrachyus*, but the great trochanter of the femur is larger in the former. The proportions are more slender in the *H. ventorum* than in the *Hyrachyus agrestis*, but it cannot be assumed that such a character will distinguish all the species of both genera.

Affinities.—Outside of the close relationships to other genera of *Lophiodontidæ*, there are resemblances to the *Anchitheriidæ*. The anterior cingular tubercle of the superior molar teeth of the family to which *Hyracotherium* is here referred, is less developed in this genus than in most of the others, while it is represented by a ledge in *Anchitherium*. The humerus femur and tibia, are much alike in the two genera; so also is the zygapophysial articulation of the lumbar vertebræ. The elongate coracoid occurs in the *Anchitherium bairdi*. The differences in the feet, the forearm, and the axis are, however, great. The genus *Triplopus* presents an entirely intermediate condition of the fore leg and foot (hind leg unknown), but the crests of the molars are continuous and transverse, a structure which indicates for that genus a position in the line of *Hyracodon* or *Rhinocerus*, rather than *Anchi-*

therium. The ancestral relation of *Hyracotherium* to *Anchitherium*, which I have proposed,[1] seems nevertheless very probable, but requires the intervention of several intermediate forms. In the *Lophiodontidæ*, one of these is *Pliolophus* (*Orotherium* Cope 1873). In the *Chalicotheriidæ* through which the line probably passes, we have *Lambdotherium*, and in the *Palæotheriidæ*, *Anchitherium*.

Species.—I am acquainted with six well-defined species of this genus from the Eocene formations of the United States, and there are probably several others. Two have been described from England. A third reputed species (*H. siderolíticum* Pict.) belongs to another genus, as observed by Kowalevsky, and, as I think likely, to the family *Palæotheriidæ.* Many of the species of *Hyracotherium* are represented by numerous individuals; this is particularly true of the *H. angustidens*, one of the smallest of them. In giving a comparative table of characters of the species, it is necessary to confine it to the madibular rami and teeth, as it is the part most frequently found, and in some species the only part known. The characters thus obtained are as follows:

Large; last molar, 0m.014 in length; ramus, 0m.018 in depth at last premolar; a strong external basal cingulum; little anterior ledge; anterior tubercles well separated. *H. craspedotum.*

Large; last molar, 0m.015; ramus, 0m.020; anterior ledges not prominent; external cingula .... *H. vasacciense.*

Last molar, 0m.012; ramus, 0m.0165; second and third premolars, 0m.0135; external cingula .... *H. ventorum.*

Last molar, 0m.011; first molar, 0m.0087; ramus, 0m.0127; second and third premolars, 0m.0139 .... *H. osbornianum.*

First molar, 0m.0065; ramus, 0m.0130; second and third premolars, 0m.0135. *H. angustidens.*

First molar, 0m.0059; ramus, 0m.0115; second and third premolars, 0m.0080 .. *H. index.*

In the following table the characters of the superior molars are given so far as they can be ascertained. I have not seen them in *H. osbornianum.* and *index.*

a. External basal cingulum weak or wanting; internal strong.
Diastema behind second premolar very short .... *H. craspedotum.*
aa. Both internal and external basal cingula well developed.

[1] Proceed. Amer. Philo. Soc., May, 1873.

[2] "*Orohippus*" *procyoninus* Cope, Ann. Report U. S. Geol. Surv. Terrs., 1872 (1873), p. 647.

Diastema behind second premolar longer .... *H. ventorum.*
Diastema? size of *H. ventorum* .... *H. vassaciense.*
aa. External cingula strong; internal rudimental.
Size less than *H. ventorum* .... *H. angustidens.*

Some names which I formerly associated with this genus are not included in the above list. Thus the *Helotherium procyoninum* must, I think, be referred to the genus *Lambdotherium*. I doubt the pertinence of the *H. cuspidatum* to this genus, though it is apparently a nearly allied form. The *Lophiotherium sylvaticum* Leidy, which I formerly placed here, is probably a *Pliolophus*, having the last inferior premolar like the first true molar. As this tooth is wanting from my specimens, I am not sure of the correctness of my former identification of it with this species, and so retain it in the above table, under the new name *H. osbornianum.*

The species range in size from that of the kit fox (*H. index*) to that of the coyote (*H. craspedotum*). Their geographical and geological distribution is as follows:

WASATCH.	WIND RIVER.	BRIDGER.
H. vassacciense.	*H. craspedotum*	
H. angustidens.	*H. ventorum.*	
H. index.	*H. angustidens.*	*H. osbornianum.*

HYRACOTHERIUM CRASPEDOTUM Cope.

American Naturalist, 1880, p. 747. Bulletin U. S. Geol. Survey Terrs., vi, 1881, p. 199.

Plate LVIII; figs. 1-2.

This species is represented by a large part of both rami of the mandible, and by a crushed skull, which contains the greater number of the molars of both sides. The upper surface of the latter specimen is injured, and all anterior to the interorbital region is wanting.

The *H. craspedotum* is of the size of the *Systemodon tapirinum*, but the tubercles of the inferior molars are not connected by cross-crests, and they all possess a strong external basal cingulum, which also extends round on the posterior base of the I and II true molars. Heel of fourth inferior premolar wide, with a diagonal ridge, two anterior cusps well separated, and no tubercle in front of them. Second premolar with narrow heel, which rises into an acute cusp on the middle of its posterior border. The anterior ledges of the true molars are rudimental. The heel of the last true molar is large and wide; a median posterior tubercle represents it on the penultimate molar. The posterior part of the last molar rises with the base of the coronoid process. Enamel wrinkled.

Measurements.

	M.
Length of inferior molar series	.056
Length of true molars	.033
Length of last molar	.014
Length of first true molar	.009
Depth of ramus at last true molar	.023
Length of diastema behind first premolar	.0035

The superior surface of the skull of the second individual shows a low sagittal crest, formed by the convergence of two gradually approaching temporal ridges, which are continuous with the posterior border of the postorbital processes. The latter are rather long, and have a broadly rounded extremity, which is quite different from the acuminate form of those of the *H. venticolum*. A slight superior angle of the malar bone marks the posterior border of the orbit below. The lateral occipital and squamosal crests are prominent, and send a branch downwards, forming the prominent edge of the posttympanic process, which is thus well distinguished from the surface of the paroccipital. The latter is the longer of the two, and is directed backwards, while the posttympanic points downwards and is separated from the paroccipital to the base. There is no osseous floor to the *neatus auditorius*. The *os petrosum* is flat, and of a suboval outline. The postglenoid process is obtuse, and has a subtriangular section. The basioccipital is posteriorly flat on each side, with a strong median keel; anteriorly the keel disappears, and the borders become more prominent. A short distance within the postglenoid process, the surface is roughened.

The infraorbital foramen opens above the anterior part of the fourth premolar. A strong ridge from the pterygoid process passes upwards and backwards over the alisphenoid canal. The foramina lacera anterius and posterius are rather large and are connected. The f. l. posterius is contracted and terminates anterior to the *f. condyloideum*. The postglenoid foramen is large, and there is no supraglenoid.

The diastema posterior to the first premolar is only as long as the diameter of one of the roots of that tooth, and much less than that anterior to it. The third and fourth premolars have the anterior external basal tubercle well developed, besides the two external cusps of the crown. The latter are subequally developed on the third premolar. The external faces of the molars have only the vertical ridge at the middle of each lobe, but on the true molars, while the anterior one retains its strength, it becomes obsolete on the second and third teeth. The result is that these cusps have, on wearing, a much more crescentic section than on species like *H. venticolum* and *H. angustidens*, where the external convexity remains. There are the merest traces of external cingula on the molars to represent the strong ledges of the other species mentioned. On the other hand, the cingula of the inner bases of the crowns are strong, excepting opposite the posterior inner tubercle, where they are wanting. Anterior and posterior cingula present. The inner lobes of the crowns of the third and fourth premolars are rather wide anteroposteriorly, and the valley between the internal tubercle and the external wall is uninterrupted. It is bounded anteriorly by a ridge connecting the points mentioned, which forms the anterior border of the crown, the anterior cingulum being absent or rudimental. There is a rudiment of the posterior median tubercle on the fourth premolar. Both intermediate tubercles are present on the true molars, the posterior forming a continuous ridge with the posterior inner tubercle, while the anterior is more distinct and is rather triangular in section. It sends a horn to the anterior angle of the anterior external cusp, which is quite distinct from the anterior basal cingulum and its cusp. The external face of the crown is wrinkled; the internal is smooth.

Measurements.

		M.
Length of superior molar series		.0565
Length of superior premolars		.0300
Length of base of first premolar		.0066
Length of base of second premolar		.0078
Width of base of second premolar		.0050
Length of base of third premolar		.0072
Width of base of third premolar		.0080
Diameters base first true molar	anteroposterior	.0088
	transverse	.0100
Diameters base third true molar	anteroposterior	.0090
	transverse	.0108
Length from last molar to and including occipital condyle		.0600
Width between inner borders of postglenoid processes		.0360
Width of basioccipital bone opposite posttympanic processes		.0130

A detailed comparison of this species with the *H. ventorum*, shows, besides the superior size, the following characters: (1) The postglenoid process is every way more robust, and is more obtuse. (2) The basioccipital bone is flatter, and the median keel does not reach the condyles as it does in *H. ventorum.* (3) The paroccipital process is longer. (4) The postorbital process is rounded and not acute.

The dentition of this species is in its dimensions and proportions intermediate between the two species of *Systemodon.* Its three premolars equal four of those of the *S. semihians*, while the molars of the two species are about equal.

The two individuals of this species were found by Mr. Wortman, from the Bad Lands of the Wind River, Wyoming.

A specimen having the proportions of the *H. craspedotum* was found by Mr. Wortman on the Big Horn, but unfortunately it does not exhibit the characteristic cingula of the two dental series. The second superior premolar, like that of *Systemodon semihians*, has no internal tubercle. It is not certain whether there is any diastema posterior to the first superior premolar. I therefore cannot yet ascertain whether this specimen represents an undescribed species of *Systemodon* or *Hyracotherium*, or a strong variety of the *H. craspedotum.* The accompanying inferior true molars are intermediate in size between those of the latter species and the *H. vasacciense.*

Hyracotherium vassacciense Cope.

Expl. Surv. W. of 100th Meridian. G. M. Wheeler, Vol. IV, Part II, p. 264. Plate LXVI, figs. 7-11.

Orohippus vasacciensis Cope, System. Cat. Vert. Eocene New Mexico, U. S. Geog. Survs. W. of 100th M., 1875, p. 21.

Lophiotherium vasacciense, Proc. Amer. Philos. Soc., 1872, p. 474.

The locality from which the typical specimen of this species was procured is near Evanston, Wyoming, in the beds of the Wasatch Epoch. They consist of two inferior molar teeth, one of which is attached to a part of the mandible. The latter tooth had suffered some erosion of one of its cusps, which led to the introduction of some abnormalities into the description. Better specimens from New Mexico furnished the fuller descriptions and the figures given in the report of the Wheeler Survey above cited. As no more perfect specimens having been received since that publication was issued, I do not repeat, but simply refer to it.

A lower jaw from the Big-Horn Basin may belong to this species, or may possibly be a new species of *Pliolophus*. The uncertainty is due to the fact that the crowns of only two molar teeth remain, one or both of which may be premolars. Supposing the anterior tooth to be the fourth premolar, it is shorter than the corresponding tooth in *H. vasacciense*.

The bases of the second and third premolars are of about equal length, and each is considerably shorter than that of the fourth premolar. The crown of the third premolar is remarkable for the size of its internal tubercle, which is more robust at the apex than the external. Its position is very little posterior to the latter. It is unlike the latter in having no ridge descending to the anterior base. There is no distinct anterior basal tubercle. The tubercle of the heel is well developed, and sends a crest forward to the middle of joined anterior tubercles. On the internal side of the heel there is a rudimental internal tubercle. No external cingulum. The fourth premolar has a strong external and posterior cingulum, the latter culminating in a median posterior tubercle.

Measurements.

	M.
Length of Pm. iii and iv	.011
Length of M. i	.007
Width of M. i posteriorly	.0045
Depth of ramus at M. i	.018

Besides the above a second specimen, a part of a lower jaw, was sent from the Big Horn.

HYRACOTHERIUM VENTICOLUM Cope.

Bulletin U. S. Geological Survey Terrs., VI, 1881, p. 198. *Hyracotherium vasacciense* Cope. American Naturalist, 1880, p. 747; not of previous publications.

Plates XLIX *a*, XLIX *b*, and XLIX *c*.

Represented by an entire skull, with many bones of the skeleton, of one individual, and the mandible of a second.

In general, this species is to be distinguished from its near ally, the *H. vasacciense*, by the slender mandibular ramus. The depth of this bone is about equal to that found in the larger varieties of the *H. angustidens*, but the teeth are much larger, having the proportions of those of the *H. vasacciense*. This remark applies especially to the last inferior molar.

The pieces which represent this species in my collection include the entire skeleton with the exception of the pelvis. They have been already mentioned in the course of the description of the genus. They are sufficiently numerous to furnish a good basis for a complete restoration. The skull is somewhat distorted by pressure, and some of the long bones are more or less crushed. In general, the condition of the specimens is good.

The form of the skull is not unlike that of a fox. It is elongate, with the orbit median in position, and the muzzle compressed. The profile is nearly horizontal from the end of the nasal bones to the summit of the occiput, allowing a little convexity for the sagittal crest. The nasal bones project to a point exactly above the anterior border of the premaxillary bone, and the border of the nostril extends as far posteriorly as a point above the posterior border of the canine tooth. The orbit is of medium size, and its anterior border is above the middle of the first true molar. The slight angle of the inner superior border of the malar is a little behind a point above the posterior border of the last molar. The postorbital process is acuminate and quite elongate. Its posterior borders pass into the anterior temporal ridges, and these soon unite to form a moderately elevated sagittal crest. The latter is quite prominent posteriorly owing to the decurvature of the brain-case. The lateral occipital crests are prominent and convex, giving a cordate outline when viewed from above. They contract above the mastoid region, and continue to the extremity of the paroccipital process, and send a branch forwards to the squamosal process of the zygoma, the latter ridge not being prominent. The occiput is concave on each side above each occipital condyle. The paroccipital process is narrow and feeble, and is separated from the posttympanic by a space as wide as the width of the *meatus auditorius externus*, which is considerably wider than the corresponding space in *H. craspedotum.* In that species, the two processes are in contact at the base. The postglenoid processes are short and pyramidal in form; less prominent than in *H. craspedotum.* The glenoid facet is perfectly flat, and has a longer diagonal-longitudinal than transverse diameter. The inner border is sharply defined from the surface which leads to the *foramen lacerum anterius*, but there is no rugose band at this point as in *H. craspedotum.* The basioccipital and basisphenoid bones are strongly convex below, and the former is angulate on the middle line for a short distance. The zygomatic arches are not expanded, and are much compressed, with flat external and internal faces. The ptyerygoid alæ are deep, and support prominent acute hamular processes. Below these there is a longitudinal ridge on the external face, and still posterior is another external posterior ridge situated near the alisphenoid bone, which sends a branch downward at nearly a right angle from its anterior extremity. The sphenoid descending ala originates opposite the middle of the glenoid facet. The maxillary bone does not project much behind the last molar tooth. The palate is injured so that the depth of the notch cannot be ascertained. The side of the face in front of the orbit is injured on both sides. There is a lachrymal tuberosity on the anterior orbital border. The ascending branch of the premaxillary grows wider upward. The borders of the nasal bones are decurved more and more, posteriorly from a flat apex. Viewed from above the apex is rounded, and the lateral outline gently convex. Their common median suture is distinct.

The infraorbital foramen is above the anterior border of the fourth premolar. The other foramina have been generally described under the head of the genus. The condyloid foramen perforates a flat part of the basioccipital behind the foramen lacerum posterius, from which it is well separated. The superior border of the foramen magnum is notched.

The elevated condyle, and full convex outline of the angle, are characteristic of the mandible of this as of other species of the genus. The coronoid process is elevated and has a convex anterior and concave posterior borders. There is no postcondylar tuberosity or crest, and the outline of the angle extends far behind the condyle.

The mandibular ramus is compressed. The ascending ramus rises almost vertically a short distance posterior to the last molar. The symphysis is narrow, and extends to below the middle of the first premolar.

The inferior canines form part of an uninterrupted series with the incisors. The superior canine is separated from the superior incisors by a diastema. The first premolar in both jaws is isolated. The second superior premolars have two cusps, and an internal ledge posteriorly. The third and fourth superior premolars are similar, the fourth displaying a little larger transverse diameter. The true molars are of subequal dimensions. Their external cusps are subconic. All the superior molars except the first and second premolars are entirely surrounded by a basal cingulum, which rises into a low cusp at the anterior external angle of the crown. The third inferior premolar has its two median cusps well separated, and a wide posterior heel. The heel of the last premolar is wider, but carries no internal cusp. The external cusps on all the teeth wear into well-defined Vs. The posterior five inferior molars have an external basal cingulum, but no other.

Measurements of skull.

		M.
Length of consecutive superior molars		.0450
Length of diastema between Pm. i and ii		.0035
Length of second premolar		.0070
Width of second premolar posteriorly		.0050
Diameters fourth premolar	anteroposterior	.0070
	transverse	.0090
Diameters second true molar	anteroposterior	.0085
	transverse	.0112
Length of entire premolar series		.0580
Length of entire inferior true molar series		.0282
Diameters first true molar	anteroposterior	.0080
	transverse	.0060
Diameters last true molar	anteroposterior	.0120
	transverse	.0065
Depth of ramus at fourth premolar		.0165
Depth of ramus at third true molar in front		.0165

Twenty-two vertebræ are more or less perfectly preserved. Three are cervicals, twelve are dorsals, and six are lumbars. The vertebrarterial canal of the atlas issues at the middle of the lower side of the transverse process, in a longitudinal oval fossa. The inferior face between the transverse processes, is slightly concave on each side, and slightly convex in the middle. There is a slight atlantal process. The neural spine of the axis has an acute border, and is elevated behind, and extends forwards over the neural canal to above the middle of the odontoid process. The atlantal facets are subtriangular, and as high as wide. The anterior orifice of the vertebrarterial canal issues between two ridges, which extend from the upper and lower borders respectively of the atlantal facet to the transverse process. Below the inferior ridge the centrum is deeply concave to a high acute keel of the middle line. Between the atlantal facets this keel is abruptly thickened, and it becomes a low obtuse hypapophysis below the posterior articular face. The latter is subcordate, and as wide as deep. The centrum of the second or third cervical is deeply excavated below the transverse processes, leaving the middle line a prominent keel. This is not produced into a hypapophysis of any prominence posteriorly. The transverse process is extended anteriorly to beyond the line of the articular surface. It has no transverse laminar expansion. The anterior face is very oblique, is moderately convex, and is deeper than wide. The neural canal is deeper than wide. The neurapophyses are concave on account of the prominence of an obtuse ridge which connects the zygapophyses. The neural spine is a keel.

The articular faces of the first dorsal vertebra are not oblique, and are moderately opisthocœlous. The centrum has a median inferior angular line, on each side of which the surface is a little concave. The prezygapophyses are inclined at 45°, and the postzygapophyses are horizontal. The tubercular facet is horizontal and projects about as far as the prezygapophysis which is directly above it. The capitular facets are large. The sides of the centra of the dorsal vertebræ which follow, are concave, and the middle line below is narrowed so as to be a ridge. The ridge becomes successively more obtuse posteriorly, becoming on the posterior dorsals an angle, which becomes quite obtuse on the lumbars. The neural canal grows smaller towards the posterior part of the column, its section being subround. On the posterior dorsals, the tubercular rib facet is near the anterior margin of the middle of the neural arch. A ridge passes from it to the posterior capitular facet. Behind the point of its disappearance this ridge continues to connect the two capitular facets. The neural spines of the dorsals are compressed, and present a simple edge posteriorly as well as anteriorly. On the posterior lumbars, the sides of the centrum below the base of the very wide diapophysis, become again concave. The articular faces of the lumbars are plane, and wider than deep.

Measurements of vertebræ.

		M.
Length of centrum of axis on inferior middle line	with odontoid	.033
	without odontoid	.025
Length of base of neural spine of axis		.024
Elevation of neural spine posteriorly		.016
Width of axis at atlantal facets		.0200

			M.
Diameters of posterior articular face	vertical		.0078
	transverse		.0097
Diameters of neural canal anteriorly	vertical		.0070
	transverse		.0082
Expanse of postzygapophyses			.0148
Length centrum of third or fourth cervical, on middle line below			.0150
Diameters of posterior articular face	vertical		.0090
	transverse		.0095
Diameters of neural canal behind	vertical		.0075
	transverse		.0052
Width, including transverse processes			.0170
Width at postzygapophyses			.0190
Diameters of first dorsal centrum	anteroposterior		.0110
	vertical-posterior		.0075
	transverse	posterior	.0075
		median	.0130
Width of first dorsal with diapophysis			.0260
Diameters of Dorsal iv or v	anteroposterior		.0140
	posterior	vertical	.0075
		transverse	.0100
Expanse of prezygapophyses of D. iv or v			.0090
Diameters of a posterior dorsal	anteroposterior		.0130
	posterior	vertical	.0070
		transverse	.0100
Diameters of neural canal behind	vertical		.0060
	transverse		.0060
Diameters of a median lumbar	anteroposterior		.0115
	posterior	vertical	.0085
		transverse	.0120
Elevation of neural arch do			.0050
Width of postzygapophyses of do			.0100

The glenoid cavity of the scapula is a wide oval, and, if completed next the coracoid, would be perfectly symmetrical. The posterior border of the lamina bends strongly backwards at a point below the middle of the length. The spine is nearest the posterior border up to this angle. Neither border of the lamina is thickened near the glenoid cavity, but the posterior becomes obsolete.

Measurements of scapula.

			M.
Diameters of glenoid extremity	transverse		.0130
	anteroposterior	with coracoid	.0200
		without coracoid	.0140
Anteroposterior width of neck			.0140

The humerus is of moderate length, and the shaft is nearly straight. The head is well decurved posteriorly, where it is also narrowed. The external border of the greater tuberosity forms a curved crest of no great prominence, and the surface of muscular insertion of the same is not wider, though much longer, than that of the lesser tuberosity. The latter surface has a subround outline, and is quite smooth, although doubtless the surface of insertion of the subscapularis muscle. The facet for the *teres minor* is a subround concavity. The bicipital groove is short, owing to the early discontinuance of its bounding ridges. Thus the deltoid crest is little prominent. At the distal part of the shaft the inner side is rather flat; the external has only an obtuse indication of the crest usual in many types, *e. g.*, *Hyrachyus agrestis*. In this respect it resembles *Triplopus cubitalis*. There are no epicondyles. The principal or internal condyle is, as usual, a trochlea with bounding ridges, of which the internal is an arc of wider diameter than the external, and is the more acute of the two. It is, however, not expanded as in many other mammals. The internal condyle has an articular surface which is flat transversely and is widest posteriorly. Its external border is obliquely directed inwards and forwards, but turns abruptly inwards at the middle of the inferior or distal curve, thus discontinuing the condyle. Its place is then occupied by an oblique muscular insertion. This external condyle, which applies to an oblique process of the head of the radius, is much more prominent than in either *Hyrachyus eximius* or *Triplopus cubitalis*.

Measurements of humerus.

			M.
Length from edge of great tuberosity			.096
Width of head with tuberosities (greatest)			.022
Diameters of articular surface of head	anteroposterior		.018
	transverse		.018
Diameters of middle of shaft	anteroposterior		.0095
	transverse		.0100
Diameters of condyles	anteroposterior	of inner ridge	.0120
		of outer ridge	.0125
	transverse	of inner condyle	.0090
		of outer condyle	.0060

The radius is about as long as the humerus; it is thus longer than in *Hyrachyus agrestis* and not nearly so long as in *Triplopus cubitalis*. As it has no acute edges, its section is everywhere a transverse, flattened oval. The diameters of the shaft are subequal, increasing a little distally. The head is flattened below and above; and the internal and external borders are gently convex. The portion that applies to the external condyle is an aliform projection with an acute external border. Its articular plane is oblique to that of the rest of the head, extending upwards and forwards. The carpal articular face is three times as long as wide, and is rather contracted. The ulnar facet is rather large, and is sessile. The ulna is rather slender, becoming more so distally. Its section is triangular, interrupted on the middle of the shaft by an angular ridge, which extends from the superior to the inferior borders on the outer side. The olecranon is compressed, deep, and truncate. Its inferior border is a little thickened, and is continuous with the external surface. The inner side is marked by a rather elongate excavation, which fits the radius. The carpal facet is small and subsemicircular in form.

Measurements of ulna and radius.

		M.
Length of ulna		.0116
Length of olecranon		.014
Depth of ulna	at end of olecranon	.012
	at head of radius	.0085
	at middle of shaft	.006
	at carpal extremity	.006
Length of radius		.093
Diameter of head of radius	vertical	.0075
	transverse	.0150
Diameters of radius at middle	vertical	.005
	transverse	.009
Diameters of distal enlargement	vettical	.0100
	transverse	.0145
Diameters of carpal surface	vertical	.0045
	transverse	.0110

Grooves for the extensor tendons are not seen on the superior side of the distal extremity of the radius. The scaphoid part of the carpal surface is very little recurved on the inferior face, and there is a mere trace of the fossa that bounds this recurved surface on the outer side or below the lunar facet in the higher *Ungulata*.

In the *carpus* the scaphoid is the largest bone of the proximal row. Its radial facet has a greater anteroposterior than transverse diameter, and there is a short tuberosity behind it. Its trapezoid facet is a little larger than that joined by the magnum. The radial facet of the lunar is a good deal wider than deep. The posterior tuberosity is large and compressed, but does not project as far posteriorly as that of the cuneiform. Its anterior face is shorter than that of *Triplopus cubitalis*, and has much the form of that of *Hyrachyus agrestis*. The cuneiform has an obliquely descending proximal articular face as in *T. cubitalis*, not a concavely excavated one as in *H. eximius*. It differs from that of the former species in having the pisiform facet largely facing posteriorly instead of proximally. The external face of the unciform is larger than that of the cuneiform, and nearly as large as that of the scaphoid, and is subquadrate in form. Its lunar facet is not large. The lunar and scaphoid facets of the magnum are subequal. It rises in a compressed arch posteriorly, and has the usual long decurved tuberosity posteriorly. The trapezoides has no posterior tuberosity, resembling in this respect the *Hyrachyus eximius*, and differing from the *Triplopus cubitalis*, where it is present. The trapezium is a small bone representing a hemisphere with one end cut off. It has about equal facets for the scaphoid and trapezoides. The pisiform is spoon-shaped with a stout neck, and the two proximal facets. It has a much thinner distal edge than either of the other species above mentioned, which is much more expanded than in the *H. eximius*.

Measurements of carpus.

		M.
Length of carpus at lunar		.0120
Width of carpus at proximal row		.0160
Diameters of scaphoid	length	.0066
	depth	.0100
Diameters of lunar	length	.0076
	median width	.0055
	depth (greatest)	.0080
Diameters of cuneiform	length	.0068
	depth	.0070
Length of pisiform		.0100
Width of pisiform distally		.0060
Diameters of trapezium	length	.0040
	depth	.0045
Diameters trapezoides	length in front	.0035
	width in front	.0035
	depth	.0060
Diameters of magnum	length in front	.0050
	length at middle	.0075
	width in front	.0050
	depth	.0096
Diameters of unciform	length	.0070
	width	.0070
	depth	.0055

The *femur* is much like that of *Pliolophus vulpiceps* figured by Owen,[1] and that of *P. vintanus* figured by myself.[2] The great trochanter is acumi-

[1] Quarterly Journ. Geolog. Soc., XIV, Pl. IV.
[2] Report U. S. G. G. Surv. W. 100th Mer., IV, Pt. II, Pl. XLV.

nate with the apex slightly curved inwards. Its anterior and interior faces are abruptly contracted, the narrowing of the former commencing at a curved transverse ridge which is on a level with the head of the femur. The anterior convexity of the shaft commences at this transverse ridge. The third trochanter is rather prominent, and is recurved. The rotular groove is rather narrow and deep, and its smooth surface is continuous with that of the condyles. There is scarcely a trace of the fossa so marked on the inferior part of the posterior side of the shaft in *Anchitherium*. A part of the shaft above the third trochanter of this femur is lost, so that full measurements cannot be given.

The *tibia* is rather slender. The internal femoral face is wider anteroposteriorly, and the external face is wider transversely. The former has a thin external edge; the latter is separated from the fibular facet by a deep vertical border. The crest terminates below at the end of the proximal third of the bone. The shaft at its middle is flattened on the inner and posterior faces, and is rounded on the external side. A bourrelet surrounds the distal extremity just proximad to the astragalar articulation. It is interrupted on the inner side by the rather wide groove for the tibialis posticus tendon. The anterior and posterior angular extremities of the trochlear crest are of equal length with the internal malleolus.

The following differences may be observed on comparing the bones of the leg above described with those of *Hyrachyus eximius* and *Anchitherium anceps*, which they considerably resemble. In the *Hyrachyus* the great trochanter is less acuminate, and the rotular crests and condyles present greater inequality between the sides. The crest of the tibia is divided in the *Hyrachyus*, entire in *Hyracotherium venticolum*. The internal malleolus is less produced in the former than in the latter. The resemblance between *H. venticolum* and the *Anchitherium* is greater. In the latter the principal difference in the femur is the greater anteroposterior extent of the great trochanter. In the tibia the principal difference is the excavation of the spine of the tibia at its summit, in which the *Anchitherium* resemblest he *Hyrachyus* rather than the *Hyracotherium*.

Measurements of posterior leg.

		M.
Length of femur below middle of third trochanter		.088
Width of femur at head		.032
Anteroposterior diameter of head		.009
Width of rotular groove at middle		.009
Expanse of condyles		.021
Length of patella		.015
Width of patella		.011
Thickness of patella		.008
Length of tibia		.122
Proximal diameter of tibia	anteroposterior	.025
	transverse	.023
Diameters of shaft at middle	anteroposterior	.009
	transverse	.009
Diameters distal end of tibia	anteroposterior	.012
	transverse	.016

The greater part of the posterior foot is preserved, the middles of the shafts of the metatarsals, and some of the phalanges being wanting. The trochlea of the astragalus is narrower and less oblique than in *Hyrachyus eximius*. The ligamentous fossa of its external side is bounded by a vertical ridge in front; the internal fossa is round, except where it runs downwards and forwards to the inferior side of the bone. The internal tuberosity of the head is insignificant. The calcaneum is not wide at the sustentaculum, and has no distal transverse processes. Its posterior distal extremity is subround. The calcaneum is a good deal like that of *Hyrachyus eximius*. The contact of the ectocuneiform and navicular is somewhat serrate, perhaps as the result of an injury. The proximal extremity of the navicular is moderately concave, and there is no tuberosity behind the astragalar facet. The latter has no marginal notch or fossa. As the mesocuneiform is shorter than the ectocuneiform, the second metatarsal commences a little proximad to the others. Its shaft, as well as that of the fourth metatarsal, is absolutely lateral to the median one. The second is narrower than the fourth, but of the same depth. The section of the third is proximally a triangle with convex base and truncate apex. Distally the section of the shaft is a flat transverse oval. The transverse fossa proximad to the condyle is sharply defined at both proximal and distal borders. About half of the distal condyle of each lateral metatarsal is produced; it is beveled so that its ectad edge is the more prominent, representing the keel of the median metatarsal. The contracted portion narrows forwards and holds a sesamoid bone. The proximal phalanges are not elongate. The distal legamentous fossæ are perfectly defined. The lateral phalanges are but little shorter than the median, and were evidently used in progression. They are rather narrower and deeper than the median.

Measurements of posterior foot.

			M.
Length of calcaneum			.036
Depth of tree extremity (vertical)			.009
Width at sustentaculum			.0125
Length of astragalus on inner side			.017
Diameters of trochlea	anteroposterior	external	.013
		median	.010
	transverse		.008
Transverse diameter of head			.010
Diameters of cuboid	longitudinal		.010
	proximal	anteroposterior	.009
		transverse	.0073
Diameters of navicular	longitudinal		.005
	proximal	anteroposterior	.010
		transverse	.0065
Length of ectocuneiform			.006
Width of ectocuneiform			.008
Length of mesocuneiform			.0040
Width of mesocuneiform			.0055
Length of entocuneiform			.0085
Width of entocuneiform			.0070
Diameters second metatarsal	proximal	anteroposterior	.0090
		transverse	.0036
	distal	anteroposterior	.0070
		transverse	.0060
Diameters third metatarsal	proximal	anteroposterior	.0090
		transverse	.0070
	distal condyle	anteroposterior	.0065
		transverse	.0075
Diameters fourth metatarsal	proximal	anteroposterior	.0075
		transverse	.0062
	distal	anteroposterior	.0070
		transverse	.0050
Diameters second proximal phalange	longitudinal		.0110
	proximal	vertical	.066
		transverse	.006
Diameters third proximal phalanges	longitudinal		.012
	proximal	vertical	.006
		transverse	.008
Diameters second median phalange	length		.005
	width		.005
Diameters third median phalange	length		.006
	width		.008

The specimen above described is the most complete representative of the genus *Hyracotherium* that has ever been discovered. I owe it to my assistant, J. L. Wortman, who found it in the bad lands of the Wind River, of Wyoming. A considerable part of the skeleton was extracted from the matrix by Mr. Wortman, and other parts were disengaged after their arrival in Philadelphia. The mandibular ramus of the second individual agrees in characters with those of the type specimen.

Hyracotherium osbornianum Cope.

Orotherium sylvaticum Cope. Annual Report U. S. Geol. Survey Terrs., 1872 (1873), p. 607. *Hyracotherium sylvaticum* Cope. U. S. Expl. Surv. W. of 100th Meridian. G. M. Wheeler, IV, II, p. 262; not *Lopiotherium sylvaticum* Leidy.

Plate XXIV; fig. 23.

This species is represented by a left mandibular ramus, which supports more or less of five molar teeth. The proportions of the jaw are about those of *H. angustidens*, but are rather more slender than the smallest of that species that are known. On the other hand, the molar teeth are of larger size than in any of the varieties of that species. This combination of characters indicates a species distinct from the *H. angustidens*, a fact consistent with the diversity of the horizons in which the two species are found.

The sides of the ramus are very flat, the outer side even a little concave along its middle, and neither side has any prominent convexity along the alveolar borders. The masseteric fossa is scarcely concave, and is only indicated by a little swelling of the external border of the ascending ramus. There is a mental foramen beneath the posterior root of the third premolar. The first true molar is the only one in which the grinding surface is well preserved. The external Vs are well defined, and their angles are directly opposite the internal cones. The latter are both simple, and have a round section. There is a strong anterior basal cingulum, which is joined at the middle by the anterior horn of the anterior V. There is an uninterrupted external basal cingulum, and also a posterior one which is joined by the posterior branch of the posterior V, close to the posterior inner cone. Enamel smooth.

Measurements.

	M.
Length of posterior four molars	.034
Length of fourth premolar	.007
Length of third true molar	.010
Width of third true molar anteriorly	.004
Depth of ramus at front of last true molar	.014

I formerly identified my specimen with the *Pliolophus sylvaticus* of Leidy, with which it nearly agrees in proportions. That species has a wide fourth premolar, while this has a narrow one, as in the species of *Hyracotherium*, although the crown is broken away so that I cannot see the structure of the crown. The crown of the first true molar is represented by Leidy as having an intermediate tubercle, and a double anterior internal cusp. Neither of these characters is seen in *H. osbornianum*.

This species is dedicated to Mr. Henry Osborne, of Princeton, New Jersey, one of the paleontologists of the College of New Jersey.

The type specimen was found by myself on Black's Fork of Green River, Wyoming Territory.

HYRACOTHERIUM ANGUSTIDENS Cope.

Orohippus angustidens Cope. **System. Cat. Vert. Eocene New Mexico, U. S. Geog. Survs. W. of 100th Meridian, 1875, p. 22.** *Hyracotherium angustidens* Cope. **Report U. S. Ex. Surv. W. of 100th Meridian, IV, Part II, p. 265.**

Plate XLIX *a*; fig. 16.

This small species was originally found by myself in the Wasatch formation of New Mexico. I have not yet heard of its occurrence in the Bridger beds, but it is abundant in the Wind River bad lands, and still more so in those of the Big Horn, which are to be referred to the Wasatch epoch. The specimens obtained are, as usual, chiefly lower jaws, but a few maxillary bones and muzzles with teeth, were also found. While these agree in general with those from New Mexico, they present variations in proportions sufficient to require notice.

In a specimen of variety C, below noted, the third and fourth premolars and the first true molar are preserved. The former have considerable anteroposterior diameter, the third relatively the greatest. Its anterior intermediate tubercle thus almost takes the position of a second internal tubercle. There is a rudiment of a posterior intermediate tubercle on the fourth premolar. On both premolars the external cusps are distinct from each other. The anterior external cingular tubercle is well developed.

In the ramus of another specimen of variety C, the anterior cusps of the fourth premolar are equal, and the angle of the outer descends to form an anterior basal cingulum. There is no trace of the inner posterior cusp which is seen in the genus *Pliolophus*. External cingulum scarcely a trace. Enamel nearly smooth. In the true molars the anterior ledge and posterior median tubercle are little developed; the oblique ridges are well developed, and the heel of the last molar elevated. The molars have a strong cingulum on the outer side, and the enamel is slightly rugose. The ramus is slender.

Measurements of a New Mexican specimen.

	M.
Length of the bases of three true molars	0.024
Length of the penultimate molar	0.007
Width of the same	0.005
Depth of the ramus at the last molar	0.0125

The variations from the above standard of measurement found in the Wind River specimens are as follows. There are three sizes which may represent different species, but this cannot be determined without better material:

A. Depth of ramus at last premolar or first true molar, .0130; length of crown of first true molar, .0070; length of last true molar, .0100. Lower jaw of one specimen.

B Depth of ramus, .0140; length of first true molar, .0065; of last molar, .0100. One lower jaw.

C. Depth of ramus, .0155; length of first true molar, .0075; of last true molar, .0100. Two individuals.

Portions of lower jaws of three other individuals in the Wind River collection are apparently referable to the *H. angustidens*. Variety A was obtained in the Big Horn basin north of the Wind River, by Mr. Wortman. The majority of the Big Horn specimens agree with the variety B, but two others occur, one a little smaller and the other a little larger than the average. The former measures: length of last molar, .0090; of first molar, .0067; depth of ramus at M. i, .0120. The dimensions of the larger variety are: length of M. iii, .110; of M. i, .0067; depth of ramus, .0165. The New Mexican forms originally described exhibit combinations of several of these measurements.

HYRACOTHERIUM INDEX Cope.

Report U. S. G. G. Expl. Surv. W. of 100th Mer., G. M. Wheeler, IV, Part II, p. 262. *Orohippus index* Cope. Systematic Catalogue Vertebrate Eocene New Mexico, 1875, p. 20. *Orotherium index* Cope. Paleontological Bulletin No. 17, p. 4, 1873.

Represented by the mandibular ramus with many of the molars in good preservation. These number P. M. 3, M. 3. The last premolar is somewhat like the first molar, but has but one posterior tubercle, and adds a cingular projection in front of the anterior pair. The first premolar has two roots; the second is compressed, and with a broad heel behind. In the molars the anterior tubercles are connected by a cross ridge; the posterior are a little more distinct from each other. The inner anterior tubercle is obtuse but not bifid, and its base is connected with the apex of the posterior outer by a diagonal ridge. There is a small median posterior tubercle on the No. 2, and a large heel on the last molar. It supports a conic tubercle, which is connected by sharp ridges with the tubercles preceding it. There is a cingulum on the outer face of the true molars, which does not extend on the base of the tubercle of the posterior pair.

Measurements.

	M.
Length of molar series	0.0350
Length of first premolar	.0032
Length of third premolar	.0055
Width of third premolar	.0040
Length of second molar	.0065
Width of second molar	.0045
Length of last molar	.0098
Depth of ramus at first premolar	.0021
Depth of ramus at second molar	.0023

I took this species from the bed on Bear River, Wyoming, from which were obtained the typical specimens of *Bathmodon radians*, *Coryphodon latipes*, and *Phenacodus primævus*. It is the typical locality of the Wasatch formation of Hayden. The type specimen having been mislaid, is not now figured.

PLIOLOPHUS Owen.

Quarterly Journal of the Geol. Soc., XIV, 1857, p. 66. *Orotherium* "Marsh" Cope. Ann. Report U. S. Geol. Surv. Terrs., 1872 (73), p. 606. Report U. S. G. G. Expl. Surv. W. 100th Mer., IV, II, p. 252: num Marshii?

This genus differs from *Hyracotherium*, so far as I am able to ascertain, in only one point. This is in the presence of two tubercles on the heel of the fourth premolar tooth, instead of one. Most of the dentition and a good deal of the skeleton of the English *P. vulpiceps* have been described by Owen, and I have described nearly as much of the *P. vintanus* from New Mexico. From these descriptions it is evident that the characters are in general those of *Hyracotherium*. Professor Owens's diagnosis of *Pliolophus* does not specify the generic distinction above mentioned, for the reason that the mandibular dentition of *Hyracotherium* was unknown at the time he wrote. He did, however, give his reasons for distinguishing the genus from the latter, and on this account the generic name must be retained. I formerly adopted for this genus a name introduced by Professor Marsh, but without certainty that he really had the present form in view in proposing it. Professor Marsh mentions the similarity between the fourth premolar and the first true molar. I regarded this statement as a definition of the genus, and therefore adopted it. An examination of the type specimen of *Pliolophus vulpiceps* in the British Museum permitted me by Professor Owen, has satisfied me that the American species I have called *Orotherium* must be placed in the same genus.

Species.—Besides the *P. vulpiceps* from the English Suessonian, four species have been described from North American formations, and a sixth is now added. Several of them are only known from mandibular rami, which are, in two of them, fragmentary. The *Lophiotherium sylvaticum* Leidy probably belongs to this genus, if not to the one to which Dr. Leidy referred it, a point which cannot be positively settled without a more complete specimen. I refer it to the immediate neighborhood of the *P. cinctus* as the most probably correct course. The species differ as follows:

Internal tubercle of third inferior premolar smaller than the external anterior, and posterior in position.

β Anterior inner tubercles of molars bifid.

Depth of ramus at M ii, .017 . . . . *P. vintanus.*

ββ Anterior inner tubercle simple.

Depth of ramus at Pm. iv, .0155; length of Pm. ii, iii and iv, .018. . . . . *P. cristonensis.*

Depth of ramus at Pm. iv, .0135; length of Pm. ii, iii and iv, .0185 (Owen). *P. vulpiceps.*

Depth of ramus at Pm. iv, .012; length of Pm. ii, iii and iv, .015 . . . . *P. lævi.*

αα Internal tubercle of third inferior premolar as large as the external, and not much posterior to it.

Depth of ramus at Pm. iv, .011; length of Pm. iv, .0065; external Vs obtuse.
P. sylvaticus.

Depth of ramus at Pm. iv, .017; length of Pm. iv, .0065; external Vs acute angled.
P. cinctus.

P. sylvaticus and *P. vulpiceps* present an intermediate tubercle in the second inferior true molar, a character not present in *P. vintanus* and *cristonensis.* The peculiar appearances seen in Prof. Owens's figure of the last inferior molar of *P. vulpiceps* are due, according to its describer, to injuries produced in extracting it from the matrix.

This genus occupies an intermediate position between *Hyracotherium* and *Lophiotherium.*

PLIOLOPHUS CINCTUS Cope.

Oligotomus cinctus Cope. Paleontological Bulletin No. 12., p. 2, March 8, 1873. Annual Report U. S. Geol. Surv. Terrs., 1872 (1873), p. 607.

Plate XXIV, fig. 26.

The only indication of the former existence of this species which I possess, is a mandibular ramus which supports the fourth premolar and the first true molar, and contains the alveoli of the other premolars, and the second true molars. There is a pretty long diastema behind the first premolar, and the second premolar was almost as large as the third. The fourth scarcely differs in any respect from the first true molar, a circumstance which led me, when I first described the species, to regard it as a true molar. Both this tooth and the first true molar are characterized by the acuteness of the external cusps and the distinctness of the oblique ridges which extend inwards from them, both anteriorly and posteriorly. The sections of these lobes form Vs, but they are distinct from the internal cusps, so that they are not identical with the molars of the species of the *Chalicotheriidæ.* The anterior inner cusp is slightly bifid. The external base is surrounded by a distinct cingulum, which passes round behind the crown and supports a median tubercle.

The ramus is rather shallow and has flat sides. The mental foramen is below the posterior limit of the diastema.

Measurements.

	M.
Length of molars from second to fifth, inclusive	.026
Length of diastema	.0037
Length of base of Pm. ii	.0050
Length of Pm. iv	.0070
Width of Pm. iv	.0050
Length of M. i	.0070
Width of M. i	.0050
Depth of ramus at Pm. ii	.0126

Found on Cottonwood Creek, Wyoming. From the Bridger epoch.

HEPTODON Cope.

American Naturalist, 1882, p. 1029. *Pachynolophus* "Pom." Cope, Proceeds. Amer. Philos. Soc., 1881, p. 381, (not Pomel).

The dental formula is, I. $\frac{?}{3}$; C. $\frac{1}{1}$; Pm. $\frac{4}{3}$; M. $\frac{3}{3}$. There is in both jaws a considerable diastema between the canine and the anterior premolar, as well as one anterior to the canine. The inferior premolars differ from the true molars, and the last of the latter has a fifth lobe or heel. This genus is in fact *Lophiodon* with seven superior molars.

The type of the genus is *H. ventorum*. It is only in this species that the maxillary dentition is known; I associate in the same genus two similar species known from the inferior dentition. I suspect that the *Hyrachyus singularis* Cope, from the Wasatch epoch of New Mexico, belongs to this genus. It is much smaller than the *P. calciculus*, which in turn is smaller than the *P. ventorum.*

I originally employed Pomel's name *Pachynolophus* for this genus. Professor Kowalevsky's figures and descriptions[1] show plainly that the genera are distinct, that genus being one of the *Chalicotheriidæ*, and allied to *Lambdotherium.*

[1] Monographie der Gutting Anthracotherium, Palaeontographica, XXII, p. 207, 1873. The external lobes of the superior molars are subequal and are separated by a vertical rib. The transverse crests are more or less complete. To this genus Professor Gaudry (Enchainements du Monde Animal), refers the *Propalæotherium* of Gervais, which was founded on *Pachynolophus isselanus*, in 1849. *Pachynolophus* bears date 1847. The European species are the following:

I. Posterior cross crest of superior molars interrupted. *P. duvali* (type); *P. sideroliticus*.

II. Posterior cross crest of superior molars continuous. *P. argentonicus*; *P. isselanus*.

HEPTODON POSTICUS Cope.

Pachynolophus posticus Cope, Paleontological Bulletin, No. 34, p. 187, Dec., 1881 (published Feb. 20, 1882).

Plate LVI; fig. 6.

Both rami of a mandible represent this large species. They are somewhat injured, and the crowns of five of the molars only can be distinctly seen. The latter display the characters seen in the *P. ventorum* and other species of the genus. The transverse crests are well characterized, and the valleys between them uninterrupted. They are closed at the inner extremity by a low ridge nearly at right angles with the cross-crest posterior to them, as in the species of *Rhinocerus*. The anterior of these bounds an anterior ledge, which is quite large on the last true molar. The latter has a rather narrow but prominent heel, which rises posteriorly. The fourth premolar has an anterior ledge, and wide heel with a diagonal crest which is median in front. The third premolar is similar, but smaller. The only cingulum is seen on the anterior part of the external side of all the true molars.

Measurements.

		M.
Length of crowns of posterior six molars		.0700
Length of crowns of true molars		.0440
Diameters Pm. iv	anteroposterior	.0095
	transverse	.0070
Diameters M. ii	anteroposterior	.0130
	transverse	.0095
Diameters M. iii	anteroposterior	.0180
	transverse anteriorly	.0092
Depth of ramus at Pm. ii		.0280
Depth of ramus at M. ii		.0310

From the Big Horn basin and Wasatch epoch; J. L. Wortman.

HEPTODON VENTORUM Cope.

Pachynolophus ventorum Cope, Bulletin U. S. Geol. Surv. Terrs., 1881, VI, p. 197. *Lophiodon ventorum* Cope, American Naturalist, 1880, p. 747.

Plate XXIX*a*; figs. 4, 5.

Portions of the crania of six individuals represent this species. These are mostly fragmentary mandibles and maxillary bones. The most characteristic specimen includes the left maxillary bone almost entire, a part of the right maxillary, and the posterior part of the left ramus of the mandible, with the third true molar in place.

This specimen demonstrates three prominent characters of the species: (1) There is no rib of the external wall of the superior molars dividing the areas that belong to the crescents respectively; (2) the transverse crests of the true molars are perfectly continuous; (3) the first premolar is two-rooted (it is represented as two-rooted, by Pictét, in the *Pachynolophus duvali*). The middle angle or summit of the anterior external crescent of the molars is a trihedral cusp, and it is well separated from the anterior cingular cusp. The transverse crests are curved obliquely backwards. On the premolars it is the posterior transverse crest that is wanting, and the width of the crown is maintained by a posterior basal cingulum not distinct in the true molars The premolars have less anteroposterior extent than the true molars, and they diminish in transverse extent anteriorly, so that the base of the crown of the first is anteroposteriorly wedge-shaped. The last three premolars have two external cusps joined together at the base, the posterior being rudimental on the second premolar. There is a considerable diastema between the canine alveolus and the first premolar.

The transverse crests of the third inferior true molar are without interruption, and the second sends a descending ridge forwards from each end. The heel is small, supporting a single cusp which rises from its posterior margin. It is larger than in the *P. calciculus*.

Measurements.

		M.
Length of maxillary bone from canine		.074
Length of diastema		.016
Length of four premolars on bases		.027
Length of bases of three true molars		.0320
Length of bases of first premolar		.0076
Diameters of second premolar	transverse	.0070
	anteroposterior	.0070
Diameters of fourth premolar	transverse	.0100
	anteroposterior	.0080
Diameters of second true molar	anteroposterior	.0110
	transverse	.0125
Diameters of third true molar	anteroposterior	.0114
	transverse	.0130
Length of heel of third inferior true molar		.0032
Width of second crest of third inferior true molar		.0074

Part of the left mandibular ramus, supporting five molar teeth, served as the first indication of this species. The last true molar is wanting. The second premolar has a short heel with a median cusp rising from its posterior border. The third premolar is injured. The fourth has a complete anterior transverse crest with a slightly concave edge. There is a strong anterior basal cingulum. The heel is wide and supports two cusps, a larger external one and a smaller internal one. The former is trihedral in section and is wrinkled on the inner face. In the other inferior molars the transverse crests have entire edges, and there is a narrow anterior basal cingulum, rather elevated in position. There is also a rudimental posterior one, but none on either internal or external side. Length of molar series, .064; of true molars, .040; of last true molar, .016; depth of ramus at second premolar, .020; at third true molar, .030.

In a third specimen of lower jaw, both of the last molars are preserved. Their cross-crests are entire and with slightly concave edges. The heel is well developed. The only cingulum is a weak one in front. Each extremity of the second cross-crest sends an angular ridge downwards and forwards. Length of last inferior molar, .0155; length of heel, .0040; width at anterior cross-crest, .0087.

HEPTODON CALCICULUS Cope.

Lophiodon calciculus Cope, American Naturalist, 1880, p. 747. *Pachynolophus calciculus* Cope, Bulletin U. S. Geol. Surv. Terrs., VI, 197, Feb. 26, 1881.

Plate XXIX *a*; fig. 6.

The mandibular rami of two individuals of this species are in my collection. One of these contains all the molars except the first (second), with most of the diastema. As the maxillary teeth are unknown, it is not yet certain whether this species belongs to the genus to which I have referred it, or to *Lophiodon*. In one of the lower jaws there is a single-rooted first premolar on the side of the crest of the diastema. It is wanting in another, and is probably an inconstant character.

This species differs from the *P. ventorum* in the possession of a single external posterior crest on the fourth premolar instead of two posterior tubercles. It is also considerably smaller, and the heel of the last inferior molar is smaller. The lateral ridges which descend anteriorly from each transverse crest are very little marked on the true molars, and there are no oblique ridges. There is a rudimental anterior basal cingulum, which is best developed on the last true molar. The heel of the latter is a low tubercle. The only lateral cingula are situated each at the base of the anterior lobe of the crown. In the third premolar the anterior cross-crest exists but is twisted backwards at the inner extremity. The heel is like that of the fourth premolar, and the anterior cingulum is better developed than in any other molar.

The ramus is compressed and rather deep for the size of the molars. The anterior border of the masseteric fossa is not well defined. The symphysis is coössified and is contracted at the diastemata. The mental foramen is below the anterior root of the first (second) premolar.

Measurements.

		M.
Length of inferior molar series		.053
Length of true molar series		.033
Length of last true molar		.014
Width of last true molar anteriorly		.008
Diameters of fourth premolar	anteroposterior	.008
	transverse	.006
Depth of ramus at diastema		.017
Depth of ramus at penultimate molar		.025
Width of symphysis at diastema of No. 2		.0115

The Wind River Eocene of Central Wyoming. J. L. Wortman.

HYRACHYUS Leidy.

Hayden's Report, Geological Survey of Wyoming, 1871, 357. Proceed. Acad. Nat. Sci., Philada., 1871, p. 229. Cope, Proceed. Amer. Philos. Soc., 1873, p. 212. Annual Report U. S. Geol. Survey, 1872 (1873), p. 594. Leidy, Report U. S. Geol. Survey Terrs., I, 1873, p. 60. Scott, Spier, and Osborne, Pal. Report Princeton Scientific Expedn., 1878, p. 49.

Dental formula: I. $\frac{3}{3}$; C. $\frac{1}{1}$; Pm. $\frac{4}{4}$; M. $\frac{3}{3}$. Transverse crests of molars and premolars continuous. Third and fourth premolars with two transverse crests; first and second with one or none. Inferior true molars with two subtransverse crests; premolars without cross-crests, excepting the fourth, which has one corresponding with the anterior crest of the true molars. Last true molar without heel. Inferior canines forming a continuous series with the incisors, but separated from the premolars by a diastema.

Nasal bones produced posteriorly and anteriorly, contracting the anterior nares. *Meatus auditorius externus* not closed below. Posttympanic and paroccipital processes united for part of their length. Mandibular condyle presenting superiorly; symphysis coössified. Postparietal and postsquamosal foramina present; mastoid small, if present; supraglenoid, none.

Atlas with a vertebrarterial canal perforating the base of the transverse process from behind, and issuing on the inferior surface. The vertebral artery then occupied a groove at the anterior base of the transverse process, and entered the neural canal by a large foramen well posterior to the occipital cotylus. Atlas and axis rather short; other cervicals more elongate. Odontoid process very robust, and a little flattened above. Atlantal facets well separated below. Other cervicals strongly opisthocœlous, and with transverse processes which are well extended anteriorly and posteriorly and but little transversely. The diapophysis and parapophysis are quite distinct in the sixth cervical; the former lateral and subcylindric, the latter flat and extended anteroposteriorly and directed downwards. The dorsal vertebræ have well-developed metapophyses. Anteriorly they rise above the diapophyses (tubercular articulations), and, as the latter diminish, stand above the prezygapophyses and increase in elevation. The prezygapophyses of the lumbars embrace the postzygapophyses on the external side, as in the *Equidæ;* more than in the *Tapiridæ*, and much more than *Rhinoceros.* They resemble those of *Hyracotherium* more than any of the types named. The neural spines are well developed on all the dorsal and lumbar vertebræ. The interspinal foramina have no spinal foramina separate from them in the lumbar vertebræ. The sacrum is long and wedge-shaped, and consists of five vertebræ, all with well-developed neural spines. The anterior sacral is articulated to the diapophysis of the last lumbar as well as to the centrum, and the diapophysis of the last lumbar in like manner articulates with the diapophysis of the penultimate. The horse and rhinoceros display similar articulations. In the *Tapirus terrestris* the sacral articulation is less distinct, and that between the last two lumbars is wanting. The centra of the lumbars are plane in *Hyrachyus.*

The anterior ribs are, as usual, wider and flatter than the posterior, and articulate with the sternum, apparently without any intervening hæmapophyses. This is to be inferred from the facts that the ribs extend to the sternal segments and that no traces of hæmopophyses are present. The latter may have, however, been very short and cartilaginous, and have been lost. There are more or less of five sternal segments preserved; they are all much compressed, the præsternum the most so; the præsternum is twice as long as any of the mesosternal segments. In general, the sternum resembles that of the *Rhinocerus* much more than that of the horse.

The scapula has a short but well defined coracoid process. The tuberosities of the humerus are well developed, and the greater is large and incurved. The bicipital groove is simple. The condyles have an inter-trochlear ridge and a small external condyle. The ulna and radius are entirely distinct, and without proximal interlocking connection. The ulna is not very slender, and its carpal articulation is not so small as in *Triplopus* or *Anchitherium.* There are eight carpal bones. The pisiform is quite large; the trapezium is small. The lunar articulates with both the magnum and unciform. There are four digits in the manus, of which the second (first) articulates proximally with the trapezoides and laterally with the magnum. The third (second) articulates proximally with the trapezoides and laterally with the unciform.

The pelvis is rather elongate. The ilium has a contracted peduncle, and the crest is separated from the sacral plate by a long neck. The ischium is without sinuses posteriorly. The femur has the great trochanter produced proximad to the head, and separated from the general surface of the shaft by a basal angle. The third trochanter is well developed and the internal rotular crest is higher than the external. The tibia has a flattened and grooved crest, bounded by a notch on the external side of the head. The fibula is distinct at both extremities, and has a slender shaft.

The calcaneum is compressed and extends distally well beyond the astragalus, thus shortening the cuboid. The trochlear part of the astragalus is wider than in *Hyracotherium*, and the head is shorter, thus resembling the tapiroid and rhinocerontic rather than the equine types. The cuboid facet is distinct. The mesocuneiform is a small bone, and the entocuneiform is large, flat, and posterior in position. There are three posterior digits without rudiment of a fourth in the species studied, *H. eximius.* The second and third metatarsals articulate with one carpal bone only, while the fourth articulates with the two inner cuneiforms, and has a small lateral contact with the ectocuneiform. The phalangeal articulations of the lateral metatarsals are but little unsymmetrical. The median ungual phalange is wide and flat, and has proximal lateral processes. The latter ungual phalanges are unsymmetrical, and have but one, an external proximal process.

The above general characters are derived from the study of one almost perfect skeleton of *H. eximius*.

The affinities they display are most rhinocerontic, then tapirine, and least equine, of the perissodactyle families. The unsymmetrical form of the external wall of the superior molars is represented among living types by the *Rhynocerontidæ* only, while the transverse crests of the inferior molars are shared by that family and the *Tapiridæ*. The form of the sternum is rhinocerontic, while the first are much like those of the tapirs. In the articulation of the lumbar vertebræ *Hyrachyus* is rather more equine.

Species.—Eight species of this genus have been distinguished by Messrs. Scott, Spier, and Osborne,[1] in a satisfactory way. I add another to the list, and copy their analytical table almost entire:

I. A long diastema between lower canines and molars.

aa Superior molars surrounded by a cingulum; enamel wrinkled.

Last molar with two external lobes in nearly the same transverse line.. *H. intermedius.*

Last molar with two external lobes in nearly the same longitudinal line.. *H. modestus.*

aa Superior molars with cingulum complete externally only.

β With anterior cingular cusp larger than anterior lobe.

Cingular cusp separate and no ridge from anterior lobe . ... *H. princeps.*

Cingular cusp united with anterior lobe by a crest .... *H. imperialis.*

ββ With anterior cingular cusp smaller than anterior lobe.

Teeth and jaws larger .... *H. eximius.*

Teeth and jaws smaller .... *H. agrarius.*

II. Inferior diastema short.

No cingula or additional lobes of superior molars; inferior molars with strong anterior cingulum .... *H. implicatus.*

II. Inferior diastema unknown.

a No internal cingulum on superior molars; the anterior crest with a distinct median lobe.

A strong internal projection of the external wall; enamel smooth; an internal ledge. *T. amarorum.*

The *Triplopus amarorum* is mentioned here for comparison with the species of this genus. It resembles them in some portions of its skeleton.

[1] Paleontological Report of the Princeton Scientific Expedition of 1877-1878, p. 49.

Six of the above species have come under my observation, together with two others of whose generic position I am not certain. One of these is quite small and is probably the *H. nanus* of Leidy, which is said by Messrs. Scott, Spier, and Osborne to have only six inferior molars. My specimens are not sufficiently complete to indicate whether the species should be referred to *Pachynolophus*, which possesses that formula, or not. The second is intermediate in size between *H. nanus* and *H. crassidens*, and I have identified it with the *Hyrachyus boöps* of Marsh. Messrs. Scott, Spier, and Osborne state that that species has no diastema, and on this account they refer it to a distinct genus for which they employ the name *Helaletes* Marsh. My specimen does not enable me to verify these references.

Hyrachyus princeps Marsh.

American Journ. Sci. and Arts, 1872, p. 125.

Plate LII; fig. 4.

This, the largest species of the genus, is represented in my collection by the superior molar tooth of an adult specimen, Plate LII, fig. 4; and by a part of the mandibular ramus of another individual of immature age. Both the specimens were found by myself on South Bitter Creek, Wyoming, in beds of the Bridger epoch, of the Washakie basin.

Hyrachyus sp.

? *Hyrachyus eximius* pt. Leidy Report U. S. Geol. Surv. Terrs. I Plate XXVI, figs. 9–10.

Portions of mandibles of two individuals are too large to be referred to the *H. eximius*, and are considerably less in proportions of the teeth than those of *H. princeps.* They may belong to the *H. imperialis* of Scott, Spier, and Osborne, although the measurement of the second inferior molar, the only tooth of that series described by those authors, somewhat exceeds that of my specimens. I give these for comparison.

Measurements.

	M. No. 1.	M. No. 2.
Length of three true inferior molars	.060	.060
Length of second true molar	.021	.022
Width of second true molar	.015	.014
Depth of ramus at second true molar	.045	.038

Specimen No. 2 is young, as the last molar is not protruded beyond the apices of the cross crests.

Both specimens were found by myself in the Bridger basin on Cottonwood Creek, Southwest Wyoming.

HYRACHYUS EXIMIUS, Leidy.

Hayden's Geol. Survey Montana, 1871, p. 361. Annual Report U. S. Geol. Survey Terrs., I, p. 66, 1873; Plate IV, figs. 19–20. Cope, Proceeds. Amer. Philos. Soc., 1873, p. 213.

Plates XXIII*a*, fig. 1; LIII, fig. 3; LIV; LV; LV*a*; LVIII *a*; figs. 5–6.

This tapiroid left numerous remains in the sediments of the lakes of the Bridger epoch. I obtained parts of numerous individuals during my exploration in the Bridger basin during the summer of 1872, but did not obtain any fragments which I could refer to it, from the Washakie basin, whose deposits are supposed to be of the same age. The most characteristic specimens are: (1) An almost entire specimen, which lacks only the muzzle and skull anterior to the glenoid cavities, one fore-leg, and the caudal vertebræ; (2) Both mandibular rami with the atlas and axis; (3) The palatal part of a cranium with worn molar teeth; (4) A ramus mandibuli with four posterior molars, all but the last in good preservation.

The skeleton first enumerated is an unusually complete fossil, and has served as a basis of estimation of the generic characters of *Hyrachyus.* It was discovered by myself standing erect in the side of a bluff at an angle in its escarpment. The entering face of the precipice cut off the nose of the specimen, which stood with regard to the topography like the winged bulls of the ruins of Nineveh. I occupied much time in removing it from its ancient position, and it is now mounted in a life-like attitude in my collection.

Cranium.—In the specimen to be described, the anterior portion from the glenoid cavities is wanting. The sagittal crest is quite elevated, and the lateral occipital quite prominent, and continuous below with the superior margin of the squamosal portion of the zygoma. Four nutritious foramina pierce the parietal bone near its middle and above the paraoccipital process, and two enter the squamosal above the postglenoid process. The paroccipital process approaches near the occipital condyle by its posterior border. I cannot discover the sutural boundaries of the mastoid bone, but that separating the paroccipital process from the post-tympanic process in front of it is distinct. The condyle of the mandible is massive, and the posterior border of the latter extends backward with a slight obliquity. The foramen magnum has prominent supero-lateral margins which are nearly straight, and unite at a right angle above.

Measurements.

	M.
Elevation of sagittal crest above foramen magnum	.045
Width of bifurcation of crest behind	.033
Width of occiput behind meatus auditorius	.070
Width between and inclusive of occipital condyles	.046
Width temporal fossa at meatus	.050
Width meatus auditorius	.012
Width condyle of mandible	.032
Depth of ramus behind	.095

Vertebræ.—In this specimen the vertebræ anterior to the tail are fully preserved. There are seven cervicals, eighteen dorsals, seven lumbars, and five sacrals. As usual the seventh has no vertebrarterial canal. The atlas is deeply incised anteriorly above. It is rather short, and its traverse processes are flat, thin, about as long as broad, and with regular convex distal margin. The arterial foramen issues some distance above and within the notch which marks the anterior base of the transverse process. It enters at the notch at the posterior base. The neural arch is quite convex, and its anterior margin is obtusely rounded. The axis is near the same length, and bears a prominent and elongate laminate neural spine. Its diaparapophysis is narrow and overlaps the parapophysis behind it three-quarters of an inch; it is pierced for the cervical artery. The centra of the third and fourth cervicals are about equal in length to that of the axis, but the remaining ones shorten successively to the seventh, which maintains a length somewhat greater than its width. The parapophyses of these, except the seventh, are flattened, and have considerable anteroposterior extent, their extremities overlapping. A short and rather narrow and stout diapophysis is present on the sixth cervical; on the seventh it is larger, and especially expanded anteroposteriorly at the base and truncate. There is no parapophysis. The fourth, fifth, sixth, and seventh have strongly opisthocœlous centra; that of the third is injured.

Measurements.

	M.
Length of the cervical series	.175
Length of atlas between articular faces	.046
Length of base of transverse process	.035
Length of transverse process	.020
Diameter neural canal in front	.021
Diameter of anterior expanse	.050
Diameter of total expanse	.099
Length axis along basis neural arch	.021
Elevation crest (rectangular) from posterior zygapophysis	.036
Length parapophysis of fifth cervical on margin	.051
Extent zygapophyses of fifth cervical on margin	.048
Expanse zygapophyses of fifth cervical on margin behind	.044
Elevation neural spine of C. vi	.056
Elevation neural spine of C. vii	.075
Length centrum below of C. vii	.028
Diameter of cup, about	.032

The measurements indicate that the neural spines of the sixth and seventh are quite elevated, the latter nearly equal to that of the first dorsal

The spines of the dorsal vertebræ are elevated in the front of the series, rising some distance above the scapulæ. They shorten and widen rapidly from the middle of the series backward. The extremities of all from the scapula posteriorly are turned forward. The metapophyses are conspicuously elevated above the diapophysis on the eleventh dorsal, and on the eighteenth their elevation is about .4 that of the neural spine. The diapophysis is extended beyond the tubercular articulation, on the eighteenth dorsal; the extension and expansion increases rapidly on the lumbars. On the fourth they are as wide at the base as .66 the length of the centrum, and maintain their width, being directed anteriorly. On the sixth and seventh they are still wider and longer, and very thin. They present a projecting transverse surface backward one-fourth the length from the base for articulation with the seventh lumbar and first sacral respectively. The centra of the lumbars are depressed and slightly opisthocœlous, except the last, which is flat. They are contracted and keeled below.

The sacrum is long and narrow, and thoroughly coössified in the specimen. The diapophysis of the first and part of that of the second give attachment to the ilium. The intervertebral foramina are rather small.

Measurements.

	M.
Length of dorsal vertebræ along middles of neural spines	0.420
Length of lumbars do	.298
Length of sacrum along centra	.170
Diameter centrum first dorsal (transverse)	.019
Diameter centrum first dorsal (vertical)	.019
Diameter centrum fifth lumbar (vertical)	.020
Diameter centrum fifth lumbar (transverse)	.0325
Length do	.039
Length diapophysis sixth do	.065
Greatest transverse width of diapophysis sixth lumbar	.030
Length centrum seventh lumbar	.034
Transverse diameter centrum first sacral	.036
Transverse expanse diapophyses do	.086
Transverse diameter end of last sacral	.020
Transverse diameter diapophysis do	.043
Elevation neural spine second dorsal	.095
Elevation neural spine seventh dorsal above scapula	.035
Elevation neural spine eighteenth dorsal (from arch behind)	.037

The *ribs* are long and slender, the first but little expanded distally and united with the præsternum a little behind its middle. They number eighteen, but as the last is quite long, there may have been another pair of shorter ones not yet exposed in the matrix.

Measurements.

	M.
Length first	0.118
Width first, distally	.018
Length eighteenth } from tubercle	.180
Length sixteenth (end broken). } from tubercle	.223

There are four *sternal segments* preserved, with a fragment of another. They are distinct, and the first is the largest. It is a longitudinal plate, placed on edge, with the anterior border strongly excavated. The inferior margins of the succeeding segments are thickened, but the compressed form remains, the section being triangular.

The *scapula* is large for the size of the animal. It has an approximately triangular form, the base being superior. The posterior angle is right, but the anterior regularly rounded. The apex supports the glenoid cavity on a neck which is contracted by a shallow excavation of the anterior margin. The latter is bounded next the glenoid cavity by the short obtuse coracoid, which stands a short distance above the articulation. The spine is long, rather elevated, with a regular convex border curved backward. It rises gradually from a low angular ridge of the middle of the external face of the neck.

Measurements.

	M.
Length of three sternal segments	0.147
Length of first sternal segments	.084
Depth of first sternal segments in front	.044
Width of first sternal segments below	.004
Width of third sternal segments	.015
Length of scapula (median)	.215
Width above (greatest)	.130
Width of neck	.036
With of glenoid cavity	.035

Humerus.—The head is directed a little inside of directly backwards. The bicipital groove is very deep, and the inner tuberosity large and directed forwards. The external tuberosity is much larger, as usual in this group of ungulates, and rises in a hook-like apex above the level of the head. The external bicipital ridge is lateral, and not very prominent, extending on one-third the length of the shaft. The shaft is moderately compressed at the middle, but is transversely flattened below. It is nearly straight. The condyles are narrow, and the inner and outer tuberosities almost wanting; their position is marked by shallow concavities. The external continues in a lateral crest which turns into the shaft below the lower third. The inner condyle is both the widest and most prominent; the external has its carina at its middle, and its external trochlear face oblique and narrow; narrowest behind. The olecranar and coronoid fossæ are deep, and produce a small supra-condylar foramen.

The *ulna* exhibits a large and obtuse olecranon, concave on the external face. Its glenoid cavity is narrowed and elevated behind; in front it widens, and there the ulna receives the transverse proximal end of the radius, which overhangs it on both sides, leaving the little elevations of the right and left coronoid processes about equal. The vertical diameters of the shaft of the ulna are about equal throughout. Its section is triangular, the base being next the radius for the proximal third. This is followed by an edge next the ulna, and the base of the section is on the outer inferior aspect, on account of the direction of an angle from a short distance beyond the outer coronoid process to the base of the ulnar epiphysis, where it disappears. Distally there are two other very obtuse ridges above this one.

The extremity bears two facets, the larger for the cuneiform, the smaller for the pisiform bone.

The *radius* is throughout its length a stouter bone than the ulna, and bears much the greater part of the carpal articulation, viz, with the scaphoid, lunar, and part of the cuneiform bones. This articulation is transverse to that of the ulna, which is thus at one side of and behind it. The head is a transverse oval in section, the narrower end outward. The articular face consists of one and a half trochleæ, the latter wider and internal. The shaft is a transvérse oval in section, with an angular ridge along the middle externally, and the distal part proximally. A broad groove marks the upper face of the epiphysis, where the shaft has a vertical inner face.

Measurements.

	M.
Length humerus (axial)	0.270
Diameter head to bicipital groove	.037
Length along crest outer tuberosity (about)	.052
Transverse diameter, distally	.046
Anteroposterior diameter of inner condyle	.042
Width olecranar fossa	.020
Length ulna	.260
Depth olecranon distally	.027
Depth at coronoid process	.025
Depth of distal end	.019
Depth at middle shaft	.019
Length radius	.200
Width of head	.036
Depth of head	.021
Width shaft at middle	.021
Width near distal end (greatest)	.037
Width distal articulation	.030

The elements of the *carpus* are distinguished for length, and for reduction of width. The anterior faces of all are considerably longer than broad, but the longest faces of the cuneiform, scaphoid, and trapezoides are anteroposterior. The facets are as usual in the carpus; scaphoid $\frac{1}{3}$; lunar $\frac{1}{2}$; cuneiform $\frac{3}{1}$; trapezium $\frac{1}{1}$; trapezoides $\frac{1}{1}$; magnum $\frac{2}{2}$; unciform $\frac{2}{3}$. The cuneiform has a rather L-shaped external face, its ulnar face being strongly excavated in correspondence with the angular extremity of the ulna. This peculiarity is not seen in *Hyracotherium venticolum* and *Triplopus cubitalis.* The pisiform has two proximal facets, and is enlarged and thickened distally; pressed inward, it reaches the scaphoid. The trapezium is a small subdiscoid bone with convex outer face. The magnum is as broad as deep in front, where its surface is swollen; it is produced behind into a spatulate decurved hook. The unciform has a narrow sub-acute hook behind, with wide base.

Measurements.

	M.
Width of carpals of first row together	0.044
Width of lunare, outer face	.016
Depth of lunare, outer face	.020
Depth of cuneiform, outer face	.020
Width of cuneiform, outer face	.020
Length of pisiform, outer face	.030
Depth distally, outer face	.014
Width three carpals of second row	.038
Width magnum, outer face	.015
Depth magnum, outer face	.014
Depth unciform, outer face	.017
Width unciform, outer face	.020
Length unciform, anteroposterior	.021
Length magnum, anteroposterior	.029
Total length of carpals	.040

The *metacarpals* are quite slender. The first only is wanting; the third is rather stouter than the others, while the fourth is considerably the most slender. Its distal extremity is oblique, with prominent median keel, which is wanting on the superior aspect. The proximal facets of these bones are respectively (2d) 2, (3d) 2, (4th) 1, (5th) 1. There is a short, shallow groove near the proximal end of No 3. The phalanges corresponding are lost in the specimen.

Measurements.

	M.
Length of fifth metacarpal	0.070
Estimated length of foot	.187
Distal diameter of fifth metacarpal	.012
Proximal diameter of metacarpal	.007
Proximal diameter of fourth metacarpal	.012
Proximal diameter of third metacarpal	.017
Proximal diameter of second metacarpal	.012

The above are taken on the articular faces transversely.

The *pelvis* is perfectly preserved. The ischium is but little over half as long as the ilium, measuring from the middle of the acetabulum. The ilium is a triradiate bone, the superior or sacral plate rather shorter and wider than that forming the "crest," which is subsimilar to the peduncular portion. The border connecting the sacral plate with the crest is more than twice as long as the crest, and is gently concave. The crest truncates this and the inferior border of the ilium at right angles. The crest expands very slightly distally forward and downward. The ischio-pubic suture is a long one, and the obturator foramen a long oval; the inferior pelvic elements do not form a transverse, but meet at an open angle.

Measurements.

	M.
Length ilium to sacral border	0.130
Length ilium to crest	.180
Width crest	.060
Width peduncle	.030
Length ischium from middle of acetabulum	.110
Width ischium posteriorly	.080
Length obturator foramen	.041
Width obturator foramen	.034
Expanse of ischia above at middle	.076

Femur.—The head projects inward on a well-marked neck. The great trochanter is strongly recurved and presents an anterior tuberosity as well. It rises to an incurved apex much elevated above the head. The prominence of the front of the femur is continued into the front of the trochanter. The outer margin of the shaft is thin, and at a point two-fifths the length from the proximal end is produced into a low third trochanter, which is curved forward and thickened on the margin. The trochlea is well elevated, the inner margin a little the more so, and is narrow. It is continuous with the surface of the inner condyle, which is the shorter and more vertical; the external is longer and divergent; its terminal face is marked by two fossæ, one in front of the other just outside the distal end of the ridge bordering the trochlea. Little trochanter moderate.

Measurements.

	M.
Total length	0.285
Proximal width of head and trochanter	.075
Width from front to edge third trochanter	.050
Width just above condyles	.035
Width of condyles	.058
Chord of outer condyle and trochlea	.060

The *tibia* has a broad prominent crest, which is remarkable in being deeply fissured longitudinally at its superior portion. The tendinous notch separates the outer portion of the crest from the spreading margin of the outer cotyloid face. The crest disappears at the proximal third, and the shaft becomes flattened in front and on the inner side. The distal articular extremity is impressed by $1\frac{2}{3}$ trochleæ, the outer being completed by the fibula. The posterior tuberosity is more nearly median than usual, hence the inner margin of the inner trochlea is low posteriorly, and the inner malleolus has a considerable beveled inferior margin. The *fibula* has a slender shaft, but little compressed. The head is expanded fore and aft, and the malleolus is quite stout.

Measurements.

	M.
Length of tibia	0.244
Diameter from outer angle of head to inner angle of crest	.065
Diameter distal end (greatest)	.035
Diameter articular face, transverse	.027
Diameter articular face, fore and aft	.026

Both *hind feet* are perfectly preserved. The calcaneum is rather elongate and compressed, the lower face truncate with two longitudinal bounding ridges, the outer of which is discontinued before reaching the heel The surface between them is striate grooved. The outer face is slightly concave. The astragaline facets are much expanded inward; the outer is transverse and strongly convex, and separated by a groove from the inner, which is longitudinal and nearly plane. The posterior edge of this and convexity of the outer facets are received into a transverse groove of the posterior part of the lower face of the astragalus. The cuboid facet is diagonal, and is bounded within by a third narrow facet for the astragalus. The astragalus has a strongly convex, deeply grooved trochlea; the convexity extends over 158°. The trochlea is nearly in the vertical, a little oblique to the longitudinal axis of the foot. The exterior malleolar facet is well marked, and bounds a lateral fossa above. The neck of the astragalus is broad and not contracted, but not wider than the trochlea. Its navicular facet is wide and concave, the cuboid narrow, with a long angle behind. The cuboid is quite elongate, and with a narrow anterior face; it has a large posterior tuberosity not projecting much posteriorly. The navicular is flat, with a sigmoid proximal face, convex on the inner side, concave on the outer. It has the three cuneiform facets below, the inner anteroposterior. The entocuneïform is flat with anteroposterior plane, and apex directed backward, and considerably oblique facet for the second metatarsal. The mesocuneïform is much the smaller, and brings the third metatarsus a short distance proximad to the fourth. The ectocuneïform is a little wider than deep. The metatarsals are three, and are rather slender. The two outer are equal in length, and the median but little wider proximally, the increased width being more obvious distally. They have no proximal grooves, and the outer has a low outer tuberosity. The facets of the second row of tarsals are $\frac{1}{1}\frac{1}{1}\frac{1}{1}$. The phalanges, including ungual, are 3, 3, 3. The proximal ones are longer than wide and contracted at the ends; the penultimate are still stouter in form. The ungues of middle line are symmetrical and broad, with the margin a segment of an ovoid, and slight contraction at the neck. The proximal articulation is bounded by a fossa on each side, which is in its turn isolated by the elongate process found in the tapir and in the horse. The margin is marked by radiating striæ separated by grooves, of which the median is the most marked. The lateral ungues are contracted on the inner side, and only possess the proximal fossa and hook on the outer side. The median distal groove is well marked.

Measurements.

	M.
Length of hind foot from heel	0.286
Length of calcaneum	.083
Length of cuboid facet of calcaneum	.024
Depth calcaneum behind	.025
Width calcaneum at astragalus	.035
Greatest axial length of astragalus	.045
Width between trochlear crests astragalus	.022
Length neck between trochlear crests, outer side	.014
Width head	.030
Width navicular	.031
Length navicular at middle	.010
Length cuboid	.022
Depth do. outside	.025
Length ectocuneiform in front	.013
Width ectocuneiform in front	.019
Width mesocuneiform in front	.019
Length mesocuneiform in front	.008
Length entocuneiform at side	.021
Depth entocuneiform at side	.015
Length of metatarsus ii	.102
Length of metatarsus iii	.107
Width of metatarsus ii proximally	.016
Width of metatarsus iii proximally	.020
Width of metatarsus ii distally within fossa	.016
Width of metatarsus iii distally within fossa	.025

	M.
Length median phalanges i	0.025
Width median phalanges ii distally	.015
Depth median phalanges iii distally	.009
Length median phalanges ii	.015
Length median phalanges unguis	.029
Width of articular facet unguis	.014
Width of neck of facet unguis	.021
Width of greatest expanse facet unguis	.029
Length phalanges of metatarsal ii	.060
Length unguis of metatarsal ii	.028
Width unguis (greatest)	.018
Length metatarsus and phalanges iv	.158

Restoration.—The following dimensions may be relied on as a basis for a restoration of this species:

		M.
Length	head	0.220
	vertebral column less tail	.063
	equal 42.1 inches	1.283
Height	of neural spines exposed	.035
	of scapula	.215
	of fore leg	.692
	total 31.05 inclusive	.947
Height	of hind leg	.770
	of elevation of ilium	.135
	total 29.7 inches	.905
Depth of body at middle manubrium		.255
Depth of body at 15th rib		.250

Allowance being made for the obliquity of the humerus, scapula, femur, and ilium, the elevation in life was—

	M.
At the withers (26.6 inches)	.872
At the rump	.762

The size of this species was, then, that of a large sheep.

Comparison of the skeleton with that of Tapirus roulini.—For the opportunity of making this comparison I am indebted to the Smithsonian Institution, which possesses a skeleton of the above species of tapir from Ecuador, presented by President Moreno.

Cranium.—In addition to the generic characters mentioned at the commencement of this description, the *H. eximius* and *T. roulini* differ as follows: In *H. eximius* there is (1) a high sagittal crest which is wanting in *T. roulini*, *T. malayanus*, and approximated in *T. terrestris;* (2) the crest of the squamosal part of the zygoma is continuous with the lateral occipital crest, which is not the case in existing tapirs.

Vertebræ.—(1) The arterial canal of the atlas is not isolated in front as in *T. roulini*, but notches the basis of the transverse process. (2) The axis is longer than in *T. roulini.* (3) The neural spines, and especially the metapophyses of the posterior dorsal vertebræ, are more elevated. (4) The ends of the centra of the lumbars are flatter and more depressed. (5) The diapophyses of the same are wider and longer and thinner, and the penultimate articulates with the last by an angular process, which is not the case in *T. roulini.*

Scapula.—(1) This bone is equal in size to that of a *T. roulini* of considerably greater general dimensions, and is hence relatively larger. (2) The spine is not angulate as in that species, has a larger base, and larger elevated margin. (3) The neck is more contracted, and (4) the coracoid is not recurved as in *T. roulini.* (5) The sinus bounded below by the latter is much shallower, and not bordered above by a recurved hook of the margin.

Humerus—(1) It is relatively smaller in *H. eximius.* (2) The internal bicipital ridge of *T. roulini* is wanting. (3) The external condyle is much shorter, whence its border is nearer its trochlear rib. The *radius* has a narrower head, (1) the external articular plane being shortened. (2) The shaft is wider with a more acute longitudinal lateral ridge medially, and more rounded distal end. The *ulna* is (1) absolutely nearly as long as in *T. roulini*, being thus relatively longer. (2) It has three weak, longitudinal ridges on a convex outer face; in *T roulini* the external face is divided by a very prominent longitudinal angle from the radial cotylus, which spreads distally, sending one angle to the upper and another to the lower base of the distal epiphysis.

Carpus.—This part is (1) absolutely and relatively smaller than in *T. roulini.* (2) The pisiform is more cylindroid distally. (3) The scaphoid is more produced backward on the inner side; the excavation of the inner side is more continued as a concavity of the outer side of the front. (4) The unciform has an acute tuberosity behind; in *T. roulini* it is short, vertical, and obtuse. (5) The trapezoides has a shorter, wider, and more swollen external face. (6) The pisiform is small and convex, instead of being larger and flat.

The *metacarpals* (1) are absolutely and relatively smaller. (2) The inner (II) has a more oblique phalangeal articulation, which is short above and with the keel prolonged upward, instead of being as in *T. roulini*, distal only.

The *pelvis* is distinguished by the much longer plate of the ilium whose extremity constitutes the crest. (1) The crest is also shorter, and more anterior. In *T. roulini* this plate does not so much exceed the sacral plate in length. (2) The pubes and ilia are not so horizontal, but meet at nearly a right angle, and (3) the ischiopubic common suture is considerably longer. (4) The obturator foramen is a more elongate oval.

The *femur* is very similar to that of *T. roulini*, being no smaller in relative size. (1) The great trochanter is wider fore and aft, and with margin more continued on the anterior aspect of the extremity of the shaft. (2) The third trochanter is nearer the middle of the length. (3) The condylar surfaces are continuous with the rotular, not isolated as in *T. roulini*. The latter also (4) lacks the two fossæ on the outer margin of the external seen in *H. eximius*. (5) The rotular groove is also narrower in the latter and not so deeply excavated as in *T. roulini*.

The *tibia* is (1) reduced in size, and especially contracted distally; the relative widths of the ends are 6 cm: 3.5; in *T. roulini* 7.5 cm to 5. (2) The crest is more prominent and is deeply fissured by a groove, which is represented by a shallow concavity in *T. roulini*. The groove (3) external to this is deeper. (4) The posterior inner tuberosity of the distal end is more median, hence the inner trochlear groove is further removed from the anterior inner malleolus, which has, therefore, a greater inner (not outer) extent.

The *tarsus* (1) is generally longer and narrower, except in the case of the cuboid bone (2), which is shorter than in *T. roulini*. (3) The astragalus has a narrower neck, which, therefore, appears more on the inner side. (4) The facet for the cuboid is smaller. The inner tuberosity of the head is more prominent. (6) The calcaneum is more slender, with larger cuboid facet, especially posteriorly. The *metatarsus* is absolutely nearly as long as in *T. roulini*, and, therefore, relatively longer and more slender. (2) The median (III) is nearly similar to the others in width; in the *T. roulini*, much larger than the lateral.

The *phalanges* of the first cross series are more contracted distally.

Thus the more important differences between the skeleton of the two species in addition to those pointed out under the head of the genus, are those of the ulna, the scapula, the lumbar vertebræ, the ilium, and the crest of the tibia. The scapula is more like that of *Tapirus terrestris*, while the ilium is approximated by that of *T. malayanus* among living species; its form leans toward the Equine series, and not to the Palæotheroid.

The following measurements of the inferior dental series are derived from the specimen enumerated at the beginning of the description of this species as No. 2:

	M.
Length of last two superior molars	0.041
Length of last	.019
Width of last	.022
Length of inferior molar series	.095
Length of premolars	.040
Length of last molar	.021
Width of last molar	.013
Depth of ramus at first true molar	.040

HYRACHYUS AGRARIUS Leidy.

Hayden's Report Geol. Surv. Wyoming, 1871, p. 357. Proceed. Academy Phila., 1871, 229; l. c. 1872, 19, 168; Hayden's Report Geol. Surv. Montana, 1872, 361. *Hyrachyus agrestis* Leidy. Hayden's Report Geol. Surv. Wyoming, 1871, 357. *Lophiodon bairdianus* Marsh, Amer. Jour. Sci. Arts, 1871, II p. 3. *Hyrachyus implicatus* Cope, Specimen 1, on some Eocene mammals, etc., Paleontological Bulletin No. 12, p. 5.

This species is nearly abundant as the *H. eximius*, from which it is chiefly distinguished by its inferior size. None of the numerous specimens which I obtained in the Bridger Basin are complete, so that I refer to Leidy for a fuller description than I can furnish. A specimen represented by both maxillary teeth with most of the molars was, with a second and more perfect specimen from the Washakie Basin, referred by me to a distinct species under the name *implicatus*. The second specimen represents a different species from the first, which is, I think, an *H. agrarius*, and I retain for it the new specific name then given.

In *H. agrarius* the superior molars have the cingular cusp smaller than the anterior external lobe, and separated from it by a ridge. A ridge or rib projects into the median valley from the external lobe, forming a loop on wearing. It is more developed in some specimens than others. There are no cingula on these molars excepting a weak one at the external base of the first true molar. The enamel is smooth.

Measurements superior molars.

	M.
Length of five molars	0.071
Length of three posterior molars	.0470
Length of last molar	.0159
Width of last molar	.0200
Width of penultimate molar	.0210
Length of penultimate molar	.0168

The measurements of a mandible of another specimen are rather smaller than those of Leidy's type.

		M.
Length of five posterior molars		.071
Length of true molars		.045
Diameters third premolar	anteroposterior	.0116
	transverse	.007
Diameters of second true molar	anteroposterior	.016
	transverse	.010
Depth of ramus at second true molar		.028

The first-named specimen came from Cottonwood Creek, and the second from Black's Fork of Green River, both in the Bridger Basin.

HYRACHYUS IMPLICATUS Cope.

Paleontological Bulletin, No. 12. (On some Eocene Mammals, etc.), March 8, 1873, p. 5; Specimen No. 2. Annual Report U. S. Geol. Surv., of the Territories, 1872 (1873), p. 605, Spec. No. 2. *Hyrachyus crassidens*, Scott, Spier, and Osborne. Paleontological Report Princeton Scientific Expedi., 877, 1878, p. 52.

Plate LVIII; figs. 6, 7.

This tapiroid species as now defined is represented by an imperfect skull in my collection. The part preserved is one side of the face with all the superior molar teeth, with the right half of the mandible supporting all the teeth. The superciliary and other orbital borders of the right side are also preserved. The specimen agrees in its proportions, and in all the details, with the description given by Princeton paleontologists of their *H. crassidens.*

The superior molar teeth have no external or internal cingula, but the anterior and posterior are present except on the first premolar on the second and third premolars, and where there is no anterior cingulum. The premolars, except the first, have two uninterrupted cross-crests, the anterior of which is the longer, and curves round the internal extremity of the posterior; to it in the fourth, and beyond it in the second and third. The cross-crests of the true molars are uninterrupted and regular. There is no tubercle between them, and the ridge from the external wall is not prominent. The anterior cingular cusp is low and is in contact with the base of the much higher anterior lobe. The second external lobe of the first true molar has a short vertical ridge near its apex. This ridge is almost as well developed as the anterior ridge on the premolars.

The inferior molars are distinguished for the width of the anterior basal cingulum, and the large anteroposterior diameter of the second and third premolars. The diastema is also very short, measuring only three-fifths the length of the bases of the first three premolars, while it equals the latter measurement in the *H. agrarius.* The canine tooth is not much larger than the third incisor, is implanted close to it, and has an antero-superior direction.

The mandibular ramus is not deep, and the symphysis is narrowed, strongly convex, and rather short. The infraorbital foramen opens above the posterior part of the third superior premolar. The orbit is entirely lateral, and its anteroposterior diameter is a little greater than its vertical. The superciliary border is rather thick, is somewhat prominent. There is no postorbital process, but the superciliary border curves round and passes with a trace of an angle into the anterior border of the temporal fossa, which for a short distance is transverse. The temporal ridges are weak in their anterior parts. The skull is broken off before their union into a sagittal crest.

Measurements.

	M.
Length of superior molar series	0.085
Length of true molars	.046
Length of penultimate	.015
Width of penultimate	.019
Length of inferior molar series	.078
Length of inferior true molars	.047
Length of penultimate	.017
Width of penultimate	.011
Width of last premolar	.008
Length of last premolar	.012
Depth of ramus at last premolar	.0235
Length of diastema	.019
Length of bases of three incisors	.018

From the Bridger Eocene of the Washakie Basin near South Bitter Creek, Wyoming

TRIPLOPODIDÆ.

Cope, American Naturalist, 1881, April (March 25), p. 340.

But one genus of this family is known at present, but the number will probably be increased when the structure of the feet of various imperfectly known species is ascertained.

TRIPLOPUS Cope.

American Naturalist, 1880, p. 383 (April 27.)

Dental formula, I. ?; C. $\frac{1}{1}$; Pm.; $\frac{4}{4}$; M. $\frac{3}{3}$; a considerable diastema anterior to the first premolar. Molars with only two vertical external ridges, viz: the anterior cingular and the approximated median of the anterior crescent. Transverse crests two, uninterrupted and rather oblique; a ? third and short crest on the posterior base of the first true molar. Premolars different from molars, the third and fourth with two transverse crests. Inferior molars with two transverse crests, as in *Lophiodon*, the last without heel.

An ossified inferior wall of the meatus auditorius externus. Posttympanic and paroccipital processes distinct from each other. No postorbital arch. Postparietal and mastoid foramina preserved; the latter large. Cervical vertebræ rather long; axis with subcylindric odontoid process. Scapula with small coronoid process. Great tuberosity of humerus long, curved. No trochlear crest on condyles of humerus; epicondyles rudimental. Ulna and radius distinct throughout their length; ulnar articulation with carpus, small. Trapezoid bone of carpus with a facet for the trapezium. Unciform with two inferior facets. Metacarpals three principal ones, and one, the fifth, rudimental; the distal extremities of the second and fifth opposite; the third a little longer.

The dentition of this genus is nearly that of *Hyrachyus*. The only exception is the possible third transverse crest of the first true molars.[1] The other portions of the skeleton known are also much like those of *Hyrachyus*, with the exception of the number of digits of the anterior foot. The en-

[1] This point is further considered in the description of the species.

tirely rudimental character of the fifth metacarpal, which, with its digit, is so well developed in *Hyrachyus*, places *Triplopus* in another family, and in another line of descent. I think that it must be regarded as one of the forms of the series connecting the lophiodonts with the rhinoceroses. The fourth digit (the fifth) was retained by the earliest type of rhinoceros in Europe, the genus *Aceratherium*, but in America it appears to have been lost earlier. None of the American rhinoceroses of the Lower Miocene, of the genus *Cœnopus* Cope, present it, and in the present genus we have an ancestral type of the Eocene period, in which the last digit is already lost. The premolars of different structure from the true molars exclude this genus from the *Rhinocerontidæ*, and with the character of the feet place it between that family and the *Lophiodontidæ*.

As yet but one species of *Triplopus* is certainly known, but a second is placed in it provisionally.

TRIPLOPUS CUBITALIS Cope.

American Naturalist, 1880, p. 383. Proceedings Amer. Philos. Soc., 1881, p. 383.
Plates LV*a*, figs. 10–12: LVI*a*.

This species is represented by a nearly entire skull with lower jaw; most of the cervical vertebræ; a left anterior limb nearly complete; a part of the left scapula, and a part of the right anterior limb; all belonging to one animal. The specimen was not quite adult, as the last superior molar is just protruding its crown through the maxillary wall, and the last two superior milk premolars still remain in place, much worn and closely pressed by the overlying successional teeth.

The *cranium* is peculiar in its wide orbital region, and short compressed muzzle; the latter is damaged in the specimen, so that the form of the nasal bones cannot be determined except at their proximal portions. The interorbital space is plane in both directions, and rises very gently posteriorly. The sagittal crest is narrow and low, until above the meatus auditorius, where it rises Above the posttympanic process it bifurcates, and each rounded lateral lobe extends posteriorly to a point above the occipital condyles. Viewed from above the head is wide between the zygomatic fossæ, and at the posterior premaxillary teeth. The top of the muzzle narrows rapidly above the latter, but does not contract below until the first premolar is reached. The zygomatic arch is not convex along its middle, and incloses a narrow fossa. The superciliary border is prominent, and nearly straight, and is bounded by a notch behind. The squamoso-occipital ridge is well marked. The posttympanic process is shorter than the paroccipital, and is separated from it by an open shallow groove, which is probably bottomed by the mastoid bone. The paroccipital process is much narrowed below and is turned a little outwards. There are two closely adjacent tubercles on the anterior border of the orbit, probably on the lachrymal bone.

Foramina.—Only a few of these are well preserved; among the lost is the *f. infraorbitale.* There are two postparietal foramina on one side, and one on the other, above the point of origin of the zygomatic process of the squamosal bone; and one in the usual posterior position. The postsquamosal has the same anterior position as the anterior postparietals, being immediately below them; I cannot discover whether there is a posterior one or not, owing to injuries to the specimen. There is apparently a fissure-like one on the parieto-squamosal suture posteriorly. The mastoid is quite large, expanding downwards and outwards; it is not so large as in a tapir, but much exceeds that in *Hyrachyus eximius.* The *meatus auditorius externus* is large, and occupies only the posterior part of the space between the postglenoid and posttympanic processes. It is inclosed anteriorly and below by the border of a wide element which may be tympanic. It incloses the petrous bone below in a bulla; as, however, the inner portion of the best preserved one is broken away, I cannot speak of its relations to the basioccipital bone. The *foramen lacerum posterius* is reduced to a jugular and perhaps another connected foramen by the close apposition of the petrous bone to the basioccipital for a considerable distance. The region of the *f. l. medius* is injured. Posterior to the *f. l. posterius* is a foramen opposite the base of the paroccipital process, anterior to the usual position of the *f. condyloideum.*

Mandible.—The angle of the lower jaw is produced posteriorly, as in some species of *Hyrachyus:* cfr., fig. 1, pl. LVI *a*, and fig. 2, pl. LIII. The coronoid process is long and is curved backwards to above the posterior border of the condyle. There is no tuberosity behind the condyle.

The symphysis is quite contracted and is short. The mental foramen is below the middle of the inferior diastema. The ramus is compressed and at the same time strong.

Dentition.—As the deciduous third and fourth premolar teeth, in a worn condition, remained in the maxillary bone, I removed them from one side, thus displaying the crowns of the corresponding permanent teeth. The first premolar may belong to the permanent dentition; the second is the deciduous. The former has two roots. The crown is cutting for a short distance anteriorly, but posteriorly it expands into a heel, much less developed than the internal lobe of the succeeding teeth The crowns of the third and fourth premolars differ externally, as well as in their crests, from those of the true molars. The median-anterior and cingular vertical ridges are not so prominent as in the latter. The external crest is not divided into two by the notch in its grinding face. The anterior cross-crest, at its inner or distal extremity, is turned shortly backwards and then inwards, giving a "pot-hook" outline to the triturating surface. The fourth deciduous premolar presents a peculiar character already ascribed to the first true molar. This consists of a crest running parallel with the posterior transverse crest and close to it, along its posterior side. It forms the border of the tooth for a short distance, but as its direction is slightly obliquely forwards as well as outwards, the posterior cingulum appears for a very short distance.

The first true molar is subquadrate in outline. The anterior transverse crest commences at the middle anterior ridge, and is first transverse, then directed a little obliquely backwards. The second crest commences at the apex of the posterior external crescent, leaving a wide posterior marginal fossa. Its internal extremity is broken off. Posterior to, and in contact with it, the posterior cingulum rises in a crest, which occupies the internal half only of the border. Its inner border is imperfect. It appears to me to be probable that the normal posterior crest is turned posteriorly on itself so as to give the "pot-hook" shape seen in the anterior crest of the fourth permanent premolar. The corresponding accessory crest in the fourth temporary premolar appears to have been distinct at its internal extremity. The second true molar has a more oblique posterior external crest, and the posterior internal is oblique and simple. It has narrow anterior and posterior basal cingula. There is no tubercle between the inner bases of the transverse crests of this or the last true molar. The latter is characterized by the rudimental character of the posterior external crescent crest, which is shortened like that of *Hyrachyus.* The transverse crests are curved backwards; the posterior is short and simple.

The canines are small, and are directed forwards. The extremity of the muzzle being broken, the relation of the incisors cannot be stated, but there was not probably any precanine diastema. An incisor preserved has the crown transversely expanded, and rather oblique.

The third and fourth inferior premolars are the deciduous ones, and are both three-lobed, but differ in the forms of the anterior lobe. In the third, it is narrow and incurved, as in the corresponding permanent teeth of some *Artiodactyla.* The transverse crests of the true molars are rather oblique, running forwards as well as outwards.

Their external extremities are bent at right angles, and there results a short descending crest running forwards and inwards; the anterior one turns inwards again, forming a transverse anterior ledge. No cingula on internal or external bases of crown; a rudimental posterior one.

Measurements of cranium.

		M.
Length from front of canine tooth to end of occipital condyles		.128
Length from same to postglenoid process		.096
Length from same to end of last molar		.069
Length from same to first premolar		.015
Length from same to line of front of orbit		.044
Width between superciliary borders		.046
Width of zygomata at orbits		.064
Width of brain-case at glenoid surface		.048
Width of occipital condyles		.023
Width of basioccipital bone between *ossa petrosa*		.006
Distance between postglenoid and post-tympanic processes		.014
Depth of occiput behind		.033
Depth of mandible from condyle		.040
Depth of mandibular ramus at third premolar		.014
Depth at diastema (axial)		.009
Least width of symphysis		.011
Diameter crown third permanent premolar	anteroposterior	.007
	transverse	.005
Anteroposterior diameter crown first premolar		.0045
Diameters crown first true molar	anteroposterior	.010
	transverse	.012
Diameters crown second inferior true molar	anteroposterior	.011
	transverse	.0075
Diameter of root of inferior canine near crown		.0035

Vertebræ.—The atlas is about as long relatively as that of the horse. Its transverse processes have more anteroposterior than transverse extent. The summit of the neural arch has a median ridge separating two grooves. The inferior surface of the centrum has a nearly median, obtuse hypapophysis The axial facets are well separated below. The vertebrarterial canal pierces the base of the transverse process behind and below, and notches it deeply anteriorly. Above this notch the usual perforation of the arch is present. The axis is not relatively quite so long as that of the horse; it is a little longer than in *Hyrachyus eximius*, but rather shorter than in *Hyracodon arcidens* (Pl. CII, Fig. 7). The atlantal facets are spread well apart, and the articulating surface of the odontoid does not connect with them. The latter is rather long, is obtuse, and slightly recurved; it has no raised borders. Between the atlantal faces the inferior surface is plane. Posterior to this the middle line bears a prominent keel. The diapophyses are long, narrow, and recurved, and each is pierced at the base by the vertebrarterial canal. The posterior articular face is but little concave, and a little oblique, and is a little wider than long.

The succeeding cervicals regularly diminish in length, and become more strongly opisthocœlous, the seventh having quite a ball in front. The sixth has a slender diapophysis directed posteriorly, and quite distinct from the wide and long parapophysis which is directed downwards and outwards. The posterior angle of the latter extends as far back as the centrum. The seventh has only a flat transverse diapophysis. The first dorsal has a very stout diapophysis excavated below for the rib tubercle. The diapophyses of the third and fourth dorsals are not so stout. The capitular fossæ are large. The centra of the anterior dorsals are flattened below; they are concealed in part by the matrix in this specimen. The neural spine of the sixth cervical is narrow, and is directed forwards. That of the seventh is vertical, and narrows rapidly from a base which is rather wide anteroposteriorly. The spines of the dorsals are wider, and are directed gently posteriorly; they are probably long, judging from the size of their bases.

Measurements of vertebræ.

	M.
Length centrum of atlas on side	.027
Length centrum of atlas below	.010

	M.
Width centrum of atlas below posteriorly	.030
Width transverse process of atlas	.010
Vertical diameter neural and odontoid canal	.015
Length axis to odontoid process	.033
Length odontoid process	.007
Diam. centrum behind, vertical, with hypapophysis	.012
Diam. centrum behind, vertical, without hypapophysis	.009
Diam. centrum behind, transverse	.0115
Length of centrum of fifth cervical	.030
Length of centrum of seventh	.017
Length of centrum of second dorsal	.014
Anteroposterior diameter of base of neural spine of second dorsal	.010
Expanse of head and tubercle of first rib	.012

Fore limb.—The greater part of the blade of the scapula is lost. The neck is stout, and the coracoid is a short aliform process. The humerus is moderately robust, most so proximally. The greater tuberosity is a strongly incurved crest, with truncate summit, which is a little elevated above the plane of the head, from which it rises rather abruptly. The bicipital ridges are not strong nor prominent. The olecranar fossa is deeper than the coronoid fossa, and they communicate by perforation. The inner part of the condyle is the largest, and forms an acute angle with the interior epicondylar surface. The exterior part of the condyle is divided by an oblique angle of the surface separating an external beveled band of the same, which narrows to extinction on the posterior side. As compared with the humerus of *Hyrachyus eximius*, that of *Triplopus cubitalis* is very similar, differing mainly in two points at the distal extremity. The olecranar fossa is smaller and is less excavated, and its lateral bounding ridges are of unequal elevation ; in *T. cubitalis* they are equal.

The ulna and radius are more than one-fourth longer than those of *H. eximius.* Although they are entirely distinct throughout, the ulna is quite slender anterior to the proximal third. The shaft is much more slender than that of *Hyrachyus eximius.* The olecranon is compressed, deep, and truncate behind. The distal epiphysis is remarkable for its length, being twice as long as that of the radius. The head of the radius is subequally divided by fossæ, the external being the shallower. The inferior or ulnar facet is regularly and gently convex downwards, and is bounded behind by a roughened ridge, which near the external border turns backwards to the humeral border. The shaft of the radius is robust and flattened. The carpal facet of the radius is contracted, and has three times the superficial area of that of the ulna. The scapholunar dividing ridge is present, but is very low. The scaphoid face is the more excavated, and then rolls backwards, forming a very narrow posterior facet, which is narrower than that found in the species of *Anchitherium*. There is no distinct fossa on its inner or posterior border, as in many ungulates. The trapezium and scaphoid are the only bones of the carpus which are wanting The latter is probably wider than long or deep, while both the lunar and cuneiform are longer than wide. The cuneiform has not its external border excavated; its proximal surface is oblique and continuous, the ulnar and pisiform facets being in line. The pisiform is large, and is enlarged distally; its proximal facets are equal. The exposed face of the trapezoides is rather larger than that of the magnum, and is nearly as large as its own face of contact with the latter. The magnum has the usual great anteroposterior extension, with elevated posterior convexity applied to the fossa of the lunar. Its posterior process is long, nearly equal to the rest of the bone, and is depressed and flattened distally. The metacarpal facet is very concave. The unciform's anterior or exposed face is a little longer than wide. Its two proximal facets are about equal. It is about as deep as wide, and extends half its length distad to the magnum. Its posterior process is rather narrow; it is narrow and abruptly decurved Distally, the facet for the fifth metacarpal is well marked, and has about half the area of that for the fourth metacarpal. The functional metacarpals are of moderate length as compared with the elongation of the ulnoradius. The third is largely in contact proximally with the unciform as well as with the magnum. The condyles are stout, and each is laterally impressed by a fossa. The second and fourth have chiefly lateral presentation, but are not much narrower in the shaft than the median metacarpal. The first phalange of the lateral digit is a little shorter than that of the median, while the seconds are of equal length. The extremity of the second digit reaches the proximal third of the length of the median ungual phalange. The fissure of the ungual phalange reaches the middle of its length. The fifth metatarsal is proximally rather stout; but it soon contracts to a thin rounded extremity at only one-fifth the length of the fourth.

Measurements of fore limb.

		M.
Anteroposterior diameter of cotyloid cavity of scapula		.015
Diameter of head of humerus	transverse	.020
	anteroposterior	.019
Diameter with greater tuberosity		.030
Length of humerus on outer side		.110
Diameter humerus at epicondyles	transverse	.021
	anteroposterior externally	.015
Length of ulna		.165
Length of radius		.141
Depth of olecranon, distally		.015
Width of ulna at coronoid		.015
Width of ulna at carpal facet (greatest)		.007
Width of radius at head		.016
Width of radius at carpal facets		.014
Width of radius at widest point distally		.016
Length of carpus at magnum		.015
Length of carpus at unciform		.018
Length of lunar		.010
Depth of lunar		.011
Length of magnum		.005
Depth of magnum		.017
Length unciform		.009
Width unciform		.009
Depth unciform (total)		.014
Depth unciform of inferior facets		.007
Length of third metacarpal		.068
Proximal diameter third metacarpal	anteroposterior	.008
	transverse	.008
Length of fifth metacarpal		.012
Length of median series of phalanges		.027
Length of first median phalange		.010
Width of first median phalange, proximally		.008
Length of second phalange		.006
Widths of median ungual phalange	proximally	.0070
	medially	.0055
	greatest	.007

The body of this animal was about the size of that of a red fox. The legs were more slender or elevated, and the head of course was shorter and thick.

The unique specimen on which our knowledge of this species rests was cut from a block of calcareous sandstone of the bed of the Washakie basin of the Bridger Epoch, near South Bitter Creek, Wyoming Territory. The bones are generally in the relation of the position in which the animal died. The neck is depressed and the left fore leg raised so as to be in contact with it, and the head is raised so as to clear the left wrist.

Triplopus amarorum Cope.

Proceeds. Amer. Philos. Soc., 1881, p. 389.

Plate LV, figs. 6–9, and LVIII *a*, fig. 2.

The characters of the fore foot of this species being unknown, it is not possible to determine its generic position. It has, however, one of the well-marked characteristics of the genus *Triplopus*, in the osseous inclosure of the *meatus auditorius externus*, through the ossification of the external prolongation of the otic bulla and tympanic cartilage. I cannot, therefore, refer it to *Hyrachyus*.

It is represented by a skull from which a large part of both maxillary bones and the mandible have been lost, and which is accompanied by parts of the ulna and radius, parts of the ilium, a femur, and tibia, and nearly all of the posterior foot of the right side. The posterior parts of both maxillary bones remain, and they support each, the last superior molar tooth from which the external wall has been broken away. The portions of molars remaining exhibit characters which lead me to suspect that the species does not belong to *Hyrachyus*. The anterior cross-crest of the molar preserved, is lobate, resembling the same ridge in the species of *Anchitherium*. The posterior cross-crest is uninterrupted. If this species possesses affinity with *Anchitherium*, it will perhaps possess three digits of the manus, in which case it will be referred to the *Triplopidæ*, in harmony with the indication furnished by the ear structure.

The *Triplopus amarorum* is much larger than the *T. cubitalis*, equaling the *Hyracodon nebrascensis*. It differs from the *T. cubitalis* in the stronger temporal ridges and more elevated sagittal crest; also in the shorter post-tympanic process. The internal lobes of the last superior molar are connected by a basal ledge, not found in the *T. cubitalis*.

The interorbital space is wide and flat, and is most expanded at the postorbital angles.

From this point the face contracts rapidly forward. From the same angle it contracts abruptly posteriorly to the rather narrow brain case. The anterior temporal ridges are nearly transverse near the postorbital processes, and then converge more gradually, uniting opposite the posterior inferior border of the zygomatic fossa. The elevated sagittal crest diverges into two lateral supraoccipital crests, which contract as they descend, and continue to the extremities of the post-tympanic processes. Although the postorbital angles are prominent, they cannot be called processes. The paroccipital processes are large, and are directed vertically downwards. They are separated by the usual concavity from the occipital condyles. The posttympanics are very short, forming only an angle projecting downwards at the anterior base of the paroccipitals, from which they are only separated by a notch. The inferior side of the tympanic bone is flat near the meatus, but opposite the stylomastoid fossa its posterior border is turned forwards, and is produced into a well-marked process. It incloses a groove in front of it, which is continuous with the pterygoid fossa. The petrous bone is not inflated, and its inferior surface is divided into two longitudinal ridges. The inner is the less prominent, and is in close contact with the basioccipital. The postglenoid processes are robust and obtuse. The basioccipital is excavated in front of each of the condyles. The inferior surface is nearly flat, with a slight median keel. The pterygoid fossa is well defined, and is long and narrow. The posterior nareal trough is elongate, the descending pterygoid processes of the sphenoid originating as far back as the apex of the *os petrosum*. This species is especially characterized by the presence of an acute keel-like ridge, which extends horizontally above the foramina sphenoörbitale and opticum, and turns upwards anterior to the latter, terminating a half inch below the inferior base of the postorbital process. All the foramina are below it, but there is a fossa above it, opposite the interspace between the *f. opticum* and *f. sphenoörbitale.*

A supraorbital foramen pierces the frontal bone a quarter of an inch within the superciliary border. There are five or six postparietal foramina, two of which are nearly on the squamosal suture There is a postsquamosal foramen, and also a not very small supraglenoid foramen. There is a small foramen anterior to the optic, and in line with the posterior part of the postfrontal angle. The *foramen opticum* is large, and is 10^{mm}. in front of the *f. sphenoörbitale.* The latter is separated by a lamina from the large and vertically oval *f. rotundum.* The latter is joined by the large alisphenoid canal, whose posterior orifice is as large as the foramen ovale. The latter is large, and is well separated from the *f. lacerum anterius.* The *f. f. lacera* are well closed up, the *posterius* being reduced to what is probably the jugular foramen. The *f. condyloideum* is large, and is an anteroposteriorly placed oval. Its anterior extremity is opposite to and well separated from the *f. jugulare.*

The nasal bones are spread out posteriorly, and their posterior extremities are truncate. The coronal suture passes downwards at the narrowest part of the cranium behind the postfrontal angles. The squamosal bone does not reach the frontal. The parietal does not extend so far posteriorly as the lateral occipital crests, except near the squamosal.

The characters of the last superior molars have already been mentioned. The posterior transverse crest is uninterrupted, but the anterior consists of closely united internal and median lobes. The division is marked on the posterior side, and on the edge of the crest; the anterior face is plane. The longitudinal external crest sends a strong protuberance into the head of the valley, which is grooved on its surface. There is a strong anterior basal cingulum which rises to an anterior cusp. On numerous surfaces the enamel is slightly rugose. The inferior canine teeth are in continuous series with the incisors, and are slightly larger than they.

Measurements of skull.

		M.
Length from line connecting anterior borders of orbits to occipital crest		0.132
Length from line connecting posterior borders of orbits to occipital crest		.100
Width between postorbital angles		.100
Width between anterior borders of orbits		.076
Elevation of occiput		.065
Width between mastoid ridges		.065
Width between *ossa petrosa* at middle		.018
Diameters third superior true molar	anteroposterior	.0200
	transverse	.0205
Diameters second superior true molar (base)	anteroposterior	.0200
	transverse	.0150

The portion of ilium remaining exhibits a rather narrow neck and a concave external face. A fragment of the femur shows a prominent third trochanter, with an obtusely rounded apex. The distal part of the fibula is not coössified with the tibia. Its shaft is exceedingly slender. The angles bounding the trochlear grooves and ridges of the tibia are of subequal lengths. The median ridge is rather wide; the inner malleolus is narrow, has no distal facets and no distinct tendinous grooves externally.

The posterior foot is both relatively and absolutely smaller than that of *Hyrachyus eximius*. The trochlea of the astragalus is narrower and more deeply grooved. The crests are obtuse, and not so narrowed as in *Anchitherium bairdi*, nor are the malleolar facets of the astragalus so sharply defined as in the latter species. The external ligamentous fossa is, however, deep, and is bounded anteriorly by a low trihedral tuberosity not found in the *A. bairdi*. The head of the astragalus is not sessile as in *A. bairdi*, and has rather the proportions of *H. eximius*. The cuboid facet is a bevel of the external side of the distal extremity, as in *H. eximius*, and is not on a produced ledge, as in *A. bairdi*. The internal tuberosity of the head is not as much developed as in either of the species named. The navicular face of the astragalus is horizontally divided by a shallow ligamentous fossa. The calcaneum is much like that of *Hyrachyus eximius*. The cuboid face is less oblique than in that species, in the anteroposterior direction, and is less crescentic in outline than in *A. bairdi*. The sustentaculum is rather more extended transversely than in *H. eximius*, but resembles that species more than the *A. bairdi*, in wanting the deep groove at its base on the inferior side, which cuts it off from the rest of the calcaneum. The remainder of the inferior surface is flat, and not grooved for a tendon as in *H. eximius*.

The remainder of the tarsus includes the usual five bones, the three cuneiforms being present. They are in general a good deal like the corresponding bone of *Hyrachyus eximius*. The navicular differs in having a low transverse ridge on its proximal face, which fits the groove of the astragalus already mentioned. The hook of the cuboid is large. The external (anterior) face of the mesocuneiform has one-third the superficial area of the anterior face of the ectocuneiform. The entocuneiform is rather large, and is flat and subsemicircular. Its position is externo-posterior. The ectocuneiform presents facets to both the second and fourth metatarsals, that with the latter the largest. The distal halves of the metatarsals are lost. At their proximal portions they are of subequal width, as in *Hyrachyus eximius*, but the lateral ones are rather narrower at the middles of the shafts.

Measurements.

	M.
Width of distal extremity of tibia	.029
Width of astragalar face of tibia	.019

	M.
Length of inner malleolus	.007
Length of astragalus on inner side	.030
Depth of trochlea on inner side	.017
Depth of head on inner side	.0145
Width of trochlea	.015
Width of navicular facet	.0195
Length of head from inner crest of trochlea	.005
Length of calcaneum	.058
Length of free part of calcaneum	.037
Distal depth of the calcaneum	.016
Diameters cuboid face calcaneum { anteroposterior	.0145
Diameters cuboid face calcaneum { transverse	.0145
Length of navicular	.008
Length of cuboid	.0145
Transverse proximal width of three metatarsals	.027
Diameters of second metatarsal { anteroposterior	.014
Diameters of second metatarsal { transverse	.007
Antero-posterior diameter of third metatarsal	.0145
Diameters of fourth metatarsal { anteroposterior	.014
Diameters of fourth metatarsal { transverse	.012

This species was obtained in 1873 from the bad lands of South Bitter Creek, Wyoming, from the Washakie basin of the Bridger formation. The locality is the same as that which furnished the *Triplopus cubitalis*, the *Achænodon insolens*, etc.

HYRACODONTIDÆ.

This family, which I characterized in 1879, includes, so far as yet known, the single genus *Hyracodon*, which is found in the Oligocene White River formation of North America. According to Marsh, the digits of this genus number three on both anterior and posterior limbs. It has a full series of incisor teeth in both jaws.

RHINOCERIDÆ.

This extensive family has left representatives in all parts of the Northern Hemisphere, and species still exist in the Old World. From the following table the range of variation of its genera can be readily seen:

I. Four anterior digits.

Incisors $\frac{2}{1}$; canine $\frac{0}{1}$; no horn; posttympanic bone distinct . . . . *Aceratherium*.

II. Three anterior digits.

a. Posttympanic process not coössified with postglenoid.

Incisors $\frac{2}{1}$; canines $\frac{0}{1}$; no dermal horn . . . . *Cænopus*.

Incisors $\frac{1}{1}$; canines $\frac{0}{1}$; no dermal horn . . . . *Aphelops*.

Incisors $\frac{0}{1}$; canines $\frac{0}{1}$; no dermal horn . . . . *Peraceras*.

Incisors $\frac{1}{1}$; canines $\frac{0}{1}$; a tuberosity for a dermal horn on each nasal bone. *Diceratherium.*
Incisors $\frac{1}{1}$; canines $\frac{0}{1}$; a median dermal-nasal horn .... *Ceratorhinus.*
Incisors $\frac{3}{2}$; canines $\frac{0}{1}$ .... *Zalabis.*
Incisors $\frac{0}{0}$; canines $\frac{0}{0}$; dermal horn median; no osseous nasal septum ...... *Atelodus.*

aa. Posttympanic process coösified with postglenoid;

Incisors $\frac{1}{1}$; canine $\frac{0}{1}$; dermal horn median; nasal septum not ossified..... *Rhinocerus.*
Incisors $\frac{0}{0}$; canine $\frac{0}{0}$; dermal horn median; nasal septum ossified .... *Cœlodonta.*

It can readily be seen that the genera above defined form a graduated series, the steps of which are measured principally by successive modifications of four different parts of the skeleton. These are, first, the reduction of the number of the toes of the anterior foot; second, the reduction in the number and development of the canine and incisor teeth; third, the degree of closure of the meatus auditorius externus below; and, fourth, in the development of the dermal horns of the nose and its supports. While these characters have that tangible and measurable quantity which renders them available for generic diagnosis, there are others which possess a similar significance, and which I have noticed in an article published in the Bulletin of the U. S. Geological Survey of the Territories for September, 1879.

This series may be represented in genealogical relation as follows:[1]

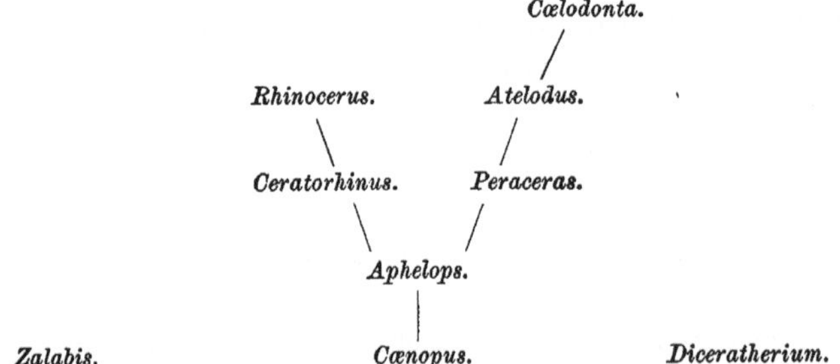

The early type, which corresponds most nearly with *Cœnopus*, and which preceded both it and the *Aceratheria* in time, is the genus *Triplopus* Cope, which has left a species in the Upper Bridger of Wyoming. Here the incisors are probably $\frac{3}{3}$ and the canines $\frac{1}{1}$. This formula is that of the Eocene tapirs, where the normal numbers $\frac{3}{3}\cdot\frac{1}{1}$ prevail. *Triplopus* further

[1] See American Naturalist, 1880, p. 611.

differs in the primitive condition of the premolars above, which, as in the *Lophiodontidæ*, differ from the molars in their greater simplicity. Thus it is probable that tapiroids, probably *Lophiodontidæ*, gave origin to the *Rhinoceridæ*, as Marsh has suggested. And it is further altogether probable that the general type of dentition presented by the *Rhinoceridæ*, *Lophiodontidæ*, etc., which I have named the palæotheriodont, took its origin from the type which is intermediate between it and the bunodont, viz, the symborodont, as I have pointed out in an essay on this subject.

The first appearance of dermal horns was apparently in a pair placed transversely on the nasal bones, in species of Eocene *Lophiodontidæ* of the genus *Colonoceras*. The same character has been observed by Marsh in species of the Lower Miocene, which probably belong to the true *Rhinoceridæ*, and which he has called *Diceratherium*. This genus appears to have terminated the line exhibiting this structure, and the family in North America remained without horn. As we have seen, the types possessing the median horn arose in Europe, in the *Ceratorhinus schleiermacheri* of the Middle Miocene, and still survives.

It may be observed, in conclusion, that a successive increase of size in the species of this line has taken place in North America with the advance of geologic time. Thus, their probable ancestors of the genus *Triplopus* were the least of all. The *Cænopoda* of the White River formation were larger; the oldest, *C. mite*, being the smallest. The *Aphelopes* of the Loup River or Upper Miocene formation were all larger, and were nearly equal to the large existing species.

TAPIRIDÆ.

The genera of this family are not numerous as yet. The oldest, *Desmatotherium*, appears in the Middle Eocene (Bridger), and *Tapirus* is first found in the Upper Miocene (Epplesheim). The recent species of the family belong to *Tapirus* L., and *Elasmognathus* (Gill). A small species, the *Tapirulus hyracinus* Gerv., is from a bed at Perreal, France, which Pictet has identified with the gypsum of Paris (Oligocene). It is sometimes referred to this family, but is not sufficiently well known to determine its position. In America a genus which has not yet been distinguished from *Tapirus* is found in the Miocenes.

The three genera are distinguished as follows:

Two superior premolars different from true molars............ ......*Desmatotherium.*
One superior premolar different from true molars; no heel of third inferior molar; nasal septum cartilaginous .... *Tapirus.*
Like *Tapirus*, but nasal septum osseous....... ....*Elasmognathus.*

CHALICOTHERIIDÆ.

Gill; Cope, American Naturalist, 1881, p. 340.

This family had numerous representatives during Eocene time, and a few species of *Chalicotherium* extended into Miocene time. The boundaries which separate the family from the *Lophiodontidæ* on the one hand and the *Menodontidæ* on the other are not always easy to determine. From the former the symmetrically developed external Vs of the superior molars, and the double Vs of the inferior molars, distinguish it. Yet in *Pachynolophus* the external Vs are not so well distinguished as in other *Chalicotheriidæ;* and the anterior cingular cusp produces a part of the asymmetry found in the *Lophiodontidæ.* The character of the double inner cusps of the superior premolars which distinguish the *Menodontidæ* is only applicable to the last premolar in *Diplacodon* of the latter, while a trace of the additional cusp of this tooth is found in the Chalicotheroid *Nestoritherium.*

In using the following table it must be borne in mind that the number of the toes has been determined in a very few of the genera. Should any of them prove to have but three digits on the anterior foot, such genera must be referred to a new family intermediate between this one and the *Palæotheriidæ.*

I. Internal cones of superior molars separate from external lobes.
A. External tubercles of superior molars separated externally by a conic cusp.
Fourth inferior premolar like first true molar ....*Ectocion.*
AA. External Vs of superior molars separated by an external ridge.
a. Incisors present.
β. No diastema in front of second inferior premolar.
Second premolar without inner lobe; last molar with one inner cone...*Leurocephalus.*
Second premolar with inner cone; last superior molar with one inner cone..*Palæosyops.*
Second premolar with inner cone; last superior molar with two inner cones..*Limnohyus.*
ββ. A diastema in front of second inferior premolar.
Two inner cones of last superior molar..............................*Lambdotherium.*
aa. Incisors absent from both jaws.
Last superior molar with one internal cone.... ....*Nestoritherium.*
II. One or both internal cusps of superior molars united with the external lobes by cross-crests.

a. External cusps of superior molars more or less conic.
An antero-external cingular cusp .... *Pachynolophus.*
aa. External lobes of superior molars, inflected Vs.
β. No crescentic inner lobes.
No intermediate lobes .... *Chalicotherium.*

The following regions have thus far furnished species of the above-mentioned genera:

Europe—*Pachynolophus, Chalicotherium.*

North America—*Ectocion, Leurocephalus, Palæosyops, Limnohyus, Lambdotherium.*

Asia—*Nestoritherium.*

Of the American genera, *Leurocephalus* S. S. and O. has been found by the Princeton Exploring Expedition of 1877 in the Bridger formation, but I have not met with it myself. *Ectocion* Cope is known from a single species found in the Wasatch formation of Wyoming.

ECTOCION Cope.

American Naturalist, June (May 20), 1882, p. 522.

Superior molars supporting eight cusps, viz, two internal, two intermediate, and two principal external, together with two which arise from the external cingulum—one opposite the space between the principal external cusps and one at the anterior external angle of the crown. Fourth superior premolar with two external and one internal cusp. The inferior molars supporting four alternating cusps, which are connected by oblique ridges, thus forming one V and more or less of a second anterior one. Fourth premolar with two posterior cusps. Third premolar with one posterior cusp. Skeleton unknown.

Until the feet of this genus are known, it will not be possible to locate it with certainty. Its dental characters have considerable resemblance to those of *Protogonia*, and should the feet have the characters of the *Condylarthra*, its position will be near that genus. It is distinguished from *Protogonia* by the two-lobed fourth superior premolar, and from *Phenacodus* by the Vs of the inferior molars.

The characters of the molars are especially interesting as displaying the type which no doubt gave origin to the prevalent type of the *Perisso-*

dactyle dentition, and demonstrates the correctness of the position which I assumed in the year 1874 on this subject. This was, in brief, that the lophodont type of molar has been derived from the bunodont by confluence of tubercles. In the family of the *Chalicotheriidæ* the genus *Pachynolophus* exhibits the transverse cross-crests and external Vs characteristic of so many genera of the order. In *Lambdotherium* these crests are partly broken up, and in *Palæosyops* entirely so. In all of these genera the external Vs are complete, though little indented in *Pachynolophus*. But in *Ectocion* these also are broken up into a median cusp representing the apex, and a lateral external one, which corresponds with the prominent lateral edge of the V. A little confluence of these elements will give the type of *Lambdotherium*. Were the median external marginal cusps wanting, the superior molars would be much like those of *Hyracotherium*, with less tendency to confluence of the median cusps than in most of the species of that genus. The superior molars are also much like those of *Acoëssus*, but the premolars are different.

But one species is known to me—

Ectocion osbornianum Cope.

Oligotomus osbornianus Cope, Paleontological Bulletin No. 34, 1881, December 20, p. 182.
Plate XXV *e*; figs. 9–10.

The true molars of both maxillary bones, with the fourth premolar of one side, are preserved more or less perfectly, with four inferior molars on two fragments of the lower jaw.

The external tubercles of the superior molars are nearly erect, and have a lenticular section. The rib which separates their external faces is prominent, and terminates in a free apex. The base of each face is marked by a strong cingulum, but the posterior one is very short. There is a strong anterior basal cingulum, but no posterior or internal one. The anterior inner tubercle is larger than the posterior. The intermediate tubercles are subround, and are anterior to the transverse line of the interior ones. They do not join the latter, excepting after very considerable wear. The external anterior cingular cusp is rather more prominent on the first than on the second true molar. The fourth superior premolar has a well-marked external anterior cingular cusp, which is, however, low; and there is no ridge dividing the external faces of the external cusps. The single inner cusp is connected with the two external by two ridges, which diverge as they extend outwards. The anterior supports a tubercle close within the anterior external. There are strong anterior and posterior basal cingula and weak external and internal ones.

The third inferior premolar has a compressed ridge on the heel. The fourth premolar is like a true molar, with the anterior inner cusp well developed and elevated, and connected with the anterior and posterior external by oblique ridges. The inner posterior cusp is less conic in form than in the true molars, and the entire crown is somewhat contracted anteriorly. The true molars are characterized by the presence of a small median tubercle on the posterior border. There is a low external basal cingulum, which is wanting opposite the posterior cusp. Enamel generally smooth.

In my original description I stated that there is a diastema in front of the second inferior premolar. This is an error based on a deceptive appearance of the jaw.

Measurements.

		M.
Length of superior true molar series		.0210
Diameters of superior M. ii	anteroposterior	.0080
	transverse	.0097
Diameters of Pm. iv	anteroposterior	.0085
	transverse	.0085
Length from inferior Pm. iii to M. ii, inclusive		.0290
Diameters of Pm. iv	anteroposterior	.0080
	transverse	.0050
Diameters of inferior M. ii	anteroposterior	.0075
	transverse	.0060
Depth of ramus between Pm. iii and Pm. iv		.0150

This species was, to judge from the size of its teeth, about the size of a red fox. The specimens of it above described were found by Mr. J. L. Wortman in the bad lands of the Big-Horn River, Wyoming. It is dedicated to my friend, Henry L. Osborne, of Princeton College, New Jersey.

PALÆOSYOPS Leidy.

Hayden's Geological Survey of Montana, 1871, p. 358 (?). Proceedings Academy Natural Sciences, Philadelphia, 1871, p. 118. Report of the U. S. Geol. Survey of the Territories, I, p. 27, Cope. Annual Report U. S. Geol. Surv. Terrs., 1872 (1873), p. 591. *Limnohyus* Marsh, American Journal Science and Arts, 1872, p. 124.

This genus has been partially described by Professor Leidy, and much light is thrown on its structure by the materials obtained by the Princeton

Scientific Expedition of 1877. As pointed out by Leidy, this genus differs from *Palæotherium* in the isolation of the internal cones of the superior molars from the external longitudinal crescentoid crests, and in the presence of but one inner tubercle on the last three premolars instead of two. There is but one internal cone on the last superior molar. Formula, I. 3, C. 1, Pm. 4, M. 3. Number of inferior molars similar. Inferior true molars, with four acute tubercles alternating in pairs and connected by oblique crests, which thus form two Vs, with their apices exterior. The last molar adds a fifth posterior tubercle. The last premolar lacks the posterior inner tubercle. The second and third have but one, the outer series of tubercles, and the first is compressed. The canines are separated by a slight interval from the premolars, and are in the continuity with the incisors. The premaxillary bones do not reach posteriorly as far as the nasals.

The molar characters are generally similar to those of *Menodus*, but the latter has two internal tubercles of the superior premolars, and only three inferior premolars.

The species originally named by Leidy *Palæosyops paludosus* belongs, according to Marsh, to the succeeding genus, *Limnohyus* Leidy. When first described, it was not characterized generically, a brief specific description only being given. He afterward included species of *Palæosyops* in it, and in so doing first characterized the genus. Hence I agree with him in retaining the name *Palæosyops* for the latter, and not the former, as is done by Marsh.

My material does not permit me to give a description of all the generic characters, since portions of skeletons in my possession cannot be referred to their species with absolute certainty. Neither Dr. Leidy nor the Princeton paleontologists have been more fortunate, judging from their writings. These authors have, however, described various bones which they believe to pertain to the genus. Neither of them gives the number of the digits, which were probably four in front and three behind. I extract the following observations from the Report of the Princeton Scientific Expedition for 1877.

"In general features it (*Palæosyops major*) strongly resembled the tapir, with stout body, slender tail, and very short neck, compensated by a proboscis of considerable length. In comparing the heads of the ancient and modern representatives of this class of *Perissodactyles* the points of contrast are the wide, stout zygomas, the deep temporal fossæ, the protruding nasals, and the narrow gaps in the dental series of the older type.

"There are also strong points of resemblance in the structure of the *Palæosyops* and the Palæothere. * * * The pelvis and particularly the ilium and acetabulum of *Palæosyops* are more palæotheroid than tapiroid. The similarity in the fore-shoulder in the two types is striking. * * * In the descending series of Palæotheres, terminating in the diminutive *P. minus* and characteristic of the upper Eocene of France, we have an interesting counterpart in the large [genus] of *Palæosyops*, of equal diversity of size."

I have obtained jaws and other parts of four different species of this genus, which may be distinguished as follows:

* Superior premolars with complete internal basal cingula.

Length of superior molar series M. .220; external cusps of superior premolars little separated; enamel smooth .... *P. vallidens.*

** Internal cingula of premolars incomplete or wanting.

Length of superior molars M. .175; external cusps of premolars well separated; enamel wrinkled .... *P. major.*

Length of superior molars M. .150; cusps of premolars separated; enamel smooth. *P. lævidens.*

Length of superior molar series M. .100; cusps of premolars separated; enamel more or less wrinkled .... *P. borealis.*

The genus ranges from the lower to the upper Eocene. It has not been found in the typical Wasatch beds of Wyoming and New Mexico, but the *P. borealis* comes from the overlying horizon, the Wind River. The *P. major* belongs to the Bridger, and the *P. lævidens* to the Bridger and allied Washakie. The *P. vallidens* has as yet only been found in the Washakie. Two other species have been proposed by Leidy on specimens from the Bridger, both of small size. These are the *P. humilis* and *P. junior.*

PALÆOSYOPS VALLIDENS Cope.

Palæontological Bulletin, No. 7, p. 1 (August 22, 1872). Proceeds. Amer. Philo. Soc., 1872, p. 487. Annual Report U. S. Geol. Survey Terrs., F. V. Hayden, 1872 (1873), p. 591.

Plates LI, fig. 1; LII, fig. 3; LIII, fig. 1.

Represented by the dentition of one maxillary bone with other bones of one individual; a portion of the same dentition of a second; with both rami of the mandible, with complete dentition, of a third. The species is distinguished by the details of the dental structure, and by the superior size. It exceeds, in this respect, the *Palæosyops major* Leidy; in which the three posterior lower molars measure M. .108 in length, while the same teeth of the present animal measure M. .145. The last superior molar of another specimen measures M. .051 in length; in a third, the first true molar is M. .038 in length, while the last inferior molar is .057 inches long. The peculiarity in the structure of the superior premolars consists in the existence of two strong transverse ridges, which connect the inner tubercle with the outer crescents, inclosing a pit between them. In the premolars is also found the peculiarity of the almost entire fusion of the outer crescents into a single ridge. These united crescents are narrower than in *P. major*, and the summits of all the crescents are relatively more elevated. The external vertical ridge is weaker than in the other species of the genus. The number of inner tubercles is the same as in that species; all the teeth have very strong basal cingula, which extends on the inner side on the premolars only. The inferior molars are relatively narrower than in *P. major*, and the posterior tubercle of the last is larger and longer, and is an elevated cone. The inner tubercles in all the lower molars have broader bases and less acumination than in that species.

The bones containing the maxillary and mandibular teeth were not found together in any instance, so that it is possible that the different series may represent different species. No other species of the genus was, however, found in the localities to which the respective parts could be referred. Should these prove not to pertain together, the lower jaws may be regarded as typical of the species.

The mandibular rami are less robust than in *P. major*, and are rather thin posteriorly. The dentary portion is compressed, and presents a longitudinal concavity on the inner side above the inferior border. There is a large mental foramen below the second premolar. The first premolar is one-rooted. The second has no anterior basal tubercle, and supports a principal cusp and a heel, which has an oblique median ridge. The fourth premolar has the two Vs, but the anterior ridge of the anterior one and the posterior ridge of the posterior one are weak. The true molars increase rapidly in size posteriorly. In the last molar the ridge connecting the posterior lobe with the median is low, only giving a connecting dentinal hand after considerable wear. The inferior canine is not very robust.

Measurements.

	M.
Length of complete ramus from anterior base of Pm. iv	0.295
Length of do. posterior to M. iii	.143
Depth of ramus at Pm. ii	.051
Depth of ramus at M. iii anteriorly	.072
Diameter of inferior canine at base of crown	.018

I found the specimens above described in the Mammoth Buttes, Southwestern Wyoming, near the headwaters of Bitter Creek. The formation is the Washakie basin of the Bridger. Since my examination of this locality the Princeton Exploring party obtained specimens of this species in the same region.

PALÆOSYOPS MAJOR Leidy.

Survey of Wyoming, 1871, p. 359. Report U. S. Geol. Survey Terrs., 1873, I, p. 45. Report of the Princeton Scientific Expedition, 1878, p. 27. *Limnohyus robustus*, Marsh, Amer. Jour. Science and Arts, 1872, p. 124.

Plate LI, fig. 2; LII, figs. 1, 2.

I found remains of this species not uncommon in the Bridger region, especially on Cottonwood Creek, Black's Fork of Green River, etc. I cannot add anything to the descriptions above cited.

PALÆOSYOPS LÆVIDENS Cope.

Annual Report U. S. Geol. Surv. Terrs., 1872 (1873), p. 591, and Report U. S. Geol. Surv. Terrs., I, 1873, p. 30; not *loc. cit.*, p. 28, nor *Limnohyus lævidens*, Cope; Proceed. Amer. Philos. Soc., 1873, published January 31. *Palæosyops paludosus*, Leidy; Hayden's Survey of Montana, 1871, p. 359; Proceed. Acad. Nat. Sci., Philada., 1870, p. 113.

Plate L, figs. 1-3.

This species is smaller than the *P. major*, but Dr. Leidy was not satisfied that it is entitled to distinct specific rank on this account. In my report to Dr. Hayden in 1872' I distinguished it from the *P. major* by the simply lobate form of the external wall of the second superior premolar. In *P. major* this tooth has two external lobes as in the succeeding premolars, the posterior lobe being smaller than the anterior. This character of the *P. lævidens* is displayed in Dr. Leidy's figures.

I have various specimens referable to a species of the size of this one, but the only one which displays the entire superior dental series is a part of a skull from the bad lands of Cottonwood Creek, Wyoming (Plate L, figs. 1, 2). It agrees in its measurements with some of the specimens described by Dr. Leidy under the name of *Palæosyops paludosus.* A second specimen (Plate L, fig. 3), from the Washakie Basin, is intermediate in dimensions between the latter and the *P. major.* I have not seen the second superior premolar. On this account, and on account of its dimensions, I cannot be sure of its specific position, but describe it under the present head. Should the skull above mentioned prove to belong to the *P. paludosus* Leidy, and the other specimen belong to another species, the latter may bear the name under which it has been and is now described.

In the cranium the prominent features are the elongate nasal bones, and the deeply excavated external nares. The former are well preserved, they are of equal width throughout, and have a flat superior surface and strongly decurved nareal borders. Their distal extremities are subtruncate, each bone slightly rounded. The nareal border is above the anterior part of the second premolar. The posterior ascending branch of the premaxillary bone disappears from the nareal border above the middle of the canine tooth, or .030 M. anterior to the posterior nareal border. The face between the nares and orbit is flat; below its middle the large infraorbital foramen interrupts the surface. The surface at the canine alveolus is very prominent, standing out much beyond the surface at the premolar teeth. The malar ridge projects immediately from the border of the infraorbital foramen. It is very prominent below the orbit, presenting a horizontal surface downwards. Its external surface slopes obliquely inwards to the border of the orbit. The anterior border of the latter is over the posterior third of the first true molar. The malar bone presents a postorbital angle upwards. It is much expanded downwards posteriorly, and gives the inferior border of the zygoma a decurved outline. The superior border of the squamosal process rises strongly posteriorly. The premaxillary bones are not prominent, and their anterior extremities are transverse and not coössified. The infraorbital foramen issues above the posterior part of the fourth premolar. The molars have the general form of those of *P. major,* but the second superior premolar has but one outer tubercle. The cingula are much less developed than in that species, those between the inner cones of the molars being entirely absent. These cones are low, and, with the rest of the crowns of all the teeth, covered with smooth and shining enamel. The anterior median small tubercle of the first true molar is wanting. The last true molar has but one interior cone. The canine tooth is powerful and bear-like, and has a rather short crown. It is wide in front, and has a cutting ridge posteriorly. The crowns of the incisors have prominent but not cingular bases.

Measurements.

	M.
Length molar series	0.140
Length true molars	.085
Length three incisors	.034
Length crown canine	.030
Length crown last molar	.039
Width crown last molar	.036
Length cranium to lateral occipital crest	.345

The teeth of the Washakie specimen are very smooth, but the basal cingula of the premolars are stronger than in the specimen above described.

Measurements of No. 2.

	M.
Length true molars (No. 2)	0.101
Length last molars (No. 2) (oblique)	.039
Width last molars (No. 2) transverse)	.038

This species was about the size of the *Anoplotherium commune*, Cuv., and intermediate between the *Palæotherium magnum* and *P. medium*. It is considerably larger than the existing tapirs, and was one of the most abundant of the quadrupeds of the Eocene of North America.

PALÆOSYOPS BOREALIS Cope.

American Naturalist, 1880, p. 746. Bulletin U. S. Geological Survey Terrs., 1881, vol. vi, p. 196.

Plate LVIII *a*, fig. 3.

At present the only parts of the skeleton of this species which are known are the following: Part of the right maxillary bone, with four molar teeth; distal portions of both radii; a lunar bone, and a basal median phalange. These fragments indicate a species much smaller than *P. lævidens* and about equal to the *Limnohyus fontinalis*.

The true molars increase rapidly in size posteriorly. The external Vs are strongly inflected, and have a low median rib. Each internal cusp sends a low ridge outwards and forwards, and the anterior intermediate tubercle is distinct. The crowns have strong anterior and posterior cingula, but no internal cingula. The ledge that connects the inner cusps in *P. major* is represented by a tubercle in this species. The posterior cingulum of the last molar rises to a prominent ridge on the inner posterior margin of the crown. In a third premolar the anterior and posterior cingula do not connect on the inner side of the crown. The external cusps have a nearly vertical external wall, and are well distinguished; the anterior one only has a median rib. It has no external cingulum. There is a weak one on the true molars. The enamel is smooth where worn; on unworn surfaces it is somewhat wrinkled.

Measurements of teeth.

		M.
Length of true molar series		0.063
Diameters of first true molar	anteroposterior	.019
	transverse	.020
Diameters of third premolar	anteroposterior	.013
	transverse	.016

The shaft of the radius is not very stout, and shows a groove for contact with the ulna below. The carpal articular face is wide and depressed, and without trace of scapholunar ridge. A small part of both surfaces is recurved inferiorly, and each part is followed by a shallow fossa: these are separated by a low tuberosity. The cuboid has a rather short posterior extension, whose superior surface makes with the transverse anterior surface an L of branches of equal length. Below, the magnum and unciform facets are strongly concave, the latter a little the larger. No posterior tuberosity. The proximal phalange is longer than wide, and stout and flat.

Measurements.

		M.
Width of carpal facet of radius		0.030
Diameter (transverse) of radius at middle		.018
Diameters of os lunare	vertical	.020
	transverse above	.017
	anteroposterior	.020
Diameters of phalange	longitudinal	.023
	transverse	.017

The above fragments were cut from a single small block of matrix.

The *P. junior*, according to Leidy, differs in the character of its superior molar teeth from the present species. That animal appears to lack the intermediate tubercles and crest, and has a much stronger external cingulum.

The *P. borealis* has been only found thus far in the Wind River Eocene by Mr. J. L. Wortman.

LIMNOHYUS Leidy.

Proceedings Academy Natural Sciences, Philadelphia, 1872, p. 242; Cope. Annual Report U. S. Geol. Survey Terrs., 1872 (1873), p. 593. *Palæosyops*, Marsh, Amer Journ. Sci. Arts, 1872, p. 122, not of Leidy, 1871.

This genus only differs from the last in possessing two conic tubercles of the inner series on the last superior molar, instead of one, a character first pointed out by Marsh.

It is yet uncertain to what genus the species originally described by Leidy as *Palæosyops paludosus* belongs. The type specimens do not include the last superior molar, according to Leidy. The only superior true molar preserved possesses two internal cusps, and is, according to its describer, the second of the series. Professor Marsh, in arranging the species of the two genera, came to the conclusion that the last superior molar of the *P. paludosus* possessed two internal cones, which character he adopted as distinctive of the genus *Palæosyops*. To the species with one inner lobe, he gave the name of *Limnohyus*. Subsequent to the original description of the *P. paludosus*, Leidy referred to it various specimens with only one internal cone of the last superior molar, and at that time defined the genus *Palæosyops* as distinguished by that character. Thus Professors Marsh and Leidy were at issue as to which of the genera should receive the name *Palæosyops*.

The decision of the question will depend on the rule adopted as to the conditions necessary to be observed in the proposal of new generic names. Those authors who deem it sufficient to establish a generic name, that it be merely printed, will follow the course adopted by Professor Marsh. In describing the original *Palæosyops paludosus* Dr. Leidy did not characterize the genus *Palæosyops*. This is a custom more to be honored in the breach than the observance, and one contrary to the rules of nomenclature. For the present writer the name *Palæosyops* did not at that time obtain a position in nomenclature. This was fully acquired at a later date, when Dr.

Leidy described both genus and species. The species then described was, however, different, according to Marsh's view, from the original *P. paludosus*. By this accident, the second species had a generic and an erroneous specific name; and the earlier described species had a specific and no generic name. Dr. Leidy supplied the latter deficiency in subsequently naming the genus with two cones *Limnohyus*, a name already used by Marsh, and synonymous with *Palæosyops*. To those who will not again use a name which has become a synonyme this course would not appear judicious; for my own part, I never saw any disadvantage to arise from such a course. As to the second species, the *Palæosyops* without name, I called it *lævidens*, thinking it very inconvenient to have two *paludosi* in two so nearly allied genera.

Besides the *L. paludosus*, there are two large species of *Limnohyus*, *L. laticeps* Marsh, and *L. diaconus* Cope. The *L. fontinalis* Cope is a smaller form.

LIMNOHYUS DIACONUS Cope.

Annual Report U. S. Geological Survey Terrs., 1872 (1873), p. 593. *Palæosyops diaconus* Cope, Palæontological Bulletin, No. 12, p. 4, March 8, 1873.

Plate LI, fig. 3.

This large species is represented by parts of the two maxillary bones, which present the crowns of the third and fourth premolars, and of the second and third true molars, with the bases of the other molars and premolars. The third true molar is injured posteriorly, but enough remains to show that it possessed two internal cones.

The jaw is as large as the *Palæosyops major* of Leidy, but differs in the relative proportions of the teeth. Thus the last three molars have the same anteroposterior length, while the space occupied by the four premolars is shorter. The anterior and posterior cingula of the true molars are very strong, but it is not well marked on the inner side, between the cones. The latter are acutely conic, and the median anterior tubercle is strongly developed. Although the wearing of the teeth indicates maturity, the enamel is coarsely and obtusely rugose. The fourth premolar differs from that of the four species of *Palæosyops* described, in its smaller size relatively and absolutely, and in the presence of a prominent vertical tubercle on the outer face, rising to the angle of the deep notch between the lobes. It has a complete interior basal cingulum. The third premolar is as wide as the fourth, and about as large as the corresponding tooth in *P. major*, but different from it in the absence of tubercle and ridge that mark its external face. The first premolar has two roots, and the canine is large and stout.

Measurements.

	M.
Length of entire molar series	0.1740
Length of true molars	.1060
Length of last molar (crown)	.0420
Width of last molar (crown)	.0473
Length second molar	.0350
Length fourth premolar	.0260
Width fourth premolar	.0260
Width third premolar	.0200
Length third premolar	.0200
Diameter of basis of canine	.0263

The *L. paludosus* Leidy is similar to this species in the rugosity of the enamel of its teeth, but appears by the measurements to be distinctly smaller. My expedition did not obtain any specimens which I can as yet certainly refer to this species. The measurements given by Professor Marsh for his *Palæosyops* (*Lymnohyus*) *laticeps* approaches very nearly to this one Thus the width of the crown of the penultimate superior molar in *L. paludosus* is M. .038; in *L. laticeps*, .036; in *L. diaconus* Cope it is .042.

In comparison with Marsh's description of his *P. laticeps*, the measurements are all larger, and the enamel is as rugose as in *L. major*, instead of smooth. The shortening of the premolar series is the same in the two species; thus, in *P. laticeps* the two sets of molars are related as 94^{mm} to 61; in the present one, as 106 : 68.

From Henry's Fork of Green River.

Limnohyus fontinalis Cope.

Palæosyops fontinalis Cope, Palæontological Bulletin, No. 11, p. 1, Jan. 31, 1873. Proceedings American Philosophical Society, 1873, p. 35. *Limnohyus fontinalis* Cope, Annual Report U. S. Geol. Survey of the Territories, 1872 (1873), p. 594.

Plate XLIX, fig. 9; Plate L, fig. 4; and Plate LVIII *a*, figs. 4-5.

This chalicotheroid is represented in my collection by a considerable part of the skull of one individual. This includes the greater part of the occipital and parietal bones, with the right squamosal almost entire; the right maxillary and malar as far forwards as the fourth premolar tooth; and the nasal and attached part of the right maxillary bones. The portions belong to a species of about the size of the *Palæosyops borealis*, and smaller than any of the other species of the two genera here described.

The nasal bones in profile have a horizontal superior surface. They are very convex in transverse section, and their extremities are emarginate. The infraorbital part of the malar is regularly convex, and overhangs the maxillary bone extensively. Its postorbital limit is marked by an angle directed at 45° inwards. The zygomatic part of the malar is flat externally. The squamosal portion of the zygoma is not much elevated. The temporal fossa is rather wide posteriorly, and the inferior temporal ridge is extended horizontally. The postglenoid process is rather narrow and elongate. The posttympanic process is robust, and is well separated from the postglenoid, leaving an especially wide *meatus auditorius*. The front is flat, and the superior temporal ridges converge gradually to an isthmus, but scarcely form a sagittal crest. The isthmus expands to the narrow lateral occipital crests.

The occipital bone sends a long process forwards on the median line, forming a half-gomphosial articulation with the parietals. The lateral suture of the two bones is considerably in advance of the posterior lateral crest. The maxillomalar suture is on the convex infraorbital ridge. There is a postparietal foramen; also a small one in the squamosal bone posterior to the position of the supraglenoid. There is above this a postsquamosal on the parietal suture.

The second true molar is abruptly larger than the first. In none of the true molars are there ridges extending outwards from either of the internal cusps, nor is there any posterior median tubercle. The anterior median tubercle is distinct. There are no external nor internal basal cingula, but the anterior and posterior are distinct. The latter, however, only connects the two cusps of the posterior side of the crown. The outer sides of the Vs have a trace of a median rib. The enamel is nearly smooth when worn, but finely rugose before protrusion.

	M.
Length of true molar series (2.75 inches)	0.067
Length of last molar	.025
Width of last molar	.026
Length of penultimate molar	.026
Width of penultimate molar	.026
Depth squamosal process	.025

Found by the writer on a bluff on Green River, near the mouth of the Big Sandy Creek, Wyoming.

LAMBDOTHERIUM Cope.

American Naturalist, 1880, p. 743, September 20. Bulletin U. S. Geological Survey Terrs., VI, 1881, p. 196. (?) *Helotherium* Cope, Palæontological Bulletin No. 2, p. 1, August 3, 1872. Proceedings American Philosoph. Society, 1872, p. 466. (*Nomen nudum.*)

This genus is as yet only known to me from teeth of both jaws and from mandibular rami. The dentition is much as in *Limnohyus*, excepting that there is a diastema in front of the second inferior premolar. Presence of first inferior premolar not ascertained. Fourth inferior premolar without posterior cusps. Superior molars with an angular ridge extending inwards from each inner cusp. Last inferior molar with heel or fifth lobe.

The inferior molars in this genus are quite as in the other genera of this family, while the superior molars show a tendency to *Anchitherium* in the well-defined tubercle and low crest extending outwards and forwards from the internal tubercles. The external Vs are separated by a strong rib. The internal cusp of the superior premolars is simple, and the external cusps are well distinguished.

If the ridges, which are rudimental in the molars of *Hyracotherium*, should be developed, and the external cusps of the superior molars be flattened externally, the result would be the dentition of *Lambdotherium*.

I am acquainted with three species of *Lambdotherium*, which differ materially in size. The larger two are from the Wind River Eocene; the smallest is from the Bridger.

LAMBDOTHERIUM BROWNIANUM Cope.

Bulletin U. S. Geological Survey Terrs., VI, 1881, p. 196.
Plate LVI *a*; fig. 10.

Considerably larger than the *L. popoagicum*, and about equal to the *Tapirus terrestris*. The greater part of a lower jaw represents the species, and on this unfortunately only one of the premolar teeth remains. The three premolars are all two-rooted, and the posterior lobe of the last true molar is well developed. The inferior part of the external side of the ramus contracts or retreats rather abruptly posteriorly below the last molar. It presents a slight external convexity below the second and third premolars.

The alveolar line rises rapidly posteriorly, so that the last true molar is quite oblique. The second (first) premolar has a considerable heel, which is narrow and elevated on the middle line. The principal cusp is large and compressed, but obtuse, and has no anterior basal tubercle.

Measurements.

		M.
Length of six molars		0.090
Length of true molars		.055
Diameters second (first) premolar	vertical	.009
	anteroposterior	.012
	transverse	.006
Length of base of first true molar		.015
Width of base of first true molar		.009
Length of base of third true molar		.023
Width of base of third true molar		.011
Depth of ramus at second premolar		.030
Depth of ramus at M. iii	at front of tooth	.039
	at end of tooth	.047

Dedicated to my friend Arthur E. Brown, superintendent of the Philadelphia Zoölogical Garden.

LAMBDOTHERIUM POPOAGICUM Cope.

American Naturalist, 1880, p. 748. Bulletin U. S. Geological Survey Terrs., VI, 1881, p. 196.

Plate XXIX *a*, fig. 7, and Plate LVIII, figs. 3–5.

This is the most abundant *Perissodactyle* of the Wind River beds, remains of at least twenty-two individuals having come under my observation. The general characters of the inferior molars are as follows: The heels of the second and third inferior premolars have a median keel; the third only has an anterior tubercle. The crest of the heel of the fourth forms an imperfect V. Heel of the last true molar small. No cingula. Enamel smooth. Length of series, .080; of true molars, .044; of last molar, .019; depth of ramus at first premolar, .021; at last molar, .031.

The crowns of the superior molars are very short; their external Vs are well distinguished by an intermediate vertical protuberance, but their external faces are less obliquely sloped inwards than in the species of the genera *Limnohyus* and *Palæosyops.* Traces of the median ribs on the external faces of the Vs are slight. The tuberosity which bounds the anterior V in front is very large, and its apex is continued into the anterior basal cingulum, as in *Lophiodon* and allies. Its size causes the external wall of the molar to be oblique to the other sides of the crown. The anterior and posterior cingula of the molars are very strong, but they do not extend quite round the inner bases of the internal cusps. The latter are rather acute, and their opposed faces are flat and meet at an angle. The intermediato tubercle is large, has a subacute summit and a triangular base. The ridge which passes forwards and inwards from the posterior internal tubercle does not reach the external lobe, and rapidly diminishes in elevation. There is no well-defined external basal cingulum. In the superior premolars the external cusps are separated by a notch, and have no external ridge between them. Each has a feeble median rib. There are posterior and anterior cingula, of which the latter terminates externally in a low tubercle. There is a low external cingulum, but no complete internal one. The superior premolars have an anterior intermediate tubercle.

Measurements of separated superior molars.

		M.
Diameters Pm. ii	anteroposterior	0.009
	transverse	.010
Diameters Pm. iii	anteroposterior	.009
	transverse	.112
Diameters Pm. iv	anteroposterior	.010
	transverse	.012
Diameters of true molar	anteroposterior	.0115
	transverse	.0160
Elevation of crown		.007

The ramus of the lower jaw is compressed and quite deep. The coronoid process rises abruptly from the alveolar border, and the masseteric fossa is little marked. There are two mental foramina, which are below the second and third premolar teeth, respectively. The posterior extremity of the symphysis is opposite the anterior root of the second premolar. This posterior position leads me to suspect that the first premolar may be wanting.

This tapiroid was about the size of the *Hyrachyus agrestis*, but less robust.

Wind River Eocene of Wyoming. Found by Jacob L. Wortman.

LAMBDOTHERIUM PROCYONINUM Cope.

Helotherium procyoninum Cope, Palæontological Bulletin No. 2, p. 1, August 3, 1872. Proceedings American Philosophical Society, 1872, p. 466. Annual Report U. S. Geol. Survey Terrs., 1872 (1873), p. 606.

Plate XXIV; fig. 22.

As this species is only known from a superior true molar, its generic position is not certainly known. The tooth in question is very much like that of the *Lambdotherium popoagicum*, but differs in easily recognizable specific characters. The crest that extends from the posterior inner tubercle reaches to a point between the external lobes, and is therefore longer than in the species named. It also shows a slight tendency to a division cutting off an intermediate tubercle, like the anterior one. The median ribs of the external Vs are more distinct than in *L. popoagicum*. The anterior and posterior cingula are distinct, and the external and internal are nearly complete. The crown is low, and the enamel nearly smooth.

Measurements.

		M.
Diameters of crown	anteroposterior	0.0070
	vertical	.0035
	transverse	.0090

The molar tooth is about the size of that of a raccoon; hence the name. Cottonwood Creek, Wyoming. Bridger Epoch.

MACRAUCHENIIDÆ.

But one genus of this family is known at the present time. The following are the dental characters of *Macrauchenia*; formula: I. $\frac{3}{3}$; C. $\frac{1}{1}$; Pm. $\frac{4}{4}$; M. $\frac{3}{3}$; forming an uninterrupted series.

The superior molars present two external Vs and two oblique transverse crests, somewhat as in *Palæotherium*. The spinous foramina pierce the neural arches of the dorsal vertebræ (Gervais). There is no intertrochlear crest of the humerus, but the carpal facets of the radius are well distinguished. The internal malleolus is small, but the fibular malleolus is coössified with the tibia at an early age, and articulates with the calcaneum. The trochlea of the astragalus is well developed. The lateral digits are large, and the distal keels of the metapodials are continued on the anterior faces of the condyles.

The position indicated by the above characters is a remarkable one. The uninterrupted dental series and the absence of intertrochlear humeral crest are primitive features among ungulate *Mammalia*. The radio-carpal articulation is faceted as in the higher ungulates, but lacks the inferior condyloid face of those types. The completeness of the metapodial distal keels is a feature of high special organization only seen in the *Equidæ* of this order. The coössification of the external malleolus is also a character peculiar to the *Equidæ* among the *Perissodactyla*. There are two other characters which are not elsewhere found in this order, viz, the articulation of the fibula with the calcaneum, and the absence of the vertebrarterial canal. The former belongs to the *Artiodactyla* generally, and to the *Proboscidea*, and the latter to the ruminant family of the *Camelidæ*. Thus the *Macraucheniidæ* stand out as one of the most distinct of the families of the *Perissodactyla*, and one to which we may anticipate considerable accessions in future. But two species of *Macrauchenia* are known, a larger *M. patachonica*, and a smaller, *M. boliviensis*, both from the Pliocene formation of South America.

MENODONTIDÆ.

The known genera of this family are not numerous. They are defined as follows:

a. Last superior premolar only with two inner tubercles.

Incisors present .... *Diplacodon*.

aa. All the superior premolars with two interior cusps.

Six inferior incisors; canines very large .... *Dæodon*.

Six inferior incisors; canines very small .... *Menodus*.

No inferior, and four small superior incisors; canines very small .... *Symborodon*.

The genera are, as yet, exclusively American. *Diplacodon*, in its simpler premolars, approaches the *Chalicotheriidæ*, and is the oldest of the American genera. It is from the Uinta or Upper Eocene. *Menodus* and *Symborodon*, which include some species of gigantic size, belong in the White River or Oligocene, while *Dæodon* has so far only been obtained from the John Day or Middle Miocene.

Up to the present time no species of this family have been described from the American Middle Eocene.

PALÆOTHERIIDÆ.

This family has been already defined. In its complex premolar teeth, which in the upper jaw resemble the molars in composition, it shows an advance over the *Chalicotheroid* and other genera of the Lower Eocene. In fact, it has not been found in the Lower Eocene, but commences in the Upper Eocene in the genera *Palæotherium* and *Paloplotherium*. Thence it extends to the very summit of the Miocene, and may even occur in the Pliocene (*Hippotherium protohippus*). Its members exhibit considerable range of variation in the details of the teeth and feet, but no striking break of family importance occurs. The most noteworthy interruption is that which is found between the *Palæotherinæ* and *Hippotheriinæ*, where there is a change in the form of the proximal extremity of the humerus from a tapiroid to a horse-like form, and a modification of similar significance in the molar teeth, by the addition of a deposit of cementum.

The characters of the genera are as follows:

I. *Palæotheriinæ*. Bicipital groove of humerus simple; teeth without cementum.

a. One or more internal tubercles of superior molars distinct.

External Vs of superior molars not well distinguished externally .... *Anchilophus*.

External Vs separated by a vertical rib; intermediate tubercles not connecting fore and aft .... *Paloplotherium*.

External Vs separated; intermediate tubercles extended fore and aft...... *Anchippus*.

aa. Internal tubercles of superior true molars continuous with the transverse ridges.

Inferior molars with two Vs only; lateral toes large .... *Palæotherium*.

Inferior molars with cusps at the inner extremities of the Vs; lateral toes small. *Anchitherium*.

II. *Hippotheriinæ*. Bicipital groove of humerus double; molars with cement in the valleys. (Intermediate tubercles connected fore and aft.)

a. One or more internal tubercles of superior molars distinct.

Inner lobes of inferior molars enlarged.............................. *Hippotherium*.

aa. Internal tubercles of molars not distinct.

Inner lobes of inferior molars enlarged .... *Protohippus*.

The genera of this family are generally of less antiquity than those of the *Chalicotheriidæ*, and they range from the Middle Eocene to the Pliocene. *Paloplotherium* is found in the Middle Eocene, and is, as might have been anticipated, more nearly allied to the *Chalicotheriidæ* than any other genus of this family. *Pachynolophus* also is not far removed from it. *Anchilophus* is Upper Eocene, and is allied to the genus just named, which connects both genera with the *Lophiodontidæ*. These early genera constitute by their similarity the bond of connection between the three families, which in their later and specialized forms are very different from each other. *Palæotherium* is chiefly found in the Upper Eocene, and *Anchitherium* commences in America in the White River or Oligocene, an age between Eocene and Miocene. In Europe it commences in the Middle Miocene, and has *Anchippus* for a cotemporary. *Hippotherium* existed only in the latter part of the Miocene epoch, consistently with the greatly specialized structure of its limbs and teeth, and the nearly allied *Protohippus* lived with it, while in Europe a species with the same type of molar teeth is found in the Pliocene epoch (Forsyth-Major). These forms were cotemporary with the *Equidæ*, which outlived them. They have many points of resemblance to that family, but nevertheless remain at a considerable interval from them in the structure of the feet.

The geographical distribution of these genera, so far as present knowledge shows, is as follows:

North America alone—*Anchippus.*

North America and Europe—*Anchitherium, Hippotherium, Protohippus.*

Europe only—*Anchilophus, Paloplotherium, Palæotherium.*

EQUIDÆ.

The two genera of this family are distinguished as follows:

Internal lobes of superior molars subequal .... *Hippidium.*
Anterior internal lobe of superior molars much larger than the posterior ...... *Equus.*

The genus *Hippidium* is extinct, and its species have been thus far only found in North and South America, in beds of Pliocene age. *Equus* made its appearance during the same period, and is represented by several existing species.

Besides the reduction in the number of digits, which is carried further here than in any other family of *Mammalia*, there are several other characteristics of specialization. Thus, in the dentition the spaces between the tubercles are filled with cementum. These valleys are generally deep, owing to the prismatic forms of the molars. The cups of the incisors are completely developed, and also filled with cementum. There are two bicipital grooves of the humerus. The preceding characters are also found in the *Hippotheriinæ* of the *Palæotheriidæ*. The *Equidæ* adds another evidence of greater specialization than the latter group in the structure of its feet, *i. e.*, the distal metapodal keels are completed forwards, as in most ruminants.

The similarity of the modifications which have supervened on the *Artiodactyle* and *Perissodactyle* lines, in attaining their most specialized extremes, has often been noticed. I repeat them here in tabular form in three columns.* These show (Table I) the modifications in which the *Equidæ* and *Bovidæ* are identical, or nearly so, which place them at the heads of their respective orders; Table II, those in which the *Equidæ* are the more specialized of the two; and Table III, those in which the *Bovidæ* display the highest differentiation.

	*Table I.	Table II.	Table III.
1	Obliteration of first premolars.	Cupping of incisor teeth.	Absence of superior incisors.
2	Selenodont and prismatic character of molars.	Deposit of cementum in teeth.	Trough-shape of odontoid.
3	Flattened odontoid process.	Double bicipital groove of humerus.	Greater involution of lumbar prezygapophyses.
4	Intertrochlear crest of humerus.	Reduction of metapodials to one.	Fibular articulation of calcaneum.
5	Reduction and coössification of ulna.		Distal ginglymus of astragalus.
6	Distal facets of radius.		
7	Involution of lumbar prezygapophyses.		
8	Elongate sacrum.		
9	Shaft of fibula atrophied and its distal extremity coössified to tibia.		
10	Reduction in number of digits.		
11	Anterior extension of carinæ of metapodial bones.		

ARTIODACTYLA.

Members of this order were very few in number in the true Eocene periods, so far as our present knowledge extends. The great subdivision of the *Ruminantia* did not exist, the only types being the Suillines and the *Eurytheria*. The characteristics of these groups will be more fully defined in Part Second of this work, which treats of the White River epoch, during which both of the divisions in question were numerously represented.

OMNIVORA.

Two, and perhaps three, genera of hogs from the Eocene beds of our Western regions have come under my notice. They differ from all existing members of the suborder in the anisognathous character of their molar dentition; that is, the superior molars have a greater transverse diameter than the inferior ones. In this they agree with allied genera already found in Europe, as *Chœropotamus* Cuv., *Anthracotherium* Blv., and show a greater resemblance to the ruminant division than do the recent hogs. The last-named genus approximates *Hyopotamus*, which is still nearer the lowest *Ruminantia.*

PANTOLESTES Cope.

Proceed. Amer. Philos. Soc., 1872, p. 467. (Separata, August 30.) Report U. S. Geol. Surv. W. of 100th Mer., iv, p. 145.

The type of this genus resembles in structural characters of the dentition of the lower jaw the *Hyopsodus* and *Sarcolemur*, already known in the collections of the different explorations of the Rocky Mountain lake basins. While it possesses the normal number of molar teeth belonging to these, it preserves a sectorial character of the premolars more posteriorly than in any of the genera named. The premolars are all two-rooted, except the first. The fourth is a simple flattened triangular cusp, with a small tubercle at the base behind, and wanting the inner cusp of other genera. The molars exhibit the usual four cusps, the external cresoentoid, the inner discoid in section, excepting the inner anterior, which is generally duplicated by an anterior twin-cusp of smaller size, closely united with it.

Six species of this genus have been described from lower jaws only, and in one only, the *M. brachystomus*, are both superior and inferior molars known. In the *P. secans* the sectorial character of the fourth inferior premolar is most strongly marked; in *P. longicaudus* that tooth has a wider heel than the other species. The species of *Pantolestes* may be distinguished as follows:

a. Fourth premolar trenchant everywhere, longer than second true molar.
Length of true molars M. .0150; second molar with but one anterior inner cusp . *P. secans.*
aa. Fourth premolar with blunt heel, not longer than second true molar.
Length of true molars .0230; no anterior cusps of premolars............. *P. etsagicus.*
Length of true molars .0160; all with double cusps...................... *P. chacensis.*
Length of true molars .0140; fourth premolar with minute anterior cusp, and long heel .... *P. longicaudus.*
Length of true molars .0130; fourth premolar large, .005, with double anterior cusp, and short heel; molars with double cusps .... *P. metsiacus.*

Length of true molars .0138; fourth premolar short, .0045; anterior cusps of true molars single .... *P. brachystomus.*
Length of true molars .0105; fourth premolar small, .0035; without anterior cusps and with two ridges on heel; true molars with double anterior inner cusps.... *P. nuptus.*

I originally arranged this genus in the *Mesodonta*, as only the dental characters were then known. These resemble also those of the suillines, where I once provisionally placed the genus *Mioclænus*, which proved later to be a creodont. The possession of the bones of *P. brachystomus* shows that that species belongs to the *Artiodactyla.*

A few bones, including metapodials, of the *P. longicaudus*, are preserved, but it is from remains of the *P. brachystomus* that the characters of the skeleton of this genus are derived.

A lumbar vertebra of the *M. brachystomus* has flat articular surfaces. The ilium has a rather long peduncle, much wider than deep, with a triangular section. The base of the triangle is rather short, and is concave. The acetubular cotyloid notch is open. Femur with prominent greater and lesser trochanters, but no third trochanter. Trochanteric fossa profound. Head with *fossa ligamenti teris* isolated from neck.

The characters of the tarsus are typically those of the order *Artiodactyla.* The astragalus exhibits a distal trochlea which is continuous with the sustentacular facet, and which articulates with both cuboid and navicular. The distal portion of the fibula is free from the tibia, and its shaft becomes very slender. It is possible that a more perfect specimen would display it as continuous. Its distal extremity articulates with the ascending tuberosity of the calcaneum. The cuboid facet of the latter is narrow. The cuboid and navicular bones are distinct from each other and from the cuneiforms. The mesocuneiform is shorter than the ectocuneiform, and *is coössified with it.* There are probably four metatarsals. The median pair are distinct, but appressed; their section together, subcircular. The lateral metatarsals are slender; the external one is wanting, but its facet on the cuboid bone is very small.

These characters are in general similar to those of the genus *Dichobune*, but Cuvier does not state whether the cuneiforms are coössified in that genus or not. They are united in *Anoplotherium*. *Pantalestes* differs from *Dichobune* in the presence of but one internal tubercle of the superior molars, and in the single external tubercle of the superior premolars. It is referable to a family to be distinguished from the *Anoplotheriidæ* by the presence of the external digits, and from that family, and from the *Dichobunidæ*, by the possession of tritubercular superior molars. It will be called the *Pantolestidæ*.

The characters of the tarsus are of much interest, and demonstrate that *Pantolestes* is the oldest type of Artiodactyle yet discovered, and that it is not altogether primitive in some of its characters. Members of this order have been found by Cuvier in the Upper Eocene (*Dichobune*, *Anoplotherium*, etc.), but none have been determined as yet from the Suessonian of America. A species represented by teeth from the Siderolithic beds of Switzerland have been referred to *Dichobune* (*D. campichii* Pict.); but dental characters alone are not sufficient to distinguish that genus from *Phenacodontidæ*. Dr. Lemoine found astragali of a small Artiodactyle in the Suessonian of Reims, and has referred them to his supposed Suilline *Lophiochœrus peroni*. I have reported an astragalus from the Wind River formation of Wyoming Territory, which is almost exactly similar to those found by Lemoine. The specimens of *P. brachystomus* enable me to characterize with some degree of completeness this interesting form, which precedes in time all the known American *Artiodactyla*.

PANTOLESTES CHACENSIS Cope.

Systematic Catal. Vert. Eocene, New Mexico, 1875, p. 15. Report U. S. Geog. Surv. W. of 100th Mer., iv, ii, p. 146, pl. xlv, fig. 17.

Plate XXIVd, fig. 5.

The largest species, represented by four mandibular rami in the Big Horn collection. It has the fourth premolar more robust and less trenchant than in *P. secans*, and shorter than the last true molar. In *P. secans* it is longer than the last true molar. For fuller description, see reference cited.

PANTOLESTES METSIACUS Cope.

Proceed. Amer. Philos. Soc., 1881, p. 149, Paleontological Bulletin, No. 34, p. 149, Feb. 20, 1882.

Plate XXIVd, fig. 6.

A small species of the size of the *P. longicaudus*, and distinguished by several peculiarities of dentition. The two cusps composing the anterior internal lobe of the molars are quite distinct, but appressed. Each one is connected with the external anterior lobe by a transverse crest, as is seen in *Esthonyx*, and these inclose between them a fossa. This fossa is closed internally by the appression of the anterior inner cusps. The fourth premolar is not so large as in *P. secans*, but resembles in proportions that of *P. chacensis*. It differs from that of *P. longicaudus* in its very short heel and its large anterior basal tubercle. The latter is double, consisting of two small cusps, one within and anterior to the other. The posterior heel is distinct on both sides of the ridge that marks the median line. The posterior external lobe is V-shaped, and the posterior inner is a small cone. Between the two is a minute median tubercle. The posterior tubercles are not so elevated as in the species of *Hyopsodus*. A weak external cingulum; enamel smooth.

Measurements.

	M.
Length Pm. iv, with M. i and ii (No. 1)	.0140
Length Pm. iv	.0048
Length M. ii	.0048
Width M. ii	.0040
Length M. iii (No. 2)	.0050
Width M. iii	.0030
Depth ramus at Pm. iv (No. 1)	.0060
Depth ramus at M. iii (No. 2)	.0070

Portions of four mandibles preserved. No. 2 is a little smaller than No. 1, and No. 4 is a little larger than No. 1.

From the Big Horn Basin. J. L. Wortman. Named from the name given by the Crow Indians to the Big Horn River.

PANTOLESTES NUPTUS Cope.

Paleontological Bulletin, No. 34, p. 150, Feb. 20, 1882. Proceed. Amer. Philos. Soc., 1881, p. 150.

Plate XXIVd, fig. 7.

This is the least species of the genus, and is represented by a portion of a right mandibular ramus which supports three molars, from the fourth to the sixth, inclusive. Besides its small size, this species is distinguished by the relatively small dimensions of the fourth premolar, which is shorter than the first true molar, instead of longer, as in all the other species. The well-developed basin of its heel, which is bounded by a ridge on each side, distinguishes it at once also from *P. secans*, and is more distinct than in *P. chacensis*. From the latter and *P. metsiacus* the entire absence of anterior basal lobes separates it. The well-developed pair of anterior inner tubercles shows that it cannot be an abnormal *Hyopsodus vicarius*, with which it agrees in size. The first anterior inner tubercle of the true molars is more widely separated from the second anterior than in any of the species of the genus, and is quite as in the species of *Pelycodus*. It is smaller than the second anterior inner, which equals in size the anterior outer. The heel is wide, inclosing a basin, which is bounded externally by an angular ridge. Its posterior inner angle supports a cusp, which is separated by a deep notch from the anterior inner cusp. External to it, on the posterior border of the crown, is a small tubercle. No basal cingula.

Measurements.

		M.
Length of true molars		0.010
Diameters M. ii	anteroposterior	.004
	transverse	.003
Depth of ramus at Pm. iv		.007

Basin of the Big Horn. J. L. Wortman.

PANTOLESTES BRACHYSTOMUS Cope.

Proceeds. Amer. Philos. Soc., 1883, p. 547. *Mioclænus brachystomus* Cope, Paleontological Bulletin No. 34, p. 187, February 20, 1882.

Plate XXIII d, figs. 16–21.

This small Artiodactyle is represented by the fragments of the skeleton of but one individual. These include the greater parts of the maxillary and mandibular bones, with the teeth; a lumbar vertebra; parts of both innominate bones, and of both femora, with the right tarsus with the attached portions of the tibia and metatarsus. The bones are about two-thirds the size of those of the Javan musk-deer (*Tragulus javanicus*). The transverse extent of the superior true molars is greater than the anteroposterior. The composition of the last molar is like that of the others. The external tubercles are lenticular in section, and the emargination which separates them is apparent on the external face of the crown. The intermediate tubercles are small, and are entirely distinct from the large external tubercle. There is a distinct cingulum which is only wanting from the inner base of the crown. The fourth superior premolar has a trilobate outline of the base of the crown, the base of the inner lobe being contracted where it joins the external part of the crown. The internal tubercle is conic, with a prolongation outwards and forwards. Intermediate tubercle not distinct. External, anterior, and posterior cingula. The third superior premolar differs much in form from that of *Mioclænus turgidus*, although having the same elemental parts. The external portion is extended anteroposteriorly, and has a cutting edge, of which the posterior tubercle forms a lobe. The internal tubercle is quite small, much less than in *M. turgidus.* A posterior and a weak external cingulum.

In the inferior true molars the external tubercles wear into crescents. The crowns increase in size posteriorly. The fifth tubercle of the last molar is rather small, but is well distinguished from the other cusps. The internal median cusp is small, the external median large. The premolars are not so much larger than the true molars in this as in the typical species of the genus. The second and third are more elongate on the base than the fourth. The latter is also less compressed than those that precede it. It has a short wide heel, and a small anterior basal tubercle. In the second and third premolars the posterior edge of the principal cusp is sharp, and descends gradually to the posterior base of the crown. Both have small acute anterior basal tubercles. The first inferior premolar is one-rooted, and has a simple crown directed somewhat forwards. It is separated from the second by a short space. The teeth anterior to this point are lost.

Measurements.

		M.
Length posterior four superior molars		.0182
Diameters Pm. iv	anteroposterior	.0040
	transverse	.0042
Diameters M. ii	anteroposterior	.0043
	transverse	.0060
Diameters M. iii	anteroposterior	.0040
	transverse	.0060
Length of inferior molars		.0330
Length of inferior premolars		.0192
Length of inferior Pm. iii		.0055
Length of inferior Pm. iv		.0045
Diameters M. i	anteroposterior	.0040
	transverse	.0033
Diameters M. iii	anteroposterior	.0052
	transverse	.0040
Depth of ramus at Pm. i		.0047
Depth of ramus at M. ii		.0090

The centrum of the lumbar vertebra is much depressed, and the base of the diapophysis occupies its entire length. The latter is slightly decurved and directed posteriorly, and its base is concave below. The inferior face of the centrum is not excavated, and is weakly keeled medially. The form of the ilium is peculiar, differing from that of most of the Artiodactyle animals with which I have compared it. The rather wide concave anterior face at the inferior part of the peduncle is due to the prominence of the anterior edge, and of a crest on the inner side. The former is prominent just above the usual position of an anterior inferior spine, and then sinks to the edge of the acetabulum. The internal face is narrower than the external, but wider than the anterior. I do not find the anterior face among any of the Ruminantia, except in a modified degree in *Antilocapra*, nor in any of the Omnivora, excepting in the *Phacochærus*, where it is not so concave. The peduncle is shorter than in *Tragulus*, and wider than in any of the Suillines or Ruminants which I am acquainted with.

The external face of the great trochanter of the femur is wide. Its apex is a little incurved, and projects further than the head. The neck is deeply constricted from above. The little trochanter is quite prominent. The head is round, and the *fossa ligamenti teris*, though behind the middle of its inner aspect, looks directly inwards. The external edge of the shaft of the femur is acute; the inner edge is rounded. The rotular groove is rather narrow, and is on an elevated base. Of the lateral crests, one is a little higher than the other near the proximal end. The internal, posterior, and external faces of the distal extremity of the tibia are flat, the last two sides separated by an angle. The internal face exhibits a shallow tendinous groove. The calcaneum has a wide longitudinal concavity of the external side. It is bounded above by a low ridge, which disappears below the fibular condyle. The cuboid face is oblique vertically and transversely, and is contracted in extent. The external face of the cuboid has a proximal concavity and a distal tuberosity running downwards and posteriorly. The external face of the fibula has a ridge which terminates abruptly a short distance above the distal extremity of the bone.

Measurements.

		M.
Diameters centrum lumbar vertebra	longitudinal	0.0070
	vertical	.0033
	transverse	.0060
Diameters of neck of ilium	anterior	.0054
	exterior	.0070
	interior	.0056
Vertical diameter of acetabulum		.0070
Width of proximal end of femur		.0137
Diameter of head of femur		.0060
Diameter of shaft of femur (transverse)		.0065
Width of rotular groove of femur		.0045
Width of tibia behind distally		.0054
Distal width of fibula		.0045
Length of astragalus		.0102
Width of trochlea behind		.0048
Diameters of cuboid	length	.0070
	width of middle	.0040

From the Wasatch formation of the Big Horn River, Wyoming. J. L. Wortman.

PANTOLESTES ETSAGICUS Cope.

Mioclænus etsagicus Cope. Paleontological Bulletin No. 34, p. 189.

Plate XXV e, fig. 21.

This, the largest species of the genus, is represented by the two rami of a mandible of an adult animal in good preservation. In their robust character the premolar teeth resemble those of *Mioclaenus turgidus*, but are not relatively so large, nor is the last true molar relatively so small, as in that species. The heel of the third premolar is obsolete, and that of the fourth is a wide cingulum. Neither exhibits an anterior basal tubercle, and in both the principal cusp is stout. The true molars widen posteriorly to the anterior part of the last molar. The latter contracts rapidly to a narrow heel. The tubercles are all subconic, and the median ones of the last molar are small. There are no cingula, and the enamel is smooth.

The ramus is not robust, and is of moderate depth. Its inferior border rises below the middle of the last molar tooth, and posteriorly. There is a "mental" foramen below the contact of the fourth premolar and first true molar.

Measurements.

	M.
Length of bases of six posterior molars	0.047
Length of bases of three premolars	.024
Length of bases of Pm. ii	.009
Length of bases of Pm. iv	.008

		M.
Length of bases of Pm. iv		.005
Diameters basis of M. ii	anteroposterior	.0075
	transverse	.0070
Diameter basis M. iii	anteroposterior	.0084
	transverse	.0070
Depth of ramus at Pm. ii		.0080
Depth of ramus at M. ii		.0140

This species is named from the Crow Indian name of the Big Horn River, *Etsagie*. Discovered by Mr. J. L. Wortman in the Wasatch beds of the Big Horn River, Wyoming.

PANTOLESTES LONGICAUDUS Cope.

Proceed. Amer. Philos. Soc., 1872, p. 467 (August 3).

Plate XXIV, figs. 13–17.

Dental formula M. 3; P. M. 4; C. 1; incisors unknown. Some of the molars in the only specimen known are so worn as to preclude exact description. They evidently possessed anterior and posterior lobes, separated by a valley, which was most expanded on the inner side. The last molar exhibits a short heel posteriorly, which probably supported a small tubercle. The last three premolars are all two-rooted and compressed in form. The fourth presents a crown composed of one large anterior, compressed cusp, and a much lower posterior heel. There is a slight cingulum in front. The canine is lost, but its alveolus indicates that it was a stout tooth.

Some caudal vertebræ found with this jaw indicate that the animal possessed a long tail.

The mandibular ramus is quite slender, and there is a large foramen below the first true molar; the masseteric fossa is pronounced. So far as the known structure goes, this species resembles the *P. chacensis*, but possesses a much larger heel of the fourth premolar.

The remains of this species were found together by the writer in the Bridger beds on Black's Fork, Wyoming.

PANTOLESTES SECANS Cope.

Bulletin U. S. Geol. Surv. Terrs., F. V. Hayden, 1881, p. 187.

Plate XXV a, fig. 6.

Represented by the adherent rami of a mandible, on both of which the posterior four molar teeth are preserved.

The species is about the size of the *P. chacensis*, and hence larger than the *P. longicaudus*. It differs from both in the proportions of its teeth, and especially in the large size and sectorial character of the fourth premolar. The length of the latter exceeds a little that of the third true molar, while in the other species it is shorter. This length is caused by the extent of the anterior basal tubercle and posterior heel. The latter is entirely surrounded by a cingulum, and its median line is elevated into a blade, which is continuous with the posterior edge of the principal cusp. The fore and aft edges of the anterior tubercle are also trenchant. The two cusps of the anterior inner tubercle of the first and third molars are well developed, but on the second molar there is but one cusp. This is probably a character to be relied on in distinguishing the species from the *P. chacensis*. No external basal cingula; enamel smooth.

Measurements.

	M.
Length of last four molars	.021
Length of fourth premolar	.0068
Elevation of fourth premolar	.0040
Length of last true molar	.0057
Depth of ramus at first true molar	.0070

As is the case with the species of *Pantolestes* already known, the *P. secans* seems to have been rare. One specimen was found by Mr. Wortman in the bad lands of the Wind River, Wyoming.

ADDENDA TO PART FIRST.

The following genera and species were received too late for insertion in their proper places in the body of the work. The title of Part First should embrace the Puerco epoch with the Wasatch and Bridger. Its fossils were not known at the time when the printing of the title and first pages of the book were in press.

PISCES.

PERCOMORPHI.

PLIOPLARCHUS Cope.

Amer. Journ. Science and Arts, 1883, May–June.

The fishes of this genus belong apparently to the Centrarchine division of the Percidæ, and although the future discovery of the structure of the ventral fins may invalidate this conclusion, I do not anticipate such a result. I am also unable to determine whether there are teeth on the vomer or not. As regards generic affinity, the species do not enter any of the genera now known from American or European tertiary formations, as will be seen from the characters about to be given. They differ from those of the recent genera of Centrarchinæ in the entire circular outline of the operculum; and from some of them in the anal fin with five spinous rays originating posterior to the line of the anterior border of the spinous dorsal fin. This new genus I have named PLIOPLARCHUS, with the following diagnosis:

Family characters, etc.—Mouth bounded above by premaxillary bone only. Branchiostegal rays seven; possibly eight. Ventral fin commencing below the base of the pectoral. Scales ctenoid.

Generic characters.—Teeth few, simple, and conic. No indication of large pharyngeal bones. Preoperculum entire posteriorly and at the angle; inferior edge unknown. Operculum rounded, entire. One dorsal fin. Anal fin commencing below middle of spinous dorsal, with five or more spinous radii. Caudal fin openly emarginate. Lateral line continuous, uninterrupted. Two species of the genus are known, both from some beds of possibly tertiary shale which lie on the Laramie beds in Western Dakota.

PLIOPLARCHUS WHITEI Cope.

Amer. Journ. Sci. Arts, 1883, May or June.

Plate XXIV g, fig. 1.

General form elongate oval, the dorsal and ventral outlines of the body about equally convex. The length of the head enters that of the head and body to the extremity of the caudal vertebræ three times, and the depth of the body at the ventral fins enters the same two and two-thirds times. The muzzle is short and obtuse, and the mouth opens obliquely upwards. The orbit is very large, and enters the length of the head to the border of the operculum three times, and is one-third of itself longer than the muzzle.

The radial formula is: D. IX – 12; C. + 17 +; A. V – 14; V. ?; P. 13. All the soft rays are fissured distally. The dorsal spines increase in length to the last one, as do also the anals. The pectoral rays reach to below the sixth dorsal spine, and beyond the extremity of the ventral fin, which does not quite reach the anal. The soft rays of the anal extend to below the extremity of the vertebral column, forming a well-developed fin. The extremity of the soft dorsal is lost. The external rays of the caudal fin are a little longer than the median. The spine of the ventral fin is not strong.

The caudal peduncle is moderately narrow. The vertebral column is convex upwards anteriorly. Its vertebræ are, caudal XVI; abdominal XII; to the edge of the operculum. A caudal vertebra preserved in place has two lateral fossæ, separated by a horizontal keel. The abdominal cavity extends posterior to the anterior spinous rays of the anal fin, so that the anterior interhæmals are directed upwards and backwards. The ribs are long. There are four interneural bones anterior to the dorsal fin. The postcoracoid is elongate.

There are seven or eight longitudinal rows of scales visible above the vertebral column, and sixteen below it, the size diminishing rapidly downwards. All the bones of the head excepting the muzzle and jaws are covered with scales. There are six rows on the cheek below the eye. The scales of the body have the basal radial grooves and ridges few and coarse. The external surface is finely but strongly rugose with tubercles or grains, with a trace of fine concentric lines near the superior and inferior edges. Marginal denticles small. The interior faces of the scales which cover part of the fossil display numerous very close and fine concentric lines, with a small triangular rough area extending from the edges toward the center.

Measurements.

	M.
Total length with caudal fin	.116
Depth at front of dorsal fin	.033
Length of caudal fin to last vertebra	.0247
Length of caudal vertebræ	.0365
Length of base of dorsal fin	.0325
Length of base of soft dorsal	.015
Length of seventh dorsal spine	.017
Length of third dorsal ray	.020
Length of fifth anal spine	.0165
Length of fourth anal ray	.022
Length of base of anal fin	.0245
Length of pectoral fin	.020
Length of ventral fin	.017
Depth of caudal preduncle	.012
Depth of head at orbit posteriorly	.026

The typical specimen of this fish is in excellent preservation. The species is dedicated to Dr. C. A. White, the distinguished geologist and paleontologist.

Plioplarchus sexspinosus Cope.

Amer. Journal Sci. Arts, 1883, May-June.

This species is represented by two specimens, both of which lack the head and body anterior to the dorsal fin. One of the specimens is accompanied by its reverse.

The differences between this species and the *P. whitei* are to be seen in the radial formula. This shows more numerous spinous and less numerous cartilaginous rays. The formula is: D. X–13; C. + 17 +; A. VI–9. The last anal radii are somewhat injured, and there may have been more than nine, but no traces of others exist, and it is clear that they were less numerous than in the *P. whitei*. There are about eighteen series of scales below the vertebral column at the front of the dorsal fin. Their external surfaces are not so rough as in the *P. whitei*, as the granules are confined to the center of the scale, and the concentric lines are much more obvious and form a wider border. Ctenoid denticles distinct. Caudal fin openly emarginate.

Measurements.

	M.
Depth of anterior edge of anal fin	.0206
Length from anterior edge of anal fin to end of caudal vertebræ	.0305
Length of caudal fin to vertebral centra	.020
Length of base of dorsal fin	.0282
Length of base of soft dorsal	.015
Length of base of anal	.017
Length of base of spinous anal	.008
Length of ninth dorsal spine	.013
Length of fifth anal spine	.015
Depth of caudal peduncle (about)	.011

REMARKS.—Among the known extinct genera of fishes it is *Mioplosus* Cope that approaches nearest this one. The former is characteristic of the Green River beds of the Lower Eocene. The genus *Plioplarchus* does not enable me to identify the horizon from which it is derived, with any of our known formations. It only permits the general statement that its age may be tertiary or upper cretaceous.

REPTILIA.

OPHIDIA.

HELAGRAS Cope.

Proceedings Amer. Philosoph. Society, 1883, p. 545.

The generic characters are drawn from vertebræ only. These display a modified form of the zygosphen articulation, as follows: The roof of the zygantrum is deeply notched on each side of the median line, so as to expose the superior lateral angles of the zygosphen. This separate median portion of the roof of the zygantrum forms a wedge-shaped body, which may be called the *episphen*. It is surmounted by a tuberosity, which constitutes the entire neural spine. The latter is thus entirely different in form from that of other serpents. Articular extremities of centrum round, the ball looking somewhat upwards. Costal articulation 8-shaped, the surfaces convex and continuous. Hypapophyses none on the two vertebræ preserved. Zygapophyses prominent. Free diapophyses none.

This genus is readily distinguished by the presence, first observed here, of the *episphen* in addition to the zygosphen, and by the peculiar form of the neural spine. We have now several vertebral articulations originally discovered in American vertebrata. These are the *episphen*, as above; the *hyposphen*, which characterizes the Opisthocœlous Dinosauria (*Sauropoda* Marsh), and the *Diadectidæ* of the Permian period; and the *zygantrapophysis*, which is present in the Diplocaulid family of Batrachia.[1]

HELAGRAS PRISCIFORMIS Cope.

Proceeds. Amer. Philosoph. Society, 1883, p. 545.
Plate XXIV g; fig. 2.

A section of the vertebra at the middle is pentagonal, the inferior side slightly convex downwards. The lateral angle is the section of the angular ridge which connects the zygapophyses. The episphen has a shallow rounded groove on its infero-posterior side, which is bounded by a projecting angle on each side at its middle. The episphen does not project so far posteriorly as the postzygapophyses, and the degree of its prominence differs in different parts of the vertebral column. In one of the two vertebræ in my possession its prominence is small. The tuberosity on its summit is a truncate oval with the long diameter anteroposterior, and equaling two-fifths the length of the arch above. It is elevated above the rest of the median line, which is roof-like, with obtuse angle. The tubercular articular facet is entirely below the prezygapophysial surface, but the free part of the prezygapophysis extends well in front of it. It is distinguished from the capitular surface by a very slight constriction. A slight ridge extends from the capitular articulation to the edge of the ball of the centrum. Below this, the surface is slightly concave, and the middle line is gently convex. The latter terminates in an obtuse angled mark just in front of the edge of the ball. This edge is also slightly free from the ball. The capitular costal surfaces do not project inferiorly quite to the line of the inferior surface of the centrum.

---

[1] Paleontological Bulletin No. 35, Nov. 11, 1882.

Measurements of a vertebra.

		M.
Length of centrum (with ball)		.0070
Diameters of ball	vertical	.0035
	transverse	.0040
Elevation of vertebra at episphen		.0085
Elevation of vertebra at middle		.0062
Width at prezygapophyses		.0120
Width of tubercular costal faces		.0105
Width of zygantrum		.0058
Vertical diameter costal faces		.0040
Transverse diameter tubercular costal face		.0028

This snake was about the size of the black snake, *Bascanium constrictor*. It is an interesting species for two reasons. First, it is the oldest serpent known from North America. Second, in the imperfection of the zygantrum we observe an approximation to the ordinary reptilian type of vertebra, from which the ophidian type was no doubt derived. In the former there is no zygosphen or zygantrum.

MAMMALIA.

POLYMASTODON Cope.

American Naturalist, August, 1882, (July), 684.

Known only from the inferior dentition. Supposed formula: I. 1; C. 0.; Pm. 0; M. 2. The first true molar is large, exceeding the second, and supports three longitudinal series of tubercles. Function of the molars masticatory.

In this genus the molar part of the dentition assumes the exclusive control of mastication, having already displayed a predominance in *Catopsalis*. The molars are similar in their general character to those of *Ptilodus* and *Catopsalis*, but the three rows of tubercles distinguish them from both.

But one species is yet known.

POLYMASTODON TAÖENSIS Cope.

American Naturalist, August, 1882, (July), 684.

Plate XXIIIc; fig. 6.

Portions of both mandibular rami of this species with the molar teeth are the only parts of this singular animal yet known to me. The inferior borders and angles of the rami are lost.

The first true molar is two-fifths of itself longer than the second molar, and viewed from above it has an oval outline, a little narrowed anteriorly, and with rounded extremities. Its tubercles are all truncate and closely packed together, so that those of the middle row have a subquadrate outline. There are eight tubercles in the internal row, twelve in the external, and nine in the median. There are no basal cingula. The second and last true molar has a pyriform outline when viewed from above, the posterior extremity being the narrow one. The contraction of the outline is regular on each side, and the posterior extremity is rounded. There are seven tubercles in the external row, five in the middle row, and only two in the internal, since the middle row forms the internal edge of more than half the length of the crown. No cingula.

Measurements.

	M.
Length of M. i	.0225
Width of M. i at middle	.0100
Length of M. ii	.0140
Width of M. ii anteriorly	.0115

Besides the three-rowed tubercles of the first molar, and the apparent absence of fourth premolars, this species differs from the *Catopsalis pollux* in the larger size, and the larger number of tubercles in each row of the molars.

The Puerco beds of Northwestern New Mexico. D. Baldwin.

CATOPSALIS Cope.

Supra, page 170.

The following additions to the characters of this genus must be made. They are derived from a second species, the *C. pollux*, described below:

Proximal caudal vertebra robust, the centrum rather short, and with biconcave extremities. The tail was evidently stout. The distal extremity of the humerus is very peculiar. There is an oval condyle on the anterior side in the position of the roller, and the internal flange is not prominent. The posterior trochlear groove is narrow, and its lateral borders are elevated, so that it resembles somewhat the rotular groove of a femur. The entire distal extremity resembles that of a flat femur, except that there is but one tuberosity resembling a condyle, instead of two. The form is almost exactly like that of the *Meniscoëssus conquistus* of the Laramie formation, which is an ally, and perhaps belongs to the same family. The arrangement described is an approach to what is seen in some of the Theromorphous saurians of the Permian. The cotylus of the ulna has a flared border adapted to the condyle of the roller.

The astragalus has a concavity to represent the trochlear groove, and the head is very small. It has a distal navicular facet, and an external lateral facet for the cuboid. It is evident that digital reduction, carried so far in the Kangaroo, has made in this genus a considerable beginning. The external two toes have been developed at the expense of the internal, which were no doubt quite small, if not so much reducéd as in *Macropus*.

The new specimen of the *C. pollux* shows that the fourth premolar is more distinct in its form from that of the genus *Ptilodus* than I had suspected. Its crown consists of a single cusp, which is compressed at the base and has a conic apex, which stands above the posterior root. The large blade of the genus *Ptilodus* is represented by the anterior edge of this cusp, and that of the anterior root with which it is continuous, at the ascent of the ramus from the concave diastema. The genealogical table of *Plagiaulacidæ* given on page 169 must then be altered. The genus *Catopsalis* must stand out of the line connecting *Ptilodus* with *Thylacoleo*, on a branch which terminates in the genus *Polymastodon*. This line will represent the diminution of the fourth premolar to probable extinction.

Catopsalis pollux Cope.

American Naturalist, August, 1882; (July), p. 685.

Plate XXIIIc; figs. 1–5.

Parts of several individuals represent this marsupial, but there are no examples of the maxillary dentition among them. The most complete ramus supports the entire dentition. Associated with this are some entire limb bones and vertebræ.

The size of this species exceeded that of the *Macropus giganteus*, and still more that of the *Catopsalis foliatus*. The ramus has the form of that of a rodent, being vertically narrowed at the diastema, and deep at the molar region. The inferior face widens and becomes flat posteriorly, and is more oblique than in the *C. foliatus*, from the greater downward extension of the external, or masseteric edge. The interior edge, on the contrary, ascends a little from the anterior inferior border, inclosing the large internal pterygoid fossa. The inferior plane commences below the anterior part of the first true molar. The symphysis is short, and was not probably strongly united, as indicated by the few rugosities of its surface. The coronoid process rises from a point opposite the posterior extremity of the first true molar.

The incisor is relatively large, and is more curved than that of a kangaroo, having the general form of that of a rodent. The acumination or bevel of the posterior face is less rapid than that of a rodent, and is perfectly gradual. The enamel band covers the anteroëxternal face as far as exposed, which is to below the anterior part of the diastema, and is gently convex in transverse section. It does not cover the entire external face, as its width is equal, while the anteroposterior diameter of the tooth increases below. The posterior face is convex, and is not much narrowed. The internal face is slightly concave, and the enamel is recurved so as to form a band on its anterior part, thus differing from most rodents. The enamel surface is delicately obsoletely line-ridged. The length of the diastema is equal to that of the combined Pm. iv and M. i. The fourth premolar is a simple tooth with a triangular transverse section, the obtuse apex of the triangle looking forwards. This edge is continued downwards by reason of the exposure of the anterior root, and is not acute. The first true molar is an elongate oval, with six tubercles on each side. These are so closely placed that their outlines are angular, and they are only separated by fissures. No cingula. The second true molar is three-fifths the length of the first, and is broadly rounded posteriorly. It supports four tubercles on the internal, and five on the external sides, and a raised edge connecting the sides posteriorly. The tubercles are appressed as in the first molar.

Measurements.

	M.
Length of ramus without incisor to a point above angle	0.094
Length from ramus to last molar, inclusive	.060
Length from ramus to fourth premolar	.025

		M.
Diameters M. i	anteroposterior	.019
	transverse	.009
Diameters M. ii	anteroposterior	.012
	transverse	.010
Diameters of incisor at base	anteroposterior	.010
	transverse	.007
Length of incisor exposed, posterior face		.020
Depth of ramus at middle of diastema		.024
Depth of ramus at middle of M. i		.033
Width of inferior face of ramus below posterior edge of M. ii		.022

An anterior caudal vertebra is short anteroposteriorly, and its long axis is oblique to the vertical axis, as in the cervical vertebræ of many vertebrates. It is flat below, and their bases indicate wide flat diapophyses The articular faces of the centra are both concave. Neural canal small; neural spine distinct.

Measurements caudal vertebra.

	M.
Length below	0.014
Depth centrum posteriorly	.016
Width centrum posteriorly	.019
Elevation of neural arch	.012

The *astragalus* is characterized by its relative width transversely, the small concavity of the trochlea, and the small head. The inner posterior angle is produced flattened and rounded, as viewed from above. The posterior and lateral internal fossæ are well marked. The external or fibular face is not longer than deep, and is nearly vertical. Its anterior inferior angle is little produced, but is separated from the edge of the trochlea by deep vertical faces both laterally and anteriorly. The navicular face is wider than deep, is convex in vertical section, and a little convex in transverse section. The cuboid facet is anteroposterior, looking outwards, and is about half the width, and the same depth, as the navicular. It is continuous inferiorly with the sustentacular facet. The latter is considerably overhung by the internal edge of the bone. From the forward divergent direction of the condyloid calcaneal facet of the astragalus, it is evident that the cuboid is a large bone, and that the external two toes are disproportionately large.

Measurements of astragalus.

		M.
Total length		0.027
Total width		.028
Diameters trochlea	transverse	.017
	anteroposterior externally	.017

	M.
Length of head	.008
Diameters navicular facet { vertical	.009
Diameters navicular facet { transverse	.012
Width of cuboid facet	.007

The Puerco beds of Northwest New Mexico. Discovered by D. Baldwin.

PTILODUS Cope.

Supra page 172.

A second species of this genus must be here recorded.

PTILODUS TROVESSARTIANUS Cope.

American Naturalist, 1882, p. 686.

Plate XXV f, fig. 19.

This species is represented by three of the characteristic fourth inferior premolars, one of which stands on a part of the ramus, giving its depth. These differ from those of the *P. mediævus* in their uniformly smaller size, and in their strongly serrate posterior edge. The number of lateral ridges is twelve, as in *P. mediævus*. Length of fourth premolar, M. .0055; elevation of do. .0040; depth of ramus at Pm. iv, .0057. Discovered by D. Baldwin, in the Puerco beds of New Mexico. Dedicated to the distinguished mammalogist Dr. E. L. Trouessart, of Angers.

BUNOTHERIA.

The name Bunotheria was proposed by me for a series of Mammalia which resemble in most technical characters the Edentata and the Rodentia. That is, they agree with these orders in having small, nearly smooth cerebral hemispheres, which leave the olfactory lobes and cerebellum entirely exposed, and in some instances the hemispheres do not cover the mesencephalum also. From the two orders in question, however, they are easily distinguished. Their enamel-covered teeth separate them from the Edentata, while the articulation of the lower jaw is different from that found in the Rodentia. It is a transverse ginglymus, with a postglenoid process in the *Bunotheria*, as distinguished from the longitudinal groove, permitting anteroposterior motion, of the Rodentia.

Such a group as is thus characterized will include two existing groups recognized as orders—the Prosimiæ and the Insectivora. The latter group has always been a crux to systematists, and when we consider the skeleton alone, as from the standpoint of the palæontologist, the difficulty is not diminished. Various extinct types discovered in latter years, chiefly in the Eocene formations, have been additions to this intermediate series of forms, giving even closer relations with the orders already adjacent; *i. e.*, the Edentata, the Rodentia, the Prosimiæ, and the Carnivora. As is known, the groups corresponding to these orders have been named respectively the Tæniodonta, Tillodonta, Mesodonta, and Creodonta. With great apparent diversity, these suborders show unmistakable gradations into each other and the two recent orders already mentioned. As such, I may mention *Psittacotherium*, which relates the Tæniodonta and Tillodonta; *Esthonyx*, which relates the Tillodonta with nearly all the other suborders; *Achænodon*, which connects Creodonta and Mesodonta, and *Cynodontomys*, which may be Mesodont or Prosimian. Then the existing *Chiromys* most certainly connects Tillodonta and Prosimiæ.

My original definitions of the suborders of the Mesodonta, given in Vol II of the United States Geological Survey under Capt. G. M. Wheeler, p. 85, omitted the Prosimiæ, and embraced a number of characters whose significance I have re-examined.[1] Thus it is impossible to characterize the Creodonta as lacking a trochlear groove of the astragalus, in view of the form of that element in *Mesonyx* and *Mioclænus*, where the groove is more or less distinct. It is impossible to distinguish the Insectivora from the Creodonta by the deficiency of canine and large development of incisor teeth. In *Rhynchocyon* the canines are large, and the superior incisors wanting, while in *Centetes* the arrangement of these teeth is precisely as in the Creodonta. As to the large *Achænodon* and other *Arctocyonidæ*, I find no characters whatever to distinguish them from the generally small Mesodonta.

In view of these inconsistencies, I have re-examined the subject, and have found the following definitions to be more nearly coincident with the natural boundaries of the divisions of this large order. The importance of

[1] See Proceedings Philadelphia Academy, 1883, p. 77, where the divisions of the Bunotheria are redefined.

the character of the tritubercular superior molar has recently impressed me (see Proceedings of the Philadelphia Academy, 1883, p. 56), as it had previously done Prof. Gill. This zoologist has already distinguished two divisions of the Insectivora (without the *Galeopithecidæ*), by the forms of the superior molar teeth. The first possesses quadritubercular molars above, the second tritubercular. That these types represent important stages in the development of the molar dentition I have no doubt. These characters far outweigh in importance those expressing the forms of the skull, matters of proportion only, with which a few systematists unnecessarily overload their diagnoses. Such characters are of little more than specific value, and serve to obscure the mind of the inquirer for a true analysis. They may be used empirically, it is true, to determine relationship when the diagnostic parts are wanting.

I proposed to transfer the Insectivora with tritubercular superior molars to the Creodonta, in spite of the fact that some of them (*Mythomys*, *Solenodon*, *Chrysochloris*) have but weakly developed canine teeth, and *Chrysochloris* has large incisors. As an extreme form, *Esthonyx* will follow, standing next the Tillodonta. It will then be necessary to transfer the Arctocyonidæ and all the Mesodonta to the Insectivora, where they will find affinity with the *Tupæidæ*. These have well-developed canines and small incisors, as in the extinct groups named. The *Chiromyidæ* must be distinguished from all of the other suborders, on account of its rodent-like incisors, combined with its lemur-like feet.

The characters of the five suborders will then be as follows:

I. Incisor teeth growing from persistent pulps:

Canines also growing from less persistent pulps, agreeing with external incisors in having molariform crowns.......................................I. *Tæniodonta.*

Canines rudimental or wanting; hallux not opposable ....II. *Tillodonta.*

Canines none; hallux opposable.. ....III. *Daubentonioidea.*

II. Incisor teeth not growing from persistent pulps:

Superior true molars quadrituberculate; hallux opposable.............IV. *Prosimiæ.*

Superior true molars quadrituberculate; hallux not opposable ....V. *Insectivora.*

Superior true molars trituberculate or bituberculate;[1] hallux not opposable. VI. *Creodonta.*

[1] The internal tubercle is wanting in the last two superior molars in *Hyænodon*. This genus, of which the osteology remains largely unknown, has been stated by Gervais to possess a brain of higher type than the Creodonta. Prof. Scott, of Princeton, is, however, of the opinion that this determination is erroneous, and that *Hyænodon* is a true Creodont in this and other respects. If so, the genus will perhaps enter the *Amblyctonidæ*.

While the above scheme defines the groups exactly, and, so far as can now be ascertained, naturally, I do not doubt but that future research among the extinct forms will add much necessary information which we do not now possess. It is possible that the group I called Mesodonta may yet be distinguished from the Insectivora by characters yet unknown. But I cannot admit any affinity between this group and any form of "Pachyderms," as suggested by Filhol, or of Suillines, as believed by Lyddeker.[1] Such suppositions are in direct opposition to what we know of the phylogeny of the Mammalia. These views are apparently suggested by the Bunodont type of teeth found in various Mesodonta, but that character gives little ground for systematic determination among Eocene Mammalia, and has deceived paleontologists from the days of Cuvier to the present time. The only connecting point where there may be doubt as to the ungulate or unguiculate type of a mammal is the family *Periptychidæ*, of the suborder *Condylarthra*. The suborder *Hyracoidea* may furnish another index of convergence.

The families included in these suborders will be the following:

TÆNIODONTA. *Calamodontidæ; Ectoganidæ.*
TILLODONTA. *Tillotheriidæ.*
DAUBENTONIOIDEA. *Chiromyidæ.*
PROSIMIÆ. *Tarsiidæ;* (?) *Anaptomorphidæ;* (?) *Mixodectidæ; Lemuridæ.*
INSECTIVORA. *Soricidæ; Erinaceidæ; Macroscelididæ; Tupæidæ; Adapidæ;*[2] *Arctocyonidæ.*
CREODONTA. *Talpidæ; Chrysochloridæ; Esthonychidæ; Centetidæ* (=*Leptictidæ* olim); *Oxyænidæ; Miacidæ; Amblyctonidæ; Mesonychidæ.*

I at one time called this order by the name Insectivora, a course which some zoologists may prefer. But a name should as nearly as possible adhere to a group to which it was first applied, and whose definition has become currently associated with it. Such an application is correct in fact, and is a material aid to the memory. There are various precedents for the

[1] Memoirs Geological Survey India, Ser. x, 1883, p. 145.

[2] Two species of *Pelycodus* must be removed from this genus and family, and be placed in the Creodonta with *Mioclænus*. They are the *P. pelvidens* and *P. angulatus*, which have the posterior inner tubercle of the superior molars, a mere projection of the cingulum. I place them in a new genus, which differs from *Mioclænus* in the possession of an internal cusp of the fourth inferior premolar, under the name of *Chriacus;* type, *C. pelvidens.*

adoption of a new general term for a group composed of subordinate divisions which have themselves already received names.

DISSACUS Cope.

Antea, p. 344.

A portion of the skeleton of *Dissacus carnifex* (antea, p. 345) includes the superior molar teeth, with some vertebræ and bones of the limbs. I take the present opportunity of making the description of that species more complete.

The general character of the superior molars is that of *Mesonyx*, but the cusps are more compressed. The last superior molar resembles the others in having a second cusp posterior to the principal one, which is not present in *Mesonyx ossifragus*. The last and penultimate molars are of the same size in this species, while the first is a little smaller, and the fourth premolar is much smaller. The last-named tooth differs from the true molars in not having a second external lobe, but instead, a large heel; also in the small size of its internal lobe. The anterior base of the crown is a little produced, but there is no tubercle. The same is true of the first true molar, but in the remaining true molars the anterior edge is recurved, so as to be a transverse low tubercle. The internal cusp is, in all the true molars, subconic; rounded on the internal face, and flattened on the external. The true molars have a posterior cingulum, which is small on the first, but strong and recurved on the second and third. There is an obtuse rib-like external cingulum on the second and third molars, but no internal cingulum. Enamel minutely roughened; smooth where used.

Measurements.

		M.
Length of last four molars		.0475
Diameters of P-m.	anteroposterior	.0105
	transverse	.0065
Diameters of M. i	anteroposterior	.011
	transverse	.0085
Diameters of M. ii	anteroposterior	.0130
	transverse	.010
Diameters of M. iii	anteroposterior	.0140
	transverse	.0125
Length of last three inferior molars		.036

A dorsal vertebra, probably posterior, has a depressed centrum nearly square in outline viewed from below, with lateral concavities and obtuse median ridge. The articular surfaces are slightly opisthocœlous. A lumbar shows subcylindric postzygapophyses as in *Mesonyx*. A femur displays a well-developed third trochanter. The rotula is strongly elevated, much more so than in the species of *Mesonyx*. Immediately above it, the section of the shaft is subtriangular, the wide base posterior.

Measurements.

		M.
Diameters of centrum of a dorsal vertebra	vertical	.008
	transverse	.015
	anteroposterior	.017
Length of a lumbar vertebra		.024
Diameters of shaft of femur below third trochanter	anteroposterior	.011
	transverse	.017
Diameters at condyles	anteroposterior	.030
	transverse	.028

SUPPLEMENT TO PART FIRST.

THE AMYZON SHALES.

In the uncertainty whether this formation belongs to the Eocene or Miocene series, I introduce it between the two.

In the American Naturalist for May, 1879, I named the strata of this epoch, that of the Amyzon beds, from the characteristic genus which it includes, and referred it to the later Eocene or early Miocene eras. Its fauna includes twelve species of fishes, distributed as follows: *Trichophanes* Cope, 3 sp.; *Amyzon* Cope, 4 sp.; *Rhineastes* Cope, 1 sp.; *Amia* L., 2 sp.; and two of birds; (*Charadrius* and *Palæospiza*). These genera have not been found represented in the fish fauna preserved in the Green River shales, which embraces eight genera and twenty-four species. But they occur in several species and specimens in the South Park of the Rocky Mountains of Colorado, associated with the genera *Rhineastes* and *Amia*, neither of which has yet been found in the Green River formation. The first named is common in the Bridger, but in a different form, and the generic identity is not yet fully established. The *Amia* is represented in the Bridger by *Pappichthys*, but in the former genus the characteristic parts have not yet been seen in the South Park specimens, so that here also the determination of the genus is not final. It however remains, that this fish fauna is different from that of the Green River beds, and the modern aspect of the genera points to an age even later than the Bridger. It is evident that the pertinence of this series of rocks to the Green River formation, asserted by King, cannot be maintained.

I published the first notice of this formation, which I examined at an exposure in the northeastern portion of Nevada, twenty-five miles northeast of Elko, on the Central Pacific Railroad. The outcrop is on the south side of the low mountain range bounding Humboldt Valley on the north. The beds are exposed in a drift and adjacent cutting, and a shaft 200 feet in depth. The strata are argillaceous and in some degree calcareous, and are very thinly laminated; so much so as to resemble thin brown or black paper in some portions of the series. They are highly carbonaceous, and burn freely; some of them with the odor of amber, which appears as a gloss on some of the laminæ Descending 60 or more feet through these shales, we reach a bed of solid argillaceous material of a dark-green color. This can be removed with the pick, but hardens on exposure to the atmosphere. It contains fresh-water shells. The first bed of coal is two and a half feet in thickness, with one or two laminæ of slate. The second bed is 12 feet deeper, and is about 3 feet in thickness. In quality both resemble cannel, but have more luster.

Masses of the laminated shales resemble the braun kohle of Bonn, Prussia; and they contain fossils disposed in the same way. These consist of multitudes of leaves, mostly of dicotyledonous plants; of mollusks, insects, and fishes; the last two often in a fine state of preservation. The mollusks present forms similar to *Planorbis*, *Viviparus*, etc. The insects are mostly diptera, and some of them are nematocera. The fishes are fresh-water forms, of which, perhaps, four species were procured. I made an examination of two of these, and found them to represent both species and genera new to science One of these is of interest, as furnishing the first evidence of the appearance of the *Catostomoid* type, now so extended in North America; the other is allied to a genus which has been discovered in the Eocene shales of Green River.

The shales are considerably less indurated in general than those of

Green River. They have been greatly disturbed by the elevation of the ranges bounding Humboldt Valley, as they dip nearly south, at an angle of forty-five degrees, at the mine.

The Tertiary shales of Florissant in the South Park of Colorado have already yielded numerous species of plants, insects, and fishes, which have been described by Messrs. Lesquereux, Scudder, and myself.* Six species of fishes have been determined, three of which pertain to the same genus of *Catostomidæ*, which I had previously ptocured from the paper coal of Osino, Nevada. On this ground, an approximation of the horizons of the two localities was made. I have recorded the occurrence of a species of the second genus found in the Osino coal, *Trichophanes*, of which the *T. hians* had been up to that time the only one known. The epochal identification of the two formations was thus confirmed.

The other species of fishes known belong to *Amia*, *Amyzon* Cope, and *Rhineastes* Cope, members of the sucker and catfish families, respectively. Both genera are nearly allied to existing forms, and the addition of the *Amia* increases the modern facies of the ichthyic fauna of the period in question. The discovery strengthens the evidence for the view that the waters inhabited by these fishes were completely isolated from access of salt or brackish water, thus differing from the beds of the Green River epoch, and occupying a later position in the scale of periods.

The regions of the John Day River and Blue Mountains furnish sections of the formations of Central Oregon. Above the Loup Fork, or Upper Miocene, there is a lava outflow, which has furnished the materials of a later lacustrine formation, which contains many vegetable remains. The material is coarse, and sometimes gravelly, and it is found on the Columbia River, and I think also in the interior basin. Professor Condon, in his unpublished notes, calls this the Dalles Group.† It is in turn overlaid by the beds of the second great volcanic outflow. Below the Loup Fork follows the Truckee Group, so rich in extinct mammalia; and below this a formation of shales. These are composed of fine material, and vary in color, from a white to a pale brown and reddish-brown. They contain vegetable remains in excellent preservation, and undeterminable fishes.

*Bulletin U. S. Geol. Surv. Terrs. 1875, p. 1, 3.

†I have published this in the Proceedings of the American Philosophical Society 1880, p. 61.

The *Taxodium* nearly resembles that from the shales at Osino, Nevada, and on various grounds I suspect that these beds form a part of the "Amyzon Group" (American Naturalist, June, 1880), with the shales of Osino and of the South Park of Colorado.

PISCES.

HALECOMORPHI.

AMIA Linn.

Two species of this genus have been found in the Amyzon shales. None of the specimens are as large as the *Amiidæ* from the Bridger formation.

AMIA SCUTATA Cope.

Bulletin U. S. Geol. Survey Terrs., 2d Series, 1875, p. 3.

Plate LX: fig. 1.

Represented by a specimen which lacks the head and body anterior to the middle of the long dorsal fin. The anal and part of the dorsal fin and the heterocercal tail are well displayed. The species differs from the existing *A. calva*, L., and its cotemporary *A. reticulata*, in the large size of its scales, of which only seven and a half longitudinal rows are visible above the vertebral column. The radii of the anal fin number nine, and the caudal vertebræ forty-six; caudal hæmapophyses number twelve.

Measurements.

	M.
Length from first caudal vertebra to end of caudal hæmapophyses	0.210
Depth of body at anal fin	.100
Length of base of anal fin	.028
Length of body of a vertebra	.005
Depth of body of a vertebra	.009

The specimen of the full size of the *A. calva*, is from near Florissant, South Park, Colorado.

AMIA DICTYOCEPHALA Cope.

Bulletin U. S. Geol. Surv. Terrs., 2d Series, 1875, p. 3.

Plate LIX; fig. 1.

Established on a number of specimens, but primarily on one in which the caudal and inferior fins are wanting, and only the posterior part of the skull remains. A second consists of the entire cranium; a third, of the tail; and a fourth, of a specimen in good condition, lacking head and tail.

The first-mentioned specimen shows that there are ten or twelve rows of scales above the vertebræ, and that the dorsal fin commences about an inch behind the line of the posterior border of the cranium. It also exhibits the strong sculpture of the surfaces of the latter to consist of narrow inosculating ridges, inclosing large and smaller pit-areas.

The specimen exhibits this sculpture to be very marked on the opercular, suborbital, parietal, frontal, and sublingual bones, the only ones where it displays the surface The branchiostegal radii number twelve, the upper large and wide. The subopercular is turned up anteriorly as in *A. calva*, and is thickened on the border of the suture with the interoperculum. The sublingual bone has much the form of that of *A. calva*, but is rather wider, and then more abruptly contracted than in a specimen of the latter before me. The orbit is smaller relatively than in *A. calva*.

It is uncertain whether this and the preceding species possessed the dentition of *Amia* or *Pappichthys* Cope, as the mandibular bone is partially and abruptly contracted near the apex, so the teeth may belong to the inner row of the true *Amia*, which is wanting in *Pappichthys*.

The fourth specimen displays the ventral fins and the characteristic femoral supports. The fins originate about an inch behind the line of origin of the dorsal fin in a specimen of $0^{m}.055$ depth of body. The scales exhibit also the dermal margin with truncate posterior outlines seen in the existing species; this character is chiefly seen on the abdominal surfaces. There are thirty-five vertebræ between vertical lines drawn from the beginning of the dorsal fin and end of the basis of the anal fin, and thirty-two dorsal radii in the same interval; anal radii, nine; vertrals, six.

Measurements.

	M.
Depth of body to vertebræ (No. 1)	0.045
Length of four dorsal vertebræ (No. 1)	.019
Depth of one dorsal vertebra (No. 1)	.010
Length of head to free border of operculum (No. 2)	.124
Depth of operculum of do	.032
Length of head on vertex	·093
Length from end of muzzle to orbit	.032
Length of orbit	.012

From South Park, Colorado.

NEMATOGNATHI.

RHINEASTES Cope.

Proceeds. American Philosophical Society, 1872, p. 486. Annual Report U. S. Geol. Survey Terrs. 1872 (1873), 638; supra, p. 60.

RHINEASTES PECTINATUS Cope.

Bulletin U. S. Geol. Survey Terrs., I, No. 2, 1874, p. 49.

Plate V, fig. 13.

This catfish is represented by a single specimen, which includes only the inferior view of the head and body anterior to the ventral fins. These exhibit characters similar in many respects to those of Amiurus, Raf, but the introperculum, the only lateral cranial bone visible, displays the dermo-ossified or sculptured surface of the Eocene genus, to which I now refer it. Other characters are those of the same genus. Thus the teeth are brush-like, and there is an inferior limb of the posttemporal bone reaching the basioccipital. The modified vertebral mass is deeply grooved below and gives off the enlarged diapophyses that extend outward and forward to the upper extremity of the clavicle. The patches of teeth are on the premaxillary, and are separated by a slight notch at the middle of the front margin. The teeth are minute. The four basihyals and the elongate anterior axial hyal are distinct; also the ceratohyal with its interlocking median suture. The number of branchiostegal radii is not determinable; three large ones are visible. The mutual sutures of the clavicles and coracoids are interlocking, and their inferior surfaces display grooves extending from the notches. The pectoral spine is rather small, and bears a row of recurved hooks on its posterior edge; there are none on its anterior face. The head is broad, short, and rounded in front, which with the uncinate character of the serration of the pectoral spine, reminds one of the existing genus Noturus.

As compared with the five species of Rhineastes, described from the Bridger Eocene, the present species is distinguished by the small size, and large uncini of the pectoral spine.

Measurements.

	M.
Length of head to clavicle (below)	0.0130
Width of head (below)	.0360
Width of scapular arch (below)	.0110
Expanse of modified diapophysis	.0200
Length of modified vertebræ	.0115
Length of pectoral spine	.0210

From the Tertiary shale of the South Park, Colorado.

PLECTOSPONDYLI.

AMYZON Cope.

Proceed. Amer. Philos. Soc., 1872, p. 480. Annual Report U. S. Geol. Surv. Terrs., 1872 (1873), p. 642.

Allied to *Bubalichthys*. An open frontoparietal fontanelle; the premaxillary forms the entire superior arch of the mouth; the pharyngeal bones are expanded behind; there are twelve to thirteen rays of the ventral fin; there is a lateral line of pores which divides the scales, piercing their margin.

Dorsal fin elongate, with a few fulcral spines in front, and the anterior jointed rays osseous for a considerable part of the length. A few short osseous rays at front of anal fin. Scales cycloid. Caudal fin emarginate. Mouth rather large, terminal.

The characters of this genus are those of the *Catostomidæ*. There are three broad branchiostegals. The vertebræ are short, and the hæmal spines for the caudal fin are distinct and rather narrow. In one specimen a pharyngeal bone is completely preserved. It is slender, and with elongate inferior limb. The teeth are arranged comb-like, are truncate, and number about thirty to forty. The dentary bone is slender and toothless, and the angular is distinct. The premaxillary appears to extend beneath the whole length of the maxillary. Should this feature be substantiated, it will indicate a resemblance to *Cyprinidæ*. The maxillary has a high expansion of its superior margin, and then contracts towards its extremity. Above it two bones descend steeply from above, which may be out of position. The preoperculum is not serrate. The superior ribs are well developed. This form approaches in its anterior mouth, the true *Cyprinidæ*, through *Bubalichthys*. It was the first extinct form of *Catostomidæ* found in this country. Species more nearly allied to the existing ones are found in the Pliocene beds of Oregon and Idaho.

AMYZON MENTALE Cope.

Proceed. Amer. Philos. Soc., 1872, p. 481. Annual Report, *l. c.*, 1872 (1873), p. 643.

Plate LIX, fig. 2, and Plate LX, fig. 2.

This fish occurs in considerable numbers in the Osino shales, and numerous specimens have been procured. Two only of these are before me at present. They are of nearly similar length, viz, M. .120 and .105. The most elevated portion of the dorsal outline is immediately in front of the dorsal fin. From this point the body contracts regularly to the caudal fin. The dorsal fin is long and elevated in front, and concave in outline, the last rays being quite short. They terminate one-half the length the fin in front of the caudal fin. The interneural bones are stout in front and weak behind. Dorsal radii III, 26, and (No. 2) (?) II, 23.

There are about twenty-three vertebræ between the first interneural spine and the end of the series in the former specimen, in which, also, there are no distinct remains of scales. In the second, scales are preserved, but no trace of the lateral line; there are six or seven longitudinal rows above the vertebral column.

The anal fin is preserved somewhat damaged; the rays are not very long, and number II, 7. The anterior interhæmal is expanded into a keel anteriorly. Ventral fins injured. The ribs and supplementaries are well developed. The inferior quadrate is a broad bone, with deep emargination for the symplectic. Depth No. 2 in front of dorsal fin, M. .025. Length basis of dorsal, .026.

AMYZON COMMUNE Cope.

Bulletin U. S. Geol. Surv. Terrs., Vol. I, No. 2, p. 50, 1874.

Plate V, fig. 21.

This is the characteristic fish of the Amyzon beds of the South Park of Colorado. In that locality it is the most abundant species.

The greatest depth of the body is just anterior to the dorsal fin, and enters the length 2.66 times the base of the caudal fin, or a little more than three times, including the caudal fin. The length of the head enters the former distance a little over 3.25 times. The general form is thus stout and the head short. The front is gently convex, and the mouth terminal. There are fifteen or sixteen rows of scales between the bases of dorsal and ventral fins. They are marked by close concentric lines, which are interrupted by the radii, of which eight to fifteen cross them on the exposed surface, forming an elegant pattern. At the center of the scale the interrupted lines inclose an areolation. The extended pectoral fin reaches the ventral, or nearly so; the latter originates beneath the anterior rays of the dorsal, or in some species a little behind that point. They do not reach the anal when appressed. The anal is rather short, and has long anterior radii. The dorsal is elevated in front, the first ray a little nearer the basis of the caudal fin than the end of muzzle. Its median and posterior rays are much shortened; the latter are continued to near the base of the anal fin. Radii D. 33; P. 14; V. 13; A. 12. The caudal is strongly emarginate and displays equal lobes.

Measurements.

	M.
Leugth of a large specimen (10.25 inches)	0.250
Length of medium specimen	.182
Depth at occiput of do.	.043
Depth at dorsal fin	.057
Depth caudal peduncle	.023
Length of head (axial)	.044
Length to D. 1 (axial)	.075
Length to end of dorsal (axial)	.131
Length to basis of caudal fin	.146
Length of basis of anal fin	.023

There are thirty-eight or thirty-nine vertebræ, of which nine are anterior to the first interneural spine, and fourteen between that point and the first caudal vertebra.

A very large number of specimens was obtained by Dr. Hayden and myself from the Tertiary shales of the Middle and South Parks, Colorado. They display but insignificant variations in all respects, and furnish a good basis of determination. They all differ from the *A. mentale* Cope, in the larger numbers of vertebræ and dorsal and anal fin radii, and greater prolongation of the dorsal fin.

AMYZON PANDATUM Cope.

Bulletin U. S. Geol. Survey Terrs., 2d series, 1875, p. 4.

Form very stout; the body deeper in relation to its length than in the known species of *Amyzon;* greatest depth just in front of dorsal fin, and two-fifths the length to basis of caudal. Length of head one-third the latter. Spines of premaxillary causing a protuberance above the end of the muzzle, as in many existing Catostomi. Mouth slightly inferior; end of muzzle obliquely truncate in profile. Dorsal fin elongate, elevated in front; radii mostly short; caudal openly emarginate; anal not very elongate in either direction; ventrals below first rays of the dorsal. Radii, D., III, 31; A., II, 11. Scales, $\frac{10.12}{10.11}$, with concentric and radiating lines well developed Vertebræ .6, 17; 10.

Measurements.

	M.
Total length	0.112
Length to basis of caudal	.093
Length to basis of anal (axial)	.071
Length to basis of ventral (axial)	.041
Depth of caudal peduncle	.015
Depth of anterior anal rays	.022
Depth at occipital crest	.030

In another rather larger specimen, which agrees with that above described, the lateral lines are well preserved.

From the South Park, Colorado.

AMYZON FUSIFORME Cope.

Bulletin U. S. Geol. Survey Terrs., 2d series, 1875, p. 5.

Represented by a very small fish, which exhibits fully ossified bones, but may be immature. It exhibits characters quite distinctive, although the caudal peduncle, anal fin, and opposite parts of dorsal, are wanting. The head is very perfectly preserved, and is of a regularly short conic form, with equal lips. The attenuated muzzle shows none of the obtuseness characteristic of the other Amyzons. Another peculiarity is seen in the ventral fins, which stand below the eighth instead of the first articulated dorsal ray. They are evidently in their normal position, and the ribs are undisturbed. The pectorals extend more than half way to the ventrals. There are seven neural spines in front of the first interneural, and sixteen between the latter and the first interhæmal. In this, as in the other species, the postclavicle is rather elongate and acute, and the parapophysial element of the anterior vertebral mass extends as far down as the line of the middle of the orbit.

Measurements.

	M.
Length of head	0.0095
Length to line of ventrals	.0180
Length to line of anal	.0239
Depth at first dorsal ray	.0105
Depth at occiput	.0070

From the South Park, Colorado.

PERCOMORPHI.

TRICHOPHANES Cope.

Proceed. Amer. Philos. Soc., 1872, p. 479 (July 29). Annual Report U. S. Geol. Surv. Terrs., 1872 (1873), p. 641.

The premaxillary bone forms all or nearly all of the superior arcade of the mouth. There are a few rows of small equal teeth *en brosse* on the dentary bone. Four rather wide branchiostegal rays are visible in the specimen. The posterior superior angle of the operculum (which is displaced in the specimen) is drawn out into an acute short spine; otherwise the bones of the head are smooth. There is a row of small teeth *en brosse* probably on the palatine or pterygoid bone. The anterior vertebræ are unmodified, and the centra are not elongate. A strong acute spine supports the dorsal fin, and a similar one the anal fin, in front. There is an elongate postclavicle on each side, which extends parallel with the femur to the base of the ventral fin. The femur is divided; the external portion is straight, and extends to the clavicle, while the other portion is curved inward and forward, reaching the apex of the corresponding bone of the opposite side. Ventral radii, 8. The dorsal originates above the ventral fin. The scales are peculiar, and characteristic of the genus. They are very thin, and without or with minute sculpture. Their borders are fringed with long, closely-set, bristle-like processes, which correspond to the teeth of the ctenoid scale.

This genus, *Amphiplaga*, and *Erismatopterus* form a group which probably belongs to the family of *Aphrodediridæ*, which is represented in modern American waters. *T. foliarum*, the only one in which the parts are large enough and sufficiently well preserved for observation, exhibits the furcate character of the femora which characterizes the family in question among Physoclystous fishes.

The first described species is the *T. hians*, from Osino, Nevada; the *T. foliarum* and *T. copei* are from the South Park of Colorado. I do not possess a specimen of the last named, which was described by the Princeton paleontologists.[1] It differs from the *T. foliarum* in its smaller scales.

[1] Pal. Report Princeton Sci. Expedition, I, 1878, p. 98.

TRICHOPHANES FOLIARUM Cope.

Bulletin U. S. Geol. Survey Terrs., IV, p. 73, 1878.

Plate LIX; fig. 4.

The scales extend on the cheeks and abdomen; there are nine or ten longitudinal rows above the vertebral column, and about sixteen below it. The head is moderately elongate, and deep behind. The mouth is subterminal, and the extremity of the premaxillary bone extended backward would reach about half way to the orbit. Ribs stout; neural spines slender. The interneurals visible number eleven, but the posterior part of the dorsal fin is wanting. These bones have thin anterior and posterior laminar expansions. The anterior interneural strikes the fifth vertebra from the head; between this one and the first interhæmal there are nine vertebræ.

Measurements.

	M.
Length of the head to first vertebra	0.028
Depth of head posteriorly	.022
Length of mandibular ramus	.013
Length to scapula	.035
Length to dorsal fin	.050
Depth at middle of dorsal fin	.023

From the Tertiary shales of Florissant, Colorado. Obtained by my friend Dr. S. H. Scudder, of Boston, collaborator of the United States Geological Survey of the Territories.

The *Trichophanes foliarum* is represented by a much larger individual than the *T. hians*, but which wants the posterior part of the body, including the caudal and part of the anal fin. The generic and family characters are, however, very clearly visible in the anterior portion of the skeleton.

TRICHOPHANES HIANS Cope.

Proceed. Amer. Philos. Soc., 1872, p. 480. Annual Report U. S. Geol. Surv. Terrs., 1872 (1873), p. 642.

Plate LIX; fig. 3.

Vertebræ, D. 9; C. 15; six between interneural spine of dorsal, and interhæmal of anal fin. Radii, D. II. (?) 6 (soft rays somewhat injured); A. II. 7; V. and P. not all preserved; caudal rays numerous, forming a deeply bifurcate fin. The ventrals reach a little over half way to the anal, and the latter about half way from its base to that of the caudal fin. The dorsal fin, laid backwards, reaches the line of the base of the first anal ray. The first dorsal ray is a little nearer the end of the muzzle than the origin of the caudal fin. The muzzle is very obtuse, and if the specimen be not distorted, not longer than the diameter of the orbit. The gape extends at least to the posterior line of the orbit. The suborbital region is deep posteriorly.

In its present somewhat distorted condition the specimen measures in—

	M.
Total length	0.059
Head	.016
Vertebræ	.029
Caudal fin	.0142
Length dorsal spine	.008
Length anal	.008
Length of hair-like fringes	.0005

The coracoid is not wide; the postclavicle has a proximal conchoidal expansion, and a long slender shaft extending to the anterior extremity of the femora. Caudal fin furcate. Interneural spines wanting anterior to dorsal fin; those of the anterior rays very strong. Interhæmals of the anterior anal rays similarly strong. Caudal fin embracing one vertebra.

From the coal shales north of Osino, Nevada; obtained by the writer.

AVES.

Two species of birds have been obtained from the Amyzon shales, which are represented by very few specimens. The best preserved specimen found up to the present time is thought by the able zoölogist, J. A. Allen, to be a passerine bird of the family *Fringillidæ*, and he accordingly names it *Palæospiza bella.** The second species was described by the writer as a plover, under the name of *Charadrius sheppardianus.*

The specimen includes three vertebræ anterior to the pelvis; the pelvis, with the vertebræ which it incloses, and the caudal vertebræ; both femora; the tibia and part of the tarsometatarse of the right leg, with the greater part of the left tibia. One-half of the tail is preserved, the feathers lying in almost undisturbed relation. There are also various light and downy

*Bulletin U. S. Geol. Surv. Terrs., IV, 1878, p. 443, Pl. 1.

feathers of the base of the tail and adjacent parts of the body lying on the block, some in place, others loose.

The characters displayed by the bones and feathers are those of a species of the order *Grallæ* and tribe *Limicolæ* (*Totanideés* A. Milne Edwards). In the absence of important parts of the skeleton, it is not possible to ascertain the family characteristics, but it is more easy to assign the species to its genus. I cannot detect any features which forbid its reference to the genus *Charadrius* in the large sense. It presents important resemblances to the species of *Totanus*, but there are some reasons, to be mentioned later, why this reference is inadmissible. It is clear that there are various genera of *Scolopacidæ* to which it cannot be referred, on account of the form of its *ossa ischii.*

GRALLÆ.

CHARADRIUS L.

CHARADRIUS SHEPPARDIANUS Cope.

Bulletin U. S. Geol. Surv. Terrs., VI, p. 83, Feb., 1881.

Plate LIX, fig. 5.

Femur one-half the length of the tibia; nine caudal vertebræ; tail gently wedge shaped, apparently without color cross-bars. The preïliac vertebræ are distinct from each other and have only moderately elongate centra. The diapophyses are of moderate length and less width, and are truncate at the extremity. Those of successive vertebræ are connected by but few osseous ligamentous spicules. The caudal vertebræ are short and wide, and have short diapophyses, except the first, which has a long narrow diapophysis. The last three are in profile, and do not display hypapophyses. The plowshare bone is an elongate triangle, considerably produced to its superior angle. The basal cotylus for articulation with the centrum immediately in front, is well excavated.

The pelvis is short and rather wide posteriorly. The fossil presents a superior view of it, with both the pubic bones turning their external faces upwards. The external borders of the anterior plates of the ilia are broken away, but enough remains of their inner portion to show their anterior extent. The postacetabular ridges diverge outwards and terminate in a prominent angulation of the posterior border, which is about equidistant between the vertebral border of the ilium and the externo-inferior border of the pubis. The posterior outline thus differs from that seen in various genera of *Scolopacidæ*, where the angle is much nearer the vertebral border, and where a second angle is produced by a notch at the point of junction of the ischium. The pelvis of *Totanus* is, however, much like that of the present species. External to the angular projection described, the border is notched, and then turns posteriorly, forming a gentle curve, which continues from the ischium to the slender pubis. The pubis is long and very narrow, and extends well posterior to the ischium. It is of uniform diameter, and is not expanded distally. The ischiopubic foramen is long and narrow, about one-seventh as wide as long. The obturator foramen is about one-third the length of the ischiopubic, and is oval. A transverse line cutting the anterior border of the acetabulum divides the pelvis between the posterior angular projection of the ilium (the true crest, *fide* Gegenbaur) and the anterior extremity into two parts of equal length.

The leg bones are quite slender, and are similar in proportions to those of several species of *Charadrius* and *Totanus*. They are more slender than in various species of *Scolopax*, *Strepsilas*, *Tringa*, &c., and less so than in *Himantopus* and *Recurvirostra*. The former is just half as long as the tibia, and seen in profile is almost straight. The crest of the tibia is very prominent, but is not produced proximally. The distal extremity of the tibia and proximal part of the tarsometatarse are so damaged as not to furnish satisfactory characters.

There are five rectrices visible in the specimen. Those which are in all probability median are the longest, while the external two are of equal length. This gives the outline a rather short wedge shape as the feathers lie closed. The expanded tail would be rounded, with a slight median angulation. The extremities of the feathers are rounded, and their whole structure is soft and delicate. The length of the longest rectrix is just about that of the tibia.

Measurements.

	M.
Length of the preïliac vertebræ	0.010
Length of centrum of first vertebræ	.0035
Length of sacrum	.021
Length of caudal vertebræ on curve	.0145
Length of plowshare bone to apex	.005
Length of ilium	.024
Length of ilium to acetabulum	.012
Length of ischium from acetabulum	.016
Length of pubis from acetabulum	.019
Width between posterior angles of ilia	.009
Length of femur	.024
Diameter of femur at middle	.003
Length of tibia	.047
Anteroposterior diameter at head	.006
Diameter of shaft at middle	.0027
Diameter of head of tarsometatarse	.004
Length of median rectrix from plowshare bone	.046
Length of external rectrix from plowshare bone	.040
Width of portion of tail preserved	.020

The strongly contrasted light and dark shades of color are not unfrequently preserved in the insects of this formation. I suspect that had the rectrices of this species originally displayed the alternating white and dark cross-bars characteristic of the *Totani*, some trace of them would be discoverable in the fossil, in spite of the fact that the entire feather is represented by carbon only. The brown tint of the specimen, both light and dark, is uninterrupted by pattern of any kind.

The tail is rather longer than in the *Tringæ*, about equal to that of many plovers and *Totani*, and shorter than that of *Actiturus*.

The *Charadrius sheppardianus* was discovered near Florissant, Colorado, by Dr. G. Hambach, a skillful naturalist. I have named it in honor of Edwin Sheppard, of Philadelphia, an excellent ornithologist and skillful artist.

PART SECOND.

THE WHITE RIVER AND JOHN DAY FAUNÆ.

PART SECOND.

THE WHITE RIVER AND JOHN DAY FAUNÆ.

REPTILIA.

No vertebrata of a rank inferior to the reptiles have yet been found in the Lacustrine Tertiary formations of the central part of North America which lies between the Upper Eocene and Loup Fork horizons. Nevertheless, as fishes are known from the Bridger, Amyzon, and Loup Fork beds, and *Batrachia* have been obtained from the Green River and the Loup Fork, it is probable that they existed in the West during the intervening time.

The *Reptilia* are represented by a good many species and a moderate number of individuals, but were much less numerous than during the epochs of the Eocene. The order of *Crocodilia* apparently became extinct; a small remnant of the tortoises only continued, and the number of species of lizards greatly diminished. The *Ophidia* only, retained their position, if we can judge by the moderate number of species discovered so far in the two formations. As to their characteristics, the Reptilia of the Miocene beds resemble those of the recent period much more closely than do those of the Eocene. All of the peculiar types of that period had disappeared, and those which came upon the scene can nearly all be arranged in existing families, and in one instance (*Testudo*), in a recent genus. The numerical and horizontal relations of these forms may be represented as follows:

Number of species.

	White River.	John Day.
Crocodilia	0	0
Testudinata	4	1
Lacertilia	8	1
Ophidia	4	1

The species are none of them of remarkable size, but coincide with the average of the species now found in the same region of the continent.

TESTUDINATA.

The species of this order which have been found in the Miocene beds were of terrestrial habits. They belong to two genera, *Testudo* and *Stylemys.* That the lakes of this period were haunted by aquatic tortoises was to have been anticipated, but as yet they have not been found.

TESTUDO Linn.

As no jaws of the Miocene species have yet been found, it is not certain whether they belong to the genus above named, or to *Xerobates* Agass., with the Loup Fork species.

Five species which I refer here, left their remains in the White River beds of Colorado. As but one species had been previously identified from the White River formation, the discovery of these was unexpected. The numerous individuals found by Dr. Hayden in the Mauvaises Terres of Dakota and Nebraska were all referred by Dr. Leidy to the *Stylemys nebrascensis.* The same species is found in Colorado, but associated with the five species here described.

The differences between these Testudos are well marked, and are best appreciated by reference to the plastron, though not wanting in the carapace also. The principal characters are the following:

I. Each half of the lip of the plastron trihedral, as deep as wide.
Marginal bones short and thick, with simple dermal grooves .... *T. cultratus.*
II. Halves of the lip of the plastron flat, thinner than wide.
a. Gular scuta square.
Lip prominent, with straight sides .... *T. quadratus.*
aa. Gular scuta triangular.
Lip prominent, narrowed forwards, border with several teeth; posterior lobe dentate; marginals wide, mucronate .... *T. laticuneus.*
Lip prominent, with parallel sides and entire margin; marginals wide, with notched border .... *T. ligonius.*
Lip not prominent, truncate, entire; marginals wide, notched medially.. *T. amphithorax.*

None of the varieties of the *Stylemys nebrascensis* present the well-defined lip of the plastron seen in these species, nor do its marginal bones ever have the projecting mucros or the step-like notches described above. In the *T. laticuneus* I have observed the tarsus and metatarsus. The latter is composed of much shorter bones than the corresponding ones in *Stylemys*, leading to the supposition that it has the short phalangeal series of the family *Testudinidæ*.[1] The positive determination of this point remains for future investigators.

TESTUDO CULTRATUS Cope.

Paleontological Bulletin No. 15, p. 6, August 20, 1873. Annual Report U. S. Geol. Surv. Terrs., 1873 (1874), p. 511.

Plate LXIII; figs. 1-3.

Parts of two individuals of this species were obtained by my expedition of 1873.

This is the smallest of the five species of *Testudo*, having about the average size of the *Stylemys nebrascensis*. This is indicated by the costal and marginal bones which accompany the portions of plastron of both specimens. The width of the lip at the base is also less than that of any of the other species, but the length and thickness are remarkable as compared with the other dimensions. The width and thickness at the base of the lip are nearly equal; both dimensions diminish to the apex, which is obtusely acuminate. The superior face of the lip is gently convex in both dimensions. The inferior surface is plane anteroposteriorly; transversely it rises to the external edge, which is subacute. The suture of the gular scutum is directed posteriorly, giving the usual triangular form. The thickness of the lip is abruptly reduced above, where the surface descends to the mesosternal bone.

The lip of one of the specimens is fissured deeply, in an eccentric manner, on both sides of central core. Whether this or the unfissured condition is characteristic of the species or not, is uncertain. It appears to be homologous with the dentation in the lip of *T. laticuneus*.

The marginal bones are robust, and are much thickened below. The edges of those of the posterior margin are acute, while those of the anterior border are obtuse, thus differing from those of *T. laticuneus*, where they are acute. At the points where the dermal sutures reach the margin, both

[1] See page 109 of this work, where the families of *Testudinata* are characterized.

specimens are unfortunately broken in every instance, so that the question of notches or mucros cannot be decided. The costal bones are moderately thick, and alternate in width, narrower and wider. The dermal sutures do not display raised margins. The anal marginal bone is wedge-shaped, with the posterior margin representing a truncate apex. Its surface and margin are convex, and the anterior sutural margin is concave.

A fragment, which is in all probability the posterior lobe of the plastron, is characteristic. It is thick, and its inferior face is somewhat recurved posteriorly. The outline of the margin presents a pronounced obtuse angle, and the edge is several times abruptly notched.

Measurements of No. 1.

			M.
Diameters of half of lip	at base	vertical	.040
		transverse	.043
	length outer edge		.056
Diameters second marginal from anal	thickness		.019
	width		.030
Length free margin of anal			.026
Width of anal above			.050
Thickness of a vertebral bone			.011
Thickness of a costal at middle			.009

Found by myself near the head of Horse Tail Creek, in Northeastern Colorado.

TESTUDO QUADRATUS Cope.

Plate LXI; fig. 5.

This tortoise was, perhaps, the largest of those of the White River beds. It is, unfortunately, only represented so far in my collection by the lip of the plastron and by an imperfect marginal bone. The former fragment indicates clearly a species quite unlike the others here described.

The lip is thick at the base, and the superior surface descends gradually to both anterior and lateral edges. The middle of the former is notched, but there are no other indentations of importance. The lateral borders do not converge, as in *T. laticuneus*, but rather diverge, as in *Hadrianus octonarius*. Its form is more like that of *T. ligonius* than any other species, but it differs from it and from all other species known to me in the absolutely transverse posterior gular suture. The suture follows an angular groove, which cuts the lip off from the inferior surface of the plastron, which lies behind it. The latter swells prominently behind the suture, but less so at the middle line than laterally. The marginal is probably the nuchal, and displays a part of the nuchal scute. The latter was at least not very narrow, and its free edge is concave. The borders of the dermal grooves are raised.

Measurements.

	M.
Length of lip from gular groove	.070
Width at base	.120
Thickness at base	.029
Length of nuchal scute	.041

Head of Horse Tail Creek, Northeastern Colorado.

TESTUDO LATICUNEUS Cope.

Paleontological Bulletin No. 15, p. 6, August 20, 1873. Annual Report U. S. Geol. Surv. Terrs., 1873 (1874), p. 511.

Plate LXI; fig. 1.

A number of fine chelonites which I obtained during my expedition of 1873 are, unfortunately, not accessible for description at present. Among these will probably be found some of the specimens of this tortoise, to which my original description refers. One individual, in a good state of preservation, must serve as the basis of the present description. It is, however, one of the best which I procured, although a little distorted in some places by pressure.

The general form is broad and depressed. The anterior and posterior outlines of the carapace are subtransverse, the former being nearly straight and the latter a little convex. The lip of the plastron projects much beyond the anterior border of the carapace, while the posterior lobe is included within that of the carapace. The free marginal bones are straight, but few of them being a little concave on their superior face. The anal bone is truncate wedge-shaped, with the free edge and the general surface gently convex. When the dermal scutal sutures reach the margin, there is a short acute mucro on both anterior and posterior marginal bones. There are no keels nor tuberosities on the carapace.

The plastron is rather flat, indicating a female animal. The anterior lip is large, and each half is rather longer than it is wide at the base of the external side. The exterior borders converge anteriorly, leaving a short margin, which presents six strong teeth, three on each half, of which the adjacent median ones are the widest. The posterior lobe is divided by a deep notch into two rounded lobes, whose border is notched three times, leaving four teeth of smaller size than those of the lip.

The lateral suture of the nuchal marginal bone is much less oblique than in *T. amphithorax* and other species. The entosternal bone has eight sides. It is as wide as long, and narrows forwards. The dermal sutures are grooves with raised edges. This character is quite marked, and cannot be looked upon as indicating immaturity, as the animal is evidently adult. The nuchal scute is as wide as long, offering a strong contrast to that of *T. ligonius*, where it is long and narrow. Its lateral sutures terminate in mucros, which leave a concave margin between them. Vertebral scuta a little longer than wide. The anal scutum has a straight superior border, and is twice as wide as long. The gular scuta have a very oblique posterior border, and each one is about as long as the external free edge of the lip. The humero-pectoral suture retreats abruptly backwards from the free humeral border before becoming transverse. At its middle it is as far anterior to the pectoro-abdominal suture as the latter is from the abdomino-anal, thus leaving a much wider pectoral scutum than is found in *T. ligonius* and *T. amphithorax*.

Measurements.

	M.
Length of carapace (axial)	.408
Width of carapace (axial)	.356
Length of plastron (greatest)	.438
Width of base of anterior lobe	.200
Width of base of lip	.085
Length of lip on external edge	.045
Greatest length of posterior lobe (axial)	.111
Width of posterior lobe at base	.200
Width of anal marginal above	.070
Width of anal marginal on margin	.037
Length of nuchal scute	.027
Length of second vertebral scute	.082
Width of second vertebral scute	.080

From the head of Horse Tail Creek, Colorado.

TESTUDO LIGONIUS Cope.

Paleontological Bulletin No. 15, p. 6, August 20, 1873. Annual Report U. S. Geol. Surv. Terrs., 1873 (1874), p. 511.
Plate LXI; figs. 2, 3.

Parts of carapace and plastron of one specimen represent this species in my collection. In the locality where I found this and most of the other specimens of tortoises from the White River beds here described, specimens were very numerous. My means of transportation being limited, I could not collect a full series, but selected diagnostic bones from the best examples for purposes of future study. Of the present individual I have the clavicle ("episternal"), nuchal vertebral, several marginals, part of pectoral, and basal part of half of posterior lobe of plastron. This is a large species, with the marginal bones and edges of plastron thickened. The lip is produced, and has parallel sides and a truncate free edge without denticulations. The angle at which the borders would meet if produced, is obliquely cut off. The inferior surface is deeply, openly grooved along the oblique line of the gular dermal suture, and is excavated on the middle sutural line just anterior to the mesosternum. The superior face of the lip slopes outwards, and is swollen laterally behind. The free marginal bones of the carapace are not recurved at the margin. There is a projecting angle or point where the dermal suture reaches the margin, which is immediately followed by a step-like notch of little depth.

The lateral suture of the gular scute on the upper side of the lip is nearly parallel with the median suture, giving thus a paralellogramnic outline. The nuchal scutum is long and narrow, and is not bounded on the edge by mucros. The pectoral scutum is very narrow.

Measurements.

	M.
Length of lip on external edge	.035
Width of half of lip at base	.046
Thickness of lip at base	.023
Width of a free posterior marginal	.082
Length of a free posterior marginal	.072
Width of posterior lobe plastron at base	.188
Anteroposterior width of pectoral scute	.018
Length of nuchal scute	.040
Width of nuchal scute	.008

Head of Horse Tail Creek, Colorado.

TESTUDO AMPHITHORAX Cope.

Paleontological Bulletin No. 15, p. 6, August 20, 1873. Annual Report U. S. Geol. Surv. Terrs., 1873 (1874), p. 511.

Plate LXI; fig. 4.

Of this large species I have before me two specimens which display the anterior lobe of the plastron, and probaby a third which does not include that diagnostic part.

The *T. amphithorax* agrees in many respects with the *T. ligonius*, but differs much in the characters of the anterior lobe of the plastron. There is scarcely any lip, but the margin is truncate, with a bounding angle on each side to represent the usual margin of the lip. In one specimen this angle is much rounded. The inferior surface differs from that of *T. ligonius* in being without impressed groove in the line of the gular suture and median fossa. In these respects it resembles *T. laticuneus*. One anterior lobe was found with the posterior lobes of two individuals, which are exactly alike. This lobe was deeply emarginate, and each sublobe is broadly rounded, wedge-shaped, with thin edge, with a notch in it at one point. The margin becomes very thick near the groin.

The relations of the inferior and superior bounding gular scuta are different in this species from what is seen in *T. ligonius*. The superior groove has the posterior position appropriate to a long lip, while the inferior leaves the margin much anterior to that point, and just posterior to the angle which bounds the very short lip which the species actually possesses. The two sutures nearly correspond in position in the *T. ligonius*. The pectoral scutum is quite narrow in the specimens, as in *T. ligonius*. The scutal sutures are impressed and do not have raised borders. The edges of the marginals are all injured in the specimens. The mesosternal bone is a little wider than long. It is more contracted anteriorly than posteriorly, but the lateral angles are about opposite the middle of the length.

The costal and vertebral bones are rather thin. A marginal bone, probably one of those in contact with the anal, has a peculiar form, such as does not exist in any other species, unless it be the *T. ligonius* and *T. quadratus*, where I have not seen it. It is recurved at the sutural edges, or is concave at the middle, so as to be openly trough-shaped. This indicates a very convex anal marginal bone, and a doubly sigmoid flexure of the free margin behind

Measurements.

No. 1.

	M.
Width of truncate lip	.080
Width of humeral scute fore and aft	.024
Length of posterior lobe on axial line of a half	.151
Width of base of posterior marginal lobe of plastron	.092
Length of axial of trough-shaped marginal (anteroposteriorly)	.072

No. 2.

	M.
Width of lip	.089
Width of mesosternum	.103
Length of mesosternum	.096
Width of anterior lobe at base	.210
Length of anterior lobe (axial)	.125
Thickness of a costal bone	.080

Head of Horse Tail Creek, northeast Colorado.

STYLEMYS Leidy.

Cope diagnosis, Transactions Amer. Philos. Society, XIV, 1866, p. 123.

This genus probably belongs to the *Emydidæ*, in the neighborhood of *Manuria*, now existing in Eastern Asia. It is characteristic of the American lacustrine Miocene, but is represented by very few species.

STYLEMYS NEBRASCENSIS Leidy.

Report U. S. Geol. Surv. Terrs., I, p. 224, 1873. Proceed. Acad. Phila., 1851, p. 172. *Testudo hemisphærica, T. oweni, T. culbersoni*, and *T. lata*, Ancient Fauna Nebraska, 1853, 105–110. Pl. XX–XXIV.

I obtained this species in Northeastern Colorado, in the White River formation, with the Tortoises already referred to the genus *Testudo*. Some of the specimens are of large size, others small. I have also numerous well-preserved chelonites from the John Day River beds of Oregon, which I refer to the same. Many of the Oregon specimens differ in some respects from the typical forms of the *S. nebrascensis*, but the characters are not constant. I observe three principal characters of the smaller and medium-sized specimens from Oregon. These are, first, the greater transverse length of the first marginal dermal scutum. This is caused by the more external position of its external bounding suture; instead of being nearly continuous with the lateral suture of the first vertebral, it is some distance external to it. This I find in all the Oregon specimens. Second, the less abrupt posterior flexure of the pectoro-humeral suture where it leaves the free margin of the anterior lobe on each side. A large Oregon specimen has the recurvature nearly as abrupt as in the Colorado and Nebraska specimens, and Leidy figures one of the latter (Ancient Fauna of Nebraska, Pl. XX), where the recurvature is as in some of the Oregon specimens. Thirdly, a convexity of the pygal and anal bones. This does not appear in a very large and a very small specimen. It may be characteristic of the male sex.

One of the Oregon specimens is much larger than the others, and has the anterior lobe a little more produced, in this respect resembling the largest Colorado specimen. I add the measurements of these two specimens:

Measurements.

No. 1.—Colorado.

	M.
Length of plastron	.390
Width of plastron at middle	.310
Width of anterior lobe at base	.200
Length of anterior lobe	.145
Elevation of carapace	.210

No. 2.—Oregon.

Length of plastron	.420
Width of plastron at middle	.345
Width of anterior lobe at base	.215
Length of anterior lobe	.157
Elevation of carapace	.170

The Oregon form has been regarded by Leidy as a distinct species, with the name *Stylemys oregonensis.** He adduces the thinness of the vertebral bones as its distinctive character. For the present I cannot admit it as more than a variety, although the skull and feet must be known before its position can be finally decided.

LACERTILIA.

Several species of this order were discovered by myself in the White River beds of Colorado in 1873, and a single species was found by one of my parties in the John Day beds of Oregon. With the exception of the *Peltosaurus granulosus* they rest upon fragmentary remains, which are in some instances not sufficient to furnish evidence of the position of the species in the system. The genera which are determinable display affinity to types now existing in the warmer parts of North America.

* Report U. S. Geol. Surv. Terrs., I, p. 225.

PELTOSAURUS Cope.

Paleontological Bulletin No. 15, p. 5, August 20, 1873. Annual Report U. S. Geol. Surv. Terrs., 1873 (1874), p. 572.

Premaxillary undivided, with spine; a zygomatic, postorbital, and parieto-quadrate arches. Parietal bones united. Teeth pleurodont, with obtuse, compressed crowns, of similar form on all the jaw-bones. Body covered with osseous scuta, which are united laterally by suture. Vertebræ depressed, without zygosphenal articulation. Median hexagonal dermal scuta on the parietal bone.

There are sufficient remains of the typical species of this genus to furnish a basis for an estimation of its affinities, a point of some interest, as this has been seldom if ever done in the case of a terrestrial lizard of the Miocene epoch. The primary group to which it is to be referred is not difficult to determine.[1]

The frontal and parietal bones are each undivided, and there is no fontanelle in either, or in their common suture.[2] There is a large postfrontal, and the usual cranial arches are present, and the quadrato-jugal absent. The frontal possesses strong lateral inferior crests, but whether they underarch the olfactory tube completely, the specimen does not show. All the usual elements of the mandibular ramus are present, but the angular is very narrow. The dentary does not extend behind the coronoid on the external face of the jaw. The coronoid is little produced either forward or backward above, but sends a process forward on the inner face of the dentary. The splenial is well developed, but becomes very slender anteriorly; it covers the meckelian groove, except for a short space distally, where it furrows the inferior aspect of the jaw. The surangular is quite peculiar; it is massive, and lacks the usual deep fossa for the pterygoid muscle, and has a broadly truncate superior margin. It is in the same vertical plane as the dentary, and not oblique or subhorizontal as in most *Gecconidæ*. The dental foramen is small, and pierces its inner face. The posterior angle of the ramus is broken off.

---

[1] See the author's Osteological Characters of the Scaled Reptiles, in Proceedings Academy Philadelphia, 1864, p. 224.

[2] What I originally thought was such is a foramen-like sinus in the posterior margin of the parietal.

The characters of the premaxillary bone, fontanelle, dentition, coronoid, dentary, splenial bones, and meckelian groove, place this genus out of the pale of the acrodont families. The parietals and vertebræ are distinct from anything known among the geccos. There is no resemblance in essentials to the *Amphisbœnia*, so that we must look for its place among the numerous pleurodont families. Here the absence of the knowledge of the periotic bones and sternum somewhat embarrasses us; but other indications are clear. The coincidence of the want of parietal fontanelle with the lateral frontal plates refers us at once to the *Leptoglossa* or *Diploglossa;* a reference confirmed by the simple frontal, and strong cranial arches. The massive form of the surangular bone, and reduction of the angular, at once distinguishes *Peltosaurus* from any known family of the tribe *Leptoglossa*, and constitutes a point of near resemblance to the *Gerrhonotidæ*. This appears to be a real affinity, which is further confirmed by the presence of a symmetrical dermal scutellation on the top of the head.

Referring *Peltosaurus*, therefore, provisionally to the *Gerrhonotidæ*, it remains to consider the generic characters. The temporal fossa was not roofed over by true bone, though the border of the postfrontal encroaches on it; and it is rather small. The orbits, on the other hand, are large, and the malar bone forms a segment of a circle. The parietal thins out behind, and its posterior border has a subround excavation. The two median dermal scuta, which left their impressions on the parietal bone, represent the interparietal and postinterparietal plates respectively; the latter especially characteristic of the *Gerrhonotidæ*, and not found in leptogloss or diplogloss families generally; those possessing it being the *Lacertidæ* in the former, and *Anguidæ* in the latter. The most prominent character which distinguishes this genus from *Gerrhonotus* is the existence of the osseous scuta which covered the body. Even the form of these is similar to the corresponding dermal scuta of the existing genus. They are rectangular and are arranged in transverse bands on the body, those of one row overlapping the bases of those of the next row posteriorly, or imbricated. The scuta of each row are joined on their long sides, by minute suture.

PELTOSAURUS GRANULOSUS Cope.

Paleontological Bulletin No. 15, p. 5, August 20, 1873. Annual Report U. S. Geol. Surv. Terrs., 1873 (1874), p. 513.

Plate LX, figs. 1-11.

I took numerous parts of the skeleton of an individual of this species from the White River chalk-bed on Cedar Creek, in Northeastern Colorado, during my expedition of 1873. The cranial fragments include the premaxillary and parts of both maxillary bones; the malar of one side; the coössified frontals and parietals separated from each other. Also the greater part of both mandibular rami. There are several vertebræ, and many dermal scuta.

The parietal is a wide element, and has a nearly plane superior surface. The granular sculpture vanishes posteriorly, leaving a wide, smooth band at the posterior border of the bone. The scutal areæ are rather narrow, each one longer than wide, and occupying just half of the length of the granular portion of the surface. The frontal suture is perfectly straight. The frontal bone is longer than wide, and is slightly convex in both directions. The greater part of it is preserved, and displays no dermal scuta, hence the frontal al scutum was large and undivided. The lateral olfactory ridges are massive and not very deep, being much thicker than in *Exostinus serratus*, and are strongly beveled on the external side; that is, to the supraorbital border. The malar and postorbital bones together form a rather slender arc of a circle, indicating a large orbit. The premaxillary bone has a wide, flat spine, which is so strongly inclined as to be nearly horizontal. No portion of it protrudes beyond the alveolar border, which supports seven teeth. There is a foramen at the base of the spine on each side. The external surface of the premaxillary is entirely smooth. On the external face of the maxillary the usual foramina occur at the usual intervals. In the mandible the post-dentary parts are extended, about equaling the dentary (the angle is broken from both rami), and are robust. The anterior fourth of the meckelian groove is exposed. The external face of the dentary is convex, and is pierced by five foramina, all in the anterior half. The superior aspect of the surangular region is truncate or flattened, with angles separating it from the lateral faces, as in other *Gerrhonotidæ*.

There are ten teeth in .10^{m} of the maxillary bone, and there are twenty-four in the dentary bone. All are closely placed. Those of the superior series are a little more robust; their inner faces are convex, giving the shaft a slight curve in profile. The apices of the teeth are not expanded, but are wedge-shaped, with a cutting edge in the long axis of the jaw. A strong bevel of the crown to the external maxillary face gives their apices an especial robustness. Under the microscope their surface is delicately parallel-wrinkled to the cutting edge.

The scuta are parallelogrammic in form, and their proximal or concealed portions are from one-fourth to one-third the entire length. The granulation is like that of the cranium, fine and without distinct pattern.

The centra of the vertebræ are moderately depressed, those of the dorsal region the most so, and with the greater part of the face of the ball looking upwards. These centra are not ridged on the middle line below, but are slightly convex in transverse section. On either side, behind the inferior extremity of the parapophysis, is an open, shallow groove looking downwards, and defining the middle region as a wide band. Above, this groove is bounded by an obtuse longitudinal ridge, and above the ridge occupying the side of the centrum is a concavity. This is bounded above by the prominent ridge that connects the zygapophyses. The transverse process is vertical and narrow; it is nearly sessile and is not subdivided. A vertebra inclosed in matrix close to the skull is probably from the anterior part of the series. Its centrum is not so depressed as in those already described, and it has a strong hypapophysial keel. Its ventral spine is rather elevated, and is grooved posteriorly.

As compared with the so-called *Placosauridæ* of the Eocene formation, this species presents various points of resemblance, which the descriptions of the former do not yet permit us fully to estimate. The presence of regular cranial scuta will distinguish *Peltosaurus* from the known forms as a genus. The granulation of the cranium and osseous scuta, is finer than in any of the Eocene species I have met with.

Measurements.

	M.
Median width of parietals	0.0140
Median width of frontals	.0080

	M.
Length of mandibular ramus to cotylus	.0400
Diameter of vertebral centrum (transverse)	.0030
Length of vertebral centrum	.0055
Length of a dorsal scutum	.0075
Width of a dorsal scutum	.0042

About the size of the *Heloderma suspectum.*

EXOSTINUS Cope.

Synopsis of New Vertebrata Colorado, 1873, p. 16. Annual Report U. S. Geol. Surv. Terrs., 1873 (1874), p. 511.

This form of lizard is represented principally by a nearly entire frontal bone. Close to it were found a zygomatic bone and a nearly complete dentary bone, with the teeth. The former is in all respects appropriate to the frontal bone, and the size of the dentary bears the usual relation of size to the same. Its dentition is appropriate to the affinities of this genus to *Peltosaurus* Cope.

The frontal bone is much narrowed between the orbits, as in recent leptogloss *Pleurodonta*, while the olfactory lobes were almost as completely underarched as in the thecagloss-type. The stout, well-developed zygomatic, with malar process, resembles the former group, and the teeth have a similar structure. These are closely placed, truly pleurodont and subcylindric. The crowns are simple, compressed, and with a convex edge. They are similar in form throughout the dentary bone. Cranial bones covered with symmetrical osseous prominences.

The sculpture of the superior surface of the frontal bone is more like that of the genus *Anolis* than any other known to me. The prominent inferior lateral olfactory crests, are, however, entirely inconsistent with any such affinity, excluding the genus from the Iguanian group altogether. It coincides with the evidence furnished by the forms of the teeth, that the genus *Exostinus* is one of the *Diploglossa*, and allied, but not very closely, to *Peltosaurus*. In the latter the frontal region is much wider, and is not covered with tubercles, and the olfactory ridges are much less prominent. In its narrow interorbital region *Exostinus* differs from any recent genus of the order known to me.

EXOSTINUS SERRATUS Cope.

Locis citatis.

Plate LX; figs. 12-14.

A series of tubercles along each supraorbital border, longitudinal at the front, and quadrate at the back part of the eyebrow. A single series of tubercles separates them. Five tubercles in a transverse row at the posterior margin of the frontal. Two series of flat tubercles on the zygomatic bone. Dentary quite convex on outer face; inner face slightly convex; 8 teeth in 0 .0050.

Measurements.

	M.
Length of frontal (nearly complete)	0.0070
Width of frontal posteriorly	.0034
Width of frontal at postorbital point	.0045
Width of frontal between orbits	.0018
Length of zygomatic	.0070
Depth of dentary at last tooth	.0030
Length of a mandibular tooth	.0018

About the size of the large males of the northern *Anolis principalis.*

ACIPRION Cope.

Synopsis New Vertebrata of Colorado, 1873, p. 17. Annual Report U. S. Geol. Surv. Terrs., 1873 (1874) p. 514.

Represented by a dentary bone, with nearly all of the teeth remaining. A groove, apparently the Meckelian, extends along the inferior border of the distal half of the bone. The teeth are truly pleurodont, closely placed, and cylindric, with compressed crowns. The latter supports a large median and two small lateral cusps, three in all.

ACIPRION FORMOSUM Cope.

Locis citatis.

Plate LX; fig. 15.

The crowns project well above the alveolar border. External face of dentary smooth, with rather distant foramina. Ten and a half teeth in $0^m.0050$.

Measurements.

	M.
Depth of dentary at middle	0.0022
Length of a median tooth	.0018
Elevation of same above alveolus	.0010

This species is about the size of our *Cnemidophori*. From all the genera of this group, *Aciprion* differs in the uniform character of the teeth, there being no simple teeth in the front of the series so far as preserved. A jaw-fragment probably represents a second species of this genus. A well-pre served lower jaw of the *Aciprion formosum* was obtained in Dakota by the Princeton Expedition of 1882.

DIACIUM Cope.

Synopsis New Vertebrata Colorado, 1873, p. 17. Annual Report U. S. Geol. Surv. Terrs., 1873 (1874), p. 514.

This genus was originally established on the sacral vertebra of a lizard which displayed the peculiarity of the absence of any trace of neural spine.

The diapophysis is subcylindric and elongate. Centrum concave below; neural arch flat above. Articulation without zygosphene or rudiment of it; zygapophyses oblique, the arch deeply excavated between the anterior ones. Obliquity of ball inferiorly.

DIACIUM QUINQUIPEDALE Cope.

Locis citatis.

Plate LX; fig. 20.

Two obscure hypapophysial tubercles below the ball. Centra slightly depressed, the cup excavated above and below. An angulation extends backwards from each anterior zygapophysis; the neural arch between them flat on the anterior half.

Measurements.

	M.
Length of centrum below	0.0100
Diameter of articular cup { transverse	.0086
Diameter of articular cup { vertical	.0064
Diameter of base of diapophysis	.0044
Width between anterior zygapophyses	.0098
Width of upper plane of neural arch	.0078

This species is as large as any of the existing species of *Iguanidæ*.

PLATYRHACHIS Cope.

Synopsis New Vertebrata Colorado, 1873, p. 19. Annual Report U. S. Geol. Surv. Terrs., 1873 (1874), 516.

This genus is known only from vertebræ, which, though abundant in the White River beds, I have been as yet unable to connect with other portions of the skeleton. The genus was founded on the supposition that the typical species, the *P. coloradoensis*, possesses the zygosphene articulation, a character which I have since been unable to verify, and do not now believe in. The subtraction of this character leaves little for the genus to stand on; but I retain the name with the following definition:

Vertebral centra with the articular faces much depressed, and without inferior or lateral carinæ. The ridge connecting the zygapophyses deeply incised. Neural spine a low keel.

The incised interzygapophysial ridge distinguishes these vertebræ from those of *Peltosaurus*.

The two best known species may be distinguished as follows:

Much smaller; neural spine not ending in a knob; costal articular face principally convex.. *P. coloradoensis*.
Much larger; neural spine ending in an obtuse apex posteriorly; costal articulation, superior part concave.. *P. rhambastes*.

A third species is only known from a sacral vertebra.

PLATYRHACHIS COLORADOENSIS Cope.

Locis citatis.

Plate LX; fig. 17.

The inferior face of the centrum in four dorsal vertebræ is plane, and is separated by an obtuse right angle from the nearly vertical lateral surfaces. Neural arch depressed, an angle connecting the zygapophyses. Neural spine a keel, projecting beyond the posterior margin in a mucro. Ball truncate below its convex face, looking slightly upward. Costal capitular surface semiglobular, directly below the anterior zygapophysis. Neural arch concave between zygapophyses.

Measurements.

	M.
Length of three dorsal vertebræ	0.0070
Length of one dorsal vertebra	.0028
Diameter of ball { transverse	.0014
Diameter of ball { vertical	.0006
Elevation of vertebra	.0019
Width between zygapophyses	.0025

This species is of small size, not exceeding the red salamander (*Spelerpes ruber*) in dimensions.

From Horse Tail Creek, Northeastern Colorado.

Platyrhachis unipedalis Cope.

Diacium unipedale Cope. Synopsis of New Vertebrata of Colorado, 1873, p. 18. *Cremastosaurus unipedalis* Cope. Annual Report U. S. Geol. Surv. Terrs., 1873 (1874), p. 516.

Plate LX; fig. 19.

Represented by a sacral vertebra of an individual much larger than any of those of the last-described species, and a little smaller than those of the species next following, characterized by the unusual protuberance of the articular ball and absence of flattening of the centrum below. Centrum depressed; plane longitudinally convex in transverse section. An annular groove round the ball. Diapophysis elongate, slightly depressed.

Measurements.

		M.
Length of centrum		0.0034
Diameter of cup	transverse	.0020
	vertical	.0018

Were this vertebra part of an individual of the *P. rhambastes*, its articular faces would have been more rather than less depressed than the dorsal vertebræ which represent it.

From Horse Tail Creek, Northeastern Colorado.

Platyrhachis rhambastes Cope.

Plate LX; fig. 18.

Established on seven dorsal vertebræ, which I formerly regarded as belonging to the *Cremastosaurus carinicollis*. (Annual Report U. S. Geol. Surv. Terrs., 1873 (1874), p. 515. They differ very much from the cervical vertebræ on which the latter genus is based; too much, I believe, to render it probable that they belong to lizards of the same genus. This also in spite of the fact that the difference is somewhat like that which prevails between cervical and dorsal vertebræ of many genera of Lacertilia. The inferior carina in the *Cremastosaurus* is not a hypapophysis of the usual kind found on only a few of the cervicals, but continues undiminished to the sixth and last I possess of the series. The articular ball at this point shows no indication of the depression characteristic of the dorsals, and which is usual at this point, even in species where the anterior cervicals have subround articular extremities.

The dorsal vertebræ have transversely-oval articular faces, and centra without inferior keel or ridge. The vertebræ are all dorsal, hence the diapophyses have the usual form in the order for costal articulation, and do not project as far inferiorly as the plane of the lower face of the centrum. It does not project beyond the anterior zygapophysis, and the lower half is especially developed as the costal condyle. The sides are separated from the wide, flat, inferior surface by an obtuse angle. Neural spine a keel extending from the front of the arch and rising into a short obtuse apex above the articular ball. There is a collar round the ball, which is faintly visible on the inferior side.

Measurements.

	M.
Length of centrum	0.0040
Width of cup	.0018
Depth of cup	.0010
Elevation of neural arch anteriorly	.0015
Elevation of neural spine and arch posteriorly	.0043
Total expanse in front	.0047

The dorsals represent several individuals.

Horse Tail Creek, Northeastern Colorado.

CREMASTOSAURUS Cope.

Synopsis New Vertebrata, Colorado, 1873, p. 18. Annual Report U. S. Geol. Surv. Terrs., 1873 (1874), p. 515.

This genus was proposed for a lizard which presents some peculiar characters of the cervical vertebræ. There are five of these preserved in a continuous series. They are quite robust and short, resembling somewhat those of the *Phrynosoma cornutum* in proportions, but have remarkably small articular ball and socket as compared with that species. Another marked feature is the rib-like hypapophysis, which is equally developed on the fifth as on the second vertebra, being bounded by a fossa or groove on each side. It is probably continued, but less distinctly, on the dorsal vertebræ.

The neural arch is capacious in the cervical region, and each neurapophysis is excavated below the posterior zygapophysis, and sending a ridge downward and backward around the centrum, continuing as a low shoulder on the inferior face. Diapophysis with a single narrow capitular articulation, extending obliquely downward and forward; that of the third vertebra smaller. Axis with an elevated neural arch, with obtuse, inferior carina. Odontoid a crescentic element, with a transverse groove on its anterior face. All the centra with an obtuse but prominent hypapophysial keel.

CREMASTOSAURUS CARINICOLLIS Cope.

Locis citatis.

Plate LX; fig. 16.

Ball of sixth cervical vertebra round. Neural arches broad, each with a low, acute keel for spine, which is elevated on the third, and produced roof-shaped backward and forward on the axis. The costal articulations are not produced below the centrum.

Measurements.

	M.
Length of cervical vertebræ II to VI	0.0140
Length of axis	.0038
Elevation of axis behind	.0047
Diameter of odontoid (in front)	.0034
Length of c. VI	.0029
Diameter of ball of same	.0011
Total elevation of same	.0030

Size of the "horned-toad", *Phrynosoma cornutum.*

OPHIDIA.

Several species of snakes have been found in the beds of the White River and John Day epochs. The characters presented by the vertebræ, the only parts as yet discovered, do not differ much from those of existing types. None of them represent species of more than the average size of the colubrine snakes now existing in the same region.

APHELOPHIS Cope.

Synopsis of New Species of Vertebrata obtained in Colorado in 1873 (Miscell. Pub. U. S. Geol. Surv. Terrs.), p. 16. Annual Report U. S. Geol. Surv. Terrs., 1873 (1874), p. 518.

The vertebræ of this genus have various well-marked characters which distinguish them. The neural spine is short and robust, and does not extend along the entire length of the neural arch as exposed in the articulated column. There is no diapophysial process below the prezygapophysis. The zygosphene exceeds the articular cup in width, and the posterior border of the neural arch is not interrupted. There is no hypapophysis, nor hypapophysial nor any other longitudinal ridges. The costal articular surface is single and uninterrupted.

These vertebræ resemble very much those of the Erycid genus *Charina*, which is found at the present time in the Pacific district of this continent. I can, in fact, detect no well-marked generic difference between them. The zygosphen of the recent species is somewhat narrower than that of the extinct one.

APHELOPHIS TALPIVORUS Cope.

Locis citatis.

Plate LX; fig. 21.

Vertebræ short and wide; the neural spine stouter and more obtuse than in any other species here described, occupying less than half the neural arch with its basis. Zygosphene wide, depressed, with nearly straight posterior margin, not sending any ridge backward from the posterior face. Articular faces of centrum depressed oval; ball looking upward, its axis making 45° with that of the centrum. Parapophysis not projecting below centrum.

Measurements.

	M.
Length of centrum	0.0026
Diameter of cup { transverse	.0018
Diameter of cup { vertical	.0012
Width between parapophyses	.0017
Depth of entire vertebra	.0034
Width of zygosphene	.0020

Represented by three vertebræ of an individual about the size of *Calamagras murivorus.*

The White River epoch of Northeastern Colorado.

OGMOPHIS Cope.

Neural spine of vertebra short and obtuse. No process below the prezygapophysis. Rib surface single, uninterrupted; rib extending posteriorly from its inferior extremity. Inferior face of centrum without hypapophysis on the dorsal region, but probably furnished with one on the cervical region.

This genus resembles *Aphelophis* in most respects, and probably has similar general affinities. It is distinguished by the ridge which extends from the parapophysis, and the groove which is included between this and the middle line of the centrum. Two species are known to me.

OGMOPHIS OREGONENSIS Cope.

Plate LVIIIa; figs. 9–11.

This snake is represented by four dorsal vertebræ and probably by a fifth from the region near the skull. The last-named vertebra has a smaller centrum than the others, as is usual with those from the anterior part of the column, but the details of its structure are much as in the others. It has an acute hypapophysial ridge which extends posteriorly into a short, subacute process. In the other centra, the median inferior ridge bounded by the lateral grooves, is obtuse, and has different forms. It is wide at both extremities and the surface is equally convex; in two others it is much narrowed medially and widened posteriorly. In all of these the posterior part is slightly angulate medially. The articular faces are nearly round, and the obliquity of the ball is only moderate. The ridge from the inferior extremity of the rib surface terminates a short distance from the articular ball. The ridge connecting the zygapophyses is very prominent. The rib articular face is one-half deeper than wide; the superior half is convex, the remainder gently concave, the inferior border projecting slightly outwards.

The only specimen of this species preserved rather exceeds in size the others here described, excepting the *Neurodromicus dorsalis*.

Measurements.

	M.
Length of centrum with ball	0.0050
Transverse diameter of ball	.0026
Total elevation of vertebra	.0060
Elevation of neural spine	.0019
Width at interzgyapophysial ridge	.0044
Vertical diameter of rib surface	.0023
Width of ball of a "cervical" centrum	.0020

From the John Day beds near the John Day River, Oregon. J. L. Wortman.

OGMOPHIS ANGULATUS Cope.

Calamagras angulatus Cope. New Vertebrata Tertiary of Colorado, 1875, p. 16. Annual Report U. S Geol. Surv. Terrs., 1873 (1874), p. 518.

Plate LVIII a; fig. 12.

This species differs from the *C. oregonensis* in the following points. The interzygapophysial ridge is not nearly so prominent but is excavated behind the prezygapophysis. The neural spine is smaller and is not split posteriorly, as is the case with a dorsal of *C. oregonensis.* The hypapophysial ridge is narrower, and is subacute in a dorsal vertebra.

The zygapophses are well expanded, and the zygosphen is wider than the cup. The superior and prominent portion of the rib articulation is narrower than the inferior flat portion.

Measurements.

		M.
Length of centrum		0.0030
Diameter of ball	transverse	.0017
	vertical	.0016
Width between parapophyses		.0024
Depth of entire vertebra		.0045

CALAMAGRAS Cope.

Locis citatis, 1873, 1874.

Neural spine small and obtuse; no process below the prezygapophysis, nor longitudinal ridge of the centrum posterior to the rib articular surfaces. The latter is undivided. Inferior median line with a keel, which is probably produced into a hypapophysis on the anterior part of the columns.

This genus belongs most probably to the same natural division as *Ogmophis* and *Aphelophis*, which when fully known may prove to be the existing family of *Erycidæ* or *Lichanuridæ*. *Calamagras* shares the characters of the two genera named, so as to stand between them in the natural system.

CALAMAGRAS MURIVORUS Cope.

New vertebrata from Colorado, 1873, p. 15. Annual Report U. S. Geol. Survey Terrs., 1873 ('74), p. 517.
Calamagras truxalis Cope l. c.

Plate LVIII a; figs. 13–15.

Parts of three vertebral columns, including thirteen vertebræ, represent this serpent.

The zygosphene is a little wider than the articular cup. The ball, viewed directly in the line of the axis of the centrum, has a wide, transversely placed oval outline. The balls of a set of three vertebræ in which the hypapophyses are best developed, and which I therefore supposed to belong to the anterior part of the series, are more nearly round. The ridge connecting the zygapophyses is deeply emarginate. In the anterior vertebræ the hypapophyses are acute keels supporting an angular prominence near the center; further posteriorly there is a slight prominence at the base of the ball. More posteriorly the hypapophysis forms a narrow but somewhat obtuse ridge, terminating posteriorly in an apical angle at the inferior side of the ball. The neural arches are separated by spaces equal to the lengths of their bases. The rib articulations present the peculiarity of other members of this group, of a single surface, of which the superior half is convex, and the inferior slightly concave.

Measurements.

No. 1.—More anterior.

	M.
Length of centrum	0.0027
Width of ball	.0016
Depth of ball	.0011
Width between parapophyses	.0020
Depth of entire vertebra	.0034

No. 2.—More posterior.

Length of centrum	0.0030
Width of ball	.0017
Depth of ball	.0013
Width between parapophyses	.0023
Depth of entire vertebra	.0040

From the beds of the White River Epoch in Northeast Colorado.

NEURODROMICUS Cope.

Synopsis of new vertebrata from the Tertiary of Colorado, 1873, p. 15. Annual Report U. S. Geological Survey Terrs., 1873 ('74), p. 516.

Centrum small, with a prominent truncate hypapophysis. Neural arch capacious, the zygantrum wider than the articular cup. Neurapophyses bounding the canal laterally below the zygosphene; its border not angulate behind. Parapophysis projecting acutely below centrum. An elevated neural spine. No process below the prezygapophysis. No prominent ridge connecting the zygapophyses.

This genus represents a different group of snakes from those included in the three genera above described. The vertebra resembles considerably those from the anterior part of the column of one of the *Crotalidæ*, but differs in the less robust hypapophysis and the absence of the process below the prezygapophysis.

NEURODROMICUS DORSALIS Cope.

Locis citatis, 1873, 1874.
Plate LVIII a; figs. 7, 8.

Articular surfaces of centrum round; the ball with a slightly upward-looking obliquity. Hypapophysis continued to cup as a prominent carina. Neural spine extending its base forward, so as to stand on the entire length of the neural arch. The rib articular process is of light form, and is separated by a considerable space from the prezygapophysis. The superior convex portion is quite small, while the inferior portion is narrow and is produced downwards. There is a trace of ridge connecting the zygapophyses, and a trace of a groove on each side of the anterior part of the base of the hypapophyses. Both the neural spine and the hypapophysis are thin edged, and all parts of the vertebra are delicate and light, in strong contrast to those of *Calamagras* and its allies.

Measurements.

		M.
Length of centrum		0.0045
Diameter of cup,	vertical	.0020
	transverse	.0021
Elevation of neural spine above centrum		.0055
Elevation of neural spine above neural arch		.0029
Length of hypapophysis below centrum		.0012
Width of hypapophysis		.0011

The zygantrum is capacious, and the whole neural arch open and light. The species was about the size of the black snake (*Bascanium constrictor*).

The White River Epoch of Northeast Colorado.

MAMMALIA.

The White River Epoch was very rich in *Mammalia*, but not more so than the preceding epochs of the Eocene proper, the Bridger and Wasatch. The composition of the fauna was very different, as a number of important groups of the earlier period were wanting, while several appear for the first time. Of the former kind may be included the Order *Amblypoda*, and the suborders *Tæniodonta*, *Tillodonta*, and the family of the *Lophiodontidæ*. The *Creodonta* and *Mesodonta* are represented by a very few remnants. The divisions which actually appear for the first time are few, for nearly all the characteristic divisions had very few and insignificant representatives during the Eocene. The *Artiodactyla ruminantia* and the *carnivora* are not known from the Eocene, while the *Artiodactyla omnivora* had very few representatives. The *Menodontidæ* and *Palæotheriidæ* chiefly belong to this epoch, with several families of *Rodentia*. The *Chiroptera*, probably the *Marsupialia*, and the rodent family of the Squirrels held over from the Eocene, while the *Proboscidea* had not yet appeared.

No species of Mammal is common to this epoch and any of those of the Eocene, and only one or two genera of Marsupial or Marsupial-like families have continued from the one to the other, so far as present information extends.

The species of the White River Epoch attained a larger average size than those of the Eocene Epochs. This is especially evident on comparison of related or corresponding types of the two periods. The *Amblypoda*, which embraces the largest Mammalia of the Eocene, became extinct. The most notable increase of size is seen in the *Perissodactyla*, where the succession is most continuous. The same is generally true of the flesh-eating series.

The number of individual mammals of this epoch in Middle North America was evidently very great. Many of the species were represented by great droves, and their bones form beds of considerable extent. A locality in Colorado, examined by the writer, embraced about forty acres of naked, soft calcareous rock, carved by erosive action into areas of various sizes. Here the surface of the rock was found to be covered with the remains of the smaller and some of the larger species of the fauna. There were innumerable rodents, and small *Artiodactyla* and *Carnivora;* numerous *Marsupialia*, *Creodonta*, and *Poëbrotherium.* There was a great abundance of *Hyracodon*, and several other rhinoceroses, with quantities of three-toed horses. No traces of the huge *Menodontidæ* were there, but at a locality some miles distant, a similar deposit of these large animals was found, mingled with rhinoceroses. Evidently the causes which overwhelmed the smaller forms did not affect the giants, which only yielded to some other and more irresistible influence.

MARSUPIALIA.

Cuvier demonstrated the existence of species of this order in the gypsum of the upper Eocene of Paris, and additional species have been made known by Aymard, Filhol, and others. In 1873 I discovered in Colorado a number of species which agree very closely with the French forms, so far as can be ascertained from the dentition. These I referred to the *Insectivora*,[1] but subsequently identified[2] them as *Marsupialia.* They are nearly related to the European forms, and whether they can be placed in a distinct genus remains to be ascertained. A species of the same group, if not the same genus, is described in Part I of this volume, from the Wind River Eocene (page 269).

Along with these species were found several others, less evidently members of the Marsupial order. I refer to the genera *Mesodectes*, *Domnina*, and *Menotherium*. Close allies of the first of these had already been referred by Leidy to the *Insectivora.* I have in the first part of this book placed the group in the suborder *Creodonta* and the family of *Leptictidæ*, but with a feeling of uncertainty whether the family may be Marsupial or not. *Domnina* is, I suspect, chiropterous *Menotherium* resembles the *Leptictidæ*, but still more the *Mesodonta*, where I originally placed it. I leave it there for the present in the immediate neighborhood of the genus *Apheliscus.* For purposes of determination from dental characters I compare these genera in a table, as follows. Some of them have been already defined in the analytical table of the genera of Creodonta, page 269.

I. Fourth inferior premolar constructed on the type of the true molars, with three anterior cusps.

Mental foramen anterior, below fourth premolar; inferior true molars subequal. *Peratherium.*

Mental foramen below first true molar; inferior molars diminishing in size posteriorly. *Domnina.*

II. Fourth premolar with three anterior cusps, as in I; but unlike true molars;

True inferior molars with only two cusps in front .... *Mesodectes.*

III. Fourth premolar simple; unlike true molars;

True inferior molars with two anterior tubercles .... *Menotherium.*

---

[1] Annual Report U. S. Geol. Surv. Terrs., 1873 ('74), p. 465.

[2] Bulletin U. S. Geol. Surv. Terrs., 1879, p. 45.

There are five species of *Peratherium* known to me, two of *Domnina*, and possibly of *Mesodectes*, and one of *Menotherium*.

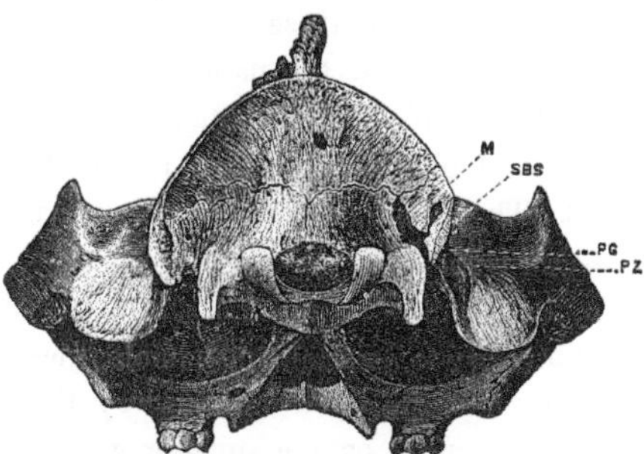

Fig. 32.—Skull of opossum (*Didelphys vriginiana*), natural size, posterior view, parts of the right mastoid and squamosal bones removed. M, mastoid foramen; SBS, subsquamosal; PG, postglenoid; PZ, postzygomatic foramen.

PERATHERIUM Aymard.

Herpetotherium Cope. Paleontological Bulletin, No. 16, August, 1873, p. 1. Synopsis of New Vertebrata from the Tertiary of Colorado (Misc. Pub. U. . Geol. Surv. Terrs.), October, 1873, p. 4. Annual Report U. S. Geol. Surv. Terrs., F. V. Hayden in charge, 1873 (1874), p. 465.

Dental formula: I. $\frac{?}{4}$; C. $\frac{1}{1}$; Pm. $\frac{3}{3}$; M. $\frac{4}{4}$. The superior canines are well developed. The premolars are compressed, with one apex, no inner lobes, and two roots. The superior molars, excepting the last, present two median **V**s, which would be termed external but for the fact that the external basal cingulum is so developed as to constitute an external crest. A single internal lobe, which is less elevated than the median. The last molar has but one median and one internal **V**. The superior incisors are unknown. The inferior incisors are subequal and closely placed; the first three are parallel, Inferior canine well developed, recurved. Inferior premolars compressed. simple unicuspid. Molars consisting of two **V**s, with apices external. The inner extremity of each branch of the anterior **V** is a pointed cusp, behind which stands, on the inner edge of the crown, a third cusp opposite to the middle of the posterior **V**. The last inferior molar is little shorter than the others, but the second **V** is narrowed so as to be only a heel. The molars do not materially enlarge anteriorly, and the posterior mental foramen is not posterior to the first true molar.

The general osteology of this genus is unknown, as the only portions of the skeleton yet discovered are mandibular rami and the cranium anterior to the middle of the orbits. From these it appears that the cranial sutures are distinct at maturity, and that the nasal bones are elongate and well developed Either the maxillary bounds the orbit in front or there is a large lachrymal bone. A pit on the anterior rim of the orbit is probably the lachrymal canal. The *foramen infraorbitale* is well anterior to the orbit. The *symphysis mandibuli* is loose.

In determining the affinities of this and other questionable genera of this epoch it is first necessary to ascertain the homologies of the cusps of the molar teeth. The opossums are characterized by the presence of three longitudinal series of tubercles on the superior molars. The homologies of these cusps are rendered clear by the character presented by the fourth superior premolar, where the anterior intermediate cusp is wanting. The external cusps are really such, and are not developed from a cingulum external to the true external cusps, as appears at first sight to be the case with such animals as the *Talpidæ*. The intermediate cusps are really such, although the posterior looks like the apex of a V-shaped external cusp. In *Peratherium* the external cusps are smaller than in *Didelphys*, and the intermediate Vs so much better developed, that the type is much like that of the *Talpidæ*, in whose neighborhood I originally referred it.

This leads to a consideration of the question of the homologies of the cusps in the genera of the old order of *Insectivora* proper, and of the *Creodonta*. Mr. St. George Mivart has briefly discussed the question so far as relates to the former group.[1] He commences with the primitive quadri-tuberculate type presented by *Gymnura* and *Erinaceus*, and believes that the external cusps occupy a successively more and more internal position, till they come to be represented by the apices of well-developed Vs, as in the ungulate types. The Vs are well developed in several families, and in *Chrysochloris* the two Vs are supposed to be united and to constitute almost the entire apex of the crown, while in *Centetes* the same kind of a V forms a still larger part of the crown.

I believe that these conclusions must be modified, in the light of the characters of various extinct genera, and of the genus *Didelphys*. In the

[1] Journal of Anatomy and Physiology, II, 138 figures.

first place, there is an inherent improbability in the supposition that the external **V**s of the superior molars of the *Insectivora* have had the same origin as those of the *Ungulata.* The movements of the jaws in the two groups are different, the one being vertical, the other partially lateral. In the one, acute apices are demanded; in the other, grinding faces and edges. We have corresponding **V**s in the inferior dental series, and we regard those as produced by the connection of alternating cusps by oblique ridges. In homologizing the superior cusps, we have, as elements, two external, two intermediate, and two internal cusps. The first are opposite the external roots, and the anterior internal is opposite the internal root.

First, as regards *Centetes* and *Chrysochloris.* Besides the strained character of the hypothesis that supposes the **V**-shaped summit of the crown to represent two **V**-s fused together, there is good evidence obtainable in support of the belief that the triangle in question is the usual one presented by the *Creodonta.* This clearly consists of the two external and the anterior internal cusps united by angular ridges. The form is quite the same as in *Leptictis* and *Ictops*[1] (Plate XXIX *a*, fig. 3), and nearly that of *Deltatherium* (Plate XXIII *d*, fig. 8 *a*), where the external cusps are present. *Centetes* and *Chrysochloris* only differ from these in that the external cusps are wanting. In addition, the latter genus presents a rudiment of the posterior inner tubercle, as is seen in *Deltatherium* (Plate XXV *a*, fig. 10). An explanation similar to this is admitted by Mr. Mivart to apply to the cusps of the inferior molar of *Centetes.* It remains to ascertain whether the apex of the **V** in *Chrysochloris* represents the internal or intermediate cusp.

Secondly, as regards the *Talpidæ* and *Soricidæ,* where the external **V**s are well marked. If we examine the external cusps in the genus *Didelphys,* we find that the posterior one becomes gradually more anterior in its position, until on the second true molar it stands largely above the interspace between the roots, instead of over the posterior root. It will also be seen that the anterior intermediate tubercle is distinct and of insignificant proportions, while the posterior intermediate is large and is related to the posterior external, as is the apex of a **V** to its anterior base. In this arrangement I conceive that we have an explanation of the **V**s of the *Talpidæ*

[1] Leidy's Extinct Fauna of Dakota and Nebraska, 351, Pl. XXVI, fig. 29.

and *Soricidæ*. The first true molar of *Scalops* is a good deal like that of *Didelphys*, but the anterior cusp is larger, and there is no anterior intermediate cusp, while the posterior external is of reduced size. The posterior **V** is better developed than in *Didelphys*, but is composed in the same way, of a posterior intermediate cusp, and a posterior external with a posterior heel. These are united by stronger ridges in *Scalops*, *Condylura*, and *Blarina* than in *Didelphys*. On the second true molar in *Scalops* a **V** represents the anterior external cusps of the first true molar. Whether this **V** has a constitution like the posterior one, *i. e.*, is composed of external and intermediate cusps joined, is difficult to determine; but it is probably so constituted. It seems to be pretty clearly the case in *Blarina*, where the fourth premolar and first true molars may be compared, with a resulting demonstration of the correctness of this view. In *Condylura* the **V**s have become more developed and the external cusps reduced, so that the analysis is more difficult.

This interpretation applied to *Urotrichus* and *Galeopithecus* gives them quadrituberculate molars, not trituberculate, as determined by Mivart. *Mystomys* is trituberculate. The intermediate tubercles are present, but are imperfectly connected with the external, so that **V**s are not developed (*vide* figures of Mivart and Allman). This genus offers as much confirmation of the homology here proposed as do the opossums, but it differs from the latter in having the anterior intermediate tubercle the larger, instead of the posterior. *Mystomys* and *Solenodon* also confirm the determination the internal angle of the crown in *Centetes*[1] is the anterior internal cusp.

In conclusion I give the following synoptic view of the constitution of the superior molar teeth in various genera of the *Bunotheria:*

External cusp. Intermediate. Two internal.	External cusp. No intermediate. Two internal.	External cusp. Intermediate. One internal.	External cusp. No intermediate. One internal.	No external cusp. No intermediate. Two internal.	No external cusp. No intermediate. One internal.
Adapidæ. Tupæidæ. Galeopithecidæ Soricidæ (with Urotrichus).	Gymnura. Erinaceus. Macroscelididæ.	Mystomyidæ. Mioclænus. Miacis. Talpidæ. (Didelphys.) (Canis.)	Mesonyx. Leptictis. Ictops. Stypolophus. Oxyæna. Chriacus. Deltatherium. Esthonyx (posterior internal rudimental).	Chrysochloris (second internal rudimental). Solenodon (second internal rudimental).	Centetes.

[1] This view was first advanced by the writer in the Annual Report of the United States Geological Survey Territories, 1873 (1874), p. 472.

In criticism of the above table, it may be added that the external cusps stand within the border in *Stypolophus* and *Didelphodus*, and are nearly confluent in the false sectorials of *Oxyæna* and *Pterodon*. One is smaller than the other in *Mesonyx*. In *Solenodon* they are rudimental.

From the above considerations it appears that the external, often minute, cusps of the teeth of *Insectivora* are the homologues of those of true external series, and do not represent an additional cingulum. Comparisons of the molars of the extinct and recent forms are thus facilitated.

Peratherium is nearly allied to *Didelphys*, but has not the inflected angle of the mandible of that genus. The difference in dentition as above pointed out consists in the elevation of the median cusps of the superior true molars into Vs, and the obsolescence of the tubercles of the external series. It is quite possible that these characters may not be sustained in a comparison of the numerous extinct and recent species. The name *Peratherium* was proposed by Aymard for some similar species from France, but his characters were derived from the inferior dentition, and were of no great importance, as remarked by Filhol. Thus he states that the third inferior premolar is larger in the extinct than in the recent species, a character not generic, and which, as Filhol shows, is not common to all the French species, and which is not found in those of America.

Animals of this genus were abundant in North America during the White River epoch of the Miocene, but I have not seen any of their bones from later deposits. Those obtained by me in Colorado represent six species, some of which I have on former occasions referred to another genus (*Embassis*). Future discovery may justify this course, but at present I suppress the name.[1]

[1] In my account of the genus *Herpetotherium*, published in the Annual Report of the Hayden Survey for 1873 (p. 465), it is stated that this genus has a greater number of molar teeth than in *Talpa*, "thus $\frac{34}{34}$ in the extinct to $\frac{22}{22}$ in the recent genus." As these figures are absurdly erroneous, it becomes necessary to explain that the numbers originally given were $\frac{3-4}{3-4}$ and $\frac{3-4}{2-3}$. The paper was printed during my absence from the East, and the editor, not understanding my meaning, allowed them to go to press in the erroneous form in which they stand.

The species are distinguished as follows:

I. Anterior triangle of inferior molars with the two inner cusps subequal and deeply separated.

a. The anterior triangle not much more elevated than the posterior.

Larger; last two molars 4.5^{mm} long; ramus 4^{mm} deep at middle............. *P. fugax.*

Smaller; last two molars 4.5^{mm}; ramus 3^{mm} .... *P. tricuspis.*

Smallest; last two molars 2.7^{mm}; ramus 2^{mm} .... *P. huntii.*

aa. The anterior triangle twice as much or more elevated than the posterior.

Larger; last two molars 4.5^{mm} .... *P. scalare.*

II. The anterior cusp of the elevated anterior triangle insignificant, the posterior inner cusp much larger.

A groove extending to the base of the anterior cusp on the inner side; heels of molars supporting two cones .... *P. marginale.*

No groove separating anterior cusps; heels of molars very low .... *P. alternans.*

Peratherium fugax Cope.

Herpetotherium fugax Cope. Paleontological Bulletin No. 15, August, 1873, p. 1. Annual Report U. S. Geol. Surv. Terrs., 1873 (1874), p. 467.

Plate LXII; figs. 1–9.

This species is represented by the anterior half of the cranium with mandibular rami, with nearly complete dentition, of one individual; and by larger and smaller parts of mandibles of fourteen others.

The portion of cranium preserved is somewhat crushed by pressure, but it was evidently wider in proportion to its length at the frontal region than in the existing *Didelphidæ* of America. The muzzle is broken away, but the inferior incisor teeth show that it was not elongate. The skull begins to expand in front of the *foramen infraorbitale*, and at the lachrymal region overhangs, by a little, the maxillary border. The orbits do not appear to have been large. The malar bone, if it existed, is lost from the specimen. The nasal bones are narrowed by the elevated maxillaries for the distal two thirds of their length. They then rather abruptly expand to double the width, and terminate in a wide slightly convex posterior border. Either the maxillo-premaxillary suture is situated well in advance of the canine tooth, or else it has become obliterated by coössification. The *foramen infraorbitale exterius* is above the third premolar tooth.

The superior canine is compressed, but neither the anterior nor posterior edges are acute. The first premolar is smaller than the others, and its crown is rather obtuse. It is separated by a short interspace from the second. The latter has a trace of a posterior basal tubercle; its edges are not acute.

The third premolar is most prominent; its posterior edge is generally worn so as to be acute. There are no external cingula on the premolars. The elevation of the external cingula of the true molars is such as to produce, on wearing with the middle cusps, two triangular areas, of which the anterior is a little smaller than the posterior. The inner lobe of these teeth is more nearly opposite the anterior than the posterior triangle, giving a straight anterior and oblique posterior outline to the crown. There is no posterior basal cingulum on the true molars. The inner lobe of the fourth true molar is compressed and acute, and sends a sharp edge to the base of the small cusp which represents the posterior external triangle of this tooth.

The fourth inferior incisor is not larger than the others, and is directed a little more upwards than they. The inferior canine is large, but smaller than the superior, though similar in form. There is no interspace between the canine and first inferior premolar; but a short one between the first and second premolars, and a shorter one between the second and third premolars. The crowns of these teeth are a simple cusp, which is mainly over the anterior root. In all the true molars the anterior triangle is not much more elevated than the posterior, and the anterior and posterior cusps of the inner side of the former are well separated from the external cusp, which much exceeds them in size. The anterior triangle of the first true molar is a little narrowed. The cusp of the internal side of the posterior triangle is about as high as the anterior cusp. In many specimens there is a small cusp behind this one; in some specimens it is very small, while in others it is wanting. This peculiarity I at one time thought characteristic of another species, the *P. tricuspis*. This I do not now believe to be the case, but find that that species possesses other and more reliable characters. The heel of the last inferior molar of the *P. fugax* is somewhat narrowed, and always presents three cusps.

There are two mental foramina—one, the larger, below the first premolar; the other below the fourth molar, counting from either end of the series.

Measurements.

	M.
Length of superior molar series plus the superior canine	.013
Length of premolar series	.0052
Length of molar series	.0063

	M.
Length of second true molar	.0016
Width of second true molar	.0020
Length of inferior molar series of a second individual	.0128
Depth of ramus at first premolar	.0016
Depth of ramus at first true molar	.0035
Depth of ramus at last true molar	.0030

These measurements indicate an animal of about the size of the common mole of this country (*Scalops aquaticus*). The teeth are relatively smaller, and the mandibular rami deeper and less robust.

PERATHERIUM TRICUSPIS Cope.

Herpetotherium tricuspis Cope. Synopsis of New Vertebrata from the Tertiary of Colorado (Miscell. pub. U. S. Geol. Survey Terrs.), October, 1873, p. 5. Annual Report U. S. Geol. Surv. Terrs., 1873 (1874), p. 466; partim.

Plate LXII; figs. 10–11.

This species as at present defined rests on a fragment of a lower jaw, which supports the second and third true molars. These teeth are similar to those of the *Peratherium tricuspis*, and possess, like some individuals of that species, a prominent posterior inner cusp of the posterior triangle. The character which distinguishes it from the typical species is the shallowness of the mandibular ramus. The depth of this bone is not over half that of some of the specimens of *P. fugax*, with teeth of equal size, and two-thirds the measurement of some others. This character is expressed in the following—

Measurements.

	M.
Length of second and third true molars	.0040
Width of third true molar	.0010
Depth of ramus at second true molar	.0025

From the same locality as the *H. fugax*.

PERATHERIUM HUNTII Cope.

Plate LXII; figs. 12–16.

Herpetotherium huntii Cope. Synopsis of New Vertebrata, Colorado, October, 1873, p. 5. Annual Rept. U. S. Geol. Surv. Terrs., 1873 (1874), p. 466. *H. stevensonii* Cope, Synopsis Vert., Colorado, p. 6. Ann. Report U. S. G. S., 1873 (1874), p. 467. *Miothen gracile*, Synopsis Vert., Col., 1873, p. 8. *Domnina gracilis* Cope, Ann. Report U. S. G. S., 1873 (1874), p. 470.

This is the smallest species of the genus known, and is readily distinguished from the *P. fugax* by other characters than its minute size. These are the uninterrupted series of premolars, which are without interspaces, and the greater acuteness and elevation of the cusp on the inner side of the posterior triangle.

The mental foramina are, as in *P. fugax*, below the first and fourth molars. The last inferior molar is as long or nearly as long as the penultimate, but has a rather narrower heel, always with three small cusps. The external anterior cusp is always much larger than the two internal, which are well apart, and not much elevated. In some specimens, as in *P. fugax*, there is a distinct third cusp on the posterior triangle, on its posterior inner border, but it is sometimes obsolete, and sometimes wanting. Hence the species, *P. stevensonii*, founded on this peculiarity, must be suppressed.

Measurements.

	M.
Length of dental series, omitting canine and Pm. 1 (No. 1)	.0072
Length of first true molar	.0013
Elevation of first true molar	.0010
Length of third true molar	.0013
Depth of ramus at third true molar	.0019
Length of last two molars	.0029
Length of third true molar	.0015
Elevation of third true molar	.0012
Depth of ramus at third true molar	.0020

I have portions of seven mandibular rami of this species, which were found at the same locality as the preceding species. It is dedicated to my friend Professor T. Sterry Hunt.

Peratherium scalare Cope.

Herpetotherium scalare Cope. Synopsis New Vertebrata from Colorado, 1874, p. 7. Ann. Report U. S. G. Survey Terrs., 1873 (1874), p. 467.

Plate LXII; figs. 17–18.

This species is represented by portions of the mandibular rami of two individuals of the size of those of the *P. fugax*. These only support some of the posterior molar teeth.

The species is readily distinguished by the increased disparity in the elevations of the anterior and posterior portions of the molar teeth, resembling in this respect the *Peratherium alternans*. It is a considerably larger species than the latter, and exhibits a distinct anterior cusp, of moderate elevation, which is separated from the median external by a deep notch. It is entirely on the inner side, and sends a cingulum to the external base of the outer median. Fourth molar largest; heel narrowed, with three tubercles.

The masseteric fossa is well defined, and is bounded anteriorly by the strong external ridge of the base of the coronoid process. This ridge is stronger, and the fossa extends further forwards than in any of the various specimens of *P. fugax* which display this region. There is no internal pterygoid fossa as in *P. alternans*.

Measurements.

	M.
Length of last three molars	.0060
Length of second true molar	.0020
Elevation of second true molar	.0022

From the same locality and horizon as the preceding species.

PERATHERIUM MARGINALE Cope.

Herpetotherium marginale Cope. Synopsis New Vertebrata, Colorado, 1873, p. 6. *Embassis marginalis* Cope, Ann. Report U. S. G. S. Terrs., 1873 (1874), p. 468.

Plate LXII, figs. 19–21.

This species has a more slender mandibular ramus than the *P. fugax*, but teeth of the same size, thus having the proportions of the *P. tricuspis*. The anterior triangle is more elevated than the posterior, and is not fissured on the inner side so as to distinguish the anterior and middle cusps, except at an elevated point. From the fissure between them a shallow groove descends inwards on the front of the tooth. The inner principal cusp is thus as large as the outer, and the triangle is transverse, or much broader than long. The species is also peculiar in the constitution of the posterior triangle or heel, which supports two elevated acute cones, which are sometimes curved forwards. The middle of the heel is concave. I at one time thought that the *P. marginale* was, with the *P. alternans*, worthy of generic separation, but so many intermediate conditions connect this form with that seen in *P. fugax*, that I no longer maintain this opinion.

There are four molars in the typical specimen, a part of the first being broken away. The last true molar is rather smaller than the others, and relatively smaller than in *P. fugax*, and is not narrowed posteriorly, supporting two opposite well-developed cusps instead of the three weak ones of the other species. It is crowded close to the base of the coronoid process, instead of standing well in front of it as in *P. huntii* and *P. fugax*, which gives the appearance of not being fully protruded, although its cusps have the same elevation as those of the other molars. The inferior border of the ramus rises strongly at this point, much more so than in the two species named, showing that the crowding of the fourth molar is not due to immaturity. It is broken off at the middle of the first molar, and does not display any mental foramen up to that point. A cingulum descends from the anterior and posterior cusps to the base of the median on the outer side. This species is about the size of *Domnina gradata*, and resembles it in the larger anterior teeth.

Measurements.

	M.
Length of last three molars	.0052
Length of third true molar	.0024
Elevation of third true molar	.0018
Depth of ramus at third true molar	.0023

Two other fragments of mandibular rami support teeth which have much the character of those of the present species as regards the form and proportions of the anterior triangle. The posterior triangle, however, is like that of some of the other species, with two low and little-developed cusps. One of the specimens includes the anterior base of the coronoid process, which does not rise abruptly as in the type. These specimens cannot be referred to the *P. marginale*, and they differ from the *P. alternans* in the anterior groove between the inner cusps of the anterior triangle.

PERATHERIUM ALTERNANS Cope.

Embassis alternans Cope. Synopsis New Vertebrata Colorado, 1873, p. 4. Annual Report U. S. Geol. Surv. Terrs., 1873 (1874), p. 468.

Plate LXII, figs. 22–24.

Our knowledge of this marsupial rests on the posterior part of the right mandibular ramus supporting the last two molars of a single individual. All these parts present characters at variance with the species already described. The anterior triangle is much elevated above the heel, and appears at first sight to have but two cusps. The heel is low and concave, and has two low cusps, neither of which is crescentic. On these accounts I formerly referred this species to a genus distinct from the *P. fugax*, which I called *Embassis*. I believe, however, that the anterior cusp of the molars exists, although in a rudimental condition, and that present evidence does not warrant generic separation.

The last molar is distinctly smaller than the penultimate, and its heel is not narrowed, as in the typical species of this genus, but is truncate and supports two cusps. There is a cingulum on the outer side of the anterior triangle of both molars. The ridges at the base of the coronoid process are stronger than in any other species. The masseteric fossa is deep, and is bounded in front by a prominent ridge which forms the external base of the coronoid process. The direction of this crest is nearly vertical, and its prominence throws the teeth to the inner side of the ramus. There is a strong horizontal ridge on the inner side of the base of the coronoid process, in line with the alveolar border, which rises posteriorly and incloses a shallow fossa with the prominent internal vertical edge of the base of the coronoid process. The last molar is situated more than half its length in front of the base of the coronoid process.

The teeth of this species are about the size of those of *P. fugax*. The depth of the mandibular ramus is unknown, as its inferior border is broken away. The *P. alternans* differs from all the species above described in the strong ridges of the coronoid region above described.

Measurements.

	M.
Length of two last molars	.0038
Width of third molar	.0020
Elevation of third molar	.0018

From the same locality and horizon as the other species of the genus.

BUNOTHERIA.

CREODONTA.

I refer provisionally to this order and suborder a few species which existed during the White River Epoch in North America. Two species of about the size of the Hedgehog were discovered in the bad lands of Southern Dakota by Dr. Hayden, and were described by Dr. Leidy as *Leptictis haydeni* and *Ictops dakotensis* in 1868.[1] A third species of about the same size, the *Mesodectes caniculus*, was discovered and described by myself in 1873, with several smaller species already referred to the genera *Peratherium* and *Domnina*. The three genera first named belong to a distinct division from those of the two genera last named, which is, perhaps, of ordinal value. But this remains uncertain on account of the incompleteness of the specimens on which they repose. The entire cranium, without mandible, of *Leptictis haydeni* is known, and part of the cranium with mandible, and part of the skeleton, of *Mesodectes caniculus* are in my possession.

[1] Extinct Mamm. Dakota and Nebraska, 1869, pp. 345–351.

MESODECTES Cope.

(Expl. Survs. W. of 100th Mer., Lt. G. M. Wheeler); April, 1875. Syst. Catal. Vert. Eocene, New Mexico, p. 30. *Isacis* Cope, Paleontological Bulletin (August, 1873) No. 16, p. 3. Bulletin U. S. Geol. Surv. Terrs., No. 1, 1874, p. 23. Ann. Rept. U. S. Geol. Surv., 1873 (1874), p. 470 (nom. præocc).

This genus embraces at present but a single species, which is known from numerous specimens discovered by the writer in the White River Miocene formation of Colorado. From these it appears that *Mesodectes* is closely allied to the *Leptictis* and *Ictops* of Leidy, occupying a position between these in the system. In *Leptictis* the last premolar is sectorial in form, consisting of a single compressed longitudinal crest, without internal tuberosity or cusp. In *Ictops* the last premolar exhibits a structure similar to that of the first true molar, viz, two exterior cusps and well-developed third on their inner side, thus giving a horizontal section of the tooth a subtriangular form. In *Mesodectes* the last premolar possesses a single acute cusp, as in *Leptictis*, with an internal cusp or heel homologous with that in *Ictops*. Such peculiarities are necessarily regarded as tangible definitions of generic groups, and are such in this case, although they separate species which have considerable resemblance in some other respects as far as known.

The *molars* of the superior series have two exterior compressed conic cusps and a stout subtriangular internal one. Behind the latter is a strong cingulum, supporting a rudimental cusp behind and within the principal one. Inferiorly there are three tubercular molars, of which the two anterior are composed of two elevated cross-crests, which form partial V's, opening to the inner side. The last premolar is larger than the true molars, and supports three anterior conic tubercles, the inner and outer equal, and a heel with a conic tubercle on the outer side. The number and character of the teeth in front of this one are unknown.

The posterior part of the *cranium* exhibits characters similar to those of *Leptictis*, but in the specimen the superior walls are wanting. The animal is not adult. The exoccipitals are distinct from the supraoccipital, displaying a wide, smooth sutural face. The mastoid is quite distinct, and is narrowed. Its superior portion enters into the posterior face of the skull, the union being formed by the posterior border of the squamosal. There are neither paroccipital nor mastoid processes, and the inferior extremity of the mastoid is grooved inwards and forwards to the inferior side of the skull. The *meatus auditorius* is wide, and has no inferior wall. There is an extensive petrous fossa inclosed between the exoccipital behind, the basisphenoid and basioccipital within, the alisphenoid in front, and the squamosal and mastoid externally. This is not overroofed on the inner side by a prolongation of the lateral border of the basisphenoid, as in *Erinaceus*. In the middle of the fossa is situated the petrous bone, which extends from the mastoid inwards and forwards to the adjacent parts of the basioccipital and basisphenoid. There is a well-marked postglenoid process, which is divided into two parts by a deep notch. The inner portion is directed obliquely somewhat forwards. Just inside of it a crest, probably the inferior border of the pterygoid process of the sphenoid, extends inwards and forwards, joining the longitudinal pterygoid. The latter terminates at the inner border of the large fossa above described. The squamosal bone has a remarkable extension behind the postglenoid process, and then rises into the oblique crest which forms the boundary of the inion. Longitudinal crests connect this with the superior extension of the postglenoid process, inclosing a fossa with a subsquamosal foramen, a postglenoid foramen between them. In this region this genus is identical with *Leptictis*, and resembles *Solenodon* more than any other living form. In that genus the mastoid is not larger, but is on the lateral instead of the posterior face of the skull. The inferior part of the squamosal in *Mystomys* also resembles considerably that of *Mesodectes*, but the mastoid is far larger, and the otic region is roofed by the basisphenoid wings. The squamosal is much shortened behind in *Erinaceus*, and there is no postglenoid process. The basioccipital and basisphenoid are continuous, as in *Solenodon*. The subsquamosal foramen is especially characteristic of the *Didelphidæ*.

The *cervical vertebræ* are short and transverse, and have well-developed lateral arterial foramina and diapophyses. The centra are depressed to a considerable degree, and are without hypapophyses. The neural arches are narrow, and without spines. The atlas is expanded, and has a very short diapophysis. The axis has a solid *processus odontoideus*. The dorsal vertebræ are smaller than the cervical in transverse diameter of the centrum, which somewhat exceeds the length; the articular faces are nearly plane.

The intervertebral foramina are quite large, and the narrow neurapophyses are almost entirely occupied by the bases of the diapophyses. These are well developed, obliquely truncate below at the end, and grooved on the under side of the shaft. The neural spines are elevated, narrow, and acute in front. The ribs are flat, and the capitular and articular faces are well developed.

The *præsternum* is shaped somewhat like the sternum of a bird. It has a prominent inferior longitudinal keel, which disappears posteriorly, leaving a vertically-oval face of articulation for the second sternal segment. The superior face is slightly concave, and the only lateral articular faces are those for the attachment of (?) first ribs, and are of considerable size, and are adjacent to the anterior extremity. The borders of the bone are but little contracted behind them. The scapula is elongate, and has an elevated crest, descending abruptly near the glenoid cavity. The latter is an elongate oval, the border at one end more produced than at the other, and terminating in a short hooked coracoid.

The *humerus* has a protuberant head and shaft, and condyles much flattened. The head is nearly 180° in arc, is posteriorly directed, and of compressed form. On the inner side is a depressed tuberosity for the pectoralis muscle, while opposite to it the greater tuberosity rises as high as the head, parallel to it. Distally, the condyles are continuous, nearly concave, and supplemented by huge inner and a smaller outer epicondyles. There is no supracondyloid foramen, but a strong arterial foramen.

The cast of almost the entire *brain* is preserved, and, as the parietal bones are wanting, the proportions are clearly traceable. The olfactory lobes are broken off. The superior face of the hemispheres and cerebellum together have a subquadrate outline, a little wider than long. The cerebellum is completely exposed behind the hemispheres, and is strongly angulate at its upper posterior border to fit the inion. The vermis is nearly as wide as each lateral lobe. The surface of the hemispheres is smooth, and the sylvian fissure distinctly indicated.

As compared with other forms, the following points may be observed: In *Chrysochloris* and *Centetes*, as above noted, the external cusps of the superior molars are wanting. In the genera in which they are present, as

Tupaia, *Talpa*, *Sorex*, &c., there are two of the middle series, as in *Peratherium*, and these add a strong internal lobe also. In *Erinaceus* and *Gymmora* they are quadrituberculate. The closest approximation is made by the genera *Mystomys* and *Solenodon*, the former African, the latter West Indian. In these the external cusps are present; there is but one well-developed median, and in the latter the internal is quite reduced. The molars of *Mesodectes* thus resemble most closely those of *Solenodon* (Brandt), but the external cusps are more developed than in that genus. Among the Eocene genera, the greatest resemblance is to be seen in the genus *Deltatherium*. The internal cusp of *Mesodectes* is connected with the external ones by oblique ridges, as in *Delthatherium;* and on one of these is a rudimental tubercle, representing an intermediate cusp. The dentition of the lower jaw is quite different in *Deltatherium* in the simple last premolar, and cutting last true molar.

In the lower series the form of the true molars is not unlike that of several diverse recent genera. It is quite unique in its large four or five cusped last premolar, which has some resemblance to a modified sectorial. The nearest approach to it which I can discover among recent genera is the Madagascar *Galeopithecus*.

In respect to the remainder of the skeleton, numerous characters distinguish it from the *Centetidæ* (which includes Solenodon) and the *Mystomyidæ*. Both of these lack the zygoma, which is present in *Septictidæ*, and have the nasals coössified, while they are distinct in these tertiary forms; *Mystomys*, further, lacks the clavicle. The presence of the zygoma without postorbital processes is a point of resemblance to *Erinaceus*, but the strongly keeled presternum and absence of cervical neural spines are found elsewhere in the *Talpidæ*. In the presence of the humeral arterial foramen, it again differs from *Erinaceus* and resembles other forms of the order.

Comparisons with the *Marsupialia* are chiefly to the *Didelphidæ*. The superior molars are much like those of *Didelphys*, the inner cusp of the third premolar being a point of difference not seen, however, in the allied *Leptictis*. The latter, according to Leidy, has only two superior incisors on each side, a wide divergence from the opossums. In the inferior molars the absence of the anterior inner cusp is a strong mark of distinction. The presence of the subsquamosal foramen is on the other hand a marsupial character. The absence of any but rudimental neural spines of the cervical vertebræ is again an important difference from *Didelphys*.

. The evidence is in favor of the *Leptictidæ*, as represented by *Mesodectes*, being a group of *Bunotheria*, and a member of the *Creodonta*, as defined by the form of the superior molar teeth.

MESODECTES CANICULUS Cope.

Isacis caniculus. Paleontological Bulletin No. 16, p. 3. Ann. Report U. S. Geol. Survey Terrs., 1873 (1874), p. 473.

Plate LXII; figs. 33–50.

This species is represented by portions of the skeletons of six individuals. All of these lack the anterior teeth of both jaws, while one includes mandibular teeth with vertebræ, ribs, humerus, scapula, presternum, a large part of the cranium, &c.

The basioccipital and sphenoid have straight lateral borders which are slightly decurved into a low ridge on each side. A low median ridge includes with these two a pair of longitudinal shallow fossæ, which fade out on the sphenoid bone. The occipital condyles are divergent, and with very small laterally-looking portion. The foramen magnum is largely transverse. The exoccipital does not extend far forwards, and the mastoid has a very short anteroposterior extent. A foramen issues from the base of the fossa inclosed by the longitudinal crests of the posterior part of the squamosal bone, and there are two other foramina above the superior of these crests. A small foramen pierces the same bone behind the inner extremity of the external half of the postglenoid process A groove follows the free posterior edge of the squamosal, separating it from the mastoid. The petrous is somewhat rectangular in its sections, and its transverse diameter exceeds its anteroposterior. The inferior face is horizontal in the outer half, and looks forwards and inwards on its inner half.

The first rib is little compressed, and becomes very robust towards and at the distal extremity, where it is truncate for articulation with the præsternum. The extremity is oval in section.

The shaft of the humerus is somewhat curved, and the anterior compressed crest does not project abruptly, but exhibits a muscular insertion some distance below the head.

The edges of the outer lobes of the first superior molar are acute; there is no external cingulum, but the diagonal crest from the median cusp passes to the posterior base of the posterior outer. There is a short but strong cingulum on the posterior base of the median lobe, which terminates in a small internal cusp. The rudimental anterior middle cusp is on the anterior diagonal ridge, which does not reach the base of the outer anterior cusp. The outer cusp of the last premolar is elongate and compressed; the inner cusp is small, acute, and opposite the posterior margin of the outer; enamel smooth. The anterior of the two prisms composing the inferior true molars is more elevated than the posterior. The crests of each form a V with the obtuse apex outward, and the anterior limb is shorter than the posterior. The last molar is a little smaller, and is produced behind by the addition of a small median lobe. In the last premolar, the conic cusps are well separated, and the inner one of the heel is insignificant. This tooth appears to have been the last one protruded; its temporary predecessor is distinguished by the obtuseness of the cusps, especially of the anterior one. Mandibular ramus deep, compressed, without inferior hook as far as opposite the basis of the coronoid process.

Measurements.

		M.
Length of sectorial and two tuberculars		0.0210
Length of sectorial alone		.0045
Width of sectorial		.0020
Width of first tubercular		.0030
Length of first tubercular		.0032
Depth of jaw at first tubercular		.0060
Width of skull behind		.026
Length from occipital condyle to postglenoid process		.013
Length of basioccipital		.006
Width of basioccipital medially		.005
Transverse extent of postglenoid process		.008
Length of two centra, third and fourth cervical vertebræ		.007
Transverse extent of fourth cervical between parapophyses		.012
Diameter of centrum behind	transverse	.006
	vertical	.003
Total elevation fourth cervical		.0066
Total length of axis		.007
Length of odontoid process		.0025
Diameter of atlas anteriorly	vertical	.0090
	transverse	.0140

		M.
Total transverse extent of atlas		.0165
Diameter centrum of anterior dorsal	longitudinal	.0040
	transverse	.0050
	vertical	.0028
Transverse extent with diapophyses		.0110
Total elevation (oblique)		.0130
Elevation of neural canal		.0028
Length of centra of four consecutive dorsal vertebræ, including spaces for intervertebral discs		.0190
Diameter of head of humerus	long	.0090
	short	.0075
Long diameter of shaft		.0045
Transverse diameter of distal extremity		.0110
Transverse diameter of distal condyles		.0070
Diameters of glenoid cavity of scapula	vertical	.0045
	transverse	.0065
Length of præsternum		.0150
Width of præsternum	anteriorly	.0110
	posteriorly	.0045
Greatest depth of præsternum		.0080

This species is about the size of the European hedge-hog. The remains representing it were found by myself at the same locality and horizon as the species of *Peratherium* above described.

GEOLABIS Cope

This genus is established to receive a species which is represented by portions of two crania which are not accompanied by either superior or inferior molar teeth. It is very distinct from *Peratherium*, but whether identical with *Domnina* or not, it is not yet possible to state. The appearances are that the molar formula is 3–3, and the number of prēmolars is doubtless as stated, while that of the true molars remains uncertain. There is no canine tooth, and there are three incisors in each premaxillary. The formula will then be 3–0–3–3. The first superior incisor is larger than the others, and is strongly decurved. The first premolar follows a quite short diastema, is two-rooted, and has a simple compressed crown.

The absence of canine tooth and the enlarged first incisor distinguish this genus from *Talpa*, and constitute a resemblance to *Scalops*. In this genus there are four teeth instead of three, posterior to the incisors and anterior to the true molars, one of which may be fairly regarded as a canine.

GEOLABIS RHYNCHÆUS Cope.

Plate LXII; figs. 30–32.

The muzzle of this species is long and narrow, with vertical sides and convex superior surface, which is nearly straight in profile. The depth is a little greater than the width, and the extremity is rather abruptly truncated downwards and forwards. The nasal orifice is exclusively anterior. The nasal bones are narrow, but widen regularly posteriorly. The palate is moderately concave. The premaxillary bone presents a rather wide external face. The cranium expands just in front of the orbits, which are well defined in front and above by a rather flat wide frontal region.

The bases of the crowns of the incisors are subround. The second premolar is separated from the first by an interspace as wide as that between the first premolar and last incisor.

Measurements.

	M.
Length muzzle of No. 1 from front of orbit	0.0100
Depth at middle	.0030
Width at middle	.0028
Length external face of premaxillary bone	.0028
Width of No. 2 between anterior borders of orbits	.0070
Interorbital width of No. 2	.0060

The cranium of this species is rather smaller than that of the *Scalops aquaticus.* The species was first described, but not named, under the head of the genus *Domnina,* in the Annual Report of the U. S. Geol. Survey of the Terrs., 1873 (1874), p. 469.

INSECTIVORA.

MENOTHERIUM Cope.

Bulletin U. S. Geological Survey Terrs., I, January, 1874, p. 22. Annual Report U. S. Geolog. Surv. Terrs., 1873 (1874), p. 510.

This genus is possibly a remnant of the lemuroid group of the *Adapidæ,* so abundantly represented during the Eocene period; but as I only possess portions of two mandibular rami with dentition, a more exact determination will be looked for with interest. It is the first indication of the existence of lemurs in the Miocene formation of the United States.

There are at least two premolars and three molars in the inferior series;

those anterior being lost in the specimens. The last premolar is somewhat sectorial in form, having a compressed but stout median cusp, a broad heel behind, and a small tubercle in front. The last molar is rather smaller than the others, and with a slight posterior or fifth tubercle. The molars support four tubercles nearly opposite, in pairs, and connected by a diagonal crest, so that when the crown is worn an S-shaped figure results. The two alveoli in front of the last premolar may have contained each a separate tooth or a single tooth longer than any of the others. The form of the true molars is as in the Mesodont genera *Apheliscus*[1] and *Anaptomorphus*, and the simple fourth premolar is also that of the former genus.

Menotherium lemurinum Cope.

Locis citatis.

Plate LXVI; figs. 34–6.

The last premolar is longer than any of the molars. There are no cingula on the molars, but the transverse crest from one of the tubercles descends to the side of that opposite to it, along the end of the crown. Enamel smooth. Ramus of the jaw rather elongate.

Measurements.

	M.
Length of bases of six molars	0.0250
Length of bases of true molars	.0120
Length of basis of first true molar	.0040
Width of basis of first true molar	.0032
Length of basis of last premolar	.0052
Width of basis of last premolar	.0030
Depth of ramus at last premolar	.0090

The animal was about as large as the domestic cat.

CHIROPTERA.

The following genus is referred to this order with some hesitation. It is founded upon mandibular rami, which are identical, so far as they go, with the corresponding parts of certain bats, but the diagnostic parts are wanting. I originally referred the genus to the same group as *Peratherium*.

[1] Report Captain Wheeler IV, pt. ii.
[2] Hujus Operis., pt. i.

DOMNINA Cope

Paleontological Bulletin No. 16, p. 1, August, 1873. Ann. Report U. S. Geol. Surv. Terrs., 1873 (1874), p. 469. ? *Miothen* Cope, Synopsis New Vert. Colorado, 1873, p. 5.

The evidence of the distinctness of this genus from *Peratherium* was originally seen in the position of the mental foramen, which is situated under the third inferior molar (counting from behind), instead of under the fourth. The deduction from this feature is that there is at least one molar tooth less in either the true or premolar series. Although a verification of this view has not yet been obtained from specimens which display the whole series, I confidently anticipate it. The true molars differ in their form from those of the species of *Peratherium*. Thus the posterior external cusp is a true crescent of the same form as the anterior one, but rather smaller. Both internal cusps are at the summits of strong subvertical ridges. The molars rapidly increase in size forwards, the third being little more than half as large as the first.

I associate with the typical species, *D. gradata*, a second one, which is represented by two mandibular rami in which the molar teeth are worn by use. They show that the posterior inner cusp had a crescentic base as in *D. gradata*, and that the last molar is even more reduced, being only half as large as the penultimate. The rami are both broken off posterior to the position of the mental foramen, so that its position cannot now be ascertained.

DOMNINA GRADATA Cope.

Paleontological Bulletin No. 16, p. 1, August, 1873. Annual Report U. S. Geol. Surv. Terrs., 1873 (1874), p. 469.

Plate LXII; figs. 25–6.

Represented by the parts of the mandibular rami of two individuals which support the three last molars. These show that the anterior base of the coronoid process is wide and oblique, and that both its external and internal borders are prominent. The line of the inner alveolar border is continued as a ridge, which projects beyond the inner face of the ramus below it.

The inner cusps of the molars are much smaller than the external cusps, the crown. The heel of the last molar is very small, and consists of only one obtuse cusp. The oblique anterior borders of the two triangles are sectorial in their character. There is a low cingulum on the external base of the crown, which bridges the interval between the lobes.

As compared with the *Peratherium fugax*, with which this species agrees in size, it differs, besides the generic characters, in the relatively larger anterior molar teeth and shallow ramus.

Measurements.

	M.
Length of last three true molars	.0060
Length of first true molar	.0024
Width of first true molar	.0013
Depth of ramus at first true molar	.0025
Length of last molar	.0016

From the same locality as the species above described.

DOMNINA CRASSIGENIS Cope.

Annual Report U. S. Geol. Survey Terrs., 1873 (1874), p. 470. *Miothen crassigenis* Cope, Synopsis New Vertebrata Colorado, 1873, p. 8.

Plate LXII; figs. 27-9.

This species is known from the posterior portion of a pair of mandibular rami of opposite sides, which look as though they might have belonged to the same individual. These show that the external border of the base of the coronoid process is prominent, but the inferior border is not, as in *Peratherium huntii*, well defined. The inner face of the rami is very flat, without the horizontal ridge behind the dental series seen in *Domnina gradata*, nor the internal pterygoid fossa of the *Peratherium alternans*. The inner face of the ramus towards the angle is gently concave. The external face of the ramus at the last two molars is convex. The last molar is only as large as the posterior triangle of the penultimate molar, and its anterior triangle is not larger than the heel. The two triangles of the penultimate are equal, and their section is strongly convex on the external side.

Measurements.

	M.
Length of last two molars	.0032
Width of penultimate molar	.0014
Length of last molar	.0012
Depth of ramus at penultimate molar	.0030

From the same locality and horizon as the preceding species.

RODENTIA.

Members of this order were very abundant during the White River and Truckee epochs in North America. They are referable to thirty-one species and eight genera. Of these genera three still exist in the regions where their fossil remains are found. These are *Sciurus*, *Hesperomys*, and *Lepus*. All of them occur in the Truckee beds, while the first named only has been found in the White River formation. All of the species belong to the three great divisions of the order which now inhabit North America, while the fourth, the *Hystricomorpha*, which is very sparingly represented on the continent, has not yet been detected in the formations in question. It appears in a single species of porcupine in the Loup Fork beds.

The four primary divisions of the order *Rodentia* are thus defined, principally after Brandt and Alston:

I. Incisor teeth $\frac{2}{2}$. Fibula not articulating with the superior condyle of the calcaneum. No intertrochlear crest of humerus.

1. Mandible with the angular portion springing from the outer side of the bony covering of the lower incisor. Fibula distinct from tibia. "Malar bone not supported below by a continuation of the maxillary zygomatic process." An interpterygoid fissure .... *Hystricomorpha*.
2. Mandible with the angle in the plane of or springing from the inferior edge of the covering of the alveolus of the inferior incisor, more or less rounded; coronoid process high, falcate. Fibula distinct from tibia. No interpterygoid fissure .... *Sciuromorpha*.
3. Mandible with the angular portion springing from the inferior edge of the sheath of the inferior incisor (except *Bathyerginæ*). Fibula coössified with the tibia. Malar short, usually supported on a maxillary process. No interpterygoid fissure (except in *Bathyerginæ*) .... *Myomorpha*.

II. Incisor teeth $\frac{4}{2}$. Fibula articulating with the condyle of the calcaneum.

4. No true alisphenoid canal; fibula ankylosed to tibia below; angle of mandible in the plane of the incisive alveolus. An intertrochlear crest of humerus. *Lagomorpha*.

These groups, as is well known, include families and genera which display adaptations to various modes of life. Some are exclusively subterranean, others are arboreal, and some live on the surface of the ground. Of the latter, some are provided with formidable spines as a protection against enemies, while others depend for their safety on their speed. Of the latter character are the *Leporidæ* of the *Lagomorpha*, and I have noted how that they have superadded to the ordinary rodent structure certain points which also characterize the most specialized *Perissodactyla* and *Artiodactyla* among ungulates. The fusion of the inferior part of the fibula with the tibia (found also in the *Myomorpha*) belongs to the higher types of these orders. The strong intertrochlear ridge of the humerus is an especial feature of the groups mentioned, distinguishing them from the lower types in all the orders. The articulation of the fibula with the calcaneum, mentioned by Mr. Alston, is a character of the *Artiodactyla*. Associated with these is the elongation of the bones of the limbs, especially the posterior one. The modification of the tarsus in *Dipus* (the jerboas) evidently has a direct relation to the projectile force transmitted through the hind legs in rapid progression by leaping. Here the metatarsals are coössified into a cannon bone, though, as there are three bones involved, the result is somewhat different from the cannon bone of the *Ruminantia*.*

The species of the American Miocenes, including Loup Fork formation, are distributed as follows:

	White River.	John Day.	Loup Fork.
HYSTRICOMORPHA.			
Hystricidæ.			
Hystricops, Leidy			1
SCIUROMORPHA.			
Mylagaulidæ.			
Mylagaulus, Cope			2
Fam.?			
Heliscomys, Cope	1		
Castoridæ.			
Eucastor, Leidy			1
Castor, L.	1	2	1
Ischyromyidæ.			
Ischyromys, Leidy	1		
Sciuridæ.			
Meniscomys, Cope		4	
Gymnoptychus, Cope	2		
Sciurus, Linn	1	2	1

*On the significance of these characters, see Bulletin U. S. Geolog. Surv. Terrs., 1881, p. 361, and American Naturalist, 1883, pp. 44–380.

	White River.	John Day.	Loup Fork.
MYOMORPHA.			
Muridæ.			
Eumys, Leidy	1	....	....
Hesperomys, Waterh	....	1	1
Paciculus, Cope	....	2	....
Geomyidæ.			
Pleurolicus, Cope	....	3	....
Entoptychus, Cope	....	5	....
LAGOMORPHA.			
Leporidæ.			
Palæolagus, Leidy	3	1	1
Panolax, Cope	....	....	1
Lepus, Linn	....	1	....

The Rodentia, like other divisions of Mammalia, present a succession of changes of structure in time, in the feet and in the teeth. The earliest known forms, as I have pointed out, are the allies of the squirrels, members of the suborder Sciuromorpha. These have the most generalized foot structure because; first, the trochlear structures of the humerus and tibia are not at all or but little developed; second, because they have five digits on the feet, and are plantigrade; and third, because the fibula is not coössified with the tibia. They are similarly primitive in the forms of the teeth, because they are rarely prismatic, and nearly always have long roots and short crowns. The cavy division, or suborder Hystricomorpha, must claim the next place, but many of its members show a decided advance in having a limited number of toes, and prismatic dentition. In the third suborder, Myomorpha (the mice, etc.), we first meet with the coössification of the fibula with the tibia. A good many genera have prismatic teeth, and some of them a restricted number of digits; and a few of them (the jerboas) even metatarsal bones coösified into a cannon bone. The rabbits have the most specialized characters in all the points mentioned, but they add another character which is most primitive, viz, the presence of four superior incisor teeth. This is probably a remnant of the primitive group from which all the Rodentia have been derived. By the law of homologous groups it is not probable that the divisions of Rodentia were descended from each other, but from corresponding groups of the primary order from which they were derived as a whole. This division may have been the suborder Tillodonta of the Eocenes, or the Rodentia may be the descendants of the Marsupialia with or without the intervention of that group.

The differentiation of the suborders of the Rodentia evidently dates from a period at least as early as the lowest Miocene. It is an important fact that the Lower Eocene (Wasatch epoch) has as yet produced nothing but the lowest type (Sciuromorpha). It is also true that the Puerco Eocene epoch has, in sixty species of Mammalia, disclosed no Rodentia at all, while Tillodonta and Tæniodonta are abundant.

The Myomorpha first appear in the White River beds (Oligocene), but none with prismatic teeth occur below the John Day epoch. The Lagomorpha, on the other hand, present us with almost all their special characters at once, in the White River. The Hystricomorpha, whose home is in South America, are unknown in North America below the Loup Fork or highest Miocene, where Leidy identified a true porcupine, *Hystricops venustus.**

Many of the above genera stand in evident genetic connection with existing forms. The Miocene Castors doubtless include the ancestor of the modern beaver. The *Ischyromys* is a primitive type of the *Sciuridæ*, and *Gymnoptychus* connects it directly with the existing forms by the character of its molar teeth. *Eumys* is the primitive form of *Hesperomys*, as *Paciculus* is of *Sigmodon*. *Entoptychus* and *Pleurolicus* are the near ancestors of the *Geomyidæ* of the Pliocene and present periods. *Palæolagus*, *Panolax*, and *Lepus* form a direct genetic line. The ancient genera all differ from their modern representatives in the same way; that is, in the greater constriction of the skull just posterior to the orbits, and accompanying absence of postorbital processes. This relation may be displayed in tabular form as follows:

Skull wider behind orbits.		Skull narrower behind orbits.	
Postorbital processes.	No postorbital processes.	Postorbital processes.	No postorbital processes.
	Castor fiber		Castor peninsulatus
Sciurus			Ischyromys
	Hesperomys		Eumys
Lepus			Palæolagus

* See American Naturalist 1883, p. 380, where these conclusions are presented.

None of this species of this fauna are of larger size than their modern representatives. In the cases of the beaver, squirrels, and rabbits, the ancient species are the smaller.

SCIUROMORPHA.

SCIURUS Linn.

In this genus the molars are $\frac{5}{4}$ or $\frac{4}{4}$, the first superior small when present. The grinding surfaces of the crowns when unworn present in the superior series a single internal cusp, which is low and anteroposterior. From this there extend to the external border of the crown two low transverse ridges, whose exterior terminations are somewhat enlarged. In the lower jaw the transverse ridges are not visible, and there is a low tubercle at each angle of the crown, between which there may be others on the border of the crown. Attrition gives the grinding surface of the latter a basin-like character. The *foramen infraorbitale* is a short, narrow fissure, situated in the inferior part of the maxillary bone in front of its tooth-bearing portion, but descending nearly to the level of the alveolar border.

The well-known characters of this genus are found in the mandibles of species which I obtained from the White River Miocene beds of Colorado and Oregon. The teeth display the subquadrate form of this genus, without any tendency to the transverse enlargement seen in *Arctomys*, *Cynomys*, and *Spermophilus*. Two of the species, *S. vortmani* and *S. relictus*, are as large as our gray and red squirrels, respectively, and the third, *S. ballovianus*, is about the size of the *Tamias quadrivittatus*.

SCIURUS VORTMANI Cope.

Paleontological Bulletin No. 31, p. 1, December 24, 1879. Proceed. Amer. Philos. Soc., 1879, p. 370. Plate LXIII; fig. 4.

Like the *S. relictus* of the Colorado White River beds, this is a rare species, being only represented by a mandibular ramus in my collection. This part is remarkable for its depth as compared with its length, and the base of the coronoid process has an anterior position. It rises opposite the posterior part of the third molar, and its anterior border descends to a point just below the posterior part of the first molar. The inferior border of the masseteric fossa is a prominent edge, which descends below the inner inferior margin of the ramus. The molars diminish regularly in size forwards. Their crowns are basin-shaped, with the anterior angle of the external border elevated, and the inner border notched medially. Incisor compressed.

Length of inferior molar series, .010; anteroposterior diameter of first molar, .0024; length of fourth molar, .003; depth of ramus at diastema, .0055; depth at third molar, .0095.

This species is considerably larger than the *S. relictus*. It is dedicated to Jacob L. Wortman, of Eugene, Oregon, who found the typical specimen in the John Day Miocene deposit of that State.

SCIURUS RELICTUS.

Annual Report U. S. Geol. Surv. Terrs., 1873 (1874), p. 475. *Paramys relictus* Cope; Synopsis New Vertebrata Colorado, October, 1875, p. 3.

Plate LXV; fig. 35.

This species was established on two mandibular rami, with all the teeth complete. It was referred to the genus *Paramys*, because I found no difference between the corresponding parts of the respective species; but as the characters of the latter are chiefly observable in the maxillary teeth, the reference was not final.

The ramus is quite robust, and has the general proportions of that of *S. hudsonius*. The diastema is much shorter than the tooth-line, and the foramen mentale is situated below its middle, a little above half way between the inferior and superior borders of the ramus. The masseteric fossa has an angular termination below the posterior border of the second molar. The ascending ramus begins opposite the posterior border of the third molar. The symphysis is not abruptly recurved, but commences in a gentle ascent of the inferior border to the incisive alveolus. The latter issues lower down than in the species of *Eumys* and *Heliscomys* described in this volume.

The inferior incisor tooth is less curved than in the species of the genera above named, and is not so much compressed. Its section is subtriangular, with the anteroposterior diameter a little greater than the transverse in front. The anterior face is gently and equally convex in transverse section, and the enamel is perfectly smooth. The inner angle is sharp, and the enamel does not pass it, but the external angle is rounded, and the enamel covers it, narrowly overlapping the side of the tooth.

The molars exhibit the basin-shaped character, with the tubercles marginal, as usual in *Sciurus*. They increase regularly in size from the front backward. The transverse crests are marginal, and terminate in cusps at the inner extremity, which are separated by a lower acute median cusp.

A longitudinal crest connects the crests just within their outer extremities; it exhibits a loop directed outward. A low ridge passes from the posterior outer butress, just in front of the posterior margin, in the last two molars. Anterior cusps of first molar contiguous.

Measurements.

	M.
Length of ramus to end of M. 4	.0120
Length of molars	.0088
Length of third molar	.0015
Width of third molar	.0018
Depth of ramus at third molar	.0050
Width of incisor	.0015

As compared with the corresponding part of the *Sciurus hudsonius*, the best preserved ramus of this species differs in the more robust incisor teeth, in the less anterior extension of the masseteric fossa, and in the rather smaller size.

SCIURUS BALLOVIANUS Cope.

Bulletin U. S. Gelog. Surv. Terrs., VI, 177, February 11, 1881.
Plate LXIII; figs. 5–6.

This squirrel is the second species of its genus supposed to occur in the John Day beds of Oregon, and the third *Sciurus* obtained thus far from the Lower Miocene or Oligocene of the West. The typical specimen fortunately includes the cranium, with both rami of the mandible, so that its reference to the genus *Sciurus* rather than to *Gymnoptychus* is assured. Like the latter genus, the infraorbital foramen is reduced to a slit, but, unlike it, there is but one internal tubercle of the crowns of the superior molars instead of two.

The skull is flat above, and the interorbital space is also flat, and is remarkably wide. Temporal ridges none. Muzzle short and narrow. Palate wide, its posterior notch extending as far forwards as the last superior molar. The ascending ramus of the mandible originates opposite the anterior part of the last inferior molar. The masseteric fossa extends to opposite the anterior border of the second inferior molar. The mental foramen is near the superior border of the posterior part of the diastema. The second and third superior molars, the only ones preserved, have two cross-crests and a strong anterior cingulum. The external extremities of the cross-crests are little elevated, and there are no other cingula. The inferior molars have basin-shaped grinding faces, with a lobe at each angle. There is a small tubercle between the lobes of the inner and outer pairs. The incisors of both jaws are much compressed, strongly convex in front, and, in the lower jaw at least, without sculpture.

Measurements.

		M.
Length of skull to orbit		.0090
Width between orbits		.0090
Width of muzzle		.0047
Width between last molars		.0040
Length of superior dental series		.0054
Diameters of second molar	anteroposterior	.0015
	transverse	.0016
Width of superior incisor		.0013
Length of mandibular ramus		.0150
Elevation of ramus at coronoid		.0080
Length of diastema		.0030
Length of inferior dental series		.0070
Depth of ramus at second molar		.0045

This species is much smaller than the *Sciurus vortmani*, from the same horizon of Oregon. The type specimen was discovered by Mr. L. S. Davis, of Mr. Wortman's party. The name is given in honor of Mr. M. H. Ballou, of Chicago, a naturalist and journalist.

GYMNOPTYCHUS Cope.

Paleontological Bulletin No. 16, p. 5 (August 20, 1873). Ann. Report U. S. Geolog. Surv. Terrs., 1873 (1874), p. 476.

Dental formula: I. $\frac{1}{1}$; C. $\frac{0}{0}$; M. $\frac{4}{4}$. Crowns of the superior molars supporting two crescents on the inner side, and two cusps on the external side opposite to them. Each cusp sends a transverse crest to the concavity of the corresponding crescent. The adjacent horns of the crescents are united, and the connecting portion sends a transverse crest into the interval between the cusps. The opposed horns of the two crescents each send a crest round the anterior and posterior sides of the crown, of which they form the borders. Incisors simple. The walls of the alveolus of the inferior incisor produced into a tuberosity on the external side of the base of the ascending ramus.

The above characters define a genus which, when fully known, will in all probability be referred to near the existing genus *Sciurus*. In confirmation of this opinion, I add that the alveolar sheath of the interior incisor is in the vertical plane of the ramus; the incisive foramen does not invade the maxillary bones, and the *foramen infraorbitale exterius* is a small fissure situated in the inferior portion of the maxillary bone, well in advance of both the orbit and first molar tooth.

As compared with the existing genera, it differs in the structure of the molar teeth. The arrangement of the tubercles and crests is more complex than in any of them, excepting *Pteromys*. Thus in all of them there is but one internal crescent of the superior molars, and but two or three cross-crests; while in the inferior molars the arrangement is unlike that of the superior teeth, the cross-crests being marginal only. In *Pteromys* (F. Cuv.) the transverse valleys of the inferior series of *Gymnoptychus* are represented by numerous isolated fossettes. The structure of the molars in the fossil genus is exactly like that which I have described as found in *Eumys*, extending even to the details. This is curious, as that genus is a Myomorph.

The protrusion of the posterior extremity of the alveolar sheath of the inferior incisor on the outer side of the ascending ramus is not exhibited by the North American *Sciuridæ*, which I have examined, nor by any of the extinct genera herein described, excepting *Castor* and the *Geomyidæ*. It is seen in a lesser degree in *Mus musculus*, *Hesperomys leucopus*, *Meriones hudsonius*, and *Arvicola riparia*—all *Muridæ*.

Whether this genus possesses a postfrontal process I have been unable to ascertain. Its absence would not in my opinion isolate it from the *Sciuridæ*, as I accord with Dr. Coues in his estimate of the value to be attached to this character.

Of other portions of the skeleton I possess incomplete humerus, ischium, femur, and tibia. Most of these are appropriate in size to the *G. minutus*, which is also the most abundant species. A fragment of a larger femur belongs perhaps to the *G. trilophus*.

The *humerus* is rather slender, and the deltoid crest does not exhibit the prominence so usual in the *Muridæ*. It is most prominent on the antero-external aspect of the shaft near its middle; an external as well as an anterior ridge diverges from it upwards. The condyles have no intertrochlear ridge, and the external trochlea is not more extended transversely than the internal, measuring from the fundus of the groove. There is a moderate internal epicondyle, and the arterial foramen is distal, and opens anteriorly below and on the external face above. The bridge is slender and moderately oblique. The external border is acute and twisted.

The *ischium* is characterized, like that of other *Sciuridæ*, by the presence of a spine or process which is wanting in North American *Muridæ*, but is present in *Perognathus*. The bone is rather short, the tuberosity is but little enlarged, and the pubic process not very wide. The proximal end of a *femur* may belong to *Eumys elegans*, but is too small according to usual data. The great trochanter is elevated as high as the head, from which a deep notch separates it; its posterior fossa is pronounced. The little trochanter is very prominent, projecting at right angles to the shaft. The shaft is broken, so that the presence of a third trochanter cannot be ascertained. The distal end of the femur is characterized by a patellar groove of moderate width, with fairly elevated ridges which are continued well posteriorly on the shaft, but not further than in existing *Sciuridæ*, and not so far as in *Palæolagus*. The distal extremity of the *tibia* displays characters of the *Sciuridæ* as distinguished from those of *Muridæ* and *Leporidæ*. The fibula is of course distinct, and the external trochlear groove opens from its fundus outwards. The internal groove is narrower, and is bounded internally by a vertical malleolus, which has no distal articular facets, and which does not project, but is flat on the inner side. The greater part of the posterior face is occupied by the bones of the wide groove for the tendon of the *flexor longus pollicis* muscle. Its inferior edge is produced downwards as far as the malleolus, from which it is separated by the deep groove for the *tibialis posticus* and *flexor longus digitorum* muscles. This groove is marked on the inner side of the distal portion of the shaft, its anterior border being especially well defined.

Two species of this genus are certainly known. They belong to the White River horizon of Colorado. They differ, so far as known, chiefly in size, and in the propörtions of the inferior premolar tooth.

Gymnoptychus minutus Cope.

Paleontological Bulletin No. 16, p. 6 (August 20, 1873). Annual Report U. S. Geol. Surv. Terrs., 1873 (1874), p. 476.

Plate LXV; figs. 19–30.

This species is about the size of the *Dipodomys philipsi*, and was abundant during the White River epoch in Western North America. The specimens which I obtained in Colorado represent twenty or more individuals. Among them are portions of four crania. One of these presents characters somewhat different from those seen in the best preserved of the others, and may belong to another species of the genus. To this supposed species I formerly gave the name of *Gymnoptychus nasutus*.

The *Gymnoptychus minutus* was originally described from mandibular rami. The cranial fragment which agrees with these in the size of its teeth is the portion anterior to the middle of the orbit, and lacks the greater part of the nasal bones. The maxillo-premaxillary suture is quite sinuous in its inferior portions, and passes in front of the infraorbital foramen, measuring one-fourth the distance from the latter to the inferior border of the incisive alveolus, or about two-fifths the distance from the last-named point to the first molar tooth. The incisive foramina extend chiefly in advance of this suture, notching the maxillary bone a little. A shallow groove entends posteriorly from each incisive foramen along the palate as far as the palatine foramina, becoming shallower posteriorly. The tuberosity behind the inferior extremity of the infraorbital foramen, usual in the *Sciuridæ*, is here represented by a scar, with an inferior angular border. A shallow and wide fossa occupies the entire side of the maxillary bone above the infraorbital foramen and behind the premaxillary suture. The side of the premaxillary anterior to the suture is also shallowly concave. The maxillary is also slightly concave in front of the inferior base of the zyomatic arch The base of the latter has considerable vertical extent, but the arch is not preserved in any of my specimens. The premaxillary bone is continued upwards and backwards to the frontal, ceasing with the posterior extremity of the nasal. Of the superior maxillary teeth the first and third are equal, while the second is a little larger. The fourth has about three-fourths the linear dimensions of the third. The outlines of all are, in horizontal section, quadrate with rounded angles. The incisors are much compressed, and the anterior face of each is strongly and equally convex. The inner angle is pronounced, while the external is rounded. The enamel is smooth, without groove or keel, and does not lap over on the inner side. It extends a short distance on the inner side, forming a narrow band.

The mandibular ramus is of rather slender proportions, much as in *Hesperomys leucopus*. The symphyseal portion leaves the inferior border at an angle of 135°, not being continuous as in *Sciurus relictus*, nor so steep as in *Eumys elegans*. The inferior border is nearly straight, with a slight contraction below the last molar. The grinding surfaces of the molars form a plane which rises a little forwards. The pterygoid fossa is deep, and is bounded below by a thickened continuation of the inferior border. The ascending ramus commences opposite to the posterior end of the third molar, and its anterior border is quite oblique. The coronoid process is a rounded laminar projection, and is quite low. A groove separates its base and that of the condylar portion from the tuberosity which contains the papilla of the inferior incisor tooth. The masseteric fossa is well defined by borders, which are not prominent, and which unite in an acute angle below the middle of the first molar, above the middle of the side of the ramus. The *foramen mentale* is opposite the middle of the diastema and close to the superior plane.

The inferior incisor is compressed like the corresponding superior tooth. The front is symmetrically convex, but rounds into the outer side, while it is separated from the inner by an angle. The enamel does not fold on to the inner side, but covers about one-third of the outer side, as seen in profile view. A feature of this tooth, where it differs from the corresponding one of the upper jaw, is seen along the line where the anterior passes into the external surface. Here is a delicate groove, which is bounded on the outer side by an equally delicate thread-like ridge. In some specimens a still more delicate groove extends along at the front of the inner angle. The second and third molars are subequal; the last molar is a little smaller, while the first is a little smaller than the fourth. All are subquadrate in horizontal outline, but the first is narrowed in front. The composition of the molars has already been described. It remains to add that the posterior cross-crest is absent from the fourth molar, and the anterior one from the first molar. There are no basal cingula on the internal or external sides of the crowns. The direction of the dental series is moderately oblique to that of the axis of the ramus, and the posterior alveolar walls do not project very far either posteriorly nor inwards. The interior face of the ramus below the last molar is quite concave. The surface of the bone of the inferior surface is sparsely punctate.

The head of the *femur*, ascribed to this species with doubt, is perfectly round in vertical section, and is not prolonged transversely. The *fossa ligamenti teris* is submedian and isolated. The neck is constricted. There is a faint rectangular connection between the posterior ridges of the great and little trochanters. The femoral condyles are well separated, and the shaft is flat above them on the posterior face. The ridge forming one of the borders of the patellar groove is continued on the shaft as a rounded angle.

The distal portion of the shaft of the *tibia* has three sides, viz, a posterior flat side, an antero-external flat side, and a rounded antero-internal side. The fibular face forms another rather elongate facet. The process which carries the tendon of the *flexor longus pollicis* is acute below, and carries an anteriorly-directed crest on the inner edge of the groove, which sinks into an angle of the shaft. The posterior border of the astragaline face of the tibia is truncate. The two trochlear grooves are separated by a wide ridge.

Measurements.

		M.
Length of mandibular ramus from incisive alveolus to dental foramen		.0110
Length from same to end of last molar		.0090
Length from same to first molar		.0040
Depth of inner face of ramus at diastema		.0030
Depth of inner face of ramus at first molar		.0036
Depth of inner face of ramus at last molar		.0032
Depth of inner face of ramus at coronoid process		.0052
Diameter of incisor	anteroposterior	.0012
	transverse	.0009
Diameter second molar	anteroposterior	.0012
	transverse	.0012
Length of superior molar series		.0043

	M.
Length of superior diastema	.0070
Length from premolar to infraorbital foramen	.0029
Depth of cranium at front of orbit	.0065
Width of cranium at front of orbit	.0084
Width of palate between premolars	.0030
Width of palate between infraorbital foramina	.0040
Width of top of muzzle above the same	.0042
Length of incisive foramina	.0028

The cranial fragment on which the *G. nasutus* was founded differs from that above described in a somewhat greater compression of the muzzle and greater prominence of the tuberosity behind the inferior part of the *foramen infraorbitale exterius*. This tuberosity is well marked, while in the muzzle of the *G. minutus* it is merely indicated. The enamel of the incisors is of a bright bay color; it is pale waxen in all those of *G. minutus*. The nasal bones are preserved; they are distinct, and were continued beyond the line of the anterior border of the incisors. They are convex anteriorly and very shallowly concave behind. They extend a little further posteriorly than the extremities ot the premaxillary bones

Measurements.

	M.
Length of superior diastema	.0078
Length from premolar to infraorbital foramen	.0035
Depth of cranium at front of orbit	.0065
Depth of palate between premolars	.0030
Depth of palate between infraorbital foramina	.0032
Width of top of muzzle above same	.0041
Length of incisive foramina	.0028

Measurements of bones.

		M.
Length of humerus below deltoid angle		.0110
Diameter of shaft above the same	anteroposterior	.0024
	transverse	.0020
Diameter of distal extremity	anteroposterior	.0020
	transverse	.0045
Length of ischium from acetabulum		.0070
Width of distal extremity		.0025
Anteroposterior diameter of acetabulum		.0025
Width of proximal end of femur at head		.0050
Width at little trochanter		.0040
Diameters of shaft from below middle	fore and aft	.0018
	transverse	.0020
Diameters of distal end femur	anteroposterior	.0038
	transverse	.0035
Diameters of distal end of tibia	anteroposterior*	.0035
	transverse	.0030

* This includes the crest of the *sulcus tendinis flexor'is longi pollicis.*

GYMNOPTYCHUS TRILOPHUS Cope.

Paleontological Bulletin No. 16, p. 6 (August 20, 1873). Annual Report U. S. Geol. Surv. Terrs., 1873 (1874), p. 476.

Plate LXV; figs. 31–34.

This squirrel is represented by mandibles only of a small number of individuals, although very similar to the species last described in most details it is constantly of larger size, and I have seen no individual proportioned so as to be intermediate between them.

Three of the inferior molars of this species have the same longitudinal extent as the four, or entire series of the *G. minutus*. The detailed structure of the molars is the same as in that species. Attrition very soon causes the union of the exterior cross-crests with those which proceed from the cusps, producing as a result only three transverse yokes, and a figure something like the Greek *ω*. The delicate groove of the inferior incisor, and the posterior protuberance of its sheath, are as in *G. minutus;* so also the form of the masseteric fossa, the position of the mental foramen, etc. The premolar is relatively a little more elongate than in the *G. minutus*, but its details are the same. The diastema is relatively shorter.

Measurements.

		M.
Length of inferior molar series		.0065
Length of inferior diastema		.0045
Depth of inner face of ramus at diastema		.0035
Depth of inner face of ramus at first molar		.0040
Depth of inner face of ramus at last molar		.0035
Depth of inner face of ramus at coronoid process		.0052
Diameter of inferior incisor	anteroposterior	.0015
	transverse	.0010
Diameter of second molar	anteroposterior	.0014
	transverse	.0015

I discovered this species in the same locality as the *G. minutus.*

MENISCOMYS Cope.

Paleontological Bulletin No. 30, p. 5, December 3, 1878. Proceed. Amer. Philos. Soc., 1878 (1879), p. 67.

The characters of this genus are derived from the dentition of both jaws, and from portions of the cranium which are preserved. The molars are rooted, and number $\frac{5}{4}$ or $\frac{2}{1}$ $\frac{3}{3}$. Those of the superior series are without enamel inflections, and the triturating surface exhibits two external and one internal crescentic sections of the investing enamel. On the second superior molar there are three external crescents, and the first molar is simply conic. Between the inner and external crescents there are the curved edges of enamel plates directed obliquely and transversely. The grinding surfaces of the inferior molars display, in the unworn condition, curved transverse crests, connected longitudinally on the median line; on wearing, the lateral emarginations of the enamel become shallower, disappearing from the inner side, but remaining on the outer. Incisor teeth not grooved. *Foramen infraorbitale anterius* small, inferior, and near the orbit. Postorbital processes; no sagittal crest.

The characters of the dentition of this genus resemble those of the genus *Pteromys*, which is now confined to Asia and the Malaysian Archipelago. The superior molars differ from those of *Pteromys* in wanting all reëntrant enamel inflection.

The general characters of the skeleton are unknown. A femur is rather slender, and a tibia rather elongate, showing that the limbs are not short.

Four species of this genus are known to me, all from the Truckee Miocene of Oregon. They differ considerably in the details of the structure of the molar teeth. Their more prominent characters may be set forth as follows:

I. Superior molars short-rooted, with the external face plane; inferior molars with a prominent median transverse crest.

Smaller; dental crests fewer, simple, not crenate.................... *M. hippodus.*

II. Superior molars long-rooted; external face of crowns inflected, forming two V's; no median transverse crest on inferior molars.

a. Crests of superior molars fewer, simple, not crenate.

Larger; crowns short.. *M. liolophus.*

aa. Crests of superior molars more numerous and much crenate.

Smaller; plications of inferior molars shallow; borders raised........ *M. cavatus.*

Larger; plications of inferior molars profound........................ *M. nitens.*

There is a suggestive resemblance between the forms of the molar teeth of the *Meniscomys hippodus* and those of the *Haplodontia rufa* now living in Oregon. The two genera have doubtless had a common origin, but the present differences are considerable. Thus, the *Haplodontia* has an extended osseous cavum tympani which does not exist in *Meniscomys*.

This genus appears to be referable to the *Sciuridæ*.

MENISCOMYS HIPPODUS Cope.

Paleontological Bulletin, No. 30, p. 5, December 3, 1878. Proceedings Amer. Philos. Soc. 1878 ('79), p. 67, Plate LXIII; figs. 7-10.

Superior molars with a vertical ridge from the points of junction of the crescents on the external side; there are thus two on the second molar, and one each on the third and fourth. Within each of the external crescents is another crescentic edge of a pair of vertical enamel plates, and the inner marginal crescent sends off a short transverse branch towards them. With attrition, all these crests unite by their extremities, inclosing four distinct lakes, which, after still further wear, disappear. The crowns of the inferior molars, in the unworn condition, terminate in two crescents; that is, in elevated anterior, external, and posterior borders, with a transverse ridge equally dividing the space thus marked out and joining the notch in the external crest. This pattern resembles somewhat that of *Anchitherium*. The first inferior molar differs from the others in its superior size, and in its having the crescents more widely separated by a deeper external emargination On attrition, the spaces bounded by the enamel crests are inclosed by the junction of the extremities of these crests on the inner side of the crown. Further attrition results in three lakes within the crown, and one notch of the external border and two notches of the internal border. The anterior molar has two lakes in its posterior area, and one large one in its anterior area. In old teeth there are, successively, one and no lakes left to interrupt the dentine. The first temporary molar (Pm. 2) has much the form of the permanent tooth, but is smaller. The inferior incisor has a wide, shallow groove or concavity on its external face. Superior incisors regularly convex.

The skull of one specimen shows that the otic bullæ are large and moderately compressed. They have no recurved lip at the meatus, and their internal surface is covered with fine inosculating septa, forming a reticulate pattern (fig. 7). On another skull I observe that the fronto-nasal and fronto-maxillary sutures are in one transverse line, crossing the skull just anterior to the anterior border of the orbit.

The maxillary bone, anterior to the molar teeth, is shorter than the premaxillary. The incisive foramina are entirely in the latter. The sides and superior aspect of the muzzle are regularly convex in transverse section. The inferior incisive alveolus is inclosed entirely in the plane of the ramus, and extends posteriorly to below the last molar tooth. The masseteric ridge is very oblique, and rises to a median point below the second molar. The coronoid process rises from the front of the last molar.

Measurements.

		M.
Length of superior molar series		.008
Diameter of second superior molar	anteroposterior	.004
	transverse	.0035
Diameter of third superior molar	anteroposterior	.0020
	transverse	.0025
Width of superior incisor		.0020
Length from base of first superior molar to base of incisor		.0065
Width between bases of first molars		.0020
Length of first inferior molar		.0033
Depth of ramus at second molar		.0050
Width of ramus below second molar		.0035

This species was evidently abundant during the early Miocene period in Oregon. I have received numerous specimens from the John Day region from Messrs. Sternberg and Wortman.

MENISCOMYS LIOLOPHUS Cope.

Bulletin U. S. Geol. Survey Terrs., VI, p. 366.

Plate LXIII; fig. 11.

This species is known from a crushed skull, which displays the second, third, and fourth superior molars and the superior incisor in good preservation.

The superior molar teeth indicate that this is the largest species of the genus. Their characters place it between the extremes of form represented by the *M. hippodus* and the *M. nitens.* The crowns of the teeth are short, and the roots are long. The external V's are distinctly inflected, and are separated by a deep notch. The external dividing ridge is not prominent, and the posterior V is smaller than the anterior. The internal lobe is a simple raised border, which would be crescentic in section. It sends a ridge inwards to the anterior intermediate tubercle. The intermediate tubercles are low cones, and the posterior is connected with the apex of the posterior V by a low, simple ridge.

The entire absence of crenations and plications of the crests and lobes of the superior molars distinguishes this species from the *M. cavatus* and *M.*

nitens, but the details differ from those of *M. hippodus* in quite as great a degree. Thus, the intermediate tubercles are lower and more conic, and the external enamel border is not produced so as to inclose lakes, as in that species.

Other portions of the dentition are not determinable, except the superior incisor. This is moderately compressed, and has a regularly convex anterior face, without grooves.

Measurements.

		M.
Length of three large anterior molar teeth		.0095
Diameters of Pm. 2	anteroposterior	.0036
	transverse	.0032
Diameters of M. 2	anteroposterior	.0030
	transverse	.0040
Width of anterior face of superior incisor		.0020

Meniscomys cavatus Cope.

Bulletin U. S. Geol. Surv. Terrs., VI, p. 366.

Plate LXIII; figs. 12–15.

A concretion contains a number of limb bones and a mutilated skull of this rodent. The latter displays the entire dentition of one side, in both jaws.

Besides the characters of dentition which distinguish this species, the ear-bullæ differ from those of the *M. hippodus*. They are, as in that species, very large and somewhat compressed, but they differ in being transversely divided by bony laminæ of no great depth, which give a sacculate appearance like that of a colon, with the difference that the external surface is not grooved at the bases of the partitions. The superior surface of the skull is flat; there is no sagittal crest, and there is a well-marked postorbital angle. The interorbital surface is plane. The occipital condyle is narrow, and is separated from the otic bulla by a space as wide as itself. The basioccipital is not excavated, but has a low median keel-angle. The superior molar teeth are all posterior to the anterior border of the orbit, excepting the anterior lobe of the second premolar.

The anterior V of the second premolar is the largest; while on the last molar the anterior V only is present, the place of the posterior one being occupied by a flat space. There is a short external rib separating the two external V-faces. On the grinding surface of the molars it is evident that the posterior intermediate tubercle is represented by two crenate ridges, which are connected with each other and with the posterior external V by a zigzag ridge. The anterior intermediate tubercle is an enlargement of a crenate ridge which connects the internal edge with the external anterior V. In the inferior series the molars have those lobes on the edge of the inner side, of which the middle one is the least prominent. There are two better defined lobes on the external margin. The crowns are basin-shaped, and although the enamel is wrinkled in a complicated manner, the wrinkles are not elevated, as in the other species of the genus. Thus the inferior molars more nearly resemble those of ordinary *Sciuridæ* than do those of the other species of *Meniscomys*. The inferior incisor is much compressed.

The inferior border of the masseteric fossa is well marked and terminates forwards below the middle of the second molar. The base of the coronoid rises opposite the posterior part of the third molar. A femur has a cylindrical shaft. The distal portion is not elevated, and the rotular groove is rather wide and with equal lateral ridges. The condyles are narrow anteroposteriorly, and are spread well apart. A broken tibia is quite elongate, and the distal part of the shaft is subcylindric. The proximal part is triangular in section. The inner side above is a little concave, the posterior side is a little more concave, while the external side is rather strongly convex. The crest is not prominent.

Measurements.

		M.
Width of skull just behind postorbital angles		.0066
Length of otic bulla		.0115
Width between otic bullæ		.0038
Length of superior molar series		.0088
Diameters second molar	anteroposterior	.0028
	transverse	.0028
Diameters last molar	anteroposterior	.0023
	transverse	.0020
Length of inferior molar series		.0090
Diameter third molar	transverse	.0018
	anteroposterior	.0025
Width of inferior incisor		.0010
Depth of ramus at front of Pm. iv		.0040
Depth of ramus at front of M. iii		.0060
Width of condyles of femur		.0060
Depth of condyles of femur		.0050
Depth of fragment of tibia		.0320
Diameters head of tibia	anteroposterior	.0060
	transverse	.0060

From the John Day River region, Oregon.

MENISCOMYS NITENS Marsh.

Allomys nitens Marsh, Amer. Journ. Sci. Arts, 1877, September, p. 253. *Meniscomys multiplicatus* Cope. Proceedings Amer. Philosoph. Society, 1878, p. 68.

Plate LXIII; figs. 16–17.

Represented in my collections by fragmentary jaws of eight individuals, two of which contain superior molar teeth. These indicate a species which has some range of size, and which may be nearly as large as the *M. liolophus.*

The characteristic peculiarity of this species is, that there are two much crenate intermediate tubercles on the posterior side of the superior molar tooth, which are connected by a ridge with the posterior **V**; and that the anterior intermediate tubercle is part of a crenate ridge which connects the inner crescent with the anterior **V**. The external face of the crown is prominent at the notches between the **V**s.

The inferior molars are bilobate on the external side, and while the inner sides of the crowns are flat, their superior edge is trilobate. The grinding-face is much complicated by enamel ridges. Four crescentic areas are discernible on the worn surfaces of the crown, of which the posterior is reduced in size on the middle two molars. The two inclosed lakes have very plicate borders which form many small loops, and sometimes they are fused into a single irregular area. The last molar is extended a little posteriorly, and all present an entrant angle between the inner columns. The coronoid process originates opposite the third molar, and the masseteric ridge ceases below the middle of the jaw below the second molar.

Measurements.

		M.
Length of Pm. iv and M. i		.0060
Diameters of Pm. iv	anteroposterior	.0040
	transverse	.0040
Diameters of Pm. I.	anteroposterior	.0020
	transverse	.0040
Probable length of inferior molar series		.0120
Length of posterior three molars		.0095
Diameter of second molar	anteroposterior	.0030
	transverse	.0025
Length of fourth molar		.0040
Depth of ramus below second molar		.0070
Width of ramus below second molar		.0050

The above measurements are taken from the largest specimens. In some of the smaller ones the length of the anterior three inferior molars is .0085, and in one instance .0080. I think that it was on a specimen of this species with the last-named measurement that Professor O. C. Marsh established his *Allomys nitens*. The species was very briefly described, but may be recognized by an accompanying figure. A new generic name was proposed, but no characters were given, and it was even made the type of a new family, but no reasons for this course were adduced.

ISCHYROMYS Leidy.

Proceed. Acad. Nat. Sci. Philad., 1856, p. 89; Extinct Fauna Dakota and Nebraska, 335.—*Colotaxis* Cope, Paleontological Bulletin, No. 15, p. 1.

Char. gen.—The essential features are, dentition, I., $\frac{1}{1}$; C., $\frac{0}{0}$; M., $\frac{5}{4}$; the molars with two crescents on the inner side above, each of which gives rise to a cross-ridge to the outer margin. In the mandibular series the crests and crescents are identical, but in a reversed relation. No cementum.

To the above characters given by Dr. Leidy, I have added the absence of postfrontal processes, and the superior position of the infraorbital foramen; also that the pterygoid fossa is large, and that its inner and outer plates are well developed, and subequal. The palate is excavated posteriorly. The acuminate anterior part of the malar bone extends as far forwards as the front of the orbit. There is no tuberosity on the side of the superior diastema near the premolar teeth. In the mandible the posterior extremity of the incisive alveolus is not distinct from the ascending ramus.

In Hayden's Bulletin, vol. vi, I described some of the characters of the remaining portions of the skeleton. Those which I possess are the distal extremities of humeri, and a tibia, astragali, and portions of the pelvis. The condyles of the humerus are narrow anteroposteriorly. The internal flange descends at once to the fundus, leaving a long external cylindrical roller without intertrochlear ridge. Anteriorly this portion is cut into for half its length by the ligamentous fossæ. There is a large internal epicondyle, which is constricted by a neck at the base, and presents a compressed edge inwards and upwards. The arterial canal opens above on the interior side of the humerus. There is no external epicondyle.

The section of the *ilium* at its base and at its middle, is triangular. There is an angle along the middle of its external face which supports a moderately prominent tuberosity, a little above the acetabulum. On the anterior margin a little higher up, is a short, compressed, rather prominent process, which probably represents the anterior inferior spine. From this point posteriorly the internal face of the ilium is deeply concave, producing an attenuation of the inner wall of the acetabulum. The ischium is rather narrow at the base.

The distal portion of the tibia is much like that of *Arctomys*, *Gymnoptychus*, and other *Sciuridæ*. The posterior median process is very large and is shallowly grooved; the usual deep tendinous groove separates it from the internal malleolus. The trochlear grooves are deep and well separated; the fibular surface is short. The head of the astragalus is horizontally oval, and is separated from the trochlear portion by a neck of moderate length. It extends obliquely inwards, so that the internal margin of the head is interior to the line of the inner margin of the trochlea. The sides of the latter are vertical. It is considerably wider than long; the trochlear carinæ are marginal, and the external is considerably more elevated than the inner. The separating groove is profound but open. The posterior inferior fossa is small and foramen-like.

Besides the very different form and position of the infraorbital foramen, this genus differs from *Gymnoptychus* in the excavation of the posterior edge of the palate.

Dr. Leidy remarks that this genus belongs to the family of the *Sciuridæ*. This is indicated by the dental characters; but in some other respects there is a greater divergence from the squirrels and marmots than is the case with the preceding genus, *Gymnoptychus*. Thus, the large *foramen infraorbitale anterius* occupies the elevated position at the origin of the zygomatic arch seen in the porcupines and cavies. There is no superciliary ridge nor postorbital process as in most *Sciuridæ*, but the front is contracted between the orbits in the same manner as, but to a less degree than, in *Fiber*, and the Eocene *Plesiarctomys*, Brav. Both the last named and *Ischyromys* present many points of resemblance to Pomel's tribe of *Protomyidæ*, but differ from any of the genera he has included in it.

This family is thus defined by Pomel:* "infraorbital foramen large as in the *Hystricidæ*, and by the position of the angular apophysis of the mandible almost in the general plane of the horizontal ramus. The jugal bone, at least in those species where we have observed it, is very much enlarged at its anterior portion, and the orbit is almost superior."

These characters apply to *Ischyromys*, excepting as regards the malar bone, which is principally unknown in the latter.

Another family, the *Ischyromyidæ*, has been proposed by E. R. Alston for the reception of this genus, to which he thinks with me† *Plesiarctomys* (=*Pseudotomus*) should be referred. He thus defines the family:‡ "Dentition as in *Sciuridæ*; skull resembling *Castoridæ*, but with the infraorbital opening large, a sagittal crest; no postorbital processes; palate broad; basioccipital keeled."

Doubtless *Ischromys* belongs to an extinct family, but which of the above names is available for it I do not yet know. I would characterize it as follows:

Dentition as in *Sciuridæ*; infraorbital foramen large, superior; pterygoid fossa large, with well-developed exterior as well as interior walls; a sagittal crest.

The superior position of the infraorbital foramen and the well-developed pterygoid laminæ are characters found in the *Muridæ*.

But one species of this genus is known.

Ischyromys typus Leidy.

Proceedings Academy Philada. 1856, p. 89. Extinct Fauna of Dakota and Nebraska, 1869, p. 335. Cope Annual Report U. S. Geol. Surv. Terrs. 1873 (1874), p. 477. *Colotaxis cristatus* Cope Paleontological Bulletin No. 15, p. 1. *Gymnoptychus chrysodon* Cope, Paleontological Bulletin No. 16, p. 5.

Plate LXVII; figs. 1-12.

The principal characters of this species have been given by Dr. Leidy as above cited. I repeat the most important of these, and several which have not been previously noticed.

In the molar teeth of both jaws there is no transverse crest arising from the connection between the adjacent horns of the crescents. The

* Catalogue Méthod. et Descr. de Vertebrés Foss. de le Bass. de la Loire, 1853, p. 32.
† Annual Report U. S. Geol. Survey Terrs. 1873 (1874,) p. 477.
‡ Proceed. Zool. Society London, 1876, p. 78.

transverse crest from the anterior horn of the anterior crescent is largely developed in both jaws, while that proceeding from the posterior horn of the posterior crescent is united at both ends with the adjacent one. On the last superior molar, the posterior of the two principal cross-crests is shortened, and the posterior marginal crest is cut off from the inner crescent by a notch. In the premolar of the inferior series, the two anterior cross-crests are wanting and are replaced by two adjacent cusps, an internal and an external.

The incisive foramina are short and rather wide. There is no tuberosity on the side of the superior diastema as in most *Sciuridæ*. There is a well-marked fossa above the first premolar, and below the *foramen infraorbitale exterius*. The anterior part of the zygomatic arch is convex above and flat below, and at right angles to the long axis of the skull. An osseous bar extends outwards and backwards from the external base of the external pterygoid ala; whether it spans an alisphenoid canal I cannot determine. The palate is excavated as far as the posterior border of the last superior molar tooth. The ascending portion of the premaxillary bone is wide.

The area of insertion of the masseter muscle on the mandible, is feebly defined, extending only to the posterior border of the second molar. The pterygoid fossa of the ramus is on the other hand, deeply excavated. The ascending ramus commences opposite the third molar. The dental foramen enters some distance behind the last molar on a level with its summit, and above the angular superior border of the pterygoid fossa. The mental foramen issues behind the middle of the length of the diastema and above the middle of the vertical diameter of the ramus at that point.

The section of the superior incisor teeth is a nearly regular oval, a little flattened on the inner side near the front. The anterior face presents a very shallow median longitudinal concavity. The enamel is nearly smooth, and is wrapped on the inner side for a moderate width. The inferior incisors are more oblique on the exterior side anteriorly, and are therefore narrower in front than the superior. The enamel is smooth, presents no sulcus, is wrapped in a narrow band on the inner side, and on half the entire width on the outer side.

There are differences in the form of the inferior premolar, some being a little wider anteriorly than others. On a specimen presenting the narrowed form the name *cristatus* was proposed. In some specimens the inferior incisors are more compressed than in others. In some of the specimens the original bright-bay color of the enamel of the superior and inferior incisors is beautifully preserved.

Measurements.

	M.
Length of inferior molars	0.0140
Length of penultimate molar	.0033
Width of penultimate molar	.0035
Width of first molar	.0030
Length of first molar	.0035
Depth of jaw at penultimate molar	.0090
Depth of incisor-tooth	.0040
Width of incisor-tooth	.0020

The olecranar fossa of the *humerus* has a small perforation. There are two small rather deep fossæ on the posterior aspect of the humerus distally; the larger is at the base of the epicondyle, the other is within and a little above it. The marginal crest from the external epicondylar region is acute and rather elongate. The shaft of the humerus above it is quite slender. In a second specimen, the inner of these two fossæ is wanting. At the middle of the shaft of the tibia, the anterior and external faces meet at a right angle; the inner and posterior faces form a nearly continuous curve. The inner crest of the posterior distal process, which bounds the *sulcus tendinis flexoris longi pollicis* is not so produced as in *Gymnoptychus minutus*, or *Arctomys monax*. The humerus bears considerable resemblance to that of both these species, but differs from both in the presence of the posterior fossæ. In *Arctomys monax* the arterial foramen is wanting.

Measurements.

		M.
Diameter distal end of humerus	anteroposterior	.0050
	transverse	.0190
Length of internal epicondyle		.0060
Diameter distal extremity tibia	anteroposterior	.0060
	transverse	.0080
Diameter of base of ilium	exterointernal	.0060
	anteroposterior	.0065
Anteroposterior diameter of acetabulum		.0065
Length of astragalus (axial to foot)		.0100
Length of astragalus obliquely		.0115
Width of trochlea of astragalus		.0060
Width of head of astragalus		.0045
Elevation of external arc of astragalus		.0045

This squirrel was about as large as the existing *Sciurus niger*. It was common during the White River epoch, specimens having been obtained in Dakota, Nebraska, Wyoming, and most abundantly in Colorado.

CASTOR Linn.

Syst. Nat. I, p. 78, 1766.—Cope, Bulletin U. S. Geol. Surv. Terrs., VI, p. 368. *Steneofiber*, E. Geoffr., Revue Encyclopédique, 1833.—"*Chalicomys* Meyer, Neues Jahrbuch, 1838, p. 404, et 1846, p. 474."—*Palæocastor* Leidy, Extinct Mammalia Dakota and Nebraska, 1869, p. 338.

The family of the *Castoridæ* differs from the *Sciuridæ* in the absence of postorbital angles or processes, and the presence of a prolonged tube of the meatus auditorius externus. In both of these points it agrees with the *Haplodontiidæ*, a family which Mr. Alston has distinguished from the *Castoridæ* on various grounds. I do not think any of his characters are tenable, excepting that drawn from the form of the mandible, which is expressed thus in Mr. Alston's diagnosis: "Angular portion of mandible much twisted" I have described this character better as follows: Angle of mandible with a transverse edge due to inflection on the one hand, and production into an apex externally; the inflection bounding a large internoposterior fossa.

Mr. Alston enumerates four genera of *Castoridæ*—*Castor*, *Diobroticus*, *Steneofiber*, and *Castoroides*. J. A. Allen has shown that the last-named genus cannot be referred to this family. The characters of *Diobroticus*, as given by Alston, are as follows: "Skull much as in *Castor*. Third upper molar and lower premolar elongate, with four enamel folds, the rest with only two; all the folds soon isolated." This diagnosis appears to separate the genus satisfactorily. The definition of *Steneofiber* is as follows: "Parietals not parallelogrammic; interparietal subhexagonal; basioccipital not concave; grinding teeth as in *Castor*, the subsidiary folds sooner isolated." The distinction from *Castor* here rests exclusively on the forms of the parietal, interparietal, and basioccipital bones. This kind of definition is always of questionable validity, as the terms "parallelogrammic," "hexagonal," &c., are not intended to be exactly used and cannot be exactly applied. The *Castor* (*Steneofiber*) *peninsulatus* illustrates this fact, for there is no striking difference in the forms of the two bones to which these terms are applied, as compared with the *Castor fiber*. The basioccipital bone differs from that of the beaver, but not so as to conform to J. W. Alston's diagnosis of the genus *Steneofiber*. Its inferior surface is concave, but doubly so, as a keel occupies the median line. In the *S. viciacensis* according to Filhol, this region is shallowly concave, without median keel. Although important as specific characters, these variations do not appear to me to require the recognition of as many genera. The possession of the epitrochlear foramen in the *S. viciacensis* is at first sight an important character. Mr. Filhol, however, informs us that of thirty-four humeri which he has studied, sixteen possess the foramen, and in eighteen it is wanting.

The *Castor tortus* was described by Leidy from the Loup Fork formation. He coined the subgeneric name *Eucastor* for it without corresponding definition. In his monograph of the *Castoridæ*, J. A. Allen referred this species[1] to a genus distinct from *Castor*, and defined it, using for it Leidy's name *Eucastor*. This genus appears to me to be valid. The three genera of *Castoridæ* will then be defined as follows:

Molars and premolars with one inner and two or three outer folds .... *Castor.*
"Inferior premolar and third superior molar elongate, with four enamel folds; the rest with only two" .... *Diobroticus.*
Superior premolar enlarged, with one inner fold; inferior molars small, with two lakes .... *Eucastor.*

Some affinity probably exists between this family and the *Mylagaulidæ*, which followed in the Loup Fork epoch.

The species of *Castor* may be distinguished as follows. I do not know the *C. nebrascensis*[2] from the White River, nor the occipital bone of the *C. pansus*[3] from the Loup Fork formation.

I. Basioccipital bone deeply concave below, simple;
  Angle of lower jaw not deflected .... *C. fiber.*
II. Basioccipital shallowly concave below:
  Angle of lower jaw deflected .... *C. viciacensis.*
III. Basioccipital concave below with a median keel; angle of lower jaw not deflected.
  Palate wider; size medium .... *C. peninsulatus.*
  Palate narrower; size small .... *C. gradatus.*

[1] Monographs of North American Rodentia, Coues and Allen, U. S. Geol. Surv. Terrs., 1877, xi, p. 450.

[2] *Steneofiber nebrascensis* Leidy, Proceed. Academy Phila., 1856, 89; 1857, 89; *Chalicomys nebrascensis* Leidy, l. c., 1857, 176; *Palæocastor nebrascensis* Leidy, Ext. Mammalia Dakota and Nebraska, 1869, p. 338, xxvi, Figs. 7-11.

[3] *Steneofiber pansus* Cope, Report Capt. G. M. Wheeler, iv, pt. ii, 1877, p. 297.

CASTOR PENINSULATUS Cope.

Bulletin U. S. Geol. Survey Terrs., 1881, VI, p. 370. American Naturalist, 1883, p. 53, figs. 9, 10. *Steneofiber ? nebrascensis* Cope, loc. cit., 1879, p. 55.

Plate LXIII; figs. 18–21.

This species is about the size of a large prairie marmot—*Cynomys ludovicianus*. It was abundant in Oregon during the period of the John Day Miocene. Leidy originally described the closely allied *C. nebrascensis* from the WhiteRiver beds of Nebraska, but I have never obtained it from that formation. Another and similar species, *C. pansus* Cope, is common in the Loup Fork beds of Nebraska and New Mexico.

Several well-preserved skulls from Oregon display characters not visible in specimens heretofore collected, and which enable me to make fuller comparisons with the European *C. viciacensis*, so fully described by M. Filhol.[1]

The postorbital constriction is much greater in this species than in the *C. viciacensis*, and greater than in the *C. nebrascensis* from the White River beds. The straight anterior temporal ridges are in line with the superciliary borders, and unite into a sagittal crest at the constriction. In the *S. viciacensis* they continue separate beyond this point one-fourth the distance to the supraoccipital crest. The zygomata are wide, and the malar ridge is very prominent anteriorly, overhanging the face, and curving rather abruptly to the base of the muzzle. The latter is rather wide, with parallel sides, and is flat above. The brain-case expands rather abruptly from the interorbital constriction, and is rather flat above. The infraorbital foramen is a narrow vertical oval and is situated low down in the vertical line with the anterior extremity of the malar angular edge. It is a little nearer the line of the first molar than the posterior border of the superior incisor. The incisive foramina are relatively larger than in the beaver, and are chiefly in the premaxillary bone. The palate between the anterior molars is as wide as the transverse diameter of the first molar. There is no distinct fossa of the maxillary bone in front of the orbit as represented by Filhol in the *C. viciacensis*. The pterygoid fossa is wide, with the inner process the longer,

[1] Étude des Mammifères Fossiles de Saint-Grand-le-Puy Allier. Bibl. de l'École des Hautes Études, XIX, Art. I, p. 44, 1879.

and reaching the otic bulla. The latter are large and obliquely placed; the meatal borders are produced into a short tube, which is not so long as that of the *C. fiber*. Its superior border is quite prominent, overhanging the inferior, and projecting more than represented by Mr. Filhol in the *C. viciacensis*. There is a strong ridge of the squamosal bone extending posteriorly from the base of the zygomatic process, which overhangs a fossa. This fossa is further defined posteriorly by the tympanic tube. This fossa is larger and deeper than in either *C. fiber* or *C. viciacensis*. Below its superior bounding ridge is a large subsquamosal foramen. The mastoid bone is distinctly exposed between the squamosal and occipital, and its surface is separated from that of the former by a groove which is not so well marked in the *C. fiber*. Its inferior angle is in contact with the bulla, and is shorter than the paroccipital process. The latter is short, not extending below the line of the condyles, and is directed downwards, not posteriorly as in *C. fiber*. The occiput is nearly vertical and flat, excepting laterally, where there are two fossæ, a superior and inferior, the latter the longer, and extending to the inferior surface.

The premaxillo-maxillary suture is just half way between the anterior molar and the superior incisor, and is vertical to opposite the middle of the incisive foramen, and then turns backwards. The frontomaxillary and frontonasal sutures are in one transverse line across the front. The malar-maxillary suture is behind the anterior border of the zygoma, thus confining the malar bone to the zygoma. The latter is much expanded in a vertical direction, but has no postorbital angle, resembling in this respect the *C. viciacensis* rather than the *C. fiber*. Its posterior portion extends well posteriorly and below almost all of the squamosal part of the zygoma. The parietal is of a parallelogrammic form; the anterior inner border cut obliquely by the frontal, and the posterior inner border cut out for the supraoccipital. The latter bone has an oval form, narrowed anteriorly and truncate posteriorly.

I describe a mandibular ramus of a second individual. It unfortunately has the coronoid and the angle broken off. The base of the latter is concave on the inner side. The external face of the ramus is everywhere convex. The base of the coronoid is separated from the molar line by a wide groove. The anterior base is opposite the second molar. The incisive alveolus is continued upwards and backwards, and ceases in a protuberance which is external to the plate which connects the condyle with the coronoid process, and is separated from it by a fossa. The condyle is subglobular, and has considerably more external than internal articular surface. The series of inferior molar teeth is quite oblique, descending posteriorly.

Dentition.—The grinding surfaces of the superior molars are none of them longer than wide, and in all but the first, the transverse diameter exceeds the anteroposterior. The dimensions diminish posteriorly in all the measurements. There is one inflection of the sheathing enamel on each side of the crown in all the molars in their present state of wear. The positions of the lakes indicate that in an earlier stage there were two external inflections. At present all the molars display a fossette external to the fundus of the internal inflection. Besides this there are two others in the first molar, and one other in the fourth. Posterior to the external inflection there is one fossette in the first and third molars, two on the fourth, and none on the second, where it is probably worn out. According to Leidy there are three in this position in all the molars in the *C. nebrascensis*, and two in the position first described. According to Filhol, there is but one in each position, in about the same stage of wear, in the *C. viciacensis.*

The inferior molars display a deep external inflection, and three transverse lakes on the inner side. These probably represent inflections at an earlier stage of wear; the median one is still continuous with the sheathing enamel on the first molar (Pm. IV). The sizes of the inferior molars increase anteriorly regularly, excepting that the first is relatively a little longer than the others.

Measurements of skull.

	M.
Length from inion to edge of nasals	.067
Length from edge of nasals to interorbital constriction	.034
Length of muzzle to preorbital angle of maxillary	.018
Width of skull at paroccipitals	.025
Width of skull at posterior edge of squamosal	.034
Width at zygomata posteriorly	.056
Width at interorbital constriction	.006

	M.
Width between anterior parts of orbits	.0275
Width at base of muzzle	.018
Length from occipital condyle to front of otic bulla, inclusive	.018
Length from bulla to last molar tooth	.008
Length from first molar to base of incisor	.023
Length of superior molar series	.0195
Diameters of first molar { anteroposterior	.0045
Diameters of first molar { transverse	.0045
Diameters of fourth molar { anteroposterior	.0030
Diameters of fourth molar { transverse	.0035

Measurements of mandible.

Length from condyle to incisor tooth	.0390
Length from incisor to Pm. iv	.0105
Length of inferior molar series	.0145
Diameters of Pm. iv { anteroposterior	.0047
Diameters of Pm. iv { transverse	.0040
Diameters of M. i { anteroposterior	.0030
Diameters of M. i { transverse	.0040
Diameters of M. ii { anteroposterior	.0030
Diameters of M. ii { transverse	.0035
Depth of ramus at diastema	.0100
Depth of ramus at Pm. iv	.0130
Depth of ramus at M. iii	.0100

A few bones accompany the mandible, all having been cut from the same fragment of matrix. The head of the femur is perfectly round, and is bounded by a well-defined neck. The great trochanter incloses a large fossa. The lesser trochanter is large; the third trochanter is not prominent as it is in *C. viciacensis*. The shaft is generally flattened, with its long diameter transverse. The condylar extremity is flattened, and the rotular groove is wide, and the condyles well separated. The epicondylar angles are distinct, but not so prominent as in *S. viciacensis* (see Filhol, *l. c.*, Pl. VI, Figs. 13, 14). The general form of the femur is robust, as in that species.

The distal extremity of the tibia resembles that of *Sciuridæ* generally, especially in the large size of the external posterior angle. Its diameters are small, and the distal part of the shaft is slender and subcylindrical. The crest extends well down from the proximal end, being much stronger than in the true squirrels, and bounds a longitudinal fossa. The fibular facet of the tibia overhangs extensively, and bounds a rather narrow proximal fossa. This continues into a narrow shallow groove on the posterior face of the shaft, which disappears near the middle of its length. The proximal half of the shaft is much compressed. The inner face is smooth and gently convex. The crest sinks rapidly into the head, and the femoral facets are well separated. The tibia resembles that of *C. viciacensis*, but is more compressed in its proximal half.

Measurements of hind leg.

	M.
Length of femur	.057
Width of femur at head	.019
Width of femur shaft below third trochanter	.009
Width of femur at epicondyles	.015
Width of condyles of femur	.013
Anteroposterior diameter at condyles	.011
Anteroposterior diameter of head	.009
Diameters of head of tibia { anteroposterior	.012
Diameters of head of tibia { transverse	.014
Diameters of distal end of tibia { anteroposterior	.0080
Diameters of distal end of tibia { transverse	.0085
Anteroposterior diameter { above middle	.010
Anteroposterior diameter { below middle	.005

I have referred to this species in former catalogues of the vertebrate fauna of Oregon as the *Castor nebrascensis* of Leidy. It is very nearly allied to that species, but I find the following differences: First, the postorbital constriction is narrower; second, there are fewer fossettes on the posterior half of the molar teeth, but one or two. Leidy figures two or three in the species described by him.

CASTOR GRADATUS Cope.

Steneofiber gradatus Cope. Paleontological Bulletin No. 30, p. 1, December 3, 1878. Proceedings American Philosophical Society, 1878 (1879), p. 63.

Plate LXIII; fig. 22.

This species is represented in my collection by parts of several individuals. One of these is a palate with both series of molars; a second is a mandibular ramus with all the molars. The best specimen is a cranium, which is nearly perfect, the principal deficiency being the absence of the mandibular rami. It is of smaller size than the *S. nebrascensis* and *S. pansus*, and differs from both these species in the relative sizes of the superior molar teeth. The first of these is the largest, and the others diminish regularly in size to the last, whose grinding face does not present more than one-third the extent of that of the first. The triturating surfaces of the second and third have their long axes transverse. In all the crowns, besides the internal and external enamel inflections, there is but one fossette, which is anterior to the external inflection. The latter has become isolated from the superficial enamel on the last three molars, by attrition. The superior incisors are flat anteriorly with the external angle rounded, and its dentine presents the transverse undulation seen in *S. pansus.*

The palate is distinctly narrower in this species than in the *C. peninsulatus.* The width between the bases of the first molars (Pm. iv) is less than the transverse diameter of each of those teeth; in *C. peninsulatus* its width is equal to that of those teeth. The temporal ridges do not unite so early in this species as in *C. peninsulatus,* but continue parallel for a considerable distance. The parietal bone is injured posteriorly from the middle, so that the union into a sagittal crest, if it takes place, cannot be seen. The middle line of the basioccipital bone is keeled, with a fossa on each side. The tympanic meatus is prolonged, and the posttympanic process is short. The space between the incisor and superior molar teeth is about two-thirds the same space in the *C. peninsulatus,* and is relatively shorter than in that species.

Measurements.

		M.
Length of skull from incisive alveolus		.0500
Width between bases of first molars		.0030
Width between bases of fourth molars		.0070
Length of molar series		.0115
Diameter of the first molar	anteroposterior	.0040
	transverse	.0045
Diameter of third molar	anteroposterior	.0028
	transverse	.0032
Diameter of fourth molar	anteroposterior	.0020
	transverse	.0024

From the above measurements it is apparent that the molar series in this species is equal in length to the anterior three molars of the *S. nebrascensis* and *S. pansus.* The posterior fossettes of the crowns seen in those species are wanting in the *S. gradatus.*

From the John Day River, Oregon.

HELISCOMYS Cope.

Synopsis of New Vertebrata from Colorado (Miss. Pub., U. S. Geol. Survey Terrs.), 1873 (October), p. 3; Annual Report U. S. Geol. Survey Terrs., 1873 (1874), p. 475.

Inferior molars four, rooted, the crowns supporting four cusps in transverse pairs. A broad ledge or cingulum projecting on the external side from base of the cusps. The inferior incisor compressed, not grooved, and with the enamel without sculpture.

This genus is only represented by a small number of specimens, which are mandibular rami exclusively. Its special affinities, therefore, cannot be ascertained, and even its general position remains somewhat doubtful. There is some probability, however, that it belongs to the *Myomorpha*, as the type of dentition is much more like that of the genera of that group than those of the *Sciuromorpha*. To the *Hystricomorpha* it does not belong.

As compared with known genera of *Myomorpha*, it is at once separated from many of them by the presence of a premolar tooth. Among recent genera of this suborder, *Sminthus* possesses this tooth in both jaws, and *Meriones* in the upper jaw only. It is present in both jaws in the *Sciuromorpha* generally. The tubercles of the teeth resemble those of the *Muridæ*, but their disposition is unlike that of any existing North American genus. A remote approximation to it is seen in the genus *Syllophodus* of the Bridger Eocene formation, where there are four subquadrate molars with tubercles; but the latter form two transverse crests, with an additional small intermediate tubercle, and the wide cingulum is absent.

But one species of *Heliscomys* is known, the *H. vetus*.

HELISCOMYS VETUS Cope.

Synopsis of New Vertebrata Colorado, 1873 (October), p. 3. Annual Report U. S. Geol. Survey Terrs., 1873 (1874), p. 475.

Plate LXV; figs. 14–18.

The mandibular ramus is rather robust in its proportions. Neither the alveolar sheaths of the molars nor that of the incisor project beyond the general surface. The anterior base of the ascending ramus commences opposite the anterior part of the last molar tooth. The masseteric fossa extends remarkably far anteriorly, its inferior border terminating close to the mental foramen. This border is moderately raised, and extends downwards and backwards, not reaching the inferior border anterior to a point below the anterior base of the ascending ramus. The symphysis rises at an angle of about 45°, and the inferior border is very little convex to below the last molar. The *foramen mentale* is situated near the superior plane of the diastema, half way between the alveolar borders of the incisor and premolar teeth.

The grinding face of the molar series rises slightly anteriorly. The sections of the middle two molars are subquadrate and are of equal size. That of the fourth is smaller by one-third; having the same longitudinal extent, but less width, thus forming a longitudinal oval. The section of the premolar is round, and its diameter is half that of the long diameter of the other molars. The premolar has two contiguous roots, and three conic cusps, one anterior and two opposite and posterior. The cusps of the other molars are separated transversely by a deep notch, and longitudinally by a fissure, which does not invade the surface of the cingulum. The latter extends a short distance round the anterior base of the crowns of the second and third molars. The crown of the fourth molar is worn or injured in my specimens, so as not to be described. The incisor is strongly compressed, as in many *Muridæ*, and the anterior face is slightly convex in section. The enamel is smooth, and is abruptly recurved in narrow borders of about equal width on both the inner and outer sides. On the inner side the surface makes a sharp right angle; on the outer side the angle is right, but is obtusely rounded. The surface of attrition is elongate, and displays a small pulp cavity.

Measurements.

	M.
Length of incisive alveolus to end of molar series	.0050
Length from incisive alveolus to base of ascending ramus	.0044
Length of diastema	.0020
Length of third molar	.0008
Width of third molar	.0008
Long diameter of incisor	.0008
Short diameter of incisor	.0004
Depth of ramus at diastema	.0018
Depth of ramus at third molar	.0021
Thickness of ramus below the same	.0014

The *Heliscomys vetus* is the least mammal of the fauna of the White River epoch. The mandibular ramus is the size of that of the *Mus musculus*, and its four teeth occupy the same length as the three of the latter species. One of the specimens indicates an individual a little larger than the one above measured.

MYOMORPHA.

EUMYS (Leidy nom.) Cope.

Annual Report of the U. S. Geological Survey of the Territories, F. V. Hayden in charge, 1873 (1874), p. 474.—*Eumys* Leidy (name only), Proceedings Academy, Philada., 1856, p. 90; loc. cit. 1857, p. 89; Extinct Mamm., Dakota and Nebraska, p. 342.

Dental formula: I. $\frac{1}{1}$; C. $\frac{0}{0}$; M. $\frac{3}{3}$. Crowns of the superior molars supporting two external cones, and two internal tubercles of crescentic section which communicate with the former by transverse ridges. Inferior molars of similar constitution, but reversed, the conic tubercles being interior and the crescentic exterior. The posterior tubercles of the posterior molars reduced, and an additional one on the anterior extremity of the first molar. Superciliary ridges none, but the supraorbital borders converging towards the middle line, and meeting above the postorbital region. No indication of postfrontal processes. Infraorbital foramen rather large above, terminating below in a vertical fissure. Incisive foramen entering the maxillary bone extensively. Incisor teeth not grooved.

I only know this genus from the cranium anterior to the pterygoid region, the mandibles, and the dentition. These parts display the characters of *Muridæ*, and in particular of the existing genus *Hesperomys*. The only character which I can find which has enabled me to distinguish *Eumys* from the latter genus is the extension upwards of the orbital fossæ so as to form an interorbital crest. In none of the Sigmodont genera of North America are the supraorbital borders contracted in this way, but the crest is seen in *Fiber* and in various degrees in the genus *Arvicola*, being as distinct in *Eumys* as in *A. xanthognathus*.[1]

A single species is certainly referable to this genus, the *E. elegans*, which was abundant during the White River Miocene epoch.

The typical species was originally described by Leidy, who gave it the generic name which I have adopted; but he at no time characterized the genus, or showed how it differed from others already known. This was first done by myself, as above cited.

[1] Report of Lieut. G. W. Wheeler, IV, p. 300, Pl. LXIX, fig. 15

EUMYS ELEGANS Leidy.

Proceedings Academy, Philada., 1856, p. 90; loc. cit. 1857, p. 89. Extinct Mammalia of Dakota and Nebraska, p. 342, Pl. XXVI; figs. 12, 13, 1869. Cope Ann. Report U. S. Geol. Surv. Terrs., 1873 (1874), p. 474.

Plate LXV; figs. 1-14.

I found remains of this rat exceedingly abundant in the shallow washes of the White River rocks in Eastern Colorado. It had previously been known from a single imperfect jaw-fragment from Dakota, found by Dr. Hayden.

The sutures of the cranium anterior to the brain-case are distinct, including those that separate the frontal and nasal bones on the median line. The suture separating the maxillary and premaxillary bones crosses the palate a little behind the line connecting the bases of the zygomata, passes upwards a little in front of the latter and then turns posteriorly, joining the nasal suture at its posterior extremity. The apices of the premaxillary and nasal bones form an acute angle above the middle of the orbit, and the frontals enter an acute notch between them. The palato-maxillary suture crosses the palate opposite the middle of the second molar tooth.

The nasal bones are strongly convex at the extremity, and extend a short distance beyond the base of the superior incisor teeth. They narrow posteriorly, become plane, and then concave on the median line. The carina of the frontals incloses a median groove. The superior portion of the maxillary bone presents a subround fossa at the entrance of the *foramen infraorbitale*. The latter is overarched by a slender bridge, which is as wide as thick. The malar bone does not appear in the base of the zygoma, whose inferior expansion presents a shallow, wide fossa, downwards and forwards. The anterior border of the base is free, inclosing a fissure which descends from the superior portion of the foramen. The posterior margin of the base is opposite to the anterior part of the first molar tooth, a point which is nearly attained by the incisive foramina. The latter are separated by a narrow septum on the median line. The free border of the palate is slightly concave, but does not advance beyond the posterior part of the third superior molar. The palatine foramen is large, and is opposite the posterior half of the second molar.

The mandible is robust, and the symphysis is very steep in profile. Its sutural face descends at an angle of 45°, and then exhibits a rugose point of union below the first molar on the inferior border of the ramus. The inferior border is subparallel with the tooth line, but presents an upward concavity below the third molar. The coronoid process originates opposite the anterior part of the third molar, is vertical and compressed. The anterior border extends obliquely backwards. The masseteric fossa has a prominent inferior border which extends upwards and forwards to below the middle of the first molar. Its surface is gently concave below, convex above. The pterygoid fossa of the ramus is deep and large. Its inferior border is a spur-like continuation of the inferior part of the ramus, from which the thin outer wall of the fossa rises. The latter is somewhat convex on the external side. The tooth basis and the incisive alveolar sheath do not project much on the inner side of the ramus. The diastema is as long as the tooth line, and the foramen mentale is below a point a little behind its middle. The foramen dentale enters a little above the base of the vertical ramus and three times the length of the last molar behind that tooth.

The superior incisor teeth are convex in front, the convexity slightly oblique to the external side. The enamel is smooth and does not overlap the sides. The crescents and cones of the molar teeth are well separated from those of the same side by vertical fissures. A narrow crest connects the base of each cone with the concave side of each crescent. The adjacent horns of the two internal crescents join, and the connecting portion sends a short crest into the interval between the external cones. The anterior and posterior horns of the two crescents are produced transversely, forming the anterior and posterior borders of the crown in the middle molar, the anterior border of the third, and posterior border of the first molars. The anterior or fifth lobe of the first molar is simple. The inner and outer lobes of the third molar are reduced in size, and close together. The last superior molar is only half as large as the first, and the middle molar is intermediate in size.

The molars of the inferior series do not differ in size so widely as those of the superior series. They are subequal, but the anterior and posterior are somewhat narrowed at their extremities. The tubercles of the first and last differ slightly from those of the superior series. The anterior odd one of the first is smaller; while the posterior outer crescent of the third is well developed, with all the transverse crests. The posterior inner cone is, however, wanting. The details of the inferior molars repeat those of the superior series, but in reversed relation, the external portions of the one being the internal of the other, and vice versa. The inferior incisors are nearly as much curved as the superior ones; they are robust and subtriangular in section. The enamel is smooth, and has a convex surface presented more obliquely outwards than the superior incisors.

Measurements.

	M.
Depth of skull at middle of orbit, without molar teeth	0.0080
Width between orbits	.0050
Length from middle of orbits	.0150
Length from front of zygoma	.0100
Length of superior diastema	.0075
Length of nasal bones	.0120
Length of superior molar series	.0060
Length of first superior molar	.0025
Length of third superior molar	.0015
Width of first superior molar	.0020
Width of third superior molar	.0018
Width of nasal bones at middle	.0020
Width of nasal bones distally	.0030
Width between bases of zygomata	.0070
Width of superior incisor tooth	.0015
Length of mandible of another individual to dental foramen	.0160
Length to end of dental series	.0115
Length to front of dental series	.0055
Length from base of vertical symphysis to front of pterygoid fossa	.0095
Depth at diastema	.0050
Depth at middle molar	.0055
Length to base of coronoid	.0090
Thickness of ramus on front border of masseteric fossa	.0040
Anteroposterior diameter of inferior incisor tooth	.0020
Anteroposterior diameter of last molar	.0020
Width of last molar	.0018

The hundred and more individuals represented by my specimens, evidently differed in various minor respects, as in size and robustness. Some, perhaps males, have the muzzle stouter in proportion to the length than others; some are more decurved than others.

With molar teeth as large as those of the Norway rat the muzzle of the *Eumys elegans* is not more than two-thirds as long; so that the species was in general proportions smaller and more robust.

HESPEROMYS Waterhouse.

This recent genus had a representative in the John Day Miocene in North America, so far as the characters of the skull and dentition may be considered to be conclusive in evidence. It is not very probable that the indications thus obtained will be invalidated by other portions of the skeleton.

The molars are $\frac{3}{3}$, and the crowns support alternating tubercles separated by shallow open transverse valleys. These are, one on the inner and two on the outer sides of the superior series, and one on the outer and two on the inner side of the inferior. In the recent species, (*H. leucopus*) there are two inflections on the inner side of the first molar, but in the species here described that tooth is constricted at the position of the anterior internal loop, and does not regain its width, but continues narrowed to the anterior extremity. The infraorbital foramen is rather large.

It is probable that there is a second species of this genus in the Loup Fork beds besides the *H. loxodon* Cope.

HESPEROMYS NEMATODON Cope.

Paleontological Bulletin No. 31, p. 1, Dec. 24, 1879. Proceedings American Philosoph. Soc., 1879–'80, p. 370.

Plate LXVI; fig. 33.

This rat is represented by a beautiful skull, discovered by Prof. Thomas Condon, of Eugene City, and by several jaws and other fragments subsequently found by Mr. Wortman.

The frontal region is not contracted as in *Eumys elegans* and *Fiber zibethicus*, but the superciliary ridges are well separated from each other, as in *Hesperomys americanus*. The frontal and posterior nasal regions are slightly concave in transverse section. The molars display tubercles on one side and crescents on the other, the former being external in the superior series. The first superior molar has an additional tubercle at its anterior extremity. The incisors have a transverse anterior face, which is divided by several delicate ridges.

Length of superior molar series, .0065; length of first superior molar, .0028; interorbital width, .0042. Length of inferior molar series (specimen

No. 2), .0064; length of first molar, .002; width of incisor, .001; depth of ramus at second molar, .004.

The *Hesperomys nematodon* is about as large as the red squirrel, *Sciurus hudsonius*, and considerably larger than the *H. loxodon*.

PACICULUS Cope.

Paleontological Bulletin No. 31, p. 2, Dec. 24, 1879. Proceedings American Philosoph. Society, 1879 (1880), p. 371.

Superior molars three, rooted. Enamel forming three entrant loops on the external face of the crown, and one on the internal face.

While the number of the superior molars of *Paciculus* is as in the *Muridæ*, the details of their structure is much as in *Dasyprocta* and *Fiber*. Two species are known.

In the *P. lockingtonianus* the cranial characters are as follows: The infraorbital foramen is very large, with a general triangular outline. The superciliary borders and temporal ridges are well separated, and there is no sagittal crest. There are no postorbital processes. The otic bullæ are large, and furnished with a very large meatus auditorius externus. The malar is a narrow bone extending to the glenoid cavity posteriorly, and resting anteriorly on a prominent peduncle composed of a maxillary bone. It probably reaches the lachrymal.

This genus is probably one of the *Muridæ*, and a near ally of *Sigmodon* and *Neotoma*. It differs from these genera in having three external inflections of the enamel in the superior molars, instead of two. It differs from *Hesperomys* as these two genera do, viz, in having deep enamel inflections instead of tubercles and valleys. It is true that the deepening and narrowing of the valleys of the molars of *Hesperomys* would result after wear in a pattern like that of *Neotoma*. The same process in *Eumys* would produce a pattern much the same as that of *Paciculus*, but that genus is further characterized by the contraction of the postorbital region and the production of a sagittal crest.

Two species of this genus are known to me, *P. insolitus*, a smaller, and *P. lockingtonianus*, a larger one. Both are from the John Day beds of Oregon. They demonstrate an early origin for the American type of *Neotoma*, as contemporaries of the first of the *Hesperomys*.

PACICULUS INSOLITUS Cope.

Locis citatis.

Plate LXVI; figs. 31, 32.

Size small. Molars regularly and rapidly diminishing in size posteriorly. Inner enamel loop turned forwards; the external straight and transverse, excepting in the first molar, where the anterior column of the tooth is extended forwards, and the anterior loop is turned backwards. No fossettes. The first molar is longer than wide, and is contracted forwards; the second and third are subquadrate in form, with the inner angles rounded. The palate is wide, considerably exceeding the long diameter of the first molar.

Measurements.

	M.
Length of superior molar series	.0060
Diameters of first molar { anteroposterior	.0021
Diameters of first molar { transverse	.0018
Length of third molar	.0010
Width of palate at first molar	.0040

The *Paciculus insolitus* is about the size of the Chipmunk, *Tamias striatus.*

PACICULUS LOCKINGTONIANUS Cope.

Eumys lockingtonianus Cope, Bulletin U. S. Geological Survey, Terrs., VI, 1881, p. 176.

Plate LXIV; fig. 10.

This rodent is represented by a nearly perfect skull, which is without lower jaw. Its specific characters separate it widely from the *Eumys elegans* Leidy and *Hesperomys nematodon* Cope. It is considerably larger than either, and the temporal ridges are very obsolete and do not unite posterior to the orbits, as in *E. elegans*, resembling in this respect the *H. nematodon.* The parietal region is wide and flat above. The interorbital region is only moderately contracted. The muzzle is rather short as compared with the total length of the skull. The interorbital region is gently convex above, and the top of the muzzle is flat. The zygoma is quite slender, and the otic bullæ are large and prominent. The notch of the palate extends as far forwards as the posterior part of the last superior molar. The infraorbital foramen is very large and round.

The anterior face of the superior incisor is nearly plane, and it is marked by a weak groove near the inner and a strong groove near the external border In *E. elegans* this face is convex and without grooves. The molars are rather small for the size of the skull; their crowns are worn by use. The third is subround in section, and its diameter is about half that of the first; the latter has the anterior odd lobe quite small.

Measurements.

	M.
Total length of skull	.0380
Length (axial) to front of orbits	.0140
Length (axial) to palatal notch	.0190
Length (axial) to first molar	.0120
Width at otic bullæ	.0205
Width at middle of zygomata	.0220
Width of interorbital space	.0060
Width between first molars	.0055
Length of molar series	.0070
Length of first molar	.0030
Width of superior incisor	.0015

This species is dedicated to my friend, W. N. Lockington, the well-known naturalist of San Francisco.

ENTOPTYCHUS Cope.

Paleontological Bulletin No. 30, p. 2, Dec. 3, 1878; Proceeds. Am. Phil. Soc., 1878-'79, p. 64.

Family *Saccomyidæ.*[1] The cranium is elongate, and presents inflated periotic bones, and slender zygoma. The foramen infraorbitale is small and anterior in position, entering the maxillary bone near its suture with the premaxillary.

Generic characters.—Molars $\frac{4}{4}$-$\frac{4}{4}$, rootless, and identical in structure. The crowns are prismatic, and in the young stage present a deep inflection of enamel from one side, the external in the superior teeth, the internal in the inferior. After a little attrition, the connection with the external enamel layer disappears, and there remains a median transverse fossette, entirely inclosed by enamel. The tooth then consists of two dentinal columns in one cylinder of enamel, separated by a transverse enamel-bordered tube. Incisors not sulcate.

[1] *Geomyidæ* Alston.

The teeth of this genus differ from those of *Perognathus* in being without distinct roots, and in having the enamel loop cut off and inclosed. In *Dipodomys*, the molars are undivided simple prisms.

The skull is compact, and does not display the vacuities or large foramina seen in some genera of *Rodentia*. The incisive foramina are rather small and posterior in position. There is a foramen on the side of the alisphenoid, which is nearly in the position of the anterior alisphenoid canal of the *Thomomys bulbivorus*. The foramen rotundum is immediately below and within the anterior part of the glenoid cavity. The foramen ovale is not distinct from the foramen lacerum anterius, and is on the external side of the apex of the petrous bone. The other *foramina lacera* are closed, so that the carotid foramen pierces the inner side of the otic bullæ. The condyloid foramen is close to the occipital condyle. The meatus auditorius externus is at the extremity of a tubular elongation of the bulla, and is separated by a space from the zygomatic process of the squamosal bone. Between the bases of these is a fossa which is bounded above by a ridge as in the genus *Castor*. Below this ridge is a subsquamosal foramen, and above it a postsquamosal. There are no postparietals nor mastoid foramina.

There are deep pterygoid fossæ, whose inner bounding laminæ unite on the middle of the palatine border, and whose external laminæ are continuous with the posterior extremity of the maxillary bone. The otic bullæ are not separated very distinctly from the mastoid. The latter looks like a continuation of the former, as in *Thomomys*, and occupies considerable space between the exoccipital and the squamosal. The latter sends downwards a process just posterior to the auricular meatus, which forms the handle to a hammer-shaped laminar bone. This is, no doubt, a dismemberment of the squamosal, as a similar process is continuous with that bone in *Thomomys*, and one somewhat different is seen in *Neotoma*, *Hesperomys*, &c. Supraoccipital distinct on superior face of skull Paroccipital process small or none. Mastoid elongate, adherent to otic tube. No postfrontal process.

A well-marked character which distinguishes the skull of this genus from *Thomomys*, *Dipodomys*, &c., is the separation of the meatal tube of the otic bulla from the zygomatic process of the squamosal bone by an interspace. There is no postsquamosal foramen in the recent genera. In *Dipodomys* the otic bulla is more largely developed, but it has the anterior bottle-neck prolongation seen in *Entoptychus.*

In the mandible the coronoid process is developed, but is not large. It is well anterior to the condyle, which it somewhat exceeds in height. The incisive alveolus forms a convexity on the outer side below the coronoid process. The angle is prominent, and is at first incurved below, and then turned outwards at the apex. The degree of obliquity of the grinding surfaces of the molar teeth vary with the species.

Parts of several skeletons are in my collection, but I cannot attach them to any cranium. They present the general characters of the genus *Thomomys* so far as they go. I describe some bones which apparently belong to one individual. The sacral vertebræ carry neural spines. There was evidently a well-developed tail. The scapula has a narrow glenoid cavity ending in a tuberosity adjacent to the coracoid hook. The spine is robust, terminating in a stout acromion. The tuberosities of the humerus are situated below the head, and are so rounded off as to be little prominent. One side of the greater is continued into a very prominent deltoid crest, which terminates abruptly below. The ilium has a narrow trilateral neck, and a projecting anterior inferior spine. The pubis is directed posteriorly at the base. The femur is not elongate. Its trochanters are well-marked, including a third. This is wanting in *Thomomys bulbivorus.* The neck rises obliquely to the rather large head. The condyles are short and spreading, and the rotular groove is short and rather wide, and with well-marked ridges. The tibia is much curved backwards at the proximal part. The crest is acute and is directed outwards, but does not project much at the head.

Individuals of this genus were very abundant in Oregon during the middle Miocene epoch. They represent several species, but how many it is difficult to determine. The most noteworthy variations are found in the development of superciliary ridges; then there are modifications in the forms of the premolar teeth, differences in the length and width of the muzzle, and some range in dimensions.

The following table represents the characters of the species so far as I can determine them at present:

A. Thickened superciliary ridges wanting; front wide.
Superciliary borders obtuse, not continued into temporal ridges; front flat, or little concave; premolars narrow.
Length of skull .046 .... *E. planifrons.*
Length of skull .038.. *E. minor.*
Superciliary borders sharp, vertical, continued into two straight temporal angles, which form a V.
Premolars narrow; size of *E. planifrons* .... *E. lambdoideus.*
AA. Thickened ridge on the superior side of each supercilium; front narrower.
Superciliary ridges soon discontinued; size of *E. planifrons*....... *E. cavifrons.*
AAA. Superciliary ridges much thickened, soon uniting, and closing the frontal groove behind. Front narrowest.
Premolar widened at the base; size of *E. planifrons*............ *E. crassiramis.*

Some differences in the form of the mastoid bone may be observed in species of this genus. Thus it is flat behind, and bears a well-marked "lateral occipital" angle in *E. planifrons* and *E. lambdoideus*, while in the remaining species it is convex, and with the angle little apparent. In some specimens the loss of the hammer-shaped squamosal dismemberment, which I will call the posttympanic bone, gives a deceptive extension forwards to the mastoid.

Parts of more than a hundred individuals of *Entoptychus* are in my collection.

ENTOPTYCHUS PLANIFRONS Cope.

Paleontological Bulletin No. 30, p. 3, December 3, 1878. Proceedings Amer. Philos. Soc., 1878 (1879), p. 65. American Naturalist, 1883, p. 170, fig. 18.

Plate LXIV; fig. 1.

A nearly perfect skull, with a portion of a second, are the only specimens that I can certainly refer to this species. They represent the largest size found in the genus.

The muzzle is long, and gently decurved anteroposteriorly, and slightly convex transversely The length from the side of the orbit at its anterior border to the extremity of the nasal bones is exactly equal to the length from the same point to the inion at one side of the middle line. The front in the more perfect skull is slightly concave, but is without lateral ridges; in the less perfect specimen it is more nearly flat. There is no trace of temporal ridges, and a delicate raised line represents the sagittal crest. The muzzle is wider above than below, and the superior and lateral faces are separated by a distinct angle, which is acute on the posterior half. In front of the superior part of the orbit is a fossa, which is directed obliquely upwards and forwards. The fundus of the orbit is not as large as in some species, giving a greater depth to the maxillary bone, which is an indication of length of the prisms of the superior molars. The palatal surface has two grooves, each of which is continued into a pterygoid fossa. The otic bullæ are rather small for the genus, and are flattened behind. Anteriorly they are continued into a neck on each side, which meets its mate of the opposite side on the middle line, resting on the basisphenoid bone. Each neck is pinched off backwards, so that the two inclose an angle between them. The tympanic or meatal tube in this species is very long, exceeding that of any other species, and equaling the long diameter of the bulla without the anterior neck. The mastoid adheres closely to the posterior side of the tube, extending to its extremity, and separated from it on the inferior side by a groove. The posterior side of the mastoid is triangular in form, and is nearly flat. Inferiorly there is no groove marking its point of separation from the bulla; superiorly it carries a strong angle from the inion to its external extremity. The superior face of the brain-case is nearly flat. The occiput is vertical, slightly concave between the points of junction with the mastoid, and with a slight median keel above. The foramen magnum is large, subquadrate, and a little wider than high. The basioccipital supports a median keel, and is concave on each side in front of the condyles.

The supraoccipital bone has a rather wide triangular exposure. The parietals diverge from the middle line anteriorly, and the suture reaches the squamosal near the posterior border of the orbit. The posterior extremity of the nasal bones reaches that of the premaxillaries, and both terminate in line with the deepest part of the preorbital fossa.

The superior molar teeth display the typical characters of the genus. Each has an external groove for part of its length, and the anterior column of the last premolar has no anterior production, but has the form of the other teeth. The sizes are, commencing with the largest, 2–1–3–4.

The inferior edge of the angular portion of the mandible is strongly inflected, and the interior face below the molars is concave. The inferior boundary of the masseteric fossa is well marked, and terminates below the anterior border of the first molar. The grinding face of the molars is very oblique anteroposteriorly, and that of the first molar is rather smaller than that of the second and third. It has no anterior expansion at the alveolus. The tuberosity of the incisive alveolus is not very prominent.

Measurements.

No. 1.

	M.
Length of skull from end of nasal bones to occipital condyles, inclusive	.046
Length from front of premaxillary bone to base of Pm. iv	.021
Length of molar series on base	.009
Length from occipital condyles to junction of otic bullæ, inclusive	.011
Length of nasal bones	.020
Length of supraoccipital bone	.008
Width of otic bulla at middle	.006
Length of tympanic tube	.005
Width of muzzle at middle	.009
Width of interorbital space	.007
Width of skull at anterior extremity of glenoid cavity	.0165
Width of skull at meatus	.0205
Width of occipital bone posteriorly	.013
Width of foramen magnum	.006
Length of inferior molars on alveoli	.0080
Elevation of first molar above alveolus	.0036
Width of crown of second molar	.0025
Depth of ramus at third molar	.0070
Width of ramus below at third molar	.0050

No. 2.

Width of interorbital space	.007
Width of muzzle at middle	.0086
Elevation of skull from second molar	.0130
Length of inferior molar series	.0072
Depth of ramus at M. 2	.0072
Width of inferior face at M. 2	.0043
Width of inferior incisor	.0018
Distance between infraorbital foramen and M. 1	.0050

Besides the characters already mentioned, the somewhat greater interorbital width distinguishes this rodent from the *E. cavifrons*.

Entoptychus lambdoideus Cope.

Bulletin U. S. Geol. Survey Terrs., IV, 380.

Plate LXIV; fig. 2.

This species is represented by two parts of crania, which include the interorbital and adjacent regions, with the molar teeth. A nearly entire skull presents some of the characters of this species and some of those of

E. cavifrons, so that its validity may be thought to be as yet not entirely beyond question.

In this species, the interorbital region is concave, but there is no ridge-like thickening of the supraorbital border. It presents, on the contrary, a subacute superior edge, flush with the inferior part of the same border. These edges leave the orbital border posteriorly, and converge in straight lines to an acute angle, forming two temporal ridges. The nasal bones do not extend as far posteriorly as the premaxillaries, which reach to the inner line of the anterior border of the orbit. The anterior molar tooth is like the others, and has no anterior basal extension.

The size is about equal to that of *E. planifrons*.

Measurements.

	M.
Interorbital width	.0064
Anteroposterior length of orbit	.0100
Depth of skull to alveolar edge	.0140
Length of series of superior molars	.0070
Width between bases of Pm. iv	.0030

The skull above referred to presents the temporal ridges of the typical specimens. It has also lateral occipital angle of the mastoid, as in *E. planifrons*, and a meatal tube nearly as long as in that species, differing in both points from the *E. cavifrons*. But it has a ridge-like thickening of the supraorbital border, as in that species, and the interorbital space has the same relative width.

Its size is that of *E. cavifrons*.

ENTOPTYCHUS MINOR Cope.

Bulletin U. S. Geol. Surv. Terrs., IV, 379.

Plate LXIV; fig. 3.

This was an abundant species. Eight more or less complete crania are in my collection, and many fragments present the characteristic measurements. It is the smallest species, and is characterized also by the perfectly flat interorbital region and the absence of temporal ridges.

The muzzle of the skull is shorter than in *E. planifrons*, as may be seen by the measurements, and it is also narrower above, as compared with the interorbital width. The preorbital fossa is not well defined in front. The otic bulla is angulate along the inferior middle line, and has a narrower form than that of any of the other species of the genus. The bottle-neck processes send out prolongations, which meet on the middle line below. The mastoid is convex posteriorly, but carries a lateral occipital angle above.

The molars of the superior series are fissured on the external side for a distance, and the anterior is like the others in both jaws. The superior incisors are slightly convex anteriorly, and are perfectly smooth, except a delicate groove close to the inner border, a character found in most of the other species.

The masseteric ridge of the mandible extends as far forward as the line of the front of the first inferior molar, and is well defined posteriorly to near the angle.

Measurements.

	M.
Length of skull from front of premaxillary bone to posterior face of mastoid	.038
Length from front of premaxillary to base of Pm. iv	.018
Length of molar series on base	.0066
Width of otic bulla at middle	.0040
Length of bulla and mastoid	.0110
Width of muzzle at middle	.0080
Width of interorbital space	.0055
Depth of mandible at M. i	.0060

John Day River, Oregon.

ENTOPTYCHUS CAVIFRONS Cope.

Paleontological Bulletin No. 30, p. 2. Proceedings American Philosophical Society, 1878 (1879), p. 64.

Plate LXIV; fig. 4.

I refer portions of six crania to individuals of this species, and seven others probably belong to it. It differs from the *E. minor* in its superior size, and in the presence of superciliary ridges. These ridges are rarely as thick and prominent as in the *Pleurolicus sulcifrons*, and do not approach the development seen in the *Entoptychus crassiramis.*

The postorbital part of the skull is subquadrate in outline and depressed in form. The interorbital region is narrowed, but the superciliary margins do not meet nor converge to form a sagittal crest. They are thickened, forming two subparallel ridges which are separated by a shallow concavity of the frontal bone. The nasal bones are very narrow, and their posterior apices just attain the line of the supero-anterior angle of the orbit. The base of the malar bone is much elevated and very oblique. The otic bulla is flattened on the inner face, as in *E. planifrons*, and is not so compressed as in the described specimen of *E minor.* The mastoid bone is convex posteriorly, and supports an obtuse keel continuously with the inion. The meatal tubes are broken off, but they were evidently not so long as in *E. planifrons.* The postsquamosal foramen is large, and is near the posterior lateral ridge of the squamosal bone. The frontal width of this species is less than that of the *E. planifrons* and *E. lambdoideus*, being just half of that of the muzzle at its middle. In most of the crania the nasal bones do not extend so far posteriorly as the premaxillaries. In one of them the sagittal crest is quite prominent; in others more or less distinct traces of it are visible.

The molar teeth are directed obliquely backwards, the alveolus of the first issuing below the anterior part of the orbit. The first superior molar is the largest, but does not differ in form from the others; and the proportions of the others diminish regularly posteriorly. The first inferior molar is a little smaller than the second and third, and is about equal to the fourth; its anterior column is contracted; while the last molar is like the second and third. The face of the inferior incisor is flat, and its enamel is smooth. The external face of the jaw is bounded below by a strong angle as far anteriorly as below the first molar.

Measurements.

	M.
Length of skull to incisive alveoli	.041
Width of skull at mastoids	.020
Width of skull between orbits	.005
Width of skull at middle of muzzle	.010
Elevation of skull from second molar	.011
Length of molar series	.007
Length of first molar	.002
Width of first molar	.002
Length of crown of last molar	.0015
Width of crown of last molar	.0015
Length from M. 1 to infraorbital foramen	.007
Depth of mandibular ramus at M. 2	.006
Width of face of inferior incisor	.0016

John Day River region of Oregon.

ENTOPTYCHUS CRASSIRAMIS Cope.

Paleontological Bulletin No. 30, p. 3. Proceeds. Am. Philo. Soc., 1878 (1879), p. 65. American Nauralist, 1883, p. 169, fig. 17.

Plate LXIV; fig. 5.

This species was originally established on mandibular rami. I find, in my later collections, rami of this character attached to crania, which enable me to fix the definition of the species with greater precision than heretofore.

The skull is about the size of that of *E. planifrons*, and has a relatively shorter muzzle. The internal orbital walls are rolled inwards at the supraorbital region so as to meet at a point opposite the posterior border of the orbital space. Opposite the anterior part of the orbit, the ridges are more widely separated, so that the interspace is a narrow wedge-shaped fossa, opening forwards. There are no temporal ridges, and only a weak sagittal crest. There is no preorbital fossa, but the preorbital ala of the maxillary bone is very prominent. The orbital fossa is small and profound, leaving a full depth to the maxillary alveoli. The otic bulla is ovate, and not keeled or flattened. Its size is medium, and its anterior bottleneck is continuous with the external pterygoid lamina. The meatal tube is elongate, but not so much so as in *E. planifrons*. The mastoid bone is not distinguished from the bulla by a groove, and it is convex both vertically and transversely. It is injured in the specimen superiorly. The superior molars display an external fissure for part of their length. The premolar has the base extended anteroposteriorly more than in any species of the genus. The premaxillo-maxillary suture is exactly half way between the bases of the incisor and first molar. It is much nearer the first molar in *E. planifrons*.

Measurements.

	M.
Length of cranium from premaxillary to mastoid, inclusive of both	.048
Length from front of premaxillary to Pm. iv	.022
Length of molar series on base	.008
Length of otic and mastoid bullæ	.015
Width of otic bulla at middle	.007
Interorbital width	.006
Width at anterior border of glenoid cavity	.017
Depth of skull at M. i	.0155

In the mandibular rami the inferior masseteric ridge extends to below the anterior border of the first molar, and is very prominent and acute. It results that both the exterior and inferior aspects of the ramus are concave to the anterior extremity of the crest, which slopes upwards. The incisive alveolus, though not prominent, as in *Hystricomorpha*, is on the inner side of the base of the ramus in front. Above the alveolar prominence the inner face of the ramus is gently concave. The anterior origin of the coronoid process is opposite the posterior border of the second molar.

In the mandible the condyle projects as far backwards as the angle, and is hence quite a distance posterior to the coronoid process. The latter is small, and is a little higher than the condyle. The angle has an acuminate apex, which is turned out. The posterior extremity of the incisive alveolus forms a prominent tuberosity, bounding a fossa of the ascending ramus on its external side. The inferior outline of the mandible follows the curve of the incisor tooth anteriorly to the middle of its length, when it reaches the line of the inferior border of the masseteric fossa. Here it commences another convexity, which is most prominent directly below the incisive tuberosity, and ascends again to the angle.

The triturating surface of the molars is not oblique, as in *E. planifrons*, but is nearly horizontal. The enamel of the incisor is smooth, and has a well-marked bounding angle at the inner edge, and another within the external edge.

Measurements.

	M.
Length of inferior molar series	.0082
Width of anterior face of inferior incisor	.0028
Depth of ramus at M. ii	.0090
Width of ramus below at M. ii	.0052

Though this species is larger than any other species excepting the *E. planifrons*, its interorbital width is the least. Its peculiar frontal groove is only approached by some specimens of the *Pleurolicus sulcifrons*. I have only two crania which undoubtedly belong to the *E. crassiramis*, while a third very probably may be referred to it. Both the former have mandibles attached.

John Day River, Oregon.

PLEUROLICUS Cope.

Paleontological Bulletin, No. 30, p. 3, Dec., 1878. Proceed. Amer. Philo. Soc., 1878 (1879), p. 66.

Family *Saccomyidæ.* Superior molars rooted and short-crowned. The crowns with a lateral fissure bordered with an inflection of the enamel sheath, extending to their bases. In the superior molars this inflection is on the external side, and does not divide the crown. Superior incisors not grooved.

This genus is curiously near to the existing *Heteromys* and *Perognathus*, the two genera of *Saccomyidæ* with rooted molars. The former differs in having the molars divided into two columns, each of which is sheathed in enamel, while *Perognathus* only differs, so far as I am aware, in having the superior incisors grooved. It is also very nearly related to *Entoptychus*, and two of the species correspond in various respects with two of those of that genus. In view of the fact that most of the specimens of the *P. sulcifrons* are old individuals with well-worn molars, the idea occurred to me that the rooted character of the molars might be common to the species of *Entoptychus*, but that it might not appear until long use had worn away most of the crown, and the protrusion had ceased. Examination of the bases of the long molars of *E. planifrons* did not reveal any roots. It is also opposed to this view that the maxillary bone of the *Pleurolici* has little depth below the orbital fossa, appropriately to the short-rooted molars, while the depth is considerable in the typical *Entoptychi*, though there is a complete gradation in this respect. But I have demonstrated satisfactorily that *Pleurolicus* is a distinct genus by observations on the *P. leptophrys*. Some of my individuals of this species are young, with the crowns of the molars little worn; yet the roots diverge immediately on entering the alveolus on all the molars. In the species of *Pleurolicus* the lateral fissure of the crown descends to its base, and hence persists longer than in the typical *Entoptychi.*

I am acquainted with two species of this genus. The posterior part of the skull of an individual represents a third species, which I refer provisionally to this genus.

The characters of the species are as follows:

I. Otic and mastoid bullæ continuous.

Temporal ridges uniting into a sagittal crest; length of skull .043; supraorbital ridges and concave front .... *P. sulcifrons.*

Temporal ridges not uniting; length of skull .035; interorbital region flat; no ridges .... *P. leptophrys.*

II. Otic and mastoid bullæ separated by a deep groove.

Temporal ridges not united; front concave; size medium; supraoccipital wide .... *P. diplophysus.*

Pleurolicus sulcifrons Cope.

Paleontological Bulletin, No. 30, p. 4. Proceedings American Philosophical Society, 1878 (1879), p. 66.

Plate LXIV; fig. 6.

Five crania represent this species in my collection; only one of these includes the posterior portions. In two of them the molar teeth are well worn; in one (the type) they present a medium degree of wear, and in two the roots only are preserved. This species resembles those of the allied genus *Entoptychus* in many respects. The superciliary borders are thickened upwards, forming two ridges, which inclose a groove between them which is more pronounced than in most specimens of the *Entoptychus cavifrons.* The muzzle is plane above and considerably wider than the interorbital space. The base of the malar is thin and oblique, and the *foramen infraorbitale exterius* is well in advance of the molar teeth and at the anterior part of the maxillary bone. A groove passes backwards from its inferior border, terminating in a small foramen which marks a point nearly half way to the first molar. This foramen is present in all the crania. Within this another shallow groove bounds the more prominent median line. The palatal surface exhibits two shallow lateral grooves, which commence opposite the posterior border of the first molar.

The otic bullæ are oval and not keeled, and have the usual bottle-neck. The mastoids are convex behind, and carry above an obtuse angle from the inion.

The grinding surfaces of the molars are transverse ovals, only interrupted by the exterior fissure. The first molar is slightly different in form, being larger, and its section, when not much worn, being nearly round. Its anterior portion extends towards the alveolus, giving an anteroposterior oval on prolonged wear. Each tooth has three roots, one interior and two exterior; in the first they may be described as two posterior and one anterior. The last molar is the smallest, the series exhibiting a regular gradation in size.

Measurements.

No. 1.		M.
Interorbital width		.0050
Width of muzzle at middle		.0080
Depth of cranium at M. ii		.0138
Length of molar series along base		.0080
Diameter of second molar	anteroposterior	.0016
	transverse	.0020
Width of face of superior incisor		.0020
No. 2.		
Length of cranium from mastoid bulla to premaxillary, inclusive		.043
Width of skull at mastoid bullæ		.021
Length of superior molar series on base		.008
Width between bases of Pm. iv		.0025

John Day River region of Oregon.

PLEUROLICUS LEPTOPHRYS Cope.

Bulletin U. S. Geol. Survey Terrs., IV, p. 381. American Naturalist, 1883, p. 167, fig. 16 a, b.

Plate LXIV; figs. 7–8.

This species is in its typical form smaller than the last, and resembles in its size and its plane interorbital region the *Entoptychus minor*. Four more or less complete crania represent it, two of which are of young and two of old animals, as indicated by the degree of attrition of the molar teeth. In only two of them is the parietal region so preserved as to show the separate temporal ridges

The interorbital region is flat and without superciliary ridges. The diastema is just twice as long as the series of molar teeth. The auditory bulla is oval and not compressed or keeled, and the mastoid bulla is very convex posteriorly, and carries a very obtuse angle from the inion above. In young individuals the enamel inflection of the Pm. iv extends entirely across the crown. When the internal groove has disappeared on wear, the grinding surface is subround. With age the protuberance of the anterior root is reached, and the form of a horizontal section of the base is pyriform. The width of the front is two-thirds that of the muzzle.

Measurements.

	M.
Length of skull, including mastoid bulla	.0340
Length from front of premaxillary to Pm. iv	.0145
Length of otic and mastoid bullæ	.0100
Interorbital width	.0050
Width of muzzle at middle	.0080
Width between the Pm. iv s	.0028

The exposure of the supraoccipital is wide and subquadrate, resembling only that of the *Pleurolicus diplophysus* among the rodents of this group. The temporal ridges converge gradually in a straight line posteriorly to the supraoccipital bone and then diverge without coming in contact.

The John Day River of Oregon.

PLEUROLICUS DIPLOPHYSUS Cope.

Bulletin U. S. Geol. Survey Terrs., IV, p. 381. American Naturalist, 1883, p. 167, fig. 16 c, d.

Plate LXIV; fig. 9.

A fragment of a skull which includes all posterior to the interorbital region is all that represents this species. The maxillary bones and teeth are lost. The interorbital region is concave, there being on each side a low angular ridge. These ridges continue into the temporal ridges, which have the same character as in the *P. leptophrys*. They are straight and converge to the anterior part of the supraoccipital bone, where they do not come in contact, but diverge to end at the inion. The supraoccipital a triangle with obtuse apex and as wide as long. The tympanic tube is quite short, the shortest found in this group. The mastoid bulla is large, a little exceeding the otic, and has a general convex external face, which is not divided into two planes, an external and a posterior, by a straight angle, as in most of the allied species. There is a small portion homologous with the external face which turns inwards and passes under the squamosal, leaving a considerable fissure-like foramen, which is wanting in most of the other species. The postsquamosal foramen is large, and the hammer-shaped bone very distinct, sending its posttympanic process to the meatus. The otic bulla is cut off from the mastoid by a deep oblique fissure. One end of the fissure is marked by the posttympanic process, and the other by the paroccipital. The bulla is compressed and flattened on both inner and external sides. A strong rib connects the occipital condyle and the paroccipital process, *above* which is the condyloid foramen.

Measurements.

	M.
Length from middle of supraorbital border to inion	.018
Width of interorbital space	.0055
Width between mastoid bullæ	.0190
Width of occipital bone	.0100
Width of foramen magnum	.0050
Length of otic bulla alone	.0100

This cranium is well preserved, and has been perfectly cleaned by weathering. It shows a good many points of difference, as compared with any of the other species of this group. It is to be regretted that the teeth are wanting.

LAGOMORPHA.

PALÆOLAGUS Leidy.

Proceedings Academy Philada., 1856, p. 89; Extinct Mamm. Dakota and Nebraska, p. 331.—Cope, Ann. Report U. S. Geol. Surv. Terrs., 1873 (1874), p. 477.

Family *Leporidæ*. Dentition: I. $\frac{2}{1}$; C. $\frac{0}{0}$; M. $\frac{6}{5}$; or, Pm. $\frac{3}{2}$, M. $\frac{3}{3}$. Superior incisors sulcate, inferior incisors not sulcate. First and last superior molars simple, intermediate ones with an enamel inflection of the inner side, which soon wears out First inferior molar of one more or less transversely divided column; other inferior molars consisting of two columns in antero-posterior relation. No postfrontal process.

The above characters approximate nearly those of the existing genus *Lepus*. The only distinction between them signalized by Dr. Leidy is the more simple first inferior molar of the extinct genus, which consists of one column more or less divided. In *Lepus* this tooth consists of two columns, the anterior of which is grooved again on the external side in the known species. I am able to reinforce this distinction by a strong character, viz, the absence of the postfrontal process in *Palæolagus*. As compared with the extinct genus *Titanomys* of Meyer,[1] the difference is well marked, as that genus has the molar teeth $\frac{4}{4}$ instead of $\frac{6}{5}$. The last inferior molar is cylin-

[1] *Amphilagus*. Catal. Méth. et Descr. Vertèbres Fossiles de la Basin de la Loire, 1853, p. 42.

dric, consisting of but one column. The first inferior molar consists of two cylinders broadly united, as in the corresponding tooth of *Palæolagus*. As compared with *Panolax* Cope,[1] which is only known from superior molar teeth, this genus may be at once recognized by the simplicity of the last tooth. In *Panolax* it consists of two columns.

Dr. Leidy's descriptions and figures, which are available for the definition of this genus, relate exclusively to the dentition. Characters drawn from the skeleton generally have been derived from my material and are now given.

The nasal bones are wide, and the suture which separates them both from the frontal is concave forwards. The median frontal suture is persistent. The ascending portion of the premaxillary, which attains the frontal, is very narrow. The superior half of the facial plate of the maxillary bone is sharply rugose with reticulate ridges, but whether perforate or not I cannot certainly determine. The *foramen infraorbitale* is small and round, and issues below the reticulate portion of the maxillary. The otic bulla is compressed globular, with very thin walls. The meatus is large and has prominent lips, which open upwards. The mastoid is coössified with the bulla, and extends with a dense surface from behind to above and in front of the meatus. The incisive foramina are very large, enter the maxillary bones deeply, and are confluent posteriorly. The palate may be said to extend to the last molar, but there is a deep though narrow median posterior emargination.

The distal extremity of the *humerus* is not so extended transversely as in *Ischyromys*, and exhibits a moderate epicondyle. The inner flange of the condyles is well developed, and on the posterior face it is supplemented by a flange of the external edge of the condyles, which is as prominent or even more so, forming an intertrochlear crest. The arterial canal is inclosed by a slender bridge, and opens on the inner side above, and anteriorly below. In an ulna supposed to belong to this genus the coronoid process is elevated. The radial facet forms a narrow transverse plane, nearly divided by a wide anterior emargination. The shaft is compressed vertico-obliquely. A radius

[1] Report Lieut. G. M. Wheeler, 4to, IV, p. 296.

exhibits a transversely oval humeral face of the head somewhat angulate at a superior and an infero-lateral extremity, which are diagonally opposed to each other. Beyond the middle the shaft becomes wider, and is flattened obliquely.

The peduncle of the *ilium* has a triangular section, the anterior face being the narrowest, and inclined at a little more than a right angle to the interior face. It expands but little at the sacral extremity, and the crest is very short. The external angle of the peduncle is very prominent and runs into the anterior extremity of the crest, from which proceeds also the more obtuse angle which is continuous with the pectineal line. A third longitudinal angle is seen on the middle of the external side of the sacral extremity, which is not continued on the peduncle. There is a prominent tuberosity on the median or first-described angle on the peduncle, which may or may not be homologous with the anterior inferior spine. There is no tuberosity on the inner bounding angle of the inner face as is seen in *Gymnoptychus.* The pubis leaves the ilium at right angles. Acetabulum nearly round.

The *femur* has well developed great and little trochanters, and a third trochanter, which rises from the shaft in line with the inferior border of the little trochanter. The fossa of the great trochanter is well marked. The head is not separated from the great trochanter by a deep emargination, and projects well within the internal face of the shaft. Its articular surface is prolonged towards the great trochanter. *Fossa ligamenti teris* isolated. The distal extremity of the femur exhibits the superiorly prolonged patellar groove characteristic of this group of rodents. The condyles are more than elsewhere produced downwards and posteriorly, and are well separated.

The spine of the *tibia* is rudimental, and the crest is very obtuse. The inferior continuation of the latter forms a prominent reverted keel on the proximal front of the shaft, which is deeply concave on its inner side. The posterior face is also concave, and is separated by a laminar external bone from the external side. The external border of the head is not deeply notched as in *Panolax.* The fibula unites with the tibia on the proximal part of the latter. The remainder of the shaft is smooth. The external malleolus is large and at right angles to the long diameter of the distal end of the bone, and its extremity is a facet for contact with the calcaneum. On its external face is a prominent process directed backwards. The external trochlear groove is deeper than the internal, and is well separated from it. The internal malleolus can scarcely be said to exist. It may be represented by a small process on the inner side of the extremity of the shaft.

The *astragalus* is elongate and flat, and the trochlear portion is oblique. The neck is elongate, and convex on the inner side; the constriction is on the inner side immediately behind the head. The long diameter of the latter makes an angle of 45° with the horizontal plane. The external trochlear are is much larger than the internal. The cotylus, which fits the external condyle of the calcaneum, possesses a peculiar impressed area on its posterior surface. The calcaneum extends nearly as far anterior to its condyle as posterior. The free portion is subcylindric or subquadrate to the end. The internal process for the astragalus is quite prominent. The cuboid facet is directed obliquely inwards, running into a short longitudinal groove. The cuboid extremity is little depressed.

The skeletal characters above enumerated were taken from the bones of *P. turgidus* and *P. haydeni*, excepting in the cases of the ulna, radius, ilium, and calcaneum, which were derived from those of *P. haydeni* only.

A cast of the cranial chamber of a specimen of *Palæolagus haydeni* displays the superficial characters of the *brain*. As in the order generally, the hemispheres are small and are contracted anteriorly. The greater part of the cast of the cerebellum is lost, but enough remains to show that it was large. The olfactory lobes are large; they are not gradually contracted to the hemispheres, but expand abruptly in front of them, being separated by a constriction only. They are wider than long, and than the anterior extremity of the hemispheres. Their cribriform surface is wide, and extends backwards on the outer sides. Traces of the three longitudinal convolutions can be observed on the hemispheres above the *lobus hippocampi*. The internal and median are continuous at both extremities, and with the external to the base of the olfactory lobes. There is no definite indication of the Sylvian fissure. The *lobus hippocampi* protrudes laterally a little beyond the border of the external convolution. Its form is depressed.

As compared with the brain of the rabbit (*Lepus cuniculus*) figured by Leuret and Gratiolet,[1] that of the *Palæolagus haydeni* is distinguished by the absolutely much smaller size of the hemispheres, and by the absolutely larger olfactory lobes, the excess being in transverse dimensions and not in the longitudinal. An important difference is also the absence of the median posterior production of the hemispheres seen in the rabbit, the prolongation in the extinct species being lateral, and extending little behind the *lobus hippocampi*. The indications of the convolutions of the superior surface are similar in the two.

As observed by Leidy, this genus presents the same number of teeth as in the existing rabbits, viz, I. $\frac{2}{1}$; C. $\frac{0}{0}$; M. $\frac{6}{5}$; and the difference consists in the fact that the first molar possesses two columns, while in *Lepus* there are three. Having collected a great number of remains of this genus, I am able to show that it is only in the immature state of the first molar that it exhibits a double column, and that in the fully adult animal it consists of a single column with a groove on its external face. The dentition undergoes other still more important changes with progressing age, so as to present the appearance of difference of species at different periods. These will be explained under the head of the *P. haydeni*, the most abundantly represented in the collections. It may be mentioned here that in neither *P. haydeni* nor *P. turgidus* is there any evidence that more than two anterior molars are preceded by deciduous teeth. The latter are present in many specimens.

Three species of this genus are known to have lived in Colorado during the White River epoch of the Miocene. Bones of two of the species have been found also in Dakota. The *P. haydeni* was probably the most abundant mammal of the fauna of that period.

Depth of ramus at penultimate molar, 9^{mm}; length of molar series, 10^{mm}; no third lobe to molars .... *P. haydeni.*

Depth of ramus at penultimate molars, 11^{mm}; length of tooth series, 14^{mm}; a third posterior lobe of the molars .... *P. triplex.*

Depth of ramus at penultimate molar, 12–14^{mm}; length of tooth series, 13–16^{mm}; no third lobe .... *P. turgidus.*

---

[1] Anatomie Comparée du Systéme Nerveux, Pl. III, Figs. 1, 2.

PALÆOLAGUS HAYDENI Leidy.

Proceedings Academy Philada., 1856, p. 89. Extinct Mamm. Dakota and Nebraska, p. 331. Cope, Ann. Report U. S. Geol. Surv. Terrs., 1873 (1874), p. 478. American Naturalist, 1883, p. 172. *Palæolagus agapetillus* Cope. Paleontological Bulletin No. 15, p. 1 (Aug., 1873). Annual Report U. S. Geol. Surv. Terrs., 1873 (1874), p. 478.

Plate LXVI; figs. 1-27.

The dentition of this rabbit has been fully described by Professor Leidy, but the material at the disposal of this distinguished naturalist did not enable him to furnish either the cranial or skeletal characters. My numerous specimens enable me to supply this deficiency. It is, however, to be observed that I cannot associate the skull with a skeleton as belonging to the same individual in any instance. The characteristic marks of the *Leporidæ* in all the bones of two species found mingled in profusion with the jaws and teeth of the *P. haydeni* and *P. turgidus*, in corresponding numbers, render their correlation sufficiently certain.

The form of the muzzle seen from above is that of an obtusely truncated wedge. The contraction in the width of the nasal bones forward is very slight; at the extremity they are strongly convex, while the posterior region is flat. The canthus rostralis is sharp, consisting of the narrow rib-like maxillary prolongation of the premaxillary bone, which at its anterior third is a little more elevated than the adjacent portion of the nasal bones. Its posterior apex does not extend quite so far posteriorly as that of the nasal bone, and the frontal sends a wedge-shaped prolongation to the outer side of it, which extends as a narrow splint anteriorly between the premaxillary and maxillary. The rugose patch of the maxillary extends more than half way towards the alveolar border of the superior incisors, and is bounded in front by the maxillo-premaxillary suture. It is separated from the orbit behind by a band of dense bone. The foramen *infraorbitale exterius* issues entirely below the rugose patch. The maxillo-premaxillary suture crosses the palate a little behind the middle point between the incisor and first molar, and on the side above the diastema bounds a long process forward. It then returns to the inferior border of the rugose patch, inclosing a notch with the inferior border of the same. The palatine bones are flat and occupy more than half the palate between the molars. Their common suture is at least as long as that of the maxillaries, and extends as far forward as the posterior border of the second molar. From this point the anterior suture extends to the posterior border of the third molar. The palatal notch is rectangular and is not wider than the palatine bone on each side of it.

The last superior molar in this species is not grooved, has a round section, and is smaller than the first molar. The first is also small, and has a deep groove on its anterior face. Of the intermediate four molars the two median are the largest. The crowns of the molars, except the first, are without enamel on the external side, and in the last there is no enamel on the posterior side also. The inner side of the four intermediate molars is deeply grooved for a short distance, which gives a fissure-like notch on attrition. This disappears after use, as does also a less profound crescentic fossa in the middle of the crown, whose concavity is directed outwards. A line connecting the external borders of the molars is strongly convex; that connecting the interior borders but slightly convex. The superior incisors describe a short arc, their triturating surfaces being directed exactly downwards. The groove of their anterior face divides the latter equally. The inner division is more prominent than the outer.

The mandible is contracted forwards, and the inferior outline rises gradually to the short symphysis. The symphyseal articular surface is oval. The ascending ramus rises abruptly opposite the middle of the last molar. Its anterior face at the base is nearly transverse, owing to the sharp prominence of the corresponding inner border of the ascending ramus. In not one of my specimens is the angular portion of the jaw preserved. The condyle is subglobular in front, with a narrow posterior prolongation, as in the *Lepus sylvaticus*. The masseteric muscular insertion is flat and has a regularly convex anterior border, which does not extend beyond the line of the posterior border of the penultimate molar. There are two mental foramina, one on the middle of the depth of the ramus, below the fissure between the second and third molars; and the other, marking the posterior third of the diastema, near its superior border. The pterygoid fossa of the ramus is deep, and its anterior border is well defined; its anterior border is pierced opposite the middle of the vertical diameter of the ramus for the dental artery and nerve. The alveolus of the inferior incisor is marked by a longitudinal convexity of the inner aspect of the ramus, which extends along its inferior border as far as the fourth molar. A shallow groove separates its posterior half from the inferior border of the ramus.

The first and last inferior molars are smaller than the others. In the intermediate ones the posterior column is smaller than the anterior one, and is in close contact with it. Both are transverse oval in section, and the anterior is a little more prolonged inwards. The inferior incisor is rounded posteriorly, and narrowed; in front the surface is very gently convex, and slightly oblique outwards. The enamel is smooth, folds over, forming a band on the outer side, and does not fold over on the inner side.

The earliest dentition of this species known to me is the presence of the two deciduous molars, the first and second in position, before the appearance of any of the permanent series. Each of these has two roots, and the crown is composed of three lobes. In the first, the first lobe is a simple cusp; the two following are divided into two cusps each; the second is similar, excepting that the simple cusp is at the posterior end of the tooth. The grooves separating the lobes descend into the alveolus on the outer side, but stop above it on the inner. The measurements at this stage are—

Measurements.

	M.
Length of two milk-molars	0.0050
Depth of ramus at No. 2	.0042
Depth of ramus at diastema	.0032

In the next stage the third permanent molar is projected, and has, like the second deciduous, a posterior simple column, whose section forms an odd cusp or lobe. The fourth true molar then follows, also with an odd fifth lobe behind. This lobed form of the molars is so different from that of the adult as to have led me to describe it as indicating peculiar species under the name of *Tricium avunculus* and *T. annæ*.

In the next stage, the fifth small molar appears in view, and the second permanent molar lifts its milk-predecessor out of the way. In a very short time, the posterior, or odd, columns entirely disappear, sinking into the shaft, and the permanent molars assume the form characteristic of the species. The last stage prior to maturity sees the first milk-molar shed, and the younger portion of the first permanent molar protruded. A specimen of this age furnished the basis of the *Palæolagus agapetillus*. Its measurements are:

Measurements.

	M.
Length of molar series	.0100
Length of penultimate molar	.0020
Depth at penultimate molar	.0070
Depth at first molar	.0050
Transverse thickness at first molar	.0037

There is the merest trace of a posterior lobe at this time, and that speedily disappears. The anterior lobe is subconical, and is entirely surrounded with enamel. By attrition, the two lobes are speedily joined by an isthmus, and for a time the tooth presents an 8-shaped section, which was supposed to be characteristic of the genus. Further protrusion brings to the surface the bottom of the groove of the inner side of the shaft, so that its section remains in adult age something like a B.

The measurements of a medium-sized adult are—

Measurements.

	M.
Length from interorbital region to end of nasal bones (Spec. No. 1)	0.0250
Length of median suture nasal bones	.0160
Width at interorbital region	.0080
Width between anterior borders of orbits	.0180
Width at front of rugose patches	.0110
Width of nasal bones at middle	.0080
Width of nasal bones at extremity	.0060
Width of anterior incisors together	.0050
Width of posterior incisors together	.0035
Width of incisive foramen at M. 1	.0040
Width of palate at M. 1	.0075
Width of palate at M. 6	.0080
Length of superior diastema	.0120
Length of superior molar series	.0110
Width of first molar	.0020
Width of third molar	.0033
Length of inferior molar series (Spec. No. 2)	.012
Length from M. 1 to end of incisor	.012
Length of diastema	.008
Length of crown of M. 1	.0029
Elevation of crown of M. 1 above alveolus	.0035
Depth at M. 1	.0070
Depth at M. 5	.0085
Inferior diameter of ramus below M. 1	.0040

The shaft of the humerus is subround at the middle. The external crest from the epicondyle extends by a long curve to the posterior side of the shaft where it ceases. The external extremity of the condyles is occupied for the anterior half by a well-defined fossa. The general form of the ilium is slender. The ilioischiatic suture traverses a concave face behind the acetabulum. The rim of the latter is pronounced. The posterior face of the femur between the little and third trochanters is nearly plane; it is strongly convex on the anterior face. The shaft below the middle is a transverse oval drawn to an obtuse angle at one extremity, on account of the upward prolongation of the angle defining the posterior face. The latter at its inferior part is divided by an obtuse longitudinal ridge which extends to the intercondylar fissure. The condyles exhibit about equal length and width, and their articulating surfaces are not distinctly cut off from that of the patellar groove. The external fossæ are very insignificant. The inner cotylus of the head of the tibia is only about half as wide as the external. The spine, which limits it towards the middle line of the head, is separated by a wide interval from the inner margin of the external glenoid cavity. The external notch of the head, though small, is well-defined. Just above the point of attachment of the distal extremity of the fibula the tibia is triangular in section, the longest side being the inner. Half-way between this point and the distal extremity, the section of the shaft is a transverse oval. As we approach the extremity, two angular lines appear on the posterior face, near the borders, which extend, the one to the external border of the external malleolus, the other to the opposite border of the inner malleolus. Each incloses a narrow groove with the margins of the tibia, which terminate on each side of the distal extremity in a small backwardly directed process. The surface between these lateral ridges is gently concave. On the anterior face the border of the distal extremity is convex forwards. Above the extremity the face is concave, and near the inner border a little further up is a short groove-like fossa separated from the border by a little ridge. Both of these concavities are differently developed in different individuals, and are sometimes obsolete.

Measurements.

	M.
Width of humerus at middle of shaft	.0025
Width of humerus at condyles	.0062
Diameter of head of radius { transverse	.0050
Diameter of head of radius { vertical	.0028
Diameter of shaft of radius distally	.0028
Length of ilium (proximally restored)	.0190

		M.
Diameter of neck of ilium	anteroposterior	.0035
	exterointernal	.0035
Long diameter of acetabulum		.0045
Anteroposterior diameter of ischium below spine		.0030
Width of femur at proximal extremity		.0090
Diameter of head	anteroposterior	.0040
	transverse	.0045
Transverse diameter at little trochanter		.0065
Transverse diameter at third trochanter		.0060
Transverse diameter at middle of shaft		.0040
Transverse diameter at condyles		.0070
Anteroposterior diameter at patellar ridge and condyles posteriorly		.0080
Diameter of head of tibia	anteroposterior	.0080
	transverse	.0085
Diameter of shaft of tibia transversely		.0030
Diameter distal end of tibia	anteroposterior	.0035
	transverse	.0065
Length of astragalus		.0070
Length of trochlea		.0035
Width of trochlea		.0035
Elevation of trochlea	externally	.0032
	internally	.0025
Diameter of head	long	.0030
	short	.0022
Length of calcaneum		.0010
Length of heel of calcaneum		.0045
Width at tibial facet		.0045
Width at cuboid facet (transverse)		.0035
Depth at cuboid facet		.0032

It is very probable that this species is the progenitor of the *Lepus sylvaticus*, which now inhabits North America in such abundance. It will then be interesting to trace the specific characters in which modification was necessary in order to effect the transition. Only the principal ones will be noticed; the few generic characters have already been pointed out.

In the recent species the muzzle is similar in size; the premaxillary bones are more produced on the extremity. The perforations of the anterior part of the maxillary bone are larger and cover a greater extent of surface than in the *Palæologus haydeni*. The posterior emargination of the palate is wider and deeper in *Lepus sylvaticus*. The otic bulla is a little larger in *P. haydeni*, and the mastoid bone is not spongy with perforations as in *L. sylvaticus*. Other parts of the skull cannot now be compared. The preserved portions of the mandibular rami of *P. haydeni* are of the same size as those of *L. sylvaticus*. The symphysis is shorter through the brevity of the inferior portion of the sutural surface in *P. haydeni*. The incisive alveolar ridge is not distinguishable from the inferior border in *L. sylvaticus*. In the same species there is but one, the usual mental foramen, in the position of the anterior one of the extinct species. The spongy condition of the rami near the symphysis in the *L. sylvaticus* is not seen in the extinct species. Excepting the first tooth, the inferior molars are of similar size and constitution in the two species.

It is probable that the *Lepus ennisianus*, described a few pages later in the present work, is the intermediate form in the line of descent between the *Palæologus haydeni* and the *Lepus sylvaticus*, or perhaps the *L. auduboni*. This species is from the John Day Miocene of Oregon, a later deposit than the true White River. The last inferior molar is like that of the *P. haydeni*, while the first molar is that of the genus *Lepus*. The postorbital process, though present, is smaller than in any existing North American species.

The *Palæologus haydeni* was excessively abundant during the White River epoch in Dakota and Colorado, as the number of its remains indicate. I find, also, quite a number of maxillary and mandibular bones with teeth, in my Oregon collections, which I cannot distinguish from the present species. I have made the identification provisionally, in anticipation of the discovery of more perfect material.

PALÆOLAGUS TRIPLEX Cope.

Paleontological Bulletin No. 16, p. 4. Annual Report U. S. Geol. Surv. Terrs., 1873 (1874), p. 479. Plate LXVI; fig. 28.

This rabbit is known from a single incomplete left mandibular ramus, which supports all the molar teeth in perfect preservation. It belonged to an adult animal, but the first inferior molar is but little worn, showing that the individual had but just attained maturity.

This species is larger than the *Palæolagus haydeni*, and is equal in size to the *P. turgidus*. It differs materially from both species in the constitution of its molar teeth. The first molar is peculiar in having an anterior as well as an external groove, the result being a trilobate instead of a bilobate section. This character would be maintained during the life of the animal, as the groove continues well into the alveolus. The four intermediate molars are characterized by the presence of a third column posterior to the second, and of smaller diameter. It is at first isolated from the second by enamel investment, but on wearing becomes connected with it. They remain distinct to an advanced age of the animal, since the grooves which bound them descend far into the alveoli before disappearing. Another characteristic of the specimen at least, is seen in the intervention of a wide isthmus between the two principal lobes of each molar, or, in other words, of a narrowed portion of the second column between its transverse portion and the anterior column. The result is that the triturating surfaces of the posterior column of the molar has a quadrilobate outline; one lobe anterior, one posterior, and two lateral.

The tuberosity of the inner side of the ramus, which incloses the incisive alveolus, extends to below the second molar. Its surface, and a portion of that above it, is roughened with small punctiform impressions. The external face of the ramus is smooth and somewhat convex anteroposteriorly and vertically. The anterior border of the masseteric fossa is not prominent, as in *P. turgidus*, is regularly convex, and extends to the line of the posterior border of the penultimate molar.

This species rests on characters which I have observed to be transitional in the *P. haydeni*, and I have attended to the possibility of the individual which has furnished them being a similarly immature *P. turgidus*. In a considerable number of specimens of the latter no approach to the present one is exhibited; the latter is a fully-grown animal, and its characters would remain after long attrition of the teeth.

Measurements.

	M.
Length of molar series	.016
Length of median three molars	.010
Width of median molar	.003
Depth of ramus at median molar	.011

This species is rather larger than the prairie-marmot (*Cynomys ludovicianus*).

PALÆOLAGUS TURGIDUS Cope.

Paleontological Bulletin No. 16, p. 4. Annual Report U. S. Geol. Surv. Terrs., 1873 (1874), p. 479.
Tricium paniense Cope, Pal. Bulletin No. 16, p. 5.

Plates LXVI, fig. 28; LXVII, figs. 13–27.

This is the largest species of the genus, and after the *P. haydeni* the most abundantly represented in collections. At the locality which furnished several hundred specimens of the former species I obtained twenty of this one, including both dental series and their supporting bones, and various portions of the skeleton, but no cranium.

The maxillary bone displays the fossa below the anterior base of the zygomatic arch much better defined than in the *P. haydeni.* The palatine bone extends to the same distance forwards, that is, to opposite the anterior border of the third molar. The portion of the bone at the side of the posterior emargination is narrower, and not so horizontal as in the *P. haydeni.* The notch has the same extent, viz, to the line of the posterior border of the fourth molar.

The superior molars are similar in general to those of the *P. haydeni,* but in none of them do I observe the fissure of the inner side of the crown. In several of them the median crescent persists, so that if the internal fissure exist at any time it must be speedily removed by attrition. In *P. haydeni* it remains after the disappearance of the crescent, as in that species there is no enamel on the external side of the crown.

The mandible is more robust than in *P. haydeni,* but has much the same form. The anterior mental foramen is behind the middle of the diastema, on its superior aspect, and the posterior is below the second molar, *below* the middle of the ramus. The dental foramen is *above* the middle line of the ramus; the last-named two foramina having relations the reverse of that seen in *P. haydeni.* The anterior border of the masseteric fossa is elevated into a prominent rough ridge in most specimens, and reaches to the penultimate molar. The tuberosity inclosing the incisive alveolus does not extend so far posteriorly as in *P. haydeni,* ceasing below some part of the second molar, varying a little in different individuals; its surface is covered with impressed punctæ. The anterior border of the pterygoid fossa of the ramus is not well defined.

There is no groove on the inner side of the first molar. The two columns of the three intermediate molars are closely appressed, and the second is lower, not so wide, but a little longer anteroposteriorly than the anterior column. The last molar is deeply grooved on the inner side; its grinding face is only half as large as that of the first molar. The inferior incisor is not deeper than wide, and is obtuse behind. The anterior face is nearly plane, and the enamel is marked by numerous approximated, faint transverse undulations in well preserved specimens.

Measurements.

No. 1.

	M.
Length of superior molar series	.0160
Length of second molar	.0025
Width of second molar	.0043
Length of fourth molar	.0030
Width of fourth molar	.0055

No. 2.

Length of molars	0.016
Length of three median molars	.010
Depth of ramus at central molars	.011
Width of central tooth	.0035

No. 3.

Depth of ramus at penultimate molar	.0130

The deciduous molars present much the same character as in *P. haydeni*, except that there is scarcely a trace of the odd posterior tubercle on the second. The posterior root of the latter extends to the bottom of the alveolus. The grooves of crown do not descend to the alveolus on either side. Measurements of such a specimen are—

Measurements.

	M.
Length of two anterior molars	0.0068
Length of first molar	.0032
Width of first molar	.0021
Depth of ramus at first molar	.0085
Depth of ramus at diastema	.0061

I have of this species portions of humeri, femora, tibiæ, and astragalus. They resemble very much the corresponding elements of the *P. haydeni*, but differ strikingly in the much larger size.

The distal extremity of the *humerus* has a greater transverse extent than that of the *P. haydeni*, chiefly because of the greater size of the internal epicondyle. The external trochlea of the condyles extends also a little further outwards beyond the external posterior trochlear flange. The corresponding internal flange is not nearly so prominent as in *P. haydeni*. The olecranar fossa is scarcely perforate; the coronoid fossa is shallow. The external marginal acute edge is prolonged well upwards. The proximal portion of the femur has the characters of that of *P. haydeni*, but the third trochanter is smaller, and its apex is the end of the truncate external surface of the great trochanter. In *P. haydeni* the two are connected by a thin edge.

The great trochanter projects considerably beyond the head, and the little trochanter is very prominent, and is directed posteriorly. The condyles of the femur are not produced so far backwards as in *P. haydeni*. In all the particulars which I have described as characterizing the distal extremity of the tibia of *P. haydeni*, *P. turgidus* agrees with it, differing only in size. The same may be said of the astragalus, excepting that the inner trochlear ridge is a little less reduced in the *P. turgidus*.

Measurements.

		M.
Diameter of distal end of humerus	anteroposterior, least	.004
	transverse	.012
Width of proximal end of femur		.013
Diameter of head	anteroposterior	.006
	transverse	.0065
Width of femur at little trochanter		.008
Width of femur at third trochanter		.007
Width of femur below third trochanter		.005
Width of condyles of femur		.010
Depth of same at posterior margin		.010
Diameter of shaft of tibia	anteroposterior	.004
	transverse	.005
Diameter of distal end of tibia	anteroposterior	.0055
	transverse	.0110
Length of astragalus		.0110
Length of trochlea		.0056
Width of trochlea		.0060
Elevation	of external ridge	.0048
	of internal ridge	.0040
Diameter of head	long	.0040
	short	.0033

White River beds of Dakota, and of Northeast Colorado.

LEPUS Linn.

Dental formula: I. $\frac{1}{1}$; C. $\frac{0}{0}$; Pm. $\frac{3}{2}$; M. $\frac{3}{3}$. First superior molar simple; first inferior molar with two external grooves; last inferior molars consisting of two cylinders. Postorbital processes present.

I am acquainted with but one extinct species of this genus, and this is from the John Day or Middle Miocene period. It proves the ancient origin of this genus, now so widely distributed over the earth. Species of *Lepus* are reported by Gervais from the Miocene (Montabuzard) and Pliocene (Montpelier) of France.

LEPUS ENNISIANUS Cope.

Bulletin U. S. Geol. Surv. Terrs., IV, 1881, p. 385. American Naturalist, 1883, p. 174, fig. 21.

Plate LXVI, page 29.

This species is abundant in the Miocene beds of the John Day River, Oregon, associated with a species which I cannot distinguish from the *Palaeolagus haydeni.* The *Lepus ennisianus* exceeds the last-named species in dimensions, being intermediate between it and the *Palaeolagus turgidus.*

The form of the skull and character of the postorbital processes refer this species to the neighborhood of the *Lepus auduboni* and *L. bachmani.*[1] The former has the general outline of that of *Lepus sylvaticus*, with which it nearly agrees in size. The postorbital processes are free and shorter and narrower than those of the *L. auduboni.* The supraorbital notch is insignificant, and is not bounded by either an angle of the border or a process. Behind the postorbital processes the cranium is narrower. The parietal region is convex in both directions. The interorbital and base of the nasal region are flat The middle of the superior part of the occipital projects table-like beyond the lateral portions, as in recent rabbits. The otic bulla is large and is flattened on the external side. The mastoid presents some subvertical grooves. The paroccipital process is rather short and is turned backwards at the apex.

The mandible has much the form of that of *L. sylvaticus*, with certain differences. A slight convexity of the anterior border of the ascending ramus is the only trace of coronoid process. The posterior border of the same projects very little behind the condyle, and is but slightly concave below that point. The inferior border of the masseteric fossa terminates below the anterior border of the base of the coronoid process, which is posterior to the corresponding position in *L. sylvaticus.* Here the masseteric fossa extends as far forwards as the line of the posterior part of the fourth inferior molar.

---

[1] See Baird, Mammalia of the U. S., Pac. R. R. Surveys, VIII, p. 574.

The superior molras have the form usual in this genus. The crowns are grooved on both the external and internal faces. The first has less transverse extent than the others (except the sixth), and has a shallow groove on the anterior face. The sixth molar is a small and simple cylinder. Of the inferior molars, the first has the greatest anteroposterior diameter, while the transverse is equal to that of the others. Of its external grooves the posterior is the strongest. The last molar is much the smallest, and its section is a figure 8, with the wider circle next to the fourth molar. The anterior column of the other molars wears so as to be higher than the posterior. Its inner edge carries a shallow groove, while the external edge is narrow and smooth, and their alveoli terminate in a swelling below the first molar (Pm. III). The groove of the superior incisors is nearer the internal than the external side. The inferior incisors are perfectly flat.

Measurements.

	M.
Length of skull from inion to above Pm. II	.048
Length from inion to base of postorbital process (axial)	.032
Width of skull at glenoid cavities	.020
Width of skull behind bases of postorbital processes	.009
Vertical diameter of orbit	.0125
Depth of skull and mandible in place, at middle of orbit	.0310
Depth of mandible at condyle	.0310
Length of mandible to exit of incisor	.044
Depth of mandible at last molar	.011
Depth of mandible at middle of diastema	.005
Length of superior molar series	.012
Width of Pm. II (above)	.002
Width of M. I	.003
Length of inferior molar series	.013
Length of inferior Pm. III	.003

A fragmentary skeleton is associated with jaws and teeth of this species, and they are presumably parts of the same animal. They resemble the corresponding parts of *Lepus sylvaticus*, but are relatively smaller. The centrum of a lumbar vertebra is much depressed. There is a prominent anterior inferior spine of the ilium. On the internal side of the distal end of the tibia the ligamentous groove is more, and its bounding process is less, distinct than in *L. sylvaticus*.

Measurements.

	M.
Width of centrum of lumbar vertebra	.0083
Depth of centrum of lumbar vertebra	.0040

		M.
Diameter of acetabulum		.0060
Diameters of head of tibia	anteroposterior	.011
	transverse	.0105
Diameter of distal end of tibia	anteroposterior	.0046
	transverse	.009
Length of free part of calcaneum		.007

From the John Day River and the north fork of the John Day River, Oregon. C. H. Sternberg.

This rabbit is the oldest species which can be referred to the genus *Lepus*. It is dedicated to my friend Prof. Jacob Ennis, of Philadelphia, the distinguished mathematician and physicist.

CARNIVORA.

This order embraces the clawed mammalia with transverse glenoid cavity of the squamosal bone, confluent scaphoid and lunar bones of the carpus, and well developed cerebral hemispheres. It is well distinguished from all others at present known, but such definition is likely to be invalidated by future discovery. Some of the Insectivora possess a united scapholunar bone, but the reduction of the cerebral hemispheres of such forms distinguishes them. The presence of the crucial fissure of the hemispheres is present under various modifications in all *Carnivora*, except one or two of the *Melinæ* (Garrod), while the parieto-occipital and calcarine fissures are absent.

The many types of existing Carnivora fall into natural groups, which are of the grade termed family in zoölogy. But the distinction of these from each other is not easily accomplished, nor is it easy to express their relations in a satisfactory manner. The primary suborders of pinnipedia and fissipedia are easily defined. Various characters have been considered in ascertaining the taxonomy of the more numerous fissiped division. The characters of the teeth, especially the sectorials, are important, as is also the number of the digits. Turner[1] has added important characters derived from the foramina at the base of the skull, and the otic bulla, which Flower[2]

[1] Proceedings Zoological Soc., London, 1848, p. 63.
[2] Loc. cit., 1869, p. 5.

has extended. Garrod[1] has pointed out the significance of the number of convolutions of the middle and posterior part of the hemispheres. I have added some characters derived from the foramina of the posterior and lateral walls of the skull.[2] Mr. Turner also defines the families by the form and relations of the paroccipital process.

In studying the extinct Carnivora of the Tertiary period it has become necessary to examine into the above definitions, in order to determine the affinities of the numerous genera which have been discovered. To take them up in order, I begin with the foramina at the base of the skull. The result of my study of these has been that their importance was not overrated by Mr. Turner, and that the divisions of secondary rank indicated by them are well founded. Secondly, as to the form and structure of the auditory bulla. Although the degree and form of inflation are characteristic of various groups of Carnivora, they cannot be used in a systematic sense, because, like all characters of proportion merely, there is no way of expressing them in a tangible form. For, if the forms in question pass into each other, the gradations are *insensible*, and not sensible, as is the case with an organ composed of distinct parts. The same objection does not apply so much to the arrangement of the septa of the bulla. The septum is absent in the Arctoidea of Flower (*Ursidæ* of Turner), small in the Cynoidea (Flower, *Canidæ* Turner), and generally large in the Æluroidea (Flower, *Felidæ* Turner). But here occurs the serious discrepancy, that in the Hyænidæ, otherwise so nearly allied to the Felidæ, the septum of the bulla is wanting. Nevertheless, the serial arrangement of the order indicated by Flower, viz, commencing with the Arctoidea, following with the Cynoidea, and ending with the Æluroidea, is generally sustained by the structure of the auditory bulla, and by the characters of the feet and dentition, as well as of the cranial foramina. Turner's arrangement in the order Ursidæ, Felidæ, and Canidæ is not sustained by his own characters, and its only support is derived from Flower's observations on the external or sylvian convolution of the hemisphere of the brain.[3] There are three simple longi-

[1] Proceedings Zool. Soc., London, 1878, p. 377.
[2] Proceedings Amer. Philosophical Society, 1880.
[3] Proceedings Zoological Society, London, 1869, p. 482.

tudinal convolutions in the raccoons; in the civets and cats the inferior convolution is fissured at the extremities, while in the dogs it is entirely divided, so that there are four longitudinal convolutions between the sylvian and median fissures.

An important set of characters hitherto overlooked confirms Flower's order. I refer to those derived from the turbinal bones. In the ursine and canine forms generally the maxilloturbinal is largely developed, and excludes the two ethmoturbinals from the anterior nareal opening. In the Feline group, as arranged by Turner, the inferior ethmoturbinal is developed at the expense of the maxilloturbinal, and occupies a part of the anterior nareal opening. These modifications are not, so far as my experience has gone, subject to the exceptions seen in the development of the otic septa and molar teeth, while they coincide with their indications. The seals possess the character of the inferior group, or Ursidæ, in a high degree.

The characters derived from the paroccipital process are of limited application, as the study of the extinct forms shows.

In view of these facts, I have proposed the following arrangement:[1]

Mr. Wortman has suggested that the Arctoidea should be distinguished as a primary division, since it differs from the Cynoidea in the articulation of the astragalus with the cuboid bone. I do not, however, find this character to be constant in the Arctoid series. It gradually disappears in the Mustelidæ, and is wanting in *Mustela pennantii* and *Procyon nasua.*

External nostril occupied by the complex maxilloturbinal bone; ethmoturbinals confined to the posterior part of the nasal fossa; the inferior ethmoturbinal of reduced size .... HYPOMYCTERI.
External nostril occupied by the inferior ethmoturbinal and the reduced maxilloturbinal .... EPIMYCTERI.

While no doubt transitional forms will be discovered, the types at present known fall very distinctly into one or the other of these divisions. The characters are readily perceived on looking into the nares of well-cleaned specimens. The *Hypomycteri* stand next to the *Pinnipedia,* since the maxilloturbinal bone has the same anterior development in that group.

In searching for definitions of the families, it is necessary to be precise

[1] Proceedings American Philosophical Society, 1882, p. 471, where a list of the genera is given.

as to the definition of terms. The meaning of the word sectorial is in this connection important, since there are so many transitional forms between the sectorial and tubercular tooth. A sectorial tooth, then, of the upper jaw is one which has at least two external tubercles, which are the homologues of the median and posterior lobes of the sectorial of the cat. By the flattening and emargination of their continuous edges, the sectorial blade is formed.[1] One or two interior and an anterior lobe may or may not exist. In the genera of the *Procyonidæ*, except in *Bassaris*, the two external tubercles do not form a blade. The inferior sectorial tooth differs from the tubercular only in having an anterior lobe or cusp, which belongs primitively to the interior side. The inferior sectorial teeth with large heels, as in Viverridæ and Canidæ, I have called tubercular sectorials. The sectorial blade is formed by the union and emargination of the edges of the anterior and the principal external cusp. This blade is not well developed in the genus *Cynogale* and still less in the *Procyonidæ* and *Ursidæ*.

In looking for causes in explanation of the modifications of structure cited, one can easily discover that there is a close relation between the arrangement of the teeth and the mechanical laws involved in the performance of their function, that of seizing an active prey and of cutting up their carcasses into pieces suitable for swallowing. It is obvious that in the latter case the flesh-teeth bear the resistance, and the masseter muscle is the power, and that the nearer these parts are together the better is the function performed. As a matter of fact, the sectorial teeth in modern *Carnivora* are placed exactly at the angle of the mouth, which is nearly the front border of the masseter muscle.

Both the muscle and the teeth have, however, moved forwards in connection with the shortening of the jaw behind. This has been due to the necessity of bringing the power (masseter) nearer to another point of resistance, viz, the canine teeth. In the early carnivores (as *Hyænodontidæ*) the long jaws supported more numerous teeth ($\frac{4-3}{4-3}$) than in any modern families, and the fissure of the mouth was probably very wide. The canine teeth were evidently very ineffective weapons. The animals probably only snapped with their jaws, and did not attempt to lacerate or hold on, as do

[1] See "On the origin of the specialized teeth of the Carnivora," American Naturalist, 1879, p. 171.

the cats. The dogs of to-day are long-jawed, and they snap in a manner quite distinct from anything seen among the cats. The only dogs that hold on are the short-jawed bulldogs.

So in the use of the canines, we have the ground of the shortening of the jaw behind and before, and the consequent change of structure, which resulted in the modern perfected *Felidæ*.

The families are then defined as follows:

HYPOMYCTERI.

I. No sectorial teeth in either jaw.

Toes 5–5 .... *Cercoleptidæ*.

II. Sectorial teeth in both jaws.

a. Toes 5–5.

β. No alisphenoid canal.

True molars $\frac{2}{2}$ .... *Procyonidæ*.

True molars $\frac{1}{2}$ .... *Mustelidæ*.

ββ. An alisphenoid canal.

Molars quadrate, $\frac{2}{2}$ .... *Aeluridæ*.

Molars longitudinal, $\frac{2}{3}$ .... *Ursidæ*.

aa. Toes 5–4 or 4–4.

Sectorials well developed, an alisphenoid canal .... *Canidæ*.

EPIMYCTERI.

I. Molars haplodont.

Toes 5–4; no alisphenoid canal .... *Protelidæ*.

II. Molars bunodont, no sectorials.

Toes 5–5; an alisphenoid canal .... *Arctictidæ*.

III. Molars bunodont, with sectorials.

a. Otic bulla with septum.

β. Alisphenoid canal and postglenoid foramen, present.

γ. True molars well developed.

Toes 5–5 .... *Viverridæ*.

Toes 5–4 .... *Cynictidæ*.

Toes 4–4 .... *Suricatidæ*.

γγ. True molars much reduced.

Toes 5–5 .... *Cryptoproctidæ*.

Toes 5–4 .... *Nimravidæ*.

ββ. No alisphenoid canal; postglenoid foramen rudimental or wanting.

Toes 5–4 .... *Felidæ*.

aa. Otic bulla without septum.

No alisphenoid canal, nor postglenoid foramen.

Toes 4–4 .... *Hyænidæ*.

Fig. 33.—*Nimravus gomphodus*, two-fifths natural size. Mus. Cope.

CANIDÆ.

Species of this family were very abundant during the Miocene period in North America as in Europe. Those of the Lower and Middle Miocene epochs belong to genera allied to, but distinct from, *Canis;* while those of the Upper Miocene (Loup Fork) and later horizons pertain to the latter genus, with a few exceptions. The characters of the Miocene genera are as follows:

I. Molar formula $\frac{4}{4}\,\frac{3}{3}$.
  Humerus with epitrochlear foramen .... *Amphicyon.*
II. Molar formula $\frac{4}{4}\,\frac{2}{3}$.
    a. No anterior lobe of superior sectorial.
   Humerus with epitrochlear foramen.
  Inferior sectorial heel trenchant .... *Temnocyon.*
  Inferior sectorial heel basin-shaped .... *Galecynus.*
   Humerus without epitrochlear foramen.
  Inferior sectorial heel basin-shaped .... *Canis.*
    aa. An anterior lobe of superior sectorial.
  Heel of lower molar not trenchant; no epitrochlear foramen .... *Ælurodon.*
III. Molar formula $\frac{3}{3}\,\frac{2}{2}$.
  Heel of inferior sectorial trenchant .... *Enhydrocyon.*
IV. Molar formula $\frac{4}{4}\,\frac{1}{2}$.
  Heel of inferior sectorial basin-shaped .... *Oligobunis.*
V. Molar formula $\frac{3}{3}\,\frac{1}{1}$.
  First inferior molar two-rooted .... *Hyænocyon.*

To these genera I refer twenty-five species of the American Miocenes.

AMPHICYON Lartet.

Bulletin Société Géologique de la France, 1836, vii, 217-220; Blainville, Comptes-Rendus, 1837, v, 434; L'Institut, 1837, v, 18-19; Blainville, Osteographie, ix, Subursus, 78-96.

Dental formula: I. $\frac{3}{3}$: C. $\frac{1}{1}$; Pm. $\frac{4}{4}$; M. $\frac{3}{3}$. The true molars of the superior series all tubercular; the last two of the inferior series also tubercular. First inferior true molar a sectorial, with an internal tubercle and a heel with a superior groove, bounded by raised borders. Humerus with an epitrochlear arterial foramen.

Much is yet to be desired in the elucidation of the characters of this genus, especially of the American forms, which are less abundant and of smaller size than those of Europe. The typical species, *Amphicyon major* Blv., was the largest, equaling a bear in size. It is derived from the Miocene of Sansan, and a smaller form of it is found, according to Pomel, at San Gerand-le-Puy. Other species are derived from the latter locality, and all are typical of the Miocene formation in Europe. In the "Mio-pliocene" of India a single species has been discovered, the *A. palæindicus* of Lydekker. Three species occur in the Lower and Middle Miocene of North America, the largest of which about equals the wolf in size. On account of the large development of the inferior tubercular teeth, I have suspected that the *Canis ursinus* Cope, from the Loup Fork group of New Mexico, would prove to be an *Amphicyon.* If so, it is the only representative of this genus in our Upper Miocene.

The three American species differ as follows: The *A. cuspigerus* is small, not exceeding the kit-fox in dimensions. The *A. hartshornianus* is about the size of a coyote, and has rather smaller tubercular molars, especially of the lower series. The *A. vetus* is a little larger, but has the tubercular molars disproportionately larger than those of the *A. hartshornianus.*

AMPHICYON VETUS Leidy.

Daphænus vetus Leidy, Proceed. Academy Philada., 1853, 393. *Amphicyon vetus* Leidy, l. c., 1854, 157, 1857, 90. Extinct Mammalia Nebraska, Dakota, 1869, p. 32 *partim.* Plate I, figs. 1, 2, and 5.

Dr. Leidy's descriptions last cited above, with the accompanying figures, cover two species, a larger and a smaller, the latter being the *A. hartshornianus.* Of the true *A. vetus* my collection includes a fragmentary skull with several teeth of both jaws. They indicate a canine animal of about the size of the *Temnocyon wallovianus;* that is, with a skull intermediate in dimensions between those of the wolf and coyote.

The superior sectorial tooth is as long as that of the *Temnocyon altigenis,* but not so thick, and with narrower blade. The anterior base of the crown is concavely emarginate. The second tubercular is more like the first of *Temnocyon coryphæus* than that of *T. altigenis,* but may be readily distinguished by the insignificance of the two external tubercles. The crown is not excavated, and it is traversed across the middle by a regularly curved low ridge, which represents the V-shaped crest of most modern dogs. The internal cingulum is wide and regularly convex. The external cingulum is prominent forwards and outwards, as in *T. coryphæus,* but is obsolete at the base of the second external tubercle.

The crowns of the third and fourth inferior premolars are preserved in a somewhat damaged state. The fourth has a posterior lobe of the border. The sectorial has lost its anterior cusp. It is rather small, and the internal tubercle is well developed, though not so high as in *Amphicyon cuspigerus.* The keel of the heel is not much elevated and is external to the middle line. There is a low inner basal cingulum, readily worn away. The first tubercular is a large tooth, nearly equal to that of *T. altigenis,* and therefore relatively much exceeding that of *A. hartshornianus* and of *T. coryphæus.* It has two opposite tubercles in front of the middle, and one external to the middle posteriorly, and no external cingulum. The alveolus of the second tubercular shows that it also was a large tooth.

Measurements.

		M.
Diameter of crown of superior canine		.011
Diameters of superior sectorial	anteroposterior	.0176
	transverse in front	.009
Diameters of second superior tubercular	anteroposterior	.013
	transverse at middle	.007
Length of inferior Pm. iii iv on base		.022
Length of heel of sectorial		.005
Diameters of first inferior tubercular	anteroposterior	.0006
	transverse	.0066
Length of alveolus of second tubercular		.004

The specimen above described agrees very nearly with the larger of those described by Leidy, and represented by his Plate I, figs. 1–2, of the extinct Mammalia of Dakota and Nebraska. It differs from all the specimens of *A. hartshornianus*, in the disproportionately larger size of the tubercular molars both above and below, and by the tendency of the heel of the inferior sectorial to form a median keel. I found the fragments lying close together without intermixture of other species, on an exposure of the White River beds in Eastern Colorado, at the same locality which furnished the specimens of *A. hartshornianus.*

AMPHICYON HARTSHORNIANUS Cope.

Bulletin U. S. Geol. Survey Terrs., vi, p. 178.

Canis hartshornianus Cope, Synopsis of New Vertebrata of Colorado, Misc. pub. U. S. Geol. Surv. Terrs., 1873, p. 9. Annual Report U. S. Geol. Surv Terrs., 1873 (1874), p. 505.

Plate LXVII*a*; fig. 4.

This was the most abundant species of the genus during the White River epoch proper. It was originally established on a specimen from Eastern Colorado, and I have subsequently obtained other specimens from the same locality. On an examination of the specimens obtained by Dr. Hayden in the Bad Lands of Dakota, now in the museum of the Philadelphia Academy, I find parts of four skulls, two of which have been figured by Dr. Leidy on his Plate I, figs. 3, 4, and 6, and referred to in the text as belonging to the *A. vetus.* I have also part of a skull with nearly complete superior dentition from the John Day Valley of Oregon, which I cannot distinguish from this species.

The original specimen consists of a portion of the mandibular ramus with the first tubercular molar and alveolus of the second. The species was nearly as large as the *Canis latrans.* The anterior molar preserved has an interrupted cingulum on the outer side, which projects considerably in front, thus interrupting the parallelogrammic outline of the crown. The outer anterior tubercle is much the larger, while the inner ones are both obsolete. In *Galecynus gregarius* Cope the tubercles are equal, and there is no cingulum. Root of tubercular molar subround in section, as in *G. gregarius.*

Measurements.

	M.
Length of bases of M. ii and iii	.0130
Length of base of crown of M. ii	.0090
Width of base of crown of M. ii	.0060
Elevation of crown of M. ii	.0050

A second specimen, which is from the same locality as the last, in Eastern Colorado, consists of the skull anterior to the orbits and the posterior nares, lacking the superior walls. The teeth are all present, excepting one first premolar and five incisors. The form is short and wide, the malar bones much expanded and the muzzle narrowed. The size of the teeth is about that of the *Temnocyon coryphæus*, but the posteriorly-expanded outline is like that of the *T. josephi.* The sectorial is a little smaller than that of the *A. vetus* from the same locality; but the principal difference becomes apparent on comparing the tubercular molars. The second of the *A. hartshornianus* has not more than half the surface area of that of the *A. vetus.*

The external incisor is short and rather stout; the canines are long, curved, and acute, and with convex inner side; the inner cutting ridge is faint, the posterior moderately distinct. They are much more slender than in the *T. coryphæus.* There is a diastema behind the canine equal to the long diameter of its alveolus. The first premolar has a long base and a short anteriorly-placed apex; it is followed by a space equal to its long diameter. The second premolar is low and is prolonged backwards at the base; it is followed by a short space. The second premolar is more symmetrical, but has no distinct basal tubercles; there is a faint trace of a posterior median tubercle. This tooth will distinguish the species from *Temnocyon coryphæus* if other parts are wanting. The sectorial is small for the size of the tuberculars; its inner anterior tubercle is well developed. The first tubercular has a well-developed external basal cingulum. The two external cusps are low; the space between them and the inner border is nearly equally divided by a low **V**-shaped ridge. This character will distinguish this tooth from those of *Temnocyon altigenis* and *T. wallovianus.* The second tubercular is much smaller and stands much within the external border of the sectorial. Its external border is very oblique, more so than that of the first. It resembles the latter in details, the points being less pronounced; and the external cingulum is obsolete. The third tubercular is opposite the inner half of the second, and its surface area is very small, rather less than that of the base of the first premolar. It is supported on a projection of the maxillary bone, which is separated from the base of the pterygoid process of the palatine by a deep notch. The posterior nareal orifice is rather narrow and its border presents a median point posteriorly, which separates two deep concavities, about as in *T. coryphæus.*

Measurements of palate, etc.

		M.
Length from premaxillary border to nares		.072
Width of dental series, including canines		.065
Length of canine tooth		.020
Long diameter of canine at base		.007
Length of premolar series		.039
Length of sectorial		.013
Width of sectorial in front		.008
Elevation of sectorial in front		.008
Diameters of first tubercular	anteroposterior	.010
	transverse	.014
Diameters of second tubercular	anteroposterior	.006
	transverse	.009
Width between bases of canines		.017
Width between sectorials posteriorly		.041
Width between third tuberculars		.023
Width of nares		.012

The third or Oregon specimen agrees closely with the one just described. The crowns of its superior incisors are preserved, so that it can be seen that they are obspatulate with angular extremities, and are not notched, as in *Galecynus geismarianus* and various living species.

The best ascertained characters of this species, then, as compared with the *A vetus*, are (1) the much smaller tubercular molars in both jaws; (2) the strongly double tubercular character of the heel of the inferior sectorial as compared with the tendency to a single keel seen in the *A. vetus.*

Dedicated to my friend Prof. Henry Hartshorne, formerly of Haverford College, Pennsylvania.

AMPHICYON CUSPIGERUS Cope.

Bulletin U. S. Geol. Survey Terrs., vi, p. 178.

Canis cuspigerus Cope, Paleontological Bulletin No. 30, p. 8, December 3, 1878. Proceedings American Philosoph. Society, 1878, p. 70. *Amphicyon entoptychi* Cope, Paleontological Bulletin No. 31, p. 3, December 24, 1879. Proceedings Amer. Philosoph. Soc., 1879, p. 372.

Plate LXVIII; figs. 1–4.

This species is considerably smaller than the *A. hartshornianus*, and intermediate in size between the *Galecyni geismarianus* and *gregarius*. It is represented in my collection by two crania, one nearly perfect and with the mandible attached; the other with its superior portions crushed and without incisive region or lower jaw.

The cranium is elongate, and the muzzle is not shortened, but is rather compressed. There are no true postorbital processes, but merely obtuse angles, from which obsolete temporal ridges converge backwards. They do not unite early, as in the *Galecynus geismarianus*, *Amphicyon vetus*, &c., but only combine to form a low sagittal keel near the middle of the parietal bone. The brain-case is rather large, and is moderately contracted behind the orbits, more than in *Vulpes cinereoargentatus* or *Canis latrans* and *C. cancrivorus.* The occipital surface is strongly convex to fit the vermis of the cerebellum. The otic bulla is small and has no tympanic prolongation. This character will distinguish the species from the *G. geismarianus*, where the bullæ are very large. The paroccipital process is short and obtuse, and is well removed from the bulla, not being even connected with it by a ridge, as in recent dogs. The posttympanic process is short and obtuse. The postglenoids are well separated from them, and have a greater transverse extent than in recent dogs. They are not overlapped posteriorly by any part of the otic bullæ. The basioccipital is wide and is marked by a shallow fossa on each side opposite each paroccipital process. This is wanting in the *G. geismarianus* and the *G. gregarius.* Between the bullæ the surface is not keeled, but is flat and slightly concave.

The mandibular rami are shallow, and their inferior margin is not stout. A gentle elevation of the latter commences below the first tubercular tooth, and the alveolar border rises but little behind. The masseteric fossa is deep and well defined.

Sutures.—The ascending branch of the premaxillary is very narrow and elongate, but fails by a little to reach the narrow acute anterior prolongation of the frontal. The nasal bones are quite narrow, and their apices are above the anterior third of the orbit. The fronto-maxillary suture is strongly arched upwards. The parietal bones extend further forwards laterally than on the superior surface of the skull, where their anterior outline is broadly truncate. Below they are extensively in contact with the alisphenoid bone. The squamosal is low and elongate.

Foramina.—The nares are rather small, and the vertical exceeds the transverse diameter. The opening of the infraorbital foramen is above the anterior part of the superior sectorial tooth. There is no postparietal foramen, but a small mastoid is visible. The *foramen magnum* is large and is subround through its extension inferiorly; its border is notched at the superior base of the condyle. The condylar foramen is small and is near the *f. lacerum posterius*, but entirely distinct from it. The latter is subround and rather large. There is no distinct *f. carotideum* visible. The *f. lacerum anterius* is much contracted. The *f. ovale* is rather large and transverse, and the *f. alisphenoidale posterius* is small and distinct. *Foramen postglenoidale* rather large.

Dentition.—The third premolar tooth in both jaws differs from the corresponding one in the *C. gregarius* and in most recent species, in lacking the lobe of the posterior cutting edge, agreeing in this (as regards the inferior series) with the *Temnocyon altigenis*. It is present in the fourth inferior premolar, which has besides, a low heel. The inferior sectorial tooth is characterized by its great robustness; the internal median tubercle is much elevated, while the principal cusp is short. The heel is wide and basin-shaped, with the inner border as much elevated as the outer. The first tubercular molar is characterized by its width as compared with its length, being nearly as wide transversely as fore and aft. It has two anterior cusps followed by a basin with elevated borders simulating two posterior cusps. There are an anterior and an exterior cingulum. The second tubercular is a miniature of the first, differing in the more robust external posterior cusp, and the absence of external basal cingulum. There are no complete cingula on the external bases of the other inferior teeth. The second superior tubercular is well developed, having two external tubercles. The anterior inner cusp of the superior sectorial is distinct and acute, and there is a cingulum along the inner base of the crown. The exserted portion of the canines is long, slender, and with an oval section narrowed behind. The enamel of all the molars is more or less rugose, a character which is only found elsewhere among our extinct dogs in the *G. geismarianus*.

Measurements.

	M.
Length of cranium to inion	.106
Length from premaxillary to condyles	.101
Length from premaxillary to postglenoid	.075
Length from premaxillary to posterior border of second tubercular	.049
Length to anterior border of orbit	.035
Width of occiput at superior border of *foramen magnum*	.023

	M.
Width between inner borders of postglenoids	.023
Width between maxillary bones	.030
Width between orbits	.020
Width of nares	.009
Length of inferior molar series	.041
Length of bases of four premolars	.023
Length of base of second premolar	.005
Elevation of crown of second premolar	.005
Length of base of fourth premolar	.0072
Elevation of crown of fourth premolar	.0055
Length of base of sectorial	.010
Elevation of principal cusp	.006
Width of heel of sectorial	.006
Diameter of first tubercular { anteroposterior	.006
Diameter of first tubercular { transverse	.005
Anteroposterior diameter second tubercular	.0037
Length of base of superior sectorial	.009
Length of bases of two tuberculars	.012
Length of base of first tubercular	.0064

The skull above described displays only two superior tubercular molars on each side, and on this account I supposed the species to belong to the genus *Canis* at the time I first described it. The second skull already mentioned, exhibits the characteristic number belonging to the genus *Amphicyon*. A careful removal of the matrix from the extremities of the maxillary bones of the specimen first described revealed the fact that they had been broken off, since traces of the alveoli of the third superior molar were found. The following characters are presented by the second skull which is the basis of the supposed species cited in the synonymy above.

The superior premolar teeth are rather short in anteroposterior diameter, while the tubercular molars are relatively large There are no posterior lobes on the former; the internal and external cingula are well developed in the first and second of the latter. The third tubercular is about as wide as the second is long. The sagittal crest is only distinct on the posterior part of the parietal region. Estimated length of skull, M. .110; length of superior molar series, .041; length of true molar series, 016; length of first tubercular, .0075; length of second tubercular, .055; width of second tubercular, .0074; length of third tubercular, .0036; width of third tubercular, .052: Length of sectorial width between anterior external angles of first tuberculars, .C30.

The teeth of this species are about half the size of those of *A. vetus* Leidy. The entire animal was probably about the size of the kit fox, *Vulpes velox*.

The specimens above described were obtained by C. H. Sternberg and J. L. Wortman, in the Bad Lands of the John Day epoch, in the John Day Valley, Oregon.

TEMNOCYON Cope.

Paleontological Bulletin, No. 30, p. 6, Dec. 3, 1878. Proceed. Amer. Phil. Soc., 1878, p. 68. Bulletin U. S. Geol. Surv. Terrs., vi, p. 179.

Dental formula: I. $\frac{3}{?3}$; C. $\frac{1}{1}$; Pm. $\frac{4}{4}$; M. $\frac{2}{3}$. Two molars in each jaw tubercular. Inferior sectorial with well-developed heel, which is keeled with a cutting edge above. An internal tubercle of the same. A postglenoid, but no postparietal foramen. Humerus with an epitrochlear arterial foramen.

The characters on which I rely at present for the discrimination of this genus from *Canis* are two. The first is the presence of a cutting edge on the superior face of the heel of the inferior sectorial, in place of a double row of tubercles surrounding a basin. When well developed these characters present a broad contrast, but indications of transitional forms are not wanting. Thus, in some extinct *Canes* the internal crest of the heel is less elevated than the external, which is the homologue of the single crest of *Temnocyon*, and in some specimens of *Temnocyon coryphæus* there is a cingulum on the inner side of the median keel, which represents the internal crest of *Canis*. Secondly, the epitrochlear foramen of the humerus, a character common to all of our Lower Miocene *Canidæ* yet known.

The keel of the sectorial, which defines this genus, is simply a repetition on that tooth of the heel which belongs to the posterior premolar teeth of many *Carnivora*. It finds resemblances in such Eocene forms as *Mesonyx* and *Palæonyctis*. Among recent *Canidæ* it is apparently unknown, and is very rare in other groups, The *Cynodictis crassirostris* Filhol, from the French Phosphorites, strongly resembles the species of *Temnocyon* in generic characters.

Three species of the genus are known to me. They may be distinguished as follows. A fourth species, *T. josephi*, is provisionally placed with these:

I. First superior tubercular molar with a wide median fossa, bounded within by a tubercle.

Length of superior molar series from canine, .070; of true molars, .0215. *T. altigenis.*

Length of molar series from canine, .067; of true molars, .014 .... *T. wallovianus.*

II. First superior tubercular molar with narrower basin, bounded within by a V-shaped crest.

Length of dental series from canine, .055; of true molars, .014 ..... *T. coryphæus.*

Length of dental series from canine, .051; of true molars, .013; muzzle narrow, zygomas wide. .... *T. josephi.*

All of the above species have been derived from the John Day Miocene beds of Oregon. I, however, anticipate the discovery of these or other species of the genus in the White River beds of Dakota and Colorado.

TEMNOCYON ALTIGENIS Cope.

Paleontological Bulletin, No. 30, p. 6 (Dec. 3, 1878). Proceed. Amer. Phil. Soc., 1878, p. 68. Bulletin U. S. Geol. Surv. Terrs., vi, p. 179.

Plate LXVIII, fig. 9; LXX, fig. 11.

This dog was about the size of the wolf, *Canis lupus*. It does not appear to have been as abundant as some other species in Oregon, as I have received portions of only three individuals. Two of these are represented by mandibular rami only, one of which is the specimen on which the species was originally established. The third is that part of a skull which is anterior to the orbits, including the mandible and dentition, and it is considerably crushed.

The first-mentioned mandibular ramus is rather deep and compressed, much more so than in the *Canis latrans*, with which it agrees in the length of the dental series. As compared with the existing species of *Canis* and *Vulpes* of North America, the sectorial tooth is relatively smaller and the premolars larger. In this respect it agrees with most other dogs of the Lower Miocene, and differs from those of the Upper Miocene (Loup Fork).[1] The posterior tubercle is wanting from the premolars, excepting the last,

[1] See Proceedings Academy Philadelphia, 1875, p. 22, where I have discussed the origin and history of the sectorial tooth.

where it is large and obtuse, differing in this respect also from most recent dogs, and from the contemporary *Galeaynus gregarius*. In the sectorial tooth the principal cusp is much elevated above the anterior, while the inner median is small, with its apex in line with the anterior. The cutting edge of the heel is not acute, and is a little external to the median line; there is a weak cingulum-like angle at its inner base. The first tubercular tooth is large, nearly equaling in anteroposterior diameter the base of the third premolar. It is parallelogrammic in transverse section, and supports two principal cusps and an anterior ledge. The cusps are pronounced, and stand exterior to the middle line; their inner side slopes to the base of the crown, where there is no cingulum. The ledge is higher on the inner than the external side. There are no basal cingula on either side of the bases of any of the teeth. The second tubercular molar is lost.

The alveolar margin of the jaw rises behind the sectorial tooth, and the inferior margin begins to ascend below the middle of the same tooth more decidedly than in *C. lupus*, *C. latrans*, or *A. cuspigerus*. The two large mental foramina are situated, the one below the second, the other below the third premolars.

Measurements.

	M.
Length of anterior six molars	.073
Length of anterior four molars	.045
Length of base of second premolar	.011
Elevation of crown of second premolar	.011
Length of base of fourth premolar	.015
Elevation of crown of fourth premolar	.014
Length of base of sectorial tooth	.0185
Elevation of principal cusp of sectorial tooth	.0160
Elevation of anterior cusp of sectorial tooth	.009
Length of heel of sectorial	.007
Elevation of heel of sectorial	.0085
Length of crown of first tubercular	.0115
Width of crown of first tubercular	.0065
Depth of ramus at Pm. ii	.024
Depth of ramus at sectorial	.028
Thickness of ramus at sectorial	.010

The second ramus contains the alveolus of the second tubercular molar.

The third specimen presents a mandibular dentition similar to that above described. The superior canine is large, being longer and more acute than in the wolf. The first and second premolars are damaged; the third does not exhibit any posterior marginal lobe. The sectorial is small for the size of the other teeth, not quite equaling that of the *Canis latrans.* Its anterior inner tubercle is prominent and acute, but of small size. The tubercular teeth are relatively large, equaling those of the wolf in transverse diameter, but not in anteroposterior. They have a wide internal cingulum, and a single low median tubercle separated by a deep valley from the two prominent external cusps. These are bounded externally by a well-marked cingulum. Enamel wrinkled where not worn.

Measurements.

		M.
Length of superior dental series on base, from canine		.070
Length of crown of canine		.030
Diameters of sectorial	anteroposterior	.019
	transverse in front	.0125
Diameters of first true molar	anteroposterior	.014
	transverse	.020
Diameters of second true molar	anteroposterior	.0075
	transverse	.014
Length of series of inferior premolars		.046

From the John Day Miocene of the John Day River, Oregon. Found by C. H. Sternberg and J. L. Wortman.

TEMNOCYON WALLOVIANUS Cope.

Bulletin U. S. Geol. Surv. Terrs., VI, p. 179, Feb. 11, 1881.

Plate LXX; fig. 10.

This species of *Temnocyon* is more nearly related to the *T. altigenis* than to any other member of the genus. Its anterior dentition is much like that of the species named, but the tubercular molars are not larger than those of the *T. coryphæus.* While they differ in details of composition from those of the latter species, they differ also from those of *T. altigenis.* The species is only known from a cranium from which all posterior to the orbits is lost. Its molar teeth, and most of the incisors, are in good preservation.

The first molar is one-rooted, and the posterior outline of its crown is quite oblique The posterior borders of the crowns of the second and third are without distinct lobes. The sectorial has the same size as that of *T. altigenis* Its blades are not very acute, nor close together. The inner anterior tubercle is small, and is set well inwards. The first tubercular has a very oblique external border, owing to the rapid contraction posteriorly of the maxillary bone. The two external cusps have the usual direction, so that it results that the anterior external angle of the crown is very protuberant, much more so than in *T. altigenis*. Moreover, the external basal cingulum is not defined at the base of the posterior cusp, as it is in *T. altigenis*. For the rest this tooth resembles the corresponding tooth in the latter species, except as to size. The middle of the crown is occupied by a deep valley, which is bounded within on the anterior border of the crown by a subtrihedral tubercle, which does not send a ridge back to the posterior border, as is usual in dogs. The internal cingulum is wide and flat. The second tubercular molar is relatively small, having about half the grinding surface of that of the corresponding tooth of *T. altigenis*. The posterior external cusp is a mere rudiment, and the cingulum is not well defined. The internal tubercle and cingulum are confounded in a uniform surface within the median concavity. The rugosity characteristic of the *T. altigenis* is little visible in this specimen.

Measurements.

		M.
Length of superior dental series on base from canines		.067
Length of bases of true molars		.014
Diameters of sectorial	anteroposterior	.017
	transverse in front	.011
Diameters of first true molar	anteroposterior	.012
	transverse at middle	.015
Diameters of second true molar	anteroposterior	.005
	transverse	.010

From the John Day Bad Lands, Oregon. J. L. Wortman.

TEMNOCYON CORYPHÆUS Cope.

Proceedings Academy, Philadelphia, 1879, p. 180. Figure 2, in Proceedings American Philosophical Society, 1880 (February). *Canis hartshornianus* Bulletin U. S. Geol. Survey Terrs., V, 1879, p. 58; not Cope Annual Report U. S. Geol. Survey Terrs., 1872 (1873), p. 506.

Plates LXXI; LXXI*a*, figs. 1–7; LXXII*a*, figs. 4–7.

That this dog was the most abundant species of the Oregon Miocene is indicated by the fact that the following material representing it is now in my collection: Seven crania, several of them with mandibles and more or less complete skeletons; seven more or less fragmentary maxillary bones with teeth, and nine more or less broken mandibles with teeth.

A nearly perfect skull displays the following characters: The orbits are entirely anterior to the vertical line dividing the skull into halves, and the muzzle is proportionately shortened. It is also narrowed anteriorly, and its median line above is shallowly grooved. The interorbital region is greatly convex to the supraorbital region, and is grooved medially. The postorbital processes are mere angles, and are flattened from below. The cranium is much constricted behind the orbits, where its diameter is not greater than the width of the premaxillary incisive border. The sagittal crest is much elevated, and forms a perfectly straight and gradually rising outline to its junction with the incisor. The borders of the latter are very prominent, extending backwards considerably beyond the brain case. The zygoma is rather slender, is elongate, and but little expanded. The otic bullæ are very large; the paroccipital processes are directed backwards, at an angle of 45°, and are rather elongate and acute; they cap the bullæ posteriorly. The lateral occipital crests bound a fossa of the occipital region near the condyles. The occipital surface is directed horizontally backwards above the foramen magnum. This part of it, and its superior portion, are divided by a median keel.

The basioccipital is keeled on the middle line below. The sphenoid is not keeled, and is concave, its borders descending on the inner side of the bullæ. The pterygoid fossa is rather narrow, and the hamular process is short. The posterior border of the palate does not extend anterior to the posterior edges of the last tubercular molar, and its middle portion projects backwards in a triangular process. The palatine fossa for the inferior sectorial is shallow. The superior surface of the postorbital region is roughened.

Foramina.—The *foramen infraorbitale exterius* is rather large, and issues above the anterior border of the sectorial tooth. The *f. incisiva* are short, not extending posterior to the middle of the canines. The *f. palatina* are opposite the posterior border of the sectorial. The *f. lachrymale* is altogether within the orbital border. The *f. opticum* is rather large. This species is peculiar in having the *f. f. spheno-orbitale, rotundum,* and *alisphenoidale anterius* united into one large external orifice. The alisphenoid canal is larger than in *Canis latrans*, and its posterior foramen small. The *f. ovale* is further removed from the *f. alisphenoidale* than in the coyote, and is exterior to and a little behind the *f. carotideum*.

The nasal bones extend to above the middles of the orbits, and contract gradually to their apex. Their combined anterior border is a regular concave, and the lateral angles at this point are produced outwards and forwards. The posterior apex of the premaxillary bone is separated from the anterior apex of the frontal by a short space. The maxillo-malar suture is deeply notched in front below, and it extends upwards to above the infraorbital foramen. A very narrow surface of the lachrymal is exposed on the external surface. The pterygoid bone is distinct, and is nearly equally bounded by the sphenoid and palatine on the outer side. The inferior suture of the orbito-sphenoid runs in a groove, which is deepest anteriorly.

Dentition.—The crowns of all the incisor teeth are narrow or compressed, and, though slightly worn, present no indication of notch. As usual, the external ones are much the largest in anteroposterior diameter. The canines have robust fangs and rapidly tapering crowns, which are but little compressed. The first superior premolar is one-rooted, and the crown is simple. The crown of the second is without posterior heel and tubercle, while the third possesses both. The sectorial is relatively short, less so than in *C. latrans*. The blades are low and obtuse as compared with recent species, and the notch separating them is quite open. The anterior external heel is small, and there is no anterior external tubercle. The first tubercular molar is large, and the crown is narrower than that of *C. latrans*. It has an obtuse external cingulum, two external conical cusps, a V-shaped median ridge, and a wide internal cingulum. This crown differs from the corresponding one of *C. latrans* in having conical instead of compressed external cusps, and a simple V-shaped crest within instead of two adjacent cusps. The second tubercular is smaller than in *C. latrans*, and its tubercles are less distinct. There are two outer tubercles, a V-shaped ridge, and an inner cingulum, all very obscure. The enamel of all these teeth is smooth.

Measurements of cranium.

	M.
Length along base of skull, including incisive border and occipital condyle	.160
Length of skull to palatal notch	.075
Length of skull to posterior border of pterygoid bone	.102
Length to front of orbit axially	.046
Width between zygomas (greatest)	.094
Width between orbits (least)	.036
Width at postorbital constriction	.021
Width between bases of canines	.017
Width between bases of second tuberculars	.027
Width between otic bullæ	.009
Width between apices of paroccipitals	.042
Width of foramen magnum	.017
Width of occiput above	.032

The cranium above described is not accompanied by a mandible, but has a cervical vertebra and a scapula associated with it in the same block of matrix.

The mandible of another skull exhibits the following characters: There is a well-developed marginal lobe of the posterior cutting edge of the third and fourth premolars as well as a low posterior heel, and a rudiment of an anterior one The heel of the sectorial is shorter than the remaining part of the tooth, and rises to a cutting edge a little external to the middle line; there is a small tubercle at its interior base. The anterior blade-cusp of the sectorial is much lower than the median, which is conical; the two diverge, diminishing the shear-like character and action of the tooth. The internal cusp is well developed. The first tubercular is of moderate size, and is a longitudinal oval in outline. The crown supports two low tubercles anterior to the middle, of which the external is the larger. The last molar has a single compressed root, and the crown is a longitudinal oval in outline. Its position is on the ascending base of the coronoid ramus, so that the crown is slightly oblique. The masseteric fossa is profound and well defined; its anterior termination is below the middle of the second tubercular tooth. The horizontal ramus is not robust, but is compressed, and rather deep.

Measurements of mandible.

	M.
Length along bases of posterior five molars	.049
Length of base of fourth premolar	.011
Elevation of crown	.008
Length of base of sectorial	.018

	M.
Elevation of crown of sectorial	.012
Length of base of first tubercular	.0075
Width of base of first tubercular	.0050
Length of base of second tubercular	.0050

While the characters of this dog do not separate it widely from the genus *Canis*, many of them are quite different from those presented by the recent species of the genus with which I am acquainted. Thus the union of the foramina spheno-orbitale and rotunda, the anterior position of the orbits, and the postorbital constriction are not seen in the wolf, domestic dog, coyote, jackal, or the North American and European foxes. The size of the brain was evidently less than in those species, and the sectorial teeth quite inferior in the efficiency of their blades. These characters may be considered in connection with the low geological position of the beds in which the species occurs.

One of the crania is accompanied by several bones of the skeleton. These are, seventh cervical and four lumbar vertebræ, commencing with the second; humerus, lacking the middle of the shaft and cuboid bone. The seventh cervical has the centrum as large as that of the coyote, but it is more depressed, and its extremities are more oblique. The opisthocœlous character is not wanting, and the inferior keel is well developed. The neural canal is as large as that of the coyote, but the roof is not so wide anteroposteriorly, and the base of the neural spine extends from one edge to the other. The lumbars are as long as those of the coyote, but the third, fourth, and fifth are narrower than in that animal. The fourth and fifth only have well developed keels. On the fifth the angle from the base of the diapophysis posteriorly is well marked. The anapophyses are invisible on the second lumbar, owing to injury; on the other vertebræ they are only a low ridge reaching the posterior edge of the neurapophysis. The parapophysial tuberosities of the anterior borders of the centra below are almost nil, as in the coyote. See Plate LXXI*a*, figs. 2–4, where they are represented two-thirds the natural size. The extremities of the centra are slightly opisthocœlous, and the articular surfaces of both extremities possess a transverse, curved, shallow groove.

Measurements of vertebræ.

			M.
Diameters of seventh cervical	longitudinal		.017
	anterior	vertical (at middle)	.009
		transverse	.0117
Diameters second lumbar	longitudinal		.023
	anterior	vertical (at middle)	.0115
		transverse	.010
Diameters fifth lumbar	longitudinal		.026
	anterior	vertical (at middle)	.012
		transverse	.018

The greater tuberosity of the humerus is of the usual straight form. As compared with both the wolf and coyote, both tuberosities are more produced, and the internal terminates in an acute edge not found in the two species named. The crest of the external tuberosity is continued farther posteriorly than in the coyote or wolf, surrounding the teres insertion space, and continuing below it on the shaft. The lesser tuberosity on the other hand graduates into the head, without the shoulder seen in *C. latrans* and *C. lupus.* The condyles have a smaller diameter and much greater transverse extent than in either of the species of *Canis* cited, or in the *Vulpes cinereoargentatus*, thus resembling the corresponding part in the Eocene *Oxyænidæ.* This resemblance is heightened by the considerable prominence of the internal epicondyle, which exceeds that seen in the recent Canidæ mentioned.

Measurements of humerus.

			M.
Anteroposterior diameter of head and greater tuberosity			.035
Anteroposterior diameter of head alone			.025
Anteroposterior diameter to bicipital groove			.021
Transverse diameter to bicipital groove			.019
Diameters of condyles	transverse		.021
	anteroposterior	external	.011
		internal	.016
Width of posterior face of condyle			.014

The cuboid is a very little longer than wide. The peroneus longus groove is profound and enters at the middle of the external border.

Measurements of cuboid.

		M.
Length		.012
Proximal diameter	anteroposterior	.007
	transverse	.011

A third cranium is accompanied by some bones, among which is a calcaneum (see Plate LXXI*a*, fig. 7). The characters of this bone distinguish it widely from that of the *Canes lupus*, *latrans*, and *cancrivorus*, and the

Vulpes cinereoargentatus. The sustentaculum is more expanded, and the process to which the external calcaneo-cuboid ligament is attached is considerably more prominent The cuboid facet is shaped much as in the gray fox, being wider than in *C cancrivorus* and *C. latrans*, and not so wide as in *C. lupus.*

Measurements of calcaneum.

	M.
Total length	.035
Length of heel	.021
Width at sustentaculum	.017
Width at cuboid facet	.015
Width of cuboid facet	.011
Depth of cuboid facet	.0076
Depth of heel at extremity	.012

This bone is but little smaller than that of an average coyote.

Restoration, etc.—The proportions of this species may be derived from the second specimen described, where the skull is accompanied by bones of the skeleton. The dimensions have been about those of the *Canis latrans*, with some parts rather more slender. The face had a very different expression, owing to its extreme shortness, as compared with the length of the head from the orbits posteriorly. The latter dimension was the same as in the *C. latrans*, but the muzzle resembled in its proportions those of a skunk or badger. The brain case is smaller than in the coyote, and the crests for muscular insertion much more elevated. As the otic bullæ are absolutely larger than in the coyote, it is fair to infer a delicate sense of hearing.

This was the most abundant carnivore in Oregon during the John Day epoch. Of the seven skulls in my possession I find but little variation in proportions. The one whose mandible is figured on Plate LXXI*a*, figs. 1, 1*a*, is the smallest.

The specimen from which the first description of this species was drawn up was obtained by Mr. J. L. Wortman Others were previously sent me by Mr. C. H. Sternberg

TEMNOCYON JOSEPHI Cope.

Bulletin U. S. Geol. Surv. Terrs., VI, p. 179, February 11, 1881.

Plate LXX, fig. 9.

The anterior portion of a cranium which supports all the teeth is all that represents this species in my collection. It is distinguished, first, by its size, being intermediate between the *Tennocyon coryphæus* and the *Galecynus geismarianus* in this respect. Secondly, it may be known by the depression of the face at the orbits and the prominence of the anterior part of the malar bone.

Laid by the side of the skull of *T. coryphæus* so that the posterior borders of the second tuberculars of this species and that one are in line, the canine of *T. josephi* originates in line with the posterior border of the first premolar of the *T. coryphæus.* Of course the molars are absolutely smaller in *T. josephi,* but the sectorial is also relatively smaller. In *T. coryphæus* the length of that tooth exceeds the space occupied by the first and second premolars; in *T. josephi* its length is considerably less. The third premolar of the *T. josephi* lacks the posterior tubercle found in the *T. coryphæus,* but a small basal heel is present. There is a narrow internal basal cingulum of the superior sectorial. The first tubercular is not so oblique as in *T. wallovianus,* and has an external cingulum. It has a strong V-shaped crest within the middle, which is well separated from both the external tubercles and the internal cingular border.

The inferior position of the orbit is indicated by the fact that the distance from its inferior border to the base of the sectorial measures one-half the distance between the anterior border of the sectorial and the posterior border of the canine. In *T. coryphæus* it is three-quarters, and in *T. geismarianus* it is three-fifths that distance.

Measurements.

		M.
Length of dental series from I 3		.062
Length of dental series from canine		.051
Length of three premolars		.025
Anteroposterior diameter of canine		.0075
Diameters of Pm. iii	anteroposterior	.009
	vertical	.006
Diameters of sectorial	anteroposterior	.013
	transverse in front	.007
Diameters of M. i	anteroposterior	.008
	transverse	.011
Diameters of M. ii	anteroposterior	.004
	transverse	.0075
Width of incisor series		.015

The reference of this species to the genus *Temnocyon* is provisional, as the character of its inferior molar teeth is unknown. It is dedicated to

Joseph, the chief of the Nez Percés Indians of Oregon, a man declared by common consent to be possessed of many noble qualities, and whose political record has been altogether creditable.

Found in the John Day beds of the John Day region, Oregon, by Mr. J. L. Wortman.

GALECYNUS Owen.

Quarterly Journal Geological Society, London, 1847, iii, 54–60. "*Cynodon* Aymard, Annales Société du Puy, 1848, xii, p. 244. *Cynodictis* Bravard et Pomel, Notice sur les Ossemens Fossiles de la Debruge, 1850, p. 5. *Cyotherium* Aymard, Ann. Soc. d'Agric. du Puy, 1850, xiv, p. 115"; Bronn.

Dental formula, I. $\frac{3}{3}$; C. $\frac{1}{1}$; Pm. $\frac{4}{4}$; M. $\frac{2}{3}$. Inferior sectorial with internal tubercle, and with a heel with raised or tubercular internal and external borders. First premolar in both jaws one-rooted. A postglenoid but no postparietal foramen. Humerus with an epitrochlear arterial foramen.

This genus, which is abundantly represented by species and individuals, existed during the Upper Eocene epoch in Europe (in the Phosphorites), and also during the White River or Oligocene in North America. As the structure of the feet of the numerous species from these epochs is not yet known, and, therefore, some doubt as to their correct generic reference may still exist, I only regard the genus as a certain inhabitant of North America during the Truckee or Middle Miocene epoch. This is indicated by the *Galecynus geismarianus*, where the number of the toes on the posterior foot has been ascertained.

All the species of the genus from Eocene and Lower Miocene beds, as well as most of those of the Loup Fork epoch, are characterized by the relatively small size of their sectorial teeth. In this they resemble the *Amphicyons*, *Temnocyons*, and other forms of *Canidæ* of the same period, and differ from such true *Canes* as *C. ursinus*, *C. sævus*, and *C. haydeni*, which display the enlarged sectorial teeth of the existing species of the genus. Of course there is every gradation in this respect between the two types. In the older species the internal tubercle of the inferior sectorial tooth is more largely developed than in the later ones, thus approaching some of the species of *Viverridæ*, where it is still more largely developed. As in other characters, there are gradations in this also, so that neither in it nor in the relative size of the sectorials do I find ground for the separation of the species in question from the genus *Canis*, as has been proposed in the case of some of the species in Europe. Through the kindness of M. Filhol, I possess jaws of a number of the species found by himself and others in the Phosphorites of Central France, including the *Canis vèlaunus*, the type of the genus *Cynodon* of Aymard. These agree very nearly with the species of dogs from the American Miocene beds as to generic characters. Professor Owen, in the paper above cited, proposed to distinguish the genus *Galecynus* on account of the greater length of the pollex as compared with that found in the existing species of *Canis*. This character appears to me to be of an unsatisfactory nature, owing to the fact that gradations in the length of a digit are difficult to express with precision in other than a specific sense; and the gradations may certainly be expected to occur. I therefore formerly regarded all these species as belonging to the genus *Canis*.

I subsequently found in the *G. geismarianus* a character which separates the genus from *Canis*, viz, the presence of the epitrochlear foramen of the humerus. In this point it agrees with *Amphicyon* and *Temnocyon*. I arrange cotemporary and generally similar species under the same generic head, as the most reasonable course in the absence of direct evidence.

The American species of *Galecynus*, then, may be arranged as follows:

I. Smaller species with little or no sagittal crest.

* Temporal ridges uniting close behind orbits; otic bullæ small.

Small; no external ridge on inferior sectorial.............. *G. gregarius* Cope,

** Temporal ridges uniting early; otic bullæ large.

Larger; no external ridge on inferior sectorial; teeth robust. *G. geismarianus* Cope.

Smaller; an external ridge on lower sectorial; teeth more robust. *G. latidens* Cope.

** Temporal ridges not uniting anteriorly; otic bullæ large.

Least; muzzle narrow; superior tuberculars wide; no external ridge on inferior sectorial.. *G. lemur* Cope.

There are three species which are only known from mandibular rami. whose positions in the above analytical table I cannot fix. These are *Canes vafer* and *temerarius* of Leidy, of the Loup Fork epoch, and the *C. lippincottianus* Cope, of the White River. As I can add nothing to Mr. Leidy's descriptions of the first two, I only mention them here. They are probably true *Canes*.

GALECYNUS GREGARIUS Cope.

Bulletin U. S. Geol. Survey Terrs., VI, p. 181.

Canis gregarius Cope. Paleontological Bulletin No. 16, p. 3, August 20, 1873. Annual Report U. S. Geol. Surv. Terrs., 1873 (1874), p. 506. Bulletin U. S. Geol. Surv. Terrs., 1879, p. 58. *Amphicyon gracilis* Leidy. Proceedings Academy Philada., 1856, p. 90; 1857, p. 90. Extinct Mammalia Dakota and Nebraska, 1869, p. 36; Pl. I, fig. 7; Pl. V, figs. 7–9. Not *Amphicyon gracilis* of Pomel.

Plate LXVII*a*, figs. 7–11; LXVIII, figs. 5–8.

In the White River beds of Colorado this species is more abundant than all the other carnivora together, and is the only one that bears due proportion to the numbers of rodentia, on which it no doubt depended for food. Slight and unimportant variations may be observed among the numerous specimens. In the Oregon John Day it is less abundant. It is a common species of the White River bad lands of Dakota and Nebraska.

Mandibles from Colorado present the following characters:

About half the size of the red fox (*Vulpes fulvus*), or equal to the *V. littoralis*, Baird, but with relatively deeper mandibular ramus than either. The premolars are in contact with each other, and the middle posterior lobe is well developed, except in the first, which is also one-rooted. Sectorial, with stout inner tubercle as high as the anterior lobe; heel rather small. First tubercular with two roots relatively smaller than in the species last described, with two anterior and one posterior tubercle. The second tubercular is very small, and has a single subcompressed or round root. It remains in very few specimens, and in a few has evidently never existed A premaxillary with part of the maxillary bone displays parts and alveoli of two incisors, one canine, and the first premolar. There is scarcely any diastema, and the canine is compressed oval in section. The exterior incisor is quite large, exceeding by several times the inner one. The premaxillary bone has but little anterior production.

Measurements.

	M.
Length of molar series	0.036
Length of premolar series	.019
Length of fourth premolar	.006
Length of sectorial	.009
Width of sectorial	.004
Height of sectorial	.006
Depth of ramus at sectorial	.010

A nearly entire skull from the John Day region of Oregon is figured on Plate LXVIII, figs. 5, 8, of the natural size. Its form partakes of the anterior abbreviation common to the *Canidæ* of the Miocene, in which that region is shorter than in recent species of *Canis* and *Vulpes*. The nearest approach to such proportions is made by the *Vulpes cinereoargentatus*, where the length anterior to the orbits enters the length of the skull two and a half times. In the *G. gregarius* the proportions of these lengths are as one to three. The muzzle descends regularly from the parietal region, without convexity or concavity of profile. The premaxillary border is not very prominent. The interorbital region is moderately transversely convex, and has a median longitudinal groove which continues to the saggital crest. The postorbital processes are not very prominent, but appear as angles, chiefly on account of the rather abrupt constriction of the cranium behind them. This constriction is not so great as in *Temnocyon oryphaeus*, and is more anterior in position, but is greater than in *Vulpes fulvus* and *cinereoargentatus*, or *Canis latrans* or *cancrivorus*. The brain-case is quite as large relatively as that of *C. latrans*. The temporal ridges extend obliquely backwards and unite to form a sagittal crest above the coronoid process of the mandible. This crest is a low ridge until it approaches the repressed inion, when it becomes more apparent. The occiput is low and broadly rounded, and its bounding crests not very prominent, resembling a good deal in this respect the gray fox. The lateral crest is continuous and quite prominent over the *meatus auditôrius externus*. On the occiput the protuberance for the *vermis cerebelli* is well marked and short and wide; it is much more prominent than in the red and gray foxes. The paroccipital process is small and obtuse, and is directed backwards. It is separated from the otic bulla by a considerable space. The posttympanic process is a rounded tuberosity which does not descend on the bulla, but is in contact with the inferior temporal crest, which is not so distinctly angulate at that point as in most recent species of *Canis*. The postglenoid crest is well marked, and is rather more extended transversely than in recent *Canes*. There is no indication of preglenoid crest The bulla is expanded, but not so much as in the other and cotemporary extinct species of the genus. In this respect the *Galecynus gregarius* nearly resembles the

Amphicyon cuspigerus. It has no reverted lip at the meatus as in most recent dogs. The basioccipital is wide and rather flat There is a faint median ridge posterior to the middle of the bullæ, which has behind, on each side, a shallow concavity, which is much less marked than in the *Amphicyon cuspigerus*, but which is not represented in *Galecynus geismarianus*. The fossa between the occipital condyle and the paramastoid process is deep. The zygomata are broken off, but enough remains to show that they are compressed, and have a lateral masseter surface, which occupies half the width of the malar process.

Sutures.—The ascending process of the premaxillary is long and slender, but does not reach the acuminate anterior process of the frontal. The nasal bones are narrow, and their posterior suture forms an acute angle above the very anterior part of the orbit.

Foramina.—The nareal orifice is oblique and about as wide as deep. The infraorbital foramen is large, and is over the anterior root of the superior sectorial. The *meatus auditorius externus* is large, and has a narrow extension downwards and forwards. The condyloid foramen is, as in recent *Canis*, anterior to the transverse ridge extending to the paroccipital process, and is small, and quite near the *foramen lacerum posterius*. The bulla is closely appressed to the sphenoid, so that the *f. l. anterius* is of very small size. The other foramina are still concealed by matrix. The posterior or nareal border of the palate forms two shallow concavities.

Dentition.—The superior incisors have simple crowns, and the median ones are very small. The crowns of the canines are slender, acute, and with an oval section. The second and third superior premolars have a posterior cutting lobe, though that of the anterior is very small. The sectorial is small and robust. It has a basal cingulum both internally and externally. The tuberculars are moderate, the anterior narrower internally than externally, the second of equal width at both ends. Their tubercles, external cingulum, median V-shaped ridge, and internal cingulum are well developed. The three posterior inferior premolars have a well-marked posterior marginal lobe. The sectorial has no external tubercle at the posterior base of the principal cusp. No external cingula on the inferior molars.

Measurements.

	M.
Length of skull to inion	.087
Length from premaxillary border to edge of palate	.040
Length from premaxillary border to postglenoid process	.060
Length from premaxillary border to paroccipital process	.077
Long diameter of otic bulla	.0135
Elevation of occiput above foramen	.0:2
Diameter of foramen magnum { transverse	.012
Diameter of foramen magnum { vertical	.009
Width between posttympanics	.032
Width between bullæ	.008
Width between pterygoids	.0065
Width of zygomata opposite postorbital angle	.043
Width of palate at posterior angles of superior sectorials	.021
Width of palate between canines	.008
Width between orbits above	.019
Length of muzzle to orbit (axial)	.028
Length of superior molar series	.031
Length of superior true molars	.009
Diameter of superior sectorial { anteroposterior	.008
Diameter of superior sectorial { transverse in front	.0045
Diameters first tubercular { anteroposterior	.006
Diameters first tubercular { transverse	.008
Diameters second tuberculars { anteroposterior	.0035
Diameters second tuberculars { transverse	.0066
Length of mandible to condyle	.059
Depth of ramus at Pm. i	.006
Depth of ramus at sectorial	.010

Besides the skull above described, fragments of three other individuals of the *Galecynus gregarius* were obtained from Oregon by Messrs. Sternberg and Wortman. None of them include any portions of the skeleton.

GALECYNUS LIPPINCOTTIANUS Cope.

Canis lippincottianus Cope, Synopsis of Vertebrata Collected in Colorado in 1873. Miscell. Pub. U. S. Geol. Survey Terrs., 1873, p. 9. Annual Report U. S. Geol. Survey Terrs., 1873 (1874), p. 506.

Plate LXVIIa; figs. 5–6.

Among the numerous remains of dogs associated with those of rodents and insectivora in the White River beds of Colorado, I have observed several portions of mandibular rami with teeth which indicate a species intermediate in size between the *Ampicyon hartshornianus* and the *C. gregarius*. Selecting one specimen as type, which contains the teeth which correspond to those which represent the species last described, I find the following peculiarities: The root of the last molar is much compressed. There is only a trace of cingulum on the penultimate, and the tubercles of the inner side of the crown are well developed. Dimensions half as large again as in *C. gregarius*, as indicated by many specimens of the latter. In it the anterior lateral tubercles are subequal.

A second specimen from the same locality is a mandibular ramus, with the alveoli of the entire molar series and the last premolar and sectorial perfectly preserved. As compared with a larger number of specimens of *C. gregarius*, the jaw is larger, but is chiefly distinguished by the relatively stouter and broader teeth. The first premolar is one-rooted.

Measurements.

	M.
Length of bases of crowns of M. II and III (No. 1)	0.0095
Length of base of crown of M. II	.0060
Width of base of crown of M. II	.0035
Elevation of crown of M. II	.0030
Length of bases of five anterior molars (No. 2)	.0320
Length of bases of four premolars	.0220
Width of sectorial at middle	.0045
Elevation of sectorial at middle	.0070
Depth of ramus at sectorial	.0130
Thickness of ramus at sectorial	.0055

A fragment of a mandible containing the last three molars found in the John Day basin of Oregon agrees in proportions with the above specimens.

Unfortunately there is not enough material in my hands to render it clear whether the specimens represent a distinct species, or a large variety of the *C. gregarius*. For the present I retain it as distinct.

Galecynus geismarianus Cope.

Canis geismarianus Cope, Paleontological Bulletin No. 30, p. 9, Dec. 3, 1879. Proceedings American Philosophical Society, 1879 (1880), p. 71.

Plates LXX, figs. 2–3; LXX*a*.

This species was about the size of the fisher, *Mustela pennantii*, and was larger than the *G. gregarius*. It is represented in my collection by the greater part of a skeleton accompanied by a skull with lower jaw complete; by a second skull, from which the end of the muzzle and the teeth have been broken; and by a fragment of a mandible which supports a sectorial tooth. On the last specimen the species was originally founded.

The characters of the cranium which demonstrate the distinctness of the species, are (1) the low and long sagittal crest; (2) the large size of the ótic bullæ; (3) the long paroccipital process directed backwards and in contact with the bulla at its base; (4) the absence of lateral fossæ of the basioccipital bone.

The skull is rather elongate, the elongation being behind the anterior border of the orbits. In profile the front is convex, and the top of the muzzle subhorizontal. The premaxillary border is only moderately prominent, and the sides of the muzzle are convex. The interorbital region is convex transversely, and the postorbital processes are small, but rather pinched, so as to be rather acute. Above the cephalic contraction behind these, the temporal ridges rapidly converge to a low sagittal ridge, which only rises into a crest when it extends backwards to join the inion. The lateral occipital crests are quite prominent and project a good deal posteriorly, giving the occiput a narrower outline than that seen in *G. gregarius.* It is not interrupted in its course into the suprameatal crest, as is the case in the coyote. The protuberance for the vermis is distinct. The posttympanic process is inconsiderable and is truncate below, inclosing a fossa between it and the bulla. The basioccipital presents a median longitudinal angle, which disappears between the middles of the bullæ. Basisphenoid nearly plane. Postglenoid processes large, the posterior border very oblique downwards and forwards. The zygomata do not exhibit much lateral convexity, and the postorbital angle is distinct. The rami of the lower jaw have been under vertical pressure, so that their inferior edges are a little flattened. The symphysis extends as far back as the middle of the second premolar. The inferior border of the jaw begins to rise gradually opposite the posterior cusps of the first inferior tubercular molar, and continues with less obliquity into the long and rather narrow angle, which is not incurved, and which extends as far posteriorly as a vertical line from the condyle. The coronoid process is rather broad, and is considerably more elevated above the condyle than the latter is above the angle. The masseteric fossa is well defined all round below the coronoid process, better at the anterior inferior angle than the wolf, coyote, or red fox.

Sutures and foramina.—The surface of the cranium is much fractured, although it preserves its general form. Hence the sutures are not easily observed. The frontomaxillary has the usual upward arch, and incloses a long strip of frontal bone with nasomaxillary sutures. The nasal bones terminate in a gradual acumination above the anterior border of the orbit. The frontal-parietal crosses the sagittal crest two-fifths the distance to the inion from the postorbital processes. The infraorbital foramen opens above the posterior border of the third superior premolar. The postglenoid is at the bottom of a groove of the postglenoid process The condyloid foramen is near the *f. lacerum posterius*, and just anterior to the anterior border of the large precondylar fossa. The *foramen magnum* is a little wider than deep, and has its superior border regularly arched, and without notches or angles.

Dentition.—The superior incisors differ from those of *Galecynus gregarius* and some other species, in having their cutting edges interrupted by two notches, excepting in the external teeth, where there is but one notch, and that on the inner side. The notches are wanting or indistinct on the inferior incisors. The canines are slender like those of the red fox, but are more curved. The enamel of the inferior canine is longitudinally wrinkled, as in the fisher; the wrinkling is less distinct on the superior tooth. The superior premolars are separated from each other and from the sectorial by short spaces. Only the third has a posterior marginal lobe. The sectorial is relatively shorter than in the *G. gregarius;* its length is less than that of the inferior sectorial by the length of the anterior lobe of the latter. Its inner lobe is smaller than usual. Both tubercular molars have well-developed external cusps, a valley, and in place of the V-shaped crest, two rather prominent juxtaposed subacute tubercles, much as in the first tubercular of the coyote. Within these is a well-defined valley bounded by a prominent internal ledge. There is a complete external cingulum on the first tubercular; it is incomplete behind on the second tubercular. There is no external cingulum on the sectorial or the other premolars.

The inferior premolars are not so widely spaced as the superior; the third and fourth have well-defined posterior marginal lobes, and small anterior basal tubercles. The inferior sectorials only display thin external faces in the specimen. They have a good deal the form of those of the *Amphicyon cuspigerus*, and probably have wide heels. The first tuberculars are quite robust. The enamel on all the molars has a tendency to wrinkling; and this is most distinct on the inferior sectorial and tuberculars. No external cingula on inferior molars.

Measurements of skull.

	M.
Length of cranium to inion	.116
Length from premaxillary border to condyles	.110
Length from premaxillary border to posttympanic	.096
Length from premaxillary border to postglenoid	.079
Length from premaxillary border to posterior border of second tubercular tooth	.053
Length to anterior border of orbit (axial)	.037
Elevation of occiput above foramen	.023
Width of occiput at foramen	.128
Width between posttympanics (outside)	.040
Width between otic bullæ	.090
Width of zygomata opposite postorbital angle	.048
Width of palate at posterior angles of sectorials	.033
Width between canines	.015
Width between orbits above	.023
Long diameter of otic bulla	.019
Length of superior dental series, including canine	.046
Length of superior premolar series	.028
Length of superior true molars	.013
Length of base of third premolar	.0065
Length of base of sectorial	.0095
Width of base of sectorial in front	.005
Diameters of first tubercular { anteroposterior	.008
Diameters of first tubercular { transverse	.0094
Diameters of second tubercular { anteroposterior	.005
Diameters of second tubercular { transverse	.0075
Length of crown of superior canine	.012
Length of crown of inferior canine	.0095
Length of inferior premolar series	.023
Length of inferior true molar series	.022
Length of base of inferior fourth premolar	.0075
Length of base of inferior sectorial	.0115
Length of base of inferior first tubercular	.0064
Length of base of inferior second tubercular	.0035
Length of ramus mandibuli to angle	.077
Depth of mandible at coronoid	.038
Depth of mandible at sectorial (partly restored)	.012
Depth of mandible second premolar	.010

Of *vertebræ* there are preserved five cervicals, two represented by parts only; ten dorsals, four lumbars, three or four sacrals, and five caudals. The cervicals have the characters of those of typical *Canidæ* and *Viverridæ*. The centra are moderately elongate, and their articular surfaces moderately obliquely opisthocœlous. They are strongly keeled inferiorly, the keel running out into a hypapophysial tuberosity on the posterior border of the centrum, where it terminates abruptly. There are opisthapophyses, or small but distinct tuberosities on the borders of the neural arches of the third and fourth cervicals. The seventh is strongly keeled below, the carina widening into a flat triangle posteriorly, and not produced downwards. The inferior side of the centrum is sharply defined laterally by an acute border extending posteriorly from the base of the diapophysis. The anterior articular face of the centrum is a transverse oval, looking a little downwards.

The capitular rib-facets of the dorsal vertebræ are quite small, making very slight emarginations of the borders, and those of the last three vertebræ are entirely on the anterior portion of each centrum. The first dorsal is not keeled below, and the second has a slight downward projection of the anterior border. All the other centra are rounded below, and their articular extremities are halves of circular disks, except those of the last, which are more depressed. The centra increase materially in length posteriorly, and the elongate ones show traces of the keel, which is well marked on the lumbars. On the antepenultimate centrum an angle of the circumference of the posterior articular face appears just below the line of the neural canal. On successive centra it becomes more prominent, and descends to the inferior plane. It is so prominent as to be a process on each side on the first lumbar. The succeeding lumbars increase in length so as to be twice as long as the anterior dorsals; indeed, nearly three times as long as the first dorsal. They have a low inferior keel, a lateral angle extending from the diapophysis, and a low keel-like neural spine extending the length of the neural arch. The diapophyses have short and narrow bases at the anterior extremity of the centrum. They commence, as usual, as lateral ridges on the dorsals. The first ridge is visible on the penultimate dorsal, extending obliquely backwards and upwards from the superior border of the capitular facet. The anapophyses are very large on the last three dorsals and the first lumbar. They are small on the antepenultimate lumbar, and absent from the last two. The metapophyses are first developed on the penultimate dorsal. On the first lumbar they are large, and have enlarged rounded summits. They are absent from the last three lumbars. The last lumbar is short, as in most carnivora, and has a rather low neural spine, which extends along the entire length of the neural arch. There are three sacral vertebræ, which have a keel for neural spine, which is most elevated on the last. Four caudal vertebræ are in place, succeeding the sacrum. They all have well-developed diapophyses, which are directed backwards and are of depressed section. The neural arch is complete on all, and not much shortened on the fourth. A more distal caudal has no neural arch, but a median ridge on the posterior half and an elevated lateral process in front. Nearly the entire length is occupied by a thin diapophysis on each side.

As compared with other *Carnivora*, the cervical vertebræ most resemble those of *Canidæ* and *Viverridæ*, as already observed. The dorsals differ from all of the former in their greater elongation, especially posteriorly. The lumbars differ from those of dogs, and from those members of the *Mustelidæ* and *Viverridæ* which are within my reach, in their greater elongation. They resemble those of *Canidæ* in the anterior position and short bases of the diapophyses. They differ from them in the long, low neural spines, and resemble those of *Putorius erminea.* The lumbars also resemble those of that species, and of the *Mangusta apiculata*, in the size and posterior persistence of the anapophyses. In the large development of the processes and arches of the caudal vertebræ the *Galecynus geismarianus* resembles the *Viverridæ.*

The vertebræ above described were found as follows: Two cervicals and parts of two others were taken from one block; one cervical and three dorsals from another; two dorsals from another. Five dorsals and one lumbar were adherent to the side of the skull. Four lumbars were attached to the block containing the sacral and four caudal vertebræ, pelvis, humerus, and fore foot.

Measurements of vertebræ.

			M.
Diameters fourth cervical	anteroposterior		.016
	anterior	vertical	.005
		transverse	.0077
Diameters neural canal of cervical	transverse		.007
	vertical		.006
Diameters seventh cervical	longitudinal		.013
	anterior	vertical	.006
		transverse	.0077
Length centrum of second dorsal			.010
Diameters fourth dorsal from last	anteroposterior		.0117
	posterior	vertical	.0057
		transverse	.011
Diameters last dorsal	anteroposterior		.0145
	posterior	vertical	.006
		transverse	.011
Diameters neural canal last dorsal	vertical		.004
	transverse		.0065.

			M.
Elevation of neural spine of last dorsal			.020
Elevation of metapophyses first lumbar			.018
Diameter antepenultimate lumbar	anteroposterior		.024
	posterior	vertical	.0067
		transverse	.012
Length of last lumbar vertebra			.014
Length of sacrum			.025
Length of four anterior caudals			.042
Expanse of diapophyses fourth caudal			.018
Vertical diameter centrum fourth caudal			.0045
Length of centrum longer caudal			.018
Vertical diameter of longer caudal			.006

The *scapula* is in a damaged condition, the coracoid process being lost. The spine is high, most so at the acromion. The glenoid cavity is lost above the line of the spine; the remaining part has the transverse greater than the vertical diameter, and is very little concave. The tuberosities of the head of the humerus are prominent, especially the great one. The latter does not extend as far posteriorly as in *Temnocyon coryphæus*, and is obliquely truncated by the teres facet. The proximal border of the great tuberosity is a ridge, but neither tuberosity develops a ridge of the shaft. The anterior face of the latter is rounded. I find this part of the humerus a good deal more like that of *Canis* than those of *Mangusta*, *Putorius*, or *Mustela*, where the tuberosities continue immediately downwards into angular ridges. The shaft of the humerus is lost. The distal portion is preserved. It resembles that of *Mangusta* rather than that of *Canis* or the *Mustelidæ*. It has greater transverse extent both as to the condyle and the epicondyles than in *Canis* sp., but rather less than in *Temnocyon coryphæus*. The internal flange of the condyle is well developed, but the radial convexity is weak. The olecranar trochlea is narrow, and the fossa is deep, rather narrow and transverse and not double as in some *Mustelidæ*. The proximal parts of the ulna and radius are preserved. The inferior edge of the ulna is turned inwards. The bicipital tuberosity of the radius is well marked, and bounds a fossa which is between it and the head. The latter presents a wide oval to the humerus. The *anterior foot* of the right side remains attached to the block which contains the pelvis, and from which the right humerus was cut. It includes the proximal halves of the metacarpals except the first, and the magnum, unciforme, and cuneiforme of the carpus.

The latter do not differ much from those of *Canis*, with one important exception. The cuneiforme and pisiforme are coössified. Whether this is an individual peculiarity, or a character of the species or of the genus, future observations only can decide. The character does not exist in my specimens of *Mangusta apiculata, Putorius erminea, Mustela pennanti*, nor in any of the *Canes* in my collection. The second metacarpal has the proximal lateral facet for contact with the first metacarpal, and the small tuberosity below it for ligamentous connection. The shafts of the metacarpals are subequal in diameters and of subround section near their middles.

The *pelvis* is preserved in position nearly entire. Its general form is more elongate and compressed than in *Canidæ* and *Mustelidæ* generally, and more compressed than in *Mangusta*. The postacetabular portion is more like that of the *Felidæ* in its narrowness and compression, but the ischiopubic symphysis is more compressed and keeled than in the latter family. The crests of the ilium are broken off, but fragments show that they reached as far forwards as the anterior margin of the last lumbar vertebra. They are moderately wide and flat, and the external face is gently concave. There is a weak anterior inferior spinous tuberosity. The spine of the ischium is a convexity of the narrow superior edge of the ischium, which descends a little to the flatter part of the border representing the lesser sacroïschiatic notch. The portion of the ischium which descends to the symphysis is compressed and subvertical; the notch the opposite sides define, is V-shaped, and deeper than in the species of *Canidæ* or other *Carnivora* above mentioned. The obturator foramina are large and are longitudinally oval.

Measurements.

	M.
Elevation of spine of scapula	.010
Transverse diameter of glenoid cavity of scapula	.0075
Diameters of head of humerus (greatest) { anteroposterior	.020
Diameters of head of humerus (greatest) { transverse	.016
Diameter of head of humerus to bicipital notch	.015
Diameters of shaft of humerus { anteroposterior	.0105
Diameters of shaft of humerus { transverse	.0065
Diameter distal end of humerus (greatest) { anteroposterior	.012
Diameter distal end of humerus (greatest) { transverse	.018
Diameters of condyles of humerus { anteroposterior, median	.006
Diameters of condyles of humerus { transverse in front	.012
Diameter of shaft of radius at middle	.0055
Width of palm without pollex	.013

	M.
Probable length of pelvis	.172
Length of ilium (inferential)	.037
Anteroposterior width of ilium at base	.013
Anteroposterior diameter of acetabulum	.013
Depth of pelvis at ischiopubic symphysis	.018
Width of obturator foramen	.011

The fragment of lower jaw above mentioned was the specimen from which the existence of this species was originally inferred. Its characters are as follows:

The mandibular ramus is robust and shallow, and quite distinct from the deep jaw of *A. hartshornianus*. The sectorial has perhaps twice the bulk of those of the *G. lippincottianus* and *A. cuspigerus*. From that of the latter it differs further in the small inner tubercle and contracted heel.

The sectorial part of the tooth is relatively small, not exceeding the heel in length, and its cusps are low. The heel is notable for the elevation of the tubercle of the inner side, which exceeds that of the outer; the latter also is contracted, standing within the external base, which is represented by a short cingulum. A weak cingulum below the sectorial blades. Surface of the enamel rugose where not exposed to friction.

Measurements.

		M.
Diameters of sectorial	vertical, anterior cusp	.006
	vertical, heel	.0038
	anteroposterior	.0115
	transverse, middle	.006
Depth of ramus at sectorial		.012
Thickness of ramus at sectorial		.007

The third specimen, a cranium, already mentioned, displays the base of the skull more distinctly than any other. The condyloid foramina are seen to be close to the *f. l. postica*, and the *f. caroticus* is not externally distinguished from the latter. The *f. alisphenoidale posticum* is small, and is well in advance of the large *f. ovale*. The *f. a. anticum* is within the *f. rotundum*. The latter is large and distinct from the *f. sphenoörbitale*. The postglenoid is rather large. The *meatus auditorius externus* is a subvertical oval; has no slit-like enlargement, nor a recurved lip. The palatine foramina are opposite the inner border of the first tubercular molar. A shallow groove extends anteriorly from each for a considerable distance. The maxillo-palatine suture crosses the middle line of the palate opposite the anterior border of the sectorial. The frontal sends a long process anteriorly on each side of the nasals, which does not meet the ascending process of the premaxillary. It also descends along the external border of the orbit to the lachrymal. The latter presents a very narrow surface external to the orbit, and supports a narrow tuberosity on the prominent border of the latter. The malar ascends to near this tuberosity in a narrow process, which is more acuminate in form than in *Canis latrans* or *Vulpes fulvus* and *V. cinereo-argentatus*. The lachrymal bone is much larger than in any of the above species or the *C. familiaris* or *C. cancrivorus*, and is especially expanded at its superior portion. The orbitosphenoid bone extends as far anteriorly as in those species, but carries no diagonal ridge, as in the latter. The orbito-nasal foramen is much more posterior than in them. The squamosal bone is long and low.

There is a slight concavity on the superior surface at the base of the postorbital process of the frontal bone, as in the recent red fox. The palatonareal border is a regular concavity. The otic bullæ are large and somewhat compressed. They are longer than deep, and as deep as wide.

Restoration.—Although the skull and pelvis of this species have about the size of those of the fisher, the vertebræ and humerus are more slender, and the anterior foot is decidedly smaller. It is probable that the *Galecynus geismarianus* resembled a large *Herpestes* in general proportions rather than a *Canis*. It stood lower on the legs than a fox, and had as slender a body as the most "vermiform" of the weasels, the elongation being most marked in the region posterior to the thorax. The tail was evidently as long as in the Ichneumons. Its carnivorous propensities were as well developed as in any of the species mentioned, although, like all other *Canidæ* of the lower Miocene period, the carnassial teeth are relatively smaller than in the recent types.

History.—The specimen above described, which includes the greatest number of bones, was obtained by my assistant, Mr. Jacob L. Wortman, from the Haystack Valley, Central Oregon, from beds of the John Day Miocene age. The skull last described was found by C. H. Sternberg in the "cove" of the John Day Valley of Oregon. The species is named in honor of Jacob Geismar, an accomplished and skillful naturalist of Philadelphia.

Galecynus latidens Cope.

Bulletin U. S. Geol. Survey Terrs., VI, p. 181, Feb. 11, 1881.

Plate LXX; figs. 4, 5.

The specimens which represent this species are the following: (1) A skull distorted by pressure and lacking the portion in front of the sectorial teeth, accompanied by a mandibular ramus anterior to the base of the coronoid process; (2) a mandibular ramus similar to that of No. 1; Nos. 3, 4, and 5, fragments of mandibles supporting sectorial teeth.

The specimens show that the *Galecynus latidens* was intermediate in size between the *G. gregarius* and the *G. geismarianus*, and differed from both in various respects. The crushed condition of the cranium renders the description of the form of the brain-case difficult. The lateral occipital crests are more produced posteriorly than in the allied species. The paroccipital process is much as in *G. geismarianus*, but is more closely appressed at the base to the otic bulla than in it. The apical portion is free from the bulla, and is directed principally downwards. The bullæ are quite large, and are oblique ovals, as wide as deep. The masseteric face of the proximal part of the malar bone is lateral, and extends above the middle of the same. The lachrymal bone is wider above than in most *Canidæ*, but is not so wide as in *G. geismarianus*.

The sectorial molar is short, and its internal tubercle is well developed. The tuberculars are distinguished for their anteroposterior width, both being as wide near their inner as at their external borders. Their external cingula are well developed; their external crests are rather low. They have a median V-shaped ridge, and a wide internal cingulum. There is a short posterior cingulum on the second tubercular.

The mandibular ramus is compressed and rather deep. The teeth are not spaced, and the fourth premolar is obliquely placed, so as to overlap externally the anterior lobe of the sectorial. Both the third and fourth premolars have a well-developed posterior marginal lobe; the second premolar is lost from all the specimens. The third premolar is moderately compressed, but the fourth is more robust in its proportions than in any other species of the genus. Both premolars have an anterior and a posterior heel, the former minute. The sectorial is distinguished by the relative length and width of its heel, in which it exceeds any other of our species; and by the possession of a narrow tubercle at the *external* base of the principal cusp. This is a little higher than the heel, but not so high as the internal tubercle. This character is seen in the *Vulpes cinereoargentatus* among recent dogs. It is present in the four well-preserved sectorials of my collection. The first tubercular is relatively large. It has four cusps, all marginal, two anterior and two posterior, and an anterior basal ledge; no other basal cingula. The second tubercular is missing from all the specimens. There are two mental foramina, a larger below the posterior part of the first premolar, and a smaller below the third premolar. One of the mandibular rami of this species I described along with the skull of *Galecynus lemur*, under the impression that it belonged to an individual of that species. This error is now corrected.

Measurements.

		M.
Length of skull, exclusive of muzzle, anterior to Pm. iii		.070
Length of skull from anterior border of orbit posteriorly		.060
Length of skull from postglenoid process		.032
Length of otic bulla		.020
Width of otic bulla		.010
Diameters of sectorial	anteroposterior	.007
	transverse in front	.0055
Diameters of first tubercular	anteroposterior	.007
	transverse	.008
Diameters of second tubercular	anteroposterior	.0045
	transverse	.005
Length of inferior dental series		.048
Length of sectorial		.008
Length of heel of sectorial		.0035
Length of inferior tubercular		.055
Depth of ramus at sectorial		.0105

The typical specimen above described was obtained by Mr. J. L. Wortman in the cove of the John Day Valley, Oregon, in the John Day Miocene formation. One of the mandibles was found by Mr. C. H. Sternberg.

Galecynus lemur Cope.

Bulletin U. S. Geol. Survey Terrs., VI, p. 181. *Canis lemur* Cope. Paleontological Bulletin No. 31, p. 2, December, 1879. Proceedings American Philosophical Society, 1879 (1880), p. 371, exclusive lower jaw.

Plate LXX; figs. 6–8.

This animal is represented in my collection by the following specimens: (1) A cranium with premaxillary region preserved, without lower jaw and with the parietal region injured; (2) a cranium without muzzle or lower jaw; (3) a cranium lacking all posterior to the frontoparietal suture; (4) a cranium mutilated like the last, and without muzzle or lower jaw; (5) a left mandibular ramus, broken off behind first tubercular; (6, 7, and 8) fragments of three mandibular rami.

This species is the smallest of the genus yet discovered in the Miocene formation of Oregon. It is characterized by the contracted proportions of the muzzle, the width of the front, and the large size of the orbits. The postorbital process is only a short angle. The superior border of the temporal fossa is traceable from the postorbital process. Those of opposite sides embrace a smooth sagittal area of an elongate urceolate form, and unite posteriorly in a very short crest. The species is further characterized by the large size of the first superior tubercular molar and the swollen otic bullæ, which exceed relatively those of any other species of the genus.

While the muzzle is shorter in the species of *Galecynus* described in the present work than in existing species of *Canidæ*, the *G. lemur* still further diminishes the size of this region by lateral contraction. The profile descends gradually and regularly, and the premaxillary bone is quite oblique and prominent below. The interorbital region is wide and is quite flat. The orbit is large, and the postorbital angle of the zygoma well posterior. The occiput is much contracted; its lateral angles are well defined but not very prominent, and become obsolete on the mastoid regions. The tuberosity for the vermis is distinct. The precondylar fossa is narrow, and is, as in other species, bounded by a transverse ridge, which extends from the paroccipital process to the basioccipital median line. The paroccipital process does not descend below the level of the occipital condyles; it is appressed to the otic bulla to near its extremity, which is free and directed downwards. Its external surface, as far as the occipitosquamosal ridge and posttympanic process, is flat. The latter process is a mere tuberosity, and does not descend on the bulla. The postglenoid is narrow and small, as in recent *Canidæ*, and its external border extends obliquely downwards and inwards. The bullæ are very large, and of a moderately compressed oval form. They do not present a prominent lip at the meatus. The zygomata are more widely expanded than in any other species of Lower Miocene

Canidæ of which this part is preserved, excepting Leidy's *Amphicyon vetus* (Pl. I, fig. 1, Extinct Mammalia Dakota and Nebraska). The nareal trough is, by comparison, narrow. The posterior palatal or nareal border is regularly concave. The palate is gently concave.

Sutures.—The premaxillaries do not reach the frontal bones. The lachrymal is wide at its inferior part and narrows upwards as in the foxes. The orbitosphenoid extends to the vertical line from the postorbital process of the frontal. The contact of the alisphenoid and parietal is extensive.

Foramina.—The infraorbital foramen opens above the posterior root of the third premolar. The *f. magnum* is wider than deep, and has its superior margin regularly arched. The *f. condyloideum* is near to the *f. lacerum posterius*, but is distinct. There is a foramen in the position of the *f. carotideum* of the *Viverridæ* on one side; it is wanting on the other, and does not occur in the second skull. I suppose, therefore, that it is abnormal. It is interesting as possibly indicating a tendency to variation in the direction of the civets. The *meatus auditorius externus* is not enlarged downwards, and is an oval directed downwards and forwards. The *f. postglenoideum* is rather large, and is cut off from the meatus by the anterior rim of the latter. The *f. ovale* is large, and is very near the *f. lacerum anterius* and well posterior to the alisphenoid canal. The *f. f. rotundum* and *sphenoorbitale* are large, and are only separated by a thin partition. The *f. orbitonasale* has the posterior position seen in other species of the genus. The *f. f. palatina* are opposite the anterior part of the internal borders of the first tubercular teeth.

Dentition.—Three premolars, two true molars, and an incisor remain in this specimen. The crown of the incisor is worn; the alveolus of the canine indicates usual size for the tooth. Crown of first premolar simple. Third premolar compressed, with a well-developed posterior cutting lobe. Sectorial robust, rather wide in front, but with the inner lobe small. The tuberculars are large, and wide anteroposteriorly, but not so wide as in *G. latidens.* The external cingulum is entire on the first, and disappears posteriorly on the second. The cusps are not elevated, and there are a median V-shaped ridge and internal cingulum, on both teeth.

Measurements (*No.* 1).

		M.
Length of skull to inion		.080
Length of muzzle to orbit		.027
Length to postglenoid process		.056
Length of skull to paroccipital process		.071
Width at middle of zygomata		.040
Width at posttympanics		.030
Width between canine teeth		.007
Width between sectorials behind		.020
Width between otic bullæ		.007
Width of foramen magnum		.009
Width of occiput at foramen		.019
Width between orbits		.0115
Elevation of occiput from foramen		.014
Length of dental series from I. 1		.038
Length of molar series		.027
Length of true molars		.009
Length of sectorial		.007
Diameters of first tubercular	anteroposterior	.005
	transverse	.0078
Diameters second tubercular	anteroposterior	.0038
	transverse	.0052

The second skull already mentioned has its superior walls entire posterior to the interorbital region. The temporal ridges are, as in various species of foxes, first convergent, then divergent, and then convergent again to a very short sagittal crest close to the inion. The measurements of this specimen are as follows:

Length of cranium to front border of orbit, M. .0525; elevation of occiput, .058; length of superior sectorial, .007; length of first tubercular, .0058; width of first tubercular, .0078; width of second tubercular, .005; length of second tubercular, .0035.

The most perfect mandibular ramus was found separately (No. 5). It agrees in all necessary respects with the crania. It is, like the muzzle of the latter, attenuated anteriorly, and the molars have corresponding proportions. The fourth premolar is compressed, and has an. acute posterior cutting lobe, besides acute anterior and posterior basal tubercles. The inner tubercle of the sectorial is very prominent; there is no trace of tubercle at the external base of the principal one as is seen in *G. latidens*. The heel is quite concave; like the first tubercular, it is narrower than the corresponding tooth of *G. latidens*. The latter tooth has a wide anterior ledge, two elevated tubercles, and a heel with raised semicircular border.

Measurements of lower jaw.

	M.
Length to end of first tubercular	.033
Length of premolar series	.017
Length of sectorial	.008
Width of sectorial posteriorly	.0036
Length of first tubercular	.0047
Width of first tubercular	.0030
Depth of ramus at sectorial	.0080
Depth of ramus at first premolar	.0050

The enamel of the inferior molars is strongly wrinkled. The mandibular ramus I originally ascribed to this species belongs to the *G. latidens*, as proven by another jaw which was found with the skull of that species.

The specimens Nos. 1, 2, and 5, described as typical of this species, was discovered in the John Day River Beds of Oregon, by J. L. Wortman; specimen No. 3 was found by Mr. Sternberg, at the same locality.

ENHYDROCYON Cope.

Bulletin U. S. Geological Survey, V, 1879, p. 56; American Naturalist, 1879, 131, 1883, p. 245.

Dental formula: I. ?; C. $\frac{1}{1}$; Pm. $\frac{3}{3}$; M. $\frac{2}{?}$. The superior premolars consist of two ordinary and one sectorial; the first and second are both compressed, two-rooted, and in the typical species with median lobe of posterior cutting edge. The two true molars are transverse and tubercular. The three inferior premolars are all two-rooted, and with posterior lobe, in the known species. The heel of the sectorial is cutting, as in *Temnocyon*, and the internal tubercle is present. There is at least one inferior tubercular tooth; specimens are injured so as not to display a second.

The dentition of this genus refers it to the *Canidæ*, but the form of the skull resembles that of *Putorius* and *Lutra*. Unfortunately no other part of the skeleton of this genus is known.

ENHYDROCYON STENOCEPHALUS Cope.

Bulletin *loc. sup. cit.*, 1879 (Feb. 28), p. 56.

Plate LXIX; figs. 3-5.

In a nearly complete cranium belonging to this species we observe the shortness of the facial part of the skull as compared with the length of the cerebral, and also the constriction of the skull behind the orbits. The zygomatic arches are robust and expanded, and the sagittal crest is high. The auditory bullæ are inflated and thin-walled. The orbits look somewhat forwards and very little upwards. The superciliary region is slightly prominent, and there is a prelachrymal concavity. The infraorbital foramen is moderate, and is situated mostly above the posterior part of the fourth premolar. The muzzle is flat above, and the nasal bones are wide, and are not emarginate above the osseous nares, as in many recent *Carnivora*. Posteriorly, the superior border of the brain-case descends, but the parietal bones maintain a gently convex outline in their high sagittal crest. The supraoccipital region is elevated, and projects posteriorly. The lateral occipital crests are prominent to the base of the paroccipital process. The latter projects strongly posteriorly, and is connected with the bulla by an osseous mass which is continuous with the inner border of the process. It is connected with the posttympanic by a longitudinal crest, which includes a deep groove between it and the otic bulla. The posttympanic is a tuberosity which extends downwards below the middle of the *meatus auditorius*, and is not in contact with the bulla. The postglenoid process has greater transverse extent than in the recent species of *Canis*. There is no trace of preglenoid crest.

The basioccipital bone is elevated on the middle line, so as to include a deep fossa on each side between it and the otic bullæ. There is no transverse ridge connecting this ridge with the paroccipital process, as in *Canis*, *Galecynus*, and *Temnocyon*, but the fossa above mentioned is continuous with the precondylar fossa. The basisphenoid is not keeled, and its lateral borders rise on the otic bullæ in a thin lamina on each side. The pterygoid fossa is narrow for the size of the skull, but is appropriate to the attenuation of the postorbital region. The posterior nareal border does not advance in front of the posterior line of the last molar tooth, and is a double concavity, the dividing angle being little prominent. The maxillary is not notched on each side of the palatine on its posterior border, as in the *Amphicyon hartshornianus*. The palate is medially flat, but its sides descend to the alveolar borders, rendering the whole convex upwards. The face is concave behind the canine alveoli. The orbits are small, and the postorbital processes are obtuse angles. The postorbital process of the zygoma is not prominent. An obtuse bridge extends upwards to the postfrontal process along the superior border of the orbitosphenoid bone, as in recent dogs.

Sutures and foramina.—As the skull is that of an old animal the sutures are obliterated. The anterior nares are small and but little oblique, though not so vertical as those of the *Lutra valetoni*, *fide* Filhol. They are about as wide as deep. The infraorbital foramen is rather small, and is widely oval in form. The palatine foramina are opposite the posterior part of the sectorial, more anterior than in *Galecynus lemur.* There are, however, two other foramina of smaller size, one opposite the anterior part of the first tubercular, the other behind the posterior margin of the second tubercular, near the posterior nostril. Lachrymal foramen rather large; orbito-nasal medium. Optic moderate; *f. f. sphenoörbitale* and *rotundum* separated by a lamina only, the latter receiving the *f. alisphenoidale anticum.* *F. a. posticum* small and in a fossa common to the rather small *f. ovale. F. lacerum anterius* small, divided into two, and bounded inwardly by an obtuse ridge of the sphenoid directed forwards from the anterior region of the bulla. No distinct carotid foramen. *F. lacerum posticum* small and linear. *F. condyloideum* rather large, situated between the anterior and posterior portions of the large precondylar fossa. *Foramen magnum* rather small, wider than deep. Mastoid very small; postparietal large; no postsquamosal. *Meatus auditorius* nearly round, without reverted lip separating the post-glenoid from it.

Dentition.—The teeth of this skull are worn with age and use. The sectorial is short and is widened anteriorly. The first tubercular is large, and has considerable transverse extent; it is a little wider externally than internally, and has much the form of the corresponding tooth in *Canis.* The second tubercular is transverse and small, not being much more than half the length of the first, and is situated in contact with it.

Measurements.

	M.
Total length of cranium	0.170
Width across zygomatic arches	0.114
Least width behind orbits	0.024
Depth of cranium with crest at otic bulla	0.070
Vertical diameter of orbit	0.025
Length from orbit to end of muzzle (axial)	0.040

	M.
Interorbital width	0.043
Width of muzzle above second premolar	0.018
Length of superior molar series	0.051
Length of fourth premolar	0.012
Length of sectorial	0.016
Length of first tubercular	0.008
Width of first tubercular	0.015
Width of second tubercular	0.0085

The dentition is better displayed by a second specimen. This consists of the middle portion of a mandibular ramus and anterior part of the maxillary of a young animal. The teeth are not fully protruded and their crowns are beautifully perfect on exposure. The principal cusps of the inferior premolars present cutting edges, as does the median posterior lobe. In both third and fourth there is a small conic heel posteriorly, but an anterior basal tubercle on the fourth only. The sectorial is large and robust, and the heel is short, with an absolutely median cutting edge. The first tubercular is longer than wide, and presents a nearly median cusp in front, which is joined to a low one on the internal border of the crown.

The superior canine has an obtuse cutting edge on the anterior and posterior borders of the inner side. The first (third) superior premolar is near to it, and is rather large, displaying a median cutting lobe and low posterior heel. The fourth is similar but larger.

Measurements of No. 2.

		M.
Anteroposterior diameter of second superior premolar		0.010
Anteroposterior diameter of third inferior premolar		0.013
Width of base of third inferior premolar		0.0065
Elevation of crown of third inferior premolar		0.010
Diameter of inferior sectorial	transverse	0.010
	anteroposterior	0.021
Width of first tubercular		0.006

Remarks.—The *Enhydrocyon stenocephalus* is an aberrant member of the family *Canidæ*. Besides the generic characters already pointed out, it presents numerous specific peculiarities. Especially to be noticed is the arrangement of the parts between the postglenoid process and the occipital condyle. These do not present striking affinities to any form outside the *Canidæ*, so far as I can ascertain.

Restoration, etc.—Without limb-bones the general form of this species cannot be inferred. It was as large as the coyote in all probability, and its facial physiognomy must have been that of a large mink or otter. Its canine teeth are formidable from their size, and the high sagittal crest and wide zygomata indicate great power in the action of the lower jaw. If anything may be derived from similarity of cranial form to the otter, its habits were aquatic. Its large otic bullæ indicate a well developed and sensitive sense of hearing. In the development of these parts it is only exceeded among the *Canidæ* here described by the *Galecynus lemur.*

OLIGOBUNIS Cope.

American Naturalist, 1881, June (May 19), p. 497; 1883, p. 246.

The dental formula is, I. $\frac{3}{3}$; C. $\frac{1}{1}$; Pm. $\frac{4}{4}$; M. $\frac{1}{2}$. The single superior tubercular molar is similar in general to that of other *Canidæ*. The inferior sectorial has an internal cusp and posterior heel, the latter with a basin and low cutting edge on one side. Inferior tubercular well developed.

The only known species of this genus I formerly referred to *Icticyon* of Lund, which it nearly approaches. It seems, however, that in that genus there are two tubercular superior molars, and the heel of the inferior sectorial is trenchant as in *Temnocyon.*

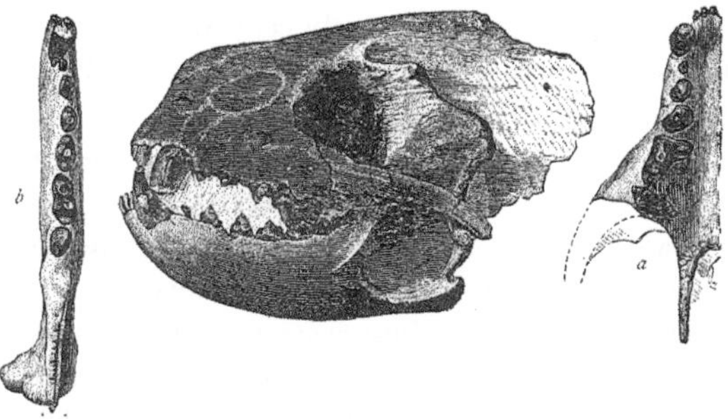

FIG. 34.—*Oligobunis crassivultus* Cope, one-half natural size. *a*, right maxillary bone with teeth from below; *b*, right mandibular ramus from above. From the John Day river of Oregon.

OLIGOBUNIS CRASSIVULTUS Cope.

American Naturalist, 1881, p. 497, and 1882, p. 246, fig. 14. *Icticyon crassivultus* Cope. Proceedings Academy Philadelphia, 1879, p. 190.

Plate LXIX; fig. 1, 2.

The specimen representing this species is a skull with both mandibular rami in place, and without the parietal and occipital regions. These parts belonged to an animal of about the size of the American badger (*Taxidea americana*).

The snout is short and robust, and the profile from the parietal region is straight and descending. The premaxillary border projects but little beyond the line of the extremity of the nasal bones. The muzzle is slightly contracted in front of the orbit and above the fundus of the canine alveoli. The latter cause a swelling on the side. The interorbital region is somewhat cracked, but appears to have been nearly flat medially; laterally it descends steeply to the supraorbital border. The orbit is not large, and the zygomatic fossa is short. The nasal bones are narrowed posteriorly, a little contracted medially, and expanded anteriorly, their lateral portions being produced along the premaxillaries. Their combined nareal border is concave, and is without the notches of some forms. The foramen infraorbitale exterius is of medium size, and issues above the interval between the sectorial tooth and the one in advance of it. The mandibular ramus is quite robust, and its inferior border is gently convex. The masseteric fossa is bounded by elevated borders especially inferiorly, and the angular hook is prominent and robust. The condyle is situated on the horizontal line of the tubercular molar, or a little above the others, and has a wide transverse extent, chiefly inwards. The coronoid process is high and wide, and is turned backwards so as to vertically overhang the condyle. Its anterior border is wide below, and becomes horizontal above.

The teeth partake of the robust character of the skull, with the exception of the incisors. Of these, the crowns of the external are long and narrow, and the median small in the premaxillaries, while those of the lower jaw are all small, and in a regular series. The canines in both jaws are quite robust, and those of the lower jaw are rather abruptly recurved. The first premolar is small, and has a simple crown and single root. The crowns of the other premolars are wide at the base and form each a simple cone, with a short posterior basal heel. The sectorial is relatively not long, but is robust, and with thick blades. The internal heel is well developed as in *Canis*, while a cingulum represents an anterior lobe. The tubercular molar is narrower in fore and aft diameter than in *Temnocyon coryphæus*, or *Canis latrans*, although it presents the same details. These are a wide obtuse external cingulum, two external tubercles, a median, an obtuse internal tubercle, and a wide internal cingulum. The premolars of the lower jaw are similar to those of the maxillary bone. The inferior sectorial is quite robust, and the internal cusp is well developed. The heel is shorter than the blades of the crown, and is wide and without tubercles in its somewhat worn condition. Its external border rises to an edge. The tubercular is wider than the corresponding tooth in the cotemporary species of *Canidæ*, although not so wide as long. Its crown rises in two low tubercles which stand transversely near the middle.

Measurements.

	M.
Length of skull to orbit (axial)	.049
Depth of skull to orbit (axial)	.042
Interorbital width	.040
Width of nares	.017
Length of superior molar series	.038
Length of bases of three premolars	.019
Length of base of sectorial	.013
Width of sectorial in front	.009
Width of first tubercular anteroposteriorly	.006
Width of first tubercular transversely	.014
Length of mandible to angle	.093
Elevation at coronoid	.051
Elevation at sectorial	.020
Length of inferior molar series	.045
Length of inferior sectorial	.014
Length of heel of inferior sectorial	.003
Length of inferior tubercular	.006
Width of inferior tubercular	.005

Van der Hoeven has given[1] descriptions and figures of the skull and dentition of the *Icticyon venaticus* of Lund, of Brazil. From these it appears that the present species differs from the latter in the greater development of the inner part of the tubercular molar of the superior series; in *I. venaticus* this part is much reduced. The tubercular molar of the lower jaw is also

[1] Overhet Gestlackt Icticyon; wis. en natuurk. vert. der Koninkl. Akademie Amsterdam, Deel III.

much smaller in the living species, the angular and coronoid processes less developed, and the condyle less extended transversely. The cranium of the *I. crassivultus* is much more robust, but not much longer than that of *I. venaticus.*

Discovered by Mr. J. L. Wortman in the John Day beds of the John Day River region of Oregon.

HYAENOCYON Cope.

Palæontological Bulletin No. 31, 1879, p. 3 (Dec. 24) (*definitione falsâ*); Proceedings American Philosophical Society, 1879, p. 372 (*definitione falsâ.* American Naturalist June (May 19), 1881, p. 497 (*definitione emendatâ*).

This genus rests on the characters furnished by two species, which are represented by but few remains. Its family position is doubtful, and my reference of it to the *Canidæ* is only provisional. It may, so far as the evidence goes, be a member of the *Mustelidæ* or even of the *Felidæ.*

Dental formula: I. $\frac{?}{3}$; C. $\frac{1}{1}$; Pm. $\frac{3}{3}$; M. $\frac{1}{II}$. Last superior molar robust, transverse, like that of the *Canidæ* generally. Inferior premolars all two-rooted, and with well-developed posterior cutting lobe. Inferior sectorial large, with heel. Probably no inferior tubercular tooth.

The characters above given agree with those of *Oligobunis* in the superior true molars, but differ in the absence of the Pm. I. and the M. II. in both jaws.

The typical species is the *Hyænocyon sectorius* (*Enhydrocyon basilatus partim* Cope olim.), from the John Day beds of Oregon.

HYAENOCYON BASILATUS Cope.

Bulletin U. S. Geol. Surv. Terrs., vi. 181, Feb., 1881. *Enhydrocyon basilatus* Cope, Bulletin U. S. Geol. Survey Territories, v. 1879, p. 57. American Naturalist, 1882, p. 246, fig. 13e.

Plate LXXV; fig. 3.

This rare species is certainly represented in my collection by parts of only one individual which is known from the greater part of both rami of a mandible, from which only the sectorial, one canine, and some incisors of one side have been lost.

These portions indicate an animal of the same general character as the *Enhydrocyon stenocephalus*, but of larger and more robust proportions, and characterized by many dental peculiarities. These will be at once pointed out. The canine is directed upwards and a little outwards, and possesses two obtuse ridges bounding the interior face. The third incisor is compressed and truncate superiorly and distally. The first (second) premolar is two-rooted, compressed, and trilobate. It consists of a principal cutting edge little elevated, and a small accessory lobe at each extremity of the crown; its base is expanded posteriorly. The principal cusp of the third premolar is more elevated, and, besides the anterior and posterior tubercles, there is a basal posterior heel, which is continued as an expansion of the inner base of the crown. In the fourth premolar, the base of the crown is expanded, especially posteriorly; the principal cusp has a nearly circular section at the base, and the posterior median lobe is a subconic tubercle standing on the middle of the heel. The sectorial is large and relatively rather narrow, but the details of its form are not ascertainable.

Measurements.

	M.
Length of dental series, including canine and sectorial	0.076
Length of the base of the sectorial	0.024
Length of the premolar series	0.037
Length of the fourth premolar	0.016
Width of the fourth premolar	0.009
Length of the third premolar	0.013
Width of the third premolar	0.008
Length of the second premolar	0.009
Width between centres of crowns of fourth premolars	0.034
Length of symphysis	0.035

This species was probably of the dimensions of the Gray Wolf. Found by Mr. Sternberg in the same region as the *E. stenocephalus.*

HYÆNOCYON SECTORIUS Cope.

American Naturalist, 1882, p. 246, fig. 13 d. *Hyænocyon basilatus* pars Cope *locis alteris citatis.*

Plate LXX; fig. 1.

This species is represented by a right maxillary bone in which the last three molars remain, with the alveolus of the first molar and the canine. I formerly supposed this piece to belong to the *H. basilatus*, and it furnished the characters of the genus *Hyænocyon* in the peculiar dental formula it presents. It belonged to too small an animal to be referable to that species.

The alveolus of the canine tooth shows that the latter is of large size, and that the root and base of the crown have a round section. The first premolar (second) follows without interspace, and all the teeth succeed each other without interruption. The second premolar is two-rooted, but the roots are close together, and the anterior is the smaller. The long axis of the tooth is oblique to that of the jaw, and the posterior root is within as well as behind the anterior. The axis of the third premolar is very little oblique. It is a robust tooth, and the crown is a little wider anteriorly than posteriorly. The apex is subconical, and there is a posterior intermediate tubercle whose section is wider than long. The heel is narrowed and not recurved. There is no anterior tubercle or cingulum.

The sectorial tooth is relatively of large size, its proportions being those of the *Felidæ* and *Hyænidæ*. Its crown is broken in front, and where not broken, is worn by use. It displays the peculiarity of a remarkably short blade posterior to the lateral external notch; and the base of the middle cusp of the crown is wide for a sectorial tooth, and was apparently subconical when complete. It is also situated remarkably posteriorly, showing either that the anterior sloping edge is long as in *Archælurus debilis*, or that there is an anterior basal lobe, as in the species of *Ælurodon* (figs. 35, 36). The internal lobe is altogether anterior, and, though worn off in the specimen, was probably small. The long axis of this tooth is that of the jaw.

The tubercular or first and only true molar is as large a tooth as in *Canidæ* generally, but its external border is more oblique to the line of the other alveoli. The posterior external cusp is well developed, though smaller than the anterior; and the anterior external angle of the crown is so prominent as to become worn by use like the true cusps. The latter are placed well within the external base of the crown, which is prominent, but rounded, not forming a distinct cingulum. The internal half of the crown is broken away. The internal root comes so near the posterior edge of the maxillary bone as to prove the non-existence of a second tubercular molar. The enamel is smooth in all these teeth, and there are no cingula.

Measurements.

		M.
Length of dental series from posterior edge of canine alveolus		.058
Length of base of Pm. ii		.008
Length of base of Pm. iii		.013
Width of base of Pm. iii		.008
Diameters of sectorial	anteroposterior	.024
	transverse in front	.013
External width of m. i (oblique)		.013
Transverse width m. i, including internal root		.021

This species, as already remarked, has some resemblance to the *Felidæ* in its very short face, and large sectorial tooth. It clearly does not closely approach any of the species of the *Nimravidæ* of contemporary age. A principal peculiarity is the large size of the tubercular molar. Such also is the robust form of the second and third premolars. These resemble somewhat those of *Enhydrocyon stenocephalus*, but are not so much compressed. They resemble those of the species of *Ælurodon* more than those of any other carnivore. Finally, the cylindric base of the canine tooth is that of the *Canidæ* rather than the *Nimravidæ*.

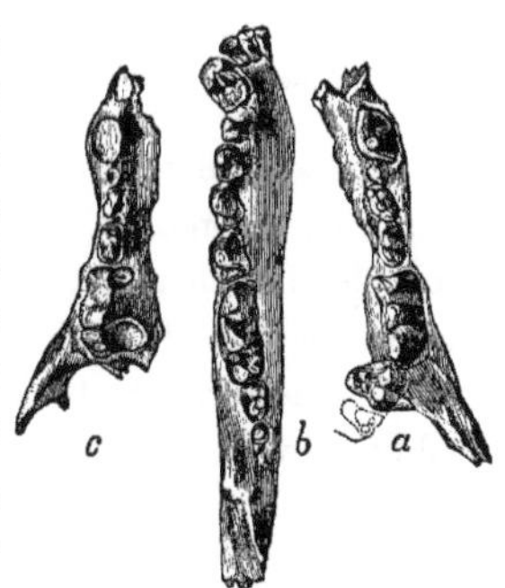

FIG. 35.—Jaws of *Ælurodons*, three-eighths natural size. Figs. *a*, *b*—*Æ. wheelerianus* Cope, from Nebraska. Fig. *c*—*Æ. hyænoides* Cope, from Nebraska. (The second true molar lost.) Both from Loup Fork beds.

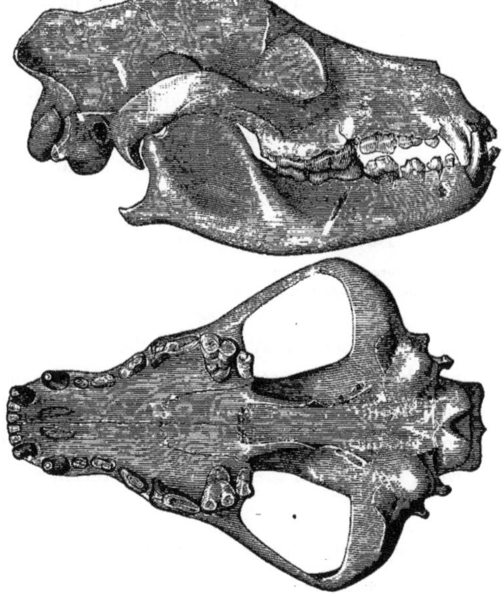

Fig. 36. *Ælurodon sævus* Leidy, from the Loup Fork beds of Nebraska, three-eighths natural size. Copied from Plate CXVIII of this work.

BUNÆLURUS Cope.

Synopsis of the New Vertebrata of Colorado, 1873, p. 8. Annual Report United States Geological Survey Territories, F. V. Hayden, 1873 (1874), p. 508.

The specimens from which this genus is known exhibit a part of the inferior dentition only, so that the number of premolars is unknown. It is probable that there are four premolars however, and there are two true molars, the first of which is sectorial.

The sectorial tooth is like that of *Putorius*. It has no internal cusp, and has a well developed blade. It has also a well developed heel, which has a trenchant median keel. Tubercular molar one, with a median cutting edge.

This genus probably belongs to the *Mustelidæ*, and is allied to *Putorius*, and perhaps to *Gulo*. It is not probable that it has the reduced number of premolars of the former genus, but is more apt to agree with the latter genus, and with *Plesiogale* Pom. in its formula. I have even referred to it under the name of *Gulo*,[1] from which *Plesiogale*[2] has never been really distinguished. The form of the tubercular tooth is, however, very peculiar, and is different from what is seen in any of the species of the genera last named. Structurally the term tubercular does not apply to it, as it is a cutting tooth, without cusps or tubercles. An approach to this form is found in *Putorius*, but in *Galictis* the tubercular molar is of the ordinary type. I know but one species of *Bunælurus*.

BUNÆLURUS LAGOPHAGUS Cope.

Synopsis New Vertebrata Colorado, 1873, p. 8. Annual Report U. S. Geol. Survey Terrs., 1873 (1874), p. 508. *Plesiogale* Cope, on plate of the present volume.

Plate LXVII*a*; figs. 13, 14.

This species is represented by parts of the lower jaws of two individuals, one adult, the other immature. The latter is the more instructive, as it presents the third and fourth premolars emerging from the jaw, and the first and second true molars fully protruded. The crown of the fourth premolar is between the roots of the deciduous sectorial, which still remains attached to the jaw.

---

[1] Bulletin U. S. Geol. Surv. Terrs., VI, p. 45. The best description of the dentition of this genus is given by Coues.—The Fur Bearing Animals of America, pp. 39, 40, 1877.

[2] See Filhol Mammiferes Fossiles de Saint Gerand le Puy, 1879, pp. 177-190, Plate 25, who refers the species to *Mustela*.

The crowns of the premolars are simple, and narrow anteroposteriorly, and without posterior lobe. They expand at the base both fore and aft, but do not present basal tubercles. In the sectorial, the median cusp is much more elevated than the anterior. Its posterior face is flattened-convex, while its anterior cutting edge is acute. The cutting edge of the anterior cutting ridge is horizontal, and is separated from that of the larger cusp, by a deep closed notch. No trace of internal cusp. The heel is shorter than the anterior lobe, and is about as wide as long. The inner side of the median keel is concave, and is bounded below by a cingulum, while the external side is convex, and has no cingulum. The second true molar has two roots, and is longer than wide. It narrows posteriorly. Its median cutting edge curves a little inwards anteriorly, and the external base forms a small angle of the opposite side. No cingula.

The deciduous sectorial differs from the permanent, in having a concave heel (perhaps partly worn), and a slight angular trace of the internal cusp.

Measurements.

		M.
Length of four last molars on bases		.0140
Length of permanent sectorial fore and aft		.0050
Length of heel of do		.0010
Elevation of anterior cusp do		.0030
Elevation of median cusp do		.0045
Diameters tubercular molar	anteroposterior	.0022
	transverse	.0013
Diameters sectorial of No. 2	anteroposterior	.0058
	vertical of median cusp	.0040
Depth of ramus mandibuli, No. 2, at sectorial		.0070

From the White River beds of Northeastern Colorado, found by myself with bones of *Peratherium*, *Eumys*, *Palæolagus*, etc. It is quite possible that the animal from the same locality which I named *Canis osorum* (see Plate LXVII a), is the same, or a second species of the genus. The typical specimen is not in a good state of preservation.

NIMRAVIDÆ.

The dental characters of the *Nimravidæ* are in general those of the *Felidæ*, the higher genera having the same dental formula. Descending the scale, the number of molar teeth increases at both ends of the series in the lower jaw, and anteriorly only in the upper, but the number of the true molars never exceeds $\frac{1}{2}$. The following table gives the definitions of the genera. I unfortunately do not know the foramina in *Pseudælurus*. I introduce *Proælurus* for comparison, although as I have pointed out,[1] it is a member of the *Cryptoproctidæ*. I do this the more readily as it is quite possible that when the number of the digits of some of the species now included under the *Nimravidæ* comes to be known that they may prove to have them five on both feet, and hence to be members of the *Cryptoproctidæ*.

I. Lateral and anterior faces of mandible continuous; no inferior flange.
   a. Inferior sectorial with a heel; canines smooth.
      Molars $\frac{4\ 1}{4\ 2}$; inferior sectorial with interior tubercle .... *Proælurus*.
      Molars $\frac{3\ 1}{3\ 1}$; inferior sectorial without interior tubercle .... *Pseudælurus*.
II. Lateral and anterior faces of mandible separated by a vertical angle; no inferior flange; incisors obspatulate.
   a. No anterior basal lobe of superior sectorial; inferior sectorial with a heel (and no internal tubercle); incisors truncate.
      Molars $\frac{4\ 1}{3\ 2}$; canine smooth .... *Archælurus*.
      Molars $\frac{3\ 1}{3\ 2}$; canine denticulate .... *Ælurogale*.
      Molars $\frac{3\ 1}{2\ 2}$; canine denticulate .... *Nimravus*.
III. Lateral and anterior faces of mandible separated by a vertical angle; an inferior flange; incisors conic; canines denticulate.[2]
   a. No anterior basal lobe of superior sectorial;[3] inferior sectorial with a heel; no posterior lobes of the crowns of the premolars.
      Molars $\frac{3\ 1}{3\ 2}$ .... *Dinictis*.
      Molars $\frac{3\ 1}{3\ 1}$ .... *Pogonodon*.
      Molars $\frac{2\text{-}3}{2}\ \frac{1}{1}$ .... *Hoplophoneus*.
      Molars $\frac{?\ ?}{1\ 1}$ .... *Eusmilus*.

It is readily perceived that the genera above enumerated form an unusually simple series, representing stages in the following modifications of parts: (1) In the reduced number of molar teeth; (2) in the enlarged size of the superior canine teeth; (3) in the diminished size of the inferior canine teeth; (4) in the conic form of the crowns of the incisors; (5) in the addition of a cutting lobe to the anterior base of the superior sectorial tooth; (6) in the obliteration of the inner tubercle of the lower sectorial, and (7) in the extinction of the heel of the same; (8) in the development of an inferior

[1] American Naturalist 1881, p. 339. *Proælurus* has also been shown by M. Filhol to be one of the *Cryptoproctidæ*. It is also uncertain whether *Pseudælurus* belongs to the *Nimravidæ*.

[2] Gervais' figures of the canines of *Eusmilus bidentatus* represent no denticulations, but the figure is not clear.

[3] Rudimental in *Hoplophoneus*.

flange and latero-anterior angle of the front of the ramus of the lower jaw; (9) in the development of cutting lobes on the posterior borders of the larger premolar teeth.

(1) The reduction in the number of molar teeth. The dental formula of *Procælurus* is that of some *Viverridæ* and *Canidæ*, and the reduction from this point to the end of the series is obvious. In *Eusmilus*, as in *Smilodon*, the number of molars is less by one in the inferior series than in *Lynx* and *Neofelis*, where the formula is the smallest known among *Felidæ* proper, viz: $\frac{2}{2}\frac{1}{1}$. (2) The enlarged size of the superior canine teeth. In *Procælurus* and *Pseudælurus* the canines of both jaws are developed as in recent *Felidæ*. In *Archælurus* the superior is the larger, but does not, relatively to the molars, exceed that of *Felis*. It is rather compressed in form and has a sharp cutting edge posteriorly. In *Nimravus* the superior canine begins to have the enlarged size of the saber-tooths, but its form is peculiar in the *N. gomphodus*, being spike-shaped rather than saber-shaped. We find the true saber shape first in the *Dinictis*, where it is compressed, and with a denticulate cutting edge on both front and rear. In *Pogonodon* it has reached a very large size, and it does not display much increase in this respect until we reach the last genus of the series, *Eusmilus*, where its proportions are enormous, almost as large as in the feline genus *Smilodon*, where they appear to have been an inconvenience to the animal. (3) The diminished size of the inferior canines becomes evident in the lower genera of the third division (supra) of the *Nimravidæ*, but is most decided in the highest genera, *Hoplophoneus* and *Eusmilus*. (4) The incisor teeth have the usual obspatulate or obovate outline in the genera of the first and second divisions of the family, including *Nimravus*. They are conic in the true saber-tooths with flared lower jaw, beginning with *Dinictis* and ending with *Eusmilus*. (5, 6, and 7) The structure of the sectorials. The presence of a heel and an inner tubercle of the lower sectorial are well-known characters of a majority of the *Carnivora*. In only the most highly organized genera are they wanting, and among them are included all those of the *Felidæ* that still exist. In the *Nimravidæ* the inferior genera have both in a reduced degree, and they soon disappear as we ascend the scale. Thus, the inner tubercle is only present in the species of *Procælurus*, *Dinictis*, and

Hoplophoneus. The heel, on the other hand, remains throughout the entire family. The anterior basal lobe of the superior sectorial has the same history, its absence being characteristic of the inferior *Carnivora*, and of all the genera of *Nimravidæ* except *Hoplophoneus*, where it is rudimental. It is well developed in *Drepanodon* as in recent *Felidæ*, and is sometimes double in *Smilodon.* (8) The development of the inferior flange and latero-anterior angle of the mandibular ramus. There is a successive advance in the development of these characters, beginning with the second group, for in the first they are wanting. The latero-anterior angle is developed in *Archælurus* and allied genera, and is merely continued on the inferior border of the ramus. In the third group it is much more acute, and is deflected downwards, forming the well-known flange of the saber-tooths. It is longest in the *Eusmilus bidentatus* Filh. (9) The highest genera of *Nimravidæ*, e. g. *Hoplophoneus*, differ from true *Felidæ* in the absence of the cutting lobes on the posterior edges of the crowns of the larger premolar teeth. But, according to Filhol, these lobes are present in the generalized genera *Proælurus* and *Pseudælurus*, which are thus brought into a relation with the *Felidæ* not possessed by the *Nimravidæ*.

A characteristic perfection of the *Felidæ* is seen in the genus *Smilodon;* that is, the vertical direction of the ungual phalanges, by which the claws become retractile. This is well displayed by the two splendid specimens of *Smilodon necator* from Buenos Ayres, which have been preserved.[1] Unfortunately, these phalanges have not yet been described in any species of the *Nimravidæ*, and it is not yet certain what their structure really was. Among the true *Felidæ* the genus *Cynælurus* displays a less degree of development in this respect than the other genera, the ungual phalanges lacking the proximal process below the articular facet. Such a condition is to be looked for among the less perfect genera of *Nimravidæ*.

The succession of genera above pointed out coincides with the order of geologic time very nearly. Those belonging to groups first and second belong to the Lower and Middle Miocene, except *Ælurogale*, which is perhaps Upper Eocene, and *Pseudælurus*, which is Middle Miocene. The genera of the third division have the same Lower Miocene age, except *Eusmi-*

[1] See American Naturalist, December, 1880, fig. 12.

lus, which has been found in the same formation (Phosphorites) as the *Ælurogale*. Among *Felidæ*, *Drepanodon* is Upper Miocene, and *Smilodon* is Pliocene.

The relations of these genera are very close, as they differ in many cases by the addition or subtraction of a single tooth from each dental series. These characters are not even always constant in the same species, so that the evidence of descent, so far as the genera are concerned, is conclusive. No fuller genealogical series exists than that which I have discovered among the extinct cats.

As to the phylogeny of this family, there are flesh-eaters of the Eocene period which may well have been the ancestors of both the *Nimravidæ* and *Felidæ*.[2] I have suggested that this position is most appropriately held by the *Oxyænidæ*, a family of several genera, which included the most formidable, rapacious mammals of that early period in both continents. The interval between them and the *Nimravidæ* is, however, great; for in the *Oxyænidæ*, when there is a superior sectorial tooth, the first true molar in the upper jaw is utilized instead of the last premolar, and the second true molar below is a sectorial as well as the first. Several intervening forms must yet be found to complete the connection, if it have ever existed. It is, however, very likely that the true *Felidæ* were derived from the genus *Proælurus*, through *Pseudælurus*, if indeed these two genera be not the primitive members of that family, for, as above remarked, the evidence of their possession of the characters of the *Nimravidæ* has not yet been obtained. There can be no reasonable doubt that the genera *Drepanodon* and *Smilodon* in the *Felidæ* are the descendants of *Hoplophoneus* and allied genera. In fact, the *Nimravidæ* and *Felidæ* are "homologous groups," having corresponding terms in the manner I foreshadowed as a general principle in 1868 (Origin of Genera).

[1] M. Filhol has shown the range of variation in the *Ælurogale intermedia* to be considerable. While the normal dental formula is as given above, molars $\frac{3}{3}\frac{1}{2}$, he believes that it may range to $\frac{3}{4}\frac{1}{2}$ in one direction, and $\frac{3}{2}\frac{1}{2}$, or $\frac{3}{2}\frac{1}{1}$, in the other direction. It never attains the formula $\frac{4}{4}\frac{1}{2}$ of *Archælurus*, while the occurrence of two premolars in the lower jaw is rare. Originally M. Filhol was inclined to believe that there were three species, and this view is confirmed by the range in size. The largest specimens measure .085 M. on the alveolar line behind the canine, and the smallest .030, or less than half as long. The discovery of various intermediate sizes led M. Filhol to combine the three species into one. I incline to think M. Filhol's belief in three species to have better foundation than his belief in but one. Were the animals living, it is probable that their characters would be much more readily defined than is possible with the jaws only. Memoires sur quelques Mammifères Fossiles des Phosphorites du Quercy, Toulouse, 1882.

[2] See On the Genera of the *Creodonta*, by E. D. Cope, Proceed. Amer. Philos. Soc. July, 1880.

The following list shows the number and distribution of the species of the *Nimravidæ*. The position of a cross on a line indicates an intermediate geological position.

	Upper Eocene.	Lower Miocene.		Upper Miocene.		Pliocene.	
	Europe.	Europe.	America.	Europe.	America.	Asia, Europe.	America.
Pseudælurus hyænoides Blv			...... +				
Pseudælurus edwardsi Filh	+						
Archælurus debilis Cope			...... +				
Ælurogale intermedia Filh	+						
Ælurogale minor Filh	+						
Ælurogale mutata Filh	+						
Nimravus gomphodus Cope			...... +				
Nimravus confertus Cope			...... +				
Dinictis felina Leidy			+				
Dinictis cyclops Cope			...... +				
Dinictis squalidens Cope			+				
Pogonodon platycopis Cope			...... +				
Pogonodon brachyops Cope			...... +				
Hoplophoneus oreodontis Cope			+				
Hoplophoneus primævus Leidy			+				
Hoplophoneus occidentalis Leidy			+				
Hoplophoneus cerebralis Cope			...... +				
Hoplophoneus strigidens Cope			...... +				
Eusmilus bidentatus Filh	+						

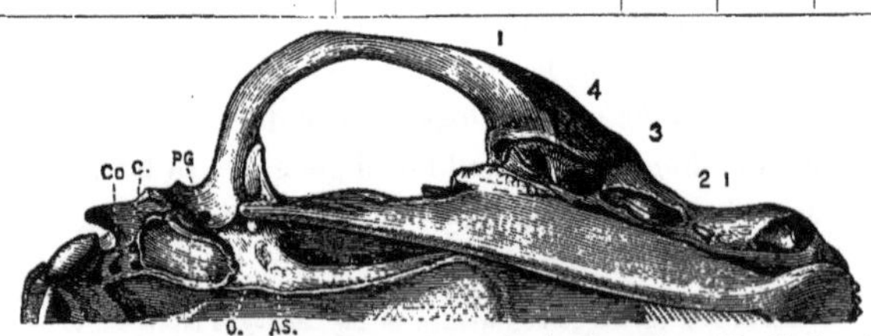

FIG. 37.—*Archælurus debilis*, one-half nat. size; inferior aspect. Foramina: AS, alisphenoid; O, ovale; PG, postglenoid; C, carotid; Co, condylar.

ARCHÆLURUS Cope.

American Naturalist, 1879, p. 798*a*. Proceed. Amer. Philos. Soc., 1879, p. 372 (Jan. 1880). Paleontological Bulletin, No. 31, p. 3, Dec. 24, 1879.

Dental formula I., $\frac{3}{3}$; C., $\frac{1}{1}$; P-m., $\frac{4}{3}$; M., $\frac{1}{2}$ Mandibular ramus without flaring inferior border anteriorly. Superior canines without cutting edge anteriorly; posterior edge not serrate. Incisors with short, slightly spatulate crowns. Postglenoid and postparietal foramina present. Superior sectorial tooth without anterior lobe. Inferior sectorial with heel, and without internal tubercle. No intermediate posterior lobes of the premolars.

This genus is of interest as completing the connection between the saber-tooth and primitive unspecialized groups of the cats, a transition also clearly indicated by the genus *Nimravus*. In dentition it adds a tooth to the number belonging to that genus in both jaws, and has a smooth-edged canine; it is otherwise identical with that genus, unless, indeed, the exostosis supporting the inferior sectorial tooth in the *A. debilis* be introduced into this category; a position I am not prepared to assume. The molar dentition only differs from that of *Dinictis* in the addition of a single tooth to the superior series, but that genus has the compressed superior canine and flared mandibular ramus of the saber-tooth. *Archælurus* also eases the passage from these genera to the *Proælurus* of Dr. Filhol, a highly interesting genus obtained by that gentleman from the phosphatic deposits of Central France. *Proælurus* has one more molar in the inferior series, and its inferior sectorial exhibits an internal tubercle similar to that seen in *Dinictis*. Like *Archælurus*, it is not a saber-tooth.

But one species of *Archælurus* is known thus far, and this one has been found in the John Day Miocene of Central Oregon.

ARCHÆLURUS DEBILIS Cope.

American Naturalist, December, 1879, p. 798 *a* (published Dec. 4). Palæontological Bulletin, No. 31, p. 3, Dec. 24, 1879. Proceed. Amer. Philos. Soc., 1879, p. 372.

Plate LXXI *a*, figs. 8–16; plate LXXII.

This species, which is about the size of the panther (*Uncia concolor*), is represented in my collection by parts of the skeletons of three individuals. First, a cranium with mandible complete, and the atlas; second, a cranium without mandible, but otherwise complete; third, a cranium anterior to the zygomatic fossa, without dentition or mandible, but accompanied by a lumbar vertebra; proximal part of scapula; both extremities of humerus; proximal part of ulna; fifth metacarpus; both extremities of tibia;

the astragalus and first metacarpal. As the cranium first mentioned furnishes the greatest number of characters, I select it for description.

The profile of the skull is convex, much as in the panther, but it appears to be somewhat less so, owing to the greater prominence of the premaxillary border. The front is rather wide at the postorbital region, and it is convex on each side of a wide shallow median longitudinal concavity. The nasal bones are rather stout, and carry their width backwards before contracting to an apex, which is above a point posterior to the lachrymal foramen. The external portion of their anterior extremity is continued as a process of moderate length along the premaxillary bone, about as far as the width of the superior border of each nostril. The superior process of the premaxillary does not reach the frontal bone by .17^{m}. The superior face of the frontal region is prolonged backwards so as to be partly diamond-shaped, owing to the gradual approximation of the anterior borders of the temporal fossæ. This form differs from that seen in the panther and leopard, and is more like the shape found in the jaguar and tiger. The extinct species differs from all of these in the very small development of the postorbital process. This is simply a slight prominence of the superciliary border; in only one of the three specimens deserving so much as the name of an angle. The brain-case is not more contracted behind the orbits than in the *Uncia concolor.* Anterior to the orbit the face is shallowly concave above the region external to the canine alveolus, which is convex. The sagittal crest is low, and the lateral crests of the inion are not very prominent. The latter do not overhang the foramen magnum so far as in the panther. The zygomatic arch is characterized by its slenderness, and the rudimental character of the postorbital process, so well developed in the existing *Felidæ.* It is less expanded posteriorly; that is, it is more nearly parallel with the axis of the skull than in the large recent cats. The occiput is not elevated. Its posterior face is divided by a short keel which descends from the inion one-third the distance to the foramen magnum. The latter is wider than deep, and its superior border is not furnished with the tuberosities so well marked in the panther, and less so in the leopard and tiger. The foramen magnum is much as in the *Lynx rufus.* The paroccipital process is moderately long, is acute, and is directed obliquely backwards. The front tympanic process is an obtuse tuberosity whose anteroposterior diameter exceeds its longitudinal. The occipital condyles are well separated below. The basioccipital is first flat below, and is then divided by a delicate median keel. The latter soon terminates, sending off a curved line to the anterior inner border of each otic bulla. The area on each side of the keel is divided into three shallow fossæ and an anterior plane. Anterior to the keel the base is plane, except a shallow fossa on each side. The posterior margin of the palate is gently concave to the line of the principal cusp of the superior sectorial. It is not so deeply concave as in the panther. The otic bullæ are large, as in existing cats.

The rami of the mandible are quite slender, in correspondence with the character of the zygomata. The inferior border of the masseteric fossa, though prominent, accompanies the inferior edge of the ramus for a short distance only, and gives it little width. The symphyseal portion rises abruptly from the rami; it is narrower below than above, and is concave in transverse section near the middle. The angle which separates the lateral from the anterior face of the ramus is rounded, and is continuous with the inferior border. This form is not seen in the species of *Felis*, *Uncia*, and *Cynœlurus*, and is a trace of the characteristic angle and flange of the sabertooths. The posterior angular process is straight, compressed, and acute, and the condyle is quite narrow. The osseous tuberosity beneath the inferior true molar teeth is a most peculiar feature. It looks as though the external alveolar border had been greatly extended, and then folded down and attached to the side of the ramus. Its free borders are a vertical and a horizontal, which are connected by a rounded angle. The ascending border passes into the base of the coronoid process by a roughened ridge. The inferior border and external face are also slightly roughened.

Foramina.—The incisive foramina are large. The *f. infraorbitale exterius* is especially large, and opens forwards. Its external border is thin, and is convex in profile; it is above the posterior edge of the last superior premolar, when the skull is resting on a plane surface. The *f. opticum* is rather large, and occupies the position usual in existing cats. The *f. sphenoörbitale* is quite large, and is round. The *f. rotundum* is smaller than the optic foramen and issues in the posterior side of the *f. sphenoörbitale*, and not separately as in the species *Uncia* at my disposal, the *Cynælurus jubatus*, and in several species of *Felis*. An alisphenoid foramen. The *f. ovale* is ransverse in direction and is well separated from the *f. lacerum anterius*. The latter is contracted. The *f. postglenoidum* is nearly as large as the *f. ovale*, and is near the anterior border of the meatus of the ear. The latter is small, and is bounded below by a distinct ossification, probably a tympanic bone, which is thickened so as to resemble somewhat a sesamoid. I find it in the two specimens where the region is preserved. There is no *foramen lacerum medius*, and the *f. carotideum* is well defined from the *f. l. posterius*, which it about equals in size. The *f. condyloideum* is quite distinct from the latter, of about the same size, and does not enter it from behind so as to have a common opening, as in the species of *Felis*, *Uncia*, *Cynælurus* *Crocuta* and *Hyæna*. Supraglenoid, postquamosal, subsquamosal, and mastoid foramina wanting. The postparietal is rather small, and is situated in the anterior inferior part of the posterior third of the parietal bone. There are two lateral and two anterior mental foramina on each side. Of the former, the posterior is below the posterior part of the second premolar, and the anterior below the first premolar.

Dentition.—There cannot be said to be any diastema in the superior dental series excepting that between the canine and the first incisor. The inferior diastema occupies the usual position, but has diminished length. The first and second superior incisors are quite small; their crowns are about as wide as their roots, and, meeting the inferior incisors squarely, are truncated by wear. The external superior incisor is several times as large as the others, and its crown is expanded outwards at the base. It presents an external worn face several times as long as its apical face. The superior canine is relatively about as long as in the panther. It is a good deal more compressed than in that species, and has a sharp posterior cutting edge which extends from near the alveolus to near the apex. This edge is not crenated. There is no corresponding anterior edge, though the wearing of the inner side of the crown by the inferior canine produces an artificial one near the apex. The front of the canine is rounded, but is not as wide as in existing cats.

The first superior premolar is one-rooted, and is of moderate size. The second is two-rooted and is small. The third is abruptly larger. The anterior face of its principal cusp is wide, and has no basal lobe; it is not trenchant posteriorly. The heel is long and rather obtuse. The fourth premolar or sectorial is relatively large. There is no anterior basal lobe, and the internal lobe is small. The cusp is robust, and not very acute posteriorly. The heel is long. The true molar is of medium size, and its external face reaches the line of the external faces of the premolars. Its grinding face is oblique upwards and backwards.

The inferior incisors are much like the superior, excepting the external one, which is only about twice as large as the others, and of the same form. The section of the inferior canine is nearly the quarter of a circle, the angle looking backwards from the inner side. It is much larger than the superior external incisor. The first inferior premolar is one-rooted, and its alveolus is as large as that of the first superior premolar, and is situated close to that of the second. The second (third) is quite large; it has an anterior basal angle, and a short posterior heel, but no posterior lobe of the principal cusp. The last premolar is larger. With the inferior sectorial it is concealed in part by the superior teeth, as the lower jaw has not been removed from its position. The posterior part of the sectorial shows a distinct, rather acute heel, like that of the *Nimravus brachyops*. The tubercular or second true molar is small and obtuse, and is directed more anteriorly than superiorly, owing to its situation on the base of the coronoid process. In all the teeth the enamel is smooth, and there are no cingula.

Measurements of skull.

	M.
Axial length from occipital condyles to premaxillary border	.180
Axial length from inion to premaxillary border	.191
Axial length from premaxillary border to canine tooth	.008
Axial length from premaxillary border to anterior border of superior sectorial	.058
Axial length from premaxillary border to palatal border	.075
Axial length from premaxillary border to end of maxillary bone	.080
Axial length from premaxillary border to postglenoid process	.135
Length of nasal bone from notch	.050
Length of sagittal crest from inion	.064
Width of premaxillary bones at canines	.034
Width of anterior nares	.027
Width of nasal bones at middle	.020
Least width between orbits	.028
Width at postorbital processes	.060
Width behind postorbital processes	.034

	M.
Width of zygomata at anterior border of orbits	.080
Width at meatus auditorius	.130
Width between paroccipital processes	.053
Width of occiput at middle	.032
Elevation of occiput above foramen	.032
Width of foramen magnum	.022
Depth of foramen magnum	.014
Width between otic bullæ	.016
Width of posterior nates	.013
Width of chin at base	.020
Width of chin at summit	.027
Length of superior canine	.026
Anteroposterior diameter at base	.013
Anteroposterior diameter at middle	.010
Transverse diameter at middle	.0065
Length of superior molar series	.057
Length of bases of Pm. i and ii	.0105
Length of base Pm. iii	.018
Elevation of crown of Pm. iii	.010
Length of base of Pm. iv	.024
Width of base in front	.010
Elevation of base in front	.012
Width of M. i (transverse)	.011
Length of base of inferior Pm. iii	.014
Length of base of inferior Pm. iv	.015
Elevation of base of inferior Pm. iii	.011

The *atlas* attached to the cranium above described is not entirely cleaned from the matrix. It displays a vertebrarterial foramen piercing the base of the transverse process posteriorly, as in various species of *Felis* and *Uncia*, and as distinguished from the lion. The base of the transverse process has a smaller anteroposterior width than in any of the species of *Felidæ* or Hyænas accessible to me. The neural arch has a similar character; a low tuberosity represents its neural spine.

The second specimen mentioned resembles the one already described so nearly as to render extended notice unnecessary. The crowns of the small first premolars have been broken off, but traces of their roots remain.

The third specimen displays the root of the first premolar with the others. The palatal surface exhibits a fossa opposite the heel of the superior sectorial to receive the apex of the inferior flesh-tooth. A last or seventh lumbar vertebra, belonging to this specimen, is represented on Plate LXXI*a*, fig. 8. It is a little smaller than that of *Uncia concolor*, and is quite similar in general proportions, especially in the form of the centrum. It differs in the following points: (1) The expanse of the postzygapophyses is a little less than the width of the centrum; in *U. concolor* it is a little greater. (2) The posterior edge of the neural spine is narrow and simple; in *U. concolor* it is wide and grooved. (3) There is no reverted process near the base of the diapophysis, as in *U. concolor*. (4) The metapophyses are more compressed, and rise higher than in *U. concolor*. As in the puma, there are no anapophyses.

Measurements of seventh lumbar vertebra.

		M.
Anterior diameters of centrum	vertical	.018
	transverse	.027
Length centrum on side		.030
Posterior diameters of centrum	vertical	.017
	transverse	.030
Expanse of prezygapophyses at middle		.025

The differences between this vertebra and the corresponding one of *Uncia concolor* are nearly those which distinguish it from the lion, leopard, and cheetah, with the following exceptions: In the lion the neural spine is thin behind, and the metapophyses thin and elevated, as in *A. debilis*. In the leopard the angle of the transverse process is not produced, and is more distal. The diapophysis in the cheetah is quite as in *A. debilis*, slender and without posterior angle.

Portions of the *scapula* and of the fore and hind limbs of this specimen are represented on Plate LXXI*a*, figs. 9–16. The glenoid cavity of the former narrows forwards to a subacute apex, which is not truncate, as in the panther. The coracoid hook, if it were present (the specimen is injured), originated farther above this apex than in the panther, leopard, or common cat. The characters may be compared with those of the recent cats, as follows:

Uncia leo. Coracoid rudimental; anterior tuberosity acute; posterior border of neck truncate; adjacent angle-ridge not continued.

Archælurus debilis. Anterior tuberosity acute; posterior border truncate and rugose for insertions; no adjacent angle-ridge.

Uncia concolor. Coracoid distinct; anterior tuberosity truncate; posterior border of neck truncate and rugose for insertions.

Uncia pardus. Coracoid distinct; anterior tuberosity acute; posterior edge not truncate; no adjacent angle-ridge.

Cynælurus jubatus. Coracoid distinct; anterior tuberosity acute; posterior edge not truncate; adjacent angle-ridge not defined.

Felis domestica. Coracoid present, not separated by notch, and continuous with anterior tuberosity; posterior edge openly grooved on account of adjacent posterior angle-ridge.

The head of the *humerus* presents some characters which distinguish it from those of the species of cats above mentioned, although it resembles them strongly in general features. While rather smaller than those of the panther and leopard, the greater tuberosity is more prominent, and the bicipital groove wider and deeper. The proximal surface of the lesser tuberosity is more distinct from the head, and is less decurved posteriorly. Viewed from the inner side its proximal surface is nearly straight and makes a right angle with the prominent posterior bicipital crest instead of being continuous with it, as in the panther and domestic cat, more nearly resembling the leopard and the lion. The tricipital fossa is larger than in any of those species, being well defined by a strong ridge posteriorly. The *fossa musculi teris* is quite small, less than in the *Uncia concolor* and *U. leo* and *Felis domestica.* The distal extremity of the humerus displays characters intermediate between those of the existing *Felidæ* and the flesh-eaters of the Eocene period. This is shown in its rather greater transverse extent; or, to particularize, in the greater length of the radial part of the trochlea and greater production of the internal epicondyle. The former has its proximal anterior border notched, and its face has a corresponding slight contraction not seen in any of the recent *Felidæ,* but not quite as strong as that of the *Crocuta brunnea* and the *Canidæ.* The distal extremity of the internal epicondyle is a truncate half-circle. Its production interior to this is withdrawn to a more proximal position. The olecranar fossa is well marked, and has a slight perforation. The radial fossa is wanting.

The *fifth metacarpal* is, relatively, smaller than in *Unciæ leo, pardus,* and *concolor, Cynælurus jubatus* and *Felis domestica.* It displays all the special characters of these species, including the keel-like interlocking process connecting it with the fourth metacarpal, so characteristic of the *Felidæ.* The proximal extremity resembles that of the lion rather than that of the panther in the plane surface of insertion of the external articular ligament.

Measurements of fore limb.

		M.
Diameters of glenoid cavity of scapula	anteroposterior	.029
	transverse at middle	.021
Diameters of head of humerus	transverse, extreme	.049
	transverse to groove	.033
	through lesser tuberosity	.036

		M.
Diameters of distal end of humerus	transverse, extreme	.044
	transverse of condyles	.033
	anteroposterior, at flange	.020
	anteroposterior at middle of condyle	.014
Length of fifth metacarpal		.039
Transverse diameters	proximally	.007
	distally (epicondyles)	.011

From the above measurements it may be seen that the fifth metacarpal bone is relatively only three-fifths as long as that of the recent *Felidæ* above mentioned.

The head of the *tibia* is about the size of that of a rather small adult *Uncia concolor* from Texas in my collection, and does not differ from it materially in characters. The anterior external tuberosity is not prominent, resembling that of *Felis domestica* rather than that of the larger cats. The external cnemial, and the popliteal fossæ are more deeply excavated than in any of the five *Felidæ* above mentioned, and the external posterior angle is acute. The distal extremity strongly resembles the corresponding part in the recent species of *Felidæ*, the external astragaline fossa being a little more expanded outwards. There is but one groove for the tendon of the peronaeus muscle instead of the two which are more or less distinct in all the species named; and its anterior bounding crest extends to the distal border of the malleolus, as in *Uncia concolor*, and does not cease above it, as in the other *Felidæ* above mentioned, and as in *Canis lupus*. The peroneal ridge is well marked.

The *astragalus* differs from that of all the *Carnivora* mentioned in this article in the shortness of its neck and the extension of the internal trochlear crest to the border of the navicular surface, from which only a narrow groove separates it. The trochlear face is a little narrower than in the species of *Felidæ* and *Hyænidæ* cited, and the inner side is more oblique outwards and downwards than in any of the species cited, excepting the *Uncia concolor*. On this side there is a well-marked tuberosity on the inferior margin posterior to the middle. Above and anterior to it the astragalus is concave. The navicular bone must have an extensive proximal external production, judging from the extent of the recurved external part of its facet on the astragalus. The trochlear groove of the astragalus is distinct, but is less profound than in the recent *Felidæ* and *Hyænidæ*. The first *metatarsus* is much shorter and more slender than that of *Uncia concolor*, *U. pardus*, *Cynœlurus jubatus*, or *Crocuta maculata*. What its length relatively to the other digits is, cannot be readily ascertained at present. Its proximal extremity is formed much as in the recent *Felidæ* and *Hyænidæ*, except that there is scarcely any external angle, from which it results that the bone is narrower when seen from the front. The shaft is subcylindric, is distally somewhat flattened, and is gently bowed forwards. The epicondyles and extremital keel are well developed, and much as in the recent species.

Measurements of posterior limb.

			M.
Head of tibia, diameters	transverse		.045
	anteroposterior at middle	greatest	.043
		least	.020
Diameters of shaft at distal fourth	transverse		.017
	anteroposterior		.016
Diameters of distal end of tibia	transverse		.030
	anteroposterior		.019
Diameters of astragalus	transverse at middle		.024
	anteroposterior		.033
Length of trochlea on external crest			.029
Width of head			.018
Length of first metacarpal			.067
Anteroposterior diameter metacarpal v	proximal		.017
	distal		.013
Width of shaft at middle			.009
Width at epicondyles			.013

Restoration.—It is probable from what has preceded, that the *Archælurus debilis* was an animal presenting much the appearance of the existing cats, and of about the size of the American panther. Omitting more technical characters, it differed from this and other species of the *Felidæ* in the greater slenderness of its feet. Its head was characterized by less breadth through the posterior part of the cheeks, and by a greater convexity of the forehead between the eyes, and a greater prolongation backwards of the same region.

Its structure plainly indicates that this species was of less sanguinary habits than the exising *Felidæ*, since its prehensile organs, both of the feet and dentition, are less robust. The slender zygomata and rami of the lower jaw show also that the impact of its bite was less powerful, although the large size and narrow form of the sectorial teeth furnish an effective cutting apparatus, which in some degree supplements the deficiency of strength. The weakness of the rami is further provided against by the curious exostosis at the base of the inferior sectorial already mentioned. This growth is symmetrical on the two sides of the skull, and is evidently normal; traces of it are seen in the species of *Nimravus*.

History.—The first description of this species was given by myself under the head of the *Nimravus brachyops* (*Machærodus brachyops*. Paleontol. Bulletin 30, p. 10, Dec., 1878), from a skull found by Mr. Sternberg, under the impression that it might belong to a female of that species. Subsequently a nearly perfect cranium, obtained by Mr. Wortman, demonstrated the distinctness of the animal, both as to species and genus. I published a wood-cut of this skull in the American Naturalist for December, 1880.

Horizon and locality.—The remains of the *Archælurus debilis* have so far been only found in the Middle Miocene formation of the John Day River, Central Oregon. Judging from the remains, it was, after the *Nimravus gomphodus*, the most abundant feline of that region.

NIMRAVUS Cope.

Proceedings of the Philadelphia Academy, 1879, p. 169.

Dental formula: I. $\frac{3}{3}$; C. $\frac{1}{1}$; Pm. $\frac{3}{2}$; M. $\frac{1}{2}$. Superior canines elongate, denticulate. Mandibular rami not flared downwards in front. Incisors with short subspatulate crowns. Superior sectorial without basal anterior lobe; inferior sectorial with well developed heel, and no internal tubercle. No intermediate posterior lobes of the premolars. Postglenoid and postparietal foramina present.

This genus has the dental formula and characters of *Hoplophoneus*, with the addition of a tubercular inferior molar tooth. It is, moreover, not a true saber-tooth, as is that genus, since it does not display the inferior anterior flange of the mandible. This is represented by an obtuse angular border, quite as in the species of *Archælurus*, in which genus *Nimravus* finds its nearest ally. The constant absence of the anterior premolars in both jaws distinguishes it sufficiently from that genus. On this account, and in view of the larger development and denticulated edge of the superior canine teeth, *Nimravus* may be considered as occupying a position between the two genera above named.

Two species are known to me, a larger and a smaller, both from the Middle Miocene formation.

NIMRAVUS GOMPHODUS Cope.

American Naturalist, 1880, p. 844. Bulletin U. S. Geolog. Surv. Terrs., 1881, p. 167.

Nimravus brachyops Cope. Proceeds. Phila. Acad., 1879, 170; not *Machærodus brachyops*. Proceeds. Amer. Philos. Soc., 1878, p. 72.

Plates LXXIII; LXXIV, figs. 1–2.

This feline is represented in my collection by parts of three individuals. The first includes a skull with one side and part of the other completely preserved, with the cervical and some dorsal vertebræ; the second is known from the posterior part of a mandibular ramus supporting three molar teeth; the third specimen embraces an entire mandibular ramus with all the teeth, and a femur, both having been found lying close together in the rock.

The skull of the first specimen has the mandible attached, with the mouth partly open, as represented on Plate LXXIII. It is as large as the large forms of the panther, and exceeds slightly the skull of the *Archælurus debilis*. It has very much the general proportions of that of the panther in the regular convexity of the profile of the frontal and nasal regions, the length anterior to the orbit, and in the degree of production of the inion. The length anterior to the glenoid cavity is relatively greater than in the panther, so that the mandible is longer. The front is moderately convex in transverse section, and is prolonged farther posteriorly than in the panther, through the more gradual convergence of the borders of the temporal fossa, resembling in this the tiger rather than the other species of *Uncia*. The obtuse postorbital angle is more prominent than in the *Archælurus debilis*; but does not deserve the name of a process, as it is in the large recent cats. The side of the face in front of the orbit is gently concave; in front of the *foramen infraorbitale exterius* it is nearly plane. The contraction behind the orbits is about as in the panther. The zygomatic arch is quite as prominent anteriorly as in the recent cats, but at the middle of its course it is less convex or flatter, as in the *Archælurus debilis*. The inion projects beyond the vertical line of the occiput, but not so far as in the tiger or the *Pogonodon platycopis*. The occiput is not higher than wide above, and it is divided by a keel which fades out on the inferior third. The lateral crest is low above the paroccipital process, where it divides into two low ridges. One of these goes to the paroccipital process and the other to the base of the posttympanic, where it turns forwards and becomes the acute supra-meatal crest. The two tuberosities on the superior border of the foramen magnum are rudimental. The paroccipital process is directed posteriorly. It is acute and rather long, is convex behind and flat in front. The post-tympanic is short, and its extremity forms an obtusely rounded tuberosity. There is an osseous tuberosity below the meatus of the ear which is doubtless the same as the piece in the *Archælurus debilis* which I identified with the tympanic bone. The basioccipital is divided by a median keel, while the basisphenoid is smooth. The palatal surface is injured in this specimen.

The ramus of the mandible is longer, deeper, and more compressed than in the recent species of *Uncia* and the *Pogonodon platycopis*. It is more robust than in the *Archælurus debilis*, and the exostosis below the last two molars is wanting. It is represented by a thickening of the external alveolar border, which is larger in some specimens than in others. The inferior border is gently convex below the teeth, and rises slightly below the anterior part of the masseteric fossa. It is continued into the hooked angular process, which is as wide as deep, and is not incurved. Anteriorly the inferior border is continued as an obtuse angle slightly everted but not in the least decurved below the diastema, and it then rises, running obliquely forwards and forming the obtusely angular border of the flattened symphyseal region. The inferior border of the symphysis descends below the level of the inferior border of the ramus. The condyle measures half the distance between the angle and the summit of the coronoid process. The latter is wide and is obtusely rounded above, where it does not extend posterior to the vertical line from the condyle The masseteric fossa is deep and wide; its narrow anterior border is reverted, while the inferior region is excavated abruptly below the general plane. There is a secondary fossa in its posterior part whose superior boundary marks half the distance between the last molar tooth and the summit of the coronoid process.

Sutures.—The premaxillary bone rises in its narrow portion to the horizontal line of the superciliary border. The maxillo-frontal suture is regularly convex upwards, and its anterior extremity falls short of the premaxillo-maxillary suture by .008 m. The nasomaxillary extends farther posteriorly, *i. e.*, to above the anterior third of the orbit, the palate and basioccipital being the horizontal base. The fronto-parietal suture is coarsely zigzag squamosal, and although it is entirely within the temporal fossa, it extends further forwards than in the recent cats. This extension is represented in the panther[1] by some curious produced laminæ, one on each side, reaching the frontal angle. The parietal and alisphenoid bones have an extensive contact. I do not find the large supraoccipital wormian bone of the recent large cats.

Foramina.—The *foramen infraorbitale anterius* is rather large, is a vertical oval in form, and is presented forwards. Its external border is above the posterior lobe of the second superior premolar. The optic foramen is large, and is separated from the much larger *f. sphenoörbitale* by a short interval. The *f. rotundum* is near the latter, but is entirely distinct from it; it is a little smaller than the *f. opticum*, and larger than the *f. ovale*. The *f. ovale* is well anterior to the *f. lacerum anterius*. An alisphenoid foramen. The *f. postglenoideum* and *f. caroticum* are well defined; the latter is separated from the jugular, and this in turn from the condyloid foramen. The *f. postparietale* is in the upper part of the inferior half of the parietal bone. There are no mastoid foramina. There are two lateral mentals, one below the anterior (third) premolar, and one below the diastema. The *foramen magnum* is large, and is wider than deep, as in *Archælurus debilis*.

Dentition.—The crowns of the superior incisors are flattened in front, where they have vertically oval outlines. The face of the external incisor is about twice as large as those of the others, and is bounded externally by a considerable face of contact for the inferior canine. The superior canine is of peculiar form, and does not exactly resemble that of any known species of the family *Felidæ*. The anterior border of the crown is a straight line from a point near the base to near the apex. The posterior outline is also nearly straight, having a slight concavity near the base and a slight convexity near the apex. As the anteroposterior diameter at the base is not large, and as the length of the crown is considerable, the resulting form is peculiarly slender and pike-shaped. The external face is strongly convex anteriorly, turning into a transverse anterior face, which is separated from the inner face by a right-angled ridge. Next to this ridge on the basal

[1] I find it in skulls which I have examined.

half of the crown, excepting near the base, the anterior face is thrown into a shallow groove. The ridge is not denticulate, while the acute posterior edge of the tooth is distinctly so from the base to the apex. The (first) second superior premolar is very small, and is two-rooted; it is a little nearer the base of the third premolar than that of the canine, leaving two diastemata. The third premolar is of large size, and consists of a triangular cusp and a short lobe-shaped acute heel. The superior sectorial is rather large. The apex of its anterior cusp is directed well backwards, and its posterior border is acute. The heel is long, and its cutting edge rises to half the height of the cusp. The external extremity of the crown of the tubercular molar is narrower than the internal; it just reaches the vertical plane of the alveolar border of the maxillary bone. Its size is moderate.

The inferior incisors are smaller than the superior, and of similar form. The external is only a little larger than the others. The inferior canine is much larger, and is nearly half the length of the superior canine. Its crown is directed somewhat outwards, and its inner face is moderately beveled above the apex of the external incisor. A section of the crown is an oval whose long axis is anteroposterior at the base, and more inwards posteriorly to an angle higher up. This angle is the posterior cutting ridge, which extends from the inner side of the base to near the apex, and is strongly denticulate. The diastema following the inferior canine is remarkably long in this species, relatively longer than in any of the large existing cats. The third inferior premolar (first present) is also remarkable for its large size, differing in this respect from true cats and saber-tooths alike. Its cusp is elevated, and the posterior basal heel is very short. There is no anterior basal tubercle. The fourth premolar has a greater longitudinal extent, and its apex is not quite so elevated as that of the third. Its anterior and posterior basal lobes are trenchant, and of equal anteroposterior extent. The inferior sectorial is large, and its principal lobes are large, and with acute cutting edges. The anterior lobe is longer than high, and the posterior higher than long, and its posterior edge is acute. The heel is small and acute. There is no trace of internal tubercle. The tubercular molar is small, and its diameters are subequal, and similar to that of the heel of the sectorial. In one of the three mandibles in my collection, represented on Plate LXXIV, fig. 1, this tooth is wanting.

The enamel is smooth, excepting on the superior canine and the anterior part of the anterior inferior premolar. Here it is roughened by minute impressed punctæ. No cingula.

Measurements of skull.

	M.
Axial length from occipital condyles to premaxillary border	.206
Axial length from inion to premaxillary border	.220
Axial length from premaxillary border to canine tooth	.017
Axial length from premaxillary border to anterior border of superior sectorial	.066
Axial length from premaxillary border to posterior extremity of maxillary bone	.097
Axial length from premaxillary border to postglenoid process	.162
Length of nasal bone from nasal notch	.065
Length of sagittal crest from inion	.082
Width of premaxillary bone (greatest)	.019
Width of each nasal bone at middle	.008
Width of each frontal bone at middle of orbit	.028
Width of each frontal bone at postfrontal angles	.035
Width of skull at anterior part of zygoma	.093
Width of zygomata at temporal fossa	.111
Width of skull at meatus auditorius	.074
Width of skull between apices of paroccipital processes	.054
Width of occiput at middle	.044
Width of foramen magnum	.024
Elevation of occiput above foramen	.036
Width of chin at base	.022
Width of chin at summit	.028
Depth of chin	.040
Depth of ramus at diastema	.027
Depth of ramus below last premolar	.031
Length of ramus	.157
Elevation of condyle	.033
Elevation of coronoid process	.071
Length of superior canine	.045
Diameters of base of superior canine { anteroposterior	.016
Diameters of base of superior canine { transverse	.008
Anteroposterior diameter at middle	.010
Distance from canine to third premolar	.016
Length of molars, including third premolar	.045
Length of base of third premolar	.018
Elevation of cusp of third premolar	.013
Length of base of sectorial	.025
Elevation of cusp of sectorial	.015
Width of tubercular	.009
Elevation of inferior canine	.024
Anteroposterior diameter of inferior canine at base	.012
Length of inferior diastema	.022
Length of inferior molar series	.063
Length of third premolar	.0175
Elevation of third premolar	.0175
Length of fourth premolar	.020
Elevation of fourth premolar	.015
Length of sectorial	.025
Elevation of median cusp of sectorial	.016

The *cervical vertebræ* have about the size and proportions of those of the leopard and the panther, differing only in minor respects. The centra are moderately depressed, and have oblique articular extremities, but the obliquity is rather less marked than in *Unciæ leo, concolor* or *pardus*, or *Felis domestica*. All are perforated by the vertebrarterial canal, excepting the sixth, as usual. The transverse processes of the atlas are about as wide as in the species named, but are more contracted at the base. The latter is pierced posteriorly, and in the middle inferiorly, by the vertebrarterial canal. The neural arch of the axis is more elevated in front than in the recent large cats, so that the interspinous foramen is larger. The axis also differs from the recent species in having the posterior borders of the arch above the zygapophyses regularly concave, and not filled (*U. leo, F. domestica*) or interrupted (*U. concolor, U. pardus*) by the opisthapophysis. The neural arch has the median posterior prolongation usual in *Felidæ*. The transverse process is acuminate, and extends as far posteriorly as the vertical line of the postzygapophyses. The hypapophysial keel of the axis is weak, but a little stronger than in the lion. The other cervicals can scarcely be said to be keeled, though traces are present on some of the centra, which on the third, fourth, and fifth, spread out posteriorly into triangular areas. Opisthapophyses are only present on the third and fourth cervicals. In the recent species I find them present, as follows:

Nimravus gomphodus, 3–4;
Uncia leo, 2–3–4–5–6–7;
Uncia concolor, 2–3–4–5;
Uncia pardus, 2–3–4–5–6–7;
Cynælurus jubatus, 2–3–4;
Felis domestica, 2–3–4–5–6–7.

All the cervicals behind the third have well-developed neural spines, that of the seventh covering the entire length of the neural arch by its base. The transverse processes present some peculiarities. The principal feature is the greater prolongation downwards of the diapophysial element, and hence less separation from the parapophysial than in the species above mentioned, especially in the fourth and succeeding vertebræ. The parapophyses have considerable expanse both transversely and fore-and-aft.

The first and second dorsal vertebræ differ from those of all the *Felidæ* above mentioned, in having the superior half of the posterior face of the neural spine grooved. The inferior half has a keel, which becomes very prominent to a point opposite the commencement of the zygapophyses and then suddenly subsides, as in *U. concolor.* The postzygapophyses are recurved outwards on the first dorsal; on the second they are plane, and slope downwards outwards. The center of both are as long as those of the last cervicals, are depressed, and have two weak angles on the inferior face.

Measurements of vertebræ.

	M.
Length of cervical vertebræ	.200
Transverse extent of atlas	.090
Transverse extent of transverse process of atlas	.025
Length of centrum of axis on side	.022
Elevation of centrum and neural canal of axis anteriorly	.029
Diameters of centrum of fourth cervical: anteroposterior	.025
Diameters of centrum of fourth cervical: vertical in front	.012
Diameters of centrum of fourth cervical: transverse in front	.020
Transverse extent with processes	.054
Diameters sixth cervical: anteroposterior	.024
Diameters sixth cervical: posterior: vertical	.013
Diameters sixth cervical: posterior: transverse	.026
Diameter with transverse processes	.050
Expanse of prezygapophyses	.035
Diameters centrum first dorsal: anteroposterior	.022
Diameters centrum first dorsal: anterior: vertical	.013
Diameters centrum first dorsal: anterior: transverse (without rib facets)	.017
Elevation of neural canal in front	.011
Elevation of neural arch and spine	.055
Expanse of postzygapophyses	.025
Expanse of postzygapophyses second dorsal	.022
Expanse of diapophyses second dorsal	.050
Anteroposterior diameter neural spine at base	.017

These vertebræ are represented in Pl. LXXIIa, fig. 3.

The second specimen of the *Nimravus gomphodus* consists of the posterior part of a mandibular ramus supporting teeth. Its dimensions are a little smaller than those of the type.

The third specimen includes a complete mandibular ramus with all the teeth, some dorsal vertebræ, and a femur. These pieces were found in juxtaposition in the same block of stone, and agree in proportions. The specimen was as large as No. 1. The *left femur* is perfectly preserved, and is represented in Plate LXXIV, fig. 2. It is rather long and slender, having nearly the size and proportions found in the leopard and panther. Its little trochanter is more prominent than in either of those species, and the trochanteric fossa descends lower down. The *fossa ligamenti teris* communicates with the neck by a groove, a character not seen in any of the five recent species at my disposal. The external *linea asper* forms a prominent ridge on the proximal half of the shaft, while the internal *linea asper* is short and only prominent proximally. Distally, the external rotular crest extends farther posteriorly than the internal. The rotular surface is a little narrower than in *Unciæ leo, concolor*, or *pardus*, and while as narrow as in *Felis domestica* and *Cynælurus jubatus*, is prolonged farther posteriorly. The lower portion of the shaft is oval in section. The intercondylar fossa is divided into three areas, viz, an anterior and posterior subtriangular separated by a deep oblique fossa. The former are the points of insertion of the internal and external crucial ligaments. The fossa between them is characteristically much deeper than in any of the recent feline animals above mentioned.

Measurements of femur.

		M.
Total length		.260
Transverse width	at proximal extremity	.052
	at distal extremity	.046
	at middle of shaft	.023
Anteroposterior width	of head	.024
	of great trochanter	.027
	of shaft	.018
	of condyles	.045
Width of rotular face at middle		.020

Restoration.—The *Nimravus gomphodus* is as large as the full-grown panther of the large varieties. It probably stood as high above the ground, but whether the body had the elongate proportions of that animal, or the more robust form of the leopard and jaguar cannot be ascertained in the absence of necessary material. Unless the animal had pendulous upper lips, a thing unknown among cats, the superior canine teeth must have been distinctly displayed on each side of the chin, their points descending entirely below the lower margin of the lower jaw when the mouth is closed. As these points are less compressed than in the true saber-tooths, they were less liable to fracture from lateral blows, but were more apt to be broken by fore-and-aft strains, owing to their slenderness.

The long canines of this species testify to bloodthirsty habits, for as weapons for penetrating wounds they are without rival among carnivorous animals. They resemble considerably the teeth of some of the *Dinosauria;* for instance, those of the Triassic *Clepsysaurus.* The sectorial apparatus is especially effective, and no tissue could long resist the combined action of the opposing blades of the two jaws. Nevertheless, this species did not, probably, attack the large *Merycochœri* of the Oregon herbivores, for their superior size and powerful tusks would generally enable them to resist an enemy of the size of this species. They were left for the two species of *Pogonodon,* who doubtless held the field in Oregon against all rivals. The compressed mandibular rami of the *Nimravus gomphodus,* though less slender than those of the *Archœlurus debilis,* are not so well calculated to resist lateral strains as the more robust jaws of the majority of the existing *Felidæ.*

History.—The first notice of this species was based on a mandible. As it is exceptionally without tubercular tooth, I referred it to the genus *Hoplophoneus;* and since its proportions are very similar to those of the *Pogonodon brachyops,* of which no mandible had been found up to that time, I identified it with that species. The reception of other specimens enabled me to distinguish the genus *Nimravus,* but it was not until some time later that I became satisfied that it was distinct from the species to which I originally referred it. The reasons for this conclusion are given under the head of *Pogonodon brachyops.*

NIMRAVUS CONFERTUS Cope.

Bulletin U. S. Geol. Survey Terrs., VI, p. 172, Feb., 1881. American Naturalist, 1880, p. 849, fig. 10.

Plate LXXI a; fig. 17.

This species is as yet represented by a mandibular ramus only. It is one-third smaller than that of *N. gomphodus.*

The inferior border of the ramus is broken off, excepting for a space below the diastema. The general form is narrow, as in *N. gomphodus,* and there is a projecting ledge along the inner base of the sectorial similar to that seen in the latter species. The angle separating the side from the front of the ramus is rather stronger than in *N. gomphodus,* but there is no indication of an inferior flare. The diastema is shorter than in the typical species, its length equaling that of the base of the third (first) premolar; in *N. gomphodus* it is half as long again. The symphysis is correspondingly shorter, ceasing a little in advance of, and at the posterior border of, the inferior canine tooth, while in *N. gomphodus* it continues for one diameter of the canine behind its posterior border.

The crown of the inferior canine tooth is directed backwards, and its serrate cutting edge is presented almost entirely inwards. The interno-anterior face of the crown is flat, and has a low shoulder at the base. The molars have the proportions of those of *N. gomphodus*, differing only in their smaller size, which is very apparent, as can be seen by the measurements. The first (third) premolar is a little longer on the base than high, has no anterior tubercle, and has a short cutting basal heel. The fourth premolar has subequal anterior and posterior basal cutting lobes, and the base is longer than the elevation of the median cusp. The sectorial tooth has a short cutting heel, but no trace of inner tubercle. The anterior lobe is as long as the median, but not so high. It overlaps the fourth premolar as far as the base of the median cusp. No incisor teeth are preserved in the specimen. Tubercular small.

Measurements of skull.

	M.
Depth of ramus at diastema	.020
Depth of chin	.027
Elevation of inferior canine	.016
Diameter of inferior canine at base	.010
Length of inferior diastema	.014
Length of inferior molar series	.053
Length of third premolar	.014
Elevation of third premolar	.010
Length of fourth premolar	.016
Elevation of fourth premolar	.013
Length of sectorial	.022
Elevation of median cusp of sectorial	.015

Although a left mandibular ramus is all that I have been able to obtain of this cat, the evidence is sufficient that it is specifically different from the others enumerated in this chapter. It is inferior in size, and peculiar in the reduced symphysial and incisive parts of the mandible. It was found by Mr. Wortman in the bad-lands of the John Day Valley, Oregon.

DINICTIS Leidy.

Proceed. Philada. Academy, 1854, p. 127. Extinct Mammalia of Dakota and Nebraska, 1869, p. 64.

With this genus we enter the group of the primitive saber-tooths, commencing with the most generalized form. The skeleton is yet unknown, but the skull and dentition are those of a true saber-tooth, and there seems to be no ground for believing the Musteline affinities suggested by Leidy.[1] It occupies the lowest position on the line of the saber-tooths, on account of its numerous and simply constructed molar teeth, and stands in immediate connection with the false saber-tooth group, having exactly the dental formula of *Ælurogale* Filh. On this account I formerly united the two genera, but now believe that the absence of the inferior flange of the mandible in *Ælurogale* is sufficient ground for maintaining them as distinct. The latter genus, in this respect, exactly resembles *Archælurus* and *Nimravus*.

Remains of this genus are quite abundant in the White River formation in Nebraska and Colorado. They principally belong to the longest known and typical species, *D. felina* Leidy. Specimens are much less numerous in the John Day beds of Oregon. Two species have been obtained from the former horizon, the *D. felina* and *D. squalidens*, and one from the latter, the *D. cyclops*. The characters are as follows:

Dental formula: I. $\frac{3}{3}$; C. $\frac{1}{1}$; Pm. $\frac{3}{3}$; M. $\frac{1}{2}$. The superior canine is long and compressed, and reposes against an inferior marginal flange of the mandible, whose surface is separated from that of the symphysis by a strong angle. Unworn incisors with wedge-shaped crowns. Superior sectorial without anterior basal lobe. Inferior sectorial with posterior cutting heel, and an inner tubercle of small size. Postglenoid and postparietal foramina present. No intermediate posterior lobes of the premolars.

The three species may be distinguished by the following characters:

First inferior molar two-rooted; first superior very small; cranium shorter behind. *D. cyclops.*

First inferior and superior molars two-rooted, the latter larger; cranium longer behind .... *D. felina.*

First inferior molar one-rooted; mandibular flange short, rounded...... *D. squalidens.*

Dinictis cyclops Cope.

Proceedings Academy Philadelphia, 1879, 176 (read July 8). American Naturalist, 1880, p. 846, fig. 8. Plate LXXV; fig. 1.

The profile of the skull is very convex, the planes of the nasal bones and sagittal crest meeting at an angle of 135°. The place of the angle is

[1] Extinct Mammalia, Dak., Nebr., p. 64.

occupied by the regularly rounded posterior frontal region. The postorbital angle is exactly half way between the inion and the anterior edge of the premaxillary bone. In the *D. felina*, according to Leidy's figure, which I find to agree with the specimen which it represents, the last-named distance is equal to the space between the postorbital angle and the postparietal foramen. Thus the region covered by the temporal fossa is shorter in the *D. cyclops*. The front is also convex transversely, excepting where it is interrupted by a median longitudinal shallow open groove. The postorbital angles are quite prominent, are subacute, and have a triangular section. They mark lateral angles equidistant between the apices of a diamond-shaped space, the posterior angle being the junction of the temporal ridges, and the anterior being theoretically situated between the anterior extremities of the frontomaxillary sutures. The nasal bones are shortened in front; their lateral angles are but little produced, and their anterior borders are concave. The premaxillaries rise high on each side of the nasals, but do not reach the frontals. The face is flat in front of the orbit, and the surface is roughened for a space just in front of its border. The zygomatic arches are strongly convex, but somewhat flattened medially, and a little more expanded posteriorly than anteriorly. They are less slender than in *Archælurus debilis* and *Nimravus gomphodus*, but their postorbital angle is not larger, being a mere angle. The brain-case is rather large, and is separated from the frontal region by a moderate constriction. The inion does not project so far posteriorly as in some of the cats, and the occiput is higher than wide, and is divided in its superior third by a weak median keel.

The paroccipital process is small and acuminate; it is directed posteriorly opposite the superior part of the occipital condyle. An obtuse ridge proceeds downwards and forwards from its base, and terminates in the posterior angle of the posttympanic process. The lateral occipital crest is continued as a delicate ridge into the external angle of the posttympanic process. The latter is well below the paroccipital; it is directed downwards, ceasing opposite the middle of the posterior face of the zygoma. Its extremity is truncate and triangular, the apex of the triangle being external. The suprameatal border is prominent and thin, and the inferior border of the meatus is the deeply set border of the bulla.

There are two areas of insertion on the basioccipital bone, separated by a low median keel, and extending anteriorly as far as the front of the otic bulla. Each is interrupted in front of the middle by a rugosity. Basisphenoid flat. The otic bullæ fit the adjacent bones closely, so as to close everything but the foramina. The pterygoid alæ are well produced downwards. The palate of *Dinictis cyclops* is flat; and is, at the widest part, as wide as long. The posterior border of the palate forms two concavities, uniting medially in a slight angle. Thus the form is that of the lynx and the leopard, and different from the panther, tiger, and jaguar. The forms of the palates of these animals are as follows:

I. Two deep excavations separated by a point: *Hoplophoneus oreodontis;* and *H. cerebralis.*
II. Two shallow excavations, separated by an angle: *Felis domestica; Uncia pardus; Lynx rufus; Hoplophoneus oreodontis; Dinictis cyclops; Proœlurus julieni.*
III. A regular shallow concavity: *Archælurus debilis; Uncia tigris.*
IV. A deeply excavated concavity: *Uncia concolor.*
V. A triple concavity; two lateral, separated by a deep median notch: *Uncia onca.*

The mandible is compressed, but is rather low at the front of the masseteric fossa. It is more robust than in the known species of *Archælurus* and *Nimravus.* The inferior anterior flange is well marked and compressed. It is short, but not so broadly rounded as in *D. squalidens.* The anterior symphysial face is flat and deep, and the incisive border elevated.

Sutures.—The frontomaxillary suture rises from the anterior border of the orbit, and then extends, with some irregularity, horizontally to the frontonasal suture, as described by Leidy in the *D. felina,* and not descending, as in *Nimravus brachyops.* The lateral nasal sutures are very little convex outwards posteriorly, and form a segment of a circle posteriorly. The frontoparietal suture is obliterated, but the superior squamosal suture does not reach it. The premaxillo-maxillary suture of the palate incloses a triangular space.

Foramina.—The *foramen infraorbitale exterius* is very large, is subtriangular, and is directed forwards. Its posterior border is above the anterior border of the superior sectorial tooth. The incisive foramina extend to a point a little behind the middles of the superior canines. The *f. f. palatina* are opposite the cusp of the third premolar. There is no alisphenoid foramen. The oval *f. lachrymale* is of medium size. The *f. opticum* is far posterior and close to the *f. sphenoörbitale*, which has twice its vertical diameter. On account of the early connection between the temporal and pterygoid parts of the sphenoid, the *f. sphenoörbitale* and *rotundum* are thrown together and forwards. They are, however, completely divided. A narrow alisphenoid canal enters the *f. rotundum* from behind, connecting it with the *f. ovale*. The *f. ovale* is large and transverse. The *f. lacerum anterius* is subround and rather large. The *f. postglenoideum* is rather small, and is in the anterior wall of the *meatus auditorius externus*. On the posterior border of the *bulla otica* are two foramina; an external larger, and an internal smaller. The latter occupies the usual position of the jugular foramen, but is probably the *f. carotideum*. The other may have carried a branch of the jugular vein, although it is in the position of the stylomastoid foramen. The anterior condylar foramen is small, and is situated far posterior to the two foramina just mentioned. A small foramen just above the paroccipital process I suspect to be the mastoid. The postparietal foramen is large, and is situated below the middle of the parietal bone. The *f. magnum* is wider than deep, and has a regularly arched superior border, without tuberosities.

Dentition.—The superior canine is quite long, and has a regularly lenticular section, without facets. Its anterior and posterior edges are denticulate. The external incisors are much larger than the internal, and have subconic crowns. The crowns of the others are subcuneiform. The first superior premolar is very small, much less developed than in *D. felina*. I originally described it as having one root, but there are indications of a second in an obsolete alveolus, filled partly by bone and partly by remains of the root. The tooth is lost from both sides of the specimen. The second premolar has a distinct anterior tubercle on the inner side, a character not seen in *D. felina;* the anterior angle of the superior sectorial is more produced than in that species. The crown of the superior tubercular looks partly inwards, is rather long, and has three roots. The inferior canines are considerably larger than the incisors. The latter are regular, and do not overlap each other. The second and third inferior premolars have well-developed basal lobes anteriorly and posteriorly. The heel of the sectorial is well developed. The tubercular is very small, and has a semiglobular crown. It is present on both sides of the specimen. The enamel of all the teeth is smooth. No cingula.

Measurements.

	M.
Length of skull on base	.140
Width of skull, measured below	.111
Length of palate	.060
Width of palate between posterior angles of sectorials	.062
Width of palate between canines	.026
Length of skull to front of orbits (axial)	.050
Vertical diameter of orbit	.031
Interorbital width (least)	.045
Elevation of inion from foramen magnum	.032
Length of inferior molar series	.050
Length of inferior sectorial	.018
Length of base of inferior first premolar	.055
Depth of ramus at sectorial	.016
Depth of ramus at first premolar	.021
Depth of ramus at flange	.026

From the Miocene beds of John Day River, Oregon. Found by J. L. Wortman.

This species was as large as the fully grown Canada lynx. Although of an inferior position in the system of *Carnivora*, its powers of destruction must have excelled those of the catamount. While the skull is generally less robust, its sectorial teeth are not smaller nor less effective than those of that animal, and the canines far excel those of the living species as instruments for cutting their prey.

Dinictis felina Leidy.

Proceedings Academy Philadelphia, 1854, 127; 1856, 91; 1857, 90. Extinct Mammalia Dakota and Nebraska, 1869, p. 64; Plate V, figs. 1-4.

This species, whose dental and cranial characters have been described by Leidy, is known from a number of crania and jaws. The former differ in their proportions from those of the *D. cyclops*, having a relatively longer cerebral and shorter facial part of the skull. The anterior premolar teeth, especially in the upper jaw, were stronger than those of *D. cyclops*.

I introduce the *D. felina* here for the purpose of mentioning its geographical range. Since the original specimens were brought from Northern Nebraska by Dr. Hayden, I have obtained a mandible from Northeastern Colorado, and the Princeton Scientific Exploring Expedition procured part of a skull from the same region. The examination necessary for this determination was kindly permitted me by Professor Guyot.

The Colorado mandible differs from that of the *D. squalidens* from the same locality in having a shallow symphysis; that is, it descends but a short distance below the inferior anterior mental foramen. The roots of the first inferior premolar are well separated. The third (fourth) has a well-developed anterior basal tubercle.

DINICTIS SQUALIDENS Cope.

Proceedings Academy Philadelphia, 1879, p. 176. *Daptophilus squalidens* Cope, No. 16, p. 2 (August 20, 1873). Annual Report U. S. Geol. Survey Terrs., 1873 (1874), p. 508.

Plate LXVIIa; figs. 15–16.

Two mandibular rami belonging to two individuals represent this species. The one on which the species was proposed is immature, with the sectorial tooth partly protruded (fig. 15), and the tubercular invisible. Having failed to find any trace of the latter in the jaw, I proposed to regard the species as typical of a genus distinct from *Dinictis*, remarking at the time that should such a tooth be ultimately found, the genus would have to be abandoned. Evidence of the existence of this tooth was afterwards obtained. Still later, another saber-tooth was found with precisely the formula supposed to characterize this discarded genus (*Daptophilus*). Under the circumstances I thought best to give the former a new name, *Pogonodon*.

In this species the first lower molar tooth has but one root, while in the others there are two. The canine tooth of the typical specimen has also a very peculiar form. The crown is short and wide, like that of a *Carcharodon* shark, or somewhat like that of the saber-tooth *Drepanodon latidens* Owen. As the first true molar tooth of this specimen was not fully protruded, it is possible that this canine belongs to the deciduous series.

The jaw of the adult specimen contains parts of the alveoli of but two inferior incisors; the existence of a third is very doubtful. The canine is large and is directed outwards. Its crown is lost. The diastema descends, and is short, not exceeding in length the length of the base of the anterior lobe of the sectorial. The alveolus of the root of the first (second) premolar is elongate anteroposteriorly, and has a very slight median constriction, indicating traces of opposite grooves on the root. The alveolar border is lost at the base of the third premolar, but from the remains of its alveoli it was not much smaller than the fourth. The fourth premolar has lost its cusp; it has a well-developed trenchant basal anterior lobe, not so large as in the known species of *Archælurus* and *Nimravus*. The base of the sectorial is absolutely and relatively shorter than in the specimens of *D. felina* figured by Leidy. The portion anterior to the heel is equal in anteroposterior extent to the fourth premolar; it is longer in the specimens above mentioned. The alveolus for the tubercular molar indicates a larger tooth than that of the *D. cyclops*.

The flange of the ramus is prominent, but short and rounded, extending but little below the symphysis, which itself is continued considerably below the inferior anterior mental foramen. The superior anterior mental foramen is below the middle of the symphysis. The anterior external mental foramen is below the anterior part of the alveolus of the first premolar. The ramus is quite robust, and is convex on both faces below the sectorial. The masseteric fossa is deep, but with sloping borders in front and below. A gradual rise towards the coronoid process begins on the inner side, at the posterior base of the sectorial.

Measurements.

	M.
Length of entire dental series	.064
Length of molar dental series	.048
Length of diastema	.008
Length of base of Pm. ii	.005
Length of base of Pm. iv	.013
Length of base of sectorial	.017
Height of anterior lobe of sectorial	.010
Long diameter of base of crown of inferior canine	.009
Depth of symphysis	.025
Depth at flange	.025
Depth at front of sectorial	.017
Width of front of sectorial	.007
Width of front of symphysis	.012

The immature specimen is instructive in displaying the last two deciduous molars, with more of the crown of the sectorial than is seen in the other specimen. The latter shows the short trenchant heel, and the internal tubercle, which is on the posterior edge of the middle lobe of the crown, about half way above the base, and a very little inside of the line of the edge of the heel. Its base is a little longer than that of the other specimen. The penultimate deciduous molar has equal anterior and posterior basal cutting lobes. The median cusp is higher than the length of its own base. The last deciduous tooth is a sectorial; its anterior lobe alone remains. The entire crown of the superior canine was found with this lower jaw (see fig. 15 *a b*). Its crown is quite short and compressed, and both edges are denticulate. It is probable that it is the temporary canine. It is much shorter than the permanent canine of any North American species.

Measurements.

	M.
Length of base of deciduous penultimate molar	.010
Elevation of cusp	.007
Elevation of anterior cusp of last deciduous molar	.007
Length of permanent sectorial	.018
Length of base of crown of superior canine	.011
Elevation of crown of superior canine	.025

The adult jaw above described differs from that of *D. felina* in the considerably shorter tooth-line, as well as in the single-rooted first premolar. As compared with Leidy's figures, it is shorter than one of them by the length of the tubercular tooth; and shorter than the other by the length of that tooth plus the heel of the sectorial.

Both specimens were found by myself in the beds of the White River formation at a single locality in Northeastern Colorado. With them were associated a multitude of jaws and bones of *Marsupialia, Carnivora, Rodentia, Ungulata, Reptilia*, etc.

POGONODON Cope.

American Naturalist, 1880, p. 143, (published January 31).

Dental formula: I. $\frac{3}{3}$; C. $\frac{1}{1}$; Pm. $\frac{3}{3}$; M. $\frac{1}{1}$. Inferior border of mandibular ramus flared downwards in front; lateral and anterior faces of ramus separated by a pronounced angle. Superior canine compressed, with anterior and posterior cutting edges. Superior sectorial without anterior basal lobe; inferior sectorial with a heel. No tubercular molar. No lobes on the edges of the crowns of the premolars. Incisors with conic crowns. Postglenoid and postparietal foramina present.

This genus represents a station on the line connecting *Dinictis* with the higher saber-tooths, being intermediate between the former genus and *Hoplophoneus*. It lacks the tubercular inferior molar of *Dinictis*, and possesses the second inferior premolar characteristic of that genus, which is wanting in *Hoplophoneus*. One species is certainly known, and a second is provisionally referred here. The two are the largest of the saber-tooths of North America, the type, *P. platycopis*, equaling in dimensions the largest species of *Drepanodon*, being only exceeded among the true saber-tooths by the species of *Smilodon*. Unfortunately, only the skull of the typical species is known. Several bones of the *P. brachyops* have been discovered.

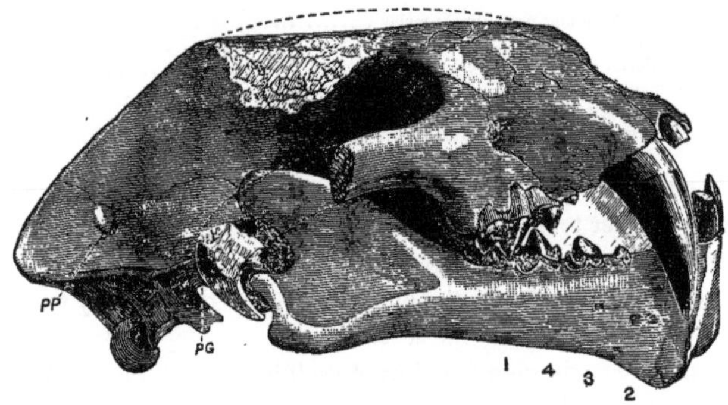

FIG. 38.—*Pogonodon platycopis*, less than two-fifths natural size. Mus. Cope.

POGONODON PLATYCOPIS Cope.

American Naturalist, 1880, p. 143. *Hoplophoneus platycopis*, American Naturalist, 1879, p. 798a, December. Proceed. Amer. Philo. Soc., 1879, p. 373 (December 24).

Plate LXXIV a.

The skull on which this species was established remains a unique. It is in good preservation, not wanting any important part. The median portions of both zygomata are wanting, and the frontal region is crushed and parts of it weathered away.

The cranium is characterized by the relative elongation anterior to the sagittal crest, the latter being a little more than one-third the length of the skull from its posterior apex to the premaxillary border, and equal the distance from the junction of the temporal ridges to the anterior rim of the orbit.

The sagittal and frontal planes make an angle with each other much as in the recent cats of large size. The brain-case is relatively smaller than in any of the *Felidæ* described in this book, and the sagittal crest and inion are more prominent. The face in front of the orbit is flat. The position of the alveolus of the superior canine is not defined posteriorly, but the surface follows it anteriorly, turning inwards to the premaxillary bone, from which it is separated by an angular groove. The free borders of the nasal bones are injured, and their posterior extremities are lost. The frontonasal suture indicates that their outline is acuminate posteriorly. The zygomata are rather short in comparison with the length of the skull. They are strongly convex and quite robust, though the vertical diameter at the orbit is rather small. This is explained by the fact that the masseter surface is mostly inferior instead of lateral. The cranium is more constricted posterior to the orbits than in any of the recent species, not excepting the *Uncia concolor*, where it is most constricted. The occiput is high and much narrowed, and its superior portion inclines at an angle of only 12° above the horizontal line passing through the occipital condyles and mandibular flanges, on which the skull rests. There are three fossæ on each side; two at the summit, separated by a keel, and two on each side, successively lower down. The inferior one is a fissure, turning upwards and outwards, only well defined by its upper border. The paroccipital process is short and acuminate, and is directed outwards and backwards. Its superior and inferior posterior borders are continued backwards, inclosing a fossa between them and the occipital condyle. The anterior border is continued as an angle which separates the lateral and inferior surfaces. The posttympanic process descends below the plane of the paroccipital process, but not so far as in the *Hoplophoneus cerebralis*. It has no inferior face, for the external and internal faces meet at an angle. The external face is an elongate triangle, with the apex upwards and outwards, and the base turned inwards. It is shallowly grooved in its length. The roof of the external meatus of the ear is prominent, and is continued as a ridge as far as a fossa which is above the space between the postglenoid and paroccipital processes. The base of the cranium being yet covered by the matrix, I cannot give its characters. The sutures of the skull are generally obliterated by coössification. The premaxillo-maxillary is sufficiently preserved to show that the premaxillary bone ascends high on each side of the nasals, so as nearly to reach the frontomaxillary.

The mandibular rami are robust, and not so high and compressed as in *Nimravus* and its allies. It deepens anteriorly to the inferior flange, which, though prominent, is not so much produced as in some of the species of later times. The anterior border of the side of the ramus is produced and acute below the superior third of its depth. The chin is deep and flat, with the median line prominent below. The masseteric fossa is very pronounced, the inferior border of the jaw at that point spreading outwards, with thick, round, horizontal border. The condyle has an appropriately large horizontal extent, but its vertical diameter is small. The angles are broken from both rami, but they were about as far below the condyle as the elevation of the coronoid process above it. The latter have an elongate anterior border, which turns backwards at the summit, and the posterior border projects as far as that of the condyle. The superior half of the coronoid process differs from that of all the other *Felidæ* here mentioned in being convex instead of concave.

Foramina.—The infraorbital foramen is large, subround, and a little wider than deep. Its posterior border is in the base of the zygoma, and above the middle of the posterior lobe of the sectorial, when the skull rests on the lower jaw. With the basicranial axis horizontal, it falls above the anterior cusp of the same tooth, so that its position is behind that observed in any of the recent or extinct cats mentioned in this book. The inferior border of the lachrymal foramen falls below the superior border of the external infraorbital foramen. The *f. postparietale* is large, and is at the border of the inferior fourth of the parietal bone, measuring from the summit of the crest. There is a small mastoid foramen. The *f. condyloideum anterius* is separate from the *f. lacerum posterius*. The *f. postglenoideum* is rather small, and is in the anterior wall of the *meatus auditorius externus*, near its inner terminus. The *foramen magnum* is rather small for the size of the skull, and is wider than high. It is not so transverse, however, as in the species already described, as its superior border forms a higher arch, and is without marginal tuberosities. There are three lateral mental foramina, one below the second (first) premolar, and the others larger and near together, anterior to it. There are three anterior mentals on each side, a superior smaller, and two inferior larger.

Dentition.—The crowns of the superior incisors are all broken off, while they remain on two of the corresponding inferior teeth. The latter are acutely, but not regularly, conical; the internal face is oblique forwards, while the external face is slightly convex from base to apex. The external incisors are twice as large as the internal, and the inferior canines are one-half larger than the external incisors in linear diameters. Their crowns are lost. The external superior incisors are a little smaller than the inferior canines, and the superior median incisors are half as large again in linear diameters as those of the inferior series. The superior diastema is rather short, or about as long as the base of the third inferior premolar. The superior canine is a very large tooth, exceeding in size that of any North American feline. It is similar in form to that of the European *Machærodus cultridens.* Both its cutting edges are denticulate, the posterior having thirteen teeth in .005^{m}. The crown, near the apex, has a regularly lenticular section, but the anterior edge turns gradually inwards until it leaves an anterior face external to it, which presents at right angles to the external face. These faces unite by a regular curve. The posterior edge is straight in the vertical plane. The second (first) superior premolar is quite small, and is probably only one-rooted, though I cannot be sure of this. Its position is near the base of the third premolar. The base of the latter is as long as the space between it and the canine. It is relatively smaller than in the species *Archælurus debilis* and *Nimravus gomphodus*, and consists of only a principal rather obtuse cusp and a moderately long cutting heel. The sectorial tooth is not as large relatively as that of the species just named, and is less compressed in form, both its cusp and heel being comparatively obtuse. The tubercular or true molar has remarkable transverse extent. Its external and posterior borders are all that can be seen in the present condition of the skull. The latter presents two concavities of the crown, separated by a low median elevation. The second inferior premolar is longer than the corresponding tooth of the superior series, and has one stout and much compressed root. The crown has a semicircular profile, and is more convex externally than internally. The third premolar is smaller than the corresponding tooth of the upper jaw, and much less than the same tooth of the known species of *Archælurus* and *Nimravus.* Its cusp is obtuse; besides, it has a short cutting heel and an anterior basal tubercle. The fourth premolar is larger, and has an anterior basal cutting tubercle, which is shorter than the cutting heel. The sectorial is small for the size of the skull, and its cusps are robust. The heel is short and thick, and has the peculiarity, entirely unique in the family, of a flat grinding surface.

Measurements of skull.

	M.
Length from inion to premaxillary border	.280
Length from inion to chin	.290
Length from premaxillary border to superior canine	.019
Length from premaxillary border to line of orbit	.090
Length from premaxillary bone axially to end of maxillary bone	.096
Length from inion to occipital condyle (horizontal)	.048
Length from inion to postglenoid process	.113
Length from inion to furcation of temporal ridges	.115
Width of zygomata (estimated)	.192
Width of occiput at middle	.037
Width of occiput at condyles	.056
Width of foramen magnum	.025
Width between posttympanic processes	.074
Width of chin at base of symphysis	.040
Length of mandibular ramus	.177
Height of chin medially	.050
Height of chin at flange	.048
Height of ramus at Pm. iii	.035
Height of ramus at M. i	.030
Height of ramus and coronoid process	.075
Length of superior dental series including canine	.090
Long diameter of base of canine	.026

Restoration and habits.—As the greater part of the skeleton of the *Pogonodon platycopis* is unknown, little can be said as to its general proportions. The skull is one-sixth shorter than that of the usual size of the tiger (*Uncia tigris*), and is equal to the largest Brazilian variety of the jaguar, and is considerably larger than the Texan form of that species.

The development of the dentition is concentrated in the canine teeth, and the powers of destruction of the animal would seem to be disproportioned to its ability to appropriate its prey as food. The molar teeth are rather small, as is the case with the earliest representatives of the canine family. The inferior sectorial is primitive and peculiar in its robust heel. We can suppose this species to have been a great destroyer of contemporary mammalian life, and that the largest ungulates of the John Day fauna were its victims.

History.—Science has hitherto had little knowledge of this species, and owes what is here recorded to a fortunate chance. The exploring party which I had sent into the John Day River valley under the direction of Mr. Jacob L. Wortman, in 1879, examined the bad lands in the locality known as The Cove. In passing the bluffs on one occasion, a member of the party saw on the summit of a pinnacle of the crag what appeared to be a skull. The large shining objects supposed to be teeth attracted his attention, and he resolved to obtain the specimen. He, however, was unable to climb the cliff, and returning to camp narrated the circumstance. The other men of the party successively attempted to reach the object, but were compelled to descend without it, and in one case, at least, the return was made at considerable peril. A later attempt, made by Leander S. Davis, of the party, an experienced collector, was more successful. By cutting notches with a pick, in the face of the rock, he scaled the pinnacle and brought down the skull, but at considerable risk to limb and life.

POGONODON BRACHYOPS Cope.

American Naturalist, 1880, p. 849, fig. 11 *Machærodus brachyops* Cope (*partim*). Paleontological Bulletin No. 30, p. 10, 1878 (December 3). Proceedings American Philosophical Society, 1878, 72.

Plates LXXIV *b*; LXXIV; figs. 3–10.

This species is the second in size of the cats of the Truckee Miocene epoch. It is represented in my collection by a fragmentary skull of one individual, and by the maxillary bone and several bones of the skeleton of a second.

The skull first mentioned includes the left side of the face, a part of the front, and the posterior regions, lacking one temporal bone. The maxillary bone is broken off at the infraorbital foramen, and the teeth are broken off. The following characters are noteworthy. The fronto-maxillary suture is transverse from the orbital border, and then turns slightly upwards rather than downwards as is usual. It most resembles the form in *Dinictis cyclops*. The face is slightly concave in front of the orbit, and is marked by an impressed fossa immediately anterior to the orbit, of about the size and form of half of the end of the human thumb. In front of the infraorbital foramen the face is gently and regularly convex to the premaxillary border. The premaxillary ascending ramus is long, but does not appear to reach the frontomaxillary suture. The fragment of the front shows that the temporal ridges converge gradually, and that the middle line is concave. The posterior part of the skull shows a high sagittal crest and inion, and a brain case of the absolute size of that of the *P. platycopis*, and therefore relatively larger. The inferior part of the lateral occipital crest is obsolete, continuing into a low rounded ridge which is continued as the prominent superior border of the *meatus auditorius externus*. A posterior ridge goes to the external superior base of the paroccipital process; another ridge extends from the internal superior base, and the space between is divided by a median ridge. The paroccipital itself is short and is directed backwards. A perpendicular ridge descends from its external base to the posterior angle of the posttympanic process. The latter has a truncate downlooking extremity of a triangular form; the longest side being anterior, and the internal and external angles of about equal length. The occipital condyles are rather small.

Foramina.—The *f. infraorbitale* is large. So is the *f. postparietale*, which is situated immediately at the base of the sagittal crest, or higher up than in any of the species here described. The *meatus auditorius externus* is contracted, owing to the more than usual approximation of the posttympanic and postglenoid processes.

Dentition.—The bases of the crowns of the superior incisors are so robust and the external so cylindric, as to render it very probable that their form is conic, as in the *P. platycopis* The superior canine is large, and has a lenticular section at the base of nearly equal acumination of front and rear. The posterior edge is denticulate. (See specimen No. 2.) The second (first) promolar is quite small, and is situated two-thirds way from the canine to the third premolar. Its root is grooved on the inner side, but I cannot say whether it is double. The third premolar is large, and is separated from the canine by a space five-sixths the diameter of the latter. There are no other teeth in the specimen.

Measurements of No. 1.

	M.
Elevation from diastema to frontomaxillary suture	.069
Width from orbit to premaxillary nateal border	.036
Length of bases of three incisors	.016

		M.
Diameters of base of canine	anteroposterior	.019
	transverse	.011
Length from canine to third premolar		.016
Oblique elevation of sagittal cusp above meatus of ear (partly estimated)		.066

The *maxillary* bone of the second specimen contains all the molars, the third and fourth somewhat mutilated. The second is plainly two-rooted, and is half way between the canine and the third, which teeth are separated by a space equal to the length of the base of the anterior lobe of the latter. This tooth has no anterior basal lobe, but the posterior heel is rather elevated, though short. The sectorial is large; its cusps are lost. The tubercular molar is small, and is especially remarkable for its small transverse extent, which is about equal to that of the *Nimravus gomphodus*, and less than half that of the *Pogonodon platycopis*. It has nevertheless three roots, the median of which is posterior. The external end of the crown is visible from the side, behind the sectorial. Parts of both canines are preserved. A fragment from the front of the crown shows that the anterior cutting edge becomes lateral, and is denticulate. The anterior face has a shallow groove next the cutting edge. Enamel slightly roughened.

The *foramen infraorbitale* is higher than wide in this species, and its external border is above the anterior border of the anterior root of the sectorial, the superior border of the molar being horizontal. The masseteric surface of the molar is vertical, not horizontal.

The remaining parts of this specimen are two dorsal and three lumbar vertebræ, portions of both femora; a patella; proximal end of tibia; both calcanea; the cuboid and navicular bones, and first metatarsal. The vertebræ are well preserved; see Plate LXXIV, figs. 3–5. To compare with these I have at hand the corresponding parts of *Uncia leo*, *U. concolor* and *U. pardus*, *Cynælurus jubatus* and *Felis domestica*. The dorsal vertebræ are shorter and a little wider than those of *U. concolor* and *U. pardus;* also flatter below, and more distinctly medially keeled. Their proportions are more as in the lion, where they are, however, regularly convex and not keeled. The anterior three lumbars are, on the other hand, of the proportions seen in *U. concolor*, *U. pardus*, and *C. jubatus;* that is, relatively longer than in the lion and shorter than in the common cat. Their centra are medially keeled below, but not so strongly as in the three species.

first named; while the lateral tuberosities are more pronounced than in any of the species excepting *U. pardus* and *F. domestica*, where they are stronger. In *U. concolor* an *U. leo* they are weak, and in *C. jubatus* wanting. As in these cats, they are below the line of the base of the diapophysis. The anapophyses are large on the last two dorsals and the first lumbar; on the second and third lumbar they are small, apparently shorter than in the *Felidæ* above mentioned excepting *Cyn. jubatus*, where they are smaller still. The posterior borders of the neural spines of the lumbars are nearly straight from their origin behind and above the postzygapophyses, and their summits are not much expanded. There is no trough-like excavation between the postzygapophyses above, as is seen in the panther, and in a less degree in the leopard and lion. The centra are not so much depressed as in the panther, leopard, and cat, or even as the lion, where they are the least depressed.

The femora are a good deal injured, and the heads and trochanters are gone. It is evident, however, that they were of the same general proportions as those of the *Nimravus gomphodus*, and not very different from those of the panther. The external *linea asper* is acute, and the rotular groove high and rather narrow. The fossa between the insertions of the crucial ligaments between the condyles is deeper than in recent cats. The head of the tibia is also injured. (See fig. 6.) The internal posterior surface for the attachment of the cruciform ligament is large. The patella is a little narrower and more convex within than that of the panther. The bones of the foot preserved show clearly that, as in the case of the *Archælurus debilis*, the extremities are relatively smaller than in recent *Felidæ* of similar general size. While the portions of the skull of the *Pogonodon brachyops* indicate an animal of rather larger size than the *Uncia concolor*, the bones of the hind foot are considerably smaller. The calcaneum is about three-fourths as long; the navicular is narrower though nearly as deep anteroposteriorly; the diameters of the cuboid are all less excepting the inferior facet anteroposteriorly; the first metatarsal is more slender, and its proximal facet is little over half as wide. One character of the calcaneum which I do not find in any of my skeletons of *Felidæ* or *Hyænidæ*, is the presence of a longitudinal fossa for the insertion of the external lateral ligament, just below the external astragaline facet. The external border at the cuboid extremity of the calcaneum is even more prominent and recurved than in the existing cats, and the groove for the internal calcaneocuboid ligament is profound. The groove of the cuboid for the tendon of the *peroneus longus* muscle is deep, and commences at the anterior face of the bone, that is, anterior to its position in *Uncia concolor* and *pardus* and *Cynælurus jubatus*, but its bounding ridge is not carried so far distally as in the panther. The small calcaneal face of the navicular is distinct from the astragaline surface. The character of the internal (free) surface of the head of the second (first) metatarsal shows that there was no hallux (first metatarsal).

Measurements of No. 2.

	M.
Length of superior dental series from canine tooth	.064
Length from canine to Pm. iii	.015
Length of base of Pm. iii	.019
Elevation of crown of Pm. iii	.013
Length of base of sectorial	.024
Width of base of tubercular	.008
Long diameter of base of canine	.017
Diameters of centrum of penultimate dorsal: anteroposterior	.028
Diameters of centrum of penultimate dorsal: anterior: vertical	.017
Diameters of centrum of penultimate dorsal: anterior: transverse	.026
Diameters of centrum of third lumbar: anteroposterior	.039
Diameters of centrum of third lumbar: anterior: vertical	.020
Diameters of centrum of third lumbar: anterior: transverse	.026
Elevation of last dorsal with spine	.055
Elevation of third lumbar with spine	.064
Diameters of middle of shaft of femur: anteroposterior	.019
Diameters of middle of shaft of femur: transverse	.025
Transverse diameter of head of tibia	.047
Length of calcaneum	.066
Width calcaneum at sustentaculum	.027
Diameters of navicular bone: longitudinal	.010
Diameters of navicular bone: horizontal: transverse	.016
Diameters of navicular bone: horizontal: fore-and-aft	.025
Diameters cuboid: longitudinal	.017
Diameters cuboid: proximal: transverse	.019
Diameters cuboid: proximal: fore-and-aft	.012

Although I do not possess a mandible of this species, I am satisfied that it is more nearly allied to *Dinictis* and the present genus than to *Nimravus*. It differs from the species of that genus and *Archælurus* in the following points: (1) the truncate triangular posttympanic process; (2) the transverse frontomaxillary suture; (3) the preorbital impressed depression; (4) the superior position of the postparietal foramen.

Restoration.—This was a most formidable animal, and its dental characters indicate a high degree of efficiency of both the lacerative and of the biting functions. While the *P. platycopis* has a larger development of the canine teeth, it is inferior in the relative size of the sectorials. In the latter respect the *P. brachyops* resembles the species of *Nimravus* and *Archælurus*, but these are furnished with smaller or more slender canines. It, however, resembled the latter in having the feet relatively smaller than in the recent cats, a character which indicates inferior prehensile power. Unfortunately, no ungual phalanges have been preserved, so that we cannot learn whether they confirm this indication by resembling those of the *Cynælurus jubatus* or the still less specialized forms of other families.

History.—This species was the first of the Oregon felines of which bones were obtained. It was first sent here by Mr. C. H. Sternberg, from the Miocene bad lands of the John Day Valley, Oregon.

HOPLOPHONEUS Cope.

Annual Report U. S. Geol. Survey Territories, F. V. Hayden, 1873 (1874), p. 509. Proceedings Academy Philadelphia, 1879, p. 170.

Dental formula: I. $\frac{3}{3}$; C. $\frac{1}{1}$; Pm. $\frac{3-2}{2}$; M. $\frac{1}{1}$. Inferior border of mandible flared downwards in front for the protection of the large double-edged superior canine tooth. Lateral and anterior faces of ramus sharply distinguished. Incisors with wedge-shaped or conic crowns. Superior sectorial without anterior basal lobe; inferior sectorial with a heel. No inferior tubercular molar.

The original diagnosis of this genus includes the ascription of a tubercular molar to the inferior dental series, which is an error due to the reference of a specimen to the *H. oreodontis* which does not belong to it.

In this genus we reach the dental formula of *Drepanodon* and the true cats, while at the same time the primitive form of the sectorials of the lower jaw remains. I have not been able to examine the skull of a *Drepanodon* (*Machærodus*), but from the silence of European authors it may be presumed that the foramina which characterize the *Nimravidæ* are absent, as in the true cats. The fact that the species have been derived from the

Pliocene and Upper Miocene formations increases the probability of the correctness of this supposition. I have ascertained that the American Pliocene saber-tooths of the genus *Smilodon* possess the characters of the *Felidæ* rather than those of the *Nimravidæ.*

Four or five species only of *Hoplophoneus* are known as yet, all from North America. We may expect, however, to find the genus in various parts of the world, wherever the beds occur which represent the time immediately preceding the epoch of the true saber-tooths. The longest known species is the *Hoplophoneus primævus* Leidy, from the White River bad lands of Dakota and Nebraska. It is about as large as the Canada lynx, and has long and slender superior canines. A larger species, the *H. occidentalis* Leidy, from the same horizon and locality, is known from two jaw fragments as large as the corresponding parts of the *Nimravus gomphodus.* Although the oldest members of the *Nimravidæ* yet known from North America, the *Drepanodon* characters of the mandible and of the superior canine tooth are well developed, much more so than in the false saber-tooth group of the later John Day epoch. In Europe, however, it must be remembered that the latter division commences still earlier; *i. e.*, in the Upper Eocene, in the genus *Ælurogale* Filhol.

As I have not obtained any parts of the *H. primævus* and *H. occidentalis* not already described by Leidy, I confine myself at present to a description of the *H. oreodontis* and *H. cerebralis.*

Hoplophoneus oreodontis Cope.

Annual Report F. V. Hayden, U. S. Geol. Survey Terrs., 1872 (1873), p. 509. *Machærodus oreodontis*, Synopsis of New Vertebrata Colorado, Misc. Pub. U. S. G. S. Terrs., 1873, p. 9.

Plate LXVII *a*, fig. 17; LXXV *a*, figs. 1, 2.

This saber-tooth was described from the fragment of a mandibular ramus supporting the temporary dentition. Subsequently I obtained at the same locality a large part of the skull, with portions of some of the limb bones of a fully grown specimen. The permānent dentition is in place, and is but little worn, while the epiphyses are not yet attached to the long bones. This specimen enables me to give a much fuller account of the species than heretofore.

The profile is convex. The face is flat in front of the orbit, but is strongly convex for the alveolus of the superior canine in front of the infra-orbital foramen. The premaxillary border is moderately produced, and the palate is generally flat. The anterior border of the posterior nares is concave on each side of the median suture, which terminates at the apex of a projecting angle. The sagittal crest is not much elevated anteriorly, but rises steeply to the inion. The lateral occipital crests are prominent and nearly parallel. The occiput is not narrowed, and it is divided by a keel on the median line. There are no lateral keels or fossæ. The suprameatal crest is thin, and the postglenoid process is not much extended transversely. The mandibular ramus is shallow, and of nearly equal depth below the molar teeth. The inferior flare is quite prominent, and the alveolar margin rises to the inferior canine. The symphysis is very deep, and the front of the chin is strongly convex at its superior part, and less so below, having a form quite different from that seen in the species of *Felidæ* already described, where it is flat and transverse. The symphyseal suture is longer above than below, and the two parts are separated by a deep sinus without contact, which almost reaches the external surface. The angle of the ramus is not as much produced as in the species above described, and it is widened horizontally inwards. The masseteric fossa is profound, and its inferior and supero-anterior borders are prominent. There is a peculiar rough, flat tuberosity on the inner side of the superior border of the ramus, opposite the posterior part of the sectorial, which I do not find in the existing cats nor in the species of *Nimravus*, and of which a trace is seen in the *Dinictis cyclops.*

Sutures.—The *frontomaxillary* is almost vertical from the orbit, and turns downwards to join the lateral nasal suture for a very short distance only. It is there widely separated from the premaxillo-maxillary. The lachrymal bone is pear-shaped, with the narrow end directed downwards and backwards, and joining the orbital plate of the palatine by a short suture. Its bounding sutures are not dentate. The anterior part of the orbitosphenoid is separated from the lachrymal by about the length of the inferior border of the latter bone. The palatal plate of the palatine bone extends as far forwards as the anterior border of the superior sectorials, and is broadly truncate anteriorly. The frontoparietal suture crosses the apex of the frontal space.

Foramina.—The infraorbital foramen is large and subtriangular. Its posterior border is directly above the middle of the anterior root of the superior sectorial. The lachrymal is small. There are two nasoörbitals perforating the orbital plate of the palatine, of which the superior is the larger. The incisive foramina are large, and the palatals are small and opposite the anterior border of the (first) third premolar. The postparietal is well below the sagittal crest. The postglenoid is small. The dental foramen is large, and its center is below the middle line of the ramus, and below the anterior border of the base of the coronoid process. There are three external mental foramina on both rami; one below the front of the fourth premolar, one anterior to the front of the third premolar, and a larger one below the middle of the surface to which the superior canine tooth is applied. There are two anterior mental foramina which mark about the thirds of the depth of the chin.

Dentition.—The width of the premaxillary region gives space for a full development of the incisors, which have robust bases. The canines, like the incisors, are broken off. Their bases show them to have been of large size, and of more compressed form than those of any of the species already described. Their long diameter considerably exceeds the length of the diastema which separates them from the third (first) premolar. There is no indication of the existence of a second premolar. As the outlines of the maxillary bones diverge strongly, the base of the third premolar is oblique. Its crown is broken off, but it was evidently not as large a tooth as in the species of *Nimravidæ* above described. Its base is one-half the length of that of the sectorial. The sectorial has its principal cusp prominent and acute, while its heel is low, but compressed and sharp. There is a rudiment of an anterior basal lobe, as is figured by Leidy in the *H. primævus*. This tooth is characterized also by the very small size of its internal heel, which is continued downwards as a perfectly straight ridge, without interruption, to near the apex of the principal cusp. The edges of the tooth are little worn, and they show at several points anterior and posterior to the principal cusp a beautiful denticulate structure. The tubercular molar is large, but not equal to that of *Pogonodon platycopis* or *Dinictis cyclops*. It has but two roots, and its crown is partly concealed from the side view by the posterior extremity of the sectorial.

The mandibular incisors are well developed, and the canine is not much larger than the external incisor. Of the molars, the first is very small, though two-rooted, and has anterior and posterior basal lobes, although the former is minute. The fourth premolar is intermediate in size between the third and the sectorial, and has anterior and posterior basal lobes, the posterior being the longer. The main cusp is not high, and its acute edges are crenulate. The sectorial has its principal cusp elevated; the heel is short, and the internal tubercle stands on the posterior edge of the main cusp two-fifths the distance from the base of the crown. The enamel of the external sides of the last two molars is slightly wrinkled.

Measurements.

	M.
Axial length from premaxillary edge to orbit	.041
Axial length from premaxillary edge to extremity of maxillary bone	.061
Axial length to middle of posterior border of palate	.060
Width of palate behind sectorials	.054
Length from premaxillary border to frontoparietal suture on profile	.095
Width of occiput at superior part	.030
Length of base of third premolar	.009
Length of base of sectorial	.020
Width of base of tubercular molar	.011
Length of mandible to angle	.101
Length of dental series from I. i	.067
Length of canine and incisor series (oblique)	.015
Length of diastema	.019
Length of base of Pm. iii	.006
Length of base of Pm. iv	.012
Length of base of sectorial	.018
Depth of chin at symphysis	.031
Depth of chin at flange	.032
Depth of ramus at M. i	.017

The mandibular ramus of the young animal already mentioned supports the temporary sectorial, and displays two of the permanent incisors inclosed in the jaw in an unworn condition (Plate LXVIIa, fig. 17). The latter have slightly recurved conic crowns, with denticulate cutting edges. The sectorial has the character of that of most *Felidæ;* there is a short acute heel, and an elevated compressed internal tubercle, which is as high as the anterior lobe of the crown. Its position is behind and a little within the median cusp of the tooth.

Remarks.—This species is nearly allied to the *Hoplophoneus primævus*, of which it may be only a regional variety. It is distinguished by its shorter and wider face and palate, a character especially seen in the shortness of the diastema, which is considerably less than in the Nebraska species. With this animal it compares much as the bull-dog does with the ordinary varieties of the genus *Canis*.

The two specimens I have described were found by myself on a denuded portion of the White River formation in Northeastern Colorado. At the same locality were multitudes of bones, mostly jaws, of fifty species of various orders of *Mammalia* and *Reptilia*, on many of which it doubtless preyed.

HOPLOPHONEUS CEREBRALIS Cope.

American Naturalist, 1880, Dec., p. 850. *Machærodus cerebralis* Cope, American Naturalist, 1880, p. 143 (January 31).

Plate LXXVa; figs. 3, 4, 5.

Probably the smallest species of the genus, and one that presents peculiar characters.

This peculiar species, the smallest of the genus, approaches nearest in dentition to the true saber-tooths (*Drepanodon*), and is represented by a skull, from which the basioccipital region, a good deal of the right side, and the lower jaw, are absent.

It differs in many respects from all the members of this family of cats heretofore discovered in North America. In almost every point in the osteology of the skull it is peculiar. There is not as much space for the temporal muscle as in most of the extinct species described, or as in the large recent *Unciæ*, but the points of origin of the muscle indicate that it was relatively stronger than in the domestic cat and the lynxes. Its single premolar is very small, so that the dentition for practical use is reduced, in the upper jaw, to the canine and sectorial. Both have been most effective instruments in the performance of their respective functions. The sectorial has a distinct anterior basal lobe. The space for the accommodation of the brain is relatively more ample than in any other feline of the formation, and the inner wall indicates that the convolutions of the hemispheres were well developed. This species, if the cranium were of usual proportions, was about the size of the red lynx (*Lynx rufus*).

The facies is quite different from that of the species heretofore described, the profile more nearly resembling that of some of the smaller species which belong to the genus *Felis*. It is much less convex than usual, for though the nasal bones descend, the sagittal crest is horizontal and the occiput vertical, when the sphenoid bone is held in a horizontal position. The side of the face is slightly concave in front of the orbit, and plane in front of the infraorbital foramen as far as the gentle curvature to the nareal border. There is a low protuberance on each frontal bone, near the maxillary suture, a little nearer the median suture than the superciliary border. From it the surface descends rapidly to the superciliary border. The postorbital process is longer and more recurved than in any of the species of the White River period, resembling more nearly that of the recent *Felidæ*. The orbit has less vertical and more lateral presentation than in many of the species, owing to the less prominence of the anterior part of the zygoma It is of a vertically oval form, owing largely to the shortness of its superciliary border. The parts of the zygoma remaining are slender, and the malar portion turns inwards below. The sagittal crest is long, but is quite narrow and low. The brain-case is relatively large, and the parietal surface of the temporal fossa is regularly convex. The occiput is quite wide, and does not project backwards as in the larger and many of the smaller cats The lateral crests are not prominent, and are quite obsolete above the mastoid region. The posterior face is impressed by a number of small fossæ and foramina. Though the occiput is somewhat imperfect, there is evidence that there was no paroccipital process; even less than the rudiment found in most of the recent *Felidæ*. Above its position, forming the external expansion of the exoccipital bone, is a low tuberosity. The posttympanic process is quite long, and is subcylindric and truncate. It does not touch the postglenoid. The lateral occipital crest is continued as a delicate ridge which divides it lengthwise externally. It does not give off the suprameatal crest, which is low, but expands into the thin posterior superior edge of the squamosal part of the zygoma. The latter is more expanded posteriorly than anteriorly. The anterior part of the basioccipital has a median groove, which is continued on the basisphenoid. The pterygoid processes of the palatine bone have rounded inferior edges. The nareal border of the palatines consists of two pronounced concavities separated by a prominent point. The border of each concavity supports two contiguous obtuse tubercles well separated from each other. The external one of these is separated from the acute deflected posterior border of the maxillary bone by a concavity of the horizontal surface. The posterior front of the palate is slightly concave on each side of the median line, and presents a deep excavation on the inner side of the sectorial, to receive the crown of a large inferior sectorial. This fossa has a straight inner border. The palate is concave between the canines; behind them on each side, a shallow groove enters the palatine foramen.

The characters of the palate described are quite peculiar among cats.

Sutures.—The premaxillomaxillary suture joins the nasal suture well above the nareal border of the nasal bones, but does not probably meet the frontomaxillary. The latter is arched upwards to a point above the anterior third of the orbit. Its nasal terminus is lost from both sides. The fronto-parietal suture crosses the sagittal crest well behind the junction of the temporal ridges. The parietosphenoid suture is quite long. The squamosal bone is not much longer than it is deep to the *meatus auditorius.* The maxillo-palatine suture extends along the inner border of the fossa for the inferior sectorial.

Foramina.—The external nares are rather small, partly curving to the prolongation of the nasal bones. The infraorbital foramen is round and small for a cat; owing to the lack of prominence of the malar bone it has a partially external opening. The lachrymal is normal and well inside the orbit. The optic is close to the sphenoörbital and is of usual size. The sphenoörbital and round foramina are united into a single large anteriorly directed opening; whether the internal perforations are united or not, the state of the specimen does not allow me to see. There is an alisphenoid canal, which connects the external opening with a common foramen with the *f. ovale.* Its diameter is small. The *f. ovale* is large, and is not widely separated from the *meatus auditorius internus.* The *f. postglenoideum* is moderate, and terminates in a groove in the anterior wall of the *meatus auditorius.* The basioccipital bone being absent, the relations of various foramina cannot be ascertained. There are two postparietals, both penetrating the inferior part of the bone, the posterior near the lateral occipital crest. The palatine foramen is rather large. It is much nearer the alveolar border than to the median suture, and is opposite the anterior border of the third premolar tooth.

Dentition.—The bases of two incisors, and both canines, are preserved with the crowns of the molars of one side. The base of the external incisor is not so much larger than that of the second, as in the cats already described. It is separated from the canine by a diastema half as long as that posterior to the canine. The latter is two-thirds as long as the long diameter of the canine. The base of the canine is of a compressed oval form, with a slight concavity on the inner side. It is much like that of the *Hoplophoneus oreodontis*, but is relatively larger. In fact, it is relatively larger than in any other American saber-tooth, excepting the *Pogonodon platycopis*. There are but three molars. The first or third premolar is quite small, but has, as usual, two roots. The crown has a low compressed anterior basal tubercle, and a small median cusp. The heel is rather long, equaling the long diameter of the base of the cusp, and is compressed. The sectorial is relatively large. It is somewhat worn by use, but its form is characteristic. Its cutting heel is very long, while the base of the principal cusp is small. There is a distinct anterior basal cutting lobe, as in the true *Drepanodontes*. The external face of the crown is concave before and behind the median cusp. The tubercular or true molar has been lost; there are alveoli for two roots, rather close together. The position of the external alveolus shows that the crown extended externally to the plane of the jaw.

Measurements.

	M.
Length from premaxillary border to inion	.117
Probable width of zygomata posteriorly	.082
Axial length from premaxillary border to orbit	.040
Axial length from premaxillary border to nares	.057
Axial length from premaxillary border to postglenoid process	.025
Width between superior canines	.020
Width between sectorials behind	.043
Width of posterior nares	.012
Width of occiput above middle	.037
Width between orbits (horizontally)	.047
Length of sagittal crest	.038
Length of postglenoid from roof of meatus	.018
Length of posttympanic (total)	.011

	M.
Length of glenoid cavity (transversely)	.017
Vertical diameter of orbit	.025
Length of base of canine	.016
Width of base of canine	.017
Length of diastema	.011
Length of molar teeth on base	.025
Length of Pm. iii	.006
Length of sectorial	.017

The unique specimen of this species was found by Mr. J. L. Wortman in the bad lands of Camp Creek, one of the head tributaries of the Crooked River, in Central Oregon.

HOPLOPHONEUS STRIGIDENS Cope.

American Naturalist, 1880, December, p. 851. *Machærodus strigidens* Cope, Paleontological Bulletin, No. 30, p. 9 (Dec. 3, 1878). Proceed. Amer. Philos. Soc., 1878, p. 71.

Plate LXXV*a*; fig. 6.

This obviously distinct species is only represented by the crown of a superior canine tooth, from which the apex has been broken. Its characters are so peculiar that I have recorded it under the above name, not knowing whether I shall have better specimens.

The tooth is long and very much compressed, much more so than in any species of the genus known to me. Its anterior and posterior edges are finely and very perfectly denticulate, without lateral flexure near the base. The center of each side of the tooth is occupied by a wide open gutter, so that the greatest transverse diameter of the crown is not at its middle. These gutters become planes towards the apex, giving an elongated hexagonal section. The size indicates an animal of the proportions of the *H. cerebralis.*

As compared with the superior canine of the *Dinictis squalidens*, which the present specimen resembles in its compression and fine denticulation, it differs in its greater relative length and in the presence of the lateral open sulci.

Measurements.

			M.
Diameters at base	anteroposterior		.0120
	transverse	greatest	.0036
		median	.0032
Length of a denticle on base			.000143

Remarks.—This tooth belonged to an animal of about the size of the *H. cerebralis*, and perhaps to that species. If so, it indicates for it a longer canine than usual, as its extremely compressed form points to a position at a considerable distance beyond the base of the crown. The probabilities are against reference to the *D. cerebralis.*

The tooth is the most elegant in form and perfect in its details yet found. As a cutting instrument it is superior to anything of human manufacture which I have seen.

Found by C. H. Sternberg, on the John Day River, Oregon, in the Truckee beds.

(The White River and John Day Faunæ are continued in Vol. IV.)

INDEX.

○

Lightning Source UK Ltd.
Milton Keynes UK
UKHW011228310119
336488UK00006B/502/P